Cooperatively Interacting Vehicles

Christoph Stiller · Matthias Althoff ·
Christoph Burger · Barbara Deml · Lutz Eckstein ·
Frank Flemisch

Editors

Cooperatively Interacting Vehicles

Methods and Effects of Automated Cooperation in Traffic

 Springer

Editors
Christoph Stiller
Institute of Measurement and Control
Systems (MRT)
Karlsruhe Institute of Technology
Karlsruhe, Baden-Württemberg, Germany

Christoph Burger
Institute of Measurement and Control
Systems (MRT)
Karlsruhe Institute of Technology
Karlsruhe, Baden-Württemberg, Germany

Lutz Eckstein
Institute for Automotive Engineering
RWTH Aachen University
Aachen, Nordrhein-Westfalen, Germany

Matthias Althoff
Department of Computer Engineering
Technical University of Munich
Garching, Germany

Barbara Deml
Institute of Human and Industrial
Engineering (ifab)
Karlsruhe Institute of Technology
Karlsruhe, Baden-Württemberg, Germany

Frank Flemisch
Institut für Arbeitswissenschaft
RWTH Aachen University
Aachen, Nordrhein-Westfalen, Germany

ISBN 978-3-031-60493-5   ISBN 978-3-031-60494-2   (eBook)
https://doi.org/10.1007/978-3-031-60494-2

This Springer imprint is published by the registered company Springer Nature Switzerland AG
The registered company address is: Gewerbestrasse 11, 6330 Cham, Switzerland

If disposing of this product, please recycle the paper.

Editorial

Mobility is a great asset to humans and our capability to move seamlessly through our world contributes to the quality of an active life. On the downside, we pay a high price for our mobility in the form of traffic accidents, environmental pollution, consumption of resources and living space, and loss of freely available time due to congestion. In this situation, two recent developments in the automotive sector are opening up attractive opportunities. Many experts expect recent advancements in automated driving research will be commercialized in the not-too-distant future. In addition, communication technology between vehicles and the infrastructure advances rapidly, so that Car2X communication systems are likely to be standard equipment in automobiles within a few years.

Used in a targeted way, these technological possibilities can lead to an innovation, if not a revolution, in our mobility. Cooperatively interacting vehicles can increase traffic efficiency and reduce accidents, economically expand public transport in peripheral areas and at off-peak times, and reduce the use of open space by traffic in cities. For people, these technologies offer comfortable, safe travel with freely available time and self-determined mobility with their own vehicle even in old age and with physical impairments.

Any driver has witnessed the benefits of cooperative behavior in traffic for safety and efficiency. Likewise, automated vehicles are expected to improve traffic safety through cooperative interaction. This book addresses research fields in cooperatively interacting vehicle technology and is structured along the information flow of automated driving.

Perception and Prediction of other road agents using on-board sensor information stands at the beginning of the perception-action cycle. The first chapters focus on body posture interpretation of cyclists and how such information can be used for prediction. It is shown how motion prediction of vulnerable road users (VRU) can benefit from information exchange between different devices carried by VRUs and sensors mounted on vehicles or in the infrastructure. Driver behavior at intersections is analyzed with an emphasis on ambiguous precedence situations.

The potential of **Explicit Communication** between traffic participants is key to enhancing the perception horizon even in adverse visibility conditions. Furthermore, proper design of V2X communication protocols can alleviate cooperative perception and maneuver negotiation.

Cooperative **Motion Planning**—whether with implicit or explicit communication—opens new abilities for automated vehicles. This book sheds light on a multitude of relevant aspects in this field. The mutual dependence of motion plans of interacting traffic participants can be framed as a Stackelberg game, where an agent plans optimal behavior considering its effects on others based on their strategy. Graph optimization techniques make it possible to assemble complex motion plans from motion primitives. Networked control forms a joint optimization task whose complexity can be reduced by a prioritized distributed model predictive control (P-DMPC) approach. Different time-variant priority assignment algorithms are investigated. V2V-communication enables negotiation for cooperative driving maneuvers. Appropriate strategies allow traffic participants to temporally reserve areas of the road for their exclusive usage. A layered architecture for motion planning is proposed to guarantee consistent and safe cooperative driving decisions. The concept of specification-compliant reachable sets makes it possible to identify and negotiate potential conflicts within a group of cooperating vehicles. Formal models for V2X communications facilitate the formation of local traffic systems whose trajectories can be verified to be free of collisions and deadlocks. Reinforcement learning has gained large attention in automated driving research. Learning reward functions from expert trajectories are proposed to mimic human driving styles. Likewise, learned cooperative maneuver policies enhance traffic efficiency and equity at mixed-traffic intersections.

Last, not least, **Human Factors** impose core design objectives for any mobility system. The combination of use cases and design/interaction patterns bears the potential to manage the complexity of future cooperative systems. SAE level 3 automated driving includes potential takeover requests to the human. For such systems, confidence horizons are proposed to predict the takeover capability based on the driver's initial orientation reaction. Alternatively, the confidence horizon concept may be instantiated with a pattern framework. Deadlock situations frequently arise in urban situations in which the right of way is not regulated. Studies on how human drivers solve such situations lead to recommendations for automated vehicles. The part closes with measures and descriptions for cooperation.

The book provides an overview of methods for the implementation of cooperatively interacting automobiles. It presents recent research results and references relevant literature in this domain. The multi-disciplinary expertise of the authors reflects the nature of the topic.

We hope the readers will find these contributions to this emerging technology fruitful and inspiring for their own work. Last, not least, the editors and authors gratefully acknowledge funding for the Focus Program SPP 1835 *Cooperatively*

Interacting Automobiles by the German Science Foundation DFG, as well as fruitful collaboration among the partners.

Karlsruhe, Germany   Christoph Stiller
October 2023   Matthias Althoff
                                                                         Christoph Burger
                                                                           Barbara Deml
                                                                          Lutz Eckstein
                                                                         Frank Flemisch

Contents

Motion Planning

Human Factors

Perception and Prediction with Implicit Communication

How Cyclists' Body Posture Can Support a Cooperative Interaction in Automated Driving

Daniel Trommler, Claudia Ackermann, Dominik Raeck, and Josef F. Krems

Abstract Automated driving is continuously evolving and will be integrated more and more into urban traffic in the future. Since urban traffic is characterized by a high number of space-sharing conflicts, the issue of an appropriate interaction with other road users, especially with pedestrians and cyclists, becomes increasingly important. This chapter provides an overview of the research project "KIRa" (Cooperative Interaction with Cyclists in automated Driving), which investigated the interaction between automated vehicles and cyclists according to four project aims. First, the investigation of body posture as a predictor of the cyclists' starting process. Second, the development of a VR cycling simulation and validation in terms of perceived criticality and experience of presence. Third, the experimental evaluation of a drift-diffusion model for vehicle deceleration detection. And fourth, the investigation of factors affecting cyclists' gap acceptance. With these research aims, it was the project's intention to contribute to a better understanding of the cyclists' perception of communication signals and to improve the ability of automated vehicles to predict cyclists' intentions. The results can provide an important contribution to the cooperative design of the interaction between automated vehicles and cyclists.

1 Introduction

While current advanced driver assistance systems (ADAS) have already improved the safety and comfort of manual driving [5], automated driving is expected to lead to even more benefits, such as reduced congestions and an increased mobility for a large number of people [18, 33]. However, there are a range of human factors issues that need to be overcome prior to launching automated vehicles [27].

D. Trommler (✉) · C. Ackermann · D. Raeck · J. F. Krems
Chemnitz University of Technology, Chemnitz, Germany
e-mail: daniel.trommler@psychologie.tu-chemnitz.de

D. Raeck
e-mail: dominik.raeck@psychologie.tu-chemnitz.de

J. F. Krems
e-mail: josef.krems@psychologie.tu-chemnitz.de

© The Author(s) 2024
C. Stiller et al. (eds.), *Cooperatively Interacting Vehicles*,
https://doi.org/10.1007/978-3-031-60494-2_1

3

When used in urban traffic, automated vehicles (AVs) need to be able to interact with vulnerable road users (VRUs), such as pedestrians and cyclists [13]. Since interaction in road traffic as a social phenomenon can be complex and can involve ambiguities [21], a detailed understanding of how human drivers and VRUs interact in road traffic is essential for the development of appropriate algorithms for AVs [16, 20]. In addition, cooperation as a collaborative effort is required for a successful interaction: In a joint action, road users share common goals and thus follow common solutions instead of interacting competitively [9, 11, 15]. Therefore, it is crucial to examine how goals and intentions of VRUs can be recognized and how to communicate AV's intentions to them. It is assumed that such a cooperative interaction between AVs and VRUs can lead to higher satisfaction, trust, acceptance, efficiency and safety in road traffic [12, 15].

While the previous project KIVI investigated the cooperative interaction between AVs and pedestrians [1, 2, 4], the current project KIRa focused on the cooperative interaction between AVs and cyclists. In both projects, researchers in traffic psychology and communications engineering jointly explored relevant questions regarding the analysis of human behavior in road traffic and the development of suitable algorithms. In the following two chapters, the results obtained in KIRa are presented, first from a psychological and then from a communications engineering perspective (see next book chapter by Raeck et al.).

1.1 Space-Sharing Conflicts Between Cyclists and AVs in Low-Speed Areas

In this project, cooperative interaction between cyclists and AVs was primarily investigated in urban low-speed areas such as parking lots or shared spaces. We assume that these low-speed areas are often characterized by (1) a shared infrastructure for vehicles and cyclists (e.g., no dedicated bike lanes), (2) less formal rules about priority (e.g., no traffic light control), (3) a higher probability that a road user will change its current behavior (e.g., starting or stopping) and (4) the (partial) occlusion of road users (e.g., due to parked vehicles).

These characteristics have the potential for space-sharing conflicts between cyclists and AVs. According to Markkula et al. (2020), a space-sharing conflict represents "an observable situation from which it can be reasonably inferred that two or more road users are intending to occupy the same region of space at the same time in the near future" (p. 736) [16]. As space-sharing conflicts between cyclists and vehicles can lead to safety-critical situations, it is necessary to either anticipate and avoid such conflicts (e.g., through recognizing the intentions of cyclists) or to handle and solve them safely (e.g., using appropriate communication cues).

1.2 Recognizing Intentions of Cyclists in Low-Speed Areas

To enable an AV to anticipate space-sharing conflicts, it can be useful to recognize the intentions of cyclists. Trajectory prediction can be used to determine the cyclist's next trajectory using the current state. Thus, potential conflicts with the AV's trajectories can be identified. In contrast, the starting of a cyclist, as a typical scenario especially in low-speed areas, cannot be determined well using trajectory prediction. Therefore, further information needs to be included. Previous research with pedestrians showed that human observers can use the body posture of pedestrians to predict their intention to cross the street, even when certain information of the body posture (e.g., head or legs) is occluded [23]. However, due to a lack of research, it is unclear whether the body posture of cyclists can contribute to the recognition of cyclists' intentions. Therefore, this project aimed to investigate how human observers use the body posture of cyclists to predict their intention to start. Further, because body parts of cyclists can be occluded in low-speed areas, it was also aimed to examine the prediction accuracy in these scenarios.

1.3 Communication Between Automated Vehicles and Cyclists in Low-Speed Areas

Implicit and explicit communication can help to solve space-sharing conflicts between AVs and cyclists safely and efficiently [10]. Implicit communication cues refer to the behavior of road users that, on the one hand, change their movement (e.g., vehicle deceleration) or perception (e.g., head turning), and, on the other hand, can be used by other road users, for example, as a sign of the willingness of a pedestrian to cross the road [16]. Explicit communication includes signals with no effect on one's own movement or perception, such as light signals to indicate intentions of an AV [16].

Several studies, however, have shown that implicit communication cues are used more frequently for interactions between vehicles and pedestrians in low-speed areas compared to explicit communication cues [7, 14]. From a reanalysis of a naturalistic cycling study, it can be assumed that priority between cyclists and vehicles in low-speed areas is similarly more likely to be negotiated using implicit communication cues [3]. For example, agents (i.e., vehicles or cyclists), who reach the conflict space earlier, often take the chance to solve the space-sharing conflict through accelerating or avoiding behavior [3]. Therefore, the present project focused on implicit communication and, in particular, investigated cyclists' gap acceptance and vehicles' deceleration maneuvers. It was examined how different time gaps, the vehicle size and speed affect the decision of cyclists to cross a street in front of a vehicle. In addition, differences in the perceptual decision-making process involved in the detection of vehicle deceleration by VRUs were further analyzed using a drift-diffusion model. The results can provide important implications for a situation-specific parameterization of vehicle deceleration maneuvers. Moreover, the results can indicate when explicit communication cues are necessary to support the decision making process.

1.4   Investigating Space-Sharing Conflicts Between Automated Vehicles and Cyclists

In the previous project KIVI, video recordings from the perspective of a pedestrian at the curb were used to investigate pedestrians' gap acceptance and deceleration detection performance [1, 2, 4]. However, this methodology had to be adapted to the perspective of a cyclist. Riding with a camera on the bicycle handlebars often resulted in blurry recordings due to the pedaling activity. In addition, in such real-world recordings, it was often difficult for the cyclist to keep a constant speed, on the one hand, and to keep the time gap to a moving vehicle, on the other hand. Therefore, this project did not use real-world recordings, but rather a VR cycling simulation which will be presented.

1.5   Aims of the Research Project "KIRa"

The following sections will provide a rough overview of the research within the project "KIRa" (Cooperative interaction with cyclists in automated driving) regarding the research topics described above:

1. Investigating the body posture as a predictor for the starting process of cyclists.
2. Development and validation of a VR cycling simulation.
3. Experimental evaluation of a drift-diffusion model for vehicle deceleration detection.
4. Investigation of factors influencing the gap acceptance of cyclists.

This chapter aims to give an overall summary of the project activities. Detailed information on the experiments can be found in the related publications at the end of each section.

2   Investigating the Body Posture as a Predictor for the Starting Process of Cyclists

A typical scenario in urban traffic, especially in shared spaces and parking lots, is the starting of a cyclist. Here, starting is understood as the process between getting on the bike and the final roll-off. It is assumed that recognizing the progress within the starting process can be crucial for an AV to decide whether it still takes priority (in early stages of the starting process) or rather yields priority (in later stages). Therefore, this project task intented to investigate how accurately the progress of the starting process can be detected based on the body posture of cyclists. Furthermore, we investigated the importance of different body parts in this rating as well as the accuracy of the ratings when body parts are masked. It is assumed that occluded parts of the cyclists' body, e.g., due to signs, parked vehicles or the bicycle frame, could be a highly relevant problem in shared spaces. The results could support the development of efficient algorithms for intention recognition and, based on human abilities, allow conclusions about how good algorithms' intention recognition needs to be at a minimum.

For the examinations, 12 cyclists were recorded while starting with their bicycle. The recordings were taken on a parking lot from two different perspectives of a car driver: The view from behind and from the side. The period of interest was the time between getting on the bike and the final roll-off. The recordings were split into four images per second to allow a better visibility of the cyclists' body posture. Further, these images were used either without masking or (after image manipulation) with masked upper or lower body. During six standardized experiments, these images were presented both in chronological (baseline condition) and random order (experimental condition). In the chronological order, participants were able to build up prior knowledge about the progress based on the images before. In the conditions with random order, prior knowledge was not available and the ratings were possible based on body posture only. For each image, participants were asked to provide ratings about the progress in the starting process using a scale between 0 and 100%. Further, the participants were asked to specify which parts of the cyclist's body were relevant to these ratings.

Surprisingly, the ratings of progress in the starting process were similar between the baseline and experimental conditions. In the conditions with random order, the ratings often increased in the order the images were originally taken. Thus, the randomly presented images of a cyclist could be rearranged well into chronological order. This could be shown for the different perspectives as well as the masked images. In particular, the ratings at the end of the starting process, shortly before the cyclist accelerated, were rated very accurately in each of these conditions. Furthermore, a lower variance could be observed in the ratings at the beginning and at the end of the starting process. It is assumed that the beginning and the end of the starting process are associated with characteristic body postures. Regarding the relevant body parts, the legs showed the highest ratings as the decisive part at the beginning. As the starting process continued, the importance of the legs decreased, while the importance of the upper body, head and feet increased. When body parts were masked, the remaining parts were substantially able to compensate the occlusion.

The results showed that the progress in the starting process can be recognized accurately based on body posture even when body parts are masked. Thus, it seems possible that the final roll-off can be detected early and ensures a safe interaction. Further analysis is required to formally describe the body posture in the starting process in order to implement these characteristics in algorithms for the intention recognition.

Related Publications:

- [30] Trommler, D., Ackermann, C., Krems, J.F.: Investigating the body posture as a predictor for the starting progress of cyclists. In: 33rd International Co-operation on Theories and Concepts in Traffic safety (ICTCT) Conference, Berlin, Germany (2021).
- Trommler, D., Krems, J.F.: Using cyclists' body posture to support a cooperative interaction in automated driving (in prep.).

3 Development and Validation of a VR Cycling Simulation

3.1 Development of a VR Cycling Simulation

While different virtual reality (VR) cycling simulations are available in the entertainment and sports sectors, these commercial products are of limited use for research purposes. For the examination of the interaction between AVs and cyclists in this project, a VR cycling simulation has to meet several requirements: (1) Accurate control of road user maneuvers, including their trajectories, speed and speed adaptations. (2) Realistic physical visualization of the environment and vehicles, including gravity and other physical forces of objects. (3) Detailed data recording. (4) Ideally, a user-friendly graphical user interface (GUI). (5) Cost-efficient development, especially regarding the hardware requirements.

For this purpose, different existing VR (driving) simulations were compared. These include CARLA [8], OpenDS [17], VICOM Editor (TÜV DEKRA arge tp 21), STISIM driving simulator platform, and Westdrive & LoopAR [19]. Considering the criteria mentioned above and seeing as it offers the opportunity to modify the VR driving simulation to a VR cycling simulation, Westdrive & LoopAR [19] was chosen. This VR implementation is based on the Unity 3D game engine. Therefore, a realistic physical behavior of all objects is ensured by the Unity3D physics engine. Due to the open-source implementation and the availability of a GUI, an individual, simple and fine-grained design of road user maneuvers is possible. Likewise, the data recording can be accurately adapted to the individual research questions. Lastly, the hardware requirements reflect the specification of a modern desktop computer.

However, Westdrive & LoopAR was originally developed for studies on automated driving from the passenger's perspective, specifically to investigate takeover requests [19]. To generate a naturalistic impression of a bicycle ride, the VR simulation was adapted to the cyclist's perspective, showing the moving bike, the handlebars and the cyclist's hands in the foreground. The VR cycling simulation can be displayed on three different monitors to provide the view to the front, right and left. In laboratory studies, these three monitors can be placed in front of a static bicycle on which participants can sit. When implemented as an online study, the VR cycling scenarios can be saved as video files with a frontal perspective.

After these adaptations, VR cycling scenarios can be implemented at relatively low cost and the setup seems to be suitable to investigate communication signals of AVs interacting with cyclists in a safe and replicable way. A disadvantage of this implementation is that the participants cannot control the behavior of the cyclist in the VR and thus, behavior of cyclists interacting with AVs, such as braking or avoiding the vehicle, cannot be studied directly.

3.2   Validation of the VR Cycling Simulation in Terms of Perceived Criticality as Well as Experience of Presence

Several validation studies were conducted using the VR cycling simulation, for example regarding the perceived criticality and experience of presence. It was aimed to investigate whether space-sharing conflicts between cyclists and vehicles with varying proximity are associated with the perceived criticality. Three typical scenarios were evaluated: (1) A vehicle exiting a parking lot and crossing the bike lane in front of the cyclist. (2) An intersection with a vehicle approaching from the left and crossing in front of the cyclist. (3) And a vehicle turning to the right and crossing the bike lane in front of the cyclist. The criticality within each space-sharing conflict was varied using the initially attempted post encroachment time (IAPT) [6]. The IAPT is defined as the time interval between one road user leaving a conflict point and another road user entering the same point, assuming no behavioral changes, such as speed changes, are initiated. Lower IAPT values are associated with a closer proximity between the two road users and thus with a higher potential for a critical outcome of the space-sharing conflict. In this validation study, the IAPT values ranged from one to three seconds for each scenario. Additionally, a baseline ride was performed for each scenario in which the crossing vehicle was absent. The perceived criticality was assessed using a scale developed by Stange et al. (2021) [26]. In addition, the experience of presence in the VR cycling simulation was evaluated using the Igroup Presence Questionnaire [24].

An online study was conducted with N = 35 participants. Each scenario with each IAPT level (including the baseline trial) was presented twice with a subsequent questionnaire on perceived criticality. At the end of the study, the experience of presence was assessed. The analysis of the perceived criticality revealed that the baseline rides were rated as significantly less critical compared to the rides with a space-sharing conflict (except for the turning scenario, which showed only a significant increase of the perceived criticality for IAPT = 1 s). Furthermore, the conditions with lower IAPT values were rated as more critical in each scenario. In addition, the results revealed that the turning scenario was perceived to be more critical compared to the intersection and parking scenarios. The analysis of the experience of presence indicated an acceptable experience of presence with a moderate score in the general presence dimension and a good score in the spatial presence dimension.

Based on these results, it is assumed that the VR cycling simulation is suitable to investigate cyclists' perceived criticality in interactions with AVs. Therefore, the simulation can support the development of safe and comfortable driving maneuvers of AVs in space-sharing conflicts. Further, an acceptable experience of presence in this VR cycling simulation can be assumed for online studies. It may be expected that the experience of presence will further increase when the cycling simulation is used in laboratory studies with a static bicycle in front of three monitors or using a VR headset.

Related Publication:

[31] Trommler, D., Bengler, P., Schmidt, H., Thirunavukkarasu, A., Krems, J.F.: Validation of a VR cycling simulation in terms of perceived criticality and experience of presence. In: Petzoldt, T., Gerike, R., Anke, J., Ringhand, M., Schröter, B. (eds.) Contributions to the 10th International Cycling Safety Conference, pp. 235–237, Dresden, Germany (2022). https://www.icsc2022.com/wp-content/uploads/icsc2022_book_of_abstracts.pdf.

4 Experimental Evaluation of a Drift-Diffusion Model for Vehicle Deceleration Detection

Vehicle deceleration can be used as an implicit communication signal to give priority to VRUs [34]. Results of the previous project KIVI showed that the detection performance of vehicle deceleration by VRUs may depend on various factors, such as deceleration rate, initial speed, age, and gender [1]. To provide a detailed understanding of the underlying differences in decision-making, these effects were further analyzed using a drift-diffusion model.

According to these models, perceptual decision-making is based on an accumulation of sensory evidence over time until a boundary is reached [22]. Several parameters are used to describe this process, which correspond to different components of the human information processing. The most important parameters are (1) drift rate, which describes the rate of evidence accumulation and is associated with the quality of evidence, (2) boundary height, which is related to the amount of evidence for a decision and reflected by the response caution in decision-making, (3) starting point, which can be positioned closer to the boundary in case of expectations towards a decision and 4) the non-decision time, which summarizes the time interval for stimulus encoding and motor response execution [22]. Using reaction times and response accuracies from empirical experiments, these parameters can be estimated after a model fitting [25].

This project task intended to investigate how deceleration rate and vehicle speed affect the parameters of a drift-diffusion model. A study was conducted with N = 62 participants which saw videos of approaching vehicles that either decelerated or not. These videos were recorded for the previous project KIVI. A detailed description of the video recordings can be found in [1]. The participants were instructed to press keys indicating whether the vehicles decelerated or not. In case of deceleration, the slowing down process was initiated immediately after the video onset. The deceleration rate (-1.5 and -3.5 m/s^2) and the vehicle speed (20 and 40 km/h) were varied as independent variables.

After the model fitting, the results showed substantial differences in the drift rate depending on the deceleration rate. This is consistent with the assumption that a higher stimulus quantity (i.e., higher deceleration rates) leads to faster evidence accumulation. Moreover, the boundary height as a measure of response caution varied slightly between the conditions with low and high vehicle speed. Higher values for the boundary height were observed for the conditions with higher vehicle speed. Additionally, there was a slight increase in non-decision time for the conditions with higher vehicle speed. This suggests that stimulus encoding needs slightly more time in the conditions with higher vehicle speed than in the conditions with lower vehicle speed. Moreover, a slight shift of the starting point towards the decision that the vehicle does not decelerate could be observed in conditions with higher vehicle speed. This finding suggests a decision bias.

In summary, a good model fit to the empirical data was achieved. The results showed that the contextual factors influenced the model parameters in a way that are in line with theoretical considerations. Further studies might investigate whether a complementary use of explicit communication signals, especially for slow decelerating and fast moving vehicles, leads to an improvement of the evidence accumulation process and thus to a higher satisfaction and perceived safety of VRUs in interaction with AVs.

Related Publication:

[29] Trommler, D., Ackermann, C., Krems, J.F.:A drift-diffusion model to explain vehicle deceleration detection of vulnerable road users. In: Stewart, T.C. (ed.) Proceedings of the 19th International Conference on Cognitive Modelling, pp. 289–294 (2021). https://acs.ist.psu.edu/papers/ICCM2021Proceedings.pdf.

5 Investigation of Factors Influencing the Gap Acceptance of Cyclists

When modelling human-like deceleration maneuvers for AVs, the deceleration rate as well as the time of a deceleration onset that VRUs expect for safe crossings need to be considered [4]. This expectation can be investigated through the VRUs' gap acceptance which defines the (time) gap that is acceptable for crossing in front of a vehicle [4]. However, previous studies on pedestrians' perspective show that the gap acceptance behavior may depend on external attributes (e.g., vehicle speed, vehicle size and time to arrival) as well as on internal attributes (e.g., gender and age of VRUs) [28]. Building on these findings for pedestrians, this project task aimed to examine the gap acceptance of cyclists. Therefore, the objective was to investigate the effects of vehicle size (car vs. truck), vehicle speed (20 vs. 40 km/h) and different levels of the time to arrival (TTA; ranging from one to five seconds).

The videos were generated in the presented VR cycling simulation with a length of approximately 10 s each. The videos were shown from the perspective of a cyclist riding towards an intersection while a vehicle is approaching from the left. Traffic signs and the study instructions indicated that the cyclist does not have priority. The TTA was measured as the time gap to the vehicle when the cyclist reaches the (theoretical) collision point at the intersection. N = 35 Participants were instructed to indicate by pressing a key, whether or not they would cross the road in front of the vehicle. As dependent variables, the crossing decision and the time of this decision before reaching the (theoretical) collision point were recorded.

The results revealed that more participants decided to cross in front of the vehicle as the TTA level increased. Further, for each TTA level, the willingness to cross was higher in conditions with faster vehicles than in conditions with slower vehicles. Within the majority of conditions, slightly more participants chose to cross in front of a truck compared to a car. Regarding the decision time, the results showed that the decision to cross or not is made approximately between two to four seconds before reaching the intersection. In conditions with faster vehicles, participants decided later (i.e., the cyclist was closer to the intersection) than in conditions with slower vehicles. In contrast, the decision was made earlier in conditions with lower TTA levels compared to conditions with higher TTA levels. Similarly, the decision was made earlier in conditions with trucks than in conditions with cars.

A further analysis focused on differences between the age of participants. For this, the sample was divided into two groups with 18 younger (18–35 years old) and 17 older (>35 years old) participants. The results indicated that participants' crossing decisions were similar, with the exception of the condition with 5 s TTA, where more younger participants than older ones expressed their intention to cross. Furthermore, older participants tended to make their crossing decisions later than younger participants.

To sum up, similar to the results for pedestrians, it is assumed that no universal parameterization is possible to design informal communication between cyclists and AVs. The study revealed that there are substantial differences in the gap acceptance of cyclists depending on vehicle size, vehicle speed and TTA. The findings also suggest the importance of considering age as a factor. Further, the results showed differences in decision time. The decision not to cross is made earlier than the decision to cross in front the vehicle. Therefore, AVs should use communication signals for giving priority early, especially when the TTA level is low and/or the AV is a truck. In future studies, additional factors, such as internal (e.g., age of the cyclist) and external attributes (e.g., time of day), need to be explored. Moreover, it could be relevant to investigate the decision-making process using drift-diffusion models as proposed in the previous section.

Related Publications:

- [32] Trommler, D., Springer-Teumer, S., Krems, J.F.: To ride or not to ride: exploring cyclists' gap acceptance in the interaction with (automated) vehicles. In: 34th

International Co-operation on Theories and Concepts in Traffic Safety (ICTCT) Conference, Györ, Hungary (2022).

- Springer-Teumer, S., Trommler, D., Krems, J.F.: How do vehicle size, speed, TTC, age and sex affect cyclists' gap acceptance when interacting with (automated) vehicles? In: 1st International Conference on Hybrid Societies, Chemnitz, Germany (2023).

6 Summary

This chapter gave an overview of the research project "KIRa", which investigated the cooperative design of the interaction between automated vehicles and cyclists according to four project aims.

First, the body posture was investigated as a predictor of the cyclists' starting process. The results showed that the progress of the starting process can be accurately detected based on body posture by human observers. This was even possible with a high accuracy when certain body parts (e.g., head or legs) were masked. Thus, it seems possible that an AV can recognize a cyclist's intention to start and can either avoid safety-critical situations with cyclists or can resolve them cooperatively at an early stage.

Second, the development of a VR cycling simulation was presented, including its validation in terms of perceived criticality and experience of presence. The findings revealed that the VR cycling simulation is suitable to investigate the cyclists' criticality perception when interacting with AVs. Different levels of proximity between a vehicle and a cyclist in three different shared-space conflicts reliably resulted in corresponding changes in the perceived criticality. This was investigated for different scenarios. Therefore, it is assumed that it is possible to investigate maneuvers of AVs interacting with cyclists in a standardized, reproducible and safe way.

Third, a drift-diffusion model for vehicle deceleration detection was empirically validated. The model parameters suggested the applicability of drift-diffusion models in applied research areas such as automated driving. This can lead to an improved understanding of the decision-making process of cyclists and further to the design of implicit and explicit communication signals adapted to humans' information processing abilities. This is expected to increase cyclists' acceptance and trust towards AVs.

And fourth, factors influencing cyclists' gap acceptance were investigated. The effects found for the gap acceptance of pedestrians could be confirmed, such as a strong dependence of the gap acceptance on the time gap, the vehicle size and vehicle speed. Furthermore, the same factors were associated with different decision time of cyclists (i.e., whether they would cross in front of the vehicle or not). These results highlight that cooperative interaction between AVs and cyclists is closely linked to context-sensitive communication.

7 Further Reading

In addition to the publications in this project, we will refer to the publications of the previous project "KIVI", which investigated the interaction between pedestrians and automated vehicles:

- Ackermann, C., Beggiato, M., Bluhm, L.F., Löw, A., Krems, J.F.: Deceleration parameters and their applicability as informal communication signal between pedestrians and automated vehicles **62**, 757–768. https://doi.org/10.1016/j.trf.2019.03.006. https://www.sciencedirect.com/science/article/pii/S1369847818306600.
- Ackermann, C., Beggiato, M., Schubert, S., Krems, J.F.: An experimental study to investigate design and assessment criteria: what is important for communication between pedestrians and automated vehicles? **75**, 272–282. https://doi.org/10.1016/j.apergo.2018.11.002. https://www.sciencedirect.com/science/article/pii/S0003687018306124.
- Beggiato, M., Witzlack, C., Springer, S., Krems, J.: The right moment for braking as informal communication signal between automated vehicles and pedestrians in crossing situations. In: Stanton, N.A. (ed.) Advances in Human Aspects of Transportation, Advances in Intelligent Systems and Computing, pp. 1072–1081. Springer International Publishing. https://doi.org/10.1007/978-3-319-60441-1_101.
- Beggiato, M., Witzlack, C., Krems, J.F.: Gap acceptance and time-to-arrival estimates as basis for informal communication between pedestrians and vehicles. In: Proceedings of the 9th International Conference on Automotive User Interfaces and Interactive Vehicular Applications, AutomotiveUI '17, pp. 50–57. Association for Computing Machinery. https://doi.org/10.1145/3122986.3122995.

Acknowledgements This project was funded within the Priority Program 1835 "Cooperative Interacting Automobiles" of the German Science Foundation DFG. The authors appreciate the fruitful collaboration with the project partners.

References

1. Ackermann, C., Beggiato, M., Bluhm, L.F., Löw, A., Krems, J.F.: Deceleration parameters and their applicability as informal communication signal between pedestrians and automated vehicles **62**, 757–768. https://doi.org/10.1016/j.trf.2019.03.006. https://www.sciencedirect.com/science/article/pii/S1369847818306600
2. Ackermann, C., Beggiato, M., Schubert, S., Krems, J.F.: An experimental study to investigate design and assessment criteria: what is important for communication between pedestrians and automated vehicles? **75**, 272–282. https://doi.org/10.1016/j.apergo.2018.11.002. https://www.sciencedirect.com/science/article/pii/S0003687018306124
3. Ackermann, C., Trommler, D., Krems, J.: Exploring cyclist-vehicle interaction - results from a naturalistic cycling study. In: Black, N.L., Neumann, W.P., Noy, I. (eds.) Proceedings of

the 21st Congress of the International Ergonomics Association (IEA 2021), Lecture Notes in Networks and Systems, pp. 533–540. Springer International Publishing

4. Beggiato, M., Witzlack, C., Springer, S., Krems, J.: The right moment for braking as informal communication signal between automated vehicles and pedestrians in crossing situations. In: Stanton, N.A. (ed.) Advances in Human Aspects of Transportation, Advances in Intelligent Systems and Computing, pp. 1072–1081. Springer International Publishing

5. Bengler, K., Dietmayer, K., Farber, B., Maurer, M., Stiller, C., Winner, H.: Three decades of driver assistance systems: review and future perspectives **6**(4), 6–22. https://doi.org/10.1109/MITS.2014.2336271. Conference Name: IEEE Intelligent Transportation Systems Magazine

6. Cunto, F.: Assessing safety performance of transportation systems using microscopic simulation. https://uwspace.uwaterloo.ca/handle/10012/4111. Accepted: 2008-11-06T15:43:28Z

7. Dey, D., Terken, J.: Pedestrian interaction with vehicles: roles of explicit and implicit communication. In: Proceedings of the 9th International Conference on Automotive User Interfaces and Interactive Vehicular Applications, AutomotiveUI '17, pp. 109–113. Association for Computing Machinery. https://doi.org/10.1145/3122986.3123009

8. Dosovitskiy, A., Ros, G., Codevilla, F., Lopez, A., Koltun, V.: CARLA: an open urban driving simulator. https://doi.org/10.48550/arXiv.1711.03938. http://arxiv.org/abs/1711.03938

9. Flemisch, F., Abbink, D., Itoh, M., Pacaux-Lemoine, M.P., Weßel, G.: Shared control is the sharp end of cooperation: towards a common framework of joint action, shared control and human machine cooperation **49**(19), 72–77. https://doi.org/10.1016/j.ifacol.2016.10.464. https://linkinghub.elsevier.com/retrieve/pii/S2405896316320547

10. Habibovic, A., Lundgren, V.M., Andersson, J., Klingegård, M., Lagström, T., Sirkka, A., Fagerlönn, J., Edgren, C., Fredriksson, R., Krupenia, S., Saluäär, D., Larsson, P.: Communicating intent of automated vehicles to pedestrians **9**. https://www.frontiersin.org/articles/10.3389/fpsyg.2018.01336

11. Hoc, J.M.: Towards a cognitive approach to human-machine cooperation in dynamic situations **54**(4), 509–540. https://doi.org/10.1006/ijhc.2000.0454. https://www.sciencedirect.com/science/article/pii/S1071581900904543

12. Kauffmann, N.: Objektivierung von kooperationsbereitschaft am beispiel eines spurwechsels im niedriggeschwindigkeitsbereich. ISBN: 9781668678602

13. Kyriakidis, M., de Winter, J.C.F., Stanton, N., Bellet, T., van Arem, B., Brookhuis, K., Martens, M.H., Bengler, K., Andersson, J., Merat, N., Reed, N., Flament, M., Hagenzieker, M., Happee, R.: A human factors perspective on automated driving **20**(3), 223–249. Publisher: Taylor & Francis. https://doi.org/10.1080/1463922X.2017.1293187

14. Lee, Y.M., Madigan, R., Giles, O., Garach-Morcillo, L., Markkula, G., Fox, C., Camara, F., Rothmueller, M., Vendelbo-Larsen, S.A., Rasmussen, P.H., Dietrich, A., Nathanael, D., Portouli, V., Schieben, A., Merat, N.: Road users rarely use explicit communication when interacting in today's traffic: implications for automated vehicles **23**(2), 367–380. https://doi.org/10.1007/s10111-020-00635-y

15. Lee, Y.M., Madigan, R., Markkula, G., Pekkanen, J., Merat, N., Avsar, H., Utesch, F., Schieben, A., Schießl, C., Dietrich, A., Boos, A., Boehm, M., Weber, F., Tango, F., Portouli, E.: interACT d. 6.1. methodologies for the evaluation and impact assessment of the interACT solutions

16. Markkula, G., Madigan, R., Nathanael, D., Portouli, E., Lee, Y.M., Dietrich, A., Billington, J., Schieben, A., Merat, N.: Defining interactions: a conceptual framework for understanding interactive behaviour in human and automated road traffic **21**(6), 728–752. Publisher: Taylor & Francis, https://doi.org/10.1080/1463922X.2020.1736686

17. Math, R., Mahr, A., Moniri, M.M., Müller, C.: OpenDS: A new open-source driving simulator for research **2**

18. Milakis, D., van Arem, B., van Wee, B.: Policy and society related implications of automated driving: a review of literature and directions for future research **21**(4), 324–348. Publisher: Taylor & Francis, https://doi.org/10.1080/15472450.2017.1291351

19. Nezami, F.N., Wächter, M.A., Maleki, N., Spaniol, P., Kühne, L.M., Haas, A., Pingel, J.M., Tiemann, L., Nienhaus, F., Keller, L., König, S.U., König, P., Pipa, G.: Westdrive x LoopAR:

An open-access virtual reality project in unity for evaluating user interaction methods during takeover requests **21**(5), 1879, https://doi.org/10.3390/s21051879. https://www.mdpi.com/1424-8220/21/5/1879. Number: 5 Publisher: Multidisciplinary Digital Publishing Institute

20. Rasouli, A., Kotseruba, I., Tsotsos, J.K.: Agreeing to cross: how drivers and pedestrians communicate. In: 2017 IEEE Intelligent Vehicles Symposium (IV), pp. 264–269. IEEE Press. https://doi.org/10.1109/IVS.2017.7995730

21. Rasouli, A., Kotseruba, I., Tsotsos, J.K.: Towards social autonomous vehicles: understanding pedestrian-driver interactions. In: 2018 21st International Conference on Intelligent Transportation Systems (ITSC), pp. 729–734. https://doi.org/10.1109/ITSC.2018.8569324. ISSN: 2153-0017

22. Ratcliff, R., McKoon, G.: The diffusion decision model: Theory and data for two-choice decision tasks **20**(4), 873–922. https://doi.org/10.1162/neco.2008.12-06-420. https://www.ncbi.nlm.nih.gov/pmc/articles/PMC2474742/

23. Schmidt, S., Färber, B.: Pedestrians at the kerb - recognising the action intentions of humans **12**(4), 300–310. https://doi.org/10.1016/j.trf.2009.02.003. https://www.sciencedirect.com/science/article/pii/S1369847809000102

24. Schubert, T., Friedmann, F., Regenbrecht, H.: The experience of presence: Factor analytic insights **10**(3), 266–281. https://doi.org/10.1162/105474601300343603

25. Shinn, M., Lam, N.H., Murray, J.D.: A flexible framework for simulating and fitting generalized drift-diffusion models **9**, e56938. https://doi.org/10.7554/eLife.56938. Publisher: eLife Sciences Publications, Ltd

26. Stange, V., Goralzik, A., Vollrath, M.: Keep your distance, automated vehicle! - configuration of automated driving behavior at an urban junction from a cyclist's perspective. In: Stanton, N. (ed.) Advances in Human Aspects of Transportation, Lecture Notes in Networks and Systems, pp. 393–402. Springer International Publishing

27. Tabone, W., de Winter, J., Ackermann, C., Bärgman, J., Baumann, M., Deb, S., Emmenegger, C., Habibovic, A., Hagenzieker, M., Hancock, P.A., Happee, R., Krems, J., Lee, J.D., Martens, M., Merat, N., Norman, D., Sheridan, T.B., Stanton, N.A.: Vulnerable road users and the coming wave of automated vehicles: expert perspectives **9**, 100293. https://doi.org/10.1016/j.trip.2020.100293. https://www.sciencedirect.com/science/article/pii/S2590198220302049

28. Tian, K., Markkula, G., Wei, C., Lee, Y.M., Madigan, R., Merat, N., Romano, R.: Explaining unsafe pedestrian road crossing behaviours using a psychophysics-based gap acceptance model https://eprints.whiterose.ac.uk/187357/. Publisher: Elsevier

29. Trommler, D., Ackermann, C., Krems, J.F.: A drift-diffusion model to explain vehicle deceleration detection of vulnerable road users. In: Stewart, T.C. (ed.) Proceedings of the 19th International Conference on Cognitive Modelling, pp. 289–294. https://acs.ist.psu.edu/papers/ICCM2021Proceedings.pdf

30. Trommler, D., Ackermann, C., Krems, J.F.: Investigating the body posture as a predictor for the starting progress of cyclists. In: Proceedings of the 33rd International Co-operation on Theories and Concepts in Traffic safety (ICTCT) conference

31. Trommler, D., Bengler, P., Schmidt, H., Thirunavukkarasu, A., Krems, J.F.: Validation of a vr cycling simulation in terms of perceived criticality and experience of presence. In: Petzoldt, T., Gerike, R., Anke, J., Ringhand, M., Schröter, B. (eds.) Contributions to the 10th International Cycling Safety Conference, pp. 235–237. https://www.icsc2022.com/wp-content/uploads/icsc2022_book_of_abstracts.pdf

32. Trommler, D., Springer-Teumer, S., Krems, J.F.: To ride or not to ride: Exploring cyclists' gap acceptance in the interaction with (automated) vehicles. In: Proceedings of the 34th International Co-operation on Theories and Concepts in Traffic Safety (ICTCT) Conference

33. Vollrath, M., Krems, J.F.: Verkehrspsychologie: Ein Lehrbuch für Psychologen, Ingenieure und Informatiker. Kohlhammer Verlag

34. Šucha, M.: Road users' strategies and communication: driver-pedestrian interaction. https://trid.trb.org/view/1327765

Prediction of Cyclists' Interaction-Aware Trajectory for Cooperative Automated Vehicles

Dominik Raeck, Timo Pech, Daniel Trommler, and Klaus Mößner

Abstract Cooperative behaviour is one of the most crucial factors for safety and comfort in shared traffic spaces. While a human driver might be able to automatically identify behavioural indicators of other traffic participants to predict their movement, an automated vehicle is not. This is especially important in interaction situations with vulnerable road users (VRU), such as cyclists. The focus of this work is to implement, evaluate and compare different possible methods of trajectory forecasts for cyclists in order to estimate their behavioural intention. With accurate trajectory information of the VRU, an automated vehicle might be able to plan a cooperative reaction ahead in time and guarantee a comfortable traffic flow. In sum, three different neural network architectures have been tested with the main focus on a CNN, which is capable of incorporating map data into the trajectory forecast. The results showed, that including external influencing factors, like the infrastructure of a traffic scene, can have a beneficial effect on the accuracy of the cyclist's predicted movement.

1 Introduction

The interaction and cooperation of road users is an integral part of urban traffic. In highly automated and connected traffic, a self-driving car must be able to identify and evaluate an interaction situation as well as perform suitable cooperative manoeuvres. This is especially important with vulnerable road users [1] to guarantee comfort and safety in space sharing conflicts (see previous book chapter by Trommler et al.). In this project called "Cooperative Interaction of Cyclists and Autonomous Vehicles" (KIRa) the behaviour of cyclists in mixed traffic was investigated. As described in the previous book chapter by Trommler et al., the focus of the research is low-speed areas, specifically urban intersection scenarios with at least one or more cyclists present

D. Raeck (✉) · T. Pech · D. Trommler · K. Mößner
Faculty of Electrical Engineering and Information Technology, Professorship of Communications Engineering, Technische Universität Chemnitz, Chemnitz, Germany
e-mail: dominik.raeck@etit.tu-chemnitz.de

T. Pech
e-mail: timo.pech@etit.tu-chemnitz.de

© The Author(s) 2024
C. Stiller et al. (eds.), *Cooperatively Interacting Vehicles*,
https://doi.org/10.1007/978-3-031-60494-2_2

19

and automated vehicles. To allow an assessment of such situations, the gathering and communication of the traffic participants' data is essential. This is especially challenging with VRUs, as technical solutions like ITS-G5 and ETSI standardized messages like cooperative awareness (CAM) [2] or manoeuvre coordination message (MCM) [3] can only be used between connected vehicles. Otherwise, the on-board sensors of an automated car could be used to detect and track VRUs, but this might not be possible at intersections, if the VRU is occluded by buildings or parking cars. Instead, a more decentralized approach was chosen: Mobile sensors on the bicycle, like GPS, accelerometers or gyroscopes can be used to gather position and movement data of a cyclists, by utilizing devices like smartphones that already broadly use these sensors. The requirement for such an approach is a connection between the road users to allow the exchange of the data.

The kinematic and position data of the traffic participants are hereby not sufficient to assess a shared space conflict properly. The behaviour of a cyclist is influenced by various factors. While the velocity and movement direction of the bicycle are decisive, the manoeuvres are also heavily affected by external conditions, for example the infrastructure or other road users. The main goal of this project was to infer and predict the cyclist's behaviour, considering and utilizing the different influencing factors. To achieve this, the identification of behavioural indicators is necessary as well as a suitable behavioural model. The main indicator for the bicycle's movement represents its trajectory, which estimates the future position and state of the bicycle. The proposed algorithm is capable to combine various types of data in order to accomplish an accurate trajectory forecast for cyclists. This information can later be used by automated vehicles, to allow an early and appropriate cooperative manoeuvre. That way a safe and comfortable traffic flow can be ensured for all road users.

2   Related Work

Trajectory forecasts for bicycles are not a very common research topic, however there is a lot of work regarding cars trajectories that were derived for this use case. Typically, there are two types of approaches: physical models and machine learning models, both with individual advantages and strengths. The research on physical approaches is usually focused around finding a suitable movement model for bicycles. As mentioned before, these can, for example, be derived from pre-existing car models [4]. While a physical model can yield high accuracies in very specific use cases, a machine learning model, like the neural network (NN) will often outperform it when evaluated in a wider array of scenarios [5]. The reason for that is the higher adaptability of a machine learning model, when it's trained with a large amount of naturalistic data. Another advantage of the NN is its capability to incorporate data of various types to allow for a more accurate trajectory forecast. This makes the approach the best choice for this project, as the aim is to combine potential influencing factors on the cyclist's behaviour. One of these influencing factors can be other road users in a given traffic scenario, especially at intersections. A proper trajectory forecast algorithm should therefore be interaction-aware and aimed to avoid conflicts, which is for example realised in the proposed algorithms of Huang [6] and Ju et al. [7]. In addition to potential interaction partners, the infrastructure of a traffic site can have a significant influence on the cyclist's behaviour. The movement of a cyclist can depend on different features of an intersection, like the lane markings, road borders and visibility of other road users [1]. This infrastructural data can be included in the prediction model, for example by converting it into rasterized maps [8].

In this project, we propose a neural network that combines and extends on the aforementioned approaches. The model can make use of different data types and sources in order to accurately predict a cyclist's future movement in interaction with AVs. For this, not only its kinematic data, but also data from all surrounding vehicles and area maps are utilized.

2.1 Datasets

Alongside the used training procedure, the quality, type and scope of the training data is crucial for the performance of the resulting model. For example, if the used position coordinates are unreliable, the accuracy of the model will be limited from the start and a meaningful evaluation will be hindered. For this reason, multiple publicly available trajectory datasets were analysed for their applicability in this project.

The criteria for a good training dataset depend on the use case of the desired model. In this work, cooperative traffic scenarios were investigated, as already outlined in Sect. 1. Given this area of application, the following characteristics are used for the selection of a suitable dataset:

1. The type of featured scenarios These should be comparable with the in Sect. 1 defined scenario, an intersection situation where multiple road users including at least one cyclist have to interact cooperatively with other vehicles. This limits the selection to recordings from urban areas. In natural data, the density of the traffic can vary a lot. A high density can make the analysis of individual interactions difficult, but offer a larger amount of potential training data.

2. The included road users While there is a multitude of recordings of cars in real traffic, the number of datasets which include the movement of the cyclists in the area is rather small. In addition to that, each type of road users must be explicitly distinguishable from each other, for example through a prior classification and a unique label.

3. The type of measurement data Since this work is focused on the trajectory of cyclists, the most important measurements are position coordinates combined with corresponding timestamps. Further interesting data are velocity, acceleration and driving direction (heading) of the vehicles. Image and video material can be helpful to improve the understanding of a traffic scenario and individual situations.

4. Frequency of the data The Frequency of the data should not be too low to ensure that even the movement of faster vehicles can be seamlessly described. Some traffic monitoring data is recorded with frequencies of 1 Hz or less and are therefore not suitable for this project. Datasets with a recording frequency of more than 2 Hz are sufficient for the analysis of low-speed scenarios in this use case.

5. Size of the dataset For the training of machine learning algorithms, especially neural networks, a large-scale and diverse dataset is required. The more complex the desired model, the more data is needed to effectively train the model. The minimum required number of training data is in most cases way smaller for a MLP than for a more complex neural networks like a LSTM or CNN, varying from a few hundred data points to several thousands. Measurement data with a high diversity will be required, if the trained model should be able to adapt to new scenarios. Alternatively, data augmentation methods can be used to enhance the model's transferability to new conditions.

Under consideration of these criteria, three public datasets were chosen: the "ApolloScape trajectories" dataset [9], the "Lankershim boulevard dataset" from the NGSIM project [10] and the "Intersection Drone Dataset" (InD) [11]. The included measurement data are best suited for the investigation of cyclists' trajectories. Nevertheless, compromises must be made when using a public dataset, which is not specifically designed for the own use case. For this reason, the mentioned datasets are further described below regarding their features, strengths and weaknesses.

The ApolloScape dataset was recorded in the city centre of Hong-Kong and includes recordings from multiple big intersections. That means it features mainly dense and complex traffic scenarios with a high amount of traffic participants. The data contains position coordinates of cars, pedestrians and cyclists, that are distinctively classified and labelled. All data were recorded with a frequency of 2 Hz, which is rather low, but sufficient in this case, since the average velocities of the road users are not very high. In conclusion, ApolloScape offers a large and accurate dataset. The downside is the complexity of the traffic scenarios, making it hardly possible to analyse individual interactions between specific traffic participants. Furthermore, a localization of the position coordinates in a map is not possible, because these are all relative to a local coordinate system and not a world coordinate system.

The second dataset is part of a collection of trajectory datasets from the NGSIM project. It contains recordings from multiple intersections along the Lankershim Boulevard in Los Angeles. The measurement data does not only consist of world coordinates from cars, trucks and cyclists, but also their velocity, acceleration and heading. The featured traffic scenarios are less dense than in ApolloScape, and allow for a better investigation of interaction situations between these road users. The biggest disadvantage of this dataset is the sparse number and therefore limited diversity of cyclist recordings.

The InD dataset was created with drones at various intersections in German cities. The high number of cyclists at these places create diverse measurement data, which are very suitable for the training of machine learning models. The observed intersections are smaller than the ones in the NGSIM dataset and not regulated by any traffic lights, which leads to very heterogeneous behaviour of the cyclists. One special feature of this dataset is, a drone image is provided for each measurement location. Later, the position coordinates of all road users can be mapped to these images, which is a decisive factor for the proposed algorithm (see Sect. 3).

The described datasets were used to develop a proof of concept for a trajectory forecasting algorithm in cooperative traffic scenarios. The large-scale ApolloScape dataset was used to design and validate a workflow for extracting, structuring and pre-processing data, as well as a first training algorithm. The NGSIM dataset could later be used to develop a method to incorporate maps into the training procedure. This approach will be further described in Sect. 3.2. The InD Dataset is best suited for the training and validation of complex neural networks because of its high diversity in cyclist-car interactions and its high amount of data.

3   Algorithm

3.1   Model Implementation

To calculate a cyclist's trajectory as an indicator for their behaviour, multiple algorithms were implemented. A trajectory forecast allows for a time discrete approximation of the future state of a bicycle through, e.g., its position and velocity. One goal of this project was to investigate different influencing parameters on VRUs' behaviours and to explore factors that can support the development of accurate predictions. A cyclist's movement in traffic is not only dependent on the kinematic of bicycles itself but can be affected, by for example, the given infrastructure or individual factors of each person. For that reason, NN have been utilized as they are capable of incorporating different types of data into the trajectory forecast. When choosing a suitable architecture for a NN, there are many possible model types, each with their own possibilities, strengths and weaknesses. In this project, three specific NN variations were implemented and compared for their applicability in trajectory forecasting. The concept of the used architectures, a multi-layer perceptron (MLP), a long-short-term-memory-NN (LSTM) and a convoluted neural network (CNN), will be described below.

1. MLP The first implemented NN was a MLP [12], as it is the most simplistic variant of the aforementioned. The input for the model is a 3s time series of the cyclist's position coordinates and the output are multiple 2D-coordinates of the cyclists' estimated future position in 0.5s intervals. This means the measurement data has to be converted and split into evenly long sequences as a preparation for the training. The focus of this first model implementation was to validate the data extraction, transformation and sequencing as well as the training procedure itself. The MLP itself consists of one input layer, three dense layers and one output layer, each connected with a ReLu activation function. The number of layers and the best suited activation function has been determined experimentally which later could be reused to design the fully connected parts of the more complex models. Despite its rather simple structure, the results showed that the MLP is already capable to predict a cyclist's trajectory, especially for straight movements.

2. LSTM The described MLP model calculated the cyclists' trajectory based on only its past position, but the behaviour of the cyclist is strongly dependent on the movement of all other road users in a scene. Therefore, the next step was to extend the model to include the cooperative aspect of a traffic scenario into the prediction to allow for an interaction-aware trajectory forecast. The chosen algorithm for this task is a LSTM based on the TrajNet++ framework [13], which was originally designed to predict movements of pedestrians in crowded areas with focus on the cooperation and interaction of the individual attendants. The pre-existing implementation was altered and adapted for the use on bicycle trajectories rather than pedestrians. The model includes a classification of potential interaction situations through given parameters like the distance between the road users, the movement direction or their field of view. The conditions for this classification were adjusted to fit the faster movement of cyclists and cars in this use case and the bigger possible distances between interaction partners. The input of the model can now not only include the cyclist's previous positions, but the movement of all road users within a specific area around the cyclist. The output of the model has been kept the same as in the MLP, a time series of estimated position coordinates for the bicycle. This ensures the comparability between the implemented models.

3. CNN After looking at the performance of the LSTM in the bicycle trajectory forecast on intersections, it is still noticeable that the model excels in predicting forward movement, but shows weaknesses when estimating tight and sudden cornering of the cyclist. Cyclists in those kinds of manoeuvres show a high amount of diversity in their actual movement, because of their ability for dynamic and fast steering. That means the greater possible range of motions for a cyclist when turning will make it harder for a model to predict the correct driving path. One way to increase the accuracy in those situations is to narrow the potential action radius of the bicycle. This is accomplished by providing infrastructural data to the model, like street markings and borders. In order to process this new kind of input data, a third NN type has been chosen. A CNN allows to extend the input for the trajectory forecast, by extracting infrastructural data from maps, but just as the LSTM it is also capable of incorporating the movement data of all road users in one given scene. How this is accomplished will be further described in the next section (see Sect. 3.2).

The main part of the CNN, the feature extractor, is based on MobileNetv2 [14], a well-established CNN with high accuracy and fast computation times. Transfer learning allows to adapt the pre-trained model for the new use case of this project, while keeping its already high precision. The fully connected part of the model is loosely based on the MLP that was described before, but experimentally optimized in this application. The output of the model is still kept the same, also to allow a comparison with the other implemented models.

Out of the three described model architectures, the CNN showed the highest potential in integrating different data types for a trajectory forecast with the highest possible accuracy. The input of the model could always be extended by additional parameters, as an indicator for the cyclist's behaviour. It will be part of future works to investigate such parameters and utilize them in the proposed algorithm.

3.2 Data Augmentation

When training the described CNN, the data pre-processing and augmentation is equally important as the model implementation itself. As mentioned earlier, the reason for using a CNN is its capability of processing images fast, in this case recordings of the infrastructure in the form of maps, satellite, or drone images. Further, the model should also be given the position coordinates and kinematic data of the cyclist and all surrounding vehicles. An algorithm is proposed to merge those different data types into one combined image, which can then be used as a training input for the CNN. This data processing will be described step by step.

The first step is to convert the cyclist's past position coordinates into local image coordinates (see Fig. 1). For consistency, the origin of the movement is always the centre of the image. The dimensions of the area that is shown in the image is a parameter of the model and was optimized for the given use case, here 100×100 m. The chosen image size must be compatible with the input layer of the CNN.

As second step the position coordinates of all surrounding vehicles in the scene were converted in the same way and added to the image (see Fig. 2). Distinct grey scale values can be used to separate individual vehicle types from each other. In this case the black line shows the cyclist and the grey line the cars.

Fig. 1 The cyclist's position coordinates (black line) added to a blank image. The global coordinates of the bicycle (UTM) were converted into a local coordinate system that is shown here

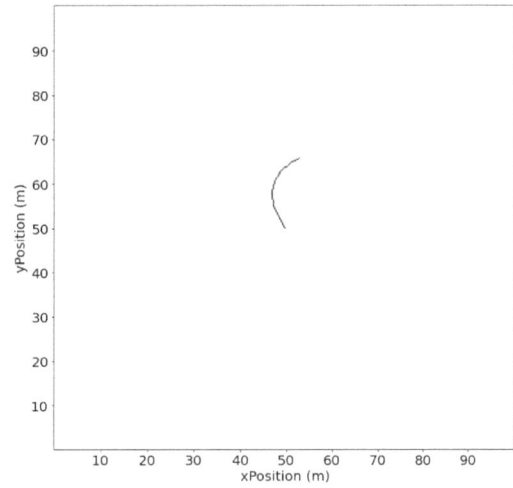

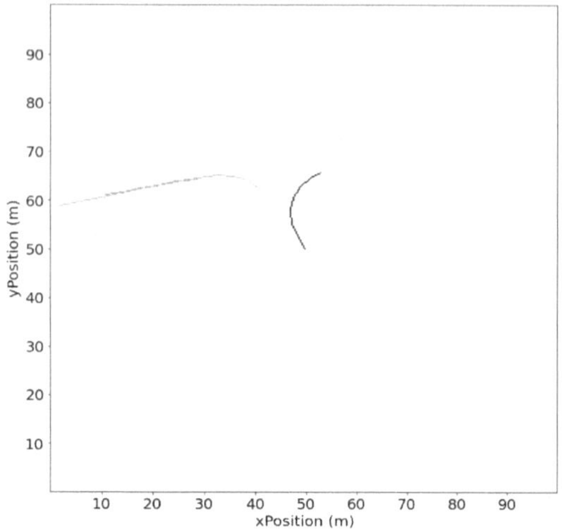

Fig. 2 All other vehicles' position coordinates in the scene are added to the same image (grey lines)

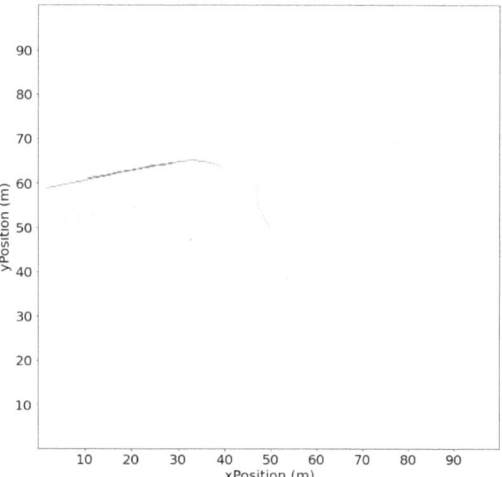

Fig. 3 Kinematic data of all vehicles and the cyclist is scaled to values between 0 and 255 (typical image color values). These values are then added to the corresponding points in the image. The cyclist is shown in blue and the car in purple

The last step of the data conversion is to include kinematic data in the image, in this case the velocity and heading value of each vehicle. For this, the given data is transformed to a scale between 0 and 255, or colour values for an image. The scaled values can then be added to one corresponding colour channel at each corresponding position of the vehicle. This means in practice that the trajectory of the vehicles is visualized in different colours depending on their velocity and heading (see Fig. 3).

Next up the infrastructural data is extracted from given images of the intersection. This can be done by utilizing satellite images or in this case drone images of the scene (see Fig. 4a). Since the color channels of the image are already used in the last step to encode kinematic data, the map is converted to grey scale. To further highlight important information, e.g., street borders, the image is processed with an

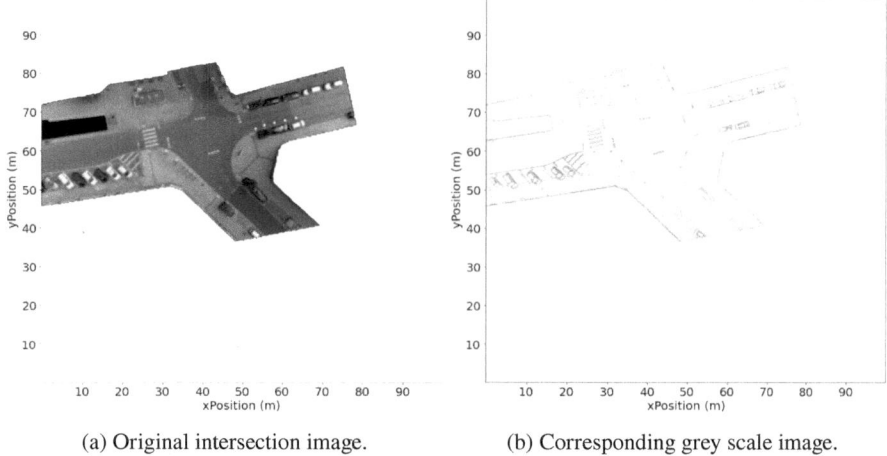

(a) Original intersection image.   (b) Corresponding grey scale image.

Fig. 4 Important features in the original image are highlighted using an edge detection algorithm

Fig. 5 Combined image
with the position and
kinematic data of the
vehicles and the
preprocessed map image of
the scene. The cyclist is
shown in blue and the car in
purple

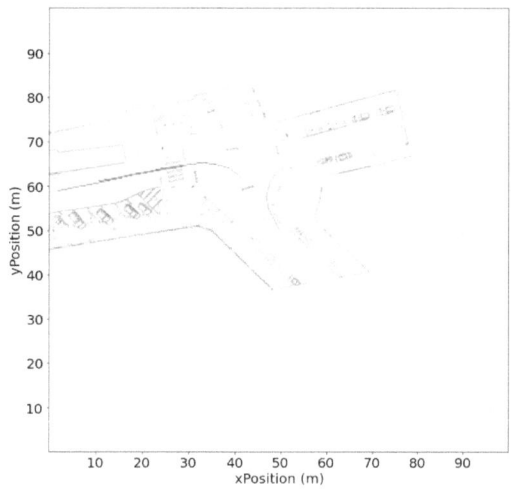

edge detection algorithm (see Fig. 4b). This new image is merged with the previously described converted vehicle data, by matching the position coordinates exactly to the cut out. Now, the generated image (see Fig. 5) contains all relevant data and can be used as input for the CNN. This algorithm is repeated for the next given time step until all measurement data are converted into images. This way a training set of more than 40,000 images could be created from the InD data, an amount that is more than sufficient to train deep neural networks.

When creating a training set with this many images, it is important that none of the possible scenarios is overrepresented, otherwise the model will over-fit on these scenes while not performing well in other conditions. In this case one specific intersection was featuring a lot more cyclists than any other location and therefore offered way more measurement data, which means that the NN could possibly be trained on this specific intersection and fail to adapt to other infrastructures. To avoid this problem, a random rotation is applied to each image that counteracts overfitting and makes the trained models usable with other scenes as well. The adaptability of the NNs is proven by utilizing a test dataset from an intersection that was completely excluded from the training.

4 Evaluation

To evaluate the prediction accuracy of the implemented models, a test dataset has been created. For that, 1,000 images have been used which were generated from a scene, that was completely excluded from the model's training. This way the model's adaptability to new scenarios and conditions can be investigated. The test dataset is the same for all used NN variations to ensure comparability between the calculated trajectory forecast precisions.

With the given test dataset, a statistical evaluation was conducted to investigate the three NN's prediction accuracy. In the field of trajectory forecasts, a specific error metric is commonly used to calculate the deviation of an estimated position from its ground truth. This error metric is called average displacement error (short ADE) and is widely used in literature to evaluate similar approaches [15]. The advantage of using an established error metric is, that it does not only allow a comparison between the proposed models but also with pre-existing ones.

$$
\begin{aligned}
dx &= \sum_{i=1}^{n} (\hat{x}_i - x_i)^2 \\
dy &= \sum_{i=1}^{n} (\hat{y}_i - y_i)^2 \\
ADE &= \sqrt{dx + dy/2n}
\end{aligned}
\tag{1}
$$

To determine the ADE, the deviation of a predicted point from the ground truth is calculated in lateral (dx) as well as longitudinal (dy) direction (see Eq. 1). Both individual values are used to evaluate a model's precision for specific manoeuvres of the bicycle. The lateral error indicates how well turning movements can be predicted and the longitudinal error shows the precision during straight driving. The combined displacement of the prediction is then calculated by the mean of these two values. These values were averaged for all predictions that correspond to each image in the test dataset in order to obtain the average error for every implemented model.

Table 1 ADE comparison of the four used models. A constant prediction time frame of 3 s has been used with all models

	KF (m)	MLP (m)	LSTM (m)	CNN (m)
ADE	1.78	1.9	1.24	0.84
dx	1.12	1.16	0.73	0.45
dy	2.44	2.64	1.75	1.23

4.1 Evaluation of Different Input Data

The main purpose of the evaluation process was to compare the influence of different data types on the trajectory prediction accuracy for the cyclist. For that, the three NN types were compared, which are used to incorporate additional data each. As mentioned in Sect. 3 algorithm the MLP uses only the cyclist's past positions for a trajectory prediction, the LSTM is capable of incorporating movements of potentially interacting vehicles, and the CNN additionally uses infrastructural data as input for the model. A Kalman filter was used to compare the proposed machine learning algorithms with a physical approach. Like the MLP, the Kalman filter uses only the cyclist's position as input and therefore both models offer a baseline accuracy for the trajectory predictions.

The table (see Table 1) shows the ADE of the three implemented NNs and the Kalman filter. A constant prediction time frame of 3 s was used for all models to make the results comparable.

The MLP revealed the highest average error of 1.9 m, followed closely by the Kalman filter with an error of 1.78 m. Both of these algorithms only use the cyclist's past position as input for the trajectory forecast, which means a lower accuracy is to be expected here. Despite the higher overall error, the lower longitudinal error indicates that a prediction of movements in driving direction is possible even with these comparatively simple models.

The LSTM column shows that the inclusion of interactions between the cyclist and vehicles in close proximity yields a significant accuracy increase. A LSTM is generally better suited for a time series forecast than the aforementioned models, which means the lower average error is caused by not only the interaction-aware prediction but also a more viable algorithm for this use case.

The best model in this comparison is the CNN with an ADE of 0.84 m. It includes all relevant vehicles' past positions and kinematic data as well as infrastructural data of the cyclist's surroundings. An accuracy increase of up to 42% can be seen in longitudinal and lateral direction compared to the LSTM. Especially the improvement in lateral movements is important, as it makes the CNN the best model to accurately predict cyclists' turning manoeuvres out of the compared approaches.

The influence of the infrastructural data on the trajectory forecasts was further investigated by implementing a variation of the CNN that is not using this data. All

Table 2 ADE comparison of two CNN variations. One with and one without the usage of map data

	CNN with map data (m)	CNN without map data (m)
ADE	0.84	1.08
dx	0.45	0.59
dy	1.23	1.57

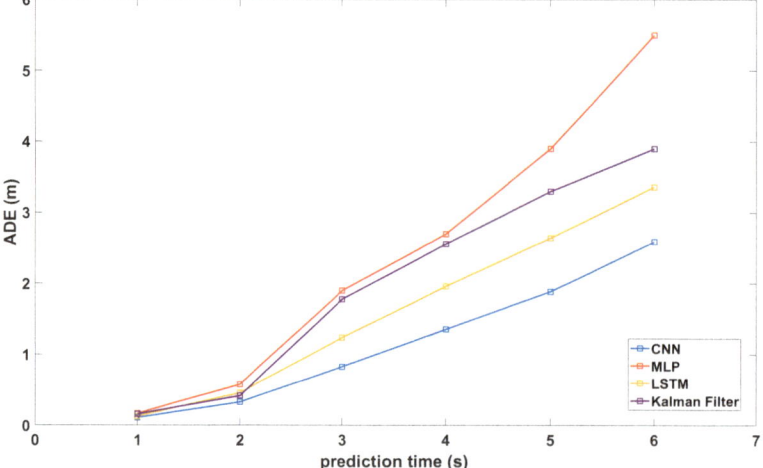

Fig. 6 Comparison of the four used trajectory forecast models with increasing prediction time horizons

other training and model parameters remained constant, which makes this version comparable to the aforementioned CNN (Table 2).

The comparison of the ADEs for both CNN variations shows a clear improvement in the forecast accuracy when using infrastructural data.

4.2   Evaluation of Prediction Time Horizons

The prediction time horizon is one of the most important parameters of any trajectory forecast model. It defines the time frame over which the prediction is supposed to happen. Naturally, the accuracy of a calculated trajectory decreases with a higher time horizon. The overall precision of a model determines how long the prediction time can be chosen before the accuracy is not sufficient anymore for the given use case. The previous results were generated using a prediction time horizon of 3 s. In the following, the influence of different time horizons on the ADE is investigated. This can be achieved by training a separate model with each of the respective prediction time frames from one to six seconds (Fig. 6).

The comparison shows that the ADE of each model increases with higher prediction times. The MLP showed the worst results at high prediction times to a point where the model is no longer trainable if the time horizon would be even higher. The best performing model is the CNN, which yields a better performance than the other models at all prediction times. All error values increase to a value where they are no longer reasonable in the use case of an intersection. The crucial difference is that the CNN could be used to predict trajectories of cyclists up to 4s length, while the Kalman Filter and the MLP are already too inaccurate with a prediction of 3 s.

In future work, possibilities of increasing the forecast time frame can be further investigated. The most important advantage of the used NN is the fact that additional types of data, e.g., behavioural or psychological parameters can be incorporated in order to allow for an even more precise and long-term trajectory prediction.

5 Discussion and Future Work

In both this book chapter and the previous one by Trommler et al., influencing factors on the behaviour of cyclists in interaction situations with cars were investigated. The analysis of naturalistic cycling studies in the book chapter by Trommler et al. showed that the cyclists' actions can be very dynamic and depend on a variety of conditions. Determining every deciding factor on a certain manoeuvre is not always possible. A neural network like the proposed model of this chapter is a well-suited method to predict such manoeuvres. Not only is this machine learning algorithm capable of utilizing input parameters of all sorts but it may also implicitly learn behaviour patterns of cyclists, by training it with a large amount of real traffic data.

The implementation and evaluation of the neural networks proved to be a viable approach for an accurate trajectory forecast of cyclists in cooperative interaction situations with cars. The dynamic and fast movement of a bicycle makes the cyclist's behaviour generally hard to anticipate. Providing information about the scenery and also the other vehicles helps to improve this estimation, by reducing the cyclist's possible range of action. Comparing the effect of additional input data showed that the cyclist's movement can be predicted better by utilizing influencing factors in its surrounding, like the infrastructure. This can be seen in the contrast between the proposed CNN's accuracy to the Kalman Filter (see Sect. 4). The advantage of a neural network in this use case is its capability to incorporate parameters that a physical model could not. In future work, this approach can be expanded to utilize even more behaviour indicators like individual factors of the cyclist.

The evaluation of the proposed CNN for cyclist trajectory prediction showed that its accuracy can vary depending on the type of movement. The amount of behaviour patterns that a machine learning algorithm can model is limited by the given scenarios that are included in the training dataset but also by the used input parameters and if they can serve as indicators for a certain manoeuvre. One example of a manoeuvre that cannot be predicted early is the starting of cyclist who was standing and waiting at an intersection. From the currently used parameters, the kinematic data of the bicycle and its past position, there is no indication for such a motion making it impossible to estimate when the cyclist is going to start. To allow for such a prediction, additional input parameters would need to be incorporated. Such potential indicators for the cyclist's starting behaviour were investigated in the previous book chapter by Trommler et al.. A body pose detection algorithm could be used to extract these features and use them as additional input for the proposed neural network. This would widen the algorithms potential field of use and will be subject to future research.

6   Conclusion

Neural networks can combine the autonomous vehicle's trajectory with infrastructural data to forecast a collision free trajectory for cyclists in interaction situations. The research in this project showed that a cyclist's behaviour in mixed traffic is dependent on many factors. A neural network, like the CNN that was discussed in this work proved to be a potent algorithm to incorporate data of various types, e.g., the kinematic data of the road users and map data, into a trajectory forecast. Combining these data types significantly increased the overall accuracy of the movement prediction. In future work this algorithm will be extended by including more potential influencing parameters on the VRUs behaviour.

Acknowledgements   This project was funded within the Priority Program 1835 "Cooperative Interacting Cars" of the German Science Foundation DFG.

References

1. Ackermann, C., Trommler, D., Krems, J.: Exploring cyclist-vehicle interaction - results from a naturalistic cycling study. In: Black, N.L., Neumann, W.P., Noy, I. (eds.) Proceedings of the 21st Congress of the International Ergonomics Association (IEA 2021), vol. 221, pp. 533–540. Springer International Publishing, Cham (2021)
2. Vehicular Communications; Basic Set of Applications; Part 2: Specification of Cooperative Awareness Basic Service. (n.d.). (302 637-2 V1.3.1 (2014))
3. Vehicular Communications; Informative report for the Maneuver Coordination Service. (n.d.). (ETSI TR 103 578 draft)
4. Polack, P., Altche, F., d'Andrea-Novel, B., La Fortelle, A. (n.d.). The kinematic bicycle model: a consistent model for planning feasible trajectories for autonomous vehicles? IEEE (2017)
5. Zernetsch, S., Kohnen, S., Goldhammer, M., Doll, K., Sick, B.: Trajectory prediction of cyclists using a physical model and an artificial neural network, pp. 833–838 (2016). https://doi.org/10.1109/IVS.2016.7535484
6. Huang, L.: A neural network based modeling and simulation of bicycle conflict avoidance behaviors at non-signalized intersections. IEEE (2014)
7. Ju, C., Wang, Z., Long, C., Zhang, X., Chang, D.E.: Interaction-aware Kalman neural networks for trajectory prediction. IEEE (2019)

8. Djuric, N., Radosavljevic, V., Cui, H., Nguyen, T., Chou, F.-C., Lin, T.-H., . . . Schneider, J.: Uncertainty-aware short-term motion prediction of traffic actors for autonomous driving. IEEE (2018)
9. Ma, Y., Zhu, X., Zhang, S., Yang, R., Wang, W., Manocha, D.: TrafficPredict: trajectory prediction for heterogeneous traffic-agents. Proc. AAAI Conf. Artif. Intell. **33**, 6120–6127 (2019)
10. Coifman, B., Li, L.: A critical evaluation of the Next Generation Simulation (NGSIM) vehicle trajectory dataset. Transp. Res. Part B: Methodol. **105**, 362–377 (2017)
11. Bock, J., Krajewski, R., Moers, T., Runde, S., Vater, L., Eckstein, L. (n.d.). The inD dataset: a drone dataset of naturalistic road user trajectories at German intersections. IEEE
12. Kawaguchi, K.: A multithreaded software model for backpropagation neural network applications. The University of Texas at El Paso (2000)
13. Kothari, P., Kreiss, S., Alahi, A.: Human trajectory forecasting in crowds: a deep learning perspective (2020)
14. Sandler, M., Howard, A., Zhu, M., Zhmoginov, A., Chen, L.-C.: MobileNetV2: inverted residuals and linear bottlenecks. IEEE (2018)
15. Twomey, J. M., Smith, A.E.: Validation and verification. Artificial Neural Networks for Civil Engineers: Fundamentals and Applications, pp. 44–64 (1997)

Detecting Intentions of Vulnerable Road Users Based on Collective Intelligence as a Basis for Automated Driving

Stefan Zernetsch, Viktor Kress, Maarten Bieshaar, Jan Schneegans, Günther Reitberger, Erich Fuchs, Bernhard Sick, and Konrad Doll

Abstract The project *Detecting Intentions of Vulnerable Road Users Based on Collective Intelligence as a Basis for Automated Driving* (DeCoInt^2) focuses on detecting the intentions of vulnerable road users (VRUs) in automated driving using cooperative technologies. Especially in urban areas, VRUs, e.g., pedestrians and cyclists, will continue to play an essential role in mixed traffic. For an accident-free and highly efficient traffic flow with automated vehicles, it is vital to perceive VRUs and their intentions and analyze them similarly to humans when driving and forecasting VRU trajectories. Doing this reliably and robustly with a multimodal sensor system (e.g., cameras, LiDARs, accelerometers, and gyroscopes in mobile devices) in real-time is a big challenge. We follow a holistic, cooperative approach to recognize humans' movements and forecast their trajectories. Heterogeneous open sets of agents, i.e., collaboratively interacting vehicles, infrastructure, and VRUs equipped with mobile

S. Zernetsch · V. Kress · K. Doll
Aschaffenburg University of Applied Sciences, Würzburger Straße 45, 63743 Aschaffenburg, Germany
e-mail: szernetsch@gmail.com

V. Kress
e-mail: viktor.kress@gmx.net

K. Doll
e-mail: kondrad.doll@th-ab.de

M. Bieshaar · J. Schneegans · B. Sick
University of Kassel, Wilhelmshöher Allee 71-73, 34121 Kassel, Germany
e-mail: mbieshaar@uni-kassel.de

J. Schneegans
e-mail: jschneegans@uni-kassel.de

B. Sick
e-mail: bsick@uni-kassel.de

G. Reitberger (✉) · E. Fuchs
University of Passau, Innstraße 41, 94032 Passau, Germany
e-mail: reitberger@forwiss.uni-passau.de

E. Fuchs
e-mail: fuchs@forwiss.uni-passau.de

© The Author(s) 2024
C. Stiller et al. (eds.), *Cooperatively Interacting Vehicles*,
https://doi.org/10.1007/978-3-031-60494-2_3

devices, exchange information to determine individual models of their surrounding environment, allowing an accurate and reliable forecast of VRU basic movements and trajectories. The collective intelligence of cooperating agents resolves occlusions, implausibilities, and inconsistencies. We developed new methods by considering and combining novel signal processing and modeling techniques with machine learning-based pattern recognition approaches. The cooperation between agents happens on several levels: the VRU perception level, the level of recognized trajectories, or the level of already detected intentions.

1 Introduction

Traffic is what moves us. Due to rapid developments in the past years in the fields of hardware, software, communication, and connectivity, it is in close reach that we do not even have to steer anymore. In special use cases, e.g., on motorways with reduced speeds, we are already there. Autonomous cars aim to combine additional comfort with exceeding efficiency when it comes to traffic jam avoidance, pathfinding, and car sharing. The most crucial aspect though is the potential to create safer traffic with fewer accidents and fatalities. This is the key issue that prevents the installation of autonomous cars right now.

The project *Detecting Intentions of Vulnerable Road Users Based on Collective Intelligence as a Basis for Automated Driving* (DeCoInt^2) aims at providing the benefits of cooperation in traffic with a focus on pedestrians and cyclists. The consortium consists of three partners who work together, contributing novel ideas and algorithms to tackle this crucial challenge in future traffic. Collaboration and cooperation are necessary for the automated domain as well. Single sources of data always lack information. The shift from forward-looking sensors to 360° perception systems is a first step to alleviate this issue, but still, a single sensor-equipped vehicle is not able to resolve occlusions or to sense behind corners. Nevertheless, this is crucial for a holistic understanding of the current situation and upcoming dangers to provide protection for VRUs and enable efficient and comfortable driving.

Autonomous cars shall not operate in isolation. Their implementation drastically influences every mobility aspect in the public space. Vulnerable road users (VRUs), i.e., micromobility users or pedestrians, are sharing parts of the same space as autonomous vehicles. Whereas cars have the capability to communicate and share information on a technical level, VRUs are not able to tell an autonomous car that they want to cross the street by establishing eye contact as humans would. Special care has to be taken to make automated traffic safe for all vulnerable traffic participants.

The vision of future traffic the DeCoInt^2 [8] project is based on is illustrated in Fig. 1. Equipped vehicles share information about their perception and predictions and thus extend their individual limits. Even VRUs themselves and static infrastructure can communicate and contribute in this local ad hoc information environment. These are the three core components in our project to perceive pedestrians and cyclists: the intersection infrastructure, the sensor-equipped vehicle, and sensor-

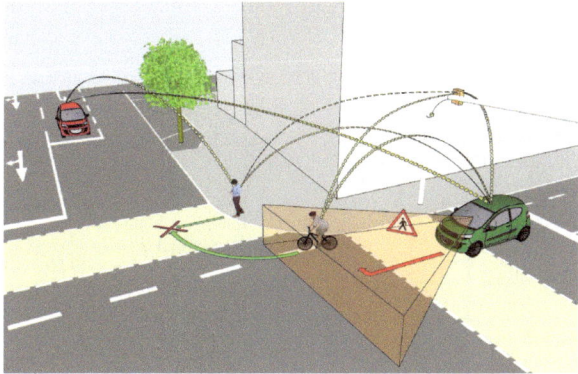

Fig. 1 Vision of a connected and cooperating world to provide safety for VRUs

equipped VRUs. Together we collect data to build a labeled ground truth database, apply existing approaches in real-world circumstances, and learn from the observed behavior by training novel models, thus pushing the state of the art of VRU protection systems in an automated and connected world. Figure 2 depicts the actual sensor setup. In the bottom, the static, wide-angle, synchronized stereo camera setup mounted at the research intersection [28] in Aschaffenburg, Germany, together with two sample images illustrating the fields of view, is shown. The area of the corner of the main road to a side road is the common field of view. Additionally, we collected data with a research vehicle [36] equipped with a LiDAR, a stereo camera, and an automotive dynamic motion analyzer (ADMA). The latter provides a self-localization ability. We created a local coordinate system with the ADMA of the research vehicle and the stereo camera setup of the intersection having the same origin. The third component are the VRUs themselves. We conducted measurement campaigns following specific scenarios involving the research vehicle and VRUs in the area of the research intersection. The VRUs are equipped with smart devices [7], as, for example, depicted in Fig. 2a. In the common field of view, labeled data can, together with the precise calibration of the stereo camera system, provide a positional ground truth. The VRU smart devices provide inertial and positional information about the VRU. Altogether, throughout the project, we collected short sequences with instructed and uninstructed VRUs capturing the movements listed in Table 1. Curated parts of the database are made publicly available [41, 44, 65, 66]. Extensive descriptions of the data collection and preparation, approaches, algorithms, and evaluations can be found in the Ph.D. theses [7, 36, 53, 64] that evolved from the project.

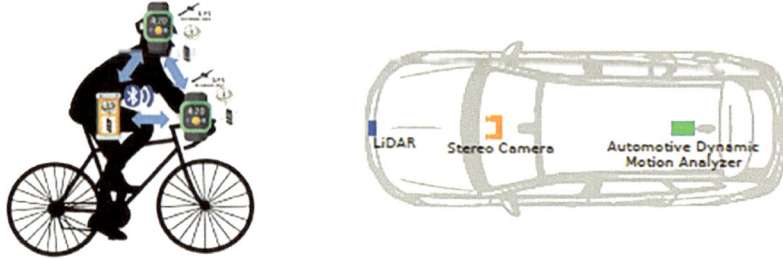

(a) Exemplary cyclist body area network of a   (b) Research vehicle with LiDAR, stereo camera,
smartphone, a smart watch, and a smart helmet   and Automotive Dynamic Motion Analyzer

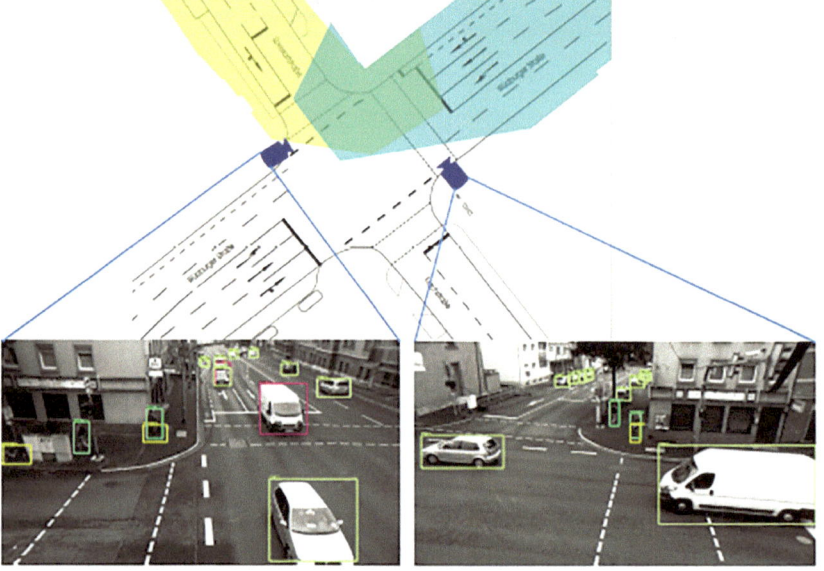

(c) Wide angle stereo camera setup at the research intersection. The individual fields of view are
marked blue and yellow, and the intersection of the fields of view is marked green.

Fig. 2 Cooperative perception and movement prediction sensor setup

1.1 Main Goals

The main focus of our approach, and consequently the DeCoInt[2] project, is the investigation of techniques for cooperative intention detection and trajectory forecasting of VRUs. Our overall goal is to detect the intentions of VRUs early and reliably using the collective intelligence of all road users. A schematic of this process is depicted in Fig. 3.

Due to the ability of VRUs to suddenly start a motion or to change the direction of motion, a dangerous situation may occur within hundreds of milliseconds. To avoid accidents, autonomous vehicles must be aware of their surroundings at all

Table 1 The gathered dataset with the number of scenes, persons, and motion primitives describing the possible motion states of VRUs

	Instructed	Uninstructed	Total
Scenes	976	672	1648
Test persons	89	≈672	761
Stop	344	189	533
Wait	348	358	706
Start	342	351	693
Straight	972	668	1640
Turn left	269	65	334
Turn right	271	130	401

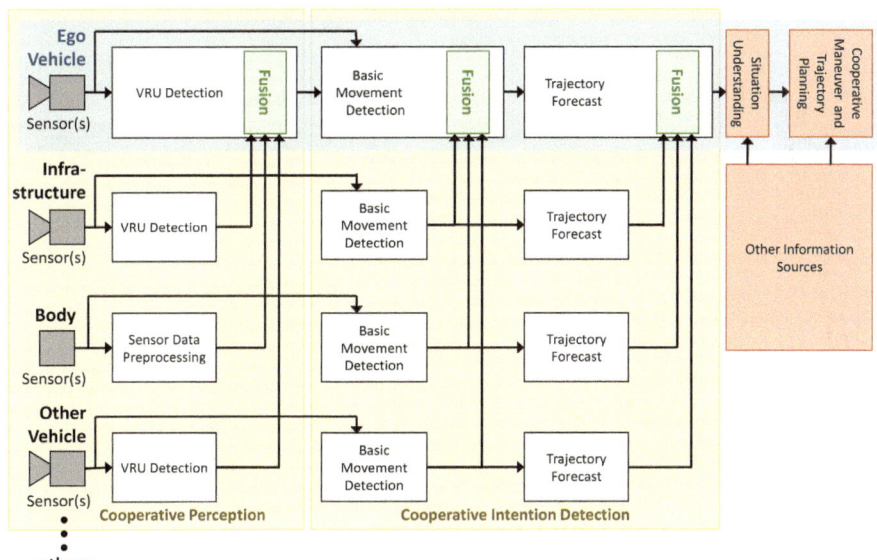

Fig. 3 Schematic representation of the overall process to cooperative perception and cooperative intention detection [8]

times. This includes not only the current but also the future positions of VRUs. Based on position forecasts, each autonomous vehicle can then plan a safe trajectory in mixed traffic. To achieve this goal, we aim to perform cooperative trajectory forecasting for VRUs. We generate forecasts over a short time horizon of 2.5 s, which is sufficient to perform emergency braking or evasive maneuvers [7, 64]. While we aim at forecasting trajectories with high positional accuracy, all predictions of VRU behavior are subject to error. This is especially true for larger forecast horizons since VRUs can change direction quickly without evidence of the behavior at the time at which the forecast is made. Therefore, we need to quantize the uncertainty of

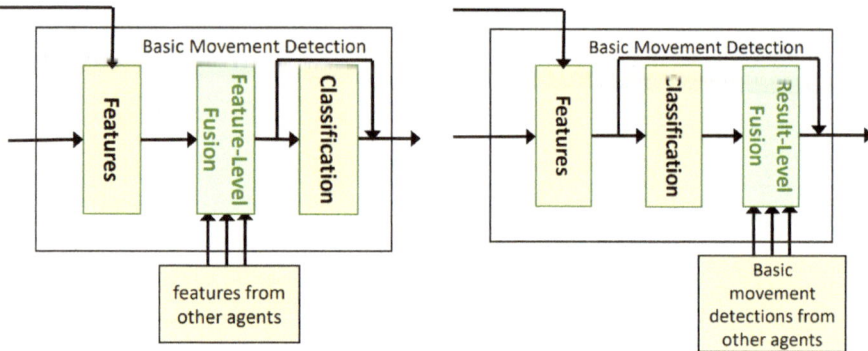

Fig. 4 Two fusion approaches for the basic movement detection [7]

our forecasts as well. This can be achieved by probabilistic trajectory forecasting, where, instead of one position for every forecast time step, regions with a certain probability are predicted. The main goal of probabilistic trajectory forecasting is to generate reliable estimates, i.e., if we estimate regions with a probability of 95%, the true position of the VRU should fall into that region in 95% of all forecasts. Another goal of probabilistic trajectory forecasting is to estimate regions that are as small as possible to allow efficient maneuver planning. Since a single sensor setup is prone to occlusion in dense traffic, our goal is to perform these forecasts cooperatively.

According to the Oxford Dictionary, cooperation is *the action of working together to the same end* [51]. Hence, the action is the process of combining information originating from different sources, i.e., vehicles, sensor-equipped infrastructure, and VRUs themselves, to increase the safety of VRUs. In the following, all involved entities are referred to as *agents*. In our work, we see the cooperative system from the perspective of an ego vehicle. All agents (including the ego vehicle) perform cyclist detection and intention detection locally. These agents exchange information via a wireless ad hoc network (i.e., V2X network). Ego-vehicle information (such as the position) and fused information of earlier stages are always available to the ego vehicle. For the sake of brevity, the corresponding arrows are not shown in Fig. 3. Perception incorporates the detection of cyclists, e.g., the detection of cyclists in camera images, RADAR, or LIDAR scans. Smart devices and other wearables detect the position using the integrated GNSS receivers, predict the VRU class, and perform intention detection (i.e., basic movement detection and trajectory forecasts) using their inertial sensors (cf. Sect. 4.2). Furthermore, we assume that the time between the agents is synchronized, e.g., via GPS time.

We conduct cooperative intention detection on the feature- and the decision-level. We can further subdivide this into the fusion of basic movements and trajectory forecasts. We depict a schematic showing the feature- and the decision-level fusion paradigms for the cyclist's basic movement detection in Fig. 4. We refer to the feature-level fusion paradigm as the fusion of sensor measurements and features from different sensors and sources. This combined information is used to detect the cyclist's intention better. In the decision-level fusion paradigm, predictions from basic movement and forecasting models of different road users are fused.

1.2 Outline

In this chapter, we describe how we solved the issues of detecting and tracking VRUs together with the follow-up process of intention detection as our main contribution. Our focus lies on the cooperative information gain based on the multimodal setting depicted in Fig. 2. In doing so, we first evaluate the weaknesses of uni-modal object detection and determine the strengths and chances of multiple different sensor sources via cooperative tracking in Sect. 2. Towards confidence estimations, we identify external factors influencing the tracking performance, i.e., context information. VRU tracks form the essential input to the intention detection process in Sect. 3, which we divide into basic movement detection and trajectory forecasting. We perform both steps separately on the three different device types we have available, i.e., the stationary cameras, the moving vehicle, and the smart devices. This is beneficial, as we point out throughout the work, due to different constraints and possibilities. Moreover, we identify additional input data derived from the sensors, such as optical flow images and poses. The basic movements form an additional input for trajectory forecasting, the essential part of our work. At first, we predict the movement of pedestrians and cyclists in a deterministic manner again separately for our three sources. Then, we show approaches of how to estimate, predict, and evaluate the confidence in the forecasts made on stationary cameras and a moving vehicle in our work on probabilistic trajectory forecasts. At this point, we emphasize the contribution of our work to the trajectory planning task of autonomous vehicles. Object detection, together with tracking and probabilistic trajectory forecast, have direct use in the planning of efficient and safe vehicle paths. In Sect. 4, we showcase the benefits of cooperatively using our three information sources in the intention detection process as we already have for the tracking stage. As shown in Fig. 2, the VRUs themselves equipped with multiple smart devices form an information source. Therefore, we show how the different devices and wearing positions contribute to specific information gains, and we can combine them beneficially. In the next step, we examine different methods for cooperative intention detection, including feature- and decision-level fusion for basic movements and trajectories. In this context, we present innovative solutions for a great variety of problems, such as delay, sensor outage and occlusion, out-of-sequence fusion, and information loops. Moreover, we investigate and compare different approaches and elaborate on the possibilities of implementing such a system utilizing current V2X protocols and standards, such as collective perception messages (CPM) and collective awareness messages (CAM).

2 Cooperative Perception and Tracking

The first step on the road to intention detection and prediction of VRUs is to detect pedestrians and cyclists and identify them throughout the scene, i.e., perform tracking. This has to be done in a precise and accurate way as it is the basis for all the following steps. Accuracy addresses the ability to ensure that a detected VRU is existing and indeed of the claimed type. If a VRU is not detected although it is in the field of view of the sensors and not occluded, the accuracy is reduced. The precision measures the distance of a detected VRU to the ground truth real-world object. Our approach is to achieve reliable and precise detection and tracking in a multimodal and multi-agent setting, i.e., in a cooperative way. We make use of a stereo camera setup at a static road site unit, a stereo camera mounted in a vehicle, and the VRUs themselves equipped with smart devices to provide superior coverage and more precise solutions than one single sensor could provide. Additionally, we will show the impact of context information on the performance of our system representative for all state of the art detection algorithms and the ability to gather context information more extensively and accurately in cooperation.

2.1 Context Dependent Detection

In this section, we explore the performance of state of the art object detection methods. On the way to fault-free and therefore safe autonomous driving, such perception techniques should be reliable in any case. For the viewing angle and lighting situation, two exemplary cases of context knowledge, we will discover significant differences to the regular performance.

2.1.1 Viewing Angle Dependent Bicycle Detection

Neural networks (NNs), in the field of image processing especially convolutional neural networks (CNNs), define the state of the art techniques for detecting objects, segmenting images, and many more tasks. Although setting the bars higher for algorithms in this domain and outperforming all previous approaches assuming the testing data to be at least similar to the training data, there is still room for improvement. In this section, we want to highlight some flaws of state of the art detection algorithms based on our data.

Figure 2c shows the viewing angles of our static stereo camera setup. There is a bike lane next to the sidewalk on the main street. The lane is directed towards the left camera. The right camera has an orthogonal view of the lane. An example of a cyclist riding on the described bike lane can be seen in Fig. 5. The figure also shows the detection boxes with classes and confidentialities created by a Faster R-CNN [56] network based on a ResNet-101 [29] backbone trained on the COCO dataset [46],

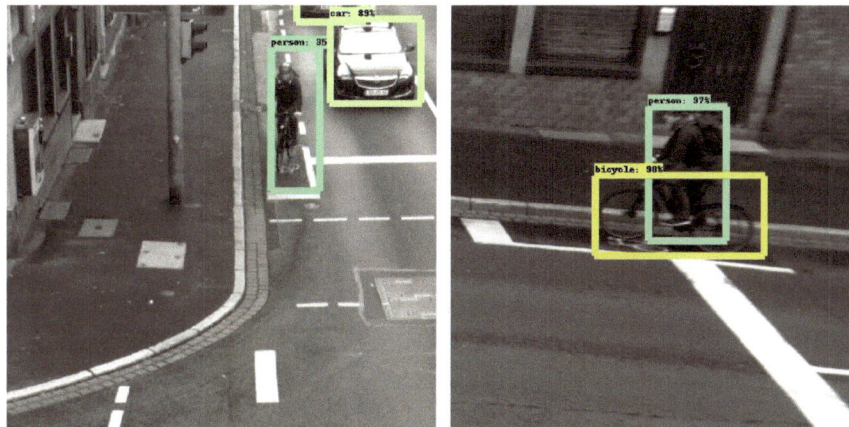

Fig. 5 Bicycle, person, and car detections from the left and right camera view

Table 2 Viewing angle dependent bicycle detection rate evaluated on 51 scenarios

View on bicycle	Average detection rate
Right, orthogonal camera	0.8993
Left, straight ahead camera	0.2476

which had a state of the art performance at the time of the evaluation considering a close to real-time execution speed. A cyclist is not labeled as a separate class. A cyclist is detected by a person bounding box and a bicycle bounding box which have an intersection over union (IoU) above a determined threshold. We experienced an IoU of 0.3 as sufficient. The detection of cyclists is a central part of our project as we want to increase the cyclists' safety by predicting their behavior. In particular, reliable bicycle detection is crucial. Figure 5 shows the bicycle detected in the right, orthogonal camera view, but not in the left, straight ahead camera view. This is indeed no exception in evaluating the bicycle detection rate with respect to the two mentioned camera angles. We evaluated 51 scenarios of cyclists riding on the bicycle lane or on the pavement next to it in the direction indicated in Fig. 5. The scenarios cover most of the possibilities cyclists can be visible in the two cameras.

Table 2 lists the detection rates with respect to all frames of all 51 scenarios. The detection rate in the right camera is 0.8993. It might be a little lower than the expected detection rate of an object detection task, but the fact that there are people sitting on the bicycles in every image and the, in some parts, low contrast with respect to the background makes the task more difficult than in the trained dataset. Moreover, the weather conditions are challenging in some scenarios. We will elaborate on that fact in more detail in the following. Nevertheless, the detection rate of 0.2476 in the left camera is significantly worse than the one on the right. The detection rate is smaller for the left camera angle in 50 of the 51 scenarios. The cause can be manifold. It might be an underrepresentation of such images in the training dataset or simply a more challenging task due to the fact that less of the bicycle is visible and more of the bicycle is hidden by the rider. Whatever the case, the consequence is that we are not able to reliably detect a cyclist in such a case if only the straight-ahead camera is available.

We, therefore, propose in our approach that at least a second camera angle is necessary to be able to track cyclists without gaps. This will be discussed in more detail in Sect. 2.3. Moreover, we created a novel algorithm to subtract the background from the foreground in static camera setups [55] that is able to identify moving objects. Regions of movements without detected objects indicate missed detections.

2.1.2   Lighting Situation Dependent Person Detection

The differences in the detection performance with respect to the viewing angle give a hint that is important to take additional information into account when we estimate our trust in the output of our perception system. We call this additional information context. It is an explicit goal of our project to investigate the influence of context information on the detection and prediction capabilities of our algorithms. Therefore, we split our database with respect to another criterion, the time of day. In Sect. 2.4, we take a look at further types of weather and lighting condition context types. Nevertheless, the main focus in the process of the creation of our dataset was to provide a basis for the development of algorithms for VRU intention detection and the corresponding evaluation. The different times of day and viewing angles are a side product of the data-capturing process. They are represented in significant amounts of scenarios to be able to make deductions. We also discovered rain or sun glare but in too few scenarios to provide statistically significant results. In the current projects of the three participating teams, we are continuing to work on extending the existing database specifically with respect to such further aspects to evaluate and address model weaknesses. To avoid the viewing angle as an influencing factor, we choose the person detection in the two camera views for the 51 scenarios already mentioned. The mean detection performance is similar for both cameras with a recall of 0.825 in the right camera view and 0.780 in the left one. There are 11 scenarios captured in the evening or during a thunderstorm which resulted in less daylight and darker images. We refer to such scenarios as *dawn_dusk* in contrast to *daytime* which is

Table 3 Lighting situation dependent person detection evaluated on 51 scenarios

Camera	Lighting situation	Average detection rate
Right	Daytime	0.8885
Right	Dawn_dusk	0.6818
Left	Daytime	0.8019
Left	Dawn_dusk	0.6910

the regular case. Table 3 shows the detection rates. The detection rate is lower for the 11 dawn_dusk scenarios for both cameras. In the left camera, the difference is about 0.11 and in the right one even 0.21. The sample size is still small such that with more data and a network training based on more dawn_dusk images the difference might not be so big anymore. Nevertheless, we have found a motivation to further work on the detection and determination of context information and to include context information in the data collection process and the assessment of detection confidentialities.

2.2 Cooperative Detection and Tracking of Cyclists

Cooperation is an integral component on the way to a comprehensive and reliable detection of VRUs in an automated traffic environment. To showcase the ability of an equipped multi-agent system to overcome the limitations of single ego-vehicles for example, we gathered data in multiple measurement campaigns in real-world environments based on the setup described in Sect. 1.

In a reduced setting, we showcase the benefit of cooperation in [54]. We concentrate on the tracking of cyclists, i.e., we assumed that the objects we want to track are cyclists. The relevant device the cyclists carry is a smartphone in the trouser pocket. Followup works described in Sect. 4.2 elaborate more on additional devices mounted at different wearing positions. The output of the VRU sensors is an estimation of the velocity, the yaw rate, and the GPS-based position. Additionally, we detect and determine the 3D positions of the cyclists with the static stereo camera setup mounted at the research intersection. The intersection sensor setting provides a positional accuracy of less than 10 cm in every direction. The accuracy of the intersection detections is superior to the ones provided by smart devices. Therefore, if both sources are available, the smart devices do not contribute to better tracking performance. Nevertheless, in cases of occlusions, e.g., a truck blocking the view of a cyclist from one camera, no 3D object detection can be performed by the stereo camera system anymore. An example of such a scenario can be seen in Fig. 6. Smart devices are

(a) Left camera view (b) Right camera view

Fig. 6 Cyclist occluded by a truck in the right camera of the static stereo camera system

Fig. 7 Comparison of static
stereo camera tracking only
(red triangles) with
cooperative tracking
including smart device data
(green circles) under a
temporal occlusion of 2 s.
The blue squares show the
ground truth trajectory

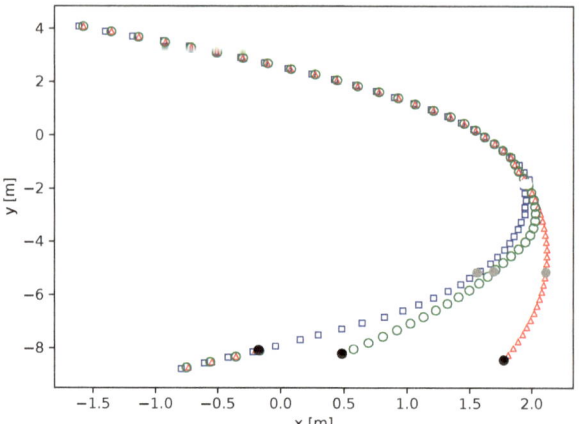

still able to communicate their measurements. In [54], we show that in more than
84% of the turning scenes under occlusion, the additional smart device information
benefits the tracking performance significantly. Figure 7 depicts a scene in which the
cooperative tracking follows the ground truth closely, whereas the tracking based on
the static camera setup only loses sight and can only predict the following positions
based on the tracking model.

The chosen model is the bicycle model tracked with an extended Kalman filter
(EKF) [3, Chap. 10]. The state transition for the state space $\mathbf{x} := [x, y, \gamma, \dot{\gamma}, v]^T$
with the positional coordinates x and y, the orientation γ, its derivation $\dot{\gamma}$, i.e., the
yaw rate, and the velocity v is given by

$$f(\mathbf{x}) := \begin{bmatrix} x + \cos(\gamma)\,a - \sin(\gamma)\,b \\ y + \sin(\gamma)\,a + \cos(\gamma)\,b \\ \gamma + \dot{\gamma}\,T \\ \dot{\gamma} \\ v \end{bmatrix} \tag{1}$$

with $a = \frac{\sin(\dot{\gamma}\,T)\,v}{\dot{\gamma}}$ and $b = \frac{(1-\cos(\dot{\gamma}\,T))\,v}{\dot{\gamma}}$ for a time step T. The z coordinate can be
determined by the stereo camera setup besides x and y, but it is not used in the
referred evaluation. The smart devices contribute with the velocity v and the yaw
rate $\dot{\gamma}$. The occlusion in the case referred to in Fig. 7 starts at the white-filled circle.
Starting at this point, the trajectories drift apart. Due to the yaw rate information by
the smart devices, the green cooperative track can follow the ground truth closely.
The grey and black-filled circles depict one and two seconds after the start of the
occlusion.

2.3 Pedestrian and Cyclist Tracking Including Class Probabilities

So far, we have mentioned the perception of cyclists that contains an intersecting detection of a bicycle and a person and the tracking of cyclists themselves. The movement model used for the cyclist tracking in Sect. 2.2 describes movements by arcs and therefore is especially suitable for cyclists but unstable in cases of sudden, not smooth, or even backward-oriented changes in the movement direction, which is in the nature of pedestrian trajectories. Therefore, a linear constant velocity or acceleration model that is independent in the lateral and longitudinal directions is more suitable for a pedestrian. We have already mentioned that being able to track VRUs is a basis for further steps in our VRU intention detection approach. Besides the choice of the tracking algorithm, a tracker tends to perform best if it is applied to an object of the class it has been designed or trained for. In our case, we have the cyclist model with the state space $[x, y, \gamma, \dot{\gamma}, v]$ and the pedestrian model with the state space $[x, \dot{x}, y, \dot{y}]$. In the following, we also develop models to predict the behavior of cyclists and pedestrians. They depend on the knowledge of the class of the VRU, too. Therefore, we want to extend the aforementioned state spaces and the tracking described in Sect. 2.2 by an additional class probability functionality. There are two information sources for the class probability. The fit of the respective model to the movement behavior observed and the object class predictions by the NN classifier. The former is a problem that is studied in the literature in the field of multiple model approaches. The idea is to have a set of possible models, and each of them is fed by the measurements, i.e., the detected object positions. In every step, it is evaluated with a probability score of how well each model fits the perception. We intend to implement the individual model tracking with Kalman filters following Sect. 2.2. Therefore, the bicycle model is implemented via an EKF and the pedestrian model via a two-dimensional constant velocity Kalman filter. The interacting multiple model (IMM) [15, 24] approach is popular, especially together with Kalman filters. It shows a robust behavior with respect to model mismatching [49]. In addition to the individual model states, the IMM holds a common mixed state and covariance estimate that form the state of the IMM model. We name the state estimates at a given point of time for the bicycle and the pedestrian model x_b and x_p, respectively. Every model is assigned a model probability μ_b and μ_p. The IMM state estimate is given by

$$x_{\text{IMM}} := x_b \mu_b + x_p \mu_p. \tag{2}$$

The covariance P_{IMM} is deduced analogously. To perform the prediction step of the IMM, mixed states $\hat{x}_j := x_b \mu^{b|j} + x_p \mu^{p|j}$ are calculated for every model j with $\mu^{i|j} := \frac{1}{\psi^j} \rho^{i,j} \mu_i$ being the conditional model probabilities for model i assuming j, ψ^j being a normalization factor, and $\rho^{i,j}$ being the respective entry in the state switching matrix ρ. The state switching matrix adds to the stability of the IMM. Initially, the probabilities of staying in a state or switching states are initialized with 0.5. With the growing age of the track, the probabilities of staying in a state iteratively grow. At every prediction step, the mixed model states $\hat{x}_b$ and $\hat{x}_p$ are propagated together with the covariances as new states to the individual models. The prediction step is performed based on the propagated states in the way defined by the individual models to gain $\tilde{x}_b$ and $\tilde{x}_p$. The update step is performed on the individual models given the incoming measurements, i.e., person or cyclist detections. The residuals r_b and r_p given by the differences of the measurements to the predicted model states define the model likelihoods λ_b and λ_p. The likelihoods are the log of the probability density function of the zero averaged normal distribution with the covariance given by the innovation covariance matrices of the Kalman filters of model b and p. The likelihoods are used to update the model probabilities by

$$[\mu_b, \mu_p] = \frac{c \cdot [\lambda_b, \lambda_p]}{c[\lambda_b, \lambda_p]^T} \quad \text{with} \quad c := [\mu_b, \mu_p]\rho$$

and '·' denoting a point-wise product. The IMM state x_{IMM} can be calculated again following formula (2). One adaptation has to be made with respect to the standard IMM algorithm described so far. The state spaces of the bicycle and the person model differ. The state space of the IMM is the union of the individual state spaces, thus $[x, \dot{x}, y, \dot{y}, \gamma, \dot{\gamma}, v]$. To make the IMM state and covariance compatible with the individual model ones in cases of propagation and update, the individual model states have to be lifted to the IMM model state space in such cases following [63].

The standalone IMM tracker is able to classify pedestrians in 38 scenarios with a precision of 0.914 by its inherent model probabilities. Nevertheless, if a cyclist is waiting at traffic lights for example, the bicycle model is unstable and does not fit the behavior very well due to small rapid movements in the process of impatient waiting for example. In the regular movements detected in our scenarios, cyclists did not follow the bicycle model enough for the IMM to classify it. The average precision in 46 scenarios with moving cyclists is 0.335. In comparison to a pedestrian, a cyclist is still classifiable on average, as a true cyclist track holds from a frame-wise perspective more cyclist classifications than a pedestrian track. Still, for a standalone classification, one would expect more from a classifier. The reason might be the tracking of the head of the cyclist that we perform. Nevertheless, by taking into account the detected class labels as well, the classification can be improved. The relative amount of assigned bike detections measured by IoU with a person detection as described in Sect. 2.1.1 with respect to the age of the track provides a sufficient feature. The classification precision is 0.970 for the pedestrian scenarios and 0.969 for the cyclist ones.

The IMM tracker extends our setup by the functionality to track two classes of VRUs simultaneously without having to decide at the level of the object detector output which measurement is assigned to what kind of tracker. In the case of the viewing angle dependent bicycle detection described in Sect. 2.1.1, the cyclist detections are unreliable. Therefore, a standalone cyclist tracker receives only a few cyclist measurements. Table 4 depicts the tracking performance averaged over the 51 cyclist scenarios already evaluated in Sect. 2.1.1 comparing the bicycle model taking only cyclist detections into account with the IMM approach based on both pedestrian and

Table 4 Comparison of cyclist tracking based on a bicycle model taking only cyclist detections into account with IMM tracking taking pedestrian and cyclist detections into account

Tracking algorithm	Average MOTP	Average MOTA
Bicycle model	0.155	0.325
IMM	0.076	0.976

cyclist detections but classifying a cyclist. The performance measures MOTP and MOTA are standard tracking metrics [4] measuring the precision and accuracy of the given track, respectively. Due to the small MOTA score, one can induce that the bicycle model is far less capable of tracking the object at all. This results from the missing detections. Because of the lower MOTP value, the IMM does not only cover the object better but is additionally capable of giving a more precise estimate of the location due to the mixed-in pedestrian component.

2.4 Cooperative Context Determination

We have already mentioned the relevance of context in the field of object detection. Moreover, we have shown that cooperation in detection and tracking can overcome the limitations of singular sensor sources and extend the tracking ability. The sources of context information can be various such as its types. In Sect. 2.1, the context information is based on external ground truth information that is able to be manually determined as the data is relatively small and the scenery is fixed. This is not possible in general. Therefore, in this section, we want to take a look at how we can extend the generation of context information and gather it in a cooperative way to aim for more reliability and better coverage.

2.4.1 Cooperative Semantic Maps

A straightforward idea that comes to mind when thinking about how to extend the available information with some extra knowledge is to use maps, more so maps that are enriched with additional annotations. We call this semantic maps. Especially in the field of prediction and motion planning, maps can help to avoid invalid impossible paths. This will be discussed in the following in more detail, especially in Sect. 3.2.2. But also the viewing angle dependent object detection evaluation Sect. 2.1.1 shows that knowledge of bike lanes with respect to the camera mounting positions and orientations contributes to a more accurate assessment of the expected detection performance.

We use a local map provided by the open street map (OSM) [48] organization to gather static map information. The amount of information and the accuracy varies depending on the contributions to the map pool by the community. In our case houses and streets are contained, cf. Fig. 8a. Considering additional annotations, high-definition maps hold precise and excessive information. Nevertheless, it is expensive to capture HD maps and thus they are only available in certain areas. To extend our initial maps, we included information about sidewalks (yellow) and parking slots (brown) in the map visible in Fig. 8b. Whereas the bare OSM map can be retrieved automatically, the extensions are made with human interaction. Still, it is expected that maps like Fig. 8b are available in close future and even already are in many locations. Another way to gain maps is to use sensor information. Using LiDAR point clouds and image-based segmentation provided by a stereo camera both mounted on our research vehicle, map Fig. 8c was created. There is more detail to it in Sect. 3.2.2. It is created and fused in multiple capturing drives. The output has the advantage that it can be captured automatically and can contain all the information the segmentation classifies. The disadvantage lies in a limited field of view and the dependency on the accuracy of the ego-positioning ability of the vehicle and the classification together

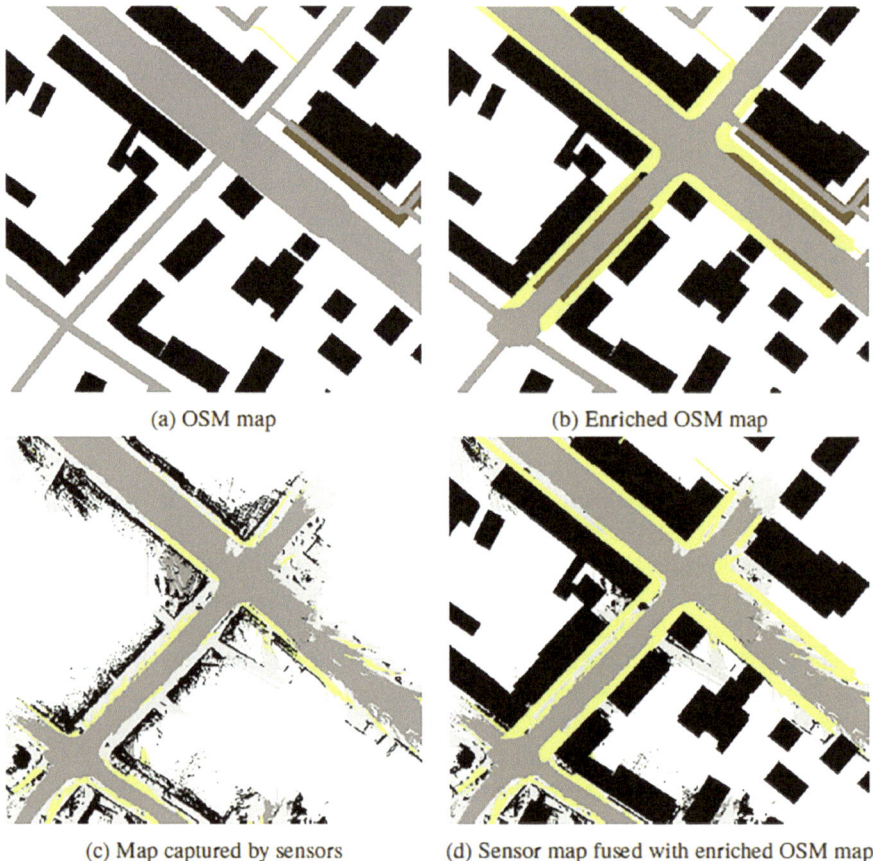

(a) OSM map   (b) Enriched OSM map

(c) Map captured by sensors   (d) Sensor map fused with enriched OSM map

Fig. 8 Semantic map fused from sensor information with enriched OSM maps

with the association to the LiDAR point cloud. The latter can have serious errors due to perspective issues. Moreover, it is difficult to create smooth and convex solutions. For example, in Fig. 8c, the holes at the edges of the pavement are visible. Therefore, it is beneficial to fuse the two information sources. Figure 8d shows the result. The houses are complete and the edges of the pavement are sharper. The benefit of such a semantic map for movement prediction is shown in Sect. 3.2.2 and the fusion benefits the accuracy in a straightforward way.

2.4.2   Cooperative Weather, Road, and Lighting Conditions

In Sect. 2.1.2 we show the effect the brightness of the daylight has on the detection rate of objects. We extend this context information to weather and conditions. The assumption is that not only the task of perception but also the behavior of traffic participants are affected, for example, by heavy rain or icy roads. We train our models to detect objects and predict their behavior under the assumption that they act the same way we have seen during training under similar conditions. It is crucial that these conditions have been met in the training phase. Otherwise, unpredictable behavior is the consequence. To support the description of conditions, context information might be useful. In this section, we describe what kind of context information we thought of being interesting and how we cooperatively detect it. As already mentioned, it was not possible to conduct enough field studies to evaluate in a statistically significant way the influence of the specific context types on the performance of our algorithms. This is one topic of the current project KI Data Tooling [33] the partners of this project also contribute to.

Table 5 lists the types of context and conditions we considered with the expected, i.e., labeled, values. Not all of them are contained in the dataset described in Sect. 1. Moreover, the list is not comprehensive and extended in the context of KI Data Tooling. As already mentioned in the evaluation in Sect. 2.1.2, the times of day 'daytime' and 'dawn_dusk' can be found in the research intersection dataset. To be able to detect the context types automatically, we trained a model for every type based on a ResNet-50V2-architecture [30]. The challenging parts are to build a good training dataset and to conduct consistent labeling. Without further knowledge, for example, it is not easy for a human spectator to detect, e.g., rain in images. Nevertheless, we labeled 28563 single images manually by ourselves. The images originate

Table 5   Types of context with the sets of possible values

Context type	Values
Precipitation	Rain, snow, nothing
Road	Dry, wet, slushy, snowy, indistinguishable
Time_of_day	Daytime, dawn_dusk, night, indistinguishable
Illumination	Natural_standard, sunglare, artificial, dark

Table 6 Number of true context detections on images of 50 scenes from the research vehicle, the left and right intersection cameras, and a fused result

Data source	Illumination	Time_of_day	Precipitation	Road
Left camera	29	39	36	41
Right camera	48	16	25	–
Vehicle	13	25	41	36
Fused	34	26	38	41

from our own dataset, the University of Passau Weather in Autonomous Robotic Driving (UPWARD) dataset containing 15566 samples and from the DENSE SeeingThroughFog dataset [13] providing another 12997 samples. We train and validate on 23028 samples (12630 UPWARD, 10398 DENSE) and test on 5535 images (2936 UPWARD, 2599 DENSE). To address the heavy class imbalance, we apply undersampling. This necessitates independent training of a separate model for each attribute.

Although the context classes are not all contained in the 50 scenarios used in Sect. 2.1—one scenario does not contain vehicle data and is therefore removed—we want to show the performance of the context detection models on them, because due to the setup we can evaluate the benefit of cooperation. We use the two cameras mounted at the research intersection and the camera mounted in the research vehicle as sensors. The mounting angle of the right intersection camera is such that the reflections of the street give, in any case, 'wet' due to the road context. This is due to the fact that the training data is gathered from lower-mounted cameras. The right camera does not contribute to a fused result as well.

The evaluation results are shown in Table 6 by the number of correctly classified scenarios. To reduce the labeling effort, one ground truth label was created for every scenario. This might not be very accurate in case of illumination for example as sun glare can be limited to a short time span and the rest of the scenario is not affected. This is also the reason why the vehicle performs much worse than the static cameras in the illumination context. The left camera is mounted in a way that allows a good detection performance with respect to the time_of_day. Precipitation is detected best by the vehicle.

Overall, we discovered two major takeaways throughout the process of cooperative context determination. Firstly, it is difficult to determine consistent labels for the specified classes and to detect them properly as a human spectator. The granularity of the labeling is also a factor that has to be covered in more detail. Secondly, the fused result does not always give the top result but does in almost every case exceed the vehicle's performance capabilities. For a single car equipped with a camera, it is not possible, at least at the state of our training data, to detect the defined classes of context with an acceptable rate. Additional information sources are necessary.

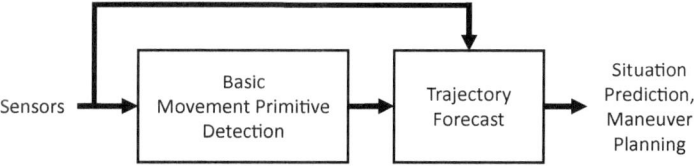

Fig. 9 Schematic of the two-stage cyclist intention detection and trajectory forecasting model [5]

3 Intention Detection

In this section, we describe our work in the field of VRU intention detection. The goal of intention detection is to create a basis for maneuver planning algorithms to be able to interact with VRUs safely. Therefore, we have to make a forecast about the VRUs' future trajectory, including uncertainties. The main focus of our project is on cooperative intention detection. In the first step, we investigate methods for intention detection in a non-cooperative way using different sensor modalities and analyze their strengths and weaknesses. These investigations are described in this section. We then select suitable methods for use in a cooperative manner and explore how much improvement can be achieved through cooperation. The investigations regarding cooperative intention detection are described in the next section.

We define intention detection as a two-stage process comprised of basic movement detection and trajectory forecasting. A schematic of the process is depicted in Fig. 9.

The first stage is basic movement detection to identify the VRUs' current state of motion, e.g., *waiting* or *starting*. As the results from basic movement detection alone do not allow to make a statement about the future VRU positions, the state estimations are used as an intermediate result within the intention detection process. Our goal is to demonstrate that the state estimations can help to improve the trajectory forecast results. Furthermore, we show that basic movement detection results can be significantly improved by incorporating video and pose information into the process. Additionally, we investigate using data from smart devices worn by VRUs as a basis for basic movement detection. Our methods for basic movement detection are described in Sect. 3.1.

The second stage of the intention detection process is trajectory forecast, which generates estimates of future VRU positions. The forecasted trajectories are the output of the intention detection process and form the basis for maneuver planning in automated vehicles. One of our goals is to include video information and basic movement detections in the forecasting process. Secondly, we aim to generate probabilistic trajectory forecasts to quantify the uncertainties of our estimates. To demonstrate the applicability of our methods, we combine our probabilistic forecasts with a maneuver planning method. Trajectory forecasting is described in Sect. 3.2.

3.1 Basic Movement Detection

Basic movement detection of VRUs has become an active field of research over the past decade. While many existing methods focus on specific scenarios or movement states often placed in a lab environment with ideal conditions [32, 52], we aim to demonstrate heuristic approaches covering all possible scenarios and states. We investigate using different sensor sources, i.e., stationary cameras mounted at an intersection, a stereo camera from within a moving vehicle, and smart devices worn by the VRUs themselves (Sect. 1). Furthermore, we examine different representations of the VRU sensor data in the form of trajectories, human poses, or video sequences. In this section, we describe different methods for basic movement detection. In Sect. 3.2, we discuss methods to incorporate basic movement detections into the forecast process.

Basic Movement Detection Using Stationary Cameras

The use of stationary cameras for intention detection leads to multiple advantages. Compared to sensors in a moving vehicle, stationary cameras can be mounted at higher positions and at an angle to each other to resolve occlusions and to reduce uncertainties of single sensors. Stationary cameras also have the advantage that the environment is known and the background is static. Furthermore, since stationary systems are less limited to space and power consumption requirements compared to systems inside vehicles, more powerful systems with regard to their computing capabilities can be used. We use these advantages by incorporating video information into our basic movement detection.

Many existing methods use a single past VRU trajectory as input data for basic movement detection [1, 27]. However, compared to the original video feed from the sensors, a lot of information about the VRU behavior is lost, e.g., movements of the upper body may signalize a starting motion, or the VRU's gaze direction can indicate a turning motion.

An approach to preserve information about the VRU's body gestures uses motion history images (MHI) [35]. To generate the MHI, the binarized silhouette of the VRU is extracted from every image. The silhouettes from the current image and past images of a certain observation period are then stacked into a single image, where the most recent silhouette has the value 1.0, and older silhouettes receive smaller values between 1.0 and 0.0 with regards to their timestamp (Fig. 10). This creates an image that encodes the past movements of the VRU, which can now be used with a simple image classifier to perform basic movement detection. However, the method strongly depends on the quality of extracted silhouettes. Also, a lot of information is lost through the binarization of the images.

To increase the level of information, more recent approaches utilize human pose trajectories for basic movement detection. Instead of using a single trajectory from an anchor point, such as the center of the VRU's head, multiple trajectories of the

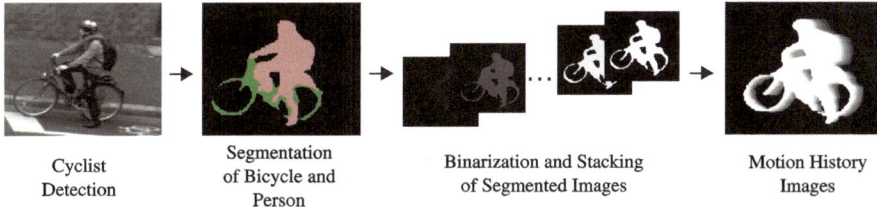

Cyclist
Detection

Segmentation
of Bicycle and
Person

Binarization and Stacking
of Segmented Images

Motion History
Images

Fig. 10 Exemplary MHI generation of a starting cyclist

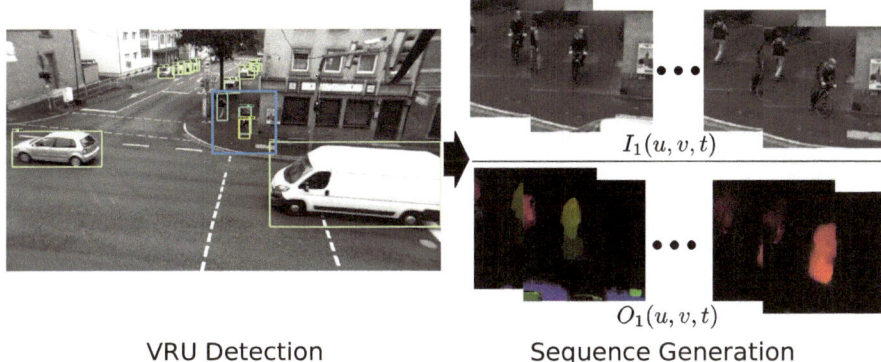

$I_1(u, v, t)$

$O_1(u, v, t)$

VRU Detection   **Sequence Generation**

Fig. 11 Extraction of a video sequence from the original video feed from camera 1. In the VRU detection step, every VRU is detected, and a region of interest is created (left). In the second step, images (right, top) and optical (right, bottom) are stacked to sequences that are used as input for our model

VRU's joints are used. This way, important features such as distinct body poses or leg movements are preserved while greatly reducing the feature size compared to the original video stream. One disadvantage of the method is that it depends strongly on the quality of the pose detection. While larger joints can be detected relatively reliably, detecting smaller features, such as the eyes, which can be used to extract the gaze direction, proves difficult. Furthermore, information about the surroundings, such as road markings or obstacles in the VRU's way, is lost.

Therefore, in our approach to basic movement detection with stationary cameras, we directly utilize video sequences for basic movement detection. Figure 11 describes the extraction of video sequences from the original video feed. In the first step, the VRU is detected in the current image. The detection window is used to create a region of interest that covers the near vicinity of the VRU, which is used to extract images from the current time step and past time steps within the observation period. In our case, the past observation horizon covers one second. The extracted images are then stacked into a short video of the VRU moving inside the region of interest, which is used as input for a three-dimensional convolutional neural network (3D-CNN). In a preliminary investigation, where we focused on detecting starting motions of a cyclist, we used these image sequences as the only input for the network [10].

However, more recent studies in the field of action recognition aside from intention detection in road traffic have shown that the use of an optical flow sequence in addition to the image sequence leads to significant improvements with regard to detection accuracy [18]. Therefore, we additionally use the optical flow sequence for our investigations. Furthermore, to reduce negative effects caused by occlusions, our movement detection is performed using inputs from both cameras of our wide-angle stereo-camera system described in Sect. 1. We investigate the use of single cameras individually, both cameras simultaneously, and the use of only image sequences or optical flow sequences, respectively. Since the past VRU trajectory is known, we also examine its additional use as input data. For our investigations of stationary systems, we use the dataset created with the wide-angle stereo-camera system described in Sect. 1. Our methods are compared to a single trajectory approach, as well as an MHI-based approach. The feature extraction from the video sequences is performed using the proposed network architecture from [18]. To evaluate the results of individual time steps, we use standard metrics used in classification. To evaluate the detection results over time, we use the segment-based approach proposed in [7, 22], allowing us to rate the detection method in terms of how often a motion state is wrongfully detected over time. We see this as an important metric since wrong detections during a motion state can lead to a false trigger of an emergency brake assistant of an automated vehicle. A detailed description of the used algorithms and the conducted experiments can be found in [71].

Our experiments regarding the user input data show that the best results are achieved by using all inputs, i.e., image and optical flow sequences from both cameras and the past trajectory. However, only slightly worse results are achieved if we omit the trajectory input. If we compare the use of input data from both cameras to only one camera, we can see a significant improvement by adding the second camera. This is partly due to the resolvement of occlusion, but we also found that some motion states are better detected using a certain camera angle. For example, starting motions are better detected when the VRU is viewed from the side. We compared our motion sequence (MS) based method to the MHI and trajectory-based methods and found that our approach outperforms both in terms of frame-based classification scores and segment scores. The inference time of the algorithm using an NVIDIA RTX 2080 Ti GPU is about 33 ms and can therefore be used in a real-time system. The detailed results of our experiments can be found in [64, 71]. While our results show that our method outperforms existing approaches, we cannot make a statement about whether or not the improvements transfer to the use in trajectory forecasting methods. This aspect will be discussed in Sect. 3.2, where we investigate the use of basic movement detections to improve trajectory forecasts.

Basic Movement Detection from a Moving Vehicle

When we compare stationary intention detection to intention detection from within a moving vehicle, the requirements for the algorithms change significantly. Since the sensors are usually mounted behind the windshield of the car or behind the radiator grill, VRUs are often occluded by other vehicles or objects at the roadside. The consequence is that we often have a significantly shorter observation period to estimate future VRU behavior. Compared to stationary cameras, we do not know the surroundings of the cameras, and we have to deal with changing backgrounds. Furthermore, the vehicle cannot accommodate large PCs, and the power consumption is limited. Due to these requirements, we investigate the use of human pose trajectories for intention detection from within a moving vehicle and compare them to single trajectory approaches. The sparsity of the representation allows us to design lightweight models that allow for real-time capability despite limited resources. At the same time, we maintain a high level of information about the VRU behavior by capturing the trajectories of the body extremities.

In the first step, we evaluate the quality of human pose estimation from within a moving vehicle. While some datasets regarding 2D pose estimation exist, e.g., [2], they are not designed for research in traffic environments. The amount of data with annotated 3D poses are quite limited. Typically, they are created in lab environments, e.g., [31], and do not include any cyclists. As a consequence, the recorded scenarios lack realism with regard to the behavior of the recorded people. Furthermore, there are too many dissimilarities compared to real-world traffic scenarios, such as the surroundings and occlusions of the VRUs. Therefore, we created a dataset recorded in real traffic. The human poses are labeled manually, and we extracted 2D and 3D poses. For the generation of reasonably good ground truth for the 3D poses, we use our wide-angle stereo-camera system at the research intersection. Using this dataset, we evaluate two methods for human pose estimation. The first method detects 2D poses in an image. The second method uses 3D lifting to estimate 3D poses, which we transfer to the world coordinate system using a stereo camera. Our investigations show that both methods perform well and can be used as a basis for vehicle-based intention detection. The detailed results can be found in [38].

Based on these results, we conduct experiments regarding the applicability of human pose trajectories for vehicle-based basic movement detection. In a preliminary investigation, we limit our traffic scenario to starting cyclists. An example scene is visualized in Fig. 12. The goal is to detect starting motions as early as possible while maintaining high detection scores. The method is compared to a single trajectory approach. The focus of the evaluation is on comparing the two approaches using different observation periods. As mentioned above, from the perspective of a moving vehicle, VRUs are often occluded, highlighting the importance of a method that functions well for small observation times. In our experiments, we evaluate observation periods between 0.12 and 1.0 s. Both methods use the same model architectures, i.e., a fully connected network (FCN). Only the inputs differ, where the input of the single trajectory model is the past head trajectory, and the pose-based model receives all joint trajectories. We find that both models show similar performance for input periods of 1.0 s. The results of the single trajectory model strongly deteriorate with smaller observation periods, while the pose-based model maintained significantly higher scores for all investigated periods. The investigations regarding observation periods for starting cyclists are published in [39].

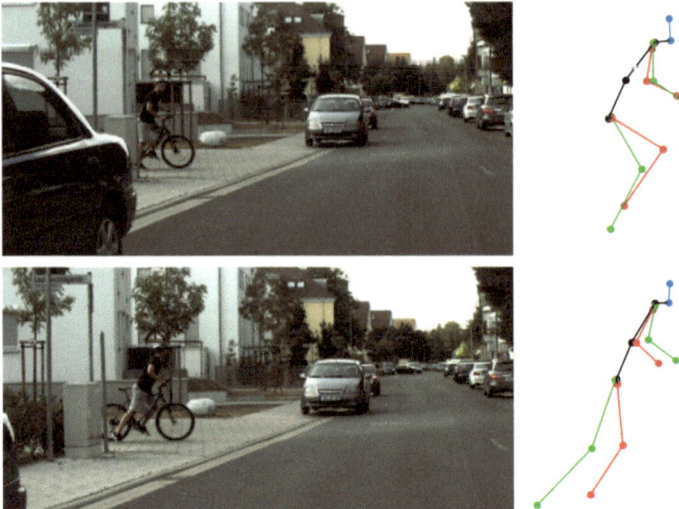

Fig. 12 Example scene of a starting cyclist recorded from within a moving vehicle. On the left, two images from the scene are visualized, showing waiting in the first image and starting in the second. The starting motion is clearly visible in the poses extracted from the images shown on the right. The cyclist's upper body is bent forward, and uses his foot to push off the ground, which is a distinct motion indicating starting process

Building on our findings on the observation period, we develop a holistic approach to pose-based basic movement detection for pedestrians and cyclists. Compared to the previous method, our investigation is not limited to a single scenario but includes all possibly occurring motion states. Furthermore, we switch our model architecture from an FCN to a recurrent neural network (RNN). The advantage is that RNNs are specifically designed to model time series and allow for variable input lengths. Our method is therefore able to estimate motion states despite short observation periods and successively improves with larger periods. As in the previous investigation, we compare our method to a single trajectory approach, where the pose-based method outperforms the single trajectory approach, especially for short observation periods. The evaluation can be found in [40].

Basic Movement Detection Using Smart Devices

In the previous sections, we described stationary and vehicle-based basic movement detection, where we used camera sensors in both cases. While both methods have different advantages and disadvantages, they are both error-prone with regard to the shortcomings of camera sensors. Camera-based approaches depend on lighting and weather conditions and are affected by occlusions. In contrast, these conditions do not affect smart devices worn by VRUs themselves. Therefore, they have great potential to serve as additional sensor sources for intention detection.

In [6, 9, 10, 12, 58], we investigate how one can use the inertial sensors of smart devices for basic movement detection. The approaches presented in the different publications are based on human activity recognition involving a machine learning classifier at its core [17]. A schematic of the six-step detection process using accelerometer and gyroscope data is shown in Fig. 13. First, the inertial sensor measurements are preprocessed (i.e., the data is transformed into a rotationally invariant coordinate system), then the signal is windowed, and features are extracted based on the windowed data. Subsequently, feature selection is performed to filter features relevant for detection. These filtered features are then used for detection. For this purpose, the detection problem is modeled as a classification problem. The classifier (e.g., an extreme gradient boosting classifier) is trained with labeled example data. Finally, a probability calibration of the detection probabilities output by the classifier is performed, and a temporal filter filters out any outliers. More details about the approach can be found in [7, 10, 12]. Regarding the early detection of cyclists' starting movements, we showed that our approach reaches an F_1 score of 67% within 0.33 s after the first movement of the bicycle wheel. Further, investigations concerning the influence of the device wearing location show that for devices worn in the trouser pocket, the detector has fewer false detections and detects starting movements faster on average. Moreover, we found that we can improve the results by training distinct classifiers for different wearing locations. In this case, we reach an F_1 score of 94% with a mean detection time of 0.34 s for the device worn in the trouser pocket.

Based on these findings, we investigate an extended smart-devices-based approach to detect longitudinal (i.e., waiting, starting, moving, and stopping) and lateral (turning left, going straight, and turning right) basic movements. Smart devices can be used very well for the detection of longitudinal basic movements; our approach achieves a macro F_1 score of 72% with an average detection time of only 0.36 s, i.e., on aver-

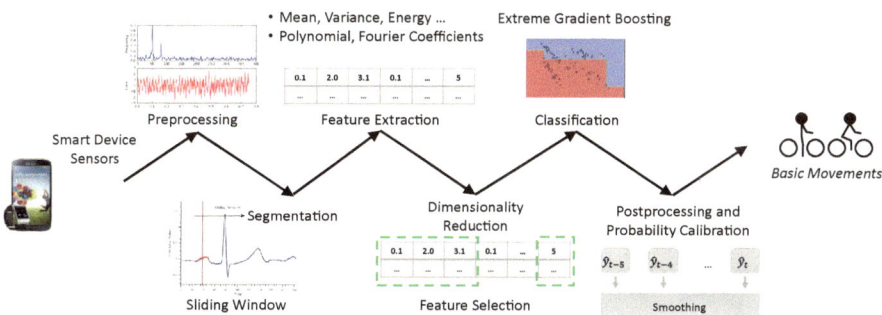

Fig. 13 Process for basic movement detection using smart devices consisting of six stages: Preprocessing, segmentation, feature extraction, feature selection, classification, as well as post-processing and probability calibration [12]

age a movement change is detected within 0.36 s. Curves or changes of direction of movement (i.e., lateral basic movements) can be detected even more reliably (F_1 score of 82%) and equally fast (mean detection time of 0.38 s). A detailed evaluation and further results can be found in [7]. In [16], we showed the successful transfer of our smart-device-based movement detection approach to the early anticipation of pedestrians' movements. Yet in [11], we moved from movement transition detection to short-term cyclist's movement transition forecasting.

3.2 Trajectory Forecasting

The goal of trajectory forecasting is to estimate future VRU positions. The forecasts build the basis for automated vehicles to safely interact with VRUs, where the forecast horizon, i.e., the time span for which the positions are estimated, depends on the application. In our case, the goal is to perform a short-time forecast for a horizon of 2.5 s, which is often named a relevant horizon to perform emergency brake maneuvers. To perform forecasting, we consider the VRU behavior we extract from video sequences. We avoid incorporating information about the traffic situation, such as traffic lights since the VRUs' disregard of such can lead to potentially dangerous situations. In the next section, we describe our methods for deterministic trajectory forecasting, where the goal is to forecast the VRU positions in the form of points. Afterward, we describe our approaches to add uncertainty estimation to our methods.

3.2.1 Deterministic Trajectory Forecasting

To perform deterministic trajectory forecasts, we utilize similar methods to the ones used for basic movement detection described in Sect. 3.1. While the same network architectures can be used, the problem is modeled as regression.

Deterministic Trajectory Forecasting Using Stationary Cameras

We investigated the incorporation of video information for trajectory forecasting using stationary cameras. We used the same representation as we used for basic movement detection, i.e., the image and optical flow sequences from both cameras and the past trajectories. We investigated the use of different inputs and compared the results to a method solely based on the past trajectory. In contrast to the results from basic movement detection, the results achieved by incorporating the optical flow sequences from both cameras and the past trajectory outperform the results achieved using all inputs. We attribute this to the fact that the optical flow sequences mainly contain information about the movement of the VRU, and excess information, such as the image background, is removed. The extraction of the optical flow is therefore comparable to an attention mechanism [62]. The positional accuracy is improved by

16.9% using the optical flow sequence compared to 8.2% when all inputs are used. We found that compared to the trajectory-based method, turning motions are better distinguished from straight motions due to a distinct head movement of the VRU towards the direction visible in the optical flow sequence. The detailed results of our investigations can be found in [64, 72].

Deterministic Trajectory Forecasting from a Moving Vehicle

From within a moving vehicle, we utilize 3D poses for trajectory forecast similar to the basic movement detection described in the previous section. In our evaluation, we perform trajectory forecasts for both pedestrians and cyclists and compare the results to a single trajectory method [36, 42]. The focus of our investigation is again on the length of the observation period, where periods between 0.2 and 1.0 s are considered. Furthermore, we compare two different variants of the poses. One variant uses joints of the entire body. In the second variant, the arms, i.e., the elbows and wrists, are not used as input. We hypothesize that the main features to forecast the future trajectory are the orientation of the VRU and the head and leg motions. We found that in the case of pedestrians and cyclists, the forecast accuracy improved by up to 6.93% for pedestrians, and 17.9% for cyclists by using poses. In both cases, no significant improvements were achieved by using the complete poses compared to the armless poses, demonstrating that the arm movements do not add additional information about the VRUs' future positions. While especially in the case of cyclists, this may seem counterintuitive since they are supposed to indicate turning motions by hand signals, we found that turns are seldom signalized. However, cyclists often perform a shoulder check before turning, demonstrating the importance of tracking head movements. In [43], we use RNNs for a pose-based trajectory forecasting of pedestrians and cyclists based on observation periods varying between 0.04 and 1 s. The use of 3D poses improves forecasting accuracy, especially for short observation periods, compared to a single trajectory method.

As discussed in the previous section, basic movement detection aims to add additional information to the trajectory forecast process. Therefore, we developed a two-stage approach to incorporate basic movement detections into the forecasting and compared the results to a single-stage approach [27, 64]. Instead of a single forecast model, we train specialized models for different VRU movements, such as *starting* or *waiting*. The forecast is generated by performing a forecast for every motion state and weighting the results with the probabilities estimated by the basic movement detection. The methods for basic movement detection, as well as trajectory forecast, are interchangeable. In our evaluation, we compare all possible combinations of the single trajectory and video-based methods. We found that forecast accuracy can be significantly improved if basic movement detection adds new information to the forecast. No improvements were achieved when the basic movement detection does not introduce new information to the model. Compared to the best video-based single-stage model, the best two-stage model did not improve the accuracy. Leading to the conclusion that in the case of deterministic trajectory forecast, incorporation of basic movement detection is only helpful if new information is introduced by the detection method, e.g., by cooperation with smart devices. While this holds for deterministic forecasts, probabilistic forecasts are a different matter, which we discuss in the next section.

Deterministic Trajectory Forecasting Using Smart Devices

Furthermore, we investigate the use of smart devices for trajectory forecasting. For this work, we focus on a single wearing position and consider a Samsung Galaxy S6 device placed in the trouser pocket. In this investigation, we do not use the two-stage intention detection process consisting of basic movement detection followed by trajectory forecasting. Instead, we focus on the realization of a trajectory forecasting module using the smart device sensors and examine the potential of this approach in principle. Since the GNSS is too inaccurate, we only forecast relative positions in the ego-frame. In doing so, we do not need absolute positioning information for forecasting. If we want to use the issued forecast with respect to a global coordinate system, we merely have to transform it back from the ego-frame. A possible use case would be, for example, that the smartphone issues a trajectory forecast in the ego-frame and transmits this forecast to an oncoming vehicle. The vehicle sees the cyclist and can determine the cyclist's position. This vehicle can now use the cyclist's position to transform the received forecast into a global or its local coordinate system. The advantage of forecasting the trajectories in the ego-frame is that the trajectory forecast is independent of the possible large absolute positioning error of the GNSS receiver integrated in the smart device. Furthermore, this approach allows us to only predict trajectories based on the inertial sensors. In the following case study, we investigate an approach to cyclist trajectory forecasting using only the smart device inertial sensors. We use a neural network for trajectory forecasting [57]. The forecasting time horizon is 2.5 s, and we have a lead time increment of 40 ms. Hence, the neural network has an output dimensionality of 126 (63×2, i.e., one for the longitudinal and one for the lateral position). The preprocessing and feature extraction of our smart device-based forecasting approach is mostly analogous to the approach for basic movement detection, i.e., we use multiple different statistical features curated from sliding windows of various sizes as input for the neural network. However, the feature selection for trajectory forecasting is more difficult because we have not only one output variable but two output variables for each forecasting lead time, i.e., 126 in total. Hence, we cannot transfer the feature selection method designed for classification tasks, i.e., basic movement detection in a straightforward way. To solve this, our approach aims to convert the multivariate regression task into a multi-class classification task. Therefore, we first perform clustering in the output domain, i.e., in the 126-dimensional target space. We use the cluster assignments to discretize the output variables into a set of 100 target classes. In this way, we reduce the multivariate regression to a classification task and may apply feature selection methods for classification tasks. Note that we only use this modeling for feature selection. We apply two feature selection approaches, a filter based on the chi-squared

Model Type	Complete	Waiting	Starting	Moving	Stopping	Straight	Turning Right	Turning Left
All Features	0.353	0.342	0.303	0.375	0.378	0.353	0.334	0.363
Selected Features	0.463	0.459	0.379	0.483	0.489	0.462	0.480	0.456

Fig. 14 The table shows the ASAEE in m/s of the respective motion types. We consider two different smart device trajectory forecasting models: One model using all features and a second using only the features selected by the feature selection procedure

statistics and a model-based approach using a gradient-boosting classifier. We union the features selected by both methods. As before, the intuition about combining two different feature selection methods is to get a diverse set of features. Subsequently, we train a neural network with these features. For this purpose, we first standardize the features. We optimize the neural network on the ASAEE [26] using the Adam optimizer [34]. The hyperparameters of the neural network, i.e., the learning rate, the number of hidden layers, the number of neurons in the hidden layers, and the number of epochs for training are determined using Bayesian optimization [61]. We use exponential linear units (ELU) as the activation function [20].

The results of our investigation are presented in Fig. 14. We compare the feature selection approach to the one where we do not reduce the number of features. As we can see, the model that uses all features has better ASAEE scores across all movement types. Additionally, we compared the smart device-based model, which uses all features, to infrastructure- and vehicle-based trajectory forecasting approaches. We observed that the smart device-based approach has worse ASAEE scores for almost all movement types than vehicle- or infrastructure-based approaches. However, there are a few exceptions, e.g., the ASAEE for starting movements is lower. Furthermore, the forecasting errors for turning, i.e., right and left, are comparable to those of the infrastructure-based approach. The smart device-based approach performs here better than the vehicle-based approach. We observe a similar result for moving cyclists. Besides, we also observe that the variance or interquartile range (IQR) is usually noticeably greater for the smart device-based approach with regard to the ASAEE. This applies to both directions, i.e., in some cases, the smart device-based approach is considerably better but sometimes also notably worse. These results show that the smart device-based approach is not yet fully competitive with the vehicle- or infrastructure-based approaches. However, the smart device-based approach performs comparably or better in some cases.

3.2.2   Probabilistic Trajectory Forecasting

Most existing methods for VRU trajectory forecasting create deterministic forecasts, i.e., estimates of the future VRU positions in the form of points (e.g., [27, 52]). Since these estimations are error-prone, methods to quantify their uncertainties are needed to create a basis for maneuver planning methods in automated vehicles. While there are few existing approaches to model uncertainties of trajectory forecasts (e.g., [1, 50]), the authors' focus is on the positional accuracy achieved by their methods.

Fig. 15 Example for the probabilistic forecast of cyclist trajectory for time steps 0.5 s (purple), 1.5 s (purple), and 2.5 s (purple) into to future. The inner regions (solid lines) describe that the cyclist will reside with a probability of 68% within the region, and the outer regions (dashed lines) with 95%, respectively. The red line describes the cyclist's head trajectory over the past second

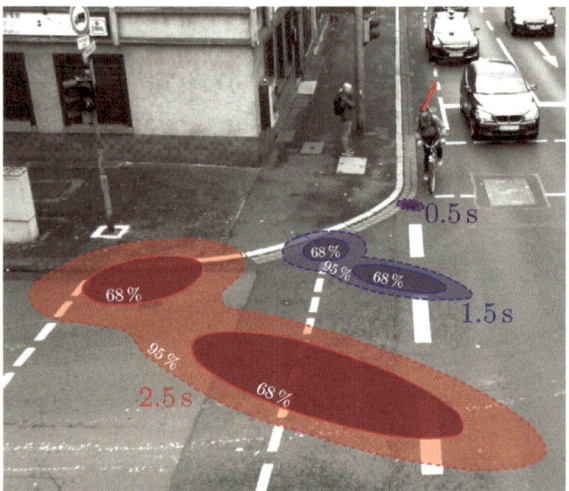

The estimated uncertainties are treated as byproducts, and no further evaluations are performed to rate the quality of the estimates. However, to use uncertainty estimates as the basis for safe maneuver planning, it is crucial to validate that the chosen methods can create reliable outputs. Furthermore, the estimated uncertainties should be kept as small as possible.

To achieve these goals, we perform probabilistic forecasts, where instead of single point estimates, we estimate confidence regions for future time steps. The regions describe an area where the VRU will reside within with a certain probability (see Fig. 15). We propose the use of three different approaches based on widely known techniques for uncertainty modeling. The first approach forecasts the parameters of probability distributions from which confidence regions can be created. The second approach extends quantile regression (QR) to multivariate targets, called quantile surfaces (QS). Both approaches are implemented using stationary cameras. The third approach is used within a moving vehicle and is based on occupancy grid maps. Furthermore, we compare standard metrics and propose novel approaches to rate the quality of our uncertainty estimates. Finally, we combine our approaches with a method for maneuver planning to demonstrate their applicability in the real world.

Probabilistic Trajectory Forecasting Using Stationary Cameras

A widely used method to add uncertainty quantification to the output of neural networks is to estimate the parameters of a probability distribution. Usually, Gaussian distributions are used. In the field of VRU trajectory forecasts, this method has been used to forecast bivariate Gaussian distributions in earlier work (e.g., [1, 50]). However, the focus of these articles is on the positional accuracy of the forecasts, i. e., only the means of the distributions are used for evaluation and no statements about the

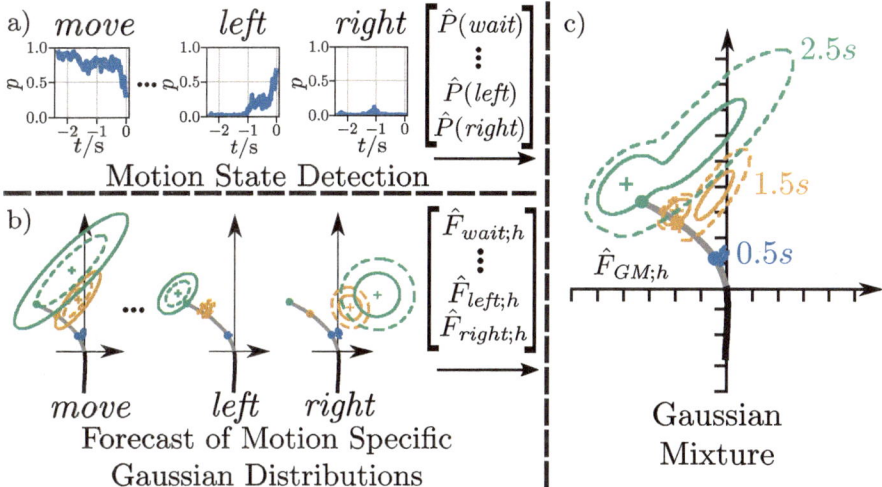

Fig. 16 Mulit-modal forecasting pipeline: **a** Motion state detection with example probabilities for states *move*, *left*, and *right*, with high probabilities for *move* and *left*. **b** Individual Gaussian forecasts for the same states for forecast horizons 0.5, 1.5, and 2.5 s. **c** Gaussian mixture generated by weighting Gaussian forecasts with motion state probabilities

quality of the uncertainty estimates are made. Therefore, we created a method that forecasts cyclist trajectories in the form of bivariate Gaussian distributions and evaluated the confidence regions generated from the estimated distributions with regards to their reliability [69]. We consider the regions to be reliable if the frequency with which the real position lies within the estimated region equals the probability of the region. For example, if we look at the 80% confidence region, the real position should fall into the region in 80% of all times. Our evaluations using our real-world dataset [67] show that the method is not able to create reliable outputs. More precisely, the method produced underconfident probabilities, meaning that the regions' probabilities are smaller than the percentage of real positions within the regions. This especially applies to waiting conditions, where an early forecast of the exact starting time is not possible, leading to the conclusion that VRU trajectories are inherently multimodal and cannot be modeled by a single Gaussian distribution.

To solve this problem, we developed a two-stage approach to forecast multimodal distributions similar to the deterministic approach from the previous section [70]. The pipeline of our approach is visualized in Fig. 16. The first stage performs basic movement detection by creating a probability for every possible VRU motion state (e.g., starting or waiting). Simultaneously, a Gaussian distribution is forecasted for every motion state using the uni-modal model from [69], where we train one specialized model for each motion state. In the second stage, the motion state probabilities are used to weigh the estimated density function of the specialized models, leading to a Gaussian mixture distribution. For detection of the current motion state, we investigate the use of the trajectory-based and image-based methods for basic movement detection described in the previous section [68, 71]. Compared to the deterministic two-stage approach, the probabilistic approach has a significant advantage. While in the deterministic case, only a weighted mean of the position is created, we add multimodality to the probabilistic approach by incorporating basic movements. Every estimated mode represents a motion state of the VRU. E.g., Fig. 15 shows an example of a cyclist beginning to make a right turn. The basic movement detection outputs high probabilities for the motion states *moving straight* and *turning right* and low probabilities for the remaining states, leading to two dominant modes. Our evaluations show that incorporating both detection methods into the probabilistic forecasts leads to reliable uncertainty estimates, solving the problem caused by the uni-modal approach. As indicated by the results of the basic movement detection, the regions estimated using the video-based method achieve a better sharpness than the trajectory-based method. The 95% confidence regions estimated by the video-based method are on average 14% smaller compared to the trajectory-based method's estimates, demonstrating that the results from basic movement detection can be applied when incorporating basic movement detection into the probabilistic trajectory forecast process. While we evaluated the method using data from the stationary cameras, the method can also be applied in a moving vehicle since the method for basic movement detection, and the architecture for trajectory forecast are interchangeable.

Our second method to forecast reliable confidence regions is based on QR. By extension of the single-output of QR to multivariate targets, we QS [7] serving the same purpose as the confidence regions created by the Gaussian mixture approach. The method consists of a two-stage model described in Fig. 17. The first stage performs deterministic point forecasting followed by the probabilistic QS estimation that uses the point estimate as the center. The method is capable of producing star-shaped estimates. While the method is based on a uni-modal approach, the star shape of the estimated regions allows us to model the uncertainties of our forecasts reliably. In contrast to the Gaussian mixtures, the method is not able to estimate multiple separate regions for a single probability, possibly leading to larger regions. However, due to the two-stage approach, any existing forecasting method that produces deterministic outputs can be extended by a probabilistic output without requiring additional detection of basic movements. This leads to a much leaner model with no need to train specialized models for every motion state, especially eliminating the need for time-consuming labeling of motion states.

Probabilistic Trajectory Forecasting from a Moving Vehicle

An approach from within a moving vehicle is described in [37]. As in the deterministic forecast, the method utilizes 3D poses. Additionally, we incorporate semantic maps to represent the surroundings of the VRUs, allowing us to prevent implausible forecasts, such as a VRU moving through an obstacle. The maps are created using 3D positions from LiDAR in combination with a semantic segmentation performed on images from a stereo camera and contain information about static obstacles, such as buildings, and dynamic obstacles, such as cars. Our forecast model is described in Fig. 18. The

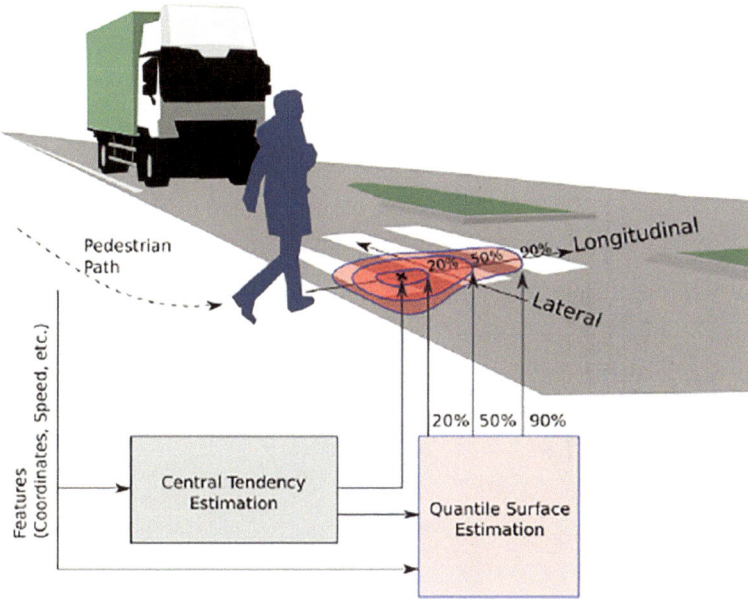

Fig. 17 Qunatile surfaces forecasting pipeline: In the first step, the central tendency estimation is performed using a classic deterministic forecasting approach. In the second step, we pass the central tendencies together with the used input features to the quantile surface estimation, which generates the probabilistic outputs for different confidences

probabilistic forecast is performed in a discrete way in the form of occupancy grids. We forecast one grid for every forecast time horizon centered at the current position of the respective VRU. Instead of a continuous probability distribution, we predict a probability for every cell within the grid. In our evaluation of the discrete method, we compare the use of only the head position to the complete pose with and without the semantic maps. Furthermore, we compare the discrete method to the single Gaussian approach from [69]. We compare the reliability, sharpness, and positional accuracy of the models. The comparison of poses with the single trajectory approach shows that the positional accuracy is improved by 9.7% in the case of the Gaussian approach and by 7.2% for the discrete method. In both cases, reliability and sharpness are also improved by using poses. While the semantic maps lead to a slight improvement in accuracy, improvements are more apparent when evaluating qualitatively, showing that fewer forecasts intersect with obstacles. Comparing the Gaussian method to the discrete method, we find both have advantages and disadvantages. While the Gaussian model overall achieves a better reliability score, only the discrete method is able to model certain motion types, e.g., *waiting*, reliably due to its ability to model multimodality.

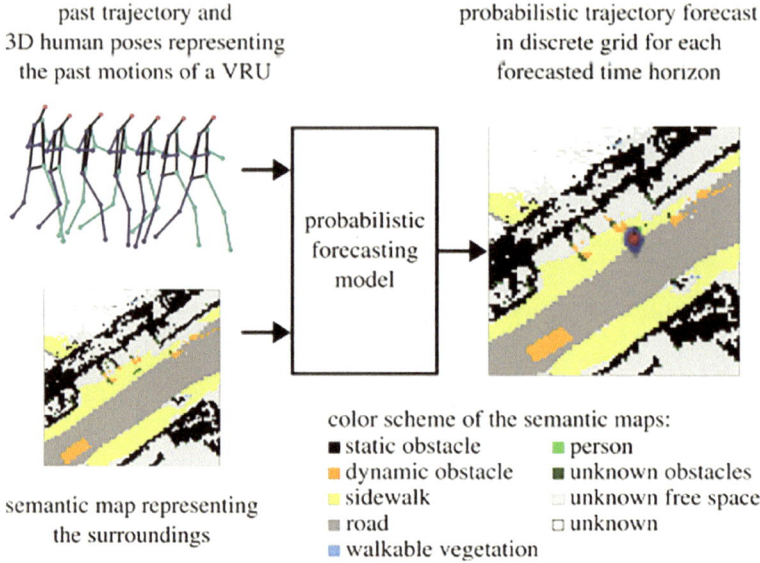

past trajectory and
3D human poses representing
the past motions of a VRU

probabilistic trajectory forecast
in discrete grid for each
forecasted time horizon

probabilistic
forecasting
model

semantic map representing
the surroundings

color scheme of the semantic maps:
■ static obstacle   ■ person
■ dynamic obstacle   ■ unknown obstacles
  sidewalk   unknown free space
  road   □ unknown
■ walkable vegetation

Fig. 18 Grid-based discrete probabilistic forecast: The model input consists of the past 3D pose trajectory of the VRU. Additionally, we use a semantic map representing the VRU's surroundings. The model outputs grid maps for each forecasted time horizon containing probabilities for every grid cell describing the likelihood of the VRU occupying the respective cell at the forecast time horizon

3.2.3   Application in Planning Algorithms for Autonomous Vehicles

To investigate whether our methods can serve as a basis for maneuver planning methods, we conducted a case study regarding an autonomous vehicle overtaking a cyclist [59] intending to safely overtake the cyclist while maintaining a lateral safety distance of at least 1.5 m. We combine our probabilistic methods with a model predictive planning (MPP) approach to achieve this goal. We simulate overtaking maneuvers based on cyclist trajectories from our real-world dataset leading to two different outcomes. Either a successful overtaking maneuver could be performed, or the vehicle stays behind the cyclist without overtaking due to larger uncertainties in the forecasted regions. While the second behavior is less desirable, it is considered safe. The MPP algorithm expects the estimated confidence region to have a convex hull in the form of a polygon. Since neither the multimodal nor the QR approach output a convex hull, we compare different approximation methods. We choose a method where a single rectangle aligned with the VRU's ego coordinate system per forecast horizon is used to approximate the region. The rectangle shape is chosen to keep the computational load of the MPP small since every edge adds to the load. For safety reasons, the rectangle over-approximates the actual region. Comparing the forecast methods showed that both methods can estimate reliable confidence regions. The multimodal approach can estimate sharper regions compared to the

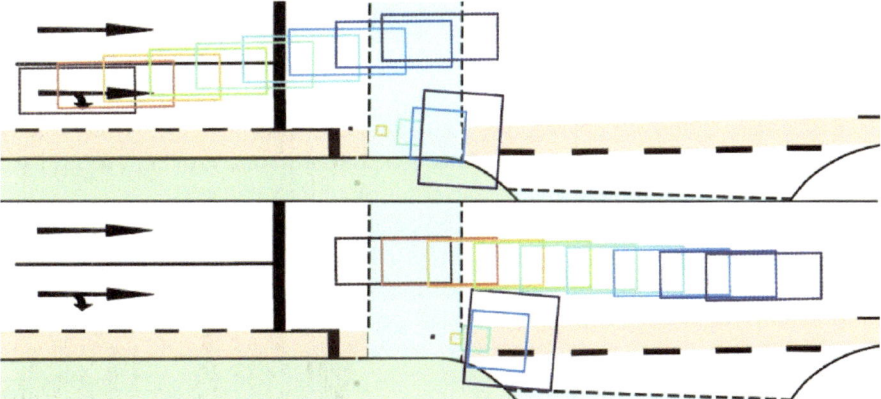

Fig. 19 Planned overtaking maneuver based on the forecasted confidence regions. The rectangles starting in the car lane represent the planned car positions. The rectangles on the bike lane are the forecasted cyclist regions. Future time steps are color coded so that the depicted boxes correspond to the same point in time

QR approach, which becomes evident, especially for larger forecast horizons. In our case study, the most desirable outcome is a successful overtaking of the cyclist. An example of a successful overtaking is displayed in Fig. 19. The less desirable yet acceptable behavior would be for the vehicle to abort the overtaking maneuver and stay behind the cyclist until overtaking is possible. The second case mainly occurred for large confidence regions. None of our tests resulted in a collision. Our results show that our methods can be used as the basis for interaction between autonomous vehicles and VRUs and highlight the importance of reliable and sharp uncertainty estimates.

4 Cooperative Intention Detection

Up till now, we focused on investigating intention detection using different sensor modalities independently. This helped us to gain an understanding of the different challenges of individual modalities. The goal of our project however is cooperative intention detection. Therefore, the following section describes our methods to combine intention detection from stationary cameras, vehicles, and smart devices into one system in order to improve the intention detection results. Before we describe our methods for cooperative intention detection, we give a short interim summary of what we have learned about the strengths and weaknesses of different sensor modalities used independently.

4.1 Interim Summary of Vehicle, Infrastructure, and Smart Device Based Intention Detection

In the previous section, we covered VRU intention detection methods. We especially showed that approaches for intention detection from within a moving vehicle face very different challenges than approaches using stationary cameras.

The process for moving vehicles is especially complicated due to occlusions from the viewpoint of the vehicle's sensors caused by other vehicles or objects on the roadside. This is challenging since many times we only have a short time frame within which we can observe the VRU's behavior to estimate future behavior. Therefore, the focus of our investigation was on finding appropriate methods that allow us to take the short observation period into account, which we achieved by the incorporation of human 3D poses into the intention detection process. We were able to improve the results for both basic movement detection and trajectory forecast, especially for short observation periods. By utilizing recurrent neural networks, we were able to consider observation periods of different lengths.

Compared to vehicle-based intention detection, stationary intention detection has many advantages. By mounting cameras at a higher elevation and using multiple cameras in a wide-angle stereo-camera system, we were able to resolve most occlusions. Furthermore, we are not as restricted to space and power consumption as we are inside a vehicle, allowing us to use a dense representation of the surroundings as a basis for our intention detection algorithms. Therefore, we investigated the direct incorporation of video sequences into our methods, leading to significant improvements compared to existing methods for basic movement detection. While stationary intention detection solves many problems of vehicle-based intention detection, it is not feasible to equip every existing road with cameras. However, stationary systems can be installed at busy traffic junctions, where many occlusions and most accidents with VRU involvement occur.

Another possibility we investigated is the use of smart device sensors for intention detection. Since smart devices are worn by the VRU directly and are not affected by occlusion at all. However, compared to camera-based intention detection we achieve far less accurate results due to sensor limitations. Therefore, we don't see smart device sensors as a feasible stand-alone solution for intention detection. However, we think that a combination of smart devices and vehicle-based intention detection can be used to improve the overall results.

4.2 Cyclists as Additional Sensors

Nowadays, almost everyone carries smart devices in form of a smartphone, smartwatch, or similar with them while taking part in traffic. Accordingly, we examine the use of smartphones and other wearable devices for the task of intention detection of vulnerable road users. These devices are equipped with a great variety of sensors, e.g., inertial measurement units or GNSS receivers. To work mobile, most smartphones are permanently online; they share their location or send the accelerometer profile to the server of the fitness application provider for further analysis. Essential for this are communication technologies such as UMTS, 4G, and 5G or, in the future, 6G, which allow us to send and receive large amounts of data within a few milliseconds. In 2010 David and Flach [21] proposed using smartphones for advanced pedestrian protection, i.e., as a sort of wireless safety belt. Many studies are investigating the usage of smartphones and other wearables for pedestrians in cooperative intelligent transport systems (C-ITS) [60]. However, cyclists have gained little attention. In contrast to vision-based approaches, smart devices also enable reliable intention detection in cases of occlusions. The position and the detected intentions, e.g., of crossing cyclists appearing from an occlusion, can then be communicated between approaching traffic participants using modern means of communication (such as 5G, V2V). Regarding our work, the utilization of smart devices worn by cyclists for the intention detection of vulnerable road users was the focus of our experimentation. We investigate various aspects, including smart device-based positioning as well as the influence of the wearing location of the smart devices [6]. We propose a novel basic movement detection approach for robust and yet fast basic movement detection using the smart device inertial sensors solely [12]. We investigate the usage of smart devices for cyclist trajectory forecasting [7]. Moreover, we propose a novel cyclist ad-hoc network involving the usage of multiple cooperating smart devices (e.g., smartphone, smartwatch, or sensor-equipped helmets) for intention detection at the same time [7, 22]. The main challenges of cooperative intention detection for cyclists are:

1. The localization of the cyclist [7]
2. The detection of the cyclist and their intention [7, 58]
3. The forecasting of the cyclists trajectory (probabilistically) [7]
4. The incorporation of multiple smart devices [7, 22].

4.3 Smart Device Cooperation for Intention Detection

Instead of a single smart device, in the future, people will carry many devices, e.g., a smartphone, smartwatch, and smart helmet. Smartwatches, for example, are already widely used today. Additionally, those may also include cloths containing sensors or helmets equipped with sensors, i.e., smart helmets. It is also likely that future bicycle generations will be equipped with intelligent assistance systems, sensors (e.g., cameras, Lidar, or Radar), and V2X communication capabilities [14]. All of these smart devices can potentially be used to anticipate cyclists' movements, to communicate them (e.g., to an oncoming vehicle), and thereby make an important contribution to improving cyclists' safety. The smart devices described previously measure different aspects of cyclist movement due to their different wearing locations or other sensor types. If these devices are connected, for example, using a kind of wireless body area network (BAN) [45] for cyclists, then the smart devices can exchange information. This information can be fused and refined and subsequently be used for cyclist intention detection. An example of this is depicted in Fig. 2a. The smart devices can communicate with each other, e.g., via Bluetooth, and the smartphone might provide communication abilities with cloud services.

The worn devices provide both redundant as well as complementary information. The smart helmet and the smartwatch, for example, might have better GNSS signal due to their wearing location, so their information should be preferably used for positioning. The smartphone, which is located, for example, in the cyclist's trouser pocket, can give information about the pedaling frequency. If we combine these two pieces of information, we could, for example, improve the positioning or the forecasting of the future trajectory. We can fuse the communicated information either in a centralized manner (e.g., on the smartphone) or in a decentralized fashion (e.g., on each device itself). This provides safe handling of a user's data in regard to privacy. Still, the data information could also be processed non-locally on a remote server through a secure cloud connection should the computational requirements exceed the capabilities of the smart devices or simply to save battery power.

In the following, we present two case studies to demonstrate the potential of a body area network incorporating the usage of multiple smart devices for a cyclist's movement anticipation. In the first case study, we investigated the use of a helmet equipped with sensors, i.e., a smart helmet. However, because off-the-shelf and ready-to-use smart helmets are not yet commonly available, we utilize a smartwatch attached to the cyclist's helmet. In the second case study, we investigate the use of multiple smart devices for longitudinal basic movement detection.

4.3.1   Combining a Smart Helmet with a Smartphone for Improved Orientation Estimation

In this section, we investigate the possibility to use of a smart helmet as an additional device connected to a smartphone. In our investigations concerning GNSS-based position, velocity, and orientation estimation, we found that especially the device placed on the helmet provides excellent velocity and orientation measurements. However, the sampling rate of 1 Hz is far too low for our intended applications, e.g., basic movement detection. Therefore, we present an approach combining inertial sensor measurements with GNSS measurements. In this case study, we combine the GNSS measurements from the smart helmet with the inertial sensors of a smartphone carried in the trouser pocket. Thereby, the utilized data comprises 48 test subjects and 257 trajectories. An implementation of our approach could be that the smart helmet sends its current GNSS measurement via Bluetooth to the cyclist's smartphone. On the smartphone, the GNSS data is now combined with the smartphone inertial sensor data to obtain an improved velocity or orientation estimate. For the orientation estimation, we use a Kalman filter running on the smartphone. The velocity estimation based on the combination of GNSS and inertial sensor data was much more difficult.

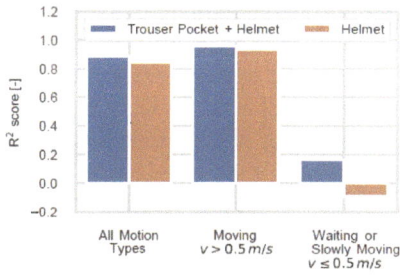
(a) R^2 score of orientation estimation

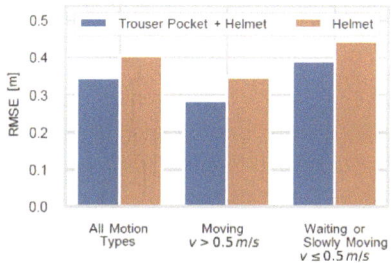
(b) RMSE of orientation estimation

Fig. 20 Performance of the cyclist's orientation estimation using a smartphone in the trouser pocket and a smart helmet, as well as the smart helmet only. The evaluation was carried out for different velocity ranges, i.e., all of the available data, moving faster than 0.5 m/s, and slower than 0.5 m/s

However, we achieved very good results using machine learning models and HAR techniques.

First, we present the results of the orientation estimation involving a smartphone and a smart helmet. We examine the following device combinations: First, GNSS from the smart helmet with gyroscope data from the smartphone in the trouser pocket and, second, only the smart helmet (i.e., GNSS and gyroscope data from the helmet). The experiments are conducted offline with real data. We do not consider any communication delays, as these are not large compared to the delay of the GNSS measurement. We tune the hyperparameters of the Kalman filter, i.e., the process- and measurement noise, using a grid search. We depict the results of our investigation in Fig. 20. The fusion of the GNSS measurements obtained from the smart helmet and the gyroscope measurements of the smartphone in the trouser pocket can greatly improve the orientation estimation. Furthermore, we observe that the orientation estimation based on the smart helmet and smartphone works differently well at different velocities. This can be explained since the cyclist might look around at slow velocities, e.g., when waiting at a traffic light, which can be mistaken as a change in orientation of the bicycle. The smartphone is less prone to such misinterpretation when it is kept in the trouser pocket. Although, the orientation of the smart helmet is a very helpful source of information to predict the intended cycling direction.

4.3.2 Inter-Device Cooperation for Basic Longitudinal Movement Detection

In this section, we present a case study for longitudinal basic movement detection using multiple smart devices. For this purpose, we consider the smartphone carried in the trouser pocket, the smartwatch at the wrist, and the device at the helmet. The results of the case study presented in the following have been published in [22]. In this case study, we restrict ourselves to data originating from the inertial sensors, i.e., we do not consider GNSS measurements. For comparison, we train classifiers for each of the three considered devices. These are our baseline models. For classification, we apply XGBoost classifiers [19], followed by an isotonic regression for probability calibration [47]. To assess the trade-off between robustness and detection time, we consider Pareto fronts. Therefore, we evaluate different hyperparameters using a randomized search with 250 trials. In this respect, we apply ten-fold cross-validation over the test subjects.

For fusing, we considered three different methods to combine the measurements of the three smart devices: (a) fusion of the feature spaces of all devices (feature stacking); (b) fusion at the decision-level of the basic movement detections (classifier stacking); (c) a hybrid approach combining the fusion of the feature spaces and the decision-level fusion. In the case under consideration, we assume that the fusion of the measurements and predictions of the smart devices is performed in a centralized manner on the smartphone. The choice of the smartphone as the point of fusion is based on the premise that today's smartphones have the necessary computing power, enabling more complex calculations to be performed here. However, this is only an example; the fusion could also be carried out on any other device. In this case study, we do not consider communication delays, i.e., we assume that the communication delay between the devices is negligible. As the devices are all worn at different locations on the body, they also measure different aspects of the motions performed by the cyclist. To prevent loss of information, we have decided against fusing these individual features (e.g., averaging) and instead decided to stack the feature spaces. We reduce the dimensionality of this feature space by applying a two-stage feature selection procedure. Based on the selected features, we then train a classifier to detect the longitudinal basic movements.

The fusion at decision-level is based on the trained classifiers of the individual smart devices. For each smart device, we train a dedicated classifier. These are referred to as base classifiers. Their outputs (i.e., predicted probabilities of the individual classes) constitute a new feature space. Subsequently, we train a new classifier based on this feature space. In literature, this approach is also known as classifier stacking or stacking ensemble [73]. We obtain the predictions of the base classifiers used for training the stacked classifiers using cross-validation.

The third and last approach is a hybrid approach. This hybrid approach uses the stacked feature space of all smart devices and, additionally, the predicted probabilities, as described before. The feature space is again reduced by applying the two-stage feature selection procedure.

Overall, when first evaluating individual smart devices, we observed that the smart helmet performs rather poorly in terms of the scores considered. Altogether, we can conclude that the classifiers based on the data from the smartwatch mounted at the wrist provide the best detection results. The results of longitudinal basic movement detection using multiple cooperating smart devices indicate that the combination of data originating from multiple smart devices leads to both faster and more robust longitudinal basic movement detection. Although, the results show that the different fusion paradigms yield considerably different results in some cases. The decision-level fusion multiple-devices classifiers have smaller detection delays than the other approaches. The detection delays are in a range from 0.194 to 0.38 s. The hybrid approach achieves detection delays between 0.24 and 0.72 s. Thus, the hybrid approach is regarding the detection speed slower but reaches higher scores. The feature stacking approach usually performs slightly worse than the hybrid approach both in terms of detection delay as well as its score. Further detailed consideration and extensive evaluation regarding the use of multiple smart devices for basic movement detection is provided in the work of Depping [22].

4.4 Cooperative Basic Movement Detection

Another approach concerns the use of cooperation to improve basic movement detection. These cooperatively determined basic movements can then be used for trajectory forecasting, i.e., for the parameterization of the forecasting models. In this regard, we examine different approaches:

Stacking of Feature Spaces: In feature space stacking, we assume that the agents exchange preprocessed features with each other. These features originating from different sensors are combined and used for basic movement detection. We realize fusion by concatenating the feature spaces of different sensors. This is, for example, the concatenation of orthogonal expansion coefficients (describing the past cyclist's trajectory) with Fourier coefficients (describing the acceleration profile derived from the smart device inertial sensors).

Stacking Ensemble: In the stacking ensemble fusion methodology, we fuse basic movement predictions employing a machine learning ensemble. These basic movement predictions, which originate from the basic movement detection models of other agents, are combined using a dedicated machine learning model. The combination of a stacking ensemble and a stacking of feature spaces is referred to as a hybrid model.

Probabilistic Fusion: Another method that we examine for cooperative basic movement detection is the independent likelihood fusion (ILP) fusion. This is a probabilistic fusion technique (similar to the Bayes filter) which is based on the assumption that the measurements of the sensors are independent of each other given the current state. It combines basic movement prediction originating from different agents.

Coopetitive Soft-Gating Ensemble (CSGE): The Coopetitive Soft Gating Ensemble (CSGE) [25] is an ensemble technique that is used to fuse forecasts of different base models. The CSGE has three different weighting aspects, i.e., global-, local-, and time-dependent, which are used to compute an overall weight for each ensemble member. We modified the original CSGE to cope with the special requirements of the task at hand, i.e., handling delayed or missing predictions.

Orthogonal Polynomials: This approach is a classifier fitted on the cooperatively acquired orthogonal expansion coefficients.

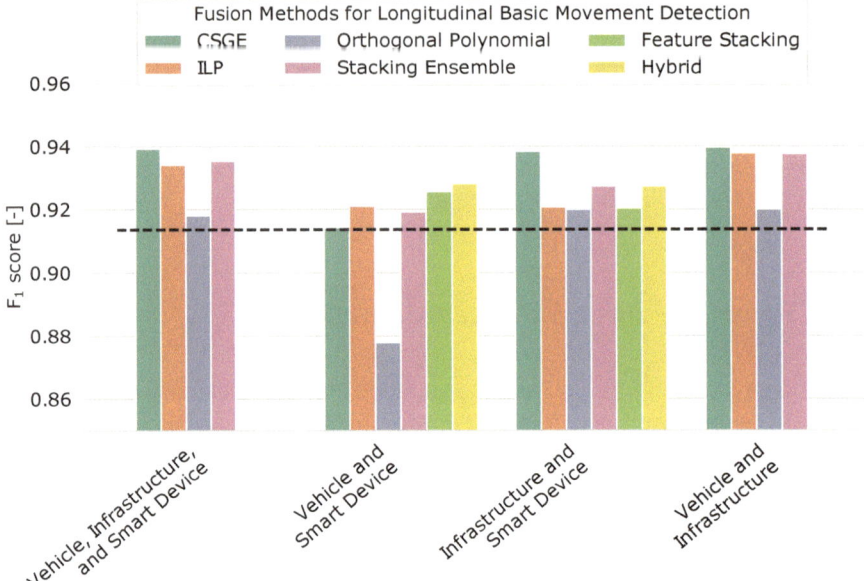

Fig. 21 Micro average F_1 score for cooperative longitudinal basic movement detection. Results of cooperative longitudinal basic movement detection for different agent configurations. The colored bars represent different fusion types. The baseline, i.e., the ego vehicle only, is given by the black, dashed line [7]

We used real data acquired at the urban research intersection in Aschaffenburg to evaluate and compare the different approaches. In this context, we examine the cooperation among three agents: a research vehicle, a sensor-equipped infrastructure, and a cyclist, i.e., a smart device carried in the cyclist's trouser pocket. We evaluate the results of the cooperative approaches from the perspective of a non-cooperatively ego vehicle. The vehicle-based approach is, therefore, our baseline against which we compare the cooperative approaches. We observe that almost all fusion methods outperform the baseline for almost all considered agent configurations.

As we can deduce from Fig. 21, cooperation is nearly always advantageous. Especially remarkable is the performance of the ILP approach. This method is almost parameter-free and performs better or at least as well as other methods with significantly more parameters. The CSGE shows the most significant improvement with up to 30% compared to the baseline. Hence, we can increase the F_1 score for basic movement detection significantly through cooperation. However, not only the detection performance is getting better, but also the mean detection time improves by up to 30% [7]. In addition, it is important to note that cooperative basic movement detection is currently the only cooperation method that effectively allows the integration of smartphones. Although this is also possible with the other methods, the use of the smartphone position often has a negative effect on the fusion result due to the poorer position estimation. The practical implementation of the cooperation techniques with current communication protocols is possible. Still, depending on the type of cooperation, it is not as straightforward to realize as with the probabilistic trajectory fusion technique.

4.5 Cooperative Trajectory Forecasting Using the CSGE

In this section, we outline an approach for cooperative cyclist trajectory forecasting using the CSGE. The underlying idea is that agents share predictions about their future trajectory. The trajectory forecasts are then combined using the CSGE. The approach described in this section fuses deterministic trajectory forecasts. The fused forecast is the starting point of the probabilistic trajectory forecast. We look at the fusion from the perspective of an ego vehicle, i.e., the fusion is conducted on the vehicle. The approach can be considered as decision-level fusion. From the perspective of sensor configuration, the approach can be classified as competitive fusion. The CSGE has three parameters, i.e., the soft gating parameters, which determine the weights of the individual ensemble members according to three influencing factors. We use the ASAEE as the target function to optimize these parameters of the CSGE. Moreover, we assume that the ensemble members are already trained. We also pretend that there is a dataset not yet used for training the ensemble members, which can be used for the CSGE training. We use ten-fold cross-validation to create this ensemble training dataset. The agents share their trajectory forecasts in the cyclist's ego-frame. The usage of this coordinate system has the advantage that errors in the absolute positioning (e.g., in the global coordinate system) of the respective agent do not influence the actual trajectory forecast. This allows us to include trajectory forecasts of agents with poor absolute positioning. This is the case, for example, with smart devices whose absolute positioning is not comparable to that of modern infrastructure- or vehicle-based approaches. Nevertheless, smart device-based trajectory forecasts can be helpful in some situations, e.g., when the field-of-view of the infrastructure or vehicle cameras is occluded. The CSGE natively supports the outage of a sensor or ensemble member. Similar to the CSGE approach for cooperative basic movement detection, we only have to re-compute the respective weights. The introduction of a new ensemble member can be handled similarly. The prerequisite for this is that the corresponding error estimates, i.e., global, local, and lead time-dependent errors, are available. However, in both cases (i.e., outage and introduction of a new ensemble member), we cannot guarantee that the soft gating parameters are still optimal.

4.5.1 Extending the CSGE for the Fusion of Delayed Trajectory Forecasts

Additionally, we proposed an extension of the CSGE for the fusion of trajectory forecasts that allows the integration of delayed trajectory forecasts. We investigate the extension in a case study considering the fusion of vehicle- and infrastructure-based trajectory forecasts. The fundamental idea of our modeling is analogous to the one used to integrate time-delayed basic movement predictions. The provider of the forecast always provides an estimate of the forecast quality, i.e., the expected error. The receiver uses this as a starting point and tries to model the increased expected error due to the delay. We distinguish three different types of expected errors, i.e., global, local, and lead time-dependent errors. Hereby, the challenge we face with delayed forecasts is that the cyclist's ego-frame changes over time. This offset is not only temporal but is also spatial, i.e., a simple temporal shift of the forecast is not sufficient. In addition to the temporal shifting, we must also simultaneously translate and rotate the cyclist's ego-frame. Hence, we cannot simply compare and fuse two trajectory forecasts of two agents without considering the time and spatial alignment of the ego-coordinate frames first. We have two possibilities for this purpose. First, the vehicle itself can estimate the change of the cyclist's ego-frame, i.e., the translation and the rotation, and apply these to the received trajectory forecast. For this purpose, the vehicle must estimate the current and the past (i.e., the time of the creation of the trajectory forecast) position and orientation of the cyclist. Subsequently, the vehicle can use these estimates to determine the translational and rotational shift. The second possibility we investigated is to use the trajectory forecast itself to estimate the change in terms of the cyclist's ego-frame and then use this estimation to translate and rotate the forecast accordingly. This method has the advantage that we can even fuse two trajectory forecasts if we cannot reconstruct exactly the past position and orientation at the time of the creation of the trajectory forecast. By artificially shifting the forecasting origin, our maximum lead time changes as well. We compensate for this by extrapolating the forecast based on its local trend and then padding it again.

4.5.2 Case-Study Delayed Trajectory Forecasts

In another case study, we examine the handling of delayed messages in the case where a vehicle receives delayed infrastructure-based trajectory forecasts and fuse these with its trajectory forecasts using the CSGE. We use the previously described modeling of the delays for the different weighting aspects of the CSGE. We assume that only the messages from the infrastructure are delayed. A delay on the side of the vehicle (e.g., due to data processing) is not considered. The results of this analysis are given in Fig. 22. We see that the improvement due to the combination of the trajectory forecasts diminishes with increasing delay. We observe a slow convergence towards the ASAEE of purely vehicle-based trajectory forecasting methodology. From this, we can conclude that the fusion of trajectory forecasts is advantageous for a maximum delay of approximately 1 s.

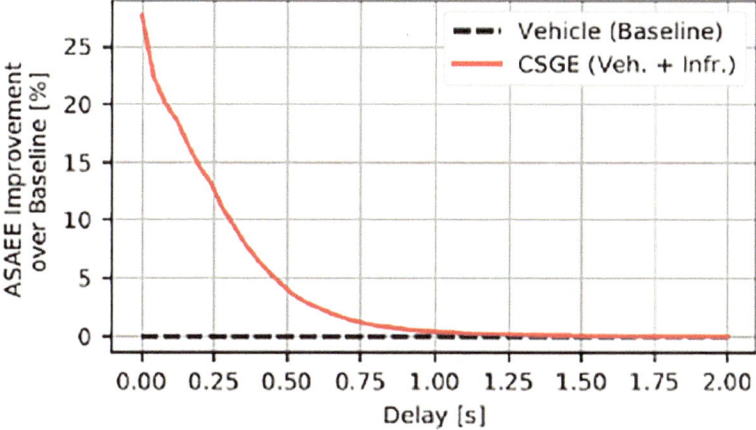

Fig. 22 CSGE forecasting performance improvements over the vehicle baseline for different delays [7]

4.5.3 Comparing Different Approaches for Cooperative Intention Detection

In the following, we compare the presented approaches to cooperative intention detection, i.e., cooperation on the data-level using the probabilistic trajectory fusion method, cooperation on the level of basic movements using various approaches, and cooperation on the level of trajectory forecasts using the CSGE. In our comparison, we examine the cyclist trajectory forecasting results of the different approaches using the example of three cooperating agents: vehicle, infrastructure, and smart devices carried by the cyclist. As a baseline, we use the forecast based on a non-cooperatively acting ego vehicle. The results of our investigation are depicted in Fig. 23. We see that the cooperative methods almost all perform better than the baseline in terms of the median ASAEE. In addition, the spread is also considerably smaller. Trajectory Fusion CSGE has the lowest ASAEE. Furthermore, the ASAEE of the infrastructure-based approach is particularly striking. This result underlines the potential of using infrastructure-based technologies for C-ITS in general and cyclist intention detection in particular.

Additionally, we performed a statistical analysis of the results to show whether there is a statistically significant difference between the performances of the cooperative methods and the baseline. The trajectory fusion CSGE approach ranks first. It is significantly better than all other approaches except the infrastructure-based trajectory forecasting approach. All cooperative approaches outperform the baseline, although the difference regarding the average rank is not statistically significant. It is not surprising that the ranks of cooperative methods for basic movements are not significantly different from the baseline. This is because the actual trajectory forecasts only use the ego vehicle data and the cooperatively determined basic movements.

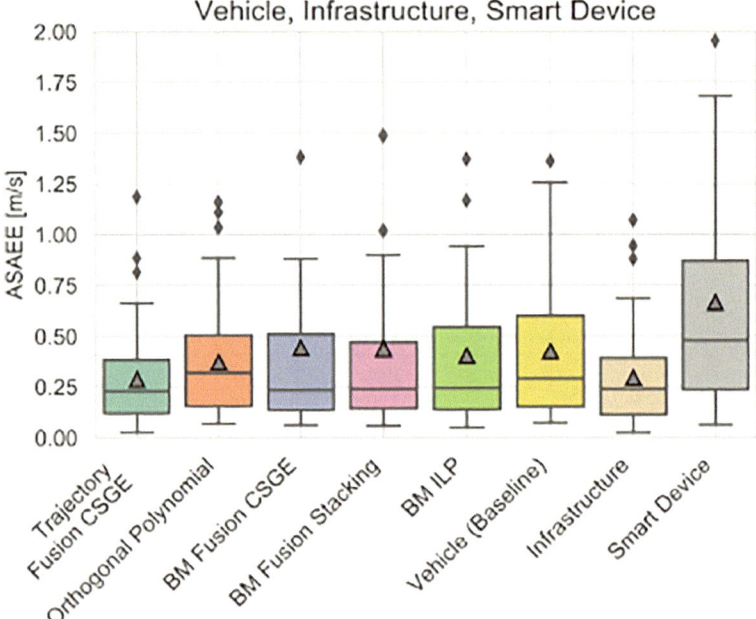

Fig. 23 Box plot showing the ASAEE for different approaches to cyclist trajectory forecasting. All cooperative approaches involve the combination of data originating from three different agents, i.e., an intelligent vehicle, sensor-equipped infrastructure, and smart device carried by the cyclist itself [7]

Nevertheless, the superior average rank shows the potential of cooperative basic movement detection. For future work, the cooperative basic movement prediction may be supplemented by cooperation based on trajectory forecasts.

4.6 Cooperative Probabilistic Trajectory Fusion Using Orthogonal Polynomials

Assuming that road users make at least partial use of the same set of features, e.g., the absolute velocity or angular velocity, the cyclist's trajectory is approximated using polynomials with orthogonal basis functions [23]. This representation is abstract, independent of the sensor's cycle time, and robust against noise due to implicit data smoothing. The feature-level fusion is realized using weighted polynomial approximation. We are exploiting specific properties of the orthogonal polynomials and the approximation technique: (1) fast incremental approximations are possible (update mechanisms are available [23]), and (2) information can be weighted individually. The former keeps the runtime short, and the latter allows us to fade out outdated

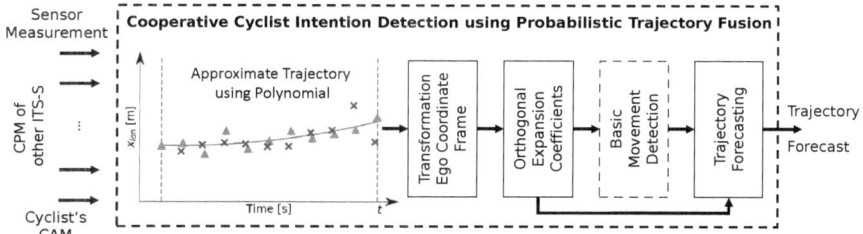

Fig. 24 Cooperative cyclist intention system using trajectory fusion from the view of a single agent, e.g., vehicle. The position and orientation estimates (indicated by the blue crosses and gray triangles) received via collective perception messages (CPM) or collective awareness messages (CAM) are fused probabilistically using a polynomial approximation with orthogonal basis polynomials. Subsequently, the orthogonal expansion coefficients are transformed into the ego-frame. These coefficients are used for basic movement detection and trajectory forecasting [7]

information or emphasize more recent information. Furthermore, by additionally modeling the posterior distribution over the polynomial coefficients in a Bayesian approach, we obtain a fully probabilistic model of the trajectory. New measurements are integrated by modeling the likelihood, i.e., implementing a sequential update methodology. We obtain the weighting of information originating from different road users through a measurement model. The measurement model describes the likelihood of an observation given the currently estimated polynomial coefficients. We derive the weight of each measurement by combining a global weight (e.g., how good is a measurement of an agent's sensor globally) and a situation-dependent weight (e.g., how good is a measurement of an agent's sensor in the current situation). Moreover, due to the usage of a polynomial approximation instead of a state-space model-based approach, e.g., a recursive Bayesian filter, we can cope with situations where, e.g., due to communication problems, the information does not arrive in the correct temporal order (out-of-sequence fusion). The coefficients of the orthogonal expansion of the approximating polynomial are optimal estimators of the average, slope, curvature, and change of curvature of the approximating trajectory in the considered time window [23]. Hence in terms of the cyclist's trajectory, the coefficients are optimal estimators of the velocity, acceleration, and jerk. As shown in [26], these are useful features for detecting the intentions of VRUs. We use these coefficients as features for basic movement detection and trajectory forecasting. A schematic of this cooperative intention detection approach is depicted in Fig. 24.

For evaluation, we utilize the data from real cyclists driving in real traffic at the research intersection. We recorded the cyclists' trajectories using a wide-angle stereo-camera system (i.e., an intelligent sensor-equipped infrastructure), a camera-equipped vehicle, and a smartphone carried by the cyclists. In the first place, we consider the evaluation of the position and orientation estimation derived from the probabilistic trajectory fusion. Therefore, we evaluate the approximating polynomial at the current time. We compare the probabilistic trajectory fusion approach to a Kalman filter for the fusion of the position measurements showing that our

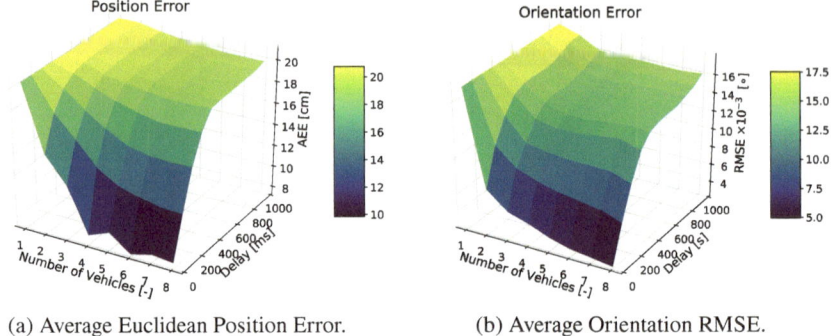

(a) Average Euclidean Position Error.   (b) Average Orientation RMSE.

Fig. 25 The average position (AEE) and orientation error (RMSE) for different numbers of vehicles and delays [7]

probabilistic approach is on par with the Kalman filter. Furthermore, we study the probabilistic trajectory fusion approach's behavior under message and measurement delays. We can show that the use of position and orientation estimates supplied by the infrastructure is beneficial, even for larger delays from an ego vehicle's perspective. The fused estimate does not worsen and up to a delay of about 0.7% always leads to an improvement. In another experiment with simulated vehicles, we showed that the approach scales well to larger vehicle collectives (cf. Fig. 25). Since this method only relies on the exchange of positions or velocities between the agents, it can be well implemented using existing standards such as CAM or CPM.

5 Prospects

We want to conclude this outline of our contribution to VRU safety by detection, tracking, basic movement detection, and trajectory forecasting with a short summary of our main findings.

First, we do not see the detection of objects as a solved problem. Despite significant improvements due to the success of data learning in the past couple of years, the resulting models still lack generality, reliability, and trustworthy confidence approximations. We introduce additional annotations of the data we collected to be able to determine types of data that cause poor results. The so-called context information is a basis for further research fields. The tasks may include a thorough determination of relevant context, concepts to gather data with respect to a specific context efficiently, and an evaluation that the model trained on the enhanced database is able to outperform the original model in any case.

Second, basic movements are an intuitive way of judging the current and short time future behavior of a VRU. More than that, they contribute greatly in a methodological way to the probabilistic trajectory prediction to reduce the future confidence regions in a multimodal approach, a single end-to-end learning approach with a single resulting distribution can not.

Third, trajectory forecasts must be made with probabilistic estimations of future VRU appearances. Only in that way is a safe and efficient coexistence of VRUs and autonomous cars possible. It will include ethics to find a way of dealing with how much risk is acceptable in the case of intersecting confidence regions.

Finally, an infrastructure to share information between traffic participants and to supply additional static knowledge is essential to exceeding the limitations of single sources, i.e., solely ego-vehicle sensors, and to be able to perform in a way that is acceptable for autonomous driving. Each data source could contribute beneficially in every processing step until trajectory prediction. Even relatively imprecise smart device data increased the tracking and trajectory forecast performance in cases of occluded infrastructure or ego-vehicle sensors. Altogether, we consider as a result of our project the proof of a concept that can estimate and predict the unsteady behavior of VRUs and thus make VRUs accessible to autonomous cars. The degree of realistic conditions and real-time performance capabilities has not been reached so far, to the best of our knowledge.

Acknowledgements This work results from the project DeCoInt[2], supported by the German Research Foundation (DFG) within the priority program SPP 1835: "Kooperativ interagierende Automobil", grant numbers DO 1186/1-1, DO 1186/1-2, SI 674/11-1, SI 674/11-2, FU 1005/1-1, and FU 1005/1-2. Additionally, the work is supported by "Zentrum Digitalisierung Bayern" and BMWI (Deutsches Bundesministerium für Wirtschaft und Energie/German Federal Ministry for Economic Affairs and Energy) within the project "KI Data Tooling—Methoden und Werkzeuge für das Generieren und Veredeln von Trainings-, Validierungs- und Absicherungsdaten für KI-Funktionen autonomer Fahrzeuge" (19A20001O).

References

1. Alahi, A., Goel, K., Ramanathan, V., Robicquet, A., Fei-Fei, L., Savarese, S.: Social lstm: human trajectory prediction in crowded spaces. In: 2016 IEEE Conference on Computer Vision and Pattern Recognition (CVPR), pp. 961–971 (2016). https://doi.org/10.1109/CVPR.2016.110
2. Andriluka, M., Pishchulin, L., Gehler, P., Schiele, B.: 2d human pose estimation: new benchmark and state of the art analysis. In: IEEE Conference on Computer Vision and Pattern Recognition (CVPR) (2014)
3. Bar-Shalom, Y., Li, X.R., Kirubarajan, T.: Estimation with Applications to Tracking and Navigation: Theory Algorithms and Software. Wiley (2001)
4. Bernardin, K., Stiefelhagen, R.: Evaluating multiple object tracking performance: the clear mot metrics. EURASIP J. Image Video Proc. **1**, 246–309 (2008)
5. Bieshaar, M.: Cooperative intention detection of vulnerable road users. In: Organic Computing: Doctoral Dissertation Colloquium, pp. 81–92. Kassel University Press (2016)
6. Bieshaar, M.: Where is my device? - Detecting the smart device's wearing location in the context of active safety for vulnerable road users. Organic Computing: Doctoral Dissertation Colloquium, pp. 27–37. Kassel University Press (2018)
7. Bieshaar, M.: Cooperative Intention Detection using Machine Learning-Advanced Cyclist Protection in the Context of Automated Driving. Intelligent Embedded Systems. Kassel University Press (2021). (Dissertation, Universität Kassel, Fachbereich Elektrotechnik/Informatik)

8. Bieshaar, M., Reitberger, G., Zernetsch, S., Sick, B., Fuchs, E., Doll, K.: Detecting intentions of vulnerable road users based on collective intelligence. In: AAET - Automatisiertes und vernetztes Fahren, pp. 67–87. Braunschweig, Deutschland (2017)

9. Bieshaar, M., Zernetsch, S., Depping, M., Sick, B., Doll, K.: Cooperative starting intention detection of cyclists based on smart devices and infrastructure. In: International Conference on Intelligent Transportation Systems (ITSC), pp. 1–8. Yokohama, Japan (2017)

10. Bieshaar, M., Zernetsch, S., Hubert, A., Sick, B., Doll, K.: Cooperative starting movement detection of cyclists using convolutional neural networks and a boosted stacking ensemble. IEEE Trans. Intell. Veh. (T-IV) **3**(4), 534–544 (2018) https://doi.org/10.1109/TIV.2018.2873900

11. Bieshaar, M., Zernetsch, S., Riepe, K., Doll, K., Sick, B.: Cyclist motion state forecasting – going beyond detection. In: Symposium Series on Computational Intelligence (SSCI), pp. 1–8. Orlando, FL, USA (2021)

12. Bieshaar M.and Depping, M., Schneegans, J., Sick, B.: Starting movement detection of cyclists using smart devices. In: International Conference on Data Science and Advanced Analytics (DSAA), pp. 313–322. Turin, Italy (2018)

13. Bijelic, M., Gruber, T., Mannan, F., Kraus, F., Ritter, W., Dietmayer, K., Heide, F.: Seeing through fog without seeing fog: deep multimodal sensor fusion in unseen adverse weather. In: The IEEE/CVF Conference on Computer Vision and Pattern Recognition (CVPR) (2020)

14. Bikes, B.: Holoscene Edge: AI-Native Bikes (2020). https://www.borealbikes.de/. Accessed: 23 June 2020

15. Blair, W., Bar-Shalom, T.: Tracking maneuvering targets with multiple sensors: does more data always mean better estimates? IEEE Trans. Aerosp. Electron. Syst. **32**(1), 450–456 (1996). https://doi.org/10.1109/7.481286

16. Botache, D., Dandan, L., Bieshaar, M., Sick, B.: Early pedestrian movement detection using smart devices based on human activity recognition. In: Workshop on ICT based Collision Avoidance for VRUs. INFORMATIK 2019, pp. 229–238. Kassel, Germany (2019)

17. Bulling, A., Blanke, U., Schiele, B.: A tutorial on human activity recognition using body-worn inertial sensors. ACM Comput. Surv. **46**(3), 1–33 (2014)

18. Carreira, J., Zisserman, A.: Quo vadis, action recognition? A new model and the kinetics dataset. In: 2017 IEEE Conference on Computer Vision and Pattern Recognition (CVPR), pp. 4724–4733 (2017). https://doi.org/10.1109/CVPR.2017.502

19. Chen, T., Guestrin, C.: Xgboost: A scalable tree boosting system. In: KDD16, pp. 785–794. San Francisco, CA (2016)

20. Clevert, D., Unterthiner, T., Hochreiter, S.: Fast and accurate deep network learning by exponential linear units (elus). In: International Conference on Learning Representations (ICLR). San Juan, Puerto Rico (2016)

21. David, K., Flach, A.: Car-2-x and pedestrian safety. IEEE Veh. Technol. Mag. **5**(1), 70–76 (2010). https://doi.org/10.1109/MVT.2009.935536

22. Depping, M.: Anfahr- und Stoppbewegungserkennung von Fahrradfahrern mittels Smart Devices. Master's thesis, University of Kassel (2018)

23. Fuchs, E., Gruber, T., Nitschke, J., Sick, B.: Online segmentation of time series based on polynomial least-squares approximations. Trans. Pattern Anal. Mach. Intell. (TPAMI) **32**(12), 2232–2245 (2010)

24. Genovese, A.: The interacting multiple model algorithm for accurate state estimation of maneuvering targets. Johns Hopkins APL Technical Digest (Appl. Phys. Lab.) **22**, 614–623 (2001)

25. Gensler, A.: Wind Power Ensemble Forecasting - Performance Measrures and Ensemble Architectures for Deteministic and Probabilistic Forecasts. Intelligent Embedded Systems. Kassel University Press (2019). (Dissertation, University of Kassel, Faculty Electrical Engineering and Computer Science)

26. Goldhammer, M.: Selbstlernende Algorithmen zur videobasierten Absichtserkennung von Fußgängern. Intelligent Embedded Systems. Kassel University Press (2016). (Dissertation, Universität Kassel, Fachbereich Elektrotechnik/Informatik)

27. Goldhammer, M., Köhler, S., Zernetsch, S., Doll, K., Sick, B., Dietmayer, K.: Intentions of vulnerable road users-detection and forecasting by means of machine learning. IEEE Trans. Intell. Transp. Syst. **21**(7), 3035–3045 (2020). https://doi.org/10.1109/TITS.2019.2923319
28. Goldhammer, M., Strigel, E., Meissner, D., Brunsmann, U., Doll, K., Dietmayer, K.: Cooperative multi sensor network for traffic safety applications at intersections. In: IEEE International Conference on Intelligent Transportation Systems (ITSC), pp. 1178–1183 (2012). https://doi.org/10.1109/ITSC.2012.6338672
29. He, K., Zhang, X., Ren, S., Sun, J.: Deep residual learning for image recognition (2015). arXiv:1512.03385
30. He, K., Zhang, X., Ren, S., Sun, J.: Identity mappings in deep residual networks. In: Leibe, B., Matas, J., Sebe, N., Welling, M. (eds.) Computer Vision - ECCV 2016, pp. 630–645. Springer International Publishing, Cham (2016)
31. Ionescu, C., Papava, D., Olaru, V., Sminchisescu, C.: Human3.6m: large scale datasets and predictive methods for 3d human sensing in natural environments. IEEE Trans. Pattern Anal. Mach. Intell. **36**(7), 1325–1339 (2014). https://doi.org/10.1109/TPAMI.2013.248
32. Keller, C.G., Gavrila, D.M.: Will the pedestrian cross? a study on pedestrian path prediction. IEEE Trans. Intell. Transp. Syst. **15**(2), 494–506 (2014). https://doi.org/10.1109/TITS.2013.2280766
33. KI Data Tooling - The Data Kit for Automotive AI (2021). https://www.ki-datatooling.de/. Accessed 03 Dec 2022
34. Kingma, D.P., Ba, J.: Adam: a method for stochastic optimization. In: International Conference for Learning Representations (ICLR). San Diego, CA (2015)
35. Köhler, S., Goldhammer, M., Zindler, K., Doll, K., Dietmeyer, K.: Stereo-vision-based pedestrian's intention detection in a moving vehicle. In: 2015 IEEE 18th International Conference on Intelligent Transportation Systems, pp. 2317–2322 (2015). https://doi.org/10.1109/ITSC.2015.374
36. Kress, V.: Posenbasierte Intentionserkennung von ungeschützten Verkehrsteinehmern aus einem Fahrzeug. Intelligent Embedded Systems. Kassel University Press (2023). (Dissertation, Universität Kassel, Fachbereich Elektrotechnik/Informatik)
37. Kress, V., Jeske, F., Zernetsch, S., Doll, K., Sick, B.: Pose and semantic map based probabilistic forecast of vulnerable road users trajectories. IEEE Trans. Intell. Veh. 1–1 (2022). https://doi.org/10.1109/TIV.2022.3149624
38. Kress, V., Jung, J., Zernetsch, S., Doll, K., Sick, B.: Human pose estimation in real traffic scenes. In: 2018 IEEE Symposium Series on Computational Intelligence (SSCI), pp. 518–523 (2018)
39. Kress, V., Jung, J., Zernetsch, S., Doll, K., Sick, B.: Pose based start intention detection of cyclists. In: 2019 IEEE Intelligent Transportation Systems Conference (ITSC), pp. 2381–2386 (2019)
40. Kress, V., Schreck, S., Zernetsch, S., Doll, K., Sick, B.: Pose based action recognition of vulnerable road users using recurrent neural networks. In: 2020 IEEE Symposium Series on Computational Intelligence (SSCI), pp. 2723–2730 (2020)
41. Kress, V., Zernetsch, S., Bieshaar, M., Reitberger, G., Fuchs, E., Doll, K., Sick, B.: Pedestrians and Cyclists in Road Traffic: Trajectories, 3D Poses and Semantic Maps (2021). https://doi.org/10.5281/zenodo.4898838
42. Kress, V., Zernetsch, S., Doll, K., Sick, B.: Pose based trajectory forecast of vulnerable road users. In: 2019 IEEE Symposium Series on Computational Intelligence (SSCI), pp. 1200–1207 (2019)
43. Kress, V., Zernetsch, S., Doll, K., Sick, B.: Pose based trajectory forecast of vulnerable road users using recurrent neural networks. In: ICPR 2021: Pattern Recognition. ICPR International Workshops and Challenges, pp. 57–71 (2021)
44. Kress, V., Zernetsch, S., Reichert, H., Hetzel, M., Bieshaar, M., Reitberger, G., Fuchs, E., Doll, K., Sick, B.: Aschaffenburg Pose Dataset (2021)
45. Lai, X., Liu, Q., Wei, X., Zhou, G., Han, G.: A survey of body sensor networks. Sensors (Basel, Switzerland) **13**, 5406–47 (2013). https://doi.org/10.3390/s130505406

46. Lin, T.Y., Maire, M., Belongie, S., Bourdev, L., Girshick, R., Hays, J., Perona, P., Ramanan, D., Zitnick, C.L., Dollár, P.: Microsoft COCO: Common Objects in Context (2014). https://doi.org/10.48550/ARXIV.1405.0312. arXiv:1405.0312
47. Niculescu-Mizil, A., Caruana, R.: Predicting good probabilities with supervised learning. In: International Conference on Machine Learning (ICML), pp. 625–632. New York, NY (2005)
48. OpenStreetMap contributors: OpenStreetMap (2017). https://www.openstreetmap.org
49. Pitre, R.R., Jilkov, V.P., Li, X.R.: A comparative study of multiple-model algorithms for maneuvering target tracking. In: Kadar, I. (ed.) Signal Processing, Sensor Fusion, and Target Recognition XIV, vol. 5809, pp. 549–560. International Society for Optics and Photonics, SPIE (2005). https://doi.org/10.1117/12.609681
50. Pool, E.A.I., Kooij, J.F.P., Gavrila, D.M.: Context-based cyclist path prediction using recurrent neural networks. In: 2019 IEEE Intelligent Vehicles Symposium (IV), pp. 824–830 (2019). https://doi.org/10.1109/IVS.2019.8813889
51. Press, O.U.: OED Online (2004). https://www.lexico.com/definition/cooperation. Accessed: 01 Oct 2020
52. Quintero, R., Parra, I., Llorca, D.F., Sotelo, M.A.: Pedestrian path prediction based on body language and action classification. In: 17th International IEEE Conference on Intelligent Transportation Systems (ITSC), pp. 679–684 (2014). https://doi.org/10.1109/ITSC.2014.6957768
53. Reitberger, G.: Detection and Tracking of Vulnerable Road Users. Ph.D. thesis, University of Passau (2023). Unpublished thesis
54. Reitberger, G., Bieshaar, M., Zernetsch, S., Doll, K., Sick, B., Fuchs, E.: Cooperative tracking of cyclists based on smart devices and infrastructure. In: 2018 21st International Conference on Intelligent Transportation Systems (ITSC), pp. 436–443 (2018)
55. Reitberger, G., Sauer, T.: Background subtraction using adaptive singular value decomposition. J. Math. Imaging Vis. **62**(8), 1159–1172 (2020). https://doi.org/10.1007/s10851-020-00967-4
56. Ren, S., He, K., Girshick, R., Sun, J.: Faster R-CNN: towards real-time object detection with region proposal networks. In: Proceedings of the 28th International Conference on Neural Information Processing Systems, NIPS'15, vol. 1, pp. 91–99. MIT Press, Cambridge, MA, USA (2015)
57. Schneegans, J.: Trajectory forecast for cyclists using smart devices and artificial neural networks. Bachelor's thesis, University of Kassel (2018)
58. Schneegans, J., Bieshaar, M.: Smart device based initial movement detection of cyclists using convolutional neuronal networks. Organic Computing: Doctoral Dissertation Colloquium, pp. 45–60. Kassel University Press (2018)
59. Schneegans, J., Eilbrecht, J., Zernetsch, S., Bieshaar, M., Doll, K., Stursberg, O., Sick, B.: Probabilistic vru trajectory forecasting for model-predictive planning – a case study: overtaking cyclists. In: 2021 IEEE Intelligent Vehicles Symposium (IV), Workshop: From Benchmarking Behavior Prediction to Socially Compatible Behavior Generation in Autonomous Driving. Nagoya, Japan (2021). Angenommen zur Veröffentlichung
60. Scholliers, J., van Sambeek, M., Moerman, K.: Integration of vulnerable road users in cooperative its systems. Eur. Trans. Res. Rev. (ETRR) **9**(2), 15 (2017)
61. Snoek, J., Larochelle, H., Adams, R.P.: Practical Bayesian optimization of machine learning algorithms. In: Conference on Neural Information Processing Systems (NIPS), pp. 2951–2959. Lake Tahoe, Nevada (2012)
62. Vaswani, A., Shazeer, N., Parmar, N., Uszkoreit, J., Jones, L., Gomez, A.N., Kaiser, L.u., Polosukhin, I.: Attention is all you need. In: Guyon, I., Luxburg, U.V., Bengio, S., Wallach, H., Fergus, R., Vishwanathan, S., Garnett, R. (eds.) Advances in Neural Information Processing Systems, vol. 30. Curran Associates, Inc. (2017)
63. Yuan, T., Bar-Shalom, Y., Willett, P., Mozeson, E., Pollak, S., Hardiman, D.: A multiple imm estimation approach with unbiased mixing for thrusting projectiles. IEEE Trans. Aerosp. Electron. Syst. **48**(4), 3250–3267 (2012). https://doi.org/10.1109/TAES.2012.6324701
64. Zernetsch, S.: Maschinelle Lernverfahren zur videobasierten Intentionserkennung von Radfahrern mit stationären Kameras. Intelligent Embedded Systems. Kassel University Press (2022). (Dissertation, Universität Kassel, Fachbereich Elektrotechnik/Informatik)

65. Zernetsch, S., Kress, V., Bieshaar, M., Reitberger, G., Fuchs, E., Doll, K., Sick, B.: Cyclist Actions: Motion History Images and Trajectories (2020)
66. Zernetsch, S., Kress, V., Bieshaar, M., Reitberger, G., Fuchs, E., Doll, K., Sick, B.: Cyclist Actions: Optical Flow Sequences and Trajectories (2020)
67. Zernetsch, S., Kress, V., Bieshaar, M., Reitberger, G., Fuchs, E., Doll, K., Sick, B.: Vru trajectory dataset (2022)
68. Zernetsch, S., Kress, V., Sick, B., Doll, K.: Early start intention detection of cyclists using motion history images and a deep residual network. In: 2018 IEEE Intelligent Vehicles Symposium (IV), pp. 1–6. Changshu, China (2018)
69. Zernetsch, S., Reichert, H., Kress, V., Doll, K., Sick, B.: Trajectory forecasts with uncertainties of vulnerable road users by means of neural networks. In: 2019 IEEE Intelligent Vehicles Symposium (IV), pp. 810–815. Paris, Frankreich (2019)
70. Zernetsch, S., Reichert, H., Kress, V., Doll, K., Sick, B.: A holistic view on probabilistic trajectory forecasting - case study. cyclist intention detection. In: 2022 IEEE Intelligent Vehicles Symposium (IV), pp. 265–272 (2022). https://doi.org/10.1109/IV51971.2022.9827220
71. Zernetsch, S., Schreck, S., Kress, V., Doll, K., Sick, B.: Image sequence based cyclist action recognition using multi-stream 3d convolution. In: 2020 25th International Conference on Pattern Recognition (ICPR), pp. 2620–2626 (2021). https://doi.org/10.1109/ICPR48806.2021.9413233
72. Zernetsch, S., Trupp, O., Kress, V., Doll, K., Sick, B.: Cyclist trajectory forecasts by incorporation of multi-view video information (2021). Zur Veröffentlichung eingereicht bei IEEE International Smart Cities Conference 2021
73. Zhou, Z.: Ensemble Methods: Foundations and Algorithms. Machine Learning & Pattern Recognition Series. Chapman & Hall/CRC, Bocan Raton, FL (2012)

Analysis and Simulation of Driving Behavior at Inner City Intersections

Hannes Weinreuter, Nadine-Rebecca Strelau, Barbara Deml, and Michael Heizmann

Abstract Inner city intersections are a challenging scenario for human drivers as well as for the development of autonomous vehicles. This is especially the case for unsignalized intersections where the *right before left* rule applies. At these intersections, ambiguous situations can arise. In this chapter, we cover two aspects of this intersection type: First, we use driving data from a field study conducted in inner city traffic to analyze the relationship between intersections and human driving behavior. For that, we describe the intersection, its surrounding environment and the traffic there by features that constitute an intersection's complexity (e.g. street width, visibility conditions, number of cooperation vehicles). With those we are able to predict features describing the driving behavior reliably. Second, we propose a decision making algorithm for unsignalized inner city T-junctions. The algorithm is modeled as a discrete event system and does not rely on any explicit communication. Instead, only the observable state is used. This includes the map, the positions and velocities of the cooperation vehicles and the driving pattern. We introduce the algorithm in detail and present results of a comprehensive simulation for validation. The algorithm is able to drive through all situations in the simulation safely.

H. Weinreuter (✉) · M. Heizmann
Institute of Industrial Information Technology (IIIT) at Karlsruhe Institute of Technology (KIT),
Karlsruhe, Germany
e-mail: hannes.weinreuter@kit.edu

M. Heizmann
e-mail: michael.heizmann@kit.edu

N.-R. Strelau · B. Deml
Institute of Human and Industrial Engineering (ifab) at Karlsruhe Institute of Technology,
Karlsruhe, Germany
e-mail: nadine-rebecca.strelau@kit.edu

B. Deml
e-mail: barbara.deml@kit.edu

© The Author(s) 2024
C. Stiller et al. (eds.), *Cooperatively Interacting Vehicles*,
https://doi.org/10.1007/978-3-031-60494-2_4

1 Introduction

The ongoing development of autonomous driving is a promising field of research. When autonomous vehicles are finally admitted onto public roads, one can expect several benefits from them. They have the potential to reduce the number and severity of traffic accidents. Additionally, it would enable people who are unable to drive for themselves access to individual mobility. There are, however, several aspects of autonomous driving that currently prohibit its introduction into real world traffic. Among them is driving through inner city traffic and especially at unsignalized intersections. This intersection type is common in Germany in areas with low or medium traffic density. At these intersections the *right before left* rule applies. It states that one has to yield to a driver approaching on the next street to one's right and that one has priority over a driver approaching from the next street to the left. Oncoming traffic has priority over turning left. This rule does not, however, provide a defined driving order in all possible scenarios. Instead, situations can occur in which each driver has to yield to at least one other driver, thus creating a deadlock at the intersection. In this case the German traffic regulations for example only state that driving before someone who has priority may only occur after the drivers communicated and thus cooperated with each other [1]. This of course is problematic for an autonomous vehicle (A-V) as it has to interpret human behavior, make a decision based on potentially unreliable predictions and still drive safely and in a way that is acceptable to both its passengers and its human interaction partners.

In this work we focus on two aspects of driving through unsignalized inner city intersections. The first aspect is how intersections influence driving behavior [42]. For that we describe an intersection by intersection complexity. We define intersection complexity based on features which describe an intersection. This includes both the static environment (e.g. visibility or the street width) and the dynamic environment, i.e. the traffic at the intersection. Driving behavior is described based on features obtained from the driven trajectory. We then predict the behavior features using the intersection features as inputs. The basis for that is data from a field study in real world traffic. The study, both the intersection and the behavior features and the prediction are described in detail in Sect. 3. The second aspect of this work focuses on the decision making at unsignalized intersections [43]. We present a decision making algorithm based on a discrete event system (DES) that is able to drive according to the traffic regulations. It is also able to cope with unclear situations like deadlocks or if a vehicle yields despite not having to. The strategy to solve these situations is based on the findings by [20]: They found that human drivers prefer not having to drive first in demanding situations such as a deadlock at a T-junction. Our approach does not require any explicit communication between the vehicles, the decisions are based only on the observable state of the cooperation vehicles, i.e. its position, velocity and acceleration. This is in line with findings from literature that state that human drivers rely on implicit communication when approaching such scenarios [19]. The algorithm, alongside a detailed validation, is presented in Sect. 4.

2 Related Work

Aspects of this work have been covered in literature before. We first present relevant publications for the behavior analysis as described in Sect. 3, and then on the behavior generation (Sect. 4). The first aspect of this work focuses on the influence of intersection complexity on the driving behavior. There are previous publications that use features describing the environment of a driving task to define complexity. [9] assume inner city scenarios as most complex and driving on a highway as least complex. The type of scenario can also be used to discriminate between complexity levels, [20] found a T-junction to be more complex than a symmetrical narrow passage. Further features that have been used before include the difference between signalized and unsignalized intersections [24], whether or not parked vehicles at the side of the road are present [8] and if a driver drove straight through an intersection or turned right or left [12]. Reference [45] uses satellite images and classifies intersections as complex if they have at least one street with multiple lanes, traffic islands, sliplanes or more than four roads leading into the intersection. Another possible feature is visual clutter [14]. All these features so far describe stationary surroundings. However, one can also consider the dynamic environment, i.e. the traffic, to describe the complexity of a situation. Reference [31] defines high complexity as situations that have high demands on both information processing and vehicle control and low complexity if there is low demand for either category. A medium complexity is assigned to scenarios that require high demand in one category and low demand in the other. Reference [21] uses the same definition but omits the medium class. Traffic density [28, 39, 44] can be considered for complexity as well as the occurrence of lane changes [39] or driving after a congestion compared to regular driving [23]. Further aspects of traffic and the environment of an intersection have also been studied, [44] included the number of vehicles from the left and whether or not a zebra crossing was present in their work. Reference [30] defines complexity by the grade of urbanization, the presence of oncoming traffic, leading traffic and the street geometry (straight road, tight corner, soft corner). Reference [4] considers a straight road as less complex than an intersection at which a stop is required or an overtaking maneuver. Reference [15] defines complexity by the number of advertisement signs, buildings, oncoming vehicles and further infrastructure while driving on a highway.

The second aspect of this work deals with decision making in the context of autonomous driving and has also been the focus of many authors. A common method for decision making at intersections and other traffic scenarios are partially observable Markov decision processes (POMDP): [26] uses a POMDP for decision making at intersections and roundabouts. Reference [18] uses a POMDP for real-time decision making where other vehicles are treated as hidden variables to adapt the driving behavior to the most likely behavior of the other drivers. Reference [38] applies a POMDP for decision making at an intersection while turning left. The autors define several critical turning points from which a turn can be executed and select the most efficient one. Additionally, one can also consider limited visibility caused by both static and dynamic objects. A possible solution for that problem is to add virtual vehicles at the edge of the obscured space [25]. Reference [2] uses POMDPs for decision making at intersections and pedestrian crossings with limited visibility. Besides POMDP, further methods for decision making have been employed as well. Reference [37] uses a mixed observability Markov decision process to predict the intention of cooperation partners and base the decision on that. Reference [29] presents a framework that combines prediction, threat detection and decision making. Using a Bayesian network the threat levels of other vehicles are classified and the decision is based on that. A decision can also be made by evaluating possible behavior policies and selecting the optimal one [5, 11]. Reference [6] selects the trajectory of an autonomous vehicle from a list of reference trajectories from human drivers during interaction with an additional vehicle. Finally, one can use a game theoretic approach by considering a game between the ego vehicle and the first oncoming vehicle [36].

All these works have in common that they do not rely on explicit communication between vehicles. Instead they rely on the vehicles' states that are observable by onboard sensors. Alternatively, decision making at intersections can also be designed to use explicit communication between the vehicles themselves or between the vehicles and a centralized coordination mechanism. Reference [27] presents an algorithm for coordination of autonomous vehicles at an intersection using model predictive control. This decentralized approach requires all vehicles to use the same algorithm and to share their current state. Reference [34] presents a centralized coordination algorithm for autonomous vehicles at unsignalized intersections. The vehicles are assigned arrival times and the problem is formulated as an absolute value problem. Reference [10] determines the driving order by centralized coordination using a mixed-integer linear problem. All vehicles transmit their state and receive their allotted time to pass the intersection. They regulate their velocity accordingly. Versions for mixed traffic and traffic lights are also suggested.

Certain aspects of inner city traffic have been modeled as DES before by using Petri-nets (PN). Reference [41] models an intersection with traffic lights using PNs for the traffic light control and to model the traffic flow. A PN can also be used to model the traffic light control mechanism at several connected intersections as well, using the largest intersection as the master control [16]. PN based traffic lights control can also be used to give arriving emergency vehicles green light at intersections [17]. Reference [7] models a city environment consisting of intersections with traffic lights and connecting streets using deterministic time-based PNs. Reference [33] controls intersections with traffic lights using deterministic and stochastic PNs. The model is adapted in case of incidents that would otherwise cause neighboring intersections to be blocked.

In this work we do not rely on explicit communication with the cooperation vehicles. Instead, the decision making is based only on the observable state of the other vehicles. We consider this to be more realistic, especially in the short term, as we cannot expect every vehicle to be equipped with such communication interfaces anytime soon. We further rely on DES as decisions by the system are easily explainable and they are made using only basic operations.

3   Intersection Complexity for Behavior Prediction

In order to autonomously drive through unsignalized inner city intersections, it is helpful to understand why human drivers drive the way they do. This is important for two reasons: Autonomous vehicles will have to interact with human drivers for the foreseeable future. An understanding of human driving behavior might make these interactions more safe and efficient. It might enable autonomous vehicles to predict the driving behavior of their interaction partners more reliably. One can secondly make such systems behave similar to human drivers, this could improve their acceptance. The evaluation of this section is based on a field study that was conducted in the inner city of Karlsruhe in Germany [42]. In that study 34 participants drove through a predefined course during which they encountered several unsignalized intersections. At one of the intersections they were confronted with instructed drivers who created a deadlock situation. In this work we are investigating the interaction with regular traffic, therefore the runs through this intersection are not part of this work. The data set includes in total 1818 runs through 13 unsignalized T-intersections and 565 runs through 4 unsignalized X-intersections. Four of the remaining T-intersections were specifically selected. This way we were able to include intersections with high and low traffic density and intersections with buildings close to and far from the street. The remaining intersections are included in the data set as they lie along the drive path between the selected intersections. The test vehicle was equipped with a 16 channel lidar, an inertial measurement unit (IMU) and two global navigation satellite system (GNSS) receivers. The data was recorded using the robotic operating system (ROS) [35] and the driven trajectory as well as the transformation of the point clouds to a global reference were generated using a simultaneous localization and mapping (SLAM) approach [13]. We then generated our data set by extracting the runs through the intersections which are included in the analysis. For that only those parts of the trajectory are included in a run that lie within a 35 m radius around the intersection center. Within the point clouds vehicles and pedestrians are detected and their trajectories are tracked. We have presented the work described in this section before in more detail [42].

3.1   Intersection and Behavior Features

From the recorded and preprocessed data we then extract several features to describe both the intersection itself and its surroundings. As we additionally need a way to describe the driving behavior of the participants, behavior features are calculated from the driven trajectories as well. The intersection features include features describing properties of the driven path, the intersection itself and features about the traffic at the intersection the participant had to interact with. The set of all features can be seen as the complexity of an intersection.

The driven path is described by the entry position and the turning direction. For the entry position p_e the T-intersection is rotated such that it resembles the letter "T". The entry position can then either take the value *left*, *bottom* or *right*. The entry position is not considered in case of the X-intersections because of their symmetry. The turning direction p_t takes one of the values *left*, *straight* or *right*. At T-intersections not all turning directions are possible depending on the entry position.

Further, we define features that describe the traffic at the intersection the participants had to interact with. For that we use the number of pedestrians n_p and the number of vehicles n_v as features. Both pedestrians and vehicles are counted if they are detected in the point clouds during the approach to the intersection. Please refer to [42] for further details on the detection and tracking. The visible vehicles are divided into further features: The number of interaction vehicles n_{vi} are those vehicles that are within 10 m from the intersection center at the same time as the test vehicle. In order to be counted their observed track has to pass the intersection center. The interaction vehicles are further analyzed if they have the right of way over the test vehicle or if they have to give way; the number of vehicles that fulfill these conditions are counted in n_{rw} and n_{gw}, respectively.

The final class of intersection features is designed to describe the static environment at the intersection. Among them is the number of trees n_t that are near the intersection and the road a participant uses to enter the intersection. To judge the occlusion of an intersection during the approach we include visibility distances. These are the distances at which reference points in the streets to the left and right of the street the vehicle enters the intersection from are visible for the first time. The reference points are placed on the center line of the streets at a distance of

$$
d_{ref} = v_{max} t_r + \frac{v_{max}^2}{2|a_b|} \tag{1}
$$

from the intersection center. This is the distance that is needed to stop when driving at the speed limit of $v_{max} = 30$ km h^{-1}. With a reaction time of $t_r = 1$ s and a braking deceleration of $a_b = 6$ ms^2, the distance of the reference points is $d_{ref} = 14.12$ m. We use two variants to calculate the visibility distance, an approach based on the point clouds and one based on object polygons. For the point clouds variant we merge the current and the two point clouds before and after to the merged point cloud $\mathbf{P}(d)$. This represents the merged point cloud at distance d from the intersection center. For that the current trajectory point is projected onto the center line of the current lane, the distance is then measured along the lane center. Within $\mathbf{P}(d)$ cylinders $C_{s,i}$ with a radius of 0.6 m are placed between the current location and the reference points i. If there is at least one point of $\mathbf{P}(d)$ within $C_{s,i}$, reference point i is considered not visible at distance d. The visibility distance $d_{v,c,i}$ to each reference point is then the distance at which the reference point is visible for the first time. Alternatively, we use polygons of the buildings and tree trunks along the intersection to determine the visibility distance. For that we draw a sight line between the current location and the reference points. If this line does not intersect with any polygon, the reference point is visible. Again, the first distance d for which this is true determines the visibility distance $d_{v,p,i}$ of a reference point. The visibility distance of an intersection is the minimum visibility distance of all its reference points: $d_{v,\cdot} = \min_i (d_{v,\cdot,i})$. To include the actual and perceived narrowness of the road leading into the intersection, we define three widths that are calculated along the normal of each point of the trajectory. The street width $w_s(d)$ is the distance from the intersection points of the normal at distance d with the street curbs and is calculated based on the map of the intersection. For the visible range the point clouds are analyzed. It describes how far a driver can see to the left and right and is supposed to model the perceived narrowness of the street. For each trajectory position the lidar data is evaluated along the normal at sensor height. The first point within $\pm 5°$ in vertical direction and $\pm 10°$ in horizontal direction determines the visual range. For the visual range $w_v(d)$ this is performed both to the left and right of the trajectory. The available width $w_a(d)$ is a combination of the previous two widths and describes the space on the street that is available to drive on. At each trajectory point the smaller one of the street width $w_s(d)$ and visual range $w_v(d)$ determines the available width. For this the calculation of the available width is adapted such that it includes all points within $\pm 15°$ in vertical direction. All three widths are averaged over the approach interval from 25 m to 7 m before the intersection center. A more detailed introduction into the features discussed here can be found in [42].

To describe the driving behavior at the intersections, we define three features based on the driven trajectory: the commit distance, the velocity drop and the minimum velocity. The commit distance is the distance from the intersection center at which, given the current velocity, stopping before the intersection center is no longer possible:

$$d_c = \max_d \left(d < v(d) \, t_r + \frac{v(d)^2}{2|a_b|} \right). \tag{2}$$

The commit distance can be interpreted as a measure for the distance at which the final decision to drive is made. The further from the intersection, the more offensive the driving behavior. The minimum velocity is the minimum velocity that the driver assumed during the approach interval of $d_s = 25$ m to $d_e = 0$ m distance to the intersection center:

$$v_{min} = \min(v(d)), \qquad d_s > d > d_e. \tag{3}$$

The final behavior feature is the velocity drop. It describes the ratio between the minimal velocity during the approach v_{min} to the mean initial approach velocity v_a in the interval from 25 m to 20 m:

$$v_d = \frac{v_{min}}{v_a}. \tag{4}$$

3.2 Prediction of Driving Behavior

Using the intersection and behavior features from above we can now predict the driving behavior. For that we train several Random Forest (RF) [3] regression models. RFs are employed because of their ease of use and because they can model non-linear dependencies [22]. Several other regression methods could be used here as well. We use the intersection features, or a subset of them, as predictors and predict the behavior features. For each combination of the three behavior features and the two intersection types (X- and T-intersections) 10 models are trained. For each of the 10 models 70 % of the runs are used as the training set, the remaining 30 % are used as the test set. In Table 1 the average and standard deviation of the 10 models are given for all variants. The performance of the RF regression models is evaluated using the root mean squared error (RMSE):

$$
\text{RMSE} = \sqrt{\frac{1}{N} \sum_{k=1}^{N} \left(\hat{y}_k - y_k \right)^2}. \tag{5}
$$

N is the number of runs in the test set, y_k is the behavior feature of the k-th run of the test set and $\hat{y}_k$ is the value of the behavior feature estimated by the regression model for the same run. A first analysis was performed using the entire feature set as introduced in Sect. 3.1. For the T-intersection models all 13 features were used. In the case of the X-intersections the entry position p_e was omitted as a feature. The results of that analysis are given in the first row of Table 1. The last row of this table contains the reference value, that is the results of a naive regression model that outputs the mean of the training set. The prediction error of the driving behavior for all three behavior features is well below the reference value with a low standard deviation for both the T-intersections and the X-intersections. The performance of this regression model is especially noteworthy given the fact that driving behavior might also be influenced by a driver's personality or mood.

Additionally, we investigate whether a dimensionality reduction of the feature set is feasible. For that we first select a subset of the most relevant complexity features. This selection is a compromise between the feature importance of all investigated model variants. The remaining features are the entry position p_e (only for the T-intersections), the turning direction p_t, both visibility distance variants $d_{v,c}$ and $d_{v,p}$, the street width w_s and the available width w_a, the number of trees n_t and the number of visible vehicles n_v. This means that there is only one feature describing the traffic. This might, at least in part, be explained by the fact that most runs did not include any cooperation partners as this study was conducted in regular traffic. The performance of the RF regression models with that feature set are given in the second row of Table 1. The regression is less accurate than with the full feature set, but the performance is very similar, indicating that these reduced complexity feature sets are sufficient to predict the driving behavior at intersections.

Table 1 Mean RMSE regression results for T-intersections and X-intersections using different feature sets and all behavior features: commit distance d_c, minimum velocity v_{min} and velocity drop v_d. The standard deviation is in brackets

	d_c in m		v_{min} in m s^{-1}		v_d	
	T-int.	X-int.	T-int.	X-int.	T-int.	X-int.
Full feature set	1.492	1.696	1.033	1.150	0.153	0.159
	(0.050)	(0.093)	(0.036)	(0.057)	(0.005)	(0.008)
Reduced feature set	1.512	1.728	1.068	1.173	0.157	0.162
	(0.049)	(0.110)	(0.036)	(0.069)	(0.006)	(0.008)
Directions feature set	1.800	2.590	1.298	1.686	0.187	0.209
	(0.068)	(0.084)	(0.041)	(0.045)	(0.006)	(0.008)
Reference	3.093	3.229	1.977	2.000	0.275	0.256
	(0.116)	(0.135)	(0.051)	(0.075)	(0.004)	(0.011)

As the entry position p_e and turning direction p_t are relevant factors to the driving behavior [42], we also train models with only these two complexity features. In case of the X-intersections we only use the turning direction p_t. The performance of these RF regression models is given in the third row of Table 1. The results show that prediction is still possible, the performance, however, decreases substantially compared to the full and reduced feature sets. This is especially true for the X-intersection. A possible explanation for the reduced performance might be that both features can only assume three distinct values each. Thus there are only six distinct value combinations possible in the case of the T-intersections and only three combinations for the X-intersections. This limits the number of possible regression values to the same numbers, thus causing a less accurate regression.

4 Behavior Generation

The second aspect of this work focuses on an approach to decide on the behavior of an A-V at a T-intersection, i.e. whether it drives first or waits for its cooperation vehicles (C-V) to pass the intersection before it. Both this high-level decision and the resulting longitudinal acceleration of the A-V is covered by our proposed decision making algorithm. There are several challenges associated with this problem: As the driving paths of the A-V and its C-Vs intersect, there oftentimes is no solution that guarantees safety from collisions in any possible scenario. This would only be possible if the A-V always waits for all other vehicles to drive first. This, however, is not a feasible option. It would firstly lead to a deadlock if there is another A-V with the same strategy. This behavior could secondly be more confusing than helpful when interacting with human drivers, especially given that human drivers prefer others to drive first in complex scenarios such as deadlocks at T-intersections [20]. In order to avoid these problems, a certain degree of risk has to be accepted. Also, another challenge is the number of possible interactions between the vehicles that are involved in the situation. If all pairwise interactions are explicitly modeled the model is dependent on the number of cooperation partners. Also, explicitly modeling all interactions would be challenging.

4.1 Basic Setup

The algorithm is modeled as a discrete event system (DES) and does not assume any communication between the vehicles. The only available information is the observable state of the C-Vs, i.e. their position, speed and acceleration and the map of the intersection. As soon as a C-V is closer than 10 m from the start of the intersection we assume that the turning direction is known, e.g. by observing the indicators or from the driven trajectory. There exist previous works from literature that support this assumption [32, 46]. In this work the vehicles follow the center line of their lane, so only the longitudinal acceleration has to be controlled. The map is a generic T-intersection with a 90° angle between the bottom street and the street going straight, see Fig. 1 for a schematic. Additionally, we consider occlusions at the intersection. For that we define two points that specify the corners of obstacles between the streets that block the direct line of sight. These points are placed on the bisecting lines between the streets and the distance from the curb is used to parameterize the visibility conditions.

To simplify the model and reduce the number of vehicles that have to be evaluated, we only consider those vehicles that are currently relevant to the A-V. Each of these vehicles is evaluated independently. With that strategy we avoid having to model the interaction between all possible pairs of vehicles as well. Each of the relevant C-Vs is assigned a virtual traffic light that is either red or green. The A-V only drives offensively if all traffic lights are green, a red light thus means that the A-V cannot drive due to that vehicle. The first relevant C-V is the vehicle that has priority (P-V) over the A-V, i.e. the vehicle closest to the intersection on the next street to the right. If the A-V will turn into the next street to the right itself, there is no P-V as the A-V does not have to yield to any vehicle in this case. Additionally, the vehicle that has to yield (Y-V) to the A-V has to be taken into consideration. The Y-V is the vehicle closest to the intersection that is approaching on the next street to the left. If its path does not intersect with the A-V's path, the vehicle behind it is evaluated. To ensure a safe passage of the intersection, two more vehicles have to be considered. The blocking vehicle (B-V) is the closest vehicle that is leaving the intersection on the same road as the A-V will and the leading vehicle (L-V) is the vehicle driving directly in front of the A-V on its path. The B-V and the L-V can be the same vehicle. All these vehicles are relevant for the decision of the A-V as either their paths intersect with the A-V's (this is the case for the P-V and the Y-V) or because they can hinder

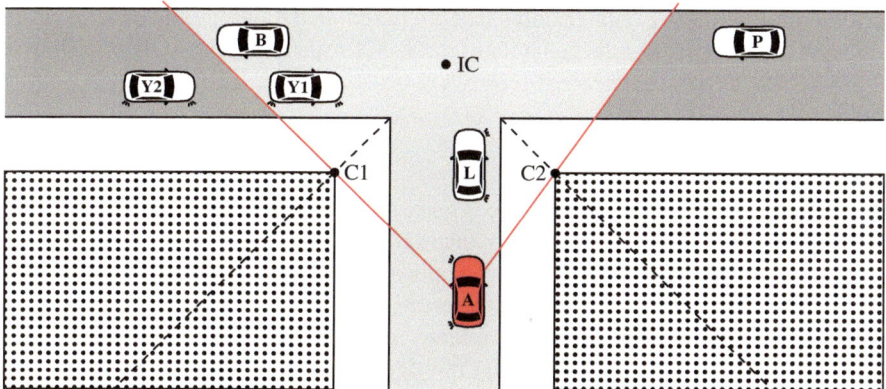

Fig. 1 Schematic representation of a scenario at a T-intersection. The visibility is determined by the visibility edges C1 and C2. These are placed on the bisecting lines between the streets originating from the intersection center IC. With that the visible street area can be calculated. In this case vehicles B, Y1 and L are visible, vehicles P and Y2 are not visible. The A-V enters the intersection from the bottom direction and turns left. It has to yield to vehicles from the right and has priority over vehicles from the left. Therefore, vehicle P is the P-V (as soon as it becomes visible). As both Y1 and Y2 are turning right, there is no Y-V. If Y2 were to drive straight it would be assigned the Y-V even before its preceding vehicle Y1 passes the intersection. Vehicle B is the B-V as it is driving on the road the A-V intends to enter and is potentially blocking this road if it is too close to the intersection. Vehicle L is driving directly in front of the A-V and is thus the L-V

the A-V from leaving the intersection right away (in the case of the B-V or the L-V). We only consider the vehicles closest to the intersection as only those are directly relevant for the decision of the A-V. A vehicle behind e.g. the P-V is irrelevant as it cannot interact with the A-V as long as the P-V is before the intersection. The same is true for the L-V: The vehicle driving in front of the L-V does not directly affect the A-V. If one of the C-Vs passes the intersection the situation is re-evaluated, the labels are assigned anew and all considerations are based on the new assignments. In the case of limited visibility the A-V might currently not be able to see some of the vehicles, despite them existing. To cope with that possibility certain non-existence is only assumed if a reference point that is placed on the road center at a radius of 25 m from the intersection center is visible. In the case of the B-V the reference point is set to a distance of 15 m and the existence of the L-V is assumed to be known in any case. If the turning direction is not yet known, the worst case is assumed. Both the vehicle assignment and the visibility is showcased in Fig. 1.

4.2 Decision Making Algorithm

As the algorithm for decision making is modeled as a DES, the vehicle is described and controlled by its current state. The state only changes if an event occurs. For the definition of these events features that are based on the observable data are used. Based on the current state the behavior of the A-V, i.e. its acceleration, is determined.

4.2.1 Features

To indicate for which vehicle a feature is calculated, it is marked by a corresponding index: $(\cdot)^x$, $x \in \{a, p, y, b, l\}$. All distances are measured along the drive path of a vehicle. The distance to scenario $d_s^x(t)$ is positive before, zero within and negative after the intersection. The begin of an intersection is defined as the point where lanes diverge and the end is the point where lanes merge. All features are calculated for the current time t. For better readability this dependence is omitted in the following.

At an intersection, the drive paths of vehicles oftentimes intersect. The area where the lanes of two vehicles overlap is referred to as the common collision zone. For the algorithm only the distances to the collision zones of the A-V with its C-Vs are needed. $d_{c,x_c,b}^x$ and $d_{c,x_c,e}^x$ are the distance of vehicle x to the beginning and the end of the collision zone of the A-V with the C-V x_c. The distance of the A-V to the beginning of the collision zone with the P-V is then $d_{c,p,b}^a$ and the distance of the P-V to the beginning of the same zone is $d_{c,p,b}^p$. Based on the distance to collision zone the time to collision zone is calculated using the current velocity v^x of vehicle x:

$$t_{c,x_c,\cdot}^x = \frac{d_{c,x_c,\cdot}^x}{v^x} . \tag{6}$$

Additionally, the distance required to brake to a complete stop assuming the velocity v_a^x and the acceleration a_a^x is used as a feature:

$$d_b^x \left(v_a^x, a_a^x \right) = \begin{cases} -\frac{(v_a^x)^2}{2a_a^x}, & a_a^x < 0 \, \text{ms}^{-2} \\ 0, & a_a^x = 0 \, \text{ms}^{-2} \wedge v_a^x = 0 \, \text{ms}^{-2} \\ \infty, & \text{otherwise} \end{cases} . \tag{7}$$

The distance to the last stopping point d_l^x is the distance to the point a vehicle has to stop to not interfere with any other driving path through the intersection. The final feature is the free distance behind the B-V. This feature measures the distance between the end of the intersection and the rear of the B-V including the distance to break in an emergency ($a_e = -7.5 \, \text{ms}^{-2}$) from the current velocity:

$$d_f^b = d_i^b - \frac{1}{2} l_v + d_b^b \left(v^b, a_e \right) , \tag{8}$$

where d_i^b is the current distance along the driven path from the end of the intersection and $l_v = 4.4$ m is the length of the vehicle.

4.2.2 Events

In our model the behavior is supposed to differ depending on the distance of the A-V to the intersection. Thus, the approach to the intersection is split into six zones. The current zone is determined by the A-V's distance to scenario d_s^a. In the first zone ($d_s^a > 40$ m) the A-V is not controlled by the decision making algorithm but drives freely. At the beginnings of the second (40 m $\geq d_s^a > 25$ m) and the third (25 m $\geq d_s^a > 10$ m) zone a single prediction of the P-V is performed and the behavior of the A-V is adapted accordingly. The A-V adapts its behavior to show its intention as early as possible. The prediction is only run twice to avoid changing the behavior too often. The fourth zone is the area just before the intersection (10 m $\geq d_s^a > 1$ m). In it the A-V constantly monitors the behavior of its C-Vs and adapts its own behavior if necessary. Zone 5 is the area within the intersection itself (1 m $\geq d_s^a \geq 0$ m). In these last two zones the final decision on the behavior has to be made and then executed accordingly. The final zone 6 is the street past the intersection where the vehicle is no longer controlled by the decision making algorithm.

The model is based on events, most events are themselves a combination of so called base events. Their meaning and definition is shown in Table 2 and the events are presented in Table 3. Each of the four relevant C-Vs has a traffic light event assigned to it. The P-V is the only vehicle that has two variants of that event. In

Table 2 Base events for the DES for decision making

Name	Description	Condition
e_{b1}	Certain non-existence of P-V	Ref. point is visible and no P-V detected
e_{b2}	No conflict expected with P-V	$t_{c,p,e}^a + \Delta t_p < t_{c,p,b}^p \wedge d_{c,p,e}^a + \Delta d_p < d_{c,p,b}^p$
e_{b3}	P-V stopped near intersection	$v^p < v_s \wedge a^p \leq 0\frac{m}{s^2} \wedge d_s^p < d_n \wedge d_{c,p,b}^p > 0$ m
e_{b4}	P-V yields	$t_w^p > t_y$
e_{b5}	Y-V inside collision zone	$d_{c,y,b}^y < 0$ m $\wedge d_{c,y,e}^y > 0$ m
e_{b6}	A-V can pass before Y-V	$t_{c,y,e}^a < t_{c,y,b}^y$
e_{b7}	Stop possible (comfort dec.)	$d_l^a > d_b^a (v_i, a_c)$
e_{b8}	Y-V stops & A-V could brake	$d_l^a > \left(d_b^a (v_i, a_h) + 0.2\,\text{m} \right) \wedge v^y < v_{sl}$ $\wedge a^y < 0\frac{m}{s^2} \wedge d_{c,y,b}^y > d_b^y (v^y, a^y)$
e_{b9}	Y-V stopped near intersection	$v^y < v_s \wedge a^y \leq 0\frac{m}{s^2} \wedge d_s^y < d_n \wedge d_{c,y,b}^y > 0$ m
e_{b10}	Certain non-existence of B-V	Ref. point is visible and no B-V detected
e_{b11}	Enough space behind B-V	$d_f^b > l_v + d_{min}$
e_{b12}	L-V does not exist	No L-V detected
e_{b13}	L-V passed intersection	$d_s^l < 0$ m
e_{b14}	Stop possible (emergency dec.)	$d_l^a > d_b^a (v^a, a_e)$
e_{b15}	Deadlock possible	A-V, P-V, Y-V: turning directions intersect
e_{b16}	A-V stopped near intersection	$v^a < v_s \wedge a^a \leq 0\,\text{m s}^{-2} \wedge < d_s^a < d_n$

Table 3 Events of the DES for decision making. Most events are a combination of base events

Definition	Description
$e_{1,p,I} = e_{b1} \vee e_{b2}$	Green light from P-V in zones 2 and 3
$e_{1,p,II} = e_{b1} \vee e_{b2} \vee (e_{b3} \wedge e_{b4})$	Green light from P-V in zones 4 and 5
$e_{1,y} = \neg e_{b5} \wedge (e_{b6} \vee e_{b7} \vee e_{b8} \vee e_{b9})$	Green light from Y-V
$e_{1,b} = e_{b10} \vee e_{b11}$	Green light from B-V
$e_{1,l} = e_{b12} \vee e_{b13}$	Green light from L-V
e_2	Entered next zone
$e_3 = e_{b14}$	Emergency stop possible
$e_4 = e_{b15}$	Deadlock possible
$e_5 = e_{b3} \wedge e_{b9} \wedge e_{b16}$	Deadlock detected
$e_6 = e_{b3} \wedge e_{b9}$	Deadlock of C-V detected

the prediction phase (zones 2 and 3) its light is green (event $e_{1,p,I}$) if the A-V is either certain that no P-V exists (base event e_{b1}) or if it does not expect a conflict with its P-V (the A-V is predicted to enter the intersection at least $\Delta t_p = 2.5$ s and $\Delta d_p = 10$ m earlier, e_{b2}). In zones 4 and 5 the light is additionally set to green ($e_{1,p,II}$) if the P-V is currently stopped close to the intersection (the velocity is below the stop threshold of $v_s = 0.15$ m s^{-1}, it does not accelerate and it is closer than the threshold $d_n = 12$ m to the start of the intersection, e_{b3}) and the wait time t_w^p has exceeded its $t_y = 2$ s limit (i.e. both vehicles stood for 2 s at the intersection and it is not due to a deadlock, e_{b4}). The parameters are either set to the authors considerations and are thus options to parameterize the model or are due to physical constraints.

The traffic light of the Y-V ($e_{1,y}$) is green if the Y-V is currently not within the common collision zone ($\neg e_{b5}$) and if at least one of these events is true: The A-V is predicted to be able to pass the collision zone before the Y-V (base event e_{b6}); the Y-V is stationary close before the intersection (e_{b9}); the distance to the last stop point of the A-V is still large enough so that it is able to stop before it without exceeding the comfort deceleration of $a_c = -2.5$ m s^{-2} and assuming a velocity within the intersection of $v_i = 6.5$ m s^{-1} if driving straight and $v_i = 4.0$ m s^{-1} if turning (e_{b7}); the Y-V is slow ($v_{sl} = 2$ m s^{-1}), it currently brakes such that it will come to a complete stop before the beginning of the collision zone and the A-V has enough space remaining for a hard stop ($a_h = -4.5$ m s^{-2}) if it should become necessary (e_{b8}). The latter two base events allow the A-V to drive despite currently not being predicted to pass the intersection before the Y-V. With these conditions we avoid unnecessarily defensive behavior. Only if the A-V is very close to the intersection and still cannot drive first safely, it yields to the Y-V.

The B-V gives green light ($e_{1,b}$) if the A-V is certain that it does not exist (base event e_{b10}) or if there is enough space (i.e. the length of a vehicle l_v and the minimum distance for a following vehicle during standstill $d_{min} = 1.5$ m) behind the B-V so that the A-V can pass the intersection without the risk of having to stop inside the intersection (e_{b11}). The L-V has a green traffic light assigned to it ($e_{1,1}$) in case it does not exist (e_{b12}) or after it has passed the intersection (e_{b13}).

Additionally, some further events are needed for the model. If the A-V enters a new zone in the current time step, event e_2 is triggered. Event e_3 is triggered if an emergency stop before the intersection is still possible. If the turning patterns of the A-V, the P-V, and the Y-V all intersect with each other, a deadlock is possible (e_4). A deadlock occurs (e_5) if both the P-V (e_{b3}) and the Y-V (e_{b9}) as well as the A-V (e_{b16}) are stopped before the intersection at the same time. If only the P-V and the Y-V are standing at the intersection, a deadlock of the C-Vs occurs (e_6).

4.2.3 DES Model

Each zone has states associated to it. The model can only be in a state that is associated with its current zone. In zones 1 and 6 there is only one state each (s_{10} and s_{60}), as the model does not influence the behavior in these states. The remaining states each have a state for offensive driving (states s_{21}, s_{31}, s_{41} and s_{51}) and defensive driving (s_{22}, s_{32}, s_{42} and s_{52}). Offensive states prepare the A-V for driving directly through the intersection or are the state in which the vehicle actually passes the intersection. The defensive states correspond with waiting before the intersection or describe the waiting state directly. State s_{53} describes offensive driving after waiting in state s_{52}. The model switches between states if certain events occur. The model and all its states and events are shown in Fig. 2.

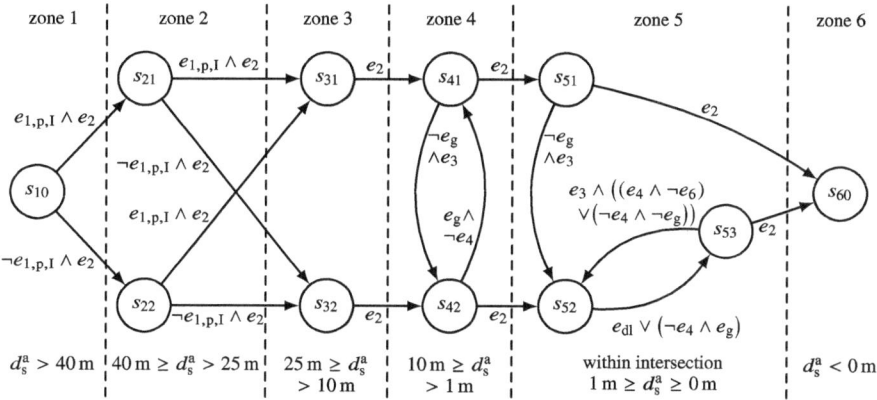

Fig. 2 DES of the A-V. If none of the events attributed to the current state occurs, the system remains in that state. These events have been omitted for better readability. The event $e_g = e_{1,p,II} \wedge e_{1,y} \wedge e_{1,b} \wedge e_{1,1}$ describes the case that the traffic lights of all four relevant C-Vs are green in zones 4 and 5. Event $e_{dl} = e_4 \wedge e_5 \wedge e_{1,b} \wedge e_{1,1}$ is true if a deadlock is possible, has occurred and both the L-V and the B-V do not obstruct the A-V from driving

During the approach the model always starts in state s_{10}. It remains there until it leaves the first zone (event e_2). When this happens, the prediction of the P-V is evaluated for the first time. In the prediction phase only the P-V is considered as the A-V only has to yield to this vehicle. In case of green light ($e_{1,p,I}$) the A-V assumes its offensive state s_{21}, otherwise it drives more defensively in state s_{22}. When it eventually enters zone 3 the same evaluation is performed again. If the evaluation leads to a green light, it enters state s_{31} that is associated with offensive behavior, otherwise it enters state s_{32} and shows defensive behavior. When the A-V leaves zone 3 there is no prediction, it transitions from state s_{31} to s_{41} or from s_{32} to s_{42}, thus keeping its offensive or defensive behavior, respectively. This can be done as the prediction is run constantly (i.e. in every time step) in zones 4 and 5.

In addition to the constant prediction, all four relevant vehicles are now considered for decision making, as the A-V is close to or within the collision zones with its C-Vs in these zones and dangerous situations can thus occur easily. If the A-V is in the defensive state s_{42} and all four lights are green (event $e_g = e_{1,p,II} \wedge e_{1,y} \wedge e_{1,b} \wedge e_{1,l}$) and if a deadlock cannot occur ($\neg e_4$), it transitions to state s_{41}. If it is in the offensive state s_{41} it switches to s_{42} if at least one of the four lights is no longer green ($\neg e_g$) and if there is still enough space for an emergency stop by the A-V (e_3). This does not pose a large risk as the parameterization for the green lights is rather conservative. Additionally, this strategy avoids a potentially dangerous stop within the intersection. If the vehicle reaches the end of zone 4 and enters zone 5 (event e_2), it progresses from s_{41} to s_{51} or from s_{42} to s_{52}, respectively. If the vehicle is in state s_{51} it remains in this offensive state unless at least one of the traffic lights is no longer green ($\neg e_g$) and there is still enough space for an emergency stop (e_3). In this case it transitions to state s_{52}. There is no transition from s_{52} to s_{51}. Instead, the A-V can only leave the waiting state s_{52} to s_{53} if all traffic lights are green again (e_g) while no deadlock is possible ($\neg e_4$) or if there is a deadlock that the A-V tries to solve ($e_{dl} = e_4 \wedge e_5 \wedge e_{1,b} \wedge e_{1,l}$). If a deadlock is detected by the A-V it always tries to drive first. An alternative strategy would be to drive after a certain waiting period. State s_{53} is an offensive state that is assumed after the A-V was defensive. From it, the A-V either progresses to s_{60} after it leaves the intersection (e_2) or it returns to the defensive state s_{52} if it can no longer drive safely. The latter is the case if an emergency stop is still possible (e_3) and either a deadlock is possible (e_4) but the cooperation vehicles are not stopped ($\neg e_6$) or a deadlock is not possible ($\neg e_4$) and not all lights are green ($\neg e_g$). State s_{60} is the only state of zone 6. This state is not controlled by the algorithm as the interaction at the intersection is now over.

4.2.4   Acceleration

So far the DES only describes the current situation of the interaction. To actually control it, the behavior of the A-V has to be set depending on the current state of the DES. For that we set a target velocity for each state (see Table 4) and control the vehicle using the intelligent driver model (IDM) [40]:

Table 4 Target velocities v_t in m s^{-1} for the states of the DES. Entries marked with an asterix are set in conjunction with a virtual vehicle to enforce stopping before the intersection

State	s_{21}	s_{22}	s_{31}	s_{32}	s_{41}, s_{51}, s_{53}	s_{42}, s_{52}
v_t straight	8.3	6.0	7.5	6.0	6.5	6.5*
v_t turning	8.3	6.0	6.0	6.0	4.0	4.0*

$$a^a = a_m \left(1 - \left(\frac{v^a}{v_t}\right)^4 - \left(\frac{d^*}{\Delta d}\right)^2\right) \quad \text{with } d^* = d_{min} + v^a \, t_{min} + \frac{v^a \, \Delta v}{2 \, \sqrt{a_m \, a_b}}. \quad (9)$$

With the maximum acceleration $a_m = 2.5\,\text{ms}^{-2}$, the braking deceleration $a_b = a_c$, the target velocity v_t as specified in Table 4, the distance along the drive path to the L-V Δd, the difference in velocity $\Delta v = v^a - v^l$ and the minimum time between following vehicles $t_{min} = 1.2$ s. The acceleration a^a by the IDM is limited to a lower threshold of $a_{min} = a_c$. If there is no L-V Δd is set to infinity and $v^l = 0\,\text{m s}^{-1}$. In states s_{42} and s_{52} the A-V is supposed to stop 1 m before the last stopping point. If this is not possible, the A-V brakes harder ($a_{min} = a_h$) to still stop at that point. If this is also no longer possible, an emergency stop with $a_{min} = a_e$ is initiated and the A-V will stop directly at the last stopping point. To ensure that the A-V stops at its stopping point, a virtual vehicle is placed such that its rear is d_{min} before the stop point. The virtual vehicle is not used if there is an L-V that is closer. v_t is set to the same value as in the offensive states s_{41} or s_{51}. This approach ensures that the A-V proceeds to its stopping point if there is no L-V before the intersection and that the A-V is able to restart after waiting in a queue to proceed to its stop point.

4.3 Simulation Results

To test and validate our proposed decision making system we implemented a simulation framework. To properly test the algorithm, also the C-Vs have to be simulated. For that a simplified version of the proposed algorithm is used because we are only interested in testing the A-V's algorithm. In it, the conditions for driving depend on fewer features and events and zones 4 and 5 of the original algorithm are merged. In this zone the decision to drive first is not revised, i.e. once the algorithm decides to drive, it continues to do so regardless of any future development of its surroundings. In case of a deadlock, the C-V waits for a random duration before it tries to resolve the situation. The C-Vs detect a deadlock before the A-V does. That way, it is also possible for the C-Vs to drive first despite the A-V driving as soon as it detects a deadlock. That way it is possible to test the behavior of the A-V's algorithm if someone else tries to resolve a deadlock. Additionally, visibility is not taken into consideration for the C-Vs, all vehicles are visible by the simplified algorithm at all times. Finally, the algorithm of the C-Vs can have some special behavior to test certain aspects of the main algorithm: They can be set to drive first despite having to yield and alternatively they can be set to wait for an arbitrary duration if they are allowed to drive first. This behavior is only shown when the relevant cooperation partner from the C-V's perspective is the A-V. With both variants we can test the A-V's behavior towards unexpected behavior. Additionally, the target velocity inside and after the intersection can be reduced. With that one can further ensure that the A-V only drives once the intersection is cleared.

Within the simulation framework, the simulation for a single run is performed as follows: First, the map for the simulation is loaded and all vehicles are initiated. Then each time step is simulated: The currently visible vehicles are determined and only the current states of these vehicles are presented to the algorithm. Then the C-Vs are identified and the features are calculated. Afterwards, the currently active events are checked and the DES is updated. Finally, the acceleration is calculated. These steps are performed for the A-V and all C-Vs.

For the simulations we used the generic map as described above, the visibility distance was set to either $d_v \in \{7\,\text{m}, 14\,\text{m}, 21\,\text{m}\}$ and there were either $n_c \in \{1, 2, 3, 4, 5, 6\}$ cooperation vehicles present in the simulation. Each of these combinations was simulated 200 times, resulting in 3600 simulations in total. In each simulation run the distances to the intersection of all vehicles and their initial velocities and turning patterns were set randomly within a certain feasible range. The special behavior and the waiting durations were set randomly as well.

None of the simulations resulted in a collision. One should note, however, that it is possible for two C-Vs to restart simultaneously after a deadlock. As the decision to drive is not revised, this would result in a collision. Such a run could safely be disregarded for evaluation as we are only interested in the performance of the A-V's algorithm. For each run we also measured the time to drive through the intersection t_d (time while the A-V was within $30\,\text{m} > d_s^a \geq 0\,\text{m}$). If we average over all runs with the same visibility distance, we get the following average durations and corresponding standard deviations: $t_d\ (d_v = 7\,\text{m}) = 12.10\,\text{s}\ (\sigma = 6.03\,\text{s})$, $t_d\ (d_v = 14\,\text{m}) = 12.14\,\text{s}\ (\sigma = 6.18\,\text{s})$ and $t_d\ (d_v = 21\,\text{m}) = 12.16\,\text{s}\ (\sigma = 6.23\,\text{s})$. As these values are very similar, we did not analyze the results separately for each visibility distance. In Table 5 the time to drive through the intersection is averaged over all runs that have the same number of P-Vs and Y-Vs. The results from that table have to be interpreted with caution as there are some aspects that are not considered, e.g. a leading vehicle that has to wait can increase the duration even though the A-V would not have had to stop. Also, there are only a few runs with more than three vehicles of a kind, the average is thus less reliable. Nonetheless, the results indicate that the algorithm results in reasonable decisions: The average time to pass the intersection increases with the number of cooperation vehicles. The increase is more pronounced for the P-Vs than for the Y-Vs. This is to be expected as one has to yield to the P-Vs instead of the interaction with Y-Vs where one should have to wait less often.

Table 5 Average time to clear the intersection by the number of P-Vs and Y-Vs for all visibility distances

	0 P-Vs	1 P-V	2 P-Vs	3 P-Vs	4 P-Vs	5 P-Vs
0 Y-Vs	9.49 s	14.02 s	19.43 s	22.49 s	24.26 s	29.39 s
1 Y-V	9.62 s	16.79 s	21.95 s	25.45 s	29.87 s	36.6 s
2 Y-Vs	13.53 s	19.83 s	25.28 s	27.31 s	36.87 s	–
3 Y-Vs	15.87 s	20.58 s	24.73 s	–	–	–
4 Y-Vs	26.87 s	15.22 s	–	–	–	–

5 Conclusion

The results from Sect. 3 show that the driving behavior of human drivers depends on the intersection. We can thus predict the driving behavior using features that describe the intersection itself, its surroundings and the traffic there. As these features can be considered as a description of an intersection's complexity, one can conclude that the complexity of an intersection has an influence on the driving behavior. We further show that it is possible to predict the driving behavior using only a subset with the most relevant features. In future work we intend to directly ask human participants for a complexity rating of such situations. With that we hope to find a dependence between the perceived complexity and the resulting behavior.

In Sect. 4 we further present a decision making algorithm that is able to reliably drive through an unsignalized T-intersection while interacting with other drivers. We validate our proposed algorithm with a simulation and the results indicate a reliable performance. Future work on this topic will include variants of this algorithm for further scenarios such as X-intersections, roundabouts or narrow passages. We further intend to run the algorithm on real world maps.

Acknowledgements This project has been funded within the priority program 1835 "Cooperatively Interacting Automobiles" by the *German Research Foundation (DFG)*. Mapping data has been provided by the City of Karlsruhe.

References

1. §11 Straßenverkehrs-Ordnung (StVO) (visited: November 23, 2022). https://www.gesetze-im-internet.de/stvo_2013/__11.html
2. Bouton, M., Nakhaei, A., Fujimura, K., Kochenderfer, M.J.: Scalable decision making with sensor occlusions for autonomous driving. In: 2018 IEEE International Conference on Robotics and Automation (ICRA), pp. 2076–2081 (2018)
3. Breiman, L.: Random forests. Mach. Learn. **45**(1), 5–32 (2001)
4. Cantin, V., Lavallière, M., Simoneau, M., Teasdale, N.: Mental workload when driving in a simulator: effects of age and driving complexity. Accident Anal. Prevent. **41**(4), 763–771 (2009)

5. Cunningham, A.G., Galceran, E., Eustice, R.M., Olson, E.: MPDM: multipolicy decision-making in dynamic, uncertain environments for autonomous driving. In: 2015 IEEE International Conference on Robotics and Automation (ICRA), pp. 1670–1677. IEEE (2015)
6. De Beaucorps, P., Streubel, T., Verroust-Blondet, A., Nashashibi, F., Bradai, B., Resende, P.: Decision-making for automated vehicles at intersections adapting human-like behavior. In: 2017 IEEE Intelligent Vehicles Symposium (IV), pp. 212–217. IEEE (2017)
7. Di Febbraro, A., Giglio, D.: On representing signalized urban areas by means of deterministic-timed Petri nets. In: Proceedings. The 7th International IEEE Conference on Intelligent Transportation Systems (IEEE Cat. No. 04TH8749), pp. 372–377. IEEE (2004)
8. Edquist, J., Rudin-Brown, C.M., Lenné, M.G.: The effects of on-street parking and road environment visual complexity on travel speed and reaction time. Accident Anal. Prevent. **45**, 759–765 (2012)
9. Faure, V., Lobjois, R., Benguigui, N.: The effects of driving environment complexity and dual tasking on drivers' mental workload and eye blink behavior. Transport. Res. F: Traffic Psychol. Behav. **40**, 78–90 (2016)
10. Fayazi, S.A., Vahidi, A.: Mixed-integer linear programming for optimal scheduling of autonomous vehicle intersection crossing. IEEE Trans. Intell. Veh. **3**(3), 287–299 (2018)
11. Galceran, E., Cunningham, A.G., Eustice, R.M., Olson, E.: Multipolicy decision-making for autonomous driving via changepoint-based behavior prediction. In: Robotics: Science and Systems, vol. 1, p. 6 (2015)
12. Hancock, P.A., Wulf, G., Thom, D., Fassnacht, P.: Driver workload during differing driving maneuvers. Accident Anal. Prevent. **22**(3), 281–290 (1990)
13. Hess, W., Kohler, D., Rapp, H., Andor, D.: Real-time loop closure in 2D LIDAR SLAM. In: 2016 IEEE International Conference on Robotics and Automation (ICRA), pp. 1271–1278. IEEE (2016)
14. Ho, G., Scialfa, C.T., Caird, J.K., Graw, T.: Visual search for traffic signs: the effects of clutter, luminance, and aging. Hum. Factors **43**(2), 194–207 (2001)
15. Horberry, T., Anderson, J., Regan, M.A., Triggs, T.J., Brown, J.: Driver distraction: The effects of concurrent in-vehicle tasks, road environment complexity and age on driving performance. Accident Anal. Prevent. **38**(1), 185–191 (2006)
16. Huang, Y.S., Weng, Y.S., Zhou, M.: Modular design of urban traffic-light control systems based on synchronized timed Petri nets. IEEE Trans. Intell. Transp. Syst. **15**(2), 530–539 (2013)
17. Huang, Y.S., Weng, Y.S., Zhou, M.: Design of traffic safety control systems for emergency vehicle preemption using timed Petri nets. IEEE Trans. Intell. Transp. Syst. **16**(4), 2113–2120 (2015)
18. Hubmann, C., Becker, M., Althoff, D., Lenz, D., Stiller, C.: Decision making for autonomous driving considering interaction and uncertain prediction of surrounding vehicles. In: 2017 IEEE Intelligent Vehicles Symposium (IV), pp. 1671–1678. IEEE (2017)
19. Imbsweiler, J., Palyafári, R., Puente León, F., Deml, B.: Untersuchung des Entscheidungsverhaltens in kooperativen Verkehrssituationen am Beispiel einer Engstelle. at-Automatisierungstechnik **65**(7), 477–488 (2017)
20. Imbsweiler, J., Ruesch, M., Weinreuter, H., Puente León, F., Deml, B.: Cooperation behaviour of road users in t-intersections during deadlock situations. Transp. Res. Part F: Traffic Psychol. Behav. **58**, 665–677 (2018)
21. Jahn, G., Oehme, A., Krems, J.F., Gelau, C.: Peripheral detection as a workload measure in driving: effects of traffic complexity and route guidance system use in a driving study. Transport. Res. F: Traffic Psychol. Behav. **8**(3), 255–275 (2005)
22. James, G., Witten, D., Hastie, T., Tibshirani, R.: An Introduction to Statistical Learning: with Applications in R, 2nd edn. Springer Texts in StatisticsSpringer eBook Collection. Springer US, New York (2021). https://doi.org/10.1007/978-1-0716-1418-1
23. Li, G., Lai, W., Sui, X., Li, X., Qu, X., Zhang, T., Li, Y.: Influence of traffic congestion on driver behavior in post-congestion driving. Accident Anal. Prevent. **141**, 105508 (2020)
24. Li, G., Wang, Y., Zhu, F., Sui, X., Wang, N., Qu, X., Green, P.: Drivers' visual scanning behavior at signalized and unsignalized intersections: A naturalistic driving study in China. J. Safety Res. **71**, 219–229 (2019)

25. Lin, X., Zhang, J., Shang, J., Wang, Y., Yu, H., Zhang, X.: Decision making through occluded intersections for autonomous driving. In: 2019 IEEE Intelligent Transportation Systems Conference (ITSC), pp. 2449–2455. IEEE (2019)
26. Liu, W., Kim, S.W., Pendleton, S., Ang, M.H.: Situation-aware decision making for autonomous driving on urban road using online POMDP. In: 2015 IEEE Intelligent Vehicles Symposium (IV), pp. 1126–1133. IEEE (2015)
27. Makarem, L., Gillet, D.: Model predictive coordination of autonomous vehicles crossing intersections. In: 16th International IEEE Conference on Intelligent Transportation Systems (ITSC 2013), pp. 1799–1804. IEEE (2013)
28. Manawadu, U.E., Kawano, T., Murata, S., Kamezaki, M., Muramatsu, J., Sugano, S.: Multiclass classification of driver perceived workload using long short-term memory based recurrent neural network. In: 2018 IEEE Intelligent Vehicles Symposium (IV), pp. 2009–2014. IEEE (2018)
29. Noh, S.: Decision-making framework for autonomous driving at road intersections safeguarding against collision overly conservative behavior and violation vehicles. IEEE Trans. Industr. Electron. **66**(4), 3275–3286 (2018)
30. Oviedo-Trespalacios, O., Haque, M.M., King, M., Washington, S.: Effects of road infrastructure and traffic complexity in speed adaptation behaviour of distracted drivers. Accident Anal. Prevent. **101**, 67–77 (2017)
31. Patten, C.J., Kircher, A., Östlund, J., Nilsson, L., Svenson, O.: Driver experience and cognitive workload in different traffic environments. Accident Anal. Prevent. **38**(5), 887–894 (2006)
32. Phillips, D.J., Wheeler, T.A., Kochenderfer, M.J.: Generalizable intention prediction of human drivers at intersections. In: 2017 IEEE Intelligent Vehicles Symposium (IV), pp. 1665–1670. IEEE (2017)
33. Qi, L., Zhou, M., Luan, W.: Modeling and control of urban road intersections with incidents via timed Petri nets. In: 2015 IEEE 12th International Conference on Networking, Sensing and Control, pp. 185–190. IEEE (2015)
34. Qian, B., Zhou, H., Lyu, F., Li, J., Ma, T., Hou, F.: Toward collision-free and efficient coordination for automated vehicles at unsignalized intersection. IEEE Internet Things J. **6**(6), 10408–10420 (2019)
35. Quigley, M., Conley, K., Gerkey, B., Faust, J., Foote, T., Leibs, J., Wheeler, R., Ng, A.Y., et al.: ROS: an open-source robot operating system. In: ICRA Workshop on Open Source Software, vol. 3, p. 5. Kobe, Japan (2009)
36. Rahmati, Y., Talebpour, A.: Towards a collaborative connected, automated driving environment: a game theory based decision framework for unprotected left turn maneuvers. In: 2017 IEEE Intelligent Vehicles Symposium (IV), pp. 1316–1321. IEEE (2017)
37. Sezer, V., Bandyopadhyay, T., Rus, D., Frazzoli, E., Hsu, D.: Towards autonomous navigation of unsignalized intersections under uncertainty of human driver intent. In: 2015 IEEE/RSJ International Conference on Intelligent Robots and Systems (IROS), pp. 3578–3585. IEEE (2015)
38. Shu, K., Yu, H., Chen, X., Li, S., Chen, L., Wang, Q., Li, L., Cao, D.: Autonomous driving at intersections: a behavior-oriented critical-turning-point approach for decision making. IEEE/ASME Trans, Mechatron (2021)
39. Teh, E., Jamson, S., Carsten, O., Jamson, H.: Temporal fluctuations in driving demand: The effect of traffic complexity on subjective measures of workload and driving performance. Transport. Res. F: Traffic Psychol. Behav. **22**, 207–217 (2014)
40. Treiber, M., Hennecke, A., Helbing, D.: Congested traffic states in empirical observations and microscopic simulations. Phys. Rev. E **62**(2), 1805 (2000)
41. Wang, J., Yan, J., Li, L.: Microscopic modeling of a signalized traffic intersection using timed Petri nets. IEEE Trans. Intell. Transp. Syst. **17**(2), 305–312 (2015)
42. Weinreuter, H., Strelau, N.R., Qiu, K., Jiang, Y., Deml, B., Heizmann, M.: Intersection complexity and its influence on human drivers. IEEE Access **10**, 74059–74070 (2022). https://doi.org/10.1109/ACCESS.2022.3189017

43. Weinreuter, H., Szigeti, B., Strelau, N.R., Deml, B., Heizmann, M.: Decision making at unsignalized inner city intersections using discrete events systems. tm - Technisches Messen **89**(2), 134–146 (2022). https://doi.org/10.1515/teme-2021-0140
44. Werneke, J., Vollrath, M.: What does the driver look at? The influence of intersection characteristics on attention allocation and driving behavior. Accident Anal. Prevent. **45**, 610–619 (2012) https://www.sciencedirect.com/science/article/pii/S0001457511002855
45. Wijnands, J.S., Zhao, H., Nice, K.A., Thompson, J., Scully, K., Guo, J., Stevenson, M.: Identifying safe intersection design through unsupervised feature extraction from satellite imagery. Computer-Aided Civil Infrastruct, Engin (2020)
46. Zyner, A., Worrall, S., Ward, J., Nebot, E.: Long short term memory for driver intent prediction. In: 2017 IEEE Intelligent Vehicles Symposium (IV), pp. 1484–1489. IEEE (2017)

Perception and Prediction with Explicit Communication

Robust Local and Cooperative Perception Under Varying Environmental Conditions

Jörg Gamerdinger, Georg Volk, Sven Teufel, Alexander von Bernuth, Stefan Müller, Dennis Hospach, and Oliver Bringmann

Abstract Robust perception of the environment under a variety of ambient conditions is crucial for autonomous driving. Convolutional Neural Networks (CNNs) achieve high accuracy for vision-based object detection, but are strongly affected by adverse weather conditions such as rain, snow, and fog, as well as soiled sensors. We propose physically correct simulations of these conditions for vision-based systems, since publicly available data sets lack scenarios with different environmental conditions. In addition, we provide a data set of real images containing adverse weather for evaluation. By training CNNs with augmented data, we achieve a significant improvement in robustness for object detection. Furthermore, we present the advantages of cooperative perception to compensate for limited sensor ranges of local perception. A key aspect of autonomous driving is safety; therefore, a robustness evaluation of the perception system is necessary, which requires an appropriate safety metric. In contrast to existing approaches, our safety metric focuses on scene semantics and the

Jörg Gamerdinger and Georg Volk are contributed equally.

J. Gamerdinger (✉) · G. Volk · S. Teufel · A. von Bernuth · S. Müller · D. Hospach · O. Bringmann
Embedded Systems, Department of Computer Science, Faculty of Science, University of Tübingen, Tübingen, Germany
e-mail: joerg.gamerdinger@uni-tuebingen.de

G. Volk
e-mail: georg.volk@uni-tuebingen.de

S. Teufel
e-mail: sven.teufel@uni-tuebingen.de

A. von Bernuth
e-mail: alexander.von-bernuth@uni-tuebingen.de

S. Müller
e-mail: stefan.mueller@uni-tuebingen.de

D. Hospach
e-mail: dennis.hospach@uni-tuebingen.de

O. Bringmann
e-mail: oliver.bringmann@uni-tuebingen.de

© The Author(s) 2024
C. Stiller et al. (eds.), *Cooperatively Interacting Vehicles*,
https://doi.org/10.1007/978-3-031-60494-2_5

113

relevance of surrounding objects. The performance of our approaches is evaluated using real-world data as well as augmented and virtual reality scenarios.

1 Introduction

Autonomous driving is one of the big challenges in society and currently of great interest in research. Autonomous vehicles are a promising approach to reduce traffic jams and the number of accidents and furthermore increase the comfort for the drivers respectively passengers. To achieve market readiness, autonomous vehicles have to be safe, this requires a complete and correct perception of the environment.

Different sensors like LiDAR, RADAR, and cameras are available for perception; of them, cameras are most frequently employed [107]. Since camera sensors are vision-based, they are affected by weather circumstances such as rain or fog [107]. The effect of adverse weather on the frequency of car crashes is not to be neglected as shown by the National Highway Traffic Safety Administration; on average over the past 10 years, adverse weather is responsible for 21% of the car accidents in the United States [106].

Therefore, these characteristics must be taken into account when developing perception algorithms in order to obtain a robust perception, which is crucial for a safe system. Car manufacturers are able to capture data using their own test vehicles, however, in research often publicly accessible data sets are utilized. Most of them contain no or only few data under adverse weather conditions. One method is to create artificial weather conditions and use those to enhance the data set.

Two major issues come with simulating weather conditions. In order to gain a benefit, it is first necessary to simulate a wide range of potential conditions, such as rain, snow, and fog. A second problem is that these simulations must be physically accurate.

Harsh weather affects not just the image itself but also neural network-based object detectors, whose performance is highly dependent on their training [108]. Hence, neural networks must be trained under different weather conditions to achieve a safe and robust perception.

Even with robust neural networks, the perception is limited due to occlusion and sensor ranges. At this point the so-called cooperative or collective perception (CP) comes into play. Using multiple distributed vehicles to perceive objects locally and share these detections with other vehicles via Vehicle-to-Everything (V2X) communication helps to acquire information even about vehicles which cannot be perceived locally.

In Sects. 1.1–1.4 we describe the concept of our work as well as the simulation framework "RESIST". Section 2 considers our physically-correct image-based weather augmentations. The following section presents a new way to evaluate safety of object perception systems. Section 4 presents optimization approaches for local and cooperative perception. Finally, in Sect. 5 we present our conclusion and give an outlook about further research topics.

1.1   Concept

A complete perception of the environment under all circumstances is crucial for autonomous driving. Hence, it must be robust against environmental influences like rain and fog as well as physical restrictions such as limited sensor ranges. This work mainly considers the environmental perception with camera sensors since this sensor type is one of the most frequently used [107]. Our goal is a safe and complete perception of the environment. An overview of our work is shown in. We consider two different ways to achieve this goal: Local perception and cooperative perception. First we consider the local perception of an ego vehicle and investigate strategies to enhance the vision-based perception. The proposed robustness enhancement is based on training CNNs for object detection on a more comprehensive data set including weather augmentations. Thus, a significant part of our research covers the physically correct simulation of the weather conditions rain, snow and fog. Realistic weather simulations allow to augment existing data sets and increase the data variety for the training of neural networks. Also, the influence of weather on object detection itself must be investigated in detail. Some findings are transferred to LiDAR sensors by our group, since these are also vision-based (see work of Teufel et al. [100]). Moreover, the robustness improvement of RADAR sensors is investigated in our group by Zlavik [95]. The work presents a noise modulated pulsed radar system which outperforms commercial state-of-the-art radar systems. Additionally, with compressive sensing the effort for signal acquisition is reduced by 70% [95]. Instead of sensor specific optimization, other approaches from our team use more generic deep learning techniques to optimize robustness. Rusak et al. [84] demonstrate that a simple

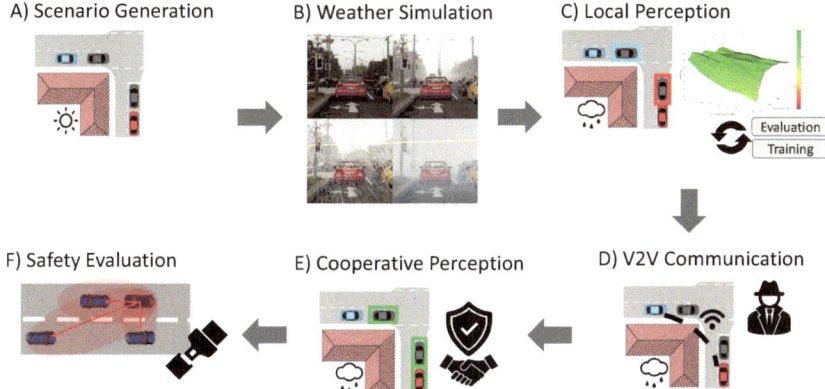

Fig. 1 Overview of the project's total concept. We focus on **b** weather simulation (see Sect. 2), the improvement of **c** local perception (see Sect. 4.2) and **e** cooperative perception (see Sect. 4.3). Furthermore, we investigate **f** how to evaluate safety for object perception (see Sect. 3.2). **a** Scenario generation is out of the scope of this work due to space limitation. **d** V2V communication is part of the workflow but not a focus of our research and covered in more detail in Chap. 6. Images from [19, 108, 110, 112]

but properly tuned training with additive Gaussian and Speckle noise generalizes surprisingly well to unseen corruptions. Michaelis et al. [64] extend the ImageNet-C robustness benchmark from image classification to object detection in order to provide an easy-to-use benchmark to assess how object detection models perform when image quality degrades.

Even with significant improvements in the local perception, a full perception of the environment is not possible due to physical restrictions regarding sensor ranges as well as covered line-of-sights through infrastructural elements or buildings. Thus, as a second improvement strategy, we investigate cooperative perception and an optimization approach to determine the validity and trustworthiness of collectively perceived information before performing a fusion. Cooperative perception aims to increase the sensor ranges through distributed perception and helps to see traffic participants that are occluded by e.g., buildings.

Since the goal is to achieve safety, we also have to consider how to evaluate safety. Therefore, we present a novel metric to evaluate safety for local and cooperative perception systems which incorporates important factors such as velocity and the object class.

1.2 Related Work

Most simulations of rain are made for computer games and only a few simulations consider physical correctness [6]. Therefore, Hospach et al. [42] proposed a realistic rain simulation based on falling, white-colored triangles. They use alpha-blending to simulate different intensities; but this approach does not consider effects like refraction. The approach of Wang et al. [115] uses ray tracing for the rendering of raindrops. Therefore, they have to know the exact position of the light source. Sato et al. [87] are using a single hemisphere in front of the camera; but this approach ignores the real distance of objects towards the camera. Furthermore, they had to use various simplifications to achieve real-time capability. Many more publications consider the rendering of realistic fluid dynamics of water droplets on different surfaces [47, 51, 52, 116]. A further study of Garg and Nayar considers the shape of falling water drops [29]. Moreover, interesting physical properties of rain can be learned from [36, 119]; both works regard the detection and removal of raindrops on images. A comprehensive work about the physical correct simulation of rain and fog was presented by Hasirlioglu [38].

For water spray of vehicles driving on wet roads (in the following: road spray) fewer works exist. The size of road spray was investigated by Kooij et al. [56]. Beginning with the work of Kamm and Wray [50] different researchers considered the movement of the road spray [31, 32, 44, 49]. Slomp et al. [96] presented a fast and efficient rendering method for water droplets based on OpenGL.

To simulate snow, a simple approach without considering depth or falling speed was introduced by Wang and Wade [117]; they produced a texture with 2D-snowflakes that surrounds the camera. Another approach is presented by Zhou and

Libaicheng [103], they propose a method to draw falling snowflakes. Since these snowflakes are visible every time, it lacks in realism regarding the rotation of flakes. While the falling snowflakes as well as how the snow covers the ground and even the process of snow melting is investigated very well [24, 69, 86, 114], the movement of the flakes while falling is considered less. This was covered by Langer and Zhang [59] who used Fourier transformations to add noise; this results in snow-like artifacts in dependence to a virtual depth.

To augment fog, very basic fog simulations are presented by Sellers et al. [91] and Aleshin et al. [3]. Further authors generate noise to simulate different fog densities [120]. Even new works only use simple light attenuation formulas such as presented by Sakaridis et al. [85]. This fog simulation was used to augment the well known Cityscapes data set [14] and create foggy Cityscapes. More realistic and advanced methods are presented by Dumont [20] as well as Jensen and Christensen [46]. Both approaches use Monte-Carlo-driven methods with multiple rays per pixel to get a realistic virtual result. Another more advanced method was presented by Biri and Michelin [9] who even integrated wind into their simulation. More realistic fog data can be produced with synthetic fog in fog chambers, such as shown by Colomb et al. [12]. They have built a 30 × 5.5 m fog chamber, which allows fog simulation for some static scenarios.

For the vision-based object detection mostly CNN-based methods are used. The effects of blurring, image compression and different types of noise on object detection were investigated by Dodge and Karam [18] and Costa et al. [15]. Both works show that noise or image corruptions lead to a lower accuracy in object detection and classification. The same result was shown by Nazaré et al. [70]. Since the accuracy of neural networks depend on the training, it is a common way to extend existing data sets by image transformations such as geometric and color transformations as proposed by Montserrat et al. [66]. The approaches in [15, 18, 70] mainly consider generic errors but no realistic environmental influences. A more realistic data set extension was presented by Hasirlioglu and Riener [39] who proposed a rain simulation to investigate the effect of weather on the detection performance. The strong effect of synthetic rain on the object detection accuracy was also shown by Müller et al. [68]. Tian et al. [102] proposed DeepTest; a methodology to evaluate neural networks for autonomous driving to detect erroneous behavior by augmenting the data. Similar to DeepTest, Pei et al. [73] proposed DeepXplore to evaluate neural networks; additionally they did an optimization with augmented data and achieved a higher detection accuracy. Further works [5, 25, 53] considered using General Adversial Networks (GANs) to augment data. Luc et al. [61] used GANs and achieved a reduction of overfitting for semantic segmentation. Karacan et al. [53] created synthetic environmental conditions through a combination of GANs and semantic image information. It is necessary to point out that for the augmentations of [5, 25, 53, 73, 102] there is no proof of realism.

As aforementioned, the local perception is not only affected by environmental conditions but also limited due to sensor ranges and occlusion. These are problems which can be addressed by cooperative perception (CP).

Two initial works to CP were presented by Rauch et al. [77, 78]. They present different approaches, how to handle and fuse information of distributed vehicles. Methods for multiple-object tracking and CP using camera and LiDAR were proposed by Obst et al. [72] and Kim et al. [54]. To evaluate the capability and advantages of CP, a correct V2V communication model must be used. An approach of modeling the reception probability and communication delay was presented by Torrent-Moreno et al. [104]. However, this approach does not consider environmental influences such as weather and buildings which cover the line-of-sight between sender and receiver. More advanced models, which are parameterizable and consider different environments such as buildings at an intersection, are proposed in [2, 62] or by Boban et al. [10]. Nowadays, the European Telecommunications Standards Institute (ETSI) works on a standard for a message format and exchange frequency for information about the ego vehicle and detected objects in cooperative perception [21, 22]. These work-in-progress standards and the rules for the message generation are reviewed by different researchers [17, 30, 101]. Since the simulations of [21] lack in realism due to missing delays and simplified sensor models. Allig and Wanielik [4] extended this simulation setup by more realistic vehicle dynamics and sensor models. Another simulation approach is presented by Schiegg et al. [89]. A real-world demonstration of the capabilities of CP was done by Shan et al. [94]. Next to simulations and real-world demonstrations there exist some analytical models for CP as presented in [45, 88].

To evaluate object detection, in common benchmarks like COCO [60] or KITTI [34] simple performance indicators like precision, accuracy, average precision (AP), and mean average precision (mAP) [16, 23, 74] are used. Since this does not satisfy the safety constraints of autonomous vehicles, this is not sufficient to evaluate object detection systems. Stiefelhagen et al. [97] proposed a slightly more comprehensive metric, using the Intersection over Union (IoU) [83] and the distance between track estimation and real position. A metric considering real-time aspects is proposed by Kim et al. [58]. The metric considers the detection time of video surveillance systems, which is also a factor for autonomous vehicles. A model to achieve safety is the Responsible-Sensitive Safety (RSS) model. Shalev-Shwartz et al. [93] proposed a guideline that mathematically describes how safety can be achieved in autonomous driving. The RSS model has become well known but does not include any metric to evaluate safety.

1.3 Data Sets

Comprehensive data is crucial for development in the field of autonomous driving. Basically, the data can be split into two groups: real-world data and simulation data. Here we present the data sources used for our experiments; therefore, it should be pointed out that it is not a complete overview over data sources for object detection in the field of autonomous driving.

As real-world data, the KITTI data set by Geiger et al. [34] was used. KITTI consists of different benchmarks for 2D- and 3D object detection as well as tracking. The KITTI data set was recorded in Karlsruhe (Germany) using a stereo-camera setup and a LiDAR sensor [34]. The data set consists mostly of inner-city recordings. The 2D object detection benchmark consists of about 15,000 images with over 80,000 labeled objects [34]. Further examples of real-world data sets are Cityscapes [14] and the Waymo data set [98]. A more comprehensive data set regarding adverse weather was presented by Bijelic et al. [8]. The presented DENSE data set contains real-world recordings including different weather conditions such as rain, snow or fog as well as recordings from a fog chamber. For a more sophisticated evaluation we created our own 2D image data set with heavy rain scenes. Therefore, we collected images of challenging rainy road scenes from our archive of self-conducted test drives and from dashcam videos on YouTube. This resulted in a very diverse data set of international road scenes. In the following, we call it *realrain data set* [112]. It contains 2062 images with 9551 labeled objects. The objects are labeled according to the KITTI label format. The realrain data set contains 7368 cars, 626 vans, 955 trucks, 395 pedestrians, 205 cyclists and one tram. The scenes are well spread from urban to freeway scenarios and contain heavy rain, mist and drops on the windshield representing challenging environmental conditions for vision-based object detection systems.

Since publicly available real-world data sets are limited, they possibly do not cover all scenarios which should be tested during development. This disadvantage can be solved by using realistic and parameterizable simulation frameworks. An exemplary commercial simulation framework is Vires VTD [1]. VTD provides different simulation scenarios such as rural road, freeway sections or an inner-city intersection. A more extensive, highly realistic (see Fig. 2) and open source simulator called *CARLA* was presented by Dosovitskiy et al. [19]. CARLA is based on the Unreal Engine and provides a set of different maps, containing many inner-city scenarios as well as rural sections and multiple freeway sections. Besides the maps, CARLA provides a wide range of different vehicles (bicycle, motorbike, truck, van, different types of cars) and pedestrians which can be spawned at different locations. Each vehicle can be equipped with different sensors such as camera, RADAR and LiDAR. More information about available sensors can be found in [19]. CARLA also includes weather variations as well as day and nighttime. Furthermore, if a specific route is to be driven, the vehicles can be controlled by a user. Scenarios can also be described using the OpenScenario standard, which CARLA can execute.

1.4 RESIST Framework and Workflow

For algorithm development and simulation a configurable and deterministic pipeline is necessary. Therefore, we use the *RESIST* framework developed in our team by Müller et al. [67] with the improvements by Volk et al. [108]. RESIST is a QT-based *C++* framework, which allows combining different plugins to a perception pipeline

Fig. 2 Example image generated with CARLA simulator [19] with the proposed weather augmentations (original, rain, snow, fog) from Sect. 2

with different inputs and an evaluation. The framework's main focus lies on local and cooperative perception with simulation of the weather conditions rain, snow, and fog as proposed in Sect. 2. RESIST can read a wide range of data sets such as KITTI [34] or Cityscapes [14]. Moreover, the simulation frameworks Vires VTD [1] and CARLA [19] can be used as input for the sensor data. This allows a comprehensive evaluation of perception algorithms using a comprehensive range of data sets. This sensor data is used to simulate the perception using realistic camera-models. Various well known vision-based object detection algorithms like Faster-RCNN [82], RRC [81] and YOLOv3 [79] are implemented in the framework, which allows a comparison between different architectures. For the object tracking, a Kalman filter [118] with different models such as constant velocity, constant acceleration or constant turn rate can be used.

RESIST is also capable to simulate cooperative perception. To simulate CP, RESIST includes a comprehensive communication channel simulation and processing delays [109]. The transmission of locally detected objects is done by V2X communication. A V2X channel simulation based on the analytical model of IEEE 802.11p by Sepulcre et al. [92] is integrated into RESIST. For the CP the focus lies in the perception and less on the V2X communication; but to gather valid results a correct communication model is necessary.

In the area of CP, different algorithms for matching and fusion are integrated. For the matching of measurements to existing tracks Hungarian matching [57], Nearest Neighbor or Expectation Maximization can be used with different cost metrics such as euclidean distance or IoU. For the Track-to-Track fusion there are also various algorithms available, such as covariance intersection [48], Kalman filter [118] or a simple mean fusion.

To evaluate algorithms a comprehensive evaluation plugin exists. This plugin allows an evaluation of a defined environment with different metrics such as precision, recall, mAP or the safety metric [110] presented in Sect. 3.2.

In conclusion RESIST is a comprehensive framework for a realistic simulation of local and cooperative object perception with physically correct vision-based weather simulation.

2 Simulation of Environmental Conditions

Simulating realistic weather influences allows extending existing data sets, which mainly consists of images with clear weather. Therefore, we present different weather augmentations for image data in this section.

2.1 Rain

The simulation of realistic rain is based on two approaches developed in our team by Hospach et al. [42] for simulation of falling rain and the simulation of raindrops on the windshield by von Bernuth et al. [6]. By combining these two steps it is possible to achieve a photorealistic simulation of rain. The rain simulation workflow is illustrated in Fig. 3. Examples of the proposed simulation are illustrated in Figs. 4 and 5. The first step is the reconstruction of the 3D scene with a depth image containing the scene depth for each pixel. Afterwards the falling rain as already introduced by Hospach et al. [42] is applied. The reconstructed 3D scene is used to distribute rain streaks in the space between camera and background, respecting the well known Marshall Palmer distribution [63]. The simulation of rain streaks respects camera parameters such as focal length, field of view, aperture, pixel size and shutter speed. Hence, the length of the simulated rain streaks varies depending on the configured shutter speed and the sharpness is depending on the aperture and the distance of the simulated rain streak from the camera. As next step, raindrops on the windshield are generated with the approach presented by von Bernuth et al. [6]. Raindrops are distributed on a virtual windshield and ray tracing is used for a physically correct rendering of these raindrops. Finally, the brightness of an image can be altered to achieve a realistic setting. This can be necessary if rain shall be simulated on a sunny image. By reducing the overall brightness of the image the simulated rain looks more realistic.

The proposed rain simulation can be parametrized with six parameters. Falling rain is parametrized by the rain intensity r_i and the rain angle r_a of the vertical rain streaks. The simulation of raindrops on the windshield uses r_i as well as the additional parameter drop count d_{count} which specifies the number of drops resting on the virtual windshield. The mean drop radius d_μ and the standard deviation d_σ

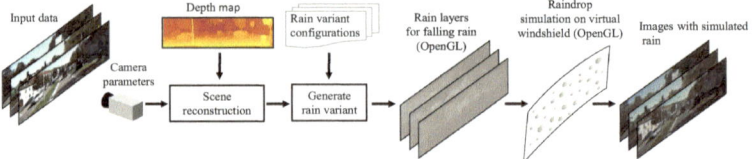

Fig. 3 Rain simulation workflow from reading the input data over the scene reconstruction to the defined rain simulation. Image from [112]

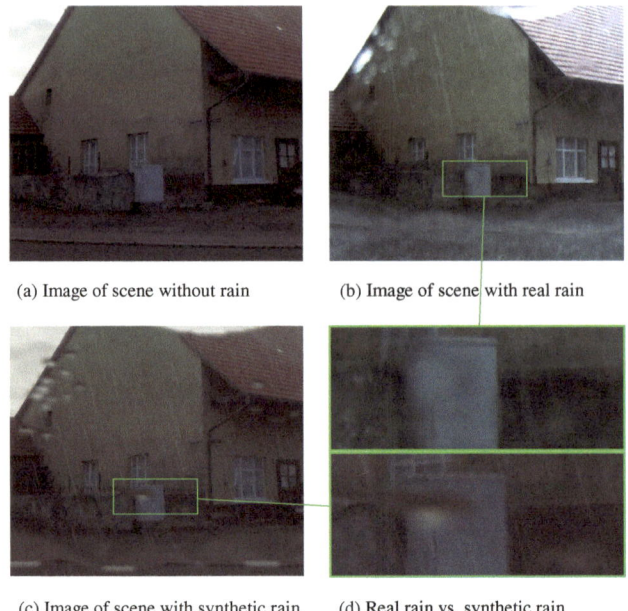

(a) Image of scene without rain   (b) Image of scene with real rain

(c) Image of scene with synthetic rain   (d) Real rain vs. synthetic rain

Fig. 4 Comparison of our synthetic rain and brightness augmentation technique against real rain. Image from [112]

specify the drop size distribution on the windshield. With parameter r_b the brightness of the image can be adapted.

Compared to other solutions such as applying a simple rainfilter mask as in [40], our approach allows a more realistic rain simulation by taking the current environment such as scene depth together with sensor characteristics into account. Additionally, our approach allows simulating variations of different rain instances by adapting the six presented parameters.

To show the visual realism of the presented synthetic rain model we compared the same scene without rain (Fig. 4a), with real rain (Fig. 4b) and with synthetic rain (Fig. 4c). The same image extract is enlarged in (Fig. 4d) for better visibility. The real rain image (Fig. 4b) as well as the synthetic rain (Fig. 4c) have identical rain streaks, blur effects and drops, showing that the used rain model produces similar optical

(a) Original image extract from KITTI dataset [34]

(b) Applied synthetic rain ($r_i = 80\,\mathrm{mm\,h^{-1}}$, $r_a = -20°$, $r_b = 75\,\%$, $d_{count} = 1985$)

Fig. 5 Synthetic rain augmentation technique on KITTI dataset [34]. Image from [112]

effects as real rain. An additional comparison of an original KITTI image compared to the same image with our synthetic rain augmentation is illustrated in Fig. 5.

In addition to the qualitative realism evaluation before, a quantitative evaluation is performed as well. Measurements of images containing real rain have shown that rain has a significant influence on basic image processing metrics like for example Harris features [42]. Edge detection based algorithms (SURF, Canny, Harris, Sobel) allow a deliberate generalization to validate the realism of this rain simulation. Therefore, these basic image processing algorithms are applied to validate our rain simulation. The influence of real rain on these features will be compared to the influence of simulated rain for the exact same scene. If simulated rain as well as real rain have similar effects on these features we state that our model is realistic. Two sets of images of a well-structured scene containing edges and corners for the algorithms to detect were recorded for validation. The first set of images was recorded under heavy real rain (RefReal). The rain intensity was averaged over the period of recording this set of images. The rain intensity of RefReal was $52\,\mathrm{mm\,h^{-1}}$. The second set of images (RefClean) has been recorded immediately after the rain had stopped. RefClean was used as input for the rain simulation with intensities of 10, 40, 70 and $100\,\mathrm{mm\,h^{-1}}$. The simulated rain will be called SimX with X specifying the simulated rain intensity. The effects of SimX and RefReal on Harris features were then compared. Therefore, the 20 best Harris features of seven randomly chosen frames of RefReal and SimX have been compared against RefClean. For RefClean 16.27 correspondences were identified correctly, while for RefReal only 15.71 correct correspondences were found. Sim40 was closest to RefReal with an average of 15.57 correct correspondences. For Sim70 and Sim100, 13.81 and 13.14 correspondences have been found respectively. Another simulation run without rain streaks, Sim0, has shown that the simulation does not produce unwanted side effects and has exactly the same value as RefClean with 16.27 correspondences. Further validation results were in close agreement to the presented example for Harris features. This shows that the presented model for simulating synthetic rain variations produces similar effects compared to real rain. For more details on validation we refer to [41].

2.2 Road Spray

In contrast to rain as presented in Sect. 2.1, road spray represents a rather locally occurring noise. It occurs behind the wheels of a vehicle driving on a surface that is covered with water. However, as road spray occurs directly behind a driving vehicle, it covers large parts of the vehicle, making it more difficult to detect by vision-based object detection algorithms. Therefore, realistic simulation of road spray is important for performance characterization.

To simulate the droplets, physical properties were used to calculate the trajectory of each droplet. When looking at the 2D case, neglecting the lateral distribution of the droplets, the drops have an initial velocity equal to the rotation velocity of the wheel [113]. After the spray is detached from the wheel, air resistance and gravity slow down the droplets until they reach the road surface again. For the 2D case the trajectory of a single droplet represents a curve given projectile motion. To transfer this simulation into 3D space, jitter was added to the droplet positions for every time step. The standard deviation of the jitter was increased the longer the time of flight of a single droplet was in simulation. A result of the 3D positioning of droplets is illustrated in Fig. 6.

As wheel positions are known and the drop positions are calculated, the droplets are rendered as spheres. The mean diameter of droplets was set to 200 μm with a standard deviation of 10 μm. This is just large enough for the droplets to influence visible light geometrically. Instead of using ray tracing for refraction and reflection calculations, reflection and refraction vectors were precalculated. Therefore, many of these vectors were calculated depending on the distance of a droplet to the camera and the location within a droplet where a ray would have hit it. With this look-up vectors the location where the reflected ray would hit the environment is the last

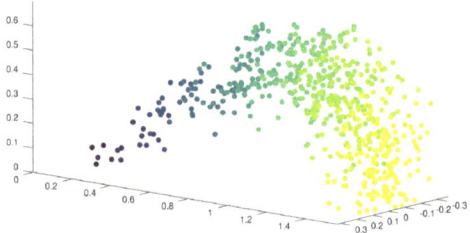

Fig. 6 Example drop distribution behind an imaginary wheel positioned at the origin. To maintain visibility, this plot reduced the number of drops. Colors indicate the longitudinal distance from the origin and aid spatial vision. The axes dimension is in [m]. Image from [113]

Fig. 7 Qualitative comparison of real spray (taken from the realrain data set [112]) on the left, and our simulated spray on the right. Because of the lack of clean spray data sets, we can only compare the occlusion of the lower end of the vehicle. Here, we can observe similar behavior: parts of the wheel are not visible, as well as part of the rear end and parts of the rear lights. The spray color blends in with the background and the color of the street; it reaches the same height as the real spray. Image from [113]

thing to be calculated for rendering. This was solved by generating an approximated cubemap of the 2D input image. Droplets too small to qualify for geometric reflection and refraction were generally considered to be fog. Instead of rendering those large number of micro droplets the sky color is assumed to be the color sampled by an up-pointing reflection vector and is mixed to the droplet color. The result of the presented rain spray simulation can be seen in Fig. 7. For more details of road spray simulation, we refer to the work from our team by von Bernuth et al. [113].

2.3 Dust

Camera sensors are affected by different types of dust throughout the year, making object detection more difficult by partially obstructing the field of view. Dirt on the windshield ranges from pollen in the spring to dirt thrown onto the windshield from the tires of vehicles in front, to tire wear particles.

Our proposed simulation of dust consists of two steps as presented by Hospach [41]. First, dust particles are distributed on a virtual windshield in front of the camera sensor. The size, number, transparency and color of the particles is configurable as well as the distance and angle of the virtual windshield. Afterwards a filter mask with the influence on each sensor pixel is calculated respecting the geometry of the particles as well as the camera parameters. In contrast to the rain simulation in Sect. 2.1 or the snow simulation in Sect. 2.4 a complete scene reconstruction is not necessary as dust is a rather static environmental influence restricted to the windshield. Hence, the filter mask can be precalculated once and applied to

Fig. 8 Original image generated with CARLA on the left and image with dust on the right with 60 simulated particles and a distance of 50 mm distance to sensor

a complete video stream, saving computation time. The calculated filter mask is then applied pixel wise to the input data in the second step. The result of the dust simulation on an image from CARLA [19] is illustrated in Fig. 8.

The dust simulation was validated comparing the influence on HARRIS and SURF features as well as the number of edge points found by Canny edge detection [41]. The experimental setup was as follows: five black, round paper particles distributed on a glass pane were recorded with a real camera. Additionally, a single particle was recorded at different distances from the glass pane to investigate different particle sizes and edge blur effects. These real dust recordings are called RefReal in the following. The same scene without particles denoted as RefClean was recorded as baseline and input for dust simulation. Afterwards the baseline image was augmented by dust simulation denoted as SimX, where X stands for the number of simulated particles. SimX is then compared against RefReal. If the influence of RefReal to basic image processing algorithms is similar to RefSim we have shown that the dust simulation produces equal effects. For evaluation based on Harris features: 20 and eight correct correspondences have been found for RefClean and RefReal, respectively. Sim10000 was closest with an average of 9.35 found correspondences. Sim20000 resulted in 5.25 correct correspondences and the simulation run with the lowest number of dust particles Sim1000 resulted in 16.6 found correspondences. This shows that the higher the number of simulated particles the lower the number of found Harris features gets. The found SURF features decrease as well with increasing amount of dust particles [41] for SimX. The results show similar effects to RefReal, which also reduces the average number of features found. Other simulation results with Canny edge detection were in close agreement. For more details on dust validation we refer to [41].

2.4 Snow

Similar to the simulation of rain (see Sect. 2.1), the first step of the snow simulation is the reconstruction of the 3D scene. Either stereo images to calculate the depth image, a camera image together with LiDAR data or simulation data from e.g., CARLA with a perfect depth image can be used for 3D scene reconstruction. After scene reconstruction, snowflakes have to be distributed in front of the camera sensor. For snow simulation the first step is to determine the number of snowflakes which shall be simulated per volume:

$$N_s = \frac{M_s}{2\,\mathrm{mg}}. \tag{1}$$

M_s represents the mass concentration in air according to Koh and Lacombe [55]:

$$M_s = 0.30 \cdot R_s, \tag{2a}$$
$$M_s = 0.47 \cdot R_s. \tag{2b}$$

Equation (2a) represents the mass concentration for dense snow such as in snow storms whereas (2b) represents the regular snow mass concentration. R_s is the snow precipitation rate in $[\mathrm{mm\ h}^{-1}]$. After having the number of snowflakes per volume specified with (1), the size of the simulated snowflakes has to be determined. With a given snowflake diameter D in [mm] and R_s, the frequency of a snowflake having diameter N_D can be calculated as follows [35, 90]:

$$N_D = N_0 \cdot e^{-\Lambda D}, \tag{3a}$$
$$N_0 = 2.50 \times 10^3 \cdot R_s^{-0.94} \qquad [\mathrm{m}^{-1}\mathrm{m}^{-3}], \tag{3b}$$
$$\Lambda = 2.29 \cdot R_s^{-0.45} \qquad [\mathrm{mm}^{-1}]. \tag{3c}$$

For each snowflake an appropriate diameter is assigned using a piece-wise defined probability distribution function weighted by N_D. Each snowflake is either represented by a flat crystal or a three-dimensional crystal constructed out of three flat ones. The orientation of each flake is randomly chosen based on velocity vectors given by gravity, the velocity of the car onto which the camera sensor is attached and additional wind speeds.

The result of the snow simulation can be seen in Fig. 9. Here, a comparison with real snow is illustrated showing the realism of the proposed simulation approach. For more details on our approach of snow simulation, we refer to [7].

2.5 Fog

Similar to rain, fog consists of little water droplets. However, the amount of water droplets per volume is extremely high (10^5 times higher than for rain), and the droplets are very small (10^3 times smaller compared to rain) [76]. Therefore, a simulation based on 3D reconstruction with trillions of particles and ray tracing would be extremely expensive considering computing power and time. Hence, the fog simulation will use light attenuation algorithms.

(a) Real Snow   (b) Simulated Snow

Fig. 9 Visual comparison of real and simulated snowflakes. The images on the left were taken during snowy weather. On the right, snow was simulated onto images of the exact same scene that were taken on days without any snow fall. Images from [7]

When light traverses fog its rays are partially scattered or absorbed when hitting the small water droplets. It can be assumed that each ray passes a fixed number of fog particles for a specific traveled distance. When passing through fog the amount of scattered or absorbed light can be described by the first term of (4), where I_i describes the incident light intensity, α_{ext} in $[m^{-1}]$ an extinction factor and d in [m] the distance the light travels through fog. Given the i-th pixel color I_i of an image and a sky color I_s, every pixel with depth d is assigned its new color [7]

$$I = I_i e^{-\alpha_{ext}d} + I_s(1 - e^{-\alpha_{ext}d}). \tag{4}$$

In Fig. 10 the resulting fog simulation on an image from Cityscapes data set is depicted. It can be seen that depending on the distance of a given pixel within the image the scattering and absorbing effects of fog differ. Distant objects are harder to spot than closer ones, as they are affected more by the fog. This results in a realistic fog simulation which takes the environment into account. For more information and results we refer to the work of von Bernuth et al. [7].

Fig. 10 The upper image is from the Cityscapes data set [14], the lower image shows the image with our fog simulation applied. Image from [7]

3 Evaluation Metrics for Object Perception

To rate and compare object detection systems, different metrics exist. These metrics consider the accuracy of the perceived bounding boxes and indicate the perception rate. An overview is given in Sect. 3.1. As aforementioned, autonomous vehicles must be safe. Since performance and safety do not always correlate, a new metric to evaluate the safety of perception systems is presented in Sect. 3.2.

3.1 Common Metrics for Perception Evaluation

In existing benchmarks like COCO [60] or KITTI [34], simple performance measures such as precision, accuracy, recall, and mean Average Precision (mAP) are used to evaluate object detection [16, 23, 74]. These metrics are calculated on the number of true positive (TP) or false positive (FP) detections. The classification of TP/FP is based on the IoU of detection and ground truth (GT) bounding box. The IoU, is a well known metric in the field of object detection [83]. For calculation the area of intersection and union of detection D and the corresponding GT G is used as described by Rezatofighi et al. [83]:

$$\text{IoU} = \frac{|D \cap G|}{|D \cup G|}. \tag{5}$$

IoU is used by object detection benchmarks like COCO [60] or Pascal VOC [23]. The threshold value to classify an object as TP can be parameterized; different threshold values like 0.5 in Pascal VOC or 0.7 in KITTI are used. The aforementioned metrics concentrate on analyzing a single frame and are applicable to both 2D and 3D bounding box-based object detection. However, none of these measures can evaluate object-tracking techniques; they only take into account tagged GT objects.

The performance metrics precision (P) and recall (R) [74] include the true negative (TN) and false negative (FN) results to describe the percentage of correct detection and how accurate the detections are:

$$P = \frac{TP}{TP + FP}, \quad R = \frac{TP}{TP + FN}. \tag{6}$$

The accuracy (A) [74] can be calculated as :

$$A = \frac{TP + TN}{TP + TN + FP + FN}. \tag{7}$$

The average precision (AP) is equal to the area of the corresponding precision recall curve (see (8)). Similarly the average accuracy (AA) is defined. The mean average precision (mAP) describes the precision averaged over all available classes.

$$AP = \int_0^1 P(R)dR \tag{8}$$

The *Classification of Events, Activities and Relationships* (CLEAR) defined different metrics to evaluate object detection,—tracking and head-pose estimation. For the detection/tracking evaluation, the Multiple-Object-Detection and Multiple-Object-Tracking precision (MODP/MOTP), and accuracy (MODA/MOTA) were defined [97].

With m_t as misses, fp_t as amount of FPs and g_t as number of GT objects at time t and the IoU of each object as well as N_t^{mapped} as number of mapped object sets at t, MODA and MODP are defined as [97]:

$$MODA(t) = 1 - \frac{\sum_t (m_t + fp_t)}{\sum_t g_t}, \quad MODP(t) = \frac{\sum_{i=1}^{N_t^{\text{mapped}}} IoU_i}{N_t^{\text{mapped}}}. \tag{9}$$

The tracking metrics include additional parameters; mme_t as number of mismatches between GT and tracking hypothesis, $d_{i,t}$ as deviation between tracking hypothesis and GT as well as c_t as number of matches. Using these parameters MOTA and MOTP are defined as [97]:

$$MOTA(t) = 1 - \frac{\sum_t (m_t + fp_t + mme_t)}{\sum_t g_t}, \quad MOTP(t) = \frac{\sum_{i,t} d_{i,t}}{\sum_t c_t}. \tag{10}$$

The CLEAR metrics are used in the KITTI Multiple-Object-Tracking benchmark [34].

Fig. 11 Exemplary scenario showing the necessity for a metric to evaluate safety

The higher level of detail in the CLEAR metrics gives them a significant edge over more fundamental performance indicators like precision and accuracy. As opposed to the binary method of computation based on TP and FP quantity, using the IoU or distance to determine the accuracy scores, allows a better statement about the precision.

3.2 Safety Metric

Since the semantics of a scenario are not taken into account by current performance measures, it is necessary to utilize a metric that assesses the real-world safety of an object perception system.

This can be shown by the scenario in Fig. 11. Based on the detections, the given perception system achieves a precision of 100% and a recall of 86% since 12 of 14 objects are correctly perceived. These results appear to be good, but the undetected vehicle in front of the ego vehicle or the one in the bottom right corner of the intersection could lead to an accident.

The goal is the development of a metric that allows to evaluate safety of various perceptual techniques in various traffic situations and weather conditions. The outcome must be a single value inside a specified range for this use. Therefore, we propose the "Comprehensive Safety Metric (CSM)".

The composition of the individual safety metric components and their relationship is presented in Fig. 12. It demonstrates the method through which our strategy integrates many factors to produce a single safety-metric score that makes it simple to compare the perception algorithms.

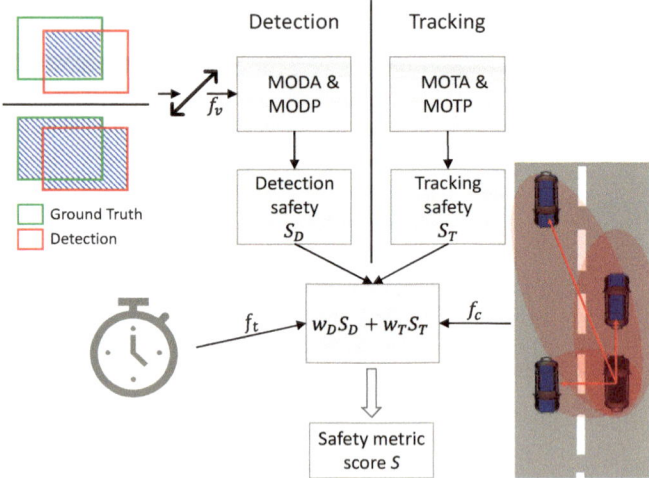

Fig. 12 Process overview of the single components and their relation to one another to determine a safety metric score S. Red areas around ego (black) indicate safety critical areas. Image from [110]

For the assessment of safety, three criteria to consider were defined:

Quality   The effectiveness of perception is crucial for subsequent activities, such as trajectory planning.

Relevance   It is important to recognize any objects that may be related to a collision. We must therefore discriminate between objects that are relevant and those that are not.

Time   Time is always an important consideration in a real-time system. Less reaction time and fewer driving maneuvers are feasible as a result of longer detection durations.

3.2.1   Basis of the Safety Metric

The accuracy of object perception is extremely important when assessing autonomous driving safety. Further activities, such as motion planning, will be carried out based on the perception. Low-quality detection or tracking may result in incorrect planning, which may put the occupants of the vehicle and other road users in danger.

Thus, perception quality is one main safety factor and will be used as basis of the CSM. To combine accuracy and precision we use the CLEAR metrics [97] (see Sect. 3.1). The choice of CLEAR metrics was based on the completeness of the metric, as it combines accuracy and precision for detection as well as tracking.

One issue with the MOTP score emerges when utilizing the CLEAR criteria to assess safety. A better tracking is indicated by a lower MOTP score. Contrary to the safety metric score, which equates a higher number to better safety, this is not the case. To invert the MOTP indication, an advanced mapping to a MOTP safety metric score $\text{MOTP}_s \in [0, 1]$ is defined. With T_u as upper and T_l as lower threshold, MOTP_s can be determined by using:

$$f_{norm}(x) = \begin{cases} 1 & x < T_l, \\ 1 - \frac{x-T_l}{T_u-T_l} & T_l \le x \le T_u, \\ 0 & \text{otherwise.} \end{cases} \tag{11}$$

For our experiments it holds that $T_l = 0.8\,\text{m}$, as this value corresponds to a step width of a vulnerable road user (VRU) to avoid a collision. By similar reasoning we set $T_u = 2.5\,\text{m}$, which roughly corresponds to a misjudgment that could lead to a collision. A linear function is used because MOTP is metrically scaled.

The threshold values of f_{norm} can be parameterized based on the application domain and the accompanying requirements. This increases the variability and makes the metric applicable for the assessment of various systems.

Precision and accuracy are equally important to us for the suggested safety measure, so we use the accuracy and precision score of detection and tracking to generate a second safety metric basis rating. The detection safety (S_D) and the tracking safety (S_T) are defined as:

$$S_D = \frac{\text{MODA} + \text{MODP}}{2}, \quad S_T = \frac{\text{MOTA} + \text{MOTP}_s}{2}. \tag{12}$$

This evaluation is just a baseline and further values must be evaluated to cover the three safety criteria, which were introduced in Sect. 3.2.

3.2.2 Distance-Based IoU Verification

A second parallel assessment is carried out before the CLEAR metrics are computed. For objects closer to the ego vehicle, the perception must be more precise. The shorter amount of time to react during motion planning is the basis for this harsher criterion for closer objects. We need to differentiate the perception quality, since these things exhibit a higher safety criticality.

The distance-based IoU verification uses the cover C_o of GT object G. For a detected object o with detection D the cover is defined as:

$$C_o = \frac{|D_o \cap G_o|}{|G_o|}. \tag{13}$$

Using C_o, a safety function f_s is defined as:

$$f_s(C_o) = \begin{cases} \frac{1+mC+(1-mC)\sin(\pi(C_o-\frac{1}{2}))}{2} & C_o \in [mC, 1], \\ 1 & C_o \in (1, oT], \\ \frac{1+\cos(\frac{\pi}{mO-oT}(C_o-oT))}{2} & C_o \in (oT, mO], \\ 0 & \text{otherwise.} \end{cases} \quad (14)$$

This function guarantees a minimum detection precision mC. Between the thresholds mC and mO, trigonometric functions are used for a smooth distance-based scaling factor depending on the precision of the detection. oT defines a threshold how much larger an object is allowed to be detected without lowering the detection precision. If C_o is larger than oT, mO represents the upper bound up to which f_s reduces the precision towards zero.

The distance-based score is calculated by function $g : [0, 1]^2 \rightarrow [-1, 1]$, where

$$g(x, y) = x - (1 - x) \cdot (1 - y). \quad (15)$$

The function $g(f_s(C_o), d_o)$ must be transformed to $[0, 1]$ to be used as a precision factor. The transformation is described by

$$f_v = \frac{g(f_s(C_o), d_o) + 1}{2}. \quad (16)$$

For each detected object o the IoU gets multiplied by f_v. This additional consideration leads to a stricter rating, which should be preferred in context of safety.

3.2.3 Consideration of the Collision Relevance

The second criteria to assess the perception safety is the relevance of an object. A possibly safety critical object has a higher importance than a non safety critical object.

First, it must be defined when an object must be considered as safety critical. An object is safety critical if its distance to the ego vehicle is less than a corresponding safety distance. To calculate the safety distance, we use the approach of the "Responsible-Sensitive Safety" (RSS) model [93]. The RSS model is an attempt to formalize the human judgment in different road scenarios in a mathematical sense. The RSS model consists of 34 definitions of different safety distances, times, and procedural rules. These rules specify how an autonomous vehicle should behave and provide a mathematical description of a safe conduct.

We use the longitudinal safety distance with same direction of movement $d_{long,s}$, with opposite direction of movement $d_{long,o}$ and the lateral safety distance d_{lat} [93, Definition 1, 2, 6].

To evaluate the collision relevance of an object, the future position must be predicted. With the ego velocity v_0 and the weather-dependent brake acceleration a, the prediction time frame t_p is defined as:

Fig. 13 Schematic identification of collision relevant objects from KITTI raw data set [33]. The right image represents the bird's eye view of the camera image on the left. Blue boxes illustrate ground truth annotations, light blue boxes represent the predicted object positions. Red filled objects are collision relevant and white ones are not. The corresponding collision relevant objects in the camera image are marked in red. Image from [110]

$$t_p = 1.1 \cdot \frac{v_0}{a}. \tag{17}$$

For each time step in the position prediction phase, it is verified whether the distance between ego and the object is higher than the corresponding safety distance. The object is marked as safety critical if this is not the case and the perception system did not perceive it.

Figure 13 shows this process schematically. The red area in the bird's eye view marks the safety critical area identified by lateral and longitudinal RSS safety distances. The collision relevant objects are marked in red. If they are not perceived, they are considered safety critical, as shown in Fig. 13.

To rate the relevance in context of safety, we need to approximate the effect of a missing detection and a hypothetical resulting collision. The first step is an approximation of the impact velocity, in case of an in fact collision.

Since safety in automated driving affects not only the vehicle occupants but also other road users, these must also be taken into account. Road users can be categorized into VRUs and road users with a crush collapsible zone, like cars, vans or trucks.

The combination of impact velocity and the road user category c of the collision relevant object leads to a collision score $s_{c,ro}$ for a relevant non-detected safety critical object ro. To assess $s_{c,ro}$, a classification of the impact velocity with four levels is defined. The level definition is based on the common accident categories used in Germany. These categories are defined by the Ministry of the Interior of the state North Rhine-Westphalia in Germany as UK 1 (fatality)—UK 3 (minor injuries only) [65]. Furthermore, an additional category UK 5 is used to include collisions with material damage only [65].

More about the effects of vehicle impact velocity in a collision can be found in the publications of Frederiksson et al. [27] and Han et al. [37].

The defined categories with their collision scores $s_{c,ro}$ are:

$$s_{c,ro} := \begin{cases} 0.9 & \text{no or almost no effect (UK5),} \\ 0.75 & \text{risk of minor injuries (UK3),} \\ 0.5 & \text{risk of serious violation (UK2),} \\ 0 & \text{high probability of fatality (UK1).} \end{cases} \tag{18}$$

In our approach, $s_{c,ro}$ is used as a factor for a single frame. A collision that is rated as having a high chance of fatalities is unacceptable and receives a score of 0. The case with almost no effect is worse than no accident, thus a factor of 0.9 is defined. $s_{c,ro}$ must not be too strict, otherwise no accurate differentiation of the final safety value would be possible.

For a single frame the worst case $s_{c,ro}$ is calculated and used as factor f_c on S_T and S_D.

3.2.4 Evaluation of Perception Time

The time is the third requirement for a safety-critical real-time perception system. The longer object identification takes, the less time there is to avert a life-threatening situation. The time requirements of the proposed safety metric is covered by the soft real-time approach of Kim et al. [58].

For the CSM, the perception time $t_{d,o}$ of object o is defined as time from falling below the safety distance (see Sect. 3.2.3) until its perception.

A weighted perception time is used to convert the detection time to a perception time factor. The introduction of the weighting was necessary, since the problem becomes more dangerous the longer it takes to identify it. The mean perception time is used to categorize long and short durations for this purpose. Let m be the number of all weights and $\overline{t_d}$ the mean perception time. The weighted perception time t_{dw} with m as number of all weights is defined as:

$$t_{dw} = \frac{1}{m} \sum_o \begin{cases} t_{d,o} & t_{d,o} \leq \overline{t_d}, \\ 2 \cdot t_{d,o} & \text{otherwise.} \end{cases} \tag{19}$$

Similar to Kim et al. [58], the CLEAR scores are mapped by a function depending on t_{dw}. The mapping of t_{dw} to f_t is done with (11). The parameter T_l is set to 0.1 s, as tolerable delay for the detection. T_u is set to ego braking time t_b. If $t_{dw} > t_b$, f_t has to be 0, since an emergency braking would not be possible anymore.

3.3 Comprehensive Safety Metric Score

The result of the CSM has the requirement of an easy comparability. Hence, the safety metric score is a single value $S \in [0, 1]$, where 1 describes the maximum

Table 1 Rating of the safety metric score. Table from [110]

$S \in$	Classification
[0.0 − 0.2]	High risk of fatality
(0.2 − 0.4]	Existing risk for serious violation
(0.4 − 0.6]	Low probability of minor injuries
(0.6 − 0.8]	Low risk UK 5 collisions
(0.8 − 1.0]	High probability of safe status

safety. Like the previous described performance metrics, S is determined for each frame of a scenario. Therefore, S_D and S_T including the evaluation of collision relevance and perception time are combined.

To achieve a high variability in the CSM, S_D and S_T can be weighted with $w_D, w_T \in [0, 1] : w_D + w_T = 1$. The safety score S is defined as:

$$S = w_D S_D + w_T S_T. \tag{20}$$

The comprehensive safety is not a percentage value, in contrast to precision or accuracy, which results in a non-intuitive interpretability. It is necessary to specify a categorization of S (see Table 1) in order to improve interpretability. The five-level defined classification is based on the evaluation of the individual CLEAR metrics values as well as the specified influences of collision relevance and detection time analysis.

This classification offers a quick and easy performance comparison safety evaluation of different test scenarios and perception systems.

3.4 Data Set Evaluation with the Safety Metric

Initially, we motivated the safety metric by the scenario shown in Fig. 11. The resulting precision of 100% and a recall of 86% indicate a very good perception. Depending on the velocity of the vehicles in an inner-city scenario, the result of the CSM would be in the range of 0.4 and therefore indicating minor to serious injuries which are far away from a safe state.

Table 2 shows the results of an image-plane object detection using YOLOv3 [79] on three VTD scenarios (freeway, crossing and rural) [1] and the KITTI raw data set [33]. As we can see, the precision is over 80% for the virtual scenarios but recall and mAP are rather low with about 30% for freeway and crossing. The significant gap between the mAP for KITTI and the simulated scenarios can be explained by the number of objects and their positioning. Multiple objects are occluded and thus cannot be perceived correctly. For the state-of-the-art performance, these results seem acceptable but the CSM has a result of 0.14/0.20 for freeway/crossing. Using

Table 2 Evaluation results for object detection with YOLOv3. Table from [110]

	KITTI	Virtual scenarios		
		Freeway	Crossing	Rural
Precision	0.59	0.82	0.86	0.96
Recall	0.60	0.23	0.36	0.60
mAP	0.51	0.21	0.35	0.60
Safety score S	0.48	0.14	0.20	0.78

the corresponding classification of Table 1, this indicates a high risk of fatality due to undetected relevant objects. For the rural scenario the safety score S is higher than the corresponding recall and mAP. Even if the recall is not perfect, we can observe that the perception is close to a safe state. This is based on the scenario of a rural road with our ego vehicle following two further vehicles. Single misdetections do not have an influence, since the distance between the objects is big enough that there is no significant risk of an accident. For the KITTI raw data set, the safety score S, the recall and the mAP are quite similar, but the interpretation of these values is quite different. While a recall or mAP of 50–60% seems good, a safety score of 0.48 indicates that some missing detections could lead to accidents with a probability of injuries, which is not acceptable.

Further results for Faster-RCNN, RCC and a Birds-Eye 3D detection can be found in [110].

4   Optimization of Object Perception

This section thematizes the optimization of local and cooperative perception. First, the need of robustness improvement is motivated by showing the influence of weather on vision-based object detection. Sections 1.3 and 4.2 present the used data sets and introduce our proposed robustness enhancement for local perception. Concluding, the advantages of cooperative perception are introduced and an environment-aware optimization approach for the data fusion in CP is presented.

4.1   Influence of Weather on Perception

Object detection relying on camera sensors is prone to adverse weather conditions such as heavy rain or difficult lighting conditions. Therefore, vision-based object detection in particular needs to be resilient to adverse and varying weather conditions. In order to determine its resilience and robustness, the capabilities of vehicle-local

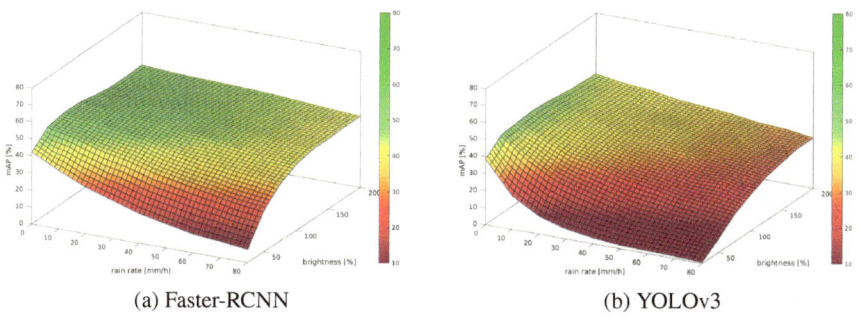

(a) Faster-RCNN   (b) YOLOv3

Fig. 14 Mean average precision depending on varying rain and brightness intensities for **a** Faster-RCNN and **b** YOLOv3. Image from [112]

perception under varying weather conditions are investigated. In the following vision-based perception will be referred to as perception.

For robustness assessment of perception, two different neural networks (Faster-RCNN [82] and YOLOv3 [79]) will be evaluated. Both networks are trained with the KITTI data set [34] and the quality of object detection is assessed with the well known average precision metric (AP) as presented in Sect. 3.1. To evaluate resilience against adverse weather conditions, a realistic synthetic rain augmentation is used to modify the KITTI data set. The augmentation consists of two steps, the generation of falling rain [43] followed by rendering raindrops on the windshield [6]. The exact process of simulating rain is explained in Sect. 2.1. The rain augmentation technique consists of various parameters to adjust the simulated rain. For evaluation, the same parameter ranges as used for the optimization from Table 3 were used. However, the ranges of rain intensity and image brightness have been adapted to cover a large variation in the evaluation phase:

- rain intensity r_i [$0 \, \text{mm} \, \text{h}^{-1}$, $80 \, \text{mm} \, \text{h}^{-1}$]
- brightness r_b [25%, 200%]

The exact parameter values were randomly chosen in between the above defined parameter ranges. The networks which were initially trained on the original and not augmented KITTI data set are then evaluated on the distinct test set of KITTI which was not used for training. The test set is augmented with synthetic rain augmentations and the perception capabilities are investigated.

The result to identify the influence of synthetic rain variations on mAP of Faster-RCNN is illustrated in Fig. 14a and b for YOLOv3 respectively. The achieved mAP is plotted for different rain intensities and brightness levels. Drop radii are implicitly included in the varying rain intensities. The angle of the falling rain is not plotted separately as it had fewer influence compared to rain rate and brightness.

Faster-RCNN achieved a mean mAP of 45.45% while YOLOv3 achieved a mean mAP of 33.74% over all rain and brightness variations. The networks were not separately trained for cars, pedestrians and cyclist only as usually done for the KITTI benchmark. We rather used all present KITTI labels for training. Hence, the mAP of 50.42% (Faster-RCNN) and 48.42% (YOLOv3) without augmentation are not to be confused with the online available results. Additionally, the online available AP values are given per class, and we average the AP over all classes regarding the number of objects per class.

Increasing rain intensities and brightness values below 100% drastically lower mAP of the investigated neural networks. For Faster-RCNN, the most critical situation was observed for 80 mm h^{-1} rain intensity, 0° rain angle and 25% brightness, which resulted in a drop by 94.21% compared to not augmented KITTI. YOLOv3 had the worst detection rates at 80 mm h^{-1} rain intensity, −30° rain angle and 25% brightness, which led to a detection drop of 99.61%. Hasirlioglu and Riener [39] found similar results in their investigation about the influence of rainy weather on the object detection performance. The investigation shows that neural networks are not robust against adverse weather conditions. Data sets such as KITTI lack weather-influenced scenarios. Therefore, it is not possible to obtain robust networks just by training on them.

4.2 Optimization of Local Perception

Vehicle-local perception is strongly affected by adverse weather conditions such as heavy rain (see Sect. 4.1). To optimize perception capabilities of vision-based object detection, we introduce a methodology that uses realistic augmentation techniques as presented in Sect. 2 to diversify existing data sets with adverse weather conditions. This makes neural networks more robust by having as diverse training data as possible. An overview of our proposed workflow is illustrated in Fig. 15.

The first step is to extend the KITTI training set [34] with augmented data. Next the training of Faster-RCNN [82] and YOLOv3 [79] is performed again on this new and diversified data set. The KITTI training set was split as before in a training set consisting of 6800 images and a test set containing 468 images. Rain augmentation is performed for the whole training set of 6800 images and added to the original training data set resulting in a training data set of 13 600 images. Hence, only half of the images from the training data are augmented while the other half are not. This prevents overfitting to adverse weather conditions and the neural networks will have still good performance on the original data set. To validate the effectiveness of the proposed data augmentation through synthetic rain and brightness variation, additional data augmentation methods were compared against our approach. Therefore, the neural networks were also trained with a data set extended by Gaussian noise (GN), Salt-and-Pepper noise (SPN) and a combination of GN and SPN.

A large variation of different augmentations (see Table 3) has been used to extend the training data set. Six parameters for synthetic rain have been chosen as in the evaluation (see Sect. 4.1). For GN two parameters, for SPN one parameter and for the combination three parameters specify the noise intensity. The selected parameter ranges were chosen as follows: The evaluation has identified that only a bright-

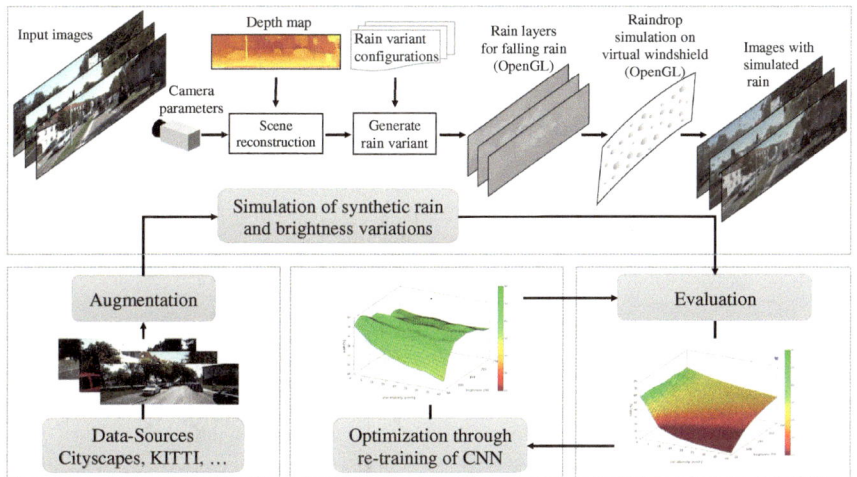

Fig. 15 Workflow of local robustness optimization and evaluation by simulating rain variations. Image from [112]

Table 3 Parameter ranges for data augmentation in the optimization phase of our workflow. Table from [112]

Case	Parameters	Value intervals
GN	μ	[10, 50]
	σ	[1, 20]
SPN	Density	[1%, 30%]
GN	μ	[10, 50]
&	σ	[1, 20]
SPN	Density	[1%, 30%]
Synthetic rain	Rain intensity r_i	[30 mm h^{-1}, 80 mm h^{-1}]
	Rain angle r_a	[$-30°$, 30°]
	Brightness r_b	[40%, 100%]
	Drop count d_{count}	[1000, 2000]
	Mean drop radius d_μ	[0.3 mm, 0.8 mm]
	Std dev of drop radius d_σ	[0.25 mm, 1.25 mm]

ness below 100% has a strong negative effect on the neural networks. For a higher brightness an increasing rain rate affects the neural networks less. The intervals for brightness augmentation and rain intensity have therefore been set to the ranges found as critical in the evaluation phase. The lower bound of the brightness augmentation was set to 40% as this has shown to be more effective compared to lower brightness values. The lower bound of rain intensity was raised to 30 mm h^{-1}, as challenging situations only occurred above this rain intensity.

(a) Examples containing rain streaks and blur

(b) Examples containing rain streaks, blur and rain drops on the windshield

Fig. 16 Comparison of GT (blue), Faster-RCNN baseline (red), optimization with GN and SPN (yellow) and our optimization with rain and brightness variations (green) on example images taken from our realrain data set. Images from [112]

Similar to the evaluation phase, the exact parameter values for every augmentation technique were randomly chosen for each image within the specified parameter ranges to generate a training set of various conditions, except for d_μ and d_σ.

d_μ and d_σ are calculated according to the randomly chosen rain intensity with equations according as introduced in Sect. 2.1. The random number generator was seeded to be able to generate reproducible results.

4.2.1 Results for Optimization of Local Perception

To evaluate the presented perception optimization approach the realrain data set (see Sect. 1.3) was used. This data set was solely used for validation and not for training. The perception capabilities in terms of AP and AA were investigated for the baseline, GN and SPN augmentation techniques and our optimization. A qualitative comparison is illustrated in Fig. 16.

Quite remarkable is the fact, that only with our optimization approach the CNN was able to detect the vehicle obstructed with raindrops in Fig. 16b. A complete overview of the results is presented in Table 4. It can be seen that our optimization performs best for YOLOv3 as well as for Faster-RCNN considering AP. With our approach, the unoptimized detection for Faster-RCNN was improved by 4.37% points (p.p.) and by 7.33 p.p. for YOLOv3. The second-best optimization in comparison achieved an improvement of 1.65 p.p. for Faster-RCNN and 2.18 for YOLOv3. Looking at AA instead of AP it can be seen that AA decreased by 1.67 p.p. for Faster-RCNN but on the other hand gets improved by 0.53 p.p. for YOLOv3. Optimization with SPN performs best for AA but worst when it comes to AP. When it comes to

Table 4 Average precision and accuracy results for Faster-RCNN, YOLOv3 and RRC on the evaluation of our realrain data set and the original KITTI test set. Table shortened from [112]

Neural network	Training method	Realrain data set		KITTI test set	
		AP in %	AA in %	AP in %	AA in %
Faster-RCNN	Baseline	7.48	36.22	50.42	41.87
	GN	10.20	38.08	**51.02**	**42.90**
	SPN	7.62	**39.93**	48.82	40.78
	GN and SPN	9.96	37.62	49.59	40.98
	Synthetic rain variations	**11.85**	34.55	49.95	42.27
YOLOv3	Baseline	5.15	37.96	**48.42**	**60.93**
	GN	10.30	42.26	45.51	58.25
	SPN	3.40	**43.95**	37.72	59.36
	GN and SPN	5.10	42.33	39.61	56.19
	Synthetic rain variations	**12.48**	38.49	47.79	59.86
RRC	Baseline	12.97	64.33	74.60	74.60

safety under adverse weather conditions not perceiving an obstacle is more severe than false positive detections which e.g., could result in additional breaking maneuvers. Therefore, the AP metric is more relevant than the AA metric for assessing perception performance because it considers recall as well as precision.

Furthermore, we compare our two optimized networks to the more robust neural network RRC [81]. RRC achieves a mean AP of 74.60% on the KITTI test set. This is a lower mean AP value compared to the online available results on the KITTI benchmark website as RRC was trained on all present KITTI labels and not separately for cars, cyclists and pedestrians. However, on the realrain data set RRC only achieves an AP of 12.97%. This shows that even more robust networks are incapable of handling adverse weather conditions such as heavy rain. Both networks which were optimized with rain variations achieve similar performance like RRC in AP on the realrain data set, although the unoptimized versions perform drastically worse.

A disadvantage of many data augmentation techniques for enlarging training data sets is the decrease of performance on the original data set. Hence, we evaluated the performance on the original KITTI data set as well. The results are present in Table 4. It can be observed that our optimization approach with synthetic rain variations almost has no negative effect on the performance on the original KITTI data set. For Faster-RCNN the AP got lowered by 0.47 p.p. and for YOLOv3 AP got decreased by 0.63 p.p. Comparing our approach with the augmentation with GN the performance got increased for Faster-RCNN and decreased for YOLOv3. The remaining augmentation techniques including SPN lowered AP slightly for Faster-RCNN but significantly for YOLOv3.

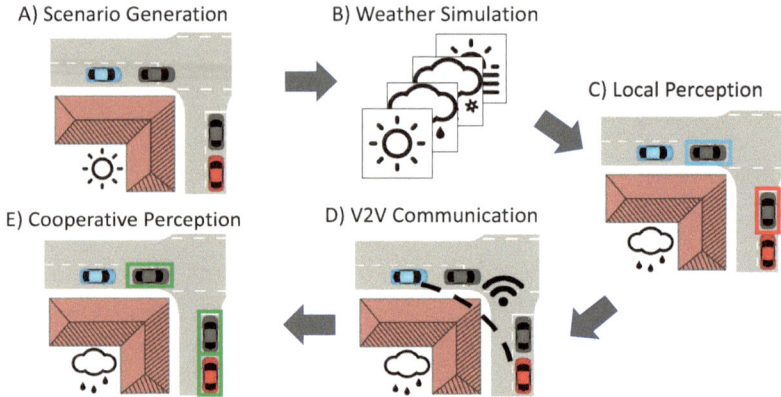

Fig. 17 Process of cooperative perception including a weather simulation (**b**). Image from [108]

The presented approach shows that using realistic synthetic rain variations to extend existing data sets for the training of neural networks can improve the robustness of these networks against adverse weather conditions. It has been shown that the performance on the completely different realrain data set could be improved while maintaining the performance on the original data set.

4.3 Cooperative Perception

Cooperative Perception describes a process in which the perception is done across multiple distributed vehicles. Information about locally perceived objects is transmitted via V2X communication between different vehicles. The ETSI defined two message formats for this purpose. The first message is the Cooperative Awareness Message (CAM) [22] which contains the state (position, velocity, orientation) of the ego vehicle. The second message type is the Collective Perception Message (CPM) [21] which contains the ego state as well as the states of the locally perceived objects. The ego must align all information of the local perception and the data from the cooperative vehicles to its ego vehicle coordinate system; afterwards all information must be matched before a fusion can be executed. The fusion is necessary to combine different information about the same object as exact one valid state per object is necessary.

The advantages of CP are manifold. The main advantage is the increase of the perception range. Local perception can be limited through weather conditions (see Sect. 4.1), limited sensor ranges and occlusion. The CP, as shown in Fig. 17, enables the perception of objects that cannot be perceived locally. The ego vehicle (blue) can only locally detect the gray vehicle in front; the other objects are occluded by a building. The cooperative vehicle (red) can detect the gray vehicle in front of it

Table 5 Comparison of mean average precision of local perception (LP) against cooperative perception (CP) over different rain rates on a rural, intersection and freeway scenario. CP40 refer to cooperative equipment rate of 40%. Table shortened from [108]

Rain-rate	Rural		Intersection		Freeway	
in mm/h	LP in %	CP in %	LP in %	CP40 in %	LP in %	CP40 in %
0	32.17	37.83	15.86	32.98	10.63	28.27
10	23.50	39.50	15.60	31.57	7.86	26.44
30	12.50	38.83	15.46	31.28	4.21	25.40
50	11.33	38.50	14.72	29.72	1.51	24.48
70	3.67	35.50	11.65	25.85	0.07	23.30
90	1.67	31.33	8.64	22.59	0.00	23.75

and send this detection together with its state to the ego. The ego now knows about the existence of two further objects behind the corner. Furthermore, CP can lead to multiple detections of the same vehicle, which allows a more precise estimation of an object's state.

The advantage of CP under different weather conditions was investigated by Volk et al. [108]. Their results can be used to quantify the above described advantage.

As shown in Table 5, they achieved remarkable results. For a freeway scenario without any rain CP could increase the mAP from 10.63 to 28.27% with 40% cooperative vehicles. At higher rain rates of about 70–90 mm h^{-1} the local perception was not able to detect any object while the CP still achieved a mAP of about 24%. Similar results could be observed for a rural and an intersection scenario. The rural scenario only consists of two vehicles except the ego; one of the further vehicles is a cooperative vehicle.

4.3.1 Optimization of Cooperative Perception

Cooperative perception complicates the measurement-to-track assignment problem, as well as data tracking and fusion. There are two basic methodologies for tracking and fusion. The first is to have a centralized tracking component that directly handles sensor data [77]. The second method, known as Track-to-Track Fusion (T2TF), employs decentralized tracking components and fuses preprocessed sensor data available as tracks (state vector and corresponding covariance/confidence). T2TF has the advantage of providing more information about object dynamics and compensating V2X transmission latencies for CP [77].

Covariance Intersection (CI) of Julier and Uhlmann [48] was one of the first fusion approaches considering unknown correlations.

The CI to determine a fused state $\hat{x}_g$ with covariance matrix (CM) P_g for two track states $\hat{x}_i$, $\hat{x}_j$ with their CMs P_i, P_j is defined by Julier and Uhlmann [48] as

Fig. 18 Process overview
from the pre-evaluation of a
local perception over the
validation of the received
tracks in the track-validation
to the T2TF fusion. Image
from [111]

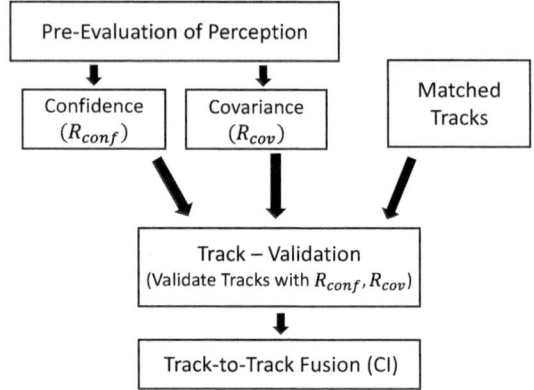

$$P_g^{-1} = \omega P_i^{-1} + (1 - \omega)P_j^{-1}, \tag{21}$$

$$\hat{x}_g = P_g(\omega P_i^{-1}\hat{x}_i + (1 - \omega)P_j^{-1}\hat{x}_j), \tag{22}$$

$$\omega = \arg\min \det P_g, \quad \omega \in [0, 1]. \tag{23}$$

Improvements of the CI regarding the sequential fusion of multiple data and the approximation of ω were presented by Cong et al. [13] as well as Niehsen [71] and Fränken and Hüpper [26].

The CI has some disadvantages; for more than two tracks it was proven by Reinhardt et al. [80] that the CI does not necessarily deliver the optimal result. Furthermore, the CI does not consider inconsistent inputs. To address the problem of inconsistent inputs, Covariance Union (CU) was presented [105]. If the deviation between two inputs exceeds a defined threshold they are considered as inconsistent [11].

In addition to CI and CU, there exist many more approaches for the T2TF. More information about T2TF can be found in [11, 75, 77, 111]

However, the CI can not fulfil the performance requirements for CP in autonomous driving. As a result, the robust but suboptimal CI must be optimized so that only accurate and trustworthy data contribute to the cooperatively perceived environmental model. A pre-evaluation analyzes the capabilities of local perception systems so that the T2TF algorithm can evaluate the trustworthiness and validity of cooperatively transmitted data before fusing it. Therefore, the assumption is made that the local perception system of each vehicle is known.

Figure 18 shows a schematic overview of our proposed optimization pipeline. The pre-evaluation is used to determine the reference data R_{conf} for the confidence and R_{cov} for the covariance. The reference data, combined with corresponding tracks used in a track-validation module; this performs the suggested validation before the CI is used for T2TF.

Pre-Evaluation of Local Perception

For our approach we assume that the local perception system l_v (sensor configuration and processing pipeline) of a cooperative vehicle v is known. Additionally, the current weather condition e including its intensity must be known. Adverse weather is considered since it has a significant influence on the perception capabilities [112]. The pre-evaluation investigates the local perception systems by their perception accuracy, measured with the CMs and the perception capabilities in terms of confidence which is measured by the recall.

The local perception is analyzed under varying weather conditions and the objects are clustered in distance bins d of size $s_{bin} = 5$ m; this approximately corresponds to the length of an average car.

The results of a local perception are analyzed to get realistic and comparable confidences and CMs, to determine if a track from a cooperative vehicle seems plausible and is considered as valid to fuse it. Based on this evaluation two weather-related lookup-tables of the local perception capabilities for each specific perception system l_v are built. These lookup-tables are $R_{conf}(l_v, e, d)$ (abbr. R_{conf}) and $R_{cov}(l_v, e, d)$ (abbr. R_{cov}).

The recall [74] at a distance bin d for l_v is used as confidence. The IoU [83] must be greater 0.5 for a classification as true positive.

A cloudy day is used as baseline for the evaluation. To include adverse weather condition, a local perception under foggy condition with different densities from $0.01\,\mu m^{-3}$ to $0.15\,\mu m^{-3}$ is performed. To achieve reliable results each weather condition is executed for 10 runs with random positioning of the vehicles. Even for 10 runs it can occur that no objects were present at specific distances. Hence, no reference data can be calculated. To avoid missing values in R_{conf} as well as R_{cov}, linear interpolation is used to determine missing values.

Optimization Strategies
Based on the pre-evaluation, our proposed approach validates collectively received tracks by comparison to R_{conf} and R_{cov}. This enables sophisticated validation of perceived data in order to improve the resilience of unoptimized data fusion methods to harsh weather conditions as well as forged data. Two different validation approaches will be investigated. First a selection of tracks to reduce the number of tracks is presented. Second an advanced filtering approach based on the pre-evaluation is investigated.

Track Selection
Reinhardt et al. [80] have proven that the CI is not necessarily optimal for more than two tracks. As a result, one optimization strategy considers reducing the number of tracks used for fusion to two. For the selection of the tracks used for fusion, two approaches based on confidence and CM are considered. The first strategy only takes the two tracks with the highest confidence into account. The second strategy uses the two estimates with the smallest trace of their CM. The two advantages of this approach are the simplicity and a reduction of noise from inaccurate estimations; but

the method only works if more than two estimations exist and can not avoid forged data.

Track Filtering

The second investigated optimization is filtering based on validation using pre-evaluated reference data. This technique addresses fusion precision as well as security; it's split into confidence-based and CM-based validation.

A received detection has an assigned confidence, describing its trustworthiness. Distant objects are perceived less precisely [108]; thus lower confidence values are expected. An attacker is interested to make sure that the forged information are considered for fusion; thus they are sent with a high confidence which can be implausible. The corresponding reference confidence from R_{conf} can be used for validation. If reference and received confidence differ more than a defined threshold the received information is either inaccurate or maybe forged and thus considered as invalid and not used for fusion. Therefore, we assume a standardized assessment for confidence values.

A similar approach is possible using the CMs of the received state estimations. The CMs of R_{cov} are used to validate the received CM by trace or element wise. The CM's main diagonal consists of the variances of the track-state items. Higher variances stand for a more inaccurate estimation such as for occluded or far distant objects. To avoid an inaccurate fusion, inaccurate estimates with a high variance must be discarded, even if their influence is small through the calculation of ω.

If the received information's trace exceeds the reference trace by a threshold t_{trace}, the received information is deemed incorrect and discarded before the fusion. However, as some variation is acceptable, the threshold should not be set too low.

Not only the trace can be used for validation but also an element wise validation on the main diagonal is possible. To do so a threshold vector t_{elem} with the size n of the main diagonal of the CM must be defined. Mathematically the validation process of the two mentioned techniques for filtering inaccurate estimations can be formulated as:

$$\text{tr}(P_s) - \text{tr}(R_{cov}) > t_{trace}, \text{ or}$$
$$P_s(i, i) - R_{cov}(i, i) > t_{elem}(i) \text{ for } i = 1, 2, \ldots, n.$$

If one of the conditions applies, the track $s(\hat{x}_s, P_s)$ is considered inaccurate and discarded.

The two advantages of the element wise approach are the higher flexibility and more detailed validation. The element wise approach allows a more specific filtering based on the requirements of the current system. Additionally, errors of single values can be detected.

To achieve an influence as high as possible an attacker would send forged data with a significant low CM. Contrary to the filtering of inaccurate estimations R_{cov} must not exceed the received information by more than t_{trace} or t_{elem}.

Mathematically this can be described with the two following conditions:

Table 6 Overall precision [%] for a cloudy day, varying fog densities and different rates of cooperative vehicles. *CR1* refer to 16.7% cooperative vehicles, *CR2* refer to 30.6% cooperative vehicles. Adapted from [111]

| | Cloudy | | Fog density [μm^{-3}] | | | | | |
| | | | 0.01 | | 0.07 | | 0.13 | |
	CR1	CR2	CR1	CR2	CR1	CR2	CR1	CR2
Baseline	61.2	56.5	58.9	54.1	67.4	57.4	67.7	68.8
2TracksConf	53.5	46.7	56.8	52.8	67.6	57.5	67.3	68.8
2TracksCov	61.0	48.7	58.4	53.7	67.6	56.9	67.3	68.3
FilterConf0.2	66.7	55.8	58.9	54.1	67.4	57.4	67.7	68.8
FilterTrace4	62.4	57.4	60.3	58.1	73.2	65.1	77.2	77.2
FilterElement	**82.4**	**73.3**	**80.6**	**75.5**	**94.7**	**85.8**	**89.4**	**91.5**

$$\text{tr}(R_{\text{cov}}) - \text{tr}(P_s) > t_{\text{trace}}, \text{ or}$$
$$R_{\text{cov}}(i, i) - P_s(i, i) > t_{\text{elem}}(i) \text{ for } i = 1, 2, \ldots, n.$$

If one of the conditions is evaluated as true, the track must be considered as possibly forged and therefore will be discarded.

Results for Optimization of Cooperative Perception

First results showed a precision increase for the detection. Table 6 shows an extract of the precision results for a Vires VTD freeway scenario with 36 vehicles in total for different optimization strategies. With 16.7% (CR1) and 30.6% (CR2), two different equipment rates for cooperative vehicles are investigated. To test the robustness of the approach under realistic environmental conditions, a fog simulation with three different densities is incorporated. *Baseline* describes the regular CI fusion using all tracks. We can observe a precision of about 55–69%. For the track selection strategies based on the confidence (*2TracksConf*) and the covariance matrix (*2TracksCov*), we can observe that the precision drops for no or low fog (0.01 μm^{-3}). For medium (0.07 μm^{-3}) and dense (0.13 μm^{-3}) fog, the precision is similar to the original CI fusion. Using a confidence deviation threshold of 0.2 for the track filtering leads to a minor increase for a cloudy day. For the different fog densities no effect on the precision can be observed. Filtering tracks by trace with $t_{\text{trace}} = 4.0$ leads to slightly better results for the cloudy day and low fog with an increase of about 2.5 p.p. For medium and dense fog the precision could be increased significantly by up to 9.5 p.p. for CR1 at dense fog. The element wise track filtering strategy with threshold $t_{\text{elem}} = \{1.5, 0.8, 0.2, 2.0, 1.0, 0.3\}$ achieved the best precision scores. For the cloudy day as well as for all fog densities, there is a significant increase of precision. Using this strategy the precision can be increased by at least 16.8 p.p. and up to 28.4 p.p. for CR2 and a fog density of 0.07 μm^{-3}.

Further details about the confidence and covariance based optimization of the covariance intersection fusion and more results can be found in Volk et al. [111].

5 Conclusion and Outlook

In this chapter we presented environment-aware approaches for robustness enhancement of local and cooperative perception. Vision-based object detection must be robust against harsh weather to ensure safety. To enhance existing data sets, which lack adverse weather scenarios, we presented physically correct image-based simulations for rain (including raindrops in the windshield and road spray of driving vehicles), snow and fog. With our proposed RESIST framework [67], a workflow to investigate different local and cooperative object perception setups exist. A wide range of possible input sources allows a comprehensive evaluation of the implemented algorithms. In addition, the realistic weather augmentations were used to study the effects of different weather conditions of varying intensity on vision-based object detection, showing a significant decrease in average accuracy as rain intensity increased. This leads to the statement that state-of-the-art neural networks are not robust against harsh weather. However, it was shown that training neural networks on data sets containing images with our proposed weather augmentations leads to an increase of the perception performance of up to 7.33 p.p. for the YOLOv3 network.

Additionally, we have shown that even with robust neural networks, the local perception is limited by different factors. To overcome these disadvantages, we considered vision-based cooperative perception. Gathering information with multiple vehicles allows perceiving objects outside the local sensor range or occluded in difficult inner-city scenarios. For different scenarios it was shown that cooperative perception can increase the mean average perception by about 18 p.p. compared to a local perception without any influence of adverse weather. Considering adverse weather it could be shown that a cooperative perception is possible to achieve a mean average precision of about 23% while the local perception was not able to detect any object. Hence, the cooperative perception increases safety.

Moreover, we have shown that state-of-the-art evaluation metrics for object perception do not necessarily satisfy the safety constraint. Hence, we considered additional factors such as velocity and object class for the evaluation of object perception systems to determine the safety with a comprehensive safety metric.

Besides all achievements some further research topics are still open.

Extend Weather Simulation to Further Sensors

The influence of weather circumstances on vision-based object detection was investigated and presented in detail. To achieve a safe system, autonomous vehicles must have a redundancy in sensors to balance the advantages and disadvantages of different sensor types. LiDAR is a promising technology for object detection because it is highly accurate and has a high sensor range. Since LiDAR sensors emit light waves, they are affected by weather as well. Rain, snow or fog can scatter the light waves such that false detections occur or the sensor range decreases. A first approach to weather simulations for LiDAR perception has been proposed by Teufel et al. [100].

Similar to the camera-based object detection, the optimization of LiDAR-based object detection has been investigated to increase robustness of LiDAR perception [99].

Safety Metric for Environment Perception

Object Perception is only one part of the perception for an autonomous vehicle. There are more subsystems such as lane detection, traffic sign recognition or motion planning. To achieve a safe autonomous vehicle all subsystems are required to be safe. Thus, the safety must be evaluated. Since lane detection and traffic sign recognition are part of the perception, the proposed safety metric can be extended to these tasks. For both tasks some requirements exist; e.g., a lane detection should at least cover the distance required for an emergency brake to enable safety.

Optimization of Cooperative Perception

An inaccurate local perception could lead to deviations of a state estimation of a cooperatively perceived object. As a worst case, the estimation error increases that much that the benefit of cooperative perception disappears. Thus, only valid and accurate information should be considered for fusion. Additionally, for cooperative perception, the communication channel should not be overloaded by transmitting erroneous information. Therefore, validation strategies at sender and receiver should be investigated to improve communication channel usage as well as fusion accuracy. Moreover, the concept of cooperative perception has been extended to lane detection by Gamerdinger et al. [28].

Acknowledgements This work has been partially funded by the German Research Foundation (DFG) under priority program 1835 grants BR2321/5-1 and BR2321/5-2. Special thanks to Patrick Schulz (FZI Forschungszentrum Informatik) for providing data from his virtual reality model of the test area autonomous driving in Karlsruhe.

References

1. Vires VTD. https://vires.mscsoftware.com/. Accessed 17 Feb 2022
2. Abbas, T., Sjöberg, K., Karedal, J., Tufvesson, F.: A measurement based shadow fading model for vehicle-to-vehicle network simulations. Int. J. Antennas Propag. **2015** (2015)
3. Aleshin, V., Afanasiev, V., Bobkov, A., Klimenko, S., Kuliev, V., Novgorodtsev, D.: Visual 3D perception of motion environment and visibility factors in virtual space. In: Transactions on Computational Science XVI, pp. 17–33. Springer (2012)
4. Allig, C., Wanielik, G.: Extending the vehicular network simulator Artery in order to generate synthetic data for collective perception. Adv. Radio Sci. **17**(F.), 189–196 (2019)
5. Arnab, A., Miksik, O., Torr, P.H.: On the robustness of semantic segmentation models to adversarial attacks. In: Proceedings of the IEEE Conference on Computer Vision and Pattern Recognition, pp. 888–897 (2018)
6. von Bernuth, A., Volk, G., Bringmann, O.: Rendering Physically Correct Raindrops on Windshields for Robustness Verification of Camera-based Object Recognition. In: 2018 IEEE Intelligent Vehicles Symposium (IV), pp. 922–927 (2018). https://doi.org/10.1109/IVS.2018.8500494

7. Bernuth, A.v., Volk, G., Bringmann, O.: Simulating photo-realistic snow and fog on existing images for enhanced CNN training and evaluation. In: 2019 IEEE Intelligent Transportation Systems Conference (ITSC), pp. 41–46 (2019). https://doi.org/10.1109/ITSC.2019.8917367

8. Bijelic, M., Gruber, T., Mannan, F., Kraus, F., Ritter, W., Dietmayer, K., Heide, F.: Seeing through fog without seeing fog: deep multimodal sensor fusion in unseen adverse weather. In: The IEEE/CVF Conference on Computer Vision and Pattern Recognition (CVPR) (2020)

9. Biri, V., Michelin, S., Arques, D.: Real-time animation of realistic fog. In: Thirteenth Eurographics Workshop on Rendering, pp. 9–16 (2002)

10. Boban, M., Barros, J., Tonguz, O.: Geometry-based vehicle-to-vehicle channel modeling for large-scale simulation. IEEE Trans. Veh. Technol. **63**(9), 4146–4164 (2014). https://doi.org/10.1109/TVT.2014.2317803

11. Castanedo, F.: A review of data fusion techniques. Sci. World J. **2013** (2013)

12. Colomb, M., Hirech, K., André, P., Boreux, J., Lacôte, P., Dufour, J.: An innovative artificial fog production device improved in the European project "FOG." Atmos. Res. **87**(3–4), 242–251 (2008)

13. Cong, J., Li, Y., Qi, G., Sheng, A.: An order insensitive sequential fast covariance intersection fusion algorithm. Inf. Sci. **367**, 28–40 (2016)

14. Cordts, M., Omran, M., Ramos, S., Rehfeld, T., Enzweiler, M., Benenson, R., Franke, U., Roth, S., Schiele, B.: The cityscapes dataset for semantic urban scene understanding. In: Proceedings of the IEEE Conference on Computer Vision and Pattern Recognition (CVPR) (2016)

15. da Costa, G.B.P., Contato, W.A., Nazaré, T.S., Neto, J.E., Ponti, M.: An empirical study on the effects of different types of noise in image classification tasks (2016). arXiv:1609.02781

16. Davis, J., Goadrich, M.: The relationship between precision-recall and ROC curves. In: ICML '06: Proceedings of the 23rd International Conference on Machine Learning, pp. 233–240 (2006)

17. Delooz, Q., Festag, A., Vinel, A.: Revisiting message generation strategies for collective perception in connected and automated driving. In: VEHICULAR 2020: The Ninth International Conference on Advances in Vehicular Systems. Technologies and Applications, Porto, Portugal, 18–22 April, 2020, pp. 46–52. International Academy, Research and Industry Association (IARIA) (2020)

18. Dodge, S., Karam, L.: Understanding how image quality affects deep neural networks. In: 2016 Eighth International Conference on Quality of Multimedia Experience (QoMEX), pp. 1–6. IEEE (2016)

19. Dosovitskiy, A., Ros, G., Codevilla, F., Lopez, A., Koltun, V.: CARLA: an open urban driving simulator. In: Conference on Robot Learning, pp. 1–16. PMLR (2017)

20. Dumont, E.: Semi-Monte Carlo light tracing applied to the study of road visibility in fog. In: Monte-Carlo and Quasi-Monte Carlo Methods 1998, pp. 177–187. Springer (2000)

21. ETSI: Intelligent Transport Systems (ITS); Vehicular Communications; Basic Set of Applications; Analysis of the Collective Perception Service (CPS); Release 2 (2019). ETSI TR 103 562 V2.1.1

22. ETSI: Intelligent Transport Systems (ITS); Vehicular Communications; Basic Set of Applications; Part 2: Specification of Cooperative Awareness Basic Service (2019). ETSI EN 302 637-2 V1.4.1

23. Everingham, M., Van Gool, L., Williams, C., Winn, J., Zisserman, A.: The pascal visual object classes (VOC) challenge. Int. J. Comput. Vis. **88**(2), 303–338 (2010)

24. Festenberg, N.v., Gumhold, S.: A geometric algorithm for snow distribution in virtual scenes. In: Eurographics Workshop on Natural Phenomena, pp. 15–25. The Eurographics Association (2009)

25. Fischer, V., Kumar, M.C., Metzen, J.H., Brox, T.: Adversarial examples for semantic image segmentation (2017). arXiv:1703.01101

26. Franken, D., Hupper, A.: Improved fast covariance intersection for distributed data fusion. In: 2005 7th International Conference on Information Fusion, vol. 1, pp. 7–pp. IEEE (2005)

27. Fredriksson, R., Rosén, E., Kullgren, A.: Priorities of pedestrian protection-a real-life study of severe injuries and car sources. Accid. Anal. & Preven. **42**(6), 1672–1681 (2010)
28. Gamerdinger, J., Teufel, S., Volk, G., Bringmann, O.: CoLD fusion: a real-time capable spline-based fusion algorithm for collective lane detection. In: 2023 IEEE Intelligent Vehicles Symposium (IV) (IEEE IV 2023), p. 8. Anchorage, USA (2023)
29. Garg, K., Nayar, S.K.: Vision and Rain. Int. J. Comput. Vis. **75**(1), 3–27 (2007)
30. Garlichs, K., Günther, H.J., Wolf, L.C.: Generation rules for the collective perception service. In: 2019 IEEE Vehicular Networking Conference (VNC), pp. 1–8. IEEE (2019)
31. Gaylard, A., Kabanovs, A., Jilesen, J., Kirwan, K., Lockerby, D.: Simulation of rear surface contamination for a simple bluff body. J. Wind Eng. Ind. Aerodyn. **165**, 13–22 (2017)
32. Gaylard, A.P.: Vehicle surface contamination, unsteady flow and aerodynamic drag. Ph.D. Thesis, University of Warwick (2019)
33. Geiger, A., Lenz, P., Stiller, C., Urtasun, R.: Vision meets robotics: the KITTI dataset. Int. J. Robot. Res. (IJRR) **32**(11), 1231–1237 (2013)
34. Geiger, A., Lenz, P., Urtasun, R.: Are we ready for autonomous driving? the KITTI vision benchmark suite. In: Conference on Computer Vision and Pattern Recognition (CVPR) (2012)
35. Gunn, K.L.S., Marshall, J.S.: The distribution with size of aggregate snowflakes. J. Meteorol. **15**(5), 452–461 (1958)
36. Halimeh, J.C., Roser, M.: Raindrop detection on car windshields using geometric-photometric environment construction and intensity-based correlation. In: 2009 IEEE Intelligent Vehicles Symposium, pp. 610–615. IEEE (2009)
37. Han, Y., Yang, J., Mizuno, K., Matsui, Y.: Effects of vehicle impact velocity, vehicle front-end shapes on pedestrian injury risk. Traffic Inj. Prev. **13**(5), 507–18 (2012)
38. Hasirlioglu, S.: A novel method for simulation-based testing and validation of automotive surround sensors under adverse weather conditions/submitted by Sinan Hasirlioglu. Ph.D. thesis, Universität Linz (2020)
39. Hasirlioglu, S., Riener, A.: Challenges in object detection under rainy weather conditions. In: First International Conference on Intelligent Transport Systems, pp. 53–65. Springer (2018)
40. Hasirlioglu, S., Riener, A.: A model-based approach to simulate rain effects on automotive surround sensor data. In: 2018 IEEE International Conference on Intelligent Transportation Systems (ITSC), pp. 2609–2615. IEEE, Maui, Hawaii, USA (2018). https://doi.org/10.1109/ITSC.2018.8569907
41. Hospach, D.: Simulation von Umgebungsbedingungen für bildbasierte Systeme. Ph.D. thesis, University of Tübingen, Germany (2020). https://d-nb.info/1221024450
42. Hospach, D., Mueller, S., Rosenstiel, W., Bringmann, O.: Simulation of falling rain for robustness testing of video-based surround sensing systems. In: 2016 Design, Automation & Test in Europe, Conference & Exhibition (DATE), pp. 233–236 (2016)
43. Hospach, D., Mueller, S., Rosenstiel, W., Bringmann, O.: Simulation of falling rain for robustness testing of video-based surround sensing systems. In: Proceedings of the 2016 Design, Automation & Test in Europe Conference & Exhibition (DATE), pp. 233–236 (2016)
44. Hu, X., Liao, L., Lei, Y., Yang, H., Fan, Q., Yang, B., Chang, J., Wang, J.: A numerical simulation of wheel spray for simplified vehicle model based on discrete phase method. Adv. Mech. Eng. **7**(7), 1687814015597190 (2015)
45. Huang, H., Li, H., Shao, C., Sun, T., Fang, W., Dang, S.: Data redundancy mitigation in V2X based collective perceptions. IEEE Access **8**, 13405–13418 (2020)
46. Jensen, H.W., Christensen, P.H.: Efficient simulation of light transport in scenes with participating media using photon maps. In: Proceedings of the 25th Annual Conference on Computer Graphics and Interactive Techniques, pp. 311–320 (1998)
47. Jonsson, M., Hast, A.: Animation of water droplet flow on structured surfaces. In: Special Effects and Rendering. Proceedings from SIGRAD 2002; Linköpings universitet; Norrköping; Sweden; November 28th and 29th; 2002, 007, pp. 17–22. Linköping University Electronic Press (2002)
48. Julier, S.J., Uhlmann, J.K.: New extension of the Kalman filter to nonlinear systems. In: Signal processing, sensor fusion, and target recognition VI, vol. 3068, pp. 182–193. International Society for Optics and Photonics (1997)

49. Kabanovs, A., Garmory, A., Passmore, M., Gaylard, A.: Investigation into the dynamics of wheel spray released from a rotating tyre of a simplified vehicle model. J. Wind Eng. Ind. Aerodyn. **184**, 228–246 (2019)
50. Kamm, I.O., Wray, G.A.: Suppression of water spray on wet roads. SAE Transactions, pp. 412–422 (1971)
51. Kaneda, K., Ikeda, S., Yamashita, H.: Animation of water droplets moving down a surface. J. Vis. Comput. Animat. **10**(1), 15–26 (1999)
52. Kaneda, K., Zuyama, Y., Yamashita, H., Nishita, T.: Animation of water droplet flow on curved surfaces. In: Proceedings of Pacific Graphics, vol. 96, pp. 50–65 (1996)
53. Karacan, L., Akata, Z., Erdem, A., Erdem, E.: Learning to generate images of outdoor scenes from attributes and semantic layouts (2016). arXiv:1612.00215
54. Kim, S.W., Qin, B., Chong, Z.J., Shen, X., Liu, W., Ang, M.H., Frazzoli, E., Rus, D.: Multivehicle cooperative driving using cooperative perception: design and experimental validation. IEEE Trans. Intell. Transp. Syst. **16**(2), 663–680 (2014)
55. Koh, G., Lacombe, J., Hutt, D.L.: Snow mass concentration and precipitation rate. Cold Reg. Sci. Technol. **15**(1), 89–92 (1988). https://doi.org/10.1016/0165-232X(88)90042-0
56. Kooij, S., Sijs, R., Denn, M.M., Villermaux, E., Bonn, D.: What determines the drop size in sprays? Phys. Rev. X **8**, 031019 (2018). https://doi.org/10.1103/PhysRevX.8.031019
57. Kuhn, H.W.: The Hungarian method for the assignment problem. Naval Res. Log. Quar. **2**(1–2), 83–97 (1955)
58. Kim, K.Y., Kim, Y., Park, J., Kim, Y.-S.: Real-Time peformance evaluation metrics for object detection and tracking of intelligent video surveillance systems. Asia Pacif. J. Contemp. Educ. Commun. Technol. **2**(2), 173–179 (2016)
59. Langer, M., Zhang, Q.: Rendering falling snow using an inverse fourier transform. ACM SIGGRAPH technical sketches program (2003)
60. Lin, T., Maire, M., Belongie, S.J., Bourdev, L.D., Girshick, R.B., Hays, J., Perona, P., Ramanan, D., Dollár, P., Zitnick, C.L.: Microsoft COCO: common objects in context (2014). arXiv:1405.0312v3
61. Luc, P., Couprie, C., Chintala, S., Verbeek, J.: Semantic segmentation using adversarial networks (2016). arXiv:1611.08408
62. Mangel, T., Klemp, O., Hartenstein, H.: A validated 5.9 GHz non-line-of-sight path-loss and fading model for inter-vehicle communication. In: 2011 11th International Conference on ITS Telecommunications, pp. 75–80. IEEE (2011)
63. Marshall, J.S., Palmer, W.M.K.: The distribution of raindrops with size. J. Meteorol. **5**(4), 165–166 (1948). https://doi.org/10.1002/qj.49707632704
64. Michaelis, C., Mitzkus, B., Geirhos, R., Rusak, E., Bringmann, O., Ecker, A.S., Bethge, M., Brendel, W.: Benchmarking robustness in object detection: autonomous driving when winter is coming. In: NeurIPS 2019 Workshop on Machine Learning for Autonomous Driving (2019)
65. Ministry of the Interior of the state Nordrhein-Westfalen: Aufgaben der Polizei bei Straßenverkehrsunfällen, Anlage 2 (2008). https://recht.nrw.de/lmi/owa/br_show_anlage?p_id=2477. Accessed 9 June 2020
66. Montserrat, D.M., Lin, Q., Allebach, J., Delp, E.J.: Training object detection and recognition CNN models using data augmentation. Electr. Imaging **2017**(10), 27–36 (2017)
67. Mueller, S., Hospach, D., Gerlach, J., Bringmann, O., Rosenstiel, W.: Framework for Varied Sensor Perception in Virtual Prototypes. In: MBMV, pp. 145–154 (2015)
68. Müller, S., Hospach, D., Bringmann, O., Gerlach, J., Rosenstiel, W.: Robustness evaluation and improvement for vision-based advanced driver assistance systems. In: 2015 IEEE 18th International Conference on Intelligent Transportation Systems, pp. 2659–2664. IEEE (2015)
69. Muraoka, K., Chiba, N.: Visual simulation of snowfall, snow cover and snowmelt. In: Proceedings Seventh International Conference on Parallel and Distributed Systems: Workshops, pp. 187–194. IEEE (2000)
70. Nazaré, T.S., Costa, G.B., Contato, W.A., Ponti, M.: Deep convolutional neural networks and noisy images. In: Iberoamerican Congress on Pattern Recognition, pp. 416–424. Springer (2017)

71. Niehsen, W.: Information fusion based on fast covariance intersection filtering. In: Proceedings of the Fifth International Conference on Information Fusion. FUSION 2002. (IEEE Cat. No. 02EX5997), vol. 2, pp. 901–904. IEEE (2002)
72. Obst, M., Hobert, L., Reisdorf, P.: Multi-sensor data fusion for checking plausibility of V2V communications by vision-based multiple-object tracking. In: 2014 IEEE Vehicular Networking Conference (VNC), pp. 143–150. IEEE (2014)
73. Pei, K., Cao, Y., Yang, J., Jana, S.: DeepXplore: automated whitebox testing of deep learning systems. In: proceedings of the 26th Symposium on Operating Systems Principles, pp. 1–18 (2017)
74. Powers, D.: Evaluation: from precision, recall and f-factor to roc, informedness, markedness & correlation. J. Mach. Learn. Technol. 2(1), 37–63 (2011)
75. Radtke, S., Li, K., Noack, B., Hanebeck, U.D.: Comparative study of track-to-track fusion methods for cooperative tracking with bearings-only measurements. In: 2019 IEEE International Conference on Industrial Cyber Physical Systems (ICPS), pp. 236–241. IEEE (2019)
76. Rasshofer, R.H., Spies, M., Spies, H.: Influences of weather phenomena on automotive laser radar systems. Adv. Radio Sci. 9(B. 2), 49–60 (2011)
77. Rauch, A.: Entwicklung von Methoden für die fahrzeugübergreifende Umfelderfassung. Ph.D. thesis, Universität Ulm (2016)
78. Rauch, A., Klanner, F., Rasshofer, R., Dietmayer, K.: Car2X-based perception in a high-level fusion architecture for cooperative perception systems. In: 2012 IEEE Intelligent Vehicles Symposium, pp. 270–275. IEEE (2012)
79. Redmon, J., Farhadi, A.: YOLOv3: an incremental improvement (2018). arXiv:1804.02767
80. Reinhardt, M., Noack, B., Arambel, P.O., Hanebeck, U.D.: Minimum covariance bounds for the fusion under unknown correlations. IEEE Signal Process. Lett. 22(9), 1210–1214 (2015). https://doi.org/10.1109/LSP.2015.2390417
81. Ren, J., Chen, X., Liu, J., Sun, W., Pang, J., Yan, Q., Tai, Y.W., Xu, L.: Accurate single stage detector using recurrent rolling convolution. In: Proceedings of the IEEE Conference on Computer Vision and Pattern Recognition, pp. 5420–5428 (2017)
82. Ren, S., He, K., Girshick, R., Sun, J.: Faster R-CNN: towards real-time object detection with region proposal networks. IEEE Trans. Pattern Anal. Mach. Intell. 39(6), 1137–1149 (2016)
83. Rezatofighi, S.H., Tsoi, N., Gwak, J., Sadeghian, A., Reid, I.D., Savarese, S.: Generalized intersection over union: a metric and a loss for bounding box regression (2019). arXiv:1902.09630
84. Rusak, E., Schott, L., Zimmermann, R.S., Bitterwolf, J., Bringmann, O., Bethge, M., Brendel, W.: A simple way to make neural networks robust against diverse image corruptions. In: European Conference on Computer Vision, pp. 53–69. Springer, Cham (2020)
85. Sakaridis, C., Dai, D., Van Gool, L.: Semantic foggy scene understanding with synthetic data. Int. J. Comput. Vision 126(9), 973–992 (2018)
86. Saltvik, I., Elster, A.C., Nagel, H.R.: Parallel methods for real-time visualization of snow. In: International Workshop on Applied Parallel Computing, pp. 218–227. Springer (2006)
87. Sato, T., Dobashi, Y., Yamamoto, T.: A method for real-time rendering of water droplets taking into account interactive depth of field effects, pp. 125–132. Springer US, Boston, MA (2003). https://doi.org/10.1007/978-0-387-35660-0_15
88. Schiegg, F.A., Bischoff, D., Krost, J.R., Llatser, I.: Analytical performance evaluation of the collective perception service in IEEE 802.11p Networks. In: 2020 IEEE Wireless Communications and Networking Conference (WCNC), pp. 1–6. IEEE (2020)
89. Schiegg, F.A., Krost, J., Jesenski, S., Frye, J.: A novel simulation framework for the design and testing of advanced driver assistance systems. In: 2019 IEEE 90th Vehicular Technology Conference (VTC2019-Fall), pp. 1–6. IEEE (2019)
90. Sekhon, R.S., Srivastava, R.C.: Snow size spectra and radar reflectivity. J. Atmos. Sci. 27, 299–307 (1970). https://doi.org/10.1175/1520-0469(1970)027<0299:SSSARR>2.0.CO;2
91. Sellers, G., Wright Jr, R.S., Haemel, N.: OpenGL SuperBible: Comprehensive Tutorial and Reference. Addison-Wesley (2013)

92. Sepulcre, M., Gonzalez-Martín, M., Gozalvez, J., Molina-Masegosa, R.: Analytical Models of the Performance of IEEE 802.11p Vehicle to Vehicle Communications (2021). ArXiv:2104.07923 [cs.NI]
93. Shalev-Shwartz, S., Shammah, S., Shashua, A.: On a Formal Model of Safe and Scalable Self-driving Cars (2017). arXiv:1708.06374
94. Shan, M., Narula, K., Wong, Y.F., Worrall, S., Khan, M., Alexander, P., Nebot, E.: Demonstrations of cooperative perception: safety and robustness in connected and automated vehicle operations. Sensors **21**(1), 200 (2020)
95. Slavik, Z.: Compressive sensing and its applications in automotive radar systems. Ph.D. thesis, Eberhard Karls Universität Tübingen (2019)
96. Slomp, M., Johnson, M., Tamaki, T., Kaneda, K.: Photorealistic real-time rendering of spherical raindrops with hierarchical reflective and refractive maps. In: Proceedings of the 2010 ACM SIGGRAPH Symposium on Interactive 3D Graphics and Games, I3D '10. Association for Computing Machinery, New York, NY, USA (2010). https://doi.org/10.1145/1730804.1730975
97. Stiefelhagen, R., Bernardin, K., Bowers, R., Garofolo, J., Mostefa, D., Soundararajan, P.: The CLEAR 2006 evaluation. In: CLEAR'06: Proceedings of the 1st International Evaluation Conference on Classification of Events, Activities and Relationships, pp. 1–44 (2006)
98. Sun, P., Kretzschmar, H., Dotiwalla, X., Chouard, A., Patnaik, V., Tsui, P., Guo, J., Zhou, Y., Chai, Y., Caine, B., Vasudevan, V., Han, W., Ngiam, J., Zhao, H., Timofeev, A., Ettinger, S., Krivokon, M., Gao, A., Joshi, A., Zhang, Y., Shlens, J., Chen, Z., Anguelov, D.: Scalability in perception for autonomous driving: waymo open dataset (2019). arXiv:1912.04838
99. Teufel, S., Gamerdinger, J., Volk, G., Gerum, C., Bringmann, O.: Enhancing robustness of LiDAR-Based perception in adverse weather using point cloud augmentations. In: 2023 IEEE Intelligent Vehicles Symposium (IV) (IEEE IV 2023), p. 6. Anchorage, USA (2023)
100. Teufel, S., Volk, G., Von Bernuth, A., Bringmann, O.: Simulating realistic rain, snow, and fog variations for comprehensive performance characterization of LiDAR perception. In: 2022 IEEE 95th Vehicular Technology Conference: (VTC2022-Spring), pp. 1–7 (2022). https://doi.org/10.1109/VTC2022-Spring54318.2022.9860868
101. Thandavarayan, G., Sepulcre, M., Gozalvez, J.: Analysis of message generation rules for collective perception in connected and automated driving. In: 2019 IEEE Intelligent Vehicles Symposium (IV), pp. 134–139. IEEE (2019)
102. Tian, Y., Pei, K., Jana, S., Ray, B.: DeepTest: automated testing of deep-neural-network-driven autonomous cars. In: Proceedings of the 40th International Conference on Software Engineering, pp. 303–314 (2018)
103. Tian, Z., Li, B.: A simulation system of snow based on particle system. In: 2015 International Conference on Intelligent Systems Research and Mechatronics Engineering, pp. 740–743. Atlantis Press (2015)
104. Torrent-Moreno, M., Jiang, D., Hartenstein, H.: Broadcast reception rates and effects of priority access in 802.11-based vehicular ad-hoc networks. In: Proceedings of the 1st ACM International Workshop on Vehicular ad Hoc Networks, pp. 10–18 (2004)
105. Uhlmann, J.K.: Covariance consistency methods for fault-tolerant distributed data fusion. Inf. Fus. **4**(3), 201–215 (2003)
106. U.S. Department of Transportation - Federal Highway Administration (FHA): How Do Weather Events Impact Roads? https://ops.fhwa.dot.gov/weather/q1_roadimpact.htm
107. Vargas, J., Alsweiss, S., Toker, O., Razdan, R., Santos, J.: An overview of autonomous vehicles sensors and their vulnerability to weather conditions. Sensors **21**(16), 5397 (2021)
108. Volk, G., von Bemuth, A., Bringmann, O.: Environment-aware development of robust vision-based cooperative perception systems. In: 2019 IEEE Intelligent Vehicles Symposium (IV), pp. 126–133. IEEE, Paris, France (2019). https://doi.org/10.1109/IVS.2019.8814148
109. Volk, G., Delooz, Q., Schiegg, F.A., Von Bernuth, A., Festag, A., Bringmann, O.: Towards realistic evaluation of collective perception for connected and automated driving. In: 2021 IEEE International Intelligent Transportation Systems Conference (ITSC), pp. 1049–1056 (2021). https://doi.org/10.1109/ITSC48978.2021.9564783

110. Volk, G., Gamerdinger, J., von Bernuth, A., Bringmann, O.: A comprehensive safety metric to evaluate perception in autonomous systems. In: 2020 IEEE 23rd International Conference on Intelligent Transportation Systems (ITSC), pp. 1–8. IEEE (2020)

111. Volk, G., Gamerdinger, J., von Bernuth, A., Teufel, S., Bringmann, O.: Environment-aware optimization of track-to-track fusion for collective perception. In: 2022 IEEE 25th International Conference on Intelligent Transportation Systems (ITSC), pp. 2385–2392 (2022). https://doi.org/10.1109/ITSC55140.2022.9922388

112. Volk, G., Müller, S., Bernuth, A.v., Hospach, D., Bringmann, O.: Towards robust CNN-based object detection through augmentation with synthetic rain variations. In: 2019 IEEE Intelligent Transportation Systems Conference (ITSC), pp. 285–292 (2019). https://doi.org/10.1109/ITSC.2019.8917269

113. Von Bernuth, A., Volk, G., Bringmann, O.: Augmenting image data sets with water spray caused by vehicles on wet roads. In: 2021 IEEE International Intelligent Transportation Systems Conference (ITSC), pp. 3055–3060 (2021). https://doi.org/10.1109/ITSC48978.2021.9564826

114. Wang, C., Wang, Z., Xia, T., Peng, Q.: Real-time snowing simulation. Vis. Comput. **22**(5), 315–323 (2006)

115. Wang, C., Yang, M., Liu, X., Yang, G.: Realistic simulation for rainy scene. J. Softw. **10**, 106–115 (2015). https://doi.org/10.17706/jsw.10.1.106-115

116. Wang, H., Mucha, P.J., Turk, G.: Water drops on surfaces. ACM Trans. Graph. (TOG) **24**(3), 921–929 (2005)

117. Wang, N., Wade, B.: Rendering falling rain and snow. In: ACM SIGGRAPH 2004 Sketches, SIGGRAPH '04, p. 14. Association for Computing Machinery, New York, NY, USA (2004). https://doi.org/10.1145/1186223.1186241

118. Welch, G., Bishop, G., et al.: An Introduction to the Kalman Filter (1995)

119. You, S., Tan, R.T., Kawakami, R., Mukaigawa, Y., Ikeuchi, K.: Raindrop detection and removal from long range trajectories. In: Asian Conference on Computer Vision, pp. 569–585. Springer (2014)

120. Zdrojewska, D.: Real Zime rendering of heterogeneous fog based on the graphics hardware acceleration. Proc. CESCG **4**, 95–101 (2004)

Design and Evaluation of V2X Communication Protocols for Cooperatively Interacting Automobiles

Quentin Delooz, Daniel Maksimovski, Andreas Festag, and Christian Facchi

Abstract This chapter studies two key communication services for the support of cooperative driving capabilities using Vehicle-to-Everything (V2X) communications: sensor data sharing and maneuver coordination. Based on the current state of the art in research and pre-standardization of V2X communications, we enhance the protocol design for both services and assess their performance by discrete-event simulations in highway and city scenarios. The first part of this chapter addresses the performance improvement of sensor data sharing by two complementary strategies. The shared sensor data are adapted to the available resources on the used channel. Furthermore, the redundancy of the transmitted information is reduced to lower the load on the wireless channel, whereas several approaches are proposed and assessed. The second part of the chapter analyzes cooperative maneuver coordination protocols. We propose a distributed approach based on the explicit exchange of V2X messages, which introduces priorities in maneuver coordination and studies several communication patterns for the negotiation and coordination of maneuvers among two and more vehicles. The results demonstrate the potential of V2X communications for automated driving, showcase several approaches for enhancements of sensor data sharing and maneuver coordination, and indicate the performance of these enhancements.

Quentin Delooz and Daniel Maksimovski contributed equally.

Q. Delooz (✉) · D. Maksimovski · A. Festag · C. Facchi
Technische Hochschule Ingolstadt, CARISSMA Institute for Electric, COnnected, and Secure Mobility (C-ECOS), Ingolstadt, Germany
e-mail: Quentin.Delooz@carissma.eu

D. Maksimovski
e-mail: Daniel.Maksimovski@carissma.eu

A. Festag
e-mail: Andreas.Festag@carissma.eu

C. Facchi
e-mail: Christian.Facchi@carissma.eu

© The Author(s) 2024
C. Stiller et al. (eds.), *Cooperatively Interacting Vehicles*,
https://doi.org/10.1007/978-3-031-60494-2_6

1 Motivation and Technical Background

In recent years, various Advanced Drivers Assistance Systems (ADAS) have been developed [1], while automated vehicles with an increasing different level of autonomy are being extensively tested and deployed on the roads. It is commonly assumed that safety, comfort, and efficiency on the road can be enhanced by the introduction of automated and fully autonomous vehicles, especially utilizing cooperative driving capabilities.

Cooperation among traffic participants is essential to reach a high level of automation. The cooperation requires interaction, which can be implicit or explicit. With implicit interaction, a vehicle infers the desired information of another traffic participant based on its behavior and actions. For example, it can recognize the intention of another vehicle by its local sensors and predict its future driving intentions when the other participant slows down. With explicit interaction, traffic participants exchange information by different means, such as by light projections in front of an automated vehicle for pedestrians. Communications can be regarded as a specific type of explicit interaction enabling an automated vehicle to warn others, exchange detailed information about the perceived environment or even negotiate maneuvers.

Vehicle-to-everything (V2X) communication enables the direct information exchange among traffic participants in an ad hoc network, as opposed to the typical communication with a communication infrastructure. It comprises communication among vehicles and with the road infrastructure as well as with pedestrians, bicyclists, etc. V2X communication operates in the 5.9 GHz frequency band, which has been specifically allocated for road safety and traffic efficiency applications in Europe and other regions of the world. Two access technologies have reached a mature status of research and development towards widespread deployment: WLAN-V2X and Cellular-V2X [37, 44]. Both facilitate an information exchange based on messages carrying application-specific content.

The development and deployment of V2X communications are commonly divided into three subsequent phases that rely on communication services with increasing complexity and requirements using dedicated, standardized message types (Fig. 1):

1. The Cooperative Awareness Message (CAM) and the Decentralized Environmental Notification Message (DENM) for the exchange of information about the vehicle state (position, speed, heading, acceleration) and about safety-critical events, respectively (Day 1);
2. The Collective Perception Message (CPM) for exchange of sensor data as lists of detected and classified objects in the perception range of a vehicle (Day 2); and
3. The Maneuver Coordination Message (MCM) for the exchange of intended maneuvers and coordination data among Connected and Automated Vehicles (CAVs) (Day 3+).

In Europe, these messages types and corresponding protocols are standardized by the European Telecommunications Standards Institute (ETSI): The specification of Day-1 message types CAM [15] and DENM is completed, CPM is close to com-

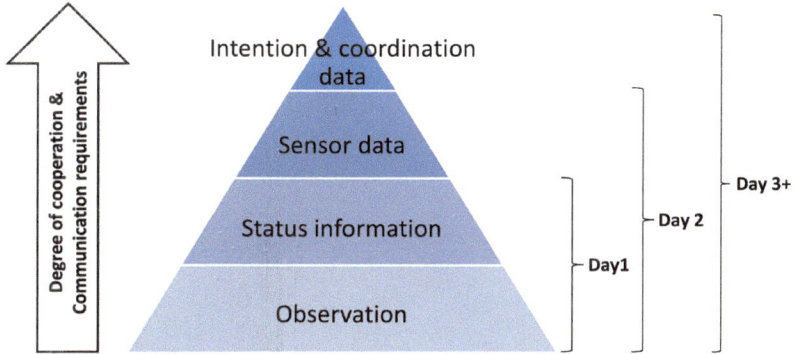

Fig. 1 Evolution of V2X communication in phases: Day 1, Day 2 and Day 3+

pletion [14], while the MCM is still very early in the research and standardization phase [17].

For performance evaluation of the studied V2X communication protocols, the discrete-event simulator Artery [42] is applied. The simulator relies on Vanetza, INET, and OMNeT++ [48] to implement the ETSI Cooperative-Intelligent Transport System (C-ITS) communication protocol stack. Furthermore, Artery realizes an environmental model and sensor models that represent vehicles to perceive objects, such as vehicles and bicycles, in their vicinity. To model node mobility, Artery is coupled with microscopic traffic simulator SUMO [31]. To model realistic traffic and vehicle movement, the traffic scenario of the city of Ingolstadt (Bavaria, Germany), referred to as InTAS [30], is chosen. The InTAS scenario lasts 24 hours long and relies on real daily data traffic from Ingolstadt. Figure 2 illustrates the road topology of the InTAS scenario.

This chapter is divided into two main parts. The first part considers the sensor data sharing based on the exchange of CPMs with lists of detected and classified objects. The second part focuses on cooperative maneuver coordination relying on the exchange of intention and coordination data among the vehicles. These parts are structured as follows:

- In the first part about sensor data sharing, an overview is provided in Sect. 2.1 along with the description of the state of the art and research questions in Sect. 2.2. Section 2.3 reviews the protocol design from research and current standardization efforts. Sections 2.4 and 2.5 present in detail the proposed changes to the current protocol design, followed by an analysis of the obtained results in Sect. 2.6.
- The maneuver coordination part consists of three sections. Section 3.1 presents an overview that also includes maneuver coordination use cases and a description of the state of the art in the field. Section 3.2 presents the Priority Maneuver (PriMa) coordination protocol design including the maneuver coordination message, communication patterns, and an example scenario. An overview of the simulation

Fig. 2 Road topology for the studied scenario of the city Ingolstadt

framework and a discussion of the results for the studied coordination scenarios are presented in Sect. 3.3.

Finally, a summary and outlook of both parts, Collective Perception (CP) and cooperative maneuver coordination are given in Sect. 4.

2 V2X Communications-Based Sensor Data Sharing

This section presents an overview of V2X communications-based sensor data sharing, i.e., CP, followed by its state of the art, protocol designs, and our proposed improvements with performance evaluations.

2.1 Overview

Sensor data sharing for V2X communications has been studied extensively during the last few years. European research and standardization activities, e.g., in ETSI, commonly refer to it as "Collective Perception (CP)". The protocol design of CP slowly reaches a stable state and is further used as a baseline. We investigate two remaining and relevant problems of CP: First, the information to be included in a

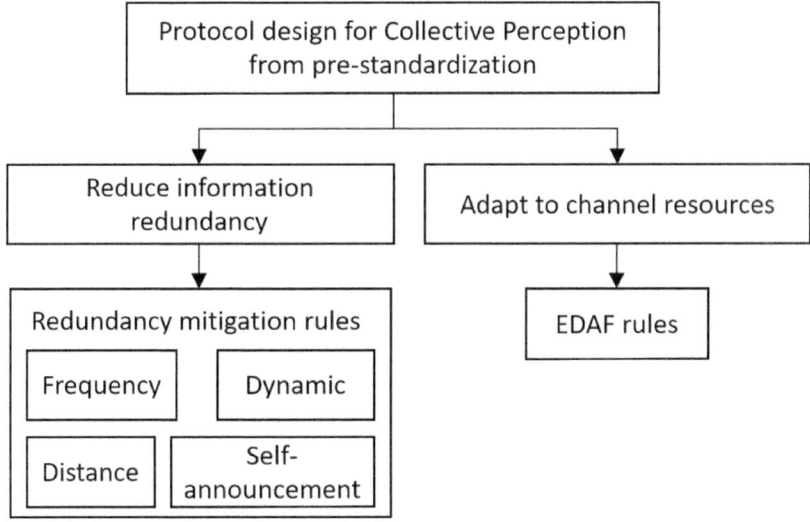

Fig. 3 Overview of the addressed research problems and studied solutions for Collective Perception

CPM—the detected and classified objects—needs to be carefully selected to avoid overloading the bandwidth-limited wireless channel. This can be achieved by applying smart filtering approaches to reduce the number of objects to transmit, called filtering rules within the rest of this chapter. The problem with the current design of these filtering rules is that they do not take into account the available channel resources, e.g., objects are unnecessarily filtered even when the channel usage is low. Our first research question to improve CP is, therefore, how and when object filtering should be modified to adapt to the available channel resources. The second problem addresses information redundancy. Currently, many vehicles can send information about the same object, unnecessarily dissipating channel resources. To diminish it, the second research question is how to address information redundancy by filtering objects considering the information received by other traffic participants. The following sections will focus on these two main research questions. Figure 3 gives a brief overview of the addressed research problems and the studied solutions (Fig. 4).

2.2 State of the Art in Collective Perception

Initial work on sensor data sharing or Collective Perception dates back to 2012 [39]. Ideas developed in [20] and others have led to standardization activities, the publication of the ETSI study item TR 103 562 [14], and draft versions of a European standard for CP in TS 103 324 [16]. Besides message format, [14] defines different design aspects of CP such as message format as well as the CPM dissemination

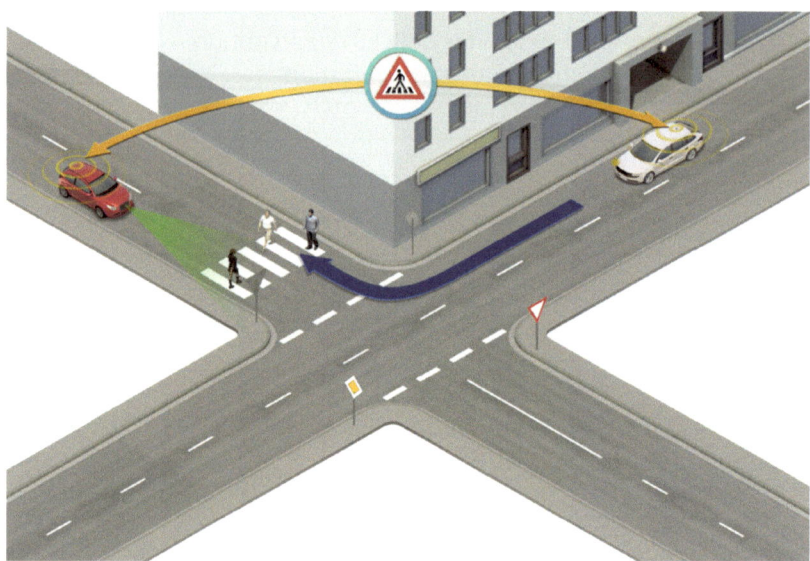

Fig. 4 Intersection use case for sensor data sharing based on V2X communication

concept to determine when to generate a CPM, i.e., the CPM generation rules. These rules determine which objects to include in a message and the triggering conditions to generate a CPM. Several publications, such as [7, 18, 46, 51], have reviewed the design for CP and elaborated on algorithms for message generation and object filtering.

The problem of information redundancy has been already considered in [14] and several redundancy mitigation approaches were defined but not evaluated. Only a few studies have addressed the redundancy problems. [6] has investigated two approaches, Dynamic and Self-Announcement redundancy mitigation rules (see Sect. 2.5), before [14] was published but applied different CPM generation rules than the currently designed ones [14, 16]. The authors of [47] focused on the redundancy mitigation approach using the object dynamics to filter objects, i.e., the dynamics-based redundancy mitigation rule (see Sect. 2.5). In [25], the authors reduce redundant object information on the wireless channel using a probabilistic object filtering approach based on the perceived density of vehicles, market penetration, and road geometry. The paper showed the efficiency of object filtering using a highway scenario and a minimal urban scenario with two roads. In [4], objects are filtered taking into account three criteria: channel load, and the number as well as the type of V2X stations that have already provided information about these objects. The main idea is to adapt the number of V2X stations that send information about the same object to the channel load. The lower the channel load is, the larger the number of stations that can send information about this object.

Congestion control algorithms in the context of safety applications have been the subject of intense research and resulted in ETSI standards such as the ETSI TS 102 687 [13] which specifies two different types of congestion control mechanism: reactive and adaptive. Both congestion control mechanisms rely on the perceived channel load to estimate available channel resources and enforce the respect of the channel access limitations imposed by the European norm [12]. These mechanisms attempt to share channel resources fairly among V2X stations by imposing constraints on the message transmission parameters, e.g., by reducing the allowed message transmission rate. The performance of CP with the constraints imposed by the reactive congestion control mechanism was initially studied in [19, 21] and later used in the evaluations realized in [14]. Our work [9] distinguishes itself by being the first to evaluate the CP performance with the adaptive approach and propose to adapt the filtering of objects based on the current channel access constraints.

In comparison to other works, we assess the performance of four redundancy mitigation approaches, the impact of their parameter settings, and the congestion control aware object filtering with the Artery framework in a complex and diverse urban scenario. We design and use novel metrics for a fair comparison, and compute the information value brought by the different message generation rules while considering the object dynamics. For an assessment of all redundancy mitigation approaches considered in the standardization process [14], we refer to our publication in [10].

2.3 Protocol Design

The published standardization document ETSI TR 103 562 [14] is considered as the baseline for the protocol design of Collective Perception. The corresponding dedicated message type CPM is composed of distinct containers with different purposes. The containers relevant for this paper are the Sensor Information Container (SIC), Perceived Object Container (POC), and the Station and Management Container (SMC) (Fig. 5). The SIC contains information about the sensing capabilities of a transmitting V2X station. The sensing capabilities are described using a list of capability descriptions of each of the sensors mounted on the vehicle, e.g., by indicating the Field of View (FOV) and the mounted position of the sensor. Since this information is static, the SIC does not need to be repeated with a high frequency and is included only once every second. The second and most relevant container is the POC, which contains all objects that a vehicle perceives with its local sensors. An object is selected for inclusion in the POC if it fulfills at least one of the following conditions [14]:

1. The object was never sent previously, i.e., it is considered new for the transmitting station.
2. The object's position changed by more than 4 m (absolute euclidian distance) since its last inclusion in a CPM of the transmitting station.
3. The object's speed changed by more than 0.5 m/s since its last inclusion.

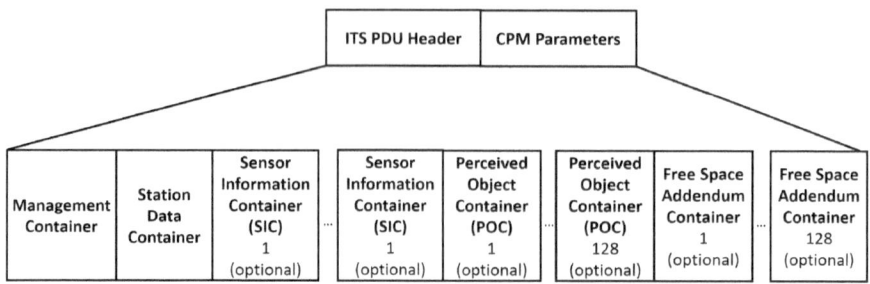

Fig. 5 CPM format and data elements as defined in [14].

4. The object's heading changed by more than 4 degrees since its last inclusion.
5. The object was previously included in a CPM of the transmitter more than one second ago.

The SMC contains information about vehicle mobility such as position and velocity.

Following the message generation rules in [14], a CPM is generated whenever the SIC needs to be transmitted, the POC contains at least one object for transmission or both. However, the CPM generation interval cannot be higher than 1 000 ms or lower than 100 ms. Following these message generation rules, the size and frequency of generated CPMs can considerably vary within the interval.

These rules were originally proposed in [18] with the idea that an object is included in a CPM whenever it would generate a CAM, presuming that the object is capable to generate messages, e.g. it is a vehicle capable of V2X communication. CP helps increase awareness of unconnected vehicles, especially in the first years of V2X deployment. However, it can overreach the goal when the V2X market penetration ratio grows over the years and it can start overusing the communication channel. On the opposite, at a low market penetration rate and when a few vehicles with V2X capabilities are within communication range, the hypothesis is that object filtering does not bring significant value (see Sect. 2.1).

Figure 6 shows an example of how to realize the CPM generation process as described in [14]. The Collective Perception Service (CPS) checks every T_{check} if a CPM has to be generated. The common value for T_{check} is 100 ms. Then, the CPS checks if any congestion control mechanism prevents the generation of a CPM (see Sect. 2.4.1), referenced as "Triggering" in Fig. 6. If allowed, the filtering rules and generation rules as described above are applied to the detected objects. Finally, the CPM is generated and passed to the lower layers for transmission.

Our contributions are the following:

1. To adapt the object filtering to the available channel resources, works have been realized in [9] and in [8]. The modification in the protocol design happens in the "Add filtered objects*" step of Fig. 6.
2. To address information redundancy, several redundancy mitigation approaches have been analyzed following the relevant approaches in ETSI TR 103 562 [14]. The resulting modifications occur in the "Filtering rules" step in Fig. 6.

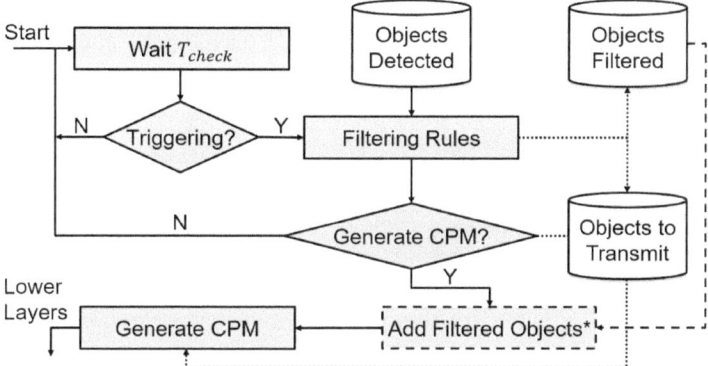

Fig. 6 CPM generation process. Figure derived from [9]

2.4 Adapting Object Filtering to the Available Channel Resources

In the 5.9 GHz bandwidth in which V2X communications are deployed, channel resources are scarce and have to be used mindfully. The default data rate for WLAN-V2X is 6 Mb/s [26]. Communication services, such as Cooperative Awareness or Collective Perception with semi-periodic message generation consume a large amount of the wireless bandwidth. Moreover, the availability of resources for a service depends on two main factors: the presence of other V2X stations transmitting messages and the number of services of a V2X station attempting to generate messages on the same channel. To adapt to these two factors, congestion control mechanisms are being considered. For WLAN-V2X in Europe, these mechanisms are referred to as Decentralized Congestion Control (DCC). We review the main DCC mechanisms in Sect. 2.4.1. To adapt the CPM generation rules to the current DCC restrictions, we propose an algorithm following a principle explained in Sect. 2.4.2. The objective is to allow more objects to be transmitted when channel resources can support it.

2.4.1 Decentralized Congestion Control (DCC)

DCC is a cross-layer functionality with interacting entities at all layers. The DCC entities at the access layer [13] provide different control for the outgoing packets. Practically, a "gatekeeper" is implemented and it realizes a first-in-first-out (FIFO) packet queuing system for each channel. A gatekeeper relies on multiple simple priority queues for the packets to be selected for transmission and acts as a single server. The non-empty queue with the highest priority is dispatched first. If a queue is full and a packet arrives or if the lifetime of a packet expires inside the queue, the packet is discarded. The gatekeeper acts as a switch by opening and closing its gate repetitively. When a packet passes the opened gate to the lower layers, the gate is closed by the gatekeeper for a period of time depending on the current channel condition and the size of the packet. To determine the closing time of the door, ETSI TS 102 687 [13] specifies two DCC access layer strategies: reactive and adaptive. Both strategies respect the following DCC-related regulatory requirements for operation in the 5.9 GHz frequency band as specified in [12]:

- $0 < T_{on} < 4\,ms$: T_{on} is the duration of a packet transmission.
- $\delta <= 3\,\%$, whereas δ is the duty cycle defined as the allowed ratio of the transmitter's total "on" time relative to 1 s. 3 % means that a station can occupy at most 3%, i.e., 30 ms, of channel time within 1 s.
- $T_{off} >= 25\,ms$: T_{off} is the duration before the gatekeeper re-opens its gate after the transmission of packet. In another word, the maximum packet transmission frequency is 40 Hz for a V2X station on a channel.
- if $CBR >= 0.62$, $T_{off} >= min(1000\,ms, T_{on}(4000 \times \frac{CBR-0.62}{CBR} - 1)$, see Sect. 2.6.1 for the definition of the Channel Busy Ratio (CBR).

LIMERIC is the adaptive DCC algorithm approach proposed in ETSI TS 102 687 [13]. Every 200 ms, it updates the duty cycle δ. Table 3 from [13] is used to parametrize LIMERIC. To improve the convergence time of LIMERIC and fairness during transition phases, [45] proposes a dual-α approach which we used for the simulations in this paper.

To enforce the allowed duty cycle δ determined by LIMERIC, the following equations from [13] are used:

$$T_{off} = min(max(\frac{T_{on_{pp}}}{\delta}, 25\,ms), 1\,s), \tag{1}$$

where $T_{on_{pp}}$ denotes the transmission time of the previous packet. If δ is updated during the T_{off} interval, the closing time of the gatekeeper needs to be updated as per $T_{off}*$ given by (2).

$$T_{off}* = min(max(\frac{T_{on_{pp}}}{\delta} \times \frac{t_{remain}}{T_{off}}, 25\,ms), 1\,s - t_{passed}) \tag{2}$$

t_{remain} is the remaining time to wait with respect to the unmodified T_{off} and t_{passed} the time since the gatekeeper closed.

An interesting feature of the gatekeeper implementation is the fact that it takes into account the size of the previously transmitted packet to determine the time to wait between two transmissions. Thus, $T_{on_{pp}}$ is also affected by the size of CPMs directly, and indirectly by the applied filtering rules. This characteristic is exploited to adapt the size of a CPM to the closing time of the gatekeeper.

2.4.2 Congestion-Aware Collective Perception

Principle: We propose to enhance the current generation rules such that the Collective Perception service avoids filtering objects when the channel resources are sufficient to transmit them. The modifications brought to the current design of the CPS [14] add a new step into the CPM generation algorithms (Fig. 6) marked as "Add filtered objects" box. This new step is motivated by two observations.

First, services such as the CPS are currently not specified to adapt to the DCC constraints. It is expected that with the Release 2 set of standards for V2X communications, mechanisms such as the one proposed here to adapt to the available channel resources and DCC restrictions will be specified or suggested. In particular for CP, when channel resources are sufficiently available, objects should not be filtered as rigorously as in congested situations.

Second, following the CPM generation rules (see Sect. 2.3), if there is no object to transmit then no CPM is generated in most of the cases. Therefore, the filtering of objects influences the message generation procedure. Moreover, the current object filtering approach relies on object dynamics, and including more objects in a message increases the probability of having none of them transmitted at the next attempt at CPM generation. The expected result is to generate a smaller number of CPMs containing more objects. The advantage of reducing the CPM generation rate is the reduction of the overhead created by the lower layer headers.

To address the two points discussed above, we worked on the following principle: if a CPM is to be generated and the addition of an object from the set of filtered objects does not directly impact the CPM generation rate, the object will be included in the CPM. This principle is possible thanks to T_{check} and the gatekeeper as explained in Sect. 2.4.1.

The following example illustrates the principle: Let's consider that T_{check} equals 100 ms. If a generated CPM just passed the gatekeeper, and the resulting T_{off} is 115 ms, the next potential CPM generation will be not earlier than 200 ms later. At the first occurrence of T_{check}, the T_{off} interval has not elapsed yet. Hence, the CP service has to postpone the message generation to the next T_{check} cycle, which is 200 ms after the preceding CPM generation. The resulting gap of 85 ms remains unused. If by adding filtered objects, the increased CPM size causes the gatekeeper to extend T_{off} by less than those 85 ms, the effective CPM generation rate would not get reduced at all. As a result, the CPM generation at an interval of 200 ms is not prevented by DCC, independently of some additional filtered objects included.

Effects on the Collective Perception Service: With consideration of the channel congestion, the previously described principle has three possible effects on the CP service:

1. The CPM includes all objects (filtered and non-filtered) without restrictions from DCC on the CPM generation rate. This will occur when channel resources are largely available. We refer to this effect as "One-For-All", i.e., all objects are included if a CPM is to be generated.

2. Part of the filtered objects is included in the generated CPM. This occurs when DCC would start delaying packet transmissions if all objects, filtered and non-filtered, would be included. As result, the CPM contains some filtered objects chosen randomly.
3. No filtered objects are included. The CPMs generated using the current ETSI rules are sufficient for DCC to restrict the generation rate.

Process: The principle elaborated in the previous section was analyzed and evaluated in [9]. We enhanced the algorithm developed in [9] by simplifying its implementation and limiting its application. This enhancement was proposed in [8]. The main modification is to avoid adding filtered objects if the DCC restrictions before the next message to generate are higher than $Toff_{thresh}$ with a default value of 100 ms, which is the minimum time to wait between two consecutive messages. In this chapter, we evaluate additional values for $Toff_{thresh}$ as shown in Table 1. Figure 7 shows the resulting process to include filtered objects considering DCC, named Enhanced DCC Aware Filtering (EDAF) rules. First, the CPM containing the objects that have passed the POC inclusion rules and other containers is created. Second, the highest $Toff_{worst}$ that DCC could impose is computed. The steps to compute $Toff_{worst}$ are not included in this book but the reader is kindly invited to find them in [9]. Then, if $Toff_{worst}$ is lower than $T_{GenCpmMin}$, one of the objects filtered is added. These steps are repeated until either there are no more objects or $Toff_{worst}$ exceeds $T_{GenCpmMin}$.

2.5   Redundancy Mitigation Rules

A problem remaining with the current CP protocol design is that many vehicles can send information about the same object without any control. Information redundancy is not bad in itself and could help to improve the perception of the surrounding. However, in the context of limited channel resources, a too-high information redundancy does not bring any benefit. Moreover, it may even decrease the perceived quality of the object by adding extra processing delay. To address this problem, different techniques were elaborated in [14] and analyzed in other documents (see Sect. 2.2). Based on [14], parts of the techniques established for the RMRs are described in the following:

The Distance-based RMR: An object is filtered if it was already received from another V2X station within the R_Redundancy range during the last time window W_Redundancy. The parameter R_Redundancy needs to be carefully tuned such that the RMR efficiently allow enough object transmissions while controlling the channel load. CAMs and CPMs are considered sources of information for this RMR. A benefit of this RMR is that it attempts to maintain the awareness range by distributing in space the sources of information.

The Dynamics-based RMR:The logic behind this RMR is inspired by the "POC inclusion rules" of the CPM and the CAM triggering rules [15]. If the last

Table 1 Summary of the simulation parameters

Parameters	Values
Protocol stack	European WLAN-V2X
Frequency band	5.9 GHz
Channel model	Two Ray Interference
Transmission power	23 dBm
DCC	Dual-alpha LIMERIC
Services	Cooperative Awareness (SCH0), Collective Perception (SCH1)
Scenarios	InTAS
MPR	$\{0.1, 0.2, ..., 1\}$
Time of simulation	9:15 a.m.
Number of vehicles	$\approx 2\,800$
Simulation time	13 s (incl. 10 s of warmup)
Number of repetitions	2
Vehicle sensor equipment	2 radars: (160 m, 35°, Front), (80 m, 325°, Back)
Area of relevance	500 m
EDAF	
$Toff_{thresh}$	$\{25, 50, 75, 100\}$ ms
Distance-based RMR	
$W\_Redundancy$	1 s
$R\_Redundancy$	$\{25, 50, 100, 200\}$ m
Dynamics-based RMR	
$P\_Redundancy$	$\{2, 4\}$ m
$S\_Redundancy$	$\{0.25, 0.5\}$ m/s
Frequency-based RMR	
$W\_Redundancy$	1 s
$N\_Redundancy$	$\{1, 3, 5, 10, 15\}$

update received time, position, or absolute speed of an object changed less than P_Redundancy meters or S_Redundancy meters per second, respectively, the object is filtered. As for the "POC inclusion rules", the advantage of this rule is that the updates of objects will depend on their dynamics, avoiding too many objects of non or slow-moving vehicles.

The Self-Announcement-based RMR:If an object is detected as capable of transmitting V2X messages (e.g., CAM or CPM), the object is filtered.

The Frequency-based RMR:An object is filtered if it is subject to a number of updates higher than N_Redundancy during last time window W_Redundancy.

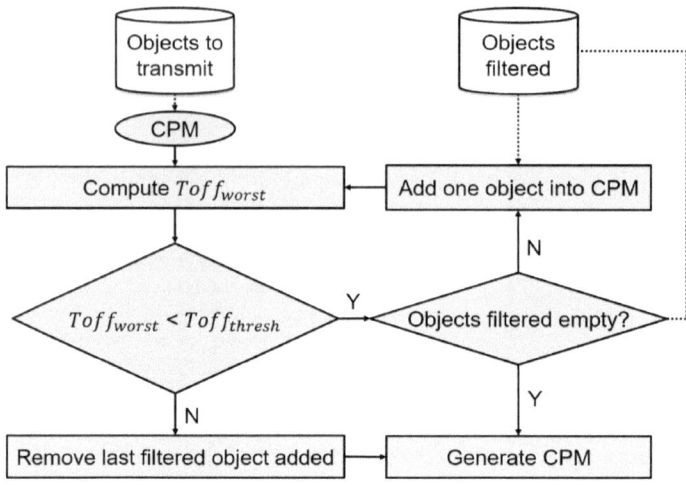

Fig. 7 The Enhanced DCC Aware Filtering (EDAF) rules

2.6 Simulation Results

The following of this section describes the evaluation framework used for the performance evaluations of the different proposed CP protocol designs.

2.6.1 Evaluation Framework

We review in this section the relevant parameters and components of our simulation framework. Table 1 summarizes the parameters used.

Communication: The insertion of V2X devices into the market is expected to increase slowly over the coming years. For our simulations, the following V2X market penetration rates (MPRs), i.e., the rate of vehicle able to generate and receive V2X messages, were investigated: $0.1, 0.2, ..., 1$.

For communication, vehicles operate a WLAN-V2X-compatible transceiver capable of working on two channels simultaneously without co-channel interference. For DCC, the adaptive approach described in Sect. 2.4.1 is deployed for all V2X capable vehicles.

Messages: CAM and CPMs are transmitted on all vehicles with enabled V2X capabilities. The CAMs are assigned to the control channel (SCH0) (or IEEE channel # 180) and generated according to ETSI EN 302 637-2 [15]. The CPMs are assigned to the SCH1 (IEEE channel # 176) and triggered following Sect. 2.3 and according to the studied filtering rules. In a CPM, the *itsPduHeader*, the *managementContainer*, and the *stationDataContainer* counts for 44 B together. A SIC is included once a second and is 12 B large. A POC in a CPM is 35 B after encoding. The *FreeSpaceAddendumContainer* is omitted.

Sensor Configuration: In Artery, object perception is assumed idealistic, i.e., there are no inaccuracies, and all the object information, such as dimensions, position, and speed, are available to the perceiving vehicle. Sensors can be allocated with a defined field of view, range, and attachment point to vehicles in the simulations. A direct line of sight between the sensor and one of the object corners is required for successful detection. Buildings and other vehicles are obstacles to the perception.

In our simulations, vehicles have two radars: one with 80 m range and 325° FOV facing backward and one with 160 m range and 35° FOV facing forwards. This configuration is inspired by the one Tesla states to use for their autopilot on its vehicles.[1]

EDAF Rules: With the sensor configuration used in our simulation, vehicles detect with their mounted sensors around 7.5 objects on average. In [9], only the configuration $Toff_{thresh}$=100ms was investigated with a different sensor configuration. In this book chapter, different $Toff_{thresh}$ values are investigated: 25, 50, 75, 100 ms. The objective is to find a threshold allowing enough filtered objects to reduce the number of CPMs to generate while avoiding high information redundancy by including too many filtered objects. The sensor-equipped configuration on the vehicle is similar to the study performed in [10] and presented in Sect. 2.5.

Metrics: The following metrics were used to assess the performance of the different object filtering techniques and CPM generation rules:

- **Channel Busy Ratio (CBR)**: Fraction of time that the radio channel is perceived as busy to the total period under observation.
- **CPM rate**: The number of CPMs generated per second.
- **CPM size**: The size in bytes of the generated CPMs.
- **Environmental Awareness Ratio (EAR)**: The ratio of vehicles known within a delimited area around a vehicle. The area considered in our simulation is a circle centered on the vehicle with a diameter of 500 m.
- **Redundancy Level (RL)**: During the last second, the number of updates received about an object is divided by the number of updates that the object would have sent if it would have generate CAMs. More details on this metric can be found in [10].
- **Score**: This metric is realized using (3). The Gompertz function G is the valuation of the RL score measured for each object and has the following parametrization: $a = 1, b = 7, c = 2.31337$. The objective is to facilitate the comparison among the CPM generation and filtering rules. More details on this metric are explained in [10].

$$Score = (1 - CBR) \times G(RL) \times EAR \tag{3}$$

---

[1] https://www.tesla.com/autopilot, last accessed: 25. Nov 2022.

2.6.2   Results for Enhanced DCC Filtering (EDAF) Rules

Results to adapt the object filtering to the available channel resources were initially presented in [9] and [8]. Because the scenario and part of the metrics used for simulation in [9] are different from the ones presented in Sect. 1, simulations were performed again to align the simulation configuration.

To evaluate the EDAF rules, we compare them to two default configurations: the "No Filtering" for which all the objects are included every 100 ms and the ETSI rules as defined in Sect. 2.3. The EDAF rules were explained in Sect. 2.4 and its parameter $Toff_{thresh}$ is configured for different values as indicated in Table 1.

Figure 8 shows the obtained results for the CBR with the different CPM generation rules. In general, the higher the MPR, the higher the CBR. This is expected as more vehicles transmit messages. Considering the greediest approach, the "No Filtering" strategy results in the highest average CBR perceived within the scenario (up to CBR around 0.55 at MPR=100 %).

In contrast, the ETSI filtering approach, which is the most conservative regarding the inclusion of objects in CPMs, shows the second lowest obtained CBR independently of the MPR. The CBR starts at around 0.04 at MPR = 0.1 and increases up to 0.42 at MPR = 1.0.

All EDAF configurations result in a lower channel usage than the No Filtering approach. With the EDAF-100 ms configuration, the CBR reaches a maximum of 0.45 at MPR = 1.0. We can observe that the EDAF-50,75,100 ms have similar CBR values up to different MPRs. For example, for MPR higher than 0.4, the EDAF-50 ms starts obtaining a lower CBR than the two other configurations. It indicates that the rules are in different phases as explained in Sect. 2.4.2. First, the three configurations are in the "One-for-All" phase. Then, as DCC starts to impose a higher closing gate

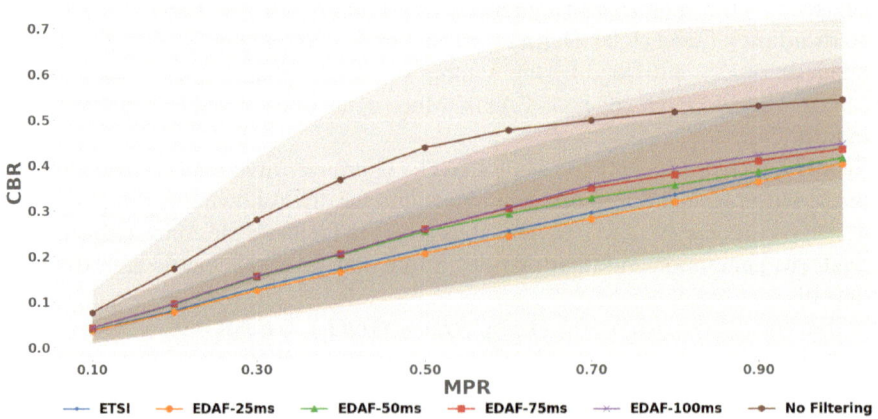

Fig. 8 Average CBR perceived by vehicles on SCH1 for different CPM generation rules

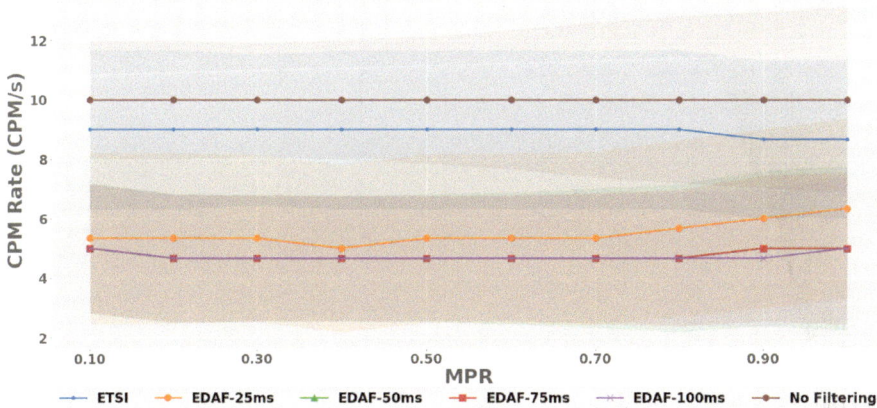

Fig. 9 Average CPM rate in [CPM/s]

delay, not all filtered objects are included in a CPM for the EDAF-50 ms. The phase change occurs at MPR = 0.4 for EDAF-50 ms and at MPR = 0.6 for EDAF-75 ms.

In contrast to the other EDAF configurations, the EDAF-25 ms results in a slightly lower CBR than the ETSI rules while being more permissive for object inclusion.

To explain this phenomenon, Figs. 9 and 10 show the resulting tradeoff between CPM size and rate obtained with both the ETSI and EDAF rules. The CPM size obtained at low MPR for the EDAF-50, −75, −100 ms is around 2.5 times higher (≈250 B) and 1.7 higher (≈170 B) for the EDAF-25 ms than for the ETSI rules (≈100 B). On the opposite, the obtained CPM rate is on average higher for the ETSI rules (≈9 CPM/s) than for the EDAF-50,75,100 ms rules (≈4.8 CPM/s) and EDAF-25 ms (≈5.3 CPM/s). This tradeoff between CPMs containing more objects but with a lower message generation rate is due to the nature of the filtering rules. For example, let's consider the scenario of a V2X vehicle perceiving two objects requiring to be transmitted every 500 ms (due to their dynamics). There are two possible scenarios of CPM generation for these objects. In the first scenario, two CPMs are transmitted every second containing both objects. In the second scenario, four CPMs are generated per second containing each one of the objects. The second scenario is what happens in many cases with the ETSI rules, as the objects are not grouped for transmission. The first scenario is an example of what happens with the EDAF rules. As there are enough channel resources, the station creates a CPM for both objects, even if one of them shouldn't have been transmitted at that time due to its last transmission. The advantage of the first scenario is to reduce the communication overhead from the lower layer headers. For the No filtering strategy, the CPM rate stays at a high value (around 10 CPM/s) with most of the included objects.

As shown b Fig. 11, the resulting EAR is similar, independently of the employed generation rules. The lowest EAR is obtained at MPR = 0.1 with around 67 % of the detected objects. At MPR = 0.3, the obtained EAR is around 92 %.

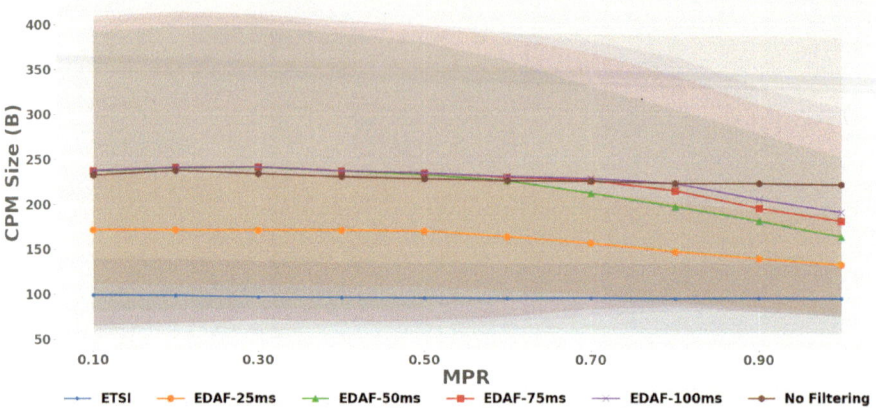

Fig. 10 Average CPM size in [B]

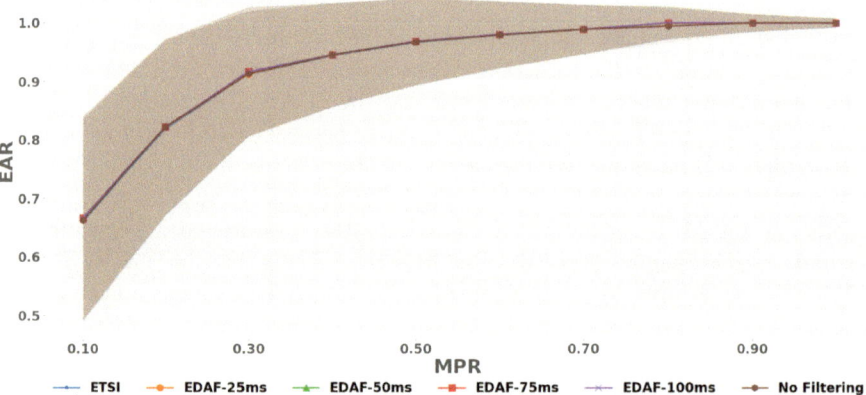

Fig. 11 EAR for the studied filtering strategies

Figure 12 for the Redundancy Level metric, the No filtering strategy results in the highest RL. Independently of the rules, the higher the MPR, the higher the RL is. Moreover, with the No Filtering rules, the RL can go up to 20 on average, i.e., a vehicle has 20 times more updates about an object than if this object would have sent CAMs following the CAM generation rules. This underlines the necessity to have RMR. We point out again that the tradeoff proposed by the EDAF-25ms creates more data redundancy, for a lower channel load, and a similar EAR than the ETSI rules.

The resulting scores depend only on RL and CBR. Indeed, the EAR does not differ significantly among the filtering rules for the same MPR. Figure 13 shows the obtained scores. At MPRs < 0.15, the EDAF-50, 75, 100 ms rules resulted in the best score of around 0.64. At the same MPRs, the No Filtering scored better (0.61) than

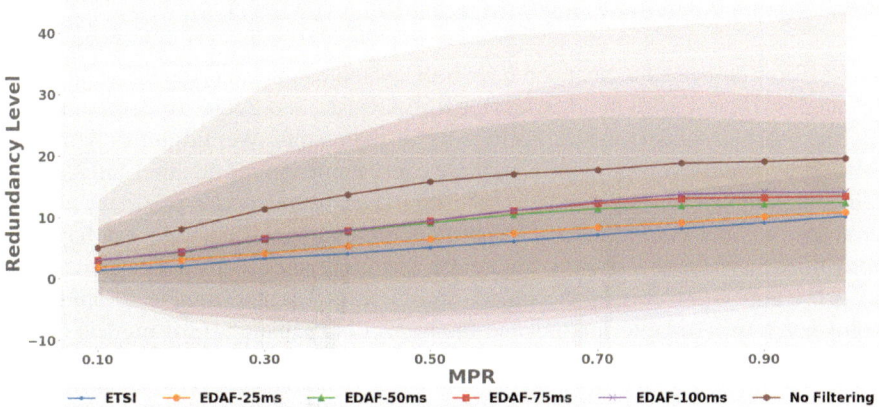

Fig. 12 Redundancy level for the different filtering strategies

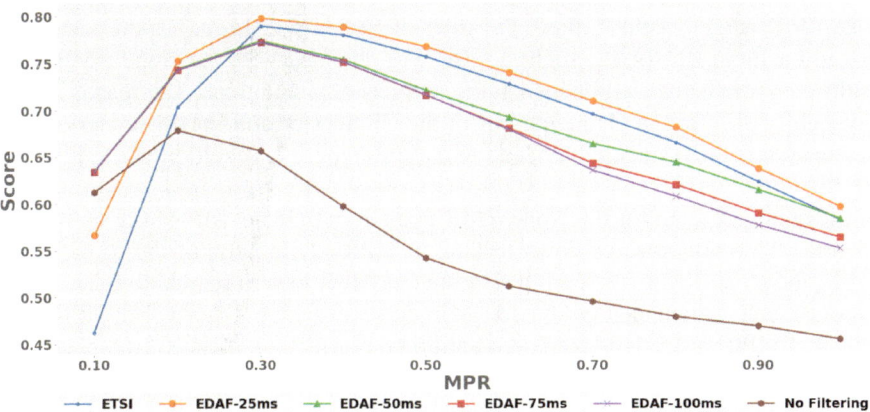

Fig. 13 Score as defined by (3) obtained for the different filtering strategies

the EDAF-25 ms (0.56) and the ETSI one (0.46). This result shows that at low MPRs, greedier approaches than the ETSI one could and should be applied. At higher MPRs, the No Filtering approach does not score well and finishes with a score of around 0.45 at MPR = 1.0. This score is expected from the high RL and CBR obtained. At MPR > 0.2, the EDAF-25 ms rule performs the best with the highest score of 0.8 obtained at MPR = 0.3. The maximum score obtained by the different filtering approaches is reached between MPR = 0.2 and MPR = 0.3. For higher MPR, the RL is too high, resulting in a reduced gain from the Gompertz function compared to the generated channel load.

2.6.3 Results for Redundancy Mitigation Rules (RMRs)

The RMRs as described in Sect. 2.5 have been evaluated in [10] using the parameters indicated in Table 1 with the metrics: CBR, EAR, RL, and Score. By lack of place, only the score is shown here as a summary of the results. We did not evaluate the RMR "confidence level" proposed in [14] because they required detailed and realistic modeling of the environmental data capturing. The current version of Artery used in our study provides detailed modeling of the communication but does not have the capabilities for realistic sensor modeling yet. Therefore, the assumed ideal perception (no delay and no errors) in the object measurement is not suitable to study the confidence level. To better understand the impact of realistic sensor models on the RMR performance, enhanced simulation models and alternative simulation tools such as [49] would need to be used.

Figure 13 shows the obtained score for the RMRs and their respective parametrizations. Similarly to the results of the EDAF rules, the score is not affected by the EAR. Consequently, the score is influenced mostly by the CBR and the RL. Note that the None RMR is equivalent to the ETSI rules of Sect. 2.6.2. In the experiments performed for these RMRs, the highest CBR was obtained with the None RMR. The results do not match exactly between the two sets of simulations. The reason is that two slightly different implementations of the CP Service and InTAS scenario were used to collect the results, which lead to the differences observed between Figs. 13 and 14.

The obtained score for the None RMR goes from 0.385 at MPR = 0.1 up to 0.77 at MPR = 0.5 and decreases down to 0.62 at MPR = 1. In comparison to the other RMRs, the None-RMR rule performs best at an MPR lower or equal to 0.25.

For the Self-Announcement-based RMR, the score evolves from 0.38 at MPR = 0.1 to 0.83 at MPR = 0.75 then decreases until 0.66 at MPR = 1. Relatively to the other rules, this rule performs as one of the best scorers up to MPR = 0.5. At a higher MPR, it starts to underperform compared to others.

The best scores relative to other RMRs obtained with the Frequency-based RMR are between MPR = 0.1 to 0.5 with N_Redundancy = 3. At higher MPRs, this rule underperforms in comparison to the best scores obtained at each MPR.

The scores obtained by the Dynamics-based RMR are lower than the best-performing RMRs at MPRs lower than 0.5. At MPR = 0.5 and higher, independently of the chosen parameter value, this RMR obtains some of the best scores from around 0.83 at MPR = 0.5 to 0.875 at MPR=1.

For the Distance-based RMR, the results differ predominantly for R_Redundancy = 200 and the other configurations. For R_Redundancy = 200, the obtained score is the same as for the Frequency-based RMR with N_Redundancy = 1, even by behaving differently. The reason is that both rules allow only a single vehicle to transmit information about an object. Still, both configurations result in some of the lowest scores obtained independently of the MPRs. With R_Redundancy = 50 m, the scores are some of the best at MPR higher and equal to 0.25, which corresponds to one of the best performing rules independently of the MPR. The other remaining configurations perform in general well but are either better at low or at high MPRs.

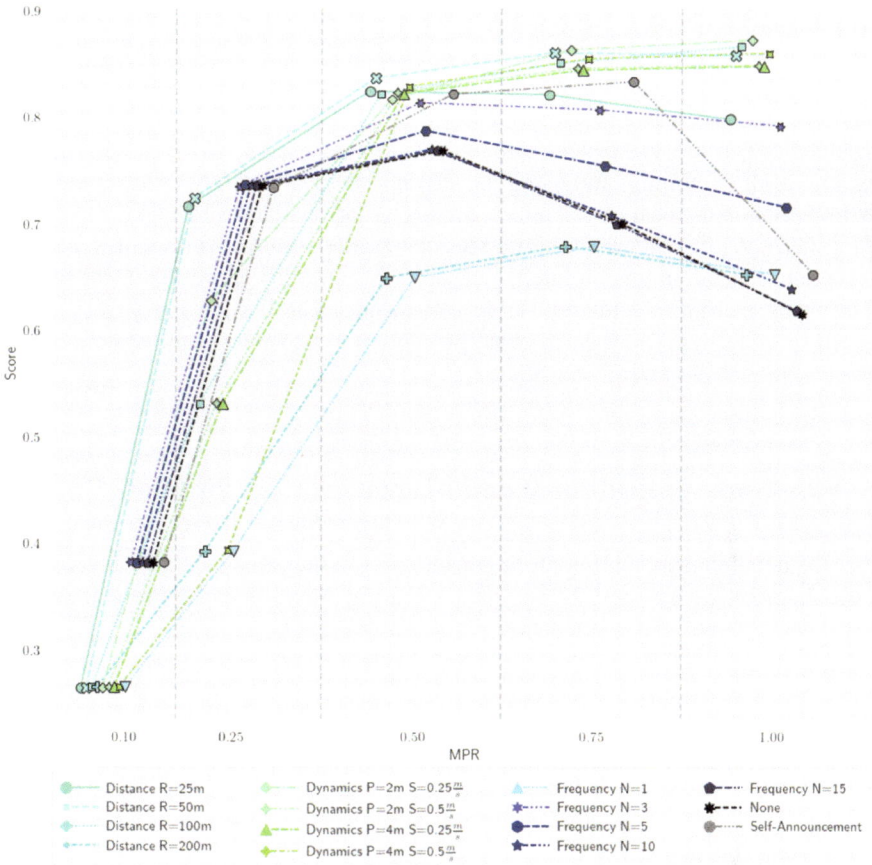

Fig. 14 Scores obtained for the different RMRs. The score is based on the obtained performance for the CBR, EAR, and RL. To ease the readability, an offset in the x axis has been applied to each line

2.6.4 Conclusion

We made proposals to address two problems of the current protocol design for CP: adapt the filtering of objects on the available resources and reduce the problem of information redundancy by applying filtering rules considering objects received by other stations.

We conclude that within this evaluation framework, the EAR is not affected by the filtering rules. The score, as currently configured, shows that at low MPR, i.e., $MPR \leq 0.25$, an RMR is not necessary. By allowing the inclusion of all or part of the filtered objects in CPMs, the EDAF rules have shown better performance than the current CP protocol design with a carefully chosen value for the parameter $Toff_{thresh}$. At higher MPRs, RMRs relying on either distance or dynamics criteria perform well to decrease the number of objects included and maintain a balance between channel usage and information redundancy. Combining EDAF rules with RMRs is the next logical step for this research and is considered future work. The end objective is to have an approach addressing both problems simultaneously. The presented research opens a clear path to the following works. We expect to derive a new approach combining and using the advantages of both EDAF and some of the RMRs to obtain the best of CP, independently of the available channel resources.

Furthermore, more work is required to analyze these filtering rules with a focus on perception. Indeed, in this research, measurement inaccuracies were not considered while being a critical factor for the development of CP. Further investigations are needed to understand how error-prone perception would impact the operation and choice of the RMRs and EDAF rules.

3   V2X Communication-Based Maneuver Coordination

After the detailed presentation of the sensor data sharing service, the following part of the chapter presents the V2X service for sharing intention and coordination data.

3.1   Overview

Cooperative Maneuver Coordination (CMC) represents a V2X-based application for exchanging of intentions and coordination data among the CAVs through V2X communication. Through this process, the vehicles can broadcast their planned maneuvers, request and negotiate a coordinated maneuver or accordingly accept a coordination offer from other CAVs which is presented by Fig. 15. Increased safety, comfort and traffic optimization are the main goals of V2X enabled CMC. The standardization process of the Maneuver Coordination Service (MCS) is still in a very early stage at ETSI [17].

CMC applications can be classified based on different criteria. One of them is to divide the applications into use case specific and generic ones where the former can

Fig. 15 Illustration of a maneuver coordination process

only be used for one traffic use case, while the latter can be applied to multiple use cases.

Another way to categorize the CMC applications is by centralized and decentralized coordination. Centralized coordination involves a central unit, mostly a Road Side Unit (RSU) that receives all the information from the involved vehicles and accordingly calculates and distributes the coordinated maneuvers to all involved vehicles. Infrastructure-controlled CMC can be applied to different use cases especially for signalized or non-signalized intersections, roundabouts and junctions, as well as cooperative merging situations through Vehicle-to-Infrastructure (V2I) communication. Decentralized CMC utilizes only Vehicle-to-Vehicle (V2V) communication among CAVs to negotiate and coordinate maneuvers.

Furthermore, the coordination can be implicit and explicit. In an implicit coordination, the vehicles broadcast messages with intention and coordination data without specifying which vehicles are included in the negotiation. This can lead to a conflicted situation if more vehicles are negotiating a maneuver. Explicit coordination involves the IDs of the negotiating and coordinating vehicles that leads to an explicit agreement among the vehicles for an acknowledged maneuver.

3.1.1 Use Cases

Various use cases exist [34] where CMC can bring certain benefits to increase the traffic safety, comfort and efficiency as well as the road capacity. Some of these use cases are presented in Fig. 16:

Cooperative-Adaptive Cruise Control (C-ACC): V2X communication enhanced ACC that enables additional information exchange between the vehicles to synchronize their velocities.

Platooning: A platoon consists of a group of vehicles driving in a stable formation, usually trucks, that can keep small distances among each other by sharing V2X messages that include their current state information. Typically, there is a master vehicle that leads and manages the platoon consisting of following vehicles.

Cooperative lane change: Two or more vehicles can cooperate to create a gap for a safe and efficient lane change maneuver.

Cooperative lane merging: Common highway situation that can be facilitated by V2X communication to allow safe and comfortable lane merging for the CAVs.

Cooperative overtaking: Another common maneuver that occurs frequently on highways and rural roads that can be utilized especially for heavy loaded trucks.

Cooperative driving at intersections, roundabouts and junctions: Such traffic use cases can often cause conflicted outcomes. By exchanging the planned and desired intentions, the involved CAVs can coordinate each other in a safe and efficient way.

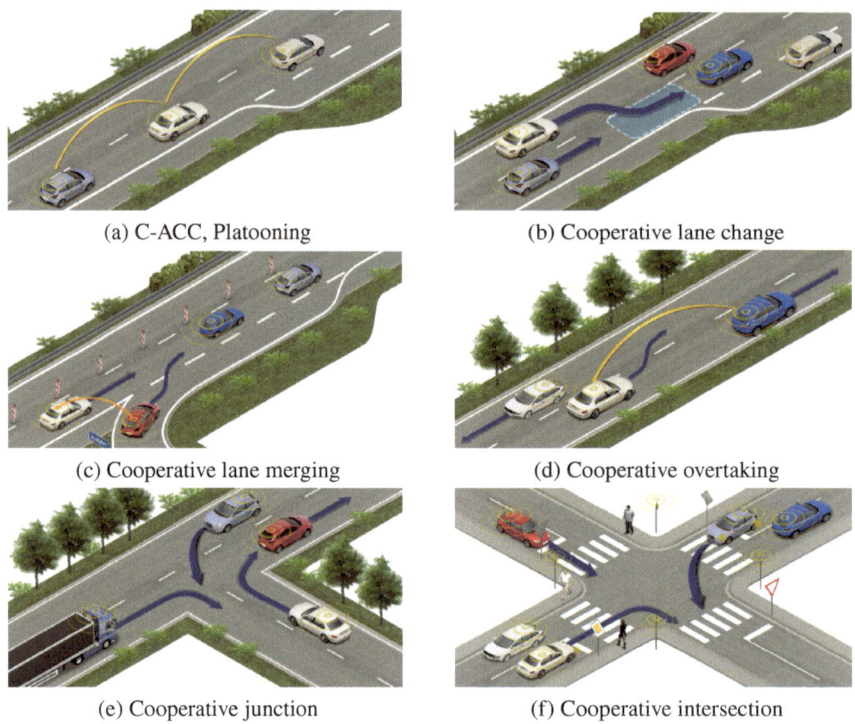

(a) C-ACC, Platooning   (b) Cooperative lane change

(c) Cooperative lane merging   (d) Cooperative overtaking

(e) Cooperative junction   (f) Cooperative intersection

Fig. 16 Cooperative driving use cases

3.1.2 State of the Art in Maneuver Coordination

The topic of cooperative maneuver coordination is a relatively new field with a lot of open research gaps. Much of the earlier research was done on use case-specific coordination, whereas in the recent years more work emerged on generic maneuver coordination. Platooning [50] and C-ACC [11] represent typical examples for use case-specific cooperative driving applications that have been more extensively researched and analyzed. These approaches utilize V2V communication to achieve coordination that is mostly focused on longitudinal acceleration and deceleration. Numerous approaches exist with different control strategies and characteristics presented in [11, 40, 50]. A research on driving in a convoy is presented in [35] where the vehicles adjust their longitudinal and lateral dynamics to keep a stable driving formation. Certain applications like lane change, merging and overtaking require coordination in both longitudinal and lateral direction in order to create the needed space. Such a lane change approach is presented in [24] consisting of three different phases: search, preparation and execution, where dedicated lane change messages are broadcast in each phase. C-ACC have been used for lane change and merging scenarios, as presented in [2]. Various message sets are utilized by distributed resource reservation protocols in [3] to analyze distributed intersection and roundabout management without infrastructure support.

Generic decentralized maneuver coordination focuses on using one protocol to achieve coordination in different traffic use cases. There are different approaches presented in the last several years, however a lot of gaps in the protocol design, testing and evaluation still exist. Maksimovski et al. already published surveys [33, 34] that analyze the up to date proposed coordination approaches and their advantages and limitations. Comparison of the approaches was performed as well. Several research gaps were highlighted focusing on the detection and decision logic, the maneuver coordination protocol and the V2X communication. The detection and decision logic part includes research questions related to: decision when to request a maneuver on the side of the requesting vehicles as well as when to accept a maneuver coordination on the side of the accepting vehicle. In the maneuver coordination protocol section, the following research questions are discussed: the message type and format, additional use case-specific information in the message, message generation rules, number of messages, number of vehicles included in a coordination, maneuver cascading, data security and privacy as well as a question related to the implementation, testing and evaluation of the maneuver coordination protocol. The V2X communication section discusses the communication requirements for CMC, the improvement of the access technologies, the communication type as well as multi-channel operation.

The early standardization work by ETSI [17] is based on the work presented in [27] that proposed a Maneuver Coordination Service that utilizes periodic broadcast of dedicated Maneuver Coordination Messages (MCMs) consisting of trajectory related data, namely planned and desired trajectories, further enhanced by an explicit coordination approach and a safety analysis [28]. An explicit approach with extended communication pattern based on [27] with three new MCMs is proposed in [52] where a lane merge scenario was also used to evaluate the approach in different situations with and without coordination among the vehicles. An implicit approach utilizing MCMs is presented in [29], where the trajectory data have cost values representing the need or willingness to cooperate with the surrounding vehicles. A space time reservation protocol (STRP) using reservations of position and time constraints has been presented in [23] and continuously upgraded and evaluated for different traffic scenarios [38]. This protocol uses an extended CAM message in an event based manner to broadcast the request in comparison to the other approaches that propose periodic MCM with trajectory data. Extended CAM message to include future vehicle trajectories in an event based manner is presented in [41] as well to be used in hazardous situations to mitigate or avoid an accident. An additional maneuver suggested container in the MCM is presented in [5] to be used by the infrastructure through V2I communication to send suggestions to the CAVs, which was also demonstrated on real world tests in [43]. The Complex Vehicular Interactions Protocol (CVIP) is another explicit approach presented in [22], utilizing four different messages in an event based manner also involving joint maneuver negotiation. The maneuvers in the message can be represented with standardized names, functions or trajectories. Maneuver Coordination Service with abstracted functions for automated driving is proposed in [36] consisting of seven messages that demonstrates reduced communication bandwidth and increases the speed of the participating vehicles in the coordination.

The Priority Maneuver Coordination (PriMa) approach is proposed in [32] which relies on decentralized and explicit exchange of MCMs introducing three levels of maneuver requests and accordingly different negotiation process among the CAVs. Different communication patterns depending on the number of included vehicles in the coordination are also proposed, including cascading scenario. The current work is based on this approach and will be presented in more details in the next section.

3.2   Protocol Design

This section presents the PriMa coordination [32] protocol design. Three different priority levels are introduced that describe the different maneuver types. Communication patterns for different cooperative driving situations depending on the number of involved vehicles are studied too. The proposed MCM format is also introduced consisting of intention and coordination data required to complete a coordination.

3.2.1   Priority Maneuver Request

The proposed PriMa coordination relies on an explicit exchange of MCMs using different communication patterns with additional three levels of maneuver requests that facilitate the decision-making process of the involved CAVs in the negotiation phase. Three different levels of priority requests were defined in the concept that are based on different metrics and costs which vary depending on the use case. The following type of maneuver requests are defined:

Low priority—desired maneuvers that the vehicle wants to execute in order to improve time efficiency.

Medium priority—necessary maneuvers that the vehicle needs to perform in order to stay on the route or significantly improve time efficiency.

High priority—critical maneuvers to avoid an emergency maneuver or accident.

A thorough analysis of cooperative driving use cases is required to determine the priority of the requested maneuvers utilizing metrics and cost values based on the road rules, velocity, acceleration, time efficiency, conflicted traffic situations, as well as potential collisions and emergency situations. Such a different level of requests will also lead to a different negotiation process between the involved participants. On the side of the accepting vehicles, similar metrics can be used to evaluate whether a request is feasible and worthy to accept because in most of the cases the vehicles that accept the maneuvers will be disadvantaged. In [32] estimated values were used for a lane change scenario where the priority level of the request was based on the time

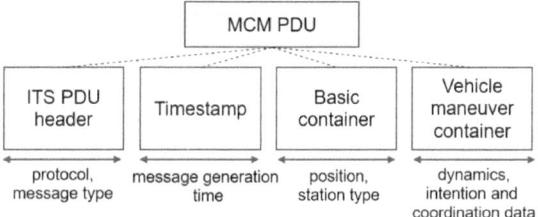

Fig. 17 Maneuver Coordination Message

gap to the front vehicle. On the other side, the accepting vehicles were evaluating the request based on the required deceleration and velocity reduction. Three different thresholds for reducing the velocity were used based on the priority level of the request. However, a high priority critical maneuver is accepted whenever the vehicle can plan and execute a conflict free trajectory.

3.2.2 Maneuver Coordination Message

The MCM consists of vehicle state information, trajectory related data that represent the planned movement of the vehicle as well as trajectories included in the negotiation process. Figure 17 presents the MCM which includes several containers and is similar to the already standardized ETSI message formats. The ITS PDU (protocol data unit) header includes the version of the protocol, the ID of the station sending the message as well as the message type, in this case the MCM subtypes which are introduced below. Along the header, the timestamp when the MCM was generated is also included in each message. The basic container consists of the current reference position and the station type that can also be a vehicle or a RSU. The main container of the message is the vehicle maneuver container that includes the required vehicle dynamics, namely the planned (PT), requested (RT) and offered (OT) trajectories. The data type that represents the trajectory is a sequence of data points that includes the vehicle pose (position and orientation), the velocity as well as the time step between the trajectory points. The pose includes the longitude and latitude values, alongside the heading of the vehicle. Furthermore, the lane ID that the vehicle is currently driving on is also included. In a request MCM, the priority level as well as the request ID are also included alongside the IDs of the potential accepting vehicles. Accordingly, during negotiation the replying vehicles also include the ID of the request they refer to. The inclusion of the ID of the request and the IDs of the vehicles involved in the negotiation makes the coordination process explicit and unambiguous.

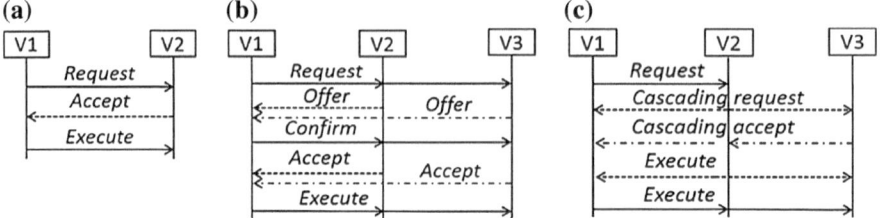

Fig. 18 Message flow in coordination between two vehicles (**a**), three vehicles (**b**) and cascading situation with three vehicles (**c**)

3.2.3 Communication Pattern

PriMa proposes three different communication patterns depicted in Fig. 18 representing coordination between two, more than two (in this case three) vehicles and a cascading situation using the following MCM subtypes explained below: *regular, request, offer, confirm, accept, reject, execute, cancel, abort, emergency, cascading request, cascading accept* and *cascading reject*.

As an intention sharing message, the vehicles *regularly* broadcast MCMs. In a coordination between two vehicles, only two MCMs are required to complete the negotiation process: *request* and *accept*. However, in a situation involving three or more vehicles, additional MCM subtypes like *offer* and *confirm* are needed in order to ensure an unambiguous and efficient negotiation process. This avoids a divergent situation that can lead to conflicted and inefficient maneuvers between the involved vehicles. In such a way, the influence of the lost message packets on the negotiation process can also be limited. The impact of the unreliable communication is also discussed in more detail in [32] and will be analyzed in the next section too. In a cascading situation, the accepting vehicle in order to accept the incoming request, has to send a *cascading request* to another vehicle which prolongs the negotiation process. PriMa proposes a limited cascading coordination involving three vehicles. Additional subtypes are *cancel, abort* and *emergency* messages that can be used to cancel a request, to abort an agreed maneuver from the requesting vehicle as well as to send an emergency MCM involving an emergency trajectory that the vehicle is going to take without a negotiation. The final *execute* message from the requesting vehicle is not needed in order to complete the negotiation, however can be useful to confirm that the vehicle is executing the request.

3.2.4 PriMa Example Scenario

Figure 19 depicts a proof of concept scenario presented in [32] that includes four CAVs. In the scenario, the ego vehicle *V1* needs to change a lane or decelerate because the vehicle *V2* in front of *V1* is stopping. In order to avoid a critical decelerating maneuver, *V1* sends a high priority *request* message to vehicle *V3*. However, at the

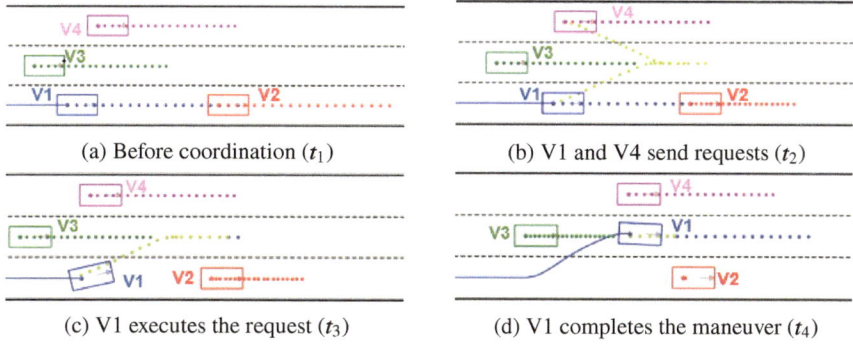

Fig. 19 PriMa coordination [32]

same time vehicle *V4* also sends a low priority lane change request to *V3* (Fig. 19b). In this situation *V3* broadcasts an *accept* message to the high priority request and allows *V1* to avoid the decelerating maneuver and perform a lane change. *V4* also receives the *accept* message, however the message includes the station ID and request ID of *V1*, therefore showing the benefit of explicit and unambiguous communication. In an implicit coordination, *V3* only sends an *accept* message that can lead to a conflicting situation where both *V1* and *V4* start to execute the lane change maneuvers. Furthermore, this scenario also demonstrates the benefit of the PriMa coordination to perform maneuvers that have a higher priority.

3.3 Simulation Results

In order to design and evaluate the proposed coordination approach, a simulation was performed for a highway lane merging scenario. The simulation is performed in the discrete-event simulator Artery [42], see Sect. 1.

3.3.1 Scenario Description and Evaluation Metrics

Scenario description: Figure 20 shows the map of the SUMO scenario taken from InTAS [30] which is modeled based on a highway in the outside parts of Ingolstadt, including the speed limits. The main highway road with a driving direction to the right has a total length of 705,78 m, of which on the first 251,86 m the speed limit is 80 km/h (22.22 m/s), while the rest of the road has a speed limit of 100 km/h (27.77 m/s). The first merging road on the bottom has a total length of 146.2 m and also has a speed limit of 80 km/h. However, the vehicles coming from this lane must give way to the vehicles on the main highway road. The merging point of the two roads is shown as well. The vehicles coming from the main road drive a total length

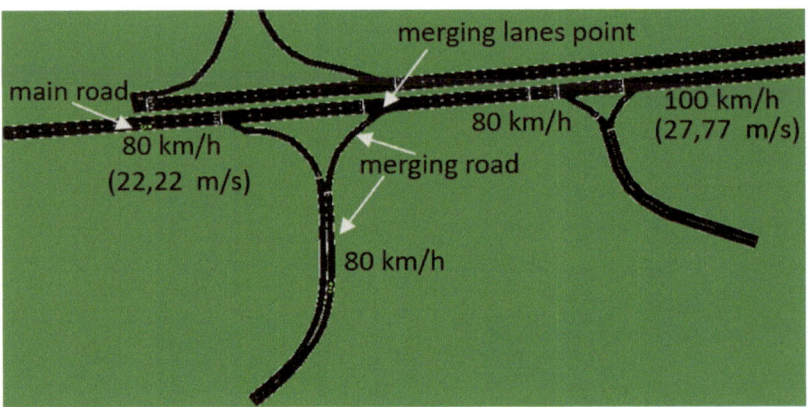

Fig. 20 SUMO scenario map

of 158,84 m until the merging point, while the vehicle coming from the first merging road drive a similar length of 145.2 m. The simulation is run for 22 s.

   Scenarios: Three different situations are considered:

- **Scenario without coordination**—SUMO simulation following the right of way rules.
- **Scenario without right of way rules**—the vehicle coming from the merging road drives without giving way to the vehicles on the main road.
- **Coordinated scenario**—the vehicle coming from the merging road coordinates a merging maneuver with the vehicles driving on the main road.

Evaluation metrics: To evaluate the coordination protocol the following metrics are considered:

- **Safety metrics**: The time gap between the negotiated and executed trajectories of the cooperative vehicles needs to be bigger than 1 s. The results also show the position of the vehicles and the distance between them during the merging maneuver.
- **Comfort metrics**: the car following models from SUMO are used in the simulation with desired acceleration of up to $3\,m/s^2$ and deceleration of up to $4\,m/s^2$ to keep the comfort values according to the ACC standard.
- **Efficiency metrics**: The results show the time loss of each cooperating vehicle in the simulation. SUMO calculates the time loss as the time that the vehicle spends in the simulation driving below the desired speed, in this case the maximum speed limit on the road.
- **Communication related metrics**—The results present the total negotiation time between the vehicles to complete the coordination. Additionally, the unreliable communication effect on the coordination is analyzed too by introducing a package loss rate of 10, 20 and 30%. The average negotiation time for different packet reception rate (PRR) is calculated based on ten simulation runs for each scenario

using random number generators with different seeds. It has to be noted that the processing delays of the motion planning system that plans a maneuver request or evaluates an incoming request are not considered, as well as the additional delay and latency that can arise due to the communication device or communication channel (Decentralized Congestion Control—DCC). More details on DCC can be found in the first part of the chapter about collective perception.

3.3.2 Coordination Protocol Implementation

The maneuver coordination protocol is implemented as a V2X service based on the ETSI ITS-G5 protocol and is analyzed for coordination involving two and three negotiating vehicles. The priority request analysis was not performed in this evaluation. MCMs are broadcast every 100 ms (broadcasting interval of 10 Hz) by each of the involved CAVs in the simulation. The broadcast trajectories include 20 points with a time step of 0.25 s between the points representing the intention of the vehicles in the next 5 s. Cartesian coordinates (x, y) are used to represent the position of the vehicle in the local coordinate system, however they can be transformed to the global coordinate system with longitude and latitude values as used in the ETSI standards. The trajectory is calculated based on the constant velocity model, in this case joining the highway with 22.22 m/s. A very important part of the coordination is the time when the ego vehicle starts to send the request. The negotiation needs to be completed at a certain distance before the merging point, to let the requesting vehicle decelerate on time with a comfortable deceleration in case if the negotiation is not successful. The required distance to stop the vehicle with the current speed and desired deceleration can be calculated using the generic braking distance formula: $d = (V_f^2 - V_i^2)/(2(-a))$, where V_f is final velocity, V_i is initial velocity, and a is the deceleration. Additionally, a timeout of 1 s is added to complete the negotiation. Therefore, the vehicle starts sending a request once it is away from the merging point at a distance equal to the braking distance + the distance required for driving 1 s with the current speed. Since the vehicle is driving with 22.22 m/s, and has a desired comfortable deceleration of 4 m/s^2, it starts sending requests around 4 s before reaching the merging point with the current speed.

3.3.3 Coordination—Two Vehicles

This simulation involves two vehicles: the ego vehicle *Vego* and *V1*. In the first scenario, *Vego* decelerates to let *V1* that has the right of way which leads to a big reduction of the velocity as shown on the velocity-time graph in Fig. 21a. The second situation involves the same vehicles without following the right of way rules, which leads to a conflict and an emergency deceleration from *V1*, which is also shown in Fig. 21b. *Vego*, on the other side, joins the road with a small deceleration that is

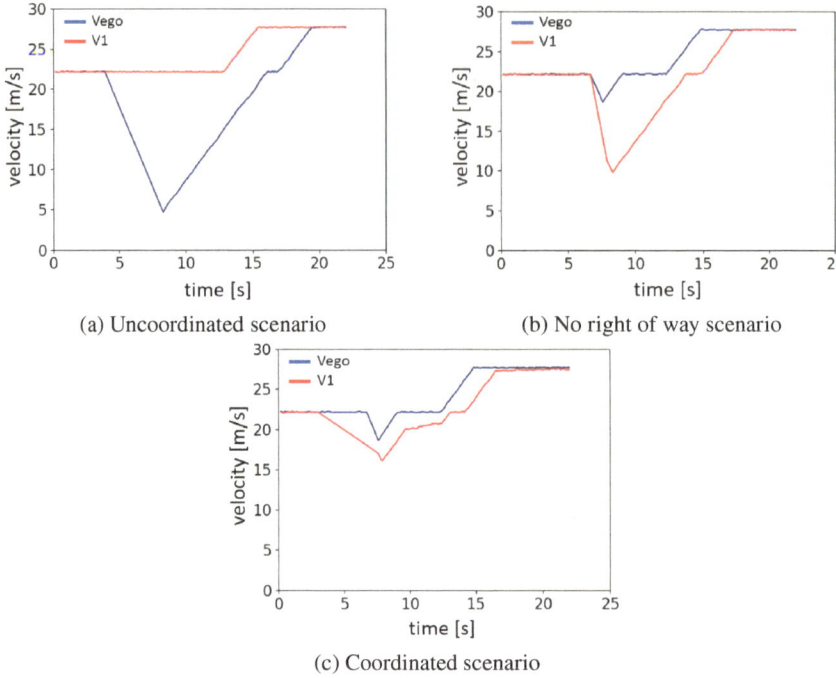

(a) Uncoordinated scenario   (b) No right of way scenario

(c) Coordinated scenario

Fig. 21 Velocity change—two coordinating vehicles

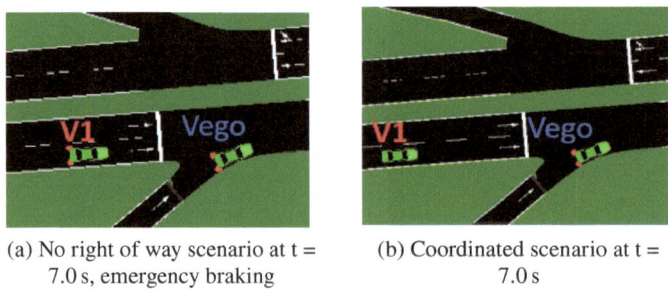

(a) No right of way scenario at t =   (b) Coordinated scenario at t =
7.0 s, emergency braking   7.0 s

Fig. 22 Coordination scenario comparison—two coordinating vehicles

required to drive in the curve. The position of the vehicles during the emergency deceleration at $t = 7.0$ s is also shown in Fig. 22a.

The third situation, involves the implemented maneuver coordination service. The message flow during the negotiation is shown in Fig. 23. Two regular messages at time steps t_1 and t_2 are included as well to show the periodic flow of the messages. Since the vehicles broadcast MCMs, *Vego* detects a conflict between its desired trajectory and the *PT* from *V1*. In order to avoid deceleration, once *Vego* arrives at the required distance before the merging point, it broadcasts a request MCM (t_3)

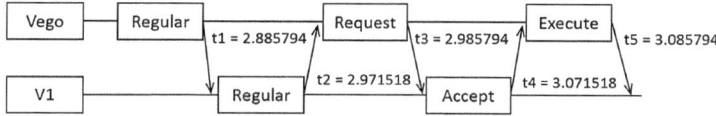

Fig. 23 Message flow in a coordination between two vehicles

Table 2 Time loss—coordination of two vehicles

Vehicle/scenario	Uncoordinated	No right of way	Coordinated
Vego	4.8 s	0.21 s	0.21 s
V1	0.3 s	2.3 s	1.53 s
Total time loss	5.1 s	2.51 s	1.74 s

including the *RT* that starts at the merging point. After performing collision check, *V1* broadcasts an *accept* message and starts adapting the new *PT* (t_4). After receiving the acceptance from *V1*, *Vego* broadcasts an *execute* message at t_5, completing the negotiation in 100 ms, in a situation with 100% communication reliability. In case of repeating the request, *Vego* accordingly updates the time when the *RT* should start, as the vehicle is getting closer to the merging point. If there is no *accept* message or if *Vego* receives a *reject* message, it will follow the right of way rules and decelerate before the merging point. After accepting the request, *V1* is reducing the speed with a small comfortable deceleration as depicted in Fig. 21c. Figure 22b shows the already coordinated merging of the vehicles at $t = 7.0$ s with a safe distance between the vehicles, in comparison with the emergency maneuver situation. The *PT* from *Vego* is calculated with constant velocity of 22.22 m/s, however it differs little bit in the execution since *Vego* requires small deceleration in the curve before merging to the highway. Because of that, it can be observed from the velocity-time graph that it leads to an additional small deceleration from *V1*, as can be seen at around 9 s. This also shows that keeping a time gap bigger than 1 s between the negotiated trajectory points also allows for safe adjustment of the these trajectories without causing further conflicts. The acceleration and deceleration values are also kept within the comfort interval as it can also be observed from the velocity change graph.

Table 2 includes the time loss of the vehicles in the different situations. It can be observed that in the uncoordinated SUMO simulation, *Vego* experienced time loss of 4.8 s, in comparison with the other two situations where the time loss is only 0.21 s, however, the second scenario leads to emergency braking from *V1*. As expected, since *V1* needs to decelerate, it will be disadvantaged and the coordination leads to a time loss of 1.53 s due to the deceleration. The total time loss is however reduced by half in comparison without coordination, equaling 2.51 s, thus making the cooperative maneuver more time efficient.

The average negotiation time was also analyzed by introducing the packet reception rate. Table 3 shows that for 70% reliability the negotiation time doubles to 210 ms, however is still well below 1 s that is set as a negotiation timeout.

Table 3 Average negotiation time—coordination of two vehicles

PRR	100%	90%	80%	70%
Negotiation time	100 ms	120 ms	150 ms	210 ms

Fig. 24 Coordinated scenario

3.3.4 Coordination—Three Vehicles

The scenario with three coordinating vehicles involves a longer negotiation process utilizing the communication pattern for a coordination with three or more vehicles, in order to ensure unambiguous and effective coordination. Figure 24 shows the coordinated scenario involving all of the vehicles in the simulation. The vehicles involved in the coordination are *Vego*, *V1* and *V2*. *V3* is included in the scenario to show the effect of the coordination on the other traffic participants but it is not participating in the coordination process. In this scenario, the ego vehicle is introduced 1.4 s after *V1* into the simulation. Vehicles *V2* and *V3* are also introduced after *V1* in the same lane and they have smaller speed than *V1* in the beginning of the simulation since they need to keep the gap with the vehicle in front: Therefore, these vehicles experience a bigger time loss while driving below the desired speed. In the uncoordinated SUMO scenario, the ego vehicle decelerates in a similar way as in the scenario with two vehicles. Hence the velocity-time graph is not shown, but this time *Vego* needs to wait longer because there are three vehicles driving on the main road. Figure 25a depicts the velocity-time graph for the second situation without the right of way. Also, the position of the vehicles at time $t = 8.3$ s is shown in Fig. 26a. *Vego* needs to perform emergency deceleration due to *V1* which is not affected by *Vego* now, since *Vego* was introduced 1.4 s later in the simulation. *V2* needs to perform emergency deceleration due to *Vego*, while *V3* also performs a high deceleration because of the emergency braking of *V2*.

The message flow in the coordination is depicted in Fig. 27. It can be observed that the negotiation time is increased to 200 ms for perfect communication conditions. Since the ego vehicle is introduced later, the request is sent at $t = 4.35$ s, once the vehicle is at the same distance from the merging point as in the previous scenario. In this situation, *Vego* requires a merging gap between *V1* and *V2*, hence the longer communication pattern is required for the negotiation process, meaning *Vego* needs to wait for the *offer* and *accept* messages from both of the vehicles in order to start the requested maneuver. After receiving the *offer* messages from *V1* and *V2* which include their offered trajectories and confirming that there is no conflict, *Vego* sends

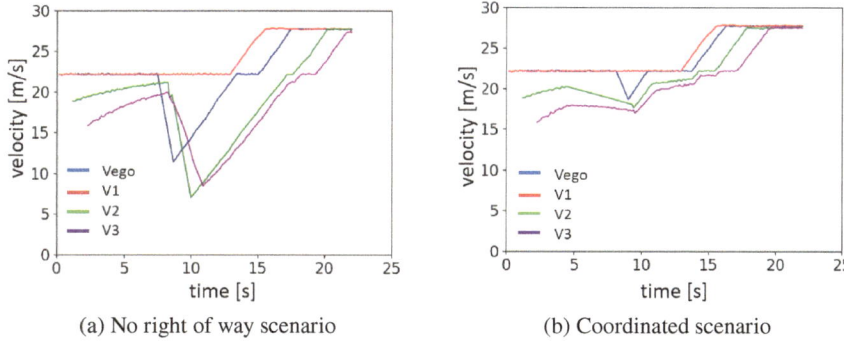

(a) No right of way scenario

(b) Coordinated scenario

Fig. 25 Velocity change—three coordinating vehicles

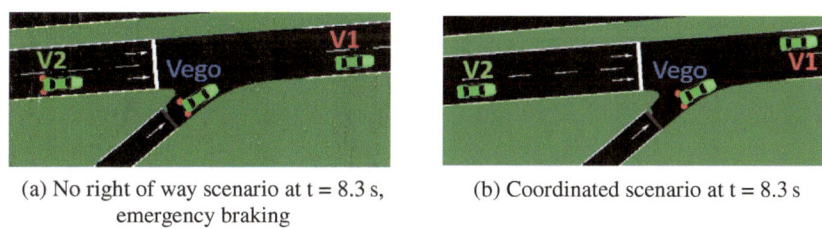

(a) No right of way scenario at t = 8.3 s,
emergency braking

(b) Coordinated scenario at t = 8.3 s

Fig. 26 Coordination scenario comparison—three coordinating vehicles

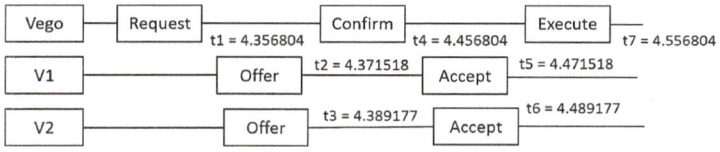

Fig. 27 Message flow in a coordination between three vehicles

the *confirm* message at t_4. However, it still needs to wait for the negotiated maneuvers to be acknowledged by both of the accepting vehicles. The maneuver negotiation process is finally over after the *accept* messages from *V1* and *V2* are received and the requested maneuver can be performed. In the end, the *execute* message is sent, but it is not needed to complete the coordination. After the negotiation, the vehicles keep sending the regular MCMs with their new *PTs*. Figures 25b and 26b present the velocity-time graph and position of the vehicles at the merging point, showing the smooth velocity change as well as the safe time gap between the involved vehicles in the coordinated maneuver.

The time loss of all included vehicles in the simulation is shown in Table 4. In an uncoordinated situation *Vego* experiences significant time loss because it has to let three vehicles pass. As mentioned, *V2* and *V3* experience higher time loss in the uncoordinated SUMO scenario due to driving with lower speeds in the beginning of the scenario. In the no right of way scenario, the vehicles experience higher time loss and additionally need to perform emergency deceleration. In the coordinated scenario, *Vego* has almost no time loss, similar as *V1* that only changes the lane in the coordinated scenario, therefore does not experience any additional time loss. *V3* is also included in the time loss table in order to show the oscillating effect of the coordination on the other traffic participants as it also has to decelerate in this situation to keep the safe time gap to the vehicle in front. However, the maneuver for *Vego* was highly optimized with no waiting time to merge into the lane, and in this situation required small deceleration from *V2*, and a lane change with no time loss from *V1*. The total time loss for the three vehicles in the coordination, as well as the one with *V3* (in brackets) is also presented in Table 4, which shows that the total time loss for all of the vehicles is significantly reduced in comparison with the uncoordinated scenario. The coordination makes the merging maneuver much more efficient without compromising the safety or comfort of driving as the vehicles also keep safe time gaps between each other and comfortable acceleration and deceleration values.

The effect of the unreliable communication on the negotiation process was also analyzed and is presented in Table 5. Since the vehicles require in total six messages to complete the negotiation process (without the *execute* message), the negotiation will be much more affected by the unreliable communication. With increased number of vehicles, the negotiation will become more complicated. However, the results show that for this traffic scenario, even with a PRR of 70%, the negotiation can still be completed under 1 s, as the average negotiation time is 500 ms and in each of the ten runs the negotiation was completed under 1 s. However, considering the processing delays of the motion planning system, the communication device and communication channel, the negotiation time will be increased and needs to be taken into account when designing the coordination protocol for real world applications. Further analysis is required to have an approximation of the negotiation time in different traffic situations.

Table 4 Time loss—coordination three vehicles

Vehicle/scenario	Uncoordinated	No right of way	Coordinated
Vego	7.74 s	1.43 s	0.21 s
V1	0.3 s	0.3 s	0.3 s
V2	0.58 s	3.99 s	1.77 s
V3	0.79 s	4.52 s	2.41 s
Total time loss (with V3)	8.62 s (9. 41 s)	5.72 s (10.24 s)	2.28 s (4.69 s)

Table 5 Average negotiation time—coordination three vehicles

PRR	100%	90%	80%	70%
Negotiation time	200 ms	290 ms	360 ms	500 ms

3.3.5 Conclusion

Cooperative maneuver coordination represents a V2X communication enabled application that has the potential to significantly improve the safety, comfort and efficiency of the CAVs on the road. By exchanging the intention and coordination data, the CAVs will be able to detect and perform cooperative maneuvers that can improve the traffic flow in different traffic situations. The proposed decentralized Priority Maneuver (PriMa) Coordination approach introduces three different levels of maneuver requests: low, medium and high priority differentiating between desired, necessary and critical maneuvers that can improve the decision making process of the cooperating vehicles. The Maneuver Coordination Message that includes trajectory related data is also presented with additional subtypes. Different communication patterns are proposed that can be utilized depending on the number of included vehicles in the coordination, whether there are two vehicles, more than two vehicles, or cascading situation. Such a designed protocol aims to ensure safe, fast, efficient and unambiguous coordination in different traffic situations.

This work shows a proof of concept and evaluation of the coordination protocol as a V2X service in the simulation framework Artery for a coordination involving two and three vehicles utilizing different communication patterns. A highway merging scenario was simulated in three different situations: uncoordinated, scenario with no right of way rules and a coordinated scenario. The evaluation is based on metrics regarding the safety, comfort, efficiency and communication. The results show the potential that the coordination offers to perform safe and comfortable maneuvers that can significantly improve the traffic flow and time efficiency of the vehicles. The impact of the unreliable communication on the coordination was evaluated for different packet reception rates as an important part in the protocol design that ensures enough time to complete the negotiation and coordination process. The presented work can be seen as a contribution to further research and development of generic decentralized maneuver coordination applications based on V2V communication.

Future work will include enhancement of the coordination protocol and implementation and evaluation for different traffic scenarios. The communication requirements will also be specified for various cooperative driving use cases. An analysis will be performed to define the threshold values for different priority maneuvers. Furthermore, the simulation results will also be verified in a real world testing environment.

4 Summary and Outlook

In this chapter, we presented, analyzed, and enhanced two key V2X communication protocols for cooperatively interacting automobiles. The first one, Collective Perception, is part of the Day 2 development of V2X applications and is expected to be deployed in the coming years. We investigated the adaption of information included in CPMs depending on the channel load and the observed information redundancy on the channel. We proposed approaches to address independently each of these problems. In future work, we plan to develop a combination of the developed solutions.

The second one, Maneuver Coordination, is part of the Day 3+ development of V2X applications. It is expected to be deployed later in the mass market and is still at an early stage of development. In this book chapter, we proposed a new maneuver coordination approach introducing three different levels of maneuver priorities. The potential of this approach was analyzed and we showed its capacity to perform safe, efficient, and comfortable maneuvers. Future works include enhancements of this protocol considering more traffic scenarios and proof of concept in a real-world testing environment.

Acknowledgements This work was gratefully supported by the German Science Foundation (DFG) by project KOALA 2 under number 273374642 within the priority program Cooperatively Interacting Automobiles (CoIn-Car, SPP 1835). The illustration toolkit from C2C-CC was used to create the figures illustrating the cooperative driving use cases (https://www.car-2-car.org).

References

1. Bengler, K., et al.: Three decades of driver assistance systems: review and future perspectives. IEEE Intell. Transp. Syst. Mag. **6**(4), 6–22 (2014). https://doi.org/10.1109/MITS.2014. 2336271
2. Bevly, D., et al.: Lane change and merge maneuvers for connected and automated vehicles: a survey. IEEE Trans. Intell. Veh. **1**(1), 105–120 (2016). https://doi.org/10.1109/TIV.2015. 2503342
3. Chen, L., Englund, C.: Cooperative intersection management: a survey. IEEE Trans. Intell. Transp. Syst. **17**(2), 570–586 (2016). https://doi.org/10.1109/TITS.2015.2471812
4. Chtourou, A., Merdrignac, P., Shagdar, O.: Context-aware content selection and message generation for collective perception services. Electronics **10**(20), 2509 (2021). https://doi.org/10. 3390/electronics10202509
5. Correa, A., et al.: Infrastructure support for cooperative maneuvers in connected and automated driving. In: IEEE Intelligent Vehicles Symposium (IV), pp. 20–25 (2019). https://doi.org/10. 1109/IVS.2019.8814044
6. Delooz, Q., Festag, A.: Network load adaptation for collective perception in V2X communications. In: 2019 IEEE International Conference on Connected Vehicles and Expo (ICCVE) (2019). https://doi.org/10.1109/ICCVE45908.2019.8964988
7. Delooz, Q., Festag, A., Vinel, A.: Revisiting message generation strategies for collective perception in connected and automated driving. In: VEHICULAR 2020 (2020)
8. Delooz, Q., Festag, A., Vinel, A.: Congestion aware objects filtering for collective perception. Electron. Commun. EASST **80** (2021)
9. Delooz, Q., Riebl, R., Festag, A., Vinel, A.: Design and performance of congestion-aware collective perception. In: 2020 IEEE Vehicular Networking Conference (VNC), pp. 1–8 (2020). https://doi.org/10.1109/VNC51378.2020.9318335
10. Delooz, Q., et al.: Analysis and evaluation of information redundancy mitigation for V2X collective perception. IEEE Access **10**, 47076–47093 (2022). https://doi.org/10.1109/ACCESS. 2022.3170029
11. Dey, K.C., et al.: A review of communication, driver characteristics, and controls aspects of cooperative adaptive cruise control (CACC). IEEE Trans. Intell. Transp. Syst. **17**(2), 491–509 (2016). https://doi.org/10.1109/TITS.2015.2483063

12. ETSI: Intelligent Transport Systems (ITS); Radiocommunications equipment operating in the 5 855 MHz to 5 925 MHz frequency band; Harmonised Standard covering the essential requirements of article 3.2 of Directive 2014/53/EU (2017). ETSI EN 302 571 V2.1.1

13. ETSI: Intelligent Transport Systems (ITS); Decentralized Congestion Control Mechanisms for Intelligent Transport Systems operating in the 5 GHz range; Access layer part (2018). ETSI TS 102 687 V1.2.1

14. ETSI: Intelligent Transport Systems (ITS); Vehicular Communications; Basic Set of Applications; Analysis of the Collective Perception Service (CPS); Release 2 (2019). ETSI TR 103 562 V2.1.1

15. ETSI: Intelligent Transport Systems (ITS); Vehicular Communications; Basic Set of Applications; Part 2: Specification of Cooperative Awareness Basic Service (2019). ETSI EN 302 637-2 V1.4.1

16. ETSI: Intelligent Transport System (ITS); Vehicular Communications; Basic Set of Applications; Specification of the Collective Perception Service (2022). ETSI TS 103 324 V0.0.45 (Draft)

17. ETSI: Intelligent Transport Systems (ITS); Vehicular Communication; Informative Report for the Maneuver Coordination Service (2022). ETSI TR 103 578 V0.0.8 (Draft)

18. Garlichs, K., Günther, H., Wolf, L.C.: Generation rules for the collective perception service. In: 2019 IEEE Vehicular Networking Conference (VNC), pp. 1–8 (2019). https://doi.org/10.1109/VNC48660.2019.9062827

19. Günther, H., Riebl, R., Wolf, L., Facchi, C.: Collective perception and decentralized congestion control in vehicular ad-hoc networks. In: IEEE VNC (2016). https://doi.org/10.1109/VNC.2016.7835931

20. Günther, H., Trauer, O., Wolf, L.: The potential of collective perception in vehicular ad-hoc networks. In: 2015 14th International Conference on ITS Telecommunications (ITST), pp. 1–5 (2015). https://doi.org/10.1109/ITST.2015.7377190

21. Günther, H.J., Riebl, R., Wolf, L., Facchi, C.: The effect of decentralized congestion control on collective perception in dense traffic scenarios. Elsevier Comput. Commun. **122** (2018). https://doi.org/10.1016/j.comcom.2018.03.009

22. Häfner, B., et al.: CVIP: a protocol for complex interactions among connected vehicles. In: IEEE Intelligent Vehicles Symposium, pp. 510–515 (2020). https://doi.org/10.1109/IV47402.2020.9304556

23. Heß, D., et al.: Fast maneuver planning for cooperative automated vehicles. In: IEEE International Conference on Intelligent Transportation Systems (ITSC), pp. 1625–1632 (2018). https://doi.org/10.1109/ITSC.2018.8569791

24. Hobert, L., Festag, A., Llatser, I., Altomare, L., Visintainer, F., Kovacs, A.: Enhancements of V2X communication in support of cooperative autonomous driving. IEEE Commun. Mag. **53**(12), 64–70 (2015). https://doi.org/10.1109/MCOM.2015.7355568

25. Huang, H., et al.: Data redundancy mitigation in V2X based collective perceptions. IEEE Access **8**, 13405–13418 (2020). https://doi.org/10.1109/ACCESS.2020.2965552

26. Jiang, D., Chen, Q., Delgrossi, L.: Optimal data rate selection for vehicle safety communications. In: Proceedings of the fifth ACM international workshop on VehiculAr InterNETworking, pp. 30–38 (2008)

27. Lehmann, B., Günther, H., Wolf, L.: A generic approach towards maneuver coordination for automated vehicles. In: IEEE International Conference on Intelligent Transportation Systems (ITSC), pp. 3333–3339 (2018). https://doi.org/10.1109/ITSC.2018.8569442

28. Lehmann, B., Wolf, L.: Safety analysis of a maneuver coordination protocol. In: IEEE Vehicular Networking Conference (VNC), p. 8 (2020). https://doi.org/10.1109/VNC51378.2020.9318359

29. Llatser, I., Michalke, T., Dolgov, M., Wildschütte, F., Fuchs, H.: Cooperative automated driving use cases for 5G V2X communication. In: IEEE 2nd 5G World Forum, pp. 120–125 (2019). https://doi.org/10.1109/5GWF.2019.8911628

30. Lobo, S., Neumeier, S., Fernandez, E., Facchi, C.: InTAS – the Ingolstadt traffic scenario for SUMO. In: SUMO User Conference. DLR, Hamburg, Germany (2020). https://www.eclipse.org/sumo/2020, Extended version in ArXiv:abs/2011.11995, GitHub repository: https://github.com/silaslobo/InTAS, retrieved Jun 2, 2022

31. Lopez, P.A., et al.: Microscopic traffic simulation using SUMO. In: 2018 21st International Conference on Intelligent Transportation Systems (ITSC), pp. 2575–2582. IEEE (2018). https://doi.org/10.1109/ITSC.2018.8569938

32. Maksimovski, D., Facchi, C., Festag, A.: Priority Maneuver (PriMa) coordination for connected and automated vehicles. In: 2021 IEEE International Intelligent Transportation Systems Conference (ITSC), pp. 1083–1089 (2021). https://doi.org/10.1109/ITSC48978.2021.9564923

33. Maksimovski, D., Facchi, C., Festag, A.: Cooperative driving: research on generic decentralized maneuver coordination for connected and automated vehicles. In: Smart Cities, Green Technologies, and Intelligent Transport Systems. VEHITS SMARTGREENS 2021. Communications in Computer and Information Science, vol 1612. Springer, Cham. (2022). https://doi.org/10.1007/978-3-031-17098-0_18

34. Maksimovski, D., Festag, A., Facchi, C.: A survey on decentralized cooperative maneuver coordination for connected and automated vehicles. In: Proceedings of the 7th International Conference on Vehicle Technology and Intelligent Transport Systems - Volume 1: VEHITS, pp. 100–111. INSTICC, SciTePress (2021). https://doi.org/10.5220/0010442501000111

35. Marjovi, A., Vasic, M., Lemaitre, J., Martinoli, A.: Distributed graph-based convoy control for networked intelligent vehicles. In: 2015 IEEE Intelligent Vehicles Symposium (IV), pp. 138–143 (2015). https://doi.org/10.1109/IVS.2015.7225676

36. Mizutani, M., Tsukada, M., Esaki, H.: Automcm: maneuver coordination service with abstracted functions for autonomous driving. In: 2021 IEEE International Intelligent Transportation Systems Conference (ITSC), pp. 1069–1076 (2021). https://doi.org/10.1109/ITSC48978.2021.9564556

37. Molina-Masegosa, R., Gozalvez, J.: LTE-V for sidelink 5G V2X vehicular communications: A new 5G technology for short-range vehicle-to-everything communications. In: IEEE Vehicular Technology Magazine, pp. 30–39 (2017). https://doi.org/10.1109/MVT.2017.2752798

38. Nichting, M., Heß, D., Schindler, J., Hesse, T., Köster, F.: Space time reservation procedure (STRP) for V2X-based maneuver coordination of cooperative automated vehicles in diverse conflict scenarios. In: 2020 IEEE Intelligent Vehicles Symposium (IV), pp. 502–509 (2020). https://doi.org/10.1109/IV47402.2020.9304769

39. Rauch, A., Klanner, F., Rasshofer, R., Dietmayer, K.: Car2X-based perception in a high-level fusion architecture for cooperative perception systems. In: 2012 IEEE Intelligent Vehicles Symposium, pp. 270–275 (2012). https://doi.org/10.1109/IVS.2012.6232130

40. Renzler, T., Stolz, M., Watzenig, D.: Decentralized dynamic platooning architecture with V2V communication tested in Omnet++. In: 2019 IEEE International Conference on Connected Vehicles and Expo (ICCVE), pp. 1–6 (2019). https://doi.org/10.1109/ICCVE45908.2019.8965224

41. Renzler, T., Stolz, M., Watzenig, D.: Looking into the path future: extending CAMs for cooperative event handling. In: 2020 IEEE 92nd Vehicular Technology Conference (VTC2020-Fall), pp. 1–5 (2020). https://doi.org/10.1109/VTC2020-Fall49728.2020.9348776

42. Riebl, R., Günther, H., Facchi, C., Wolf, L.: Artery: Extending Veins for VANET applications. In: MT-ITS 2015, pp. 450–456 (2015). https://doi.org/10.1109/MTITS.2015.7223293.

43. Schindler, J., Coll-Perales, B., Zhang, X., Rondinone, M., Thandavarayan, G.: Infrastructure-supported cooperative automated driving in transition areas. In: IEEE Vehicular Networking Conference (VNC), p. 8 (2020). https://doi.org/10.1109/VNC51378.2020.9318392

44. Sjöberg, K., Andres, P., Buburuzan, T., Brakemeier, A.: Cooperative intelligent transport systems in Europe: current deployment status and outlook. IEEE Veh. Technol. Mag. **12**(2), 89–97 (2017). https://doi.org/10.1109/MVT.2017.2670018

45. Soto, I., Amador, O., Urueña, M., Calderon, M.: Strengths and weaknesses of the ETSI adaptive DCC algorithm: a proposal for improvement. IEEE Commun. Lett. **23**(5), 802–805 (2019). https://doi.org/10.1109/LCOMM.2019.2906178

46. Thandavarayan, G., Sepulcre, M., Gozalvez, J.: Analysis of message generation rules for collective perception in connected and automated driving. In: 2019 IEEE Intelligent Vehicles Symposium (IV), pp. 134–139 (2019). https://doi.org/10.1109/IVS.2019.8813806

47. Thandavarayan, G., Sepulcre, M., Gozalvez, J.: Redundancy mitigation in cooperative perception for connected and automated vehicles. In: 2020 IEEE 91st Vehicular Technology Conference (VTC2020-Spring), pp. 1–5. IEEE (2020). https://doi.org/10.1109/VTC2020-Spring48590.2020.9129445

48. Varga, A.: Omnet++. In: Modeling and Tools for Network Simulation, pp. 35–59. Springer (2010)

49. Volk, G., et al.: Towards realistic evaluation of collective perception for connected and automated driving. In: 2021 IEEE International Intelligent Transportation Systems Conference (ITSC), pp. 1049–1056. IEEE (2021). https://doi.org/10.1109/ITSC48978.2021.9564783

50. Vukadinovic, V., et al.: 3GPP C-V2X and IEEE 802.11p for vehicle-to-vehicle communications in highway platooning scenarios. Elsevier Ad Hoc Netw. **74**, 17 – 29 (2018). https://doi.org/10.1016/j.adhoc.2018.03.004

51. Willecke, A., Garlichs, K., Schulze, F., Wolf, L.C.: Vulnerable road users are important as well: Persons in the collective perception service. In: 2021 IEEE Vehicular Networking Conference (VNC), pp. 24–31. IEEE (2021). https://doi.org/10.1109/VNC52810.2021.9644669

52. Xu, W., Willecke, A., Wegner, M., Wolf, L., Kapitza, R.: Autonomous maneuver coordination via vehicular communication. In: IEEE/IFIP International Conference on Dependable Systems and Networks Workshops (DSN-W), pp. 70–77 (2019). https://doi.org/10.1109/DSN-W.2019.00022

Motion Planning

Interaction-Aware Motion Planning
as a Game

Christoph Burger, Shengchao Yan, Wolfram Burgard, and Christoph Stiller

Abstract Motion planning for automated vehicles (AVs) in mixed traffic, where AVs share the road with human-driven vehicles, is a challenging task. To reduce the complexity, state-of-the-art planning approaches often assume that the future motion of surrounding vehicles can be predicted independently of the AV's plan. This separation can lead to suboptimal, overly conservative behavior especially in highly interactive traffic situations. In this work, we introduce a motion planning algorithm to generate interaction-aware behavior for highly interactive scenarios. The presented algorithm is based upon a reformulation of a bi-level optimization problem, which frames interactions between a human driver and a AV as a Stackelberg game. In contrast to existing works, the algorithm can account for general nonlinear state and input constraints. Further, we introduce mechanisms to integrate cooperation and courtesy into motion planning to prevent overly aggressive driving behavior.

1 Introduction

When automated vehicles (AV)s first enter traffic, they will not drive in isolation but share the road with predominantly human drivers. Thus, interacting with them is crucial for smooth and efficient operation. This is especially important in interactive situations where the actions of multiple vehicles are tightly coupled. For instance,

C. Burger (✉) · C. Stiller
Karlsruhe Institute of Technology (KIT), Institute of Measurement and Control Systems (MRT), Karlsruhe, Germany
e-mail: christoph.burger@kit.edu

C. Stiller
e-mail: stiller@kit.edu

S. Yan
University of Freiburg, Autonomous Intelligent Systems,Breisgau, Germany
e-mail: yan@informatik.uni-freiburg.de

W. Burgard
Department of Engineering, University of Technology Nuremberg,Nuremberg, Germany
e-mail: wolfram.burgard@utn.de

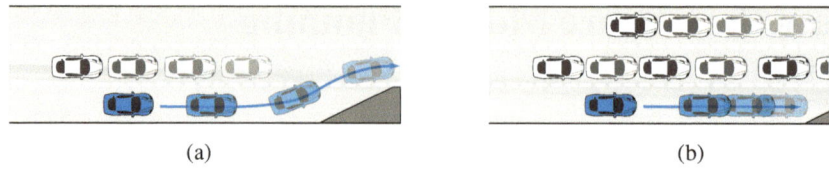

(a)   (b)

Fig. 1 Illustrated are the results of a planner following a predict-then-plan structure for a merge scenario in (**a**) low traffic and (**b**) high traffic. While in low traffic, separating prediction and planning is a useful simplification, in high traffic it can lead to suboptimal, overly conservative driving behavior of the AV

a driver on a highway might decide to slow down so that another driver can merge, or a driver might start to nudge into the adjacent lane, hoping that the driver behind will slow down and open a gap.

A key aspect to master such scenarios with AVs is to consider interactions with human drivers. However, to reduce the computational complexity of motion planning, most state-of-the-art planners follow a structure that overlooks these mechanisms. In particular, they follow a *predict-then-plan* scheme, where the motion planning is separated into a prediction step, where the future motion of surrounding drivers is predicted, and a subsequent planning step, where the motion of the AV is determined. During the planning, surrounding vehicles are treated as moving objects with an immutable trajectory.

While this separation poses a useful simplification for many traffic scenarios, it can lead to situations similar to the frozen robot problem [34], a state in which the predictions of other traffic participants block all paths, and thus the planner is not able to find a solution to its goal anymore. Fig. 1 illustrates this issue for a merge scenario. Following a predict-then-plan structure, the AV, in blue, first predicts the future motion of surrounding vehicles and plans its trajectory in a subsequent step. In Fig. 1(a) the result in low traffic is shown. Following the same principle in high traffic, shown in Fig. 1(b), the planner is unable to find a collision-free trajectory onto the highway and the AV stops at the end of the lane.

Some approaches are already able to overcome the structural limitation of planners following a predict-then-plan scheme by solving the prediction and planning task simultaneously. These planners can be categorized into the following three classes: Forward simulation methods, multi-agent methods, and game-theoretic methods.

Forward Simulation Methods: One technique to generate interaction-aware behavior is via forward simulation. Here, the current traffic scene is simulated for different actions of the AV. Transition models are used to describe how the environment changes due to the actions of the AV and further how other drivers react to these changes in the environment. We refer to such techniques as forward simulation methods. Most sampling-based planning methods that consider interactions can be associated with this category [11, 27]. An important group among the sampling-based planning methods are methods based on partially observable Markov decision process (POMDP), e.g. as presented in [17].

The behavior of other agents is often modeled with specific driver models such as the Intelligent Driver Model (IDM) [35] or the Minimum Overall Braking Induced by Lane change (MOBIL) model [18]. An example where the IDM is used to determine the reaction of other vehicles is presented in [11]. Here, to generate the behavior for the AV, multiple candidate trajectories are simulated and evaluated based on the effect imposed on others. In forward simulation methods the influence the AV exerts on others is not explicitly given, but must be determined by trying out several actions and subsequent forward simulations of the traffic scene.

Multi-Agent Methods: In multi-agent methods, the separate prediction of other vehicles is replaced by planning coupled trajectories. Therefore, the traffic scene is modeled as a multi-agent planning problem with the underlying assumption that all traffic participants behave towards optimizing a joint objective [2, 4, 5, 8, 19, 22, 32]. The AV then solves the multi-agent problem assuming that other agents will also roughly follow their part of the plan. Varying weights can be used to model different levels of cooperation or incorporate asymmetries in the traffic scene [4, 7]. To cope with uncertainties in the behavior of humans, these methods are combined with tracking approaches to estimate if humans roughly follow the same model [5, 32].

Game-Theoretic Methods: In real traffic, the assumption that each driver is behaving towards optimizing a common objective might not be valid, since some drivers are only interested in optimizing their own driving. To model interactions among agents with different objectives, a game-theoretic perspective might be more suitable. Several game-theoretic methods have already successfully been used, e.g., for lane change, merge, intersection, round-about, and overtaking scenarios [6, 10, 12, 13, 21, 29–31, 38, 39]. E.g., in [30], human-like driving behavior, e.g., slowing down before intersections or nudging into the adjacent lane while doing a lane change, could be generated.

Apart from these driving applications, game-theoretic methods have been used for agile maneuvering of multiple ground vehicles in close proximity [40], and automated car racing [23, 25, 37, 39], where it is shown that game-theoretic planners yield complex strategies such as blocking and faking and significantly outperform baseline MPC planners.

In game-theoretic formulations, there is no optimal solution in the traditional sense, but depending on the game's structure, different solutions are possible, also referred to as equilibria. Therefore, an important feature to categorize game-theoretic methods is the type of solution they are solving for. In literature, it is distinguished between Nash and Stackelberg equilibria. A Nash equilibrium describes a set of strategies where no individual agent can benefit from unilaterally changing its strategy, given that all other agents will stick to their strategy. This type of equilibrium has been investigated, e.g., in [3, 10, 14, 20, 37, 40].

Compared to a Nash equilibrium, a Stackelberg equilibrium involves turn taking and, therefore, an asymmetry in the decision-making process. It is typically modeled for a two-player game, where one player is the leader, and the other is the follower. The leader chooses its strategy first, and the follower then optimizes its strategy as the best response to the leader's strategy. In contrast, the Nash solution can be seen as the best response from everyone to everyone else without hierarchical turn-taking. Stackelberg equilibria are considered in [6, 12, 23, 29, 30, 33, 41, 42]

In this work, we present a model based on a game-theoretic formulation that directly captures interactions between a AV and a human driver (HD) as a Stackelberg game. This algorithm enables AVs to be aware of how their actions influence other drivers and thereby allows generating interaction-aware driving behavior. In contrast to existing works, the algorithm can account for general nonlinear state and input constraints. Additionally, we present mechanisms to integrate cooperation and courtesy into interaction-aware methods to prevent overly aggressive driving behavior, which has been reported as an issue of existing approaches.

2　Problem Statement

To derive a model to directly capture interactions, we consider a system with one AV, referred to as the leader L, and one HD, referred to as the follower F. The system's state at time t is given by the leader's and follower's state $\mathbf{x}_t^L, \mathbf{x}_t^F \in \mathcal{X}$, where $\mathcal{X}$ is the set that contains all possible states. The leader's and follower's actions are described by their trajectories $\xi_L(t), \xi_F(t) : [0, \mathcal{T}] \to \mathcal{X}$. Further, each agent has its individual objective function denoted by J_L and J_F.

The objective is minimized subject to the vehicle's initial state $\xi(0) = \mathbf{x}_0$ and the evolution of the state described by the trajectory, which is only allowed to pass through the set of feasible states $\mathcal{X}_{\text{feasible}}(t) \subseteq \mathcal{X}$. $\mathcal{X}_{\text{feasible}}(t)$ encodes, for instance, collision avoidance. Additionally, system dynamics and bound constraints can be enforced by $D(\xi(t), \dot{\xi}(t), \ddot{\xi}(t), \dots) = 0$. The set of all feasible trajectories $\xi(t)$ is denoted by Ξ.

In contrast to traditional multi-agent systems, we assume a turn-taking structure, where the follower optimizes its trajectory as a response to the leader's trajectory. To do so, the follower predicts the leader's future motion $\tilde{\xi}_L$ and then plans by minimizing its objective function J_F considering these predictions. Therefore, the follower's optimal trajectory can be described as:

$$\underset{\xi_F \in \Xi_F}{\arg \min} \, J_F(\mathbf{x}_0^L, \mathbf{x}_0^F, \tilde{\xi}_L, \xi_F) \tag{1}$$

For simplicity, we assume that for short time horizons, a human can predict the trajectory of the AV sufficiently well, such that the prediction $\tilde{\xi}_L$ can be assumed to be the actual trajectory ξ_L of the AV. Hence, the optimal trajectory of the follower as a function of the leader's trajectory ξ_L is given as:

$$\xi_F^*(\mathbf{x}_0^L, \mathbf{x}_0^F, \xi_L) = \underset{\xi_F \in \Xi_F}{\arg \min} \, J_F(\mathbf{x}_0^L, \mathbf{x}_0^F, \xi_L, \xi_F) \tag{2}$$

With this link between the leader's actions and the follower's actions the optimal trajectory for the AV can be stated as:

$$\xi_L^* = \underset{\xi_L \in \Xi_L}{\arg\min} \, J_L \left(\mathbf{x}_0^L, \mathbf{x}_0^F, \xi_L, \xi_F^*(\mathbf{x}_0^L, \mathbf{x}_0^F, \xi_L) \right) \tag{3}$$

Equation (3) gives the leader the ability to reason about how its actions will influence the follower's response and is therefore the fundamental model which enables interaction-aware planning.

3 Bi-level Formulation

The derived model, in (3), describes a Stackelberg game, where the leader decides on its behavior first and the follower optimizes its behavior given the decision of the leader (Fig. 2). If the follower's best response to the leader's actions can be stated in closed form, (3) can be solved as a standard optimal control problem (OCP). However, this is, in general, not the case since ξ_F^* is the outcome of an OCP itself. This results in a nested or bi-level optimization problem. Further, solving the underlying Stackelberg game would require planning until $\mathcal{T}$, which is the end of an interaction. However, the end of an interaction is not trivial to determine and requires the consideration of a varying time horizon.

In the following, we propose an approximate solution to (3) based on model predictive control (MPC), where we solve the problem on a receding horizon with a fixed length T, execute the first action and then replan. We utilize multiple shooting methods and discretize the time horizon $t \in [0, T]$ into $N = T/\tau$ intervals, where τ denotes the duration of one time step. To improve readability, we subsume the state and input sequences of the leader and follower as $\mathbf{x} := [\mathbf{x}_0, \ldots, \mathbf{x}_N]^T$ and $\mathbf{u} := [\mathbf{u}_0, \ldots, \mathbf{u}_{N-1}]^T$. In the following, the resulting nonlinear programs (NLP)s of the follower and leader are stated. The equality constraints $\mathbf{h}$ can be used to represent constraints imposed by the system dynamics while the inequality constraints $\mathbf{g}$ collect bound constraints, collision constraints, and dynamic constraints.

3.1 NLP of the Follower

The follower's NLP is parametrized by the leader's states and inputs $(\mathbf{x}^L, \mathbf{u}^L)$ and can be formulated as:

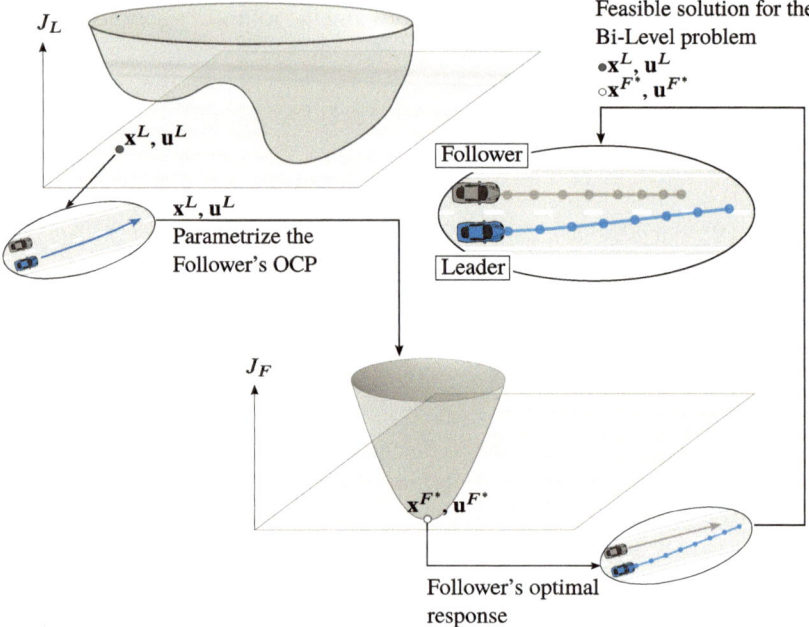

Fig. 2 Structure of a bi-level optimization problem. Here, the follower optimizes its objective function as a response to the given actions of the leader

$$\underset{\mathbf{x}^F, \mathbf{u}^F}{\arg\min} \quad J_F(\mathbf{x}^L, \mathbf{x}^F, \mathbf{u}^F) \tag{4a}$$

$$\text{s.t.} \quad \mathbf{h}_F(\mathbf{x}^F, \mathbf{u}^F) = 0, \tag{4b}$$

$$\mathbf{g}_F(\mathbf{x}^L, \mathbf{x}^F, \mathbf{u}^F) \leq 0 \tag{4c}$$

3.2 NLP of the Leader

The leader's bi-level optimization problem can be stated as:

$$\underset{\mathbf{x}^L, \mathbf{x}^F, \mathbf{u}^L, \mathbf{u}^F}{\arg\min} \quad J_L(\mathbf{x}^L, \mathbf{x}^F, \mathbf{u}^L) \tag{5a}$$

$$\text{s.t.} \quad \mathbf{h}_L(\mathbf{x}^L, \mathbf{u}^L) = 0, \tag{5b}$$

$$\mathbf{g}_L(\mathbf{x}^L, \mathbf{x}^F, \mathbf{u}^L) \leq 0, \tag{5c}$$

$$(\mathbf{x}^F, \mathbf{u}^F) \in \underset{\mathbf{x}^F, \mathbf{u}^F}{\arg\min}\{J_F(\mathbf{x}^L, \mathbf{x}^F, \mathbf{u}^F) : \mathbf{h}_F = 0, \mathbf{g}_F \leq 0\} \tag{5d}$$

Formulating the follower's optimization problem as a constraint, (5d), ensures that only optimal solutions for the follower are considered feasible solutions for the leader.

4 Single-Level Representation

To efficiently solve (5), we need to reformulate the bi-level optimization problem into a regular, single-level problem. Therefore, we assume that the follower will act optimally with respect to its own objective function (4). With this assumption, we can replace the inner optimization problem with its necessary conditions for optimality.

If the follower's problem is convex, the Karush Kuhn Tucker (KKT) conditions are necessary and sufficient for optimality. However, due to the combinatorial nature of driving it is, in general, non-convex, e.g., due to non-linear collision avoidance constraints or a non-convex cost function. To obtain locally optimal solutions, we convexity the follower's problem around an initial guess, which at the same time encodes the considered homotopy class. For the convexification, the constraints are linearized, and the cost function is approximated by a 2. order Tailer expansion.

By replacing the follower's optimization problem with its KKT conditions in (5), we obtain the following single-level optimization problem:

$$\underset{\mathbf{x}^L,\mathbf{x}^F,\mathbf{u}^L,\mathbf{u}^F,\boldsymbol{\lambda},\boldsymbol{\mu}}{\arg\min} \quad J_L(\mathbf{x}^L, \mathbf{x}^F, \mathbf{u}^L) \tag{6a}$$

$$\text{s.t.} \quad \mathbf{h}_L(\mathbf{x}^L, \mathbf{u}^L) = 0, \tag{6b}$$

$$\mathbf{g}_L(\mathbf{x}^L, \mathbf{x}^F, \mathbf{u}^L) \le 0, \tag{6c}$$

$$\nabla_{(\mathbf{x}^F,\mathbf{u}^F)} L(\mathbf{x}^L, \mathbf{x}^F, \mathbf{u}^F, \boldsymbol{\lambda}, \boldsymbol{\mu}) = 0, \tag{6d}$$

$$\mathbf{h}_{F_{\text{lin}}}(\mathbf{x}^F, \mathbf{u}^F) = 0, \tag{6e}$$

$$\mathbf{g}_{F_{\text{lin}}}(\mathbf{x}^L, \mathbf{x}^F, \mathbf{u}^F) \le 0, \tag{6f}$$

$$\boldsymbol{\mu} \ge 0, \tag{6g}$$

$$\boldsymbol{\mu} \perp \mathbf{g}_{F_{\text{lin}}}(\mathbf{x}^L, \mathbf{x}^F, \mathbf{u}^F) \tag{6h}$$

with the Lagrangian

$$L(\mathbf{x}^L, \mathbf{x}^F, \mathbf{u}^F, \boldsymbol{\lambda}, \boldsymbol{\mu}) = J_{F_{\text{con}}}(\mathbf{x}^L, \mathbf{x}^F, \mathbf{u}^F)$$
$$+ \boldsymbol{\lambda}^T \mathbf{h}_{F_{\text{lin}}}(\mathbf{x}^F, \mathbf{u}^F) + \boldsymbol{\mu}^T \mathbf{g}_{F_{\text{lin}}}(\mathbf{x}^L, \mathbf{x}^F, \mathbf{u}^F)$$

Here, λ and μ are the KKT multipliers and $\mathbf{h}_{F_{\text{lin}}}$, $\mathbf{g}_{F_{\text{lin}}}$ and $J_{F_{\text{con}}}$ are the constraints and objective after the convexification. For the reformulation we assume sufficient regularity of the follower's NLP, differentiability of $\mathbf{h}_F$ and $\mathbf{g}_F$, and the cost function J_F to be twice differentiable.

4.1 Solving the Complementarity Constraints

The leader's NLP in (6) forms an instance of a mathematical program with complementarity constraints (MPCC). Due to the complementarity constraints $\mu \perp \mathbf{g}_{F_{\text{lin}}}$, MPCCs are non-smooth and non-convex. MPCC are particularly challenging to solve because at every feasible point, ordinary constraint qualifiers (CQ) such as LICQ or Mangasarian-Fromovitz CQ are violated [9]. Therefore, to solve (6), we reformulate the complementarity constraints using relaxation methods [16] as shown in (7).

$$ -\epsilon \le \mu^T \mathbf{g}_F. \tag{7} $$

With $\epsilon > 0$ a regularized NLP is obtained, and CQ can be satisfied again. The smaller ϵ is chosen, the closer any feasible solution is to achieving complementarity. However, if ϵ is chosen too small, the problem may be numerically unstable and the solver will fail to find a feasible solution at all.

5 Application to Motion Planning for AVs

So far the derived model represents a formulation of how interactions between a robot and a human can be considered during motion planning or decision-making for robots in general. In the following, we present a modeling to apply the bi-level algorithm to motion planning for AVs. The section starts by stating the OCP used for trajectory optimization of an AV. This OCP contains the system dynamics, bound constraints, as well as an objective function to encode desirable driving behavior. For the purpose of the evaluation in Sect. 6, we assume that a good approximation of a human objective function is provided. Such a function could be obtained, e.g., via inverse reinforcement learning.

5.1 Trajectory Optimization for AVs

The OCP used for trajectory optimization can be stated as:

$$\arg\min_{\mathbf{x},\mathbf{u}} J_{\text{base}} = J_{\mathbf{x}} + J_{\mathbf{u}} + J_{\dot{\mathbf{u}}} \tag{8a}$$

s.t. :

$$\mathbf{x}_{k+1} = \mathbf{f}(\mathbf{x}_k, \mathbf{u}_k) \qquad\qquad k = 0, \dots, N-1 \tag{8b}$$

$$\mathbf{x}_0 = \hat{\mathbf{x}} \tag{8c}$$

$$\underline{\mathbf{g}_{\text{dyn}}} \leq \mathbf{g}_{\text{dyn}}(\mathbf{x}_k) \leq \overline{\mathbf{g}_{\text{dyn}}} \qquad k = 1, \dots, N \tag{8d}$$

$$\mathbf{g}_{\text{col}}(\mathbf{x}_k^F, \mathbf{x}_k^L) \leq 0 \qquad\qquad k = 1, \dots, N \tag{8e}$$

$$\mathbf{g}_{\text{obs}}(\mathbf{x}_k) \leq 0 \qquad\qquad k = 1, \dots, N \tag{8f}$$

$$\underline{\mathbf{x}} \leq \mathbf{x}_k \leq \overline{\mathbf{x}} \qquad\qquad k = 1, \dots, N \tag{8g}$$

$$\underline{\mathbf{u}} \leq \mathbf{u}_k \leq \overline{\mathbf{u}} \qquad\qquad k = 0, \dots, N-1 \tag{8h}$$

The objective function J_{base} is used to generate a desirable driving behavior. The equality constraints (8b) enforces the vehicle dynamics. Further, (8c) ensures that the trajectory is planed from the current state $\hat{\mathbf{x}}$. The inequality constraints (8d)–(8h) are used for collision avoidance and to account for physical limitations of the real system.

5.1.1 Vehicle Model

To describe the dynamics of the vehicle (8b), the kinematic single-track model is used. The vehicle state at time k, $\mathbf{x}_k = [x_k, y_k, \psi_k, v_k]^T$, is described by the lateral and longitudinal position (x, y) of the vehicle's center of gravity, the orientation ψ, and the absolute velocity v. Together with the input $\mathbf{u}_k = [\delta_k, a_k]^T$ consisting of steering angle δ and acceleration a, the dynamics of a vehicle are given by (Fig. 3):

Here, β is the slip angle which is given by $\beta = \arctan\left(\frac{l_r}{l}\tan(\delta)\right)$.

Further, l is the wheelbase, and l_r is the distance between the center of gravity and the rear axis. To obtain the discrete dynamics model $\mathbf{x}_{k+1} = f(\mathbf{x}_k, \mathbf{u}_k)$ in (8b) we use a fourth-order Runge-Kutta method.

To ensure the validity of the kinematic single-track model [28], $\mathbf{g}_{\text{dyn}}$ (8d) are introduced to limit the lateral acceleration as follows:

$$|v_k \dot{\psi}_k| = \left|\frac{v_k^2}{l}\tan(\delta_k)\cos(\beta_k)\right| \leq a_{\text{lat,max}} = 4\,\frac{\text{m}}{\text{s}^2} \tag{9}$$

To also limit the jerk, the following constraints on the acceleration change are introduced:

$$j_{\min} \leq \frac{a_k - a_{k-1}}{\tau} \leq j_{\max} \tag{10}$$

Here, $j_{\min}$ and $j_{\max}$ are the minimum and maximum allowed jerk values.

5.1.2 Collision Avoidance

The collision avoidance constraints (8e) are formulated pairwise between vehicles. Hereby, the shape of one vehicle is approximated by a finite number of circles and the shape of the second vehicle is approximated with a superellipses, as illustrated in Fig. 4. Compared to regular ellipses, superellipses provide a more accurate approximation of the vehicle's rectangular shape [24].

Collision avoidance between a point $\mathbf{p} = [x, y]^T$ and a superellipse defined by the semi-major a, the semi-minor b, and order $n \in \mathbb{N}$ can be formulated as:

$$\sqrt[n]{\left(\frac{x}{a}\right)^n + \left(\frac{y}{b}\right)^n} \geq 1 \tag{11}$$

$$\dot{\mathbf{x}}_k = \begin{pmatrix} \dot{x}_k \\ \dot{y}_k \\ \dot{\psi}_k \\ \dot{v}_k \end{pmatrix} = \begin{pmatrix} v_k \cos(\psi_k + \beta_k) \\ v_k \sin(\psi_k + \beta_k) \\ \frac{v_k}{l}\tan(\delta_k)\cos(\beta_k) \\ a_k \end{pmatrix}$$

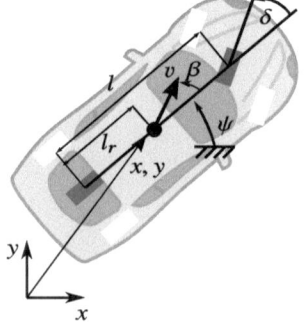

Fig. 3 Kinematic bicycle model

Similarly, collision avoidance between a circle with radius r and a superellipse can be formulated as a point mass constraint on the center point of the circle $\mathbf{p_c} = [x_c, y_c]^T$. Therefore, $\mathbf{p_c}$ needs to be outside the Minkowski sum of the superellipse and a circle with radius $\frac{r}{2}$, see Fig. 5.

To maintain an efficient formulation, the Minkowsi sum is approximated by an enlarged superellipse. In case of a superellipse of order $n = 4$, enlarging the semi-major and semi-minor by the radius r is a sufficient over approximation, see Fig. 5.

Henceforth, the collision avoidance constraints can be stated as:

$$\sqrt[n]{\left(\frac{x_c}{a+r}\right)^n + \left(\frac{y_c}{b+r}\right)^n} \geq 1 \tag{12}$$

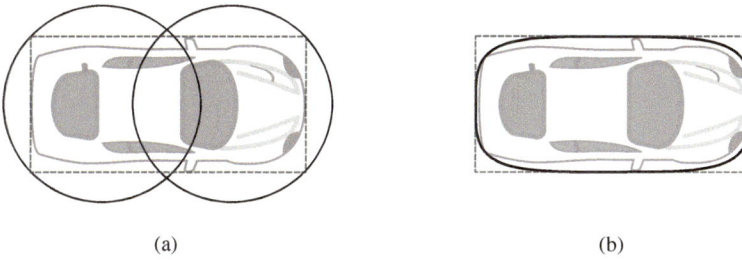

(a)   (b)

Fig. 4 Illustrated are the shape approximations by (**a**) multiple circles and (**b**) a superellipse of order $n = 4$

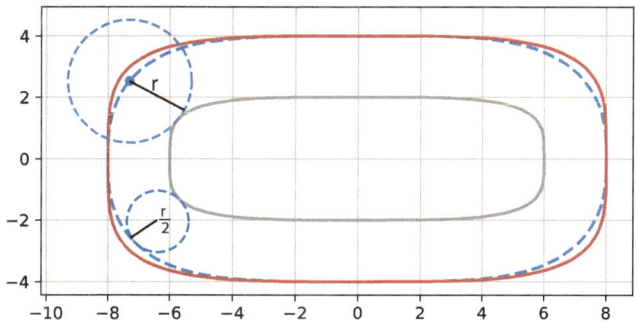

Fig. 5 Comparison of the Minkowski sum, shown in blue, and a superellipse with the semi-major and semi-minor enlarged by r, shown in red. The original superellipse is shown in grey

5.1.3 Objective Function

The objective function (8a) consists of three components, $J_{\mathbf{x}}$, $J_{\mathbf{u}}$, $J_{\dot{\mathbf{u}}}$, penalizing deviations from a desired state $\mathbf{x}_{\text{ref}} = [x_{k,\text{ref}}, y_{k,\text{ref}}, \psi_{k,\text{ref}}, v_{k,\text{ref}}]^T$, any control effort, and any changes in control, respectively. The function can be stated as:

$$J_{\text{base}}(\mathbf{x}, \mathbf{u}) = J_{\mathbf{x}} + J_{\mathbf{u}} + J_{\dot{\mathbf{u}}} \tag{13a}$$

$$= \sum_{k=1}^{N} \begin{pmatrix} x_k - x_{k,\text{ref}} \\ y_k - y_{k,\text{ref}} \\ \psi_k - \psi_{k,\text{ref}} \\ \Delta v \end{pmatrix}^T Q \begin{pmatrix} x_k - x_{k,\text{ref}} \\ y_k - y_{k,\text{ref}} \\ \psi_k - \psi_{k,\text{ref}} \\ \Delta v \end{pmatrix} \tag{13b}$$

$$+ \sum_{k=0}^{N-1} \mathbf{u}_k^T R_u \mathbf{u}_k \tag{13c}$$

$$+ \sum_{k=1}^{N-1} (\mathbf{u}_k - \mathbf{u}_{k-1})^T R_{\dot{u}} (\mathbf{u}_k - \mathbf{u}_{k-1}) \tag{13d}$$

$$+ (\mathbf{u}_0 - \hat{\mathbf{u}})^T R_{\dot{u}} (\mathbf{u}_0 - \hat{\mathbf{u}}) \tag{13e}$$

With the velocity vector $\mathbf{v} = [v \cos(\psi + \beta), v \sin(\psi + \beta)]^T$ and the road tangential unit vector $\mathbf{t}$, $\Delta v = \mathbf{v} \cdot \mathbf{t} - v_{\text{ref}}$ measures the difference between the current velocity along the road and the reference velocity v_{ref}. Further, $\hat{\mathbf{u}}$ is the control input from the previous step. Finally, R_u, $R_{\dot{u}}$ and Q are weighting matrices used to model the desired driving behavior.

6 Evaluation

The efficacy of the proposed bi-level algorithm is evaluated in two different settings. First, the ability of the AV to deliberately influence the HD's state through its driving behavior is investigated. These experiments make use of the direct link between the AV's actions and the HD's response which the bi-level approach provides.

Since in real driving applications, the goal of the AV is to drive efficiently and comfortably rather than to influence the state of other vehicles, the focus of the second part, is to demonstrate how the approach can be used to plan interaction-aware, cooperative driving behavior.

Apart from the efficacy, the algorithm's runtime is analyzed followed by a discussion highlighting the advantages and limitations of the algorithm.

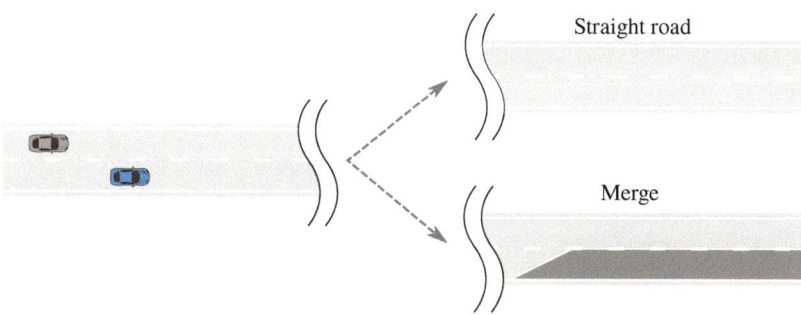

Fig. 6 Depending on the experiment, either a multi-lane or a merging scenario, where the right lane ends, is considered

Table 1 MPC parameters

Parameter	Value
N	30
T	6 s
Q	diag(0, 1, 0, 100)
R_u	diag(1, 1)
$R_{\dot{u}}$	diag(10000, 1000)
v_{min}, v_{max}	$0\,\frac{m}{s}$, $30\,\frac{m}{s}$
δ_{max}	$30°$
a_{min}, a_{max}	$-8\,\frac{m}{s^2}$, $3\,\frac{m}{s^2}$
j_{min}, j_{max}	$-10\,\frac{m}{s^3}$, $6\,\frac{m}{s^3}$
l	4 m
l_r	2 m

6.1 Base Scenario

We evaluate our approach in multi-lane scenarios as shown in Fig. 6, where the AV is depicted in blue and the HD is depicted in gray. For the purpose of these experiments, the AV is considered the leader, and the HD is considered the follower. In the following, we will use the terms leader and AV as well as follower and HD interchangeably.

Both vehicles have a width of 2.0 m and a length of 4.0 m. Collision avoidance is implemented using a superellipse of order $n = 4$ for the leader and two circles for the follower. Further parameters are given in Table 1.

The follower directly uses the cost function (13) for its trajectory optimization with the weights and vehicle characteristics given in Table 1. The leader's NLP is also based on (13) but additionally considers the KKT conditions of the follower's NLP as constraints, as stated in (6). Further, the leader's objective function is augmented with additional cost terms to set scenario-specific incentives.

Table 2 Leader's and follower's initial and reference states

Parameter	Value
$\mathbf{x}_0^L$	$[12.0\,\text{m}, 3.0\,\text{m}, 0°, 10.0\,\frac{\text{m}}{\text{s}}]^T$
$\mathbf{x}_0^F$	$[2.0\,\text{m}, 5.0\,\text{m}, 0°, 10.0\,\frac{\text{m}}{\text{s}}]^T$
$\mathbf{x}_{\text{ref}}^L = \mathbf{x}_{\text{ref}}^F$	$[0.0\,\text{m}, 5.0\,\text{m}, 0°, 10.0\,\frac{\text{m}}{\text{s}}]^T$

If not stated otherwise, the initial and reference states listed in Table 2 are used for the leader and follower.

6.2 Influence the Human's State

The following two experiments investigate the leader's ability to influence the follower's state. To provide the appropriate incentives, the leader's objective function is augmented with $J_{\text{influence}}$. The leader's objective is, therefore, the following weighted sum:

$$J_L = w_L J_{\text{base}} + w_{\text{influence}} J_{\text{influence}} \tag{14}$$

Henceforth, a ratio of $\frac{w_{\text{influence}}}{w_L} = 10^7$ is used.

6.2.1 Slow Down the Human

In this experiment, the leader's goal is to slow down the follower to a certain velocity, v_{ref}^F. To incentivize this behavior, deviations of the follower's velocity to v_{ref}^F are penalized. The scenario-specific $J_{\text{influence}}$ is therefore set to

$$J_{\text{influence}} = \sum_{k=1}^{N} (\mathbf{v}_k^F \cdot \mathbf{t}_k - v_{\text{ref}}^F)^2 \tag{15}$$

with $\mathbf{v} = [v\cos(\psi + \beta), v\sin(\psi + \beta)]^T$ and $\mathbf{t}$ as the road tangential unit vector.

The results for a desired velocity of $v_{\text{ref}}^F = 5.\frac{\text{m}}{\text{s}}$ are illustrated in Fig. 7. As can be seen, the leader changes to the left lane to get in front of the follower. Despite its interest in driving fast, the leader starts to brake, forcing the follower to slow down. To prevent the follower from overtaking, the leader drives close to the center of the road.

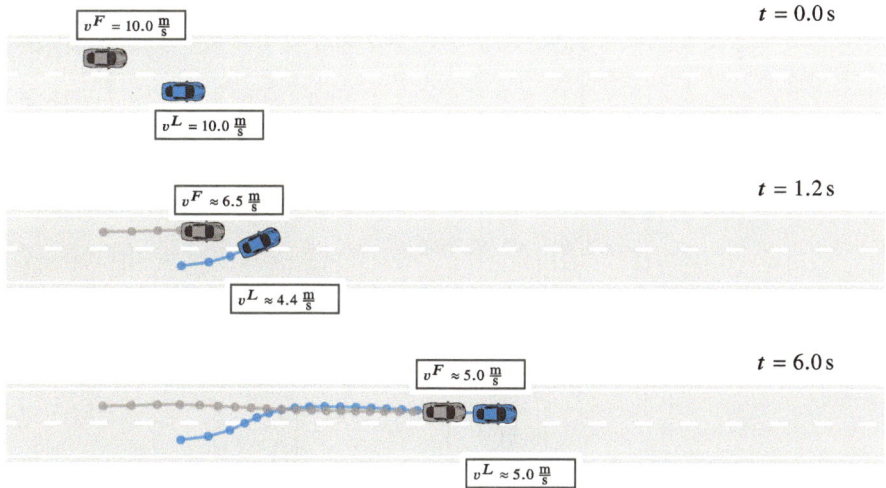

Fig. 7 By sole penalizing the follower's velocity the bi-level approach yields an intuitive solution; The leader has to change the lane and needs to brake to slow down the follower

6.2.2 Push the Human to the Adjacent Lane

In this experiment, the ability to also influence the follower in the lateral direction is investigated. Therefore, a 3-lane road is considered, see Fig. 8.

The leader's goal is to enforce a lane change of the human to the adjacent left lane. This incentive is encoded by setting $J_{\text{influence}}$ to penalize deviations of the follower's lateral position to a reference y_{ref}^F as:

$$J_{\text{influence}} = \sum_{k=1}^{N} (y_k^F - y_{\text{ref}}^F)^2 \tag{16}$$

Figure 8 shows the behavior for $y_{\text{ref}}^F = 8.5$ m, which corresponds the center of the leftmost lane. To push the follower to the left, the leader changes lanes and slows down, almost coming to a full stop. The leader thereby blocks the middle lane, which forces the follower to also slow down to avoid a collision. To continue, the follower starts an overtaking maneuver. At the same time, the leader accelerates again to stay next to the follower, blocking him from changing back to his original lane.

6.3 Interaction-Aware Trajectory Optimization

In real traffic, the primary goal of the AV is to drive comfortably and efficiently rather than to change the state of surrounding vehicles in a certain way. Therefore, the

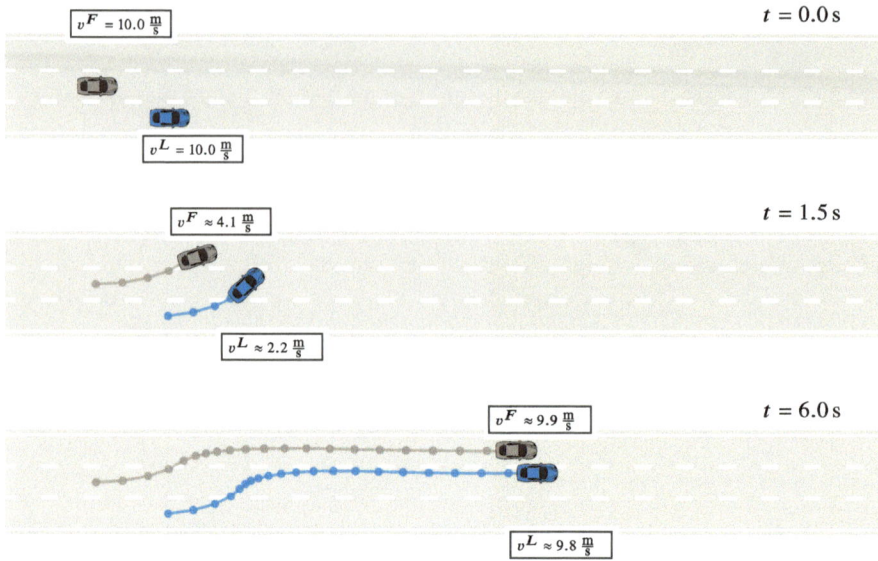

Fig. 8 The leader changes lanes and brakes harshly to enforce an overtaking maneuver of the follower. As soon as the follower tries to overtake, the leader accelerates again, blocking the follower from changing back to the middle lane

generated behavior when planning trajectories with the proposed interaction-aware algorithm in different lane change scenarios is investigated next. To better show the effect of the planned behavior, the desired velocity of the follower is increased to $v_{\text{ref}}^F = 15.\frac{m}{s}$. Throughout the scenarios, the leader aims to perform a lane change to the left.

6.3.1 Efficient Planning

We start by formulating the leader's objective in an egocentric way, similar to how it is formulated for planners following a predict-then-plan scheme. Here, the leader solely considers attributes of its own trajectory, formulated by only optimizing J_{base}.

The resulting trajectories are shown in Fig. 9. As can be seen, the leader plans a very efficient lane change without any acceleration. However, as a response, the follower has to brake harshly to avoid a collision, see Fig. 10. This aggressive cut in is a result of the leader knowing that the follower will react, which the leader then exploits to further optimize its own driving behavior.

This example shows that interactive behavior not only occurs when the leader is incentivized to alter the state of the follower but also emerges out of efficiency.

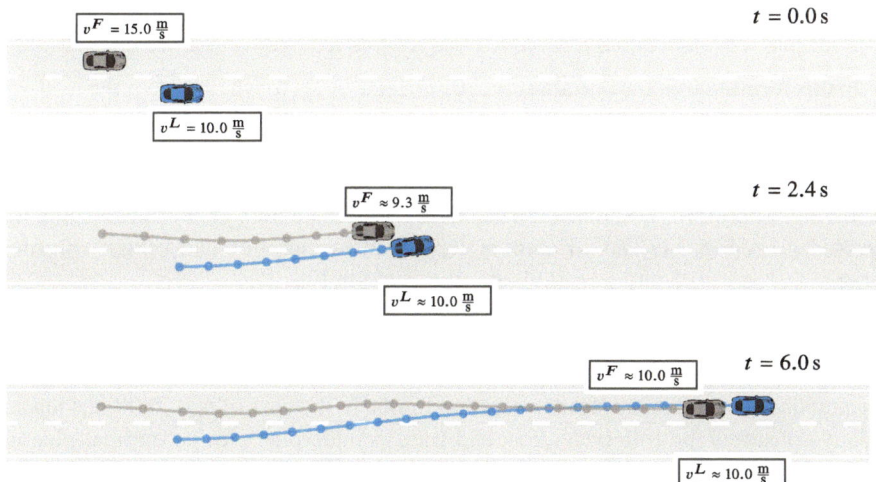

Fig. 9 When only considering its own costs, the leader performs an aggressive lane change

6.3.2   Cooperative Interaction-Aware Planning

The proposed interaction-aware model gives the leader the ability to anticipate the follower's reaction. When naively using an egocentric objective function, the leader exploits the follower's response and generates an overly aggressive behavior, as demonstrated in the previous example.

To mitigate this effect, the impact imposed on others must be considered in the objective function of the leader. Therefore, a formulation base on a cooperative cost function that includes the leader's and followers's cost in the leader's objective is considered in the following:

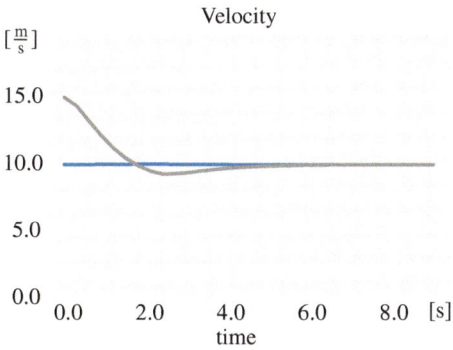

Fig. 10 While the leader can perform a smooth lane change without accelerating, the follower has to brake harshly to avoid a collision

$$J_{\text{cooperative}} = \alpha J_{F,\text{base}} + (1 - \alpha) J_{L,\text{base}} \qquad (17)$$

In this formulation, the variable $\alpha \in [0, 1]$ determines to which extent the leader's and the follower's cost are considered. Therefore, α provides a way to design different driving behaviors, ranging from overly aggressive to overly conservative.

The impact the parameter α has on the generated behavior is investigated in the following. Therefore, we consider a scenario including a mandatory lane change for the leader, see Fig. 11. The different α-dependent acceleration and velocity profiles for $\alpha = 0.0$, $\alpha = 0.5$ and $\alpha = 0.99$ are illustrated in Fig. 12.

In detail, for $\alpha = 0.0$, the leader does not accelerate, and all the discomfort has to be carried out by the follower. This represents the aggressive, egocentric behavior presented in the previous experiment. With a larger α, the leader increases its acceleration until reaching the acceleration limits. In the case of $\alpha = 0.99$, the leader mostly considers the follower's cost and tries to intervene with its optimal plan as little as possible. This value of α generates a very conservative behavior similar to a predict-then-plan approach. With $\alpha = 0.5$, the leader's and the follower's cost are

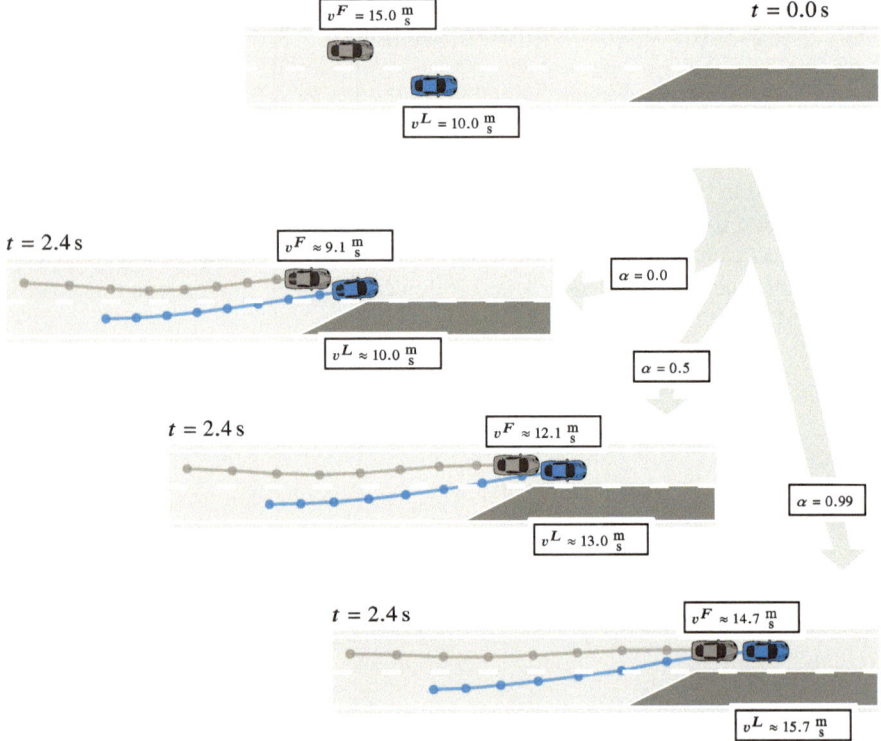

Fig. 11 Illustrated is a scenario where the leader has to perform a lane change to the left. Depending on the value of α, different behaviors are generated, ranging from overly aggressive to overly conservative

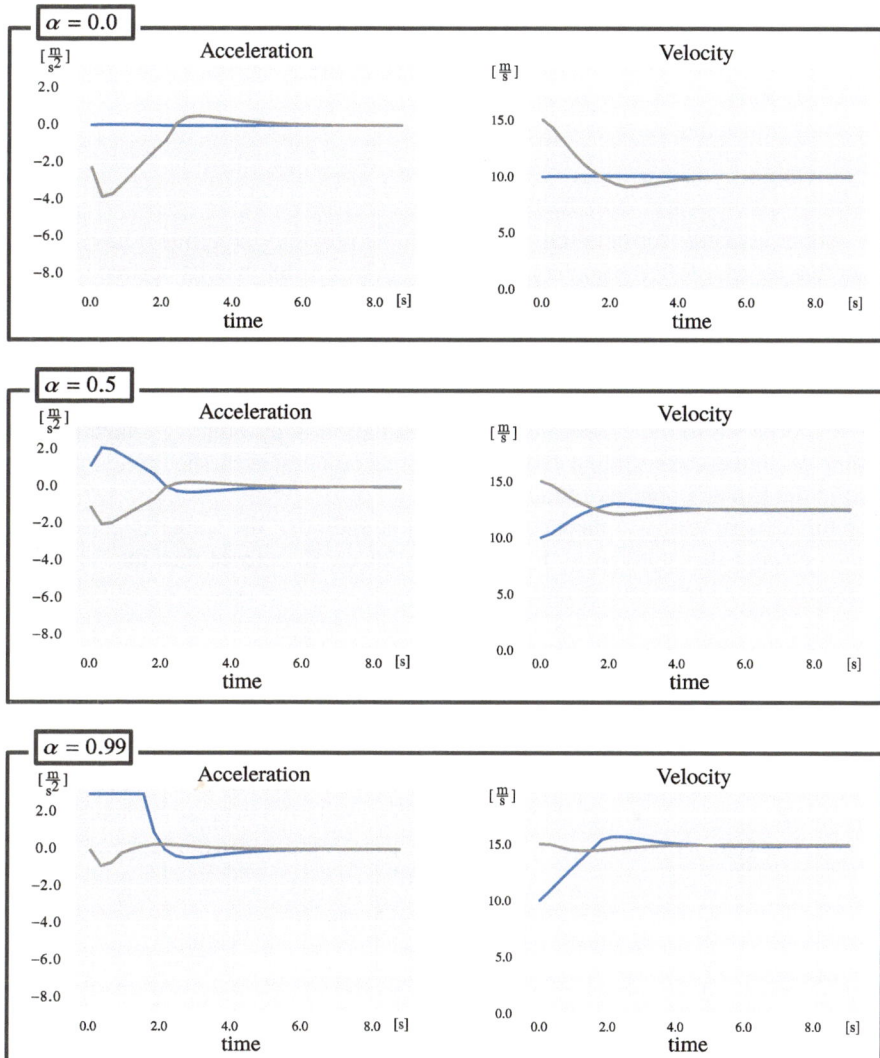

Fig. 12 Depending on the value of α different acceleration and velocity profiles are obtained. Thereby, the langer α is, the more discomfort the leader accepts. Further, with different α the vehicles approach different stationary velocities which might significantly differ from their desired velocities

considered equaly, which leads to an approximately equal distribution of discomfort. Note however, that, besides adjusting the acceleration during the lane change, the leader also adapts its stationary velocity depending on α.

6.3.3 Courtesy Constraints

The cooperative cost formulation presented in the previous experiment has the side effect that for $\alpha > 0.0$, the leader permanently drives faster than its desired velocity v_{ref}^L. For some scenarios, e.g., overtaking a slow-moving truck on the highway, a temporal increased velocity might be acceptable or even desirable for traffic efficiency. However, in most situations, a vehicle in front does not adapt its velocity to the desires of rear traffic.

An alternative to the cooperative cost formulation is introducing *courtesy* constraints. With these constraints, the leader's impact on others can be limited without altering the leader's objective function.

In this experiment, we introduce a constraint such that the leader is allowed to, at max, cause a deceleration of a_{limit} to the follower. To enforce this, the following constraints are added to the leader's NLP:

$$\mathbf{g}_{\text{courtesy},k} = a_k^F - a_{\text{limit}} \geq 0 \tag{18}$$

Here, a_k^F is the acceleration of the follower.

The effect of the courtesy constraint with $a_{\text{limit}} = -2.0$ on the considered merging scenario is illustrated in Fig. 13. By introducing the constraint, the leader accelerates during the lane change, which successfully limits the induced deceleration to $-2.0 \frac{m}{s^2}$. The velocity profiles are shown in Fig. 13b. Compared to the cooperative cost

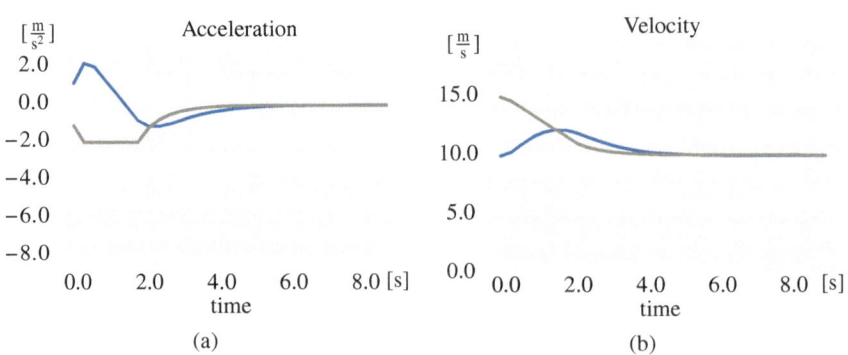

Fig. 13 Shown are the acceleration (**a**) and velocity (**b**) profiles when planning with the courtesy constraints. Introducing these constraints into the leader's NLP generates a behavior that successfully limit the follower's deceleration to $a_{\text{limit}} = -2.0$. Further, after the merge is completed, the leader returns to its desired velocity

formulation, the leader returns to its desired velocity of $10.0 \frac{m}{s}$ after the successful merge.

Both, the cooperative cost and the courtesy constraint method have traffic scenarios where they are particularly suited. E.g., when overtaking a slower driving vehicle on the highway, the cooperative cost formulation might be more suitable as it leads to a temporal increase in velocity for the duration of the overtaking maneuver. In contrast, for a merging scenario or a permanent lane change, the courteous constraint method might be the better choice since the leader returns to its desired velocity after the merge is completed.

6.4 Runtime Experiments

The presented method for interaction-aware trajectory optimization computes an open-loop solution for the AV. More precisely, the control inputs are functions of time and not of the state. To adapt to unforeseen changes in the environment, the algorithm needs to run in an MPC fashion. For MPC, a sufficiently high update rate is crucial. Therefore, we analyze the performance of the algorithm with a proof of concept MPC implementation.

The MPC was implemented in Python. All necessary derivatives were calculated using the open-source software CasADi [1]. CasADi utilizes automatic differentiation methods to accurately calculate the derivatives. Compared to, finite difference methods, automatic differentiation is faster and more accurate. Further, IPOPT [36] was used to solve the formulated NLP. IPOPT is a general-purpose solver for large-scale nonlinear problems. We cold started the IPOPT solver with a feasible solution of the desired driving maneuver, which we obtained by sequentially solving a single vehicle NLP, as in (8), first for the leader and then for the follower. This initialization was only performed for the very first iteration of the planner. All subsequent iterations were warm started with the solution of the previous iteration. To get a better initial guess, the previous solution was shifted by the duration between the planning iterations.

The timing results were obtained by considering a merging scenario, with the two most relevant methods for the application to real traffic, namely, the cooperative cost function method, with $\alpha = 0.5$ and the courtesy constraints method, with $a_{\text{limit}} = -2.0$. We simulate each method for $9.0\,$s. A horizon length of $N = 30$ steps is considered for the MPC. Further parameters were taken from Table 1. The runtime results are obtained by running the MPC implementation 100 times on the merging scenario with both methods. The mean solve time over the 100 simulation runs are shown in the histogram in Fig. 14. Additionally, the mean and standard deviation of the mean solve times are listed in Table 3. All timing results were obtained on an Intel Core i7-8565U CPU with a clock rate of 1.80GHz.

Even though the experiments were conducted with an MPC implementation that leaves great potential for improvements, we could already demonstrate our algorithm's real-time capability with mean solve times of 96.82 ms and 83.85 ms, respec-

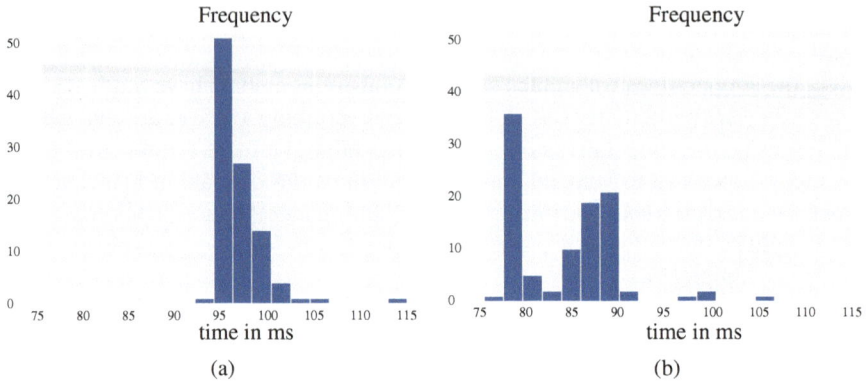

Fig. 14 Mean solve times obtained by running the MPC implementation 100 times on the merging scenario with (**a**) the cooperative cost and (**b**) the courtesy constraint method

Table 3 Mean and standard deviation of the mean solve times obtained by running the MPC implementation 100 times

Method	μ	σ
cooperative cost	96.82 ms	2.61 ms
courtesy constraints	83.85 ms	5.50 ms

tively. The presented results can be considered a conservative estimate of the achievable performance. However, in the future, this could be greatly improved by utilizing tailored solvers and implementing the approach in a high-performance programming language, e.g., C++.

7 Algorithm Discussion

A core assumption that we made to obtain the model for interaction-aware planning, stated in (3), is that the human does not try to influence the AV but rather reacts to its actions. According to [30, 31], this is a valid assumption for a wide range of interactive scenarios. Further, compared to a Nash equilibrium, it might even be the better model for how humans act in interactive situations since humans typically do not solve games in their everyday lives when they are not playing chess [15].

The formulated NLP (6), is a non-convex and non-smooth problem. As such, one can not expect to find globally optimal solutions. However, we use derivative-based optimization methods to find local optima. These methods require an initial guess, which sets the considered homotopy, as solutions of local methods are typically in the same homotopy as the initial guess. In the context of automated driving, homotopies are often thought of as maneuvers. Thus, we use the initial guess to encode the desired driving maneuver. Via the experiments, we empirically observe that initializing with a rough, but feasible initial guess of the desired maneuver is sufficient to reliably solves the problem. To take multiple maneuvers into consideration, it is advisable to combine the presented approach with a global method. E.g., a higher abstraction behavior planner based on an arbitration scheme as in [26] could be used to generate good initial guesses.

The focus of the experiments was to analyze the capabilities and the performance of the proposed bi-level planner. As such, the algorithm was evaluated in a tailored simulation environment, where one important modeling assumption was that the human driver is always attentive. However, in real traffic, this is not the case, and human drivers are sometimes distracted and do not respond to the actions of the AV. Therefore, the presented algorithm needs to be combined with an intention estimation, e.g, as presented in [5], to cope with unattentive drivers.

8 Conclusion

In this chapter, we presented an algorithm that is able to generate interaction-aware trajectories for AV. The interaction between a HD and an AV is modeled as a Stackelberg game, where the human responds rationally to the AV's actions optimizing its own objective. This leads to a nested optimization problem which we approximate by MPC based on a bi-level optimization formulation. To solve this, we reformulated the problem into a single-level representation, exploiting the assumption that the human will act optimally with respect to its objective function. We solve the obtained NLP using derivative-based optimization methods. The presented algorithm is able to solve the interaction-aware trajectory optimization problem in a continuous state and input space. Further, in contrast to existing methods, general nonlinear state and input constraints can be considered, which allows for an accurate dynamics model.

The algorithm enables the AV to anticipate how surrounding HD will react to its actions. This gives the AV the possibility to deliberately influence the state of the human. Here, simply encoding the desired effect into the AV's objective function is enough to generate complex, interaction rich behavior, without the need for hand designed decision heuristic. Further, interactive behavior does not only occur if incentivized in the AV's objective function, but also emerges out of optimizing the AV's behavior.

However, care must be taken to avoid that the AV exploits interactions to further optimize its own objective, and thereby generates an overly aggressive driving behavior. To prevent such an aggressive behavior, the AV's objective is extended to also consider the costs of the HD.

As an alternative to modifying the AV's objective function, we presented a strategy to establish courtesy in the planning algorithm via additional constraints. These constraints allow a motion planner to utilizes an egocentric objective function, provided that the negative impact imposed on other vehicles is limited.

The experiments demonstrated the efficacy of our algorithm and suggest that the algorithm can be used in challenging interactive driving scenarios. Further, we could achieve real-time performance even with an unoptimized proof-of-concept implementation.

Acknowledgements The authors thank the German Research Foundation (DFG) for being funded within the German collaborative research center "SPP 1835—Cooperative Interacting Automobiles" (CoInCar).

References

1. Andersson, J.A.E. et al.: CasADi: a software framework for nonlinear optimization and optimal control. Math. Program. Comput. **11**(1), 1–36 (2019). ISSN: 1867-2949, 1867-2957. https://doi.org/10.1007/s12532-018-0139-4
2. Bansal, S., et al.: Collaborative planning for mixed-autonomy lane merging. In: 2018 IEEE/RSJ International Conference on Intelligent Robots and Systems (IROS). Madrid, Spain, Oct. 2018, pp. 4449–4455 (2018). https://doi.org/10.1109/IROS.2018.8594197
3. Britzelmeier, A., Dreves, A., Gerdts, M.: Numerical solution of potential games arising in the control of cooperative automatic vehicles. In: 2019 Proceedings of the Conference on Control and Its Applications 24 Christoph Burger, Shengchao Yan, Wolfram Burgard and Christoph Stiller (CT). Proceedings. Society for Industrial and Applied Mathematics, Philadelphia, Jan. 1, 2019, pp. 38–45 (2019). https://doi.org/10.1137/1.9781611975758.7
4. Burger, C., Lauer, M.: Cooperative multiple vehicle trajectory planning using MIQP. In: 2018 21st International Conference on Intelligent Transportation Systems (ITSC). Maui, Hawaii, USA, Nov. 2018, pp. 602–607 (2018). https://doi.org/10.1109/ITSC.2018.8569776
5. Burger, C., Schneider, T., Lauer, M.: Interaction aware cooperative trajectory planning for lane change maneuvers in dense traffic. In: 2020 IEEE 23rd International Conference on Intelligent Transportation Systems (ITSC). Rhodes, Greece, Sept. 2020, pp. 1–8 (2020). https://doi.org/10.1109/ITSC45102.2020.9294638
6. Burger, C., et al.: Interaction-aware game-theoretic motion planning for automated vehicles using bi-level optimization. In: 2022 IEEE 25th International Conference on Intelligent Transportation Systems (ITSC). Macau, China, Oct. 2022, pp. 1–6 (2022)
7. Burger, C., et al.: Rating cooperative driving: a scheme for behavior assessment. In: 2017 IEEE 20th International Conference on Intelligent Transportation Systems (ITSC). Yokohama, Japan, Oct. 2017, pp. 1–6 (2017). https://doi.org/10.1109/ITSC.2017.8317794
8. de Campos, G.R., Falcone, P., Sjöberg, J.: Autonomous cooperative driving: a velocity-based negotiation approach for intersection crossing. In: 16th International IEEE Conference on Intelligent Transportation Systems (ITSC 2013). The Hague, Netherlands, Oct. 2013, pp. 1456–1461 (2013). https://doi.org/10.1109/ITSC.2013.6728435
9. Chen, H., Kremling, H., Allgöwer, F.: Nonlinear predictive control of a benchmark CSTR. In: Proceedings of the 3rd European Control Conference, Rome,Italy. (Jan. 1, 1995), pp. 3247–3252 (1995)
10. Dreves, A., Gerdts, M.: A generalized nash equilibrium approach for optimal control problems of autonomous cars. Optim. Control Appl. Methods **39**(1), 326–342 (2018). ISSN: 1099-1514. https://doi.org/10.1002/oca.2348. https://onlinelibrary.wiley.com/doi/abs/10.1002/oca.2348 (visited on 12/30/2022)
11. Evestedt, N., et al.: Interaction aware trajectory planning for merge scenarios in congested traffic situations. In: 2016 IEEE 19th International Conference on Intelligent Transportation Systems (ITSC). Rio de Janeiro, Brazil, Nov. 2016, pp. 465–472 (2016). https://doi.org/10.1109/ITSC.2016.7795596

12. Fisac, J.F., et al.: Hierarchical game-theoretic planning for autonomous vehicles. Oct. 12, 2018. arXiv:1810.05766 [cs, math]. http://arxiv.org/abs/1810.05766 (visited on 12/30/2022)

13. Fridovich-Keil, D., et al.: Efficient iterative linear-quadratic approximations for nonlinear multi-player general-sum differential games (2020). arXiv:1909.04694 [cs, eess]. http://arxiv.org/abs/1909.04694 (visited on 12/30/2022)

14. Fridovich-Keil, D., et al.: Efficient iterative linear-quadratic approximations for nonlinear multi-player general-sum differential games. In: Interaction-Aware Motion Planning as a Game 25 2020 IEEE International Conference on Robotics and Automation (ICRA). Paris, France, May 2020, pp. 1475–1481. https://doi.org/10.1109/ICRA40945.2020.9197129

15. Hedden, T., Zhang, J.: What do you think i think you think?: Strategic reasoning in matrix games. Cognition **85**(1), 1–36 (2002). ISSN: 0010-0277. https://doi.org/10.1016/S0010-0277(02)00054-9

16. Hoheisel, T., Kanzow, C., Schwartz, A.: Theoretical and numerical comparison of relaxation methods for mathematical programs with complementarity constraints. Math. Program. **137**(1), 257–288 (2013). ISSN: 1436-4646. https://doi.org/10.1007/s10107-011-0488-5

17. Hubmann, C., et al.: A belief state planner for interactive merge maneuvers in congested traffic. In: 2018 21st International Conference on Intelligent Transportation Systems (ITSC). Maui, USA, Nov. 2018, pp. 1617–1624 (2018). https://doi.org/10.1109/ITSC.2018.8569729

18. Kesting, A., Treiber, M., Helbing, D.: General lane-changing model MOBIL for car-following models. Transp. Res. Rec. J. Transp. Res. Board 1999 86–94 (2007)

19. Kretzschmar, H., et al.: Socially compliant mobile robot navigation via inverse reinforcement learning. Int. J. Robot. Res. **35**(11), 1289–1307 (2016). ISSN: 0278-3649, 1741-3176. https://doi.org/10.1177/0278364915619772

20. Le Cleac'h, S., Schwager, M., Manchester, Z.: ALGAMES: a fast augmented lagrangian solver for constrained dynamic games. Autonom. Robot. **46**(1), 201–215 (2022). ISSN: 1573-7527. https://doi.org/10.1007/s10514-021-10024-7

21. Le Cleac'h, S., Schwager, M., Manchester, Z.: LUCIDGames: online unscented inverse dynamic games for adaptive trajectory prediction and planning. In: IEEE Robotics and Automation Letters, p. 1 (2021). ISSN: 2377-3766, 2377-3774. https://doi.org/10.1109/LRA.2021.3074880. https://ieeexplore.ieee.org/document/9410364/ (visited on 12/30/2022)

22. Lenz, D., Kessler, T., Knoll, A.: Tactical cooperative planning for autonomous highway driving using Monte-Carlo tree search. In: 2016 IEEE Intelligent Vehicles Symposium (IV). Gothenburg, Sweden, June 2016, pp. 447–453. https://doi.org/10.1109/IVS.2016.7535424

23. Liniger, A., Lygeros, J.: A noncooperative game approach to autonomous racing. IEEE Trans. Control Syst. Technol. **28**(3), 884–897 (2020). ISSN: 1558-0865. https://doi.org/10.1109/TCST.2019.2895282

24. Ne, S., et al.: Implementing conditional inequality constraints for optimal collision avoidance. J. Aeronaut. Aerospace Engin. **06**(03) (2017). ISSN: 21689792. https://doi.org/10.4172/2168-9792.1000195

25. Notomista, G., et al.: Enhancing game-theoretic autonomous car racing using control barrier functions. In: 2020 IEEE International Confer- 26 Christoph Burger, Shengchao Yan, Wolfram Burgard and Christoph Stiller ence on Robotics and Automation (ICRA). Paris, France: IEEE, May 2020, pp. 5393–5399 (2020). isbn: 978-1-72817-395-5. https://doi.org/10.1109/ICRA40945.2020.9196757

26. Orzechowski, P.F., Burger, C., Lauer, M.: Decision-making for automated vehicles using a hierarchical behavior-based arbitration scheme. In: IEEE Intelligent Vehicles Symposium (IV). Las Vegas, NV, USA, Oct. 2020, pp. 767–774 (2020). https://doi.org/10.1109/IV47402.2020.9304723

27. Paden, B., et al.: A survey of motion planning and control techniques for self-driving urban vehicles. IEEE Trans. Intell. Veh. **1**(1), 33–55 (2016). ISSN: 2379-8858. https://doi.org/10.1109/TIV.2016.2578706

28. Polack, P., et al.: The kinematic bicycle model: a consistent model for planning feasible trajectories for autonomous vehicles? In: 2017 IEEE Intelligent Vehicles Symposium (IV). Los Angeles, USA, June 2017, pp. 812–818 (2017). https://doi.org/10.1109/IVS.2017.7995816

29. Sadigh, D., et al.: Information gathering actions over human internal state. In: 2016 IEEE/RSJ International Conference on Intelligent Robots and Systems (IROS). 2016 IEEE/RSJ International Conference on Intelligent Robots and Systems (IROS). Oct. 2016, pp. 66–73 (2016). https://doi.org/10.1109/IROS.2016.7759036
30. Sadigh, D., et al.: Planning for autonomous cars that leverage effects on human actions. In: Robotics: Science and Systems XII. Robotics: Science and Systems 2016. Robotics: Science and Systems Foundation (2016). ISBN: 978-0-9923747-2-3. https://doi.org/10.15607/RSS.2016.XII.029. http://www.roboticsproceedings.org/rss12/p29.pdf (visited on 12/30/2022)
31. Sadigh, D., et al.: Planning for cars that coordinate with people: leveraging effects on human actions for planning and active information gathering over human internal state. Autonom. Robot. **42**(7), 1405–1426 (2018). ISSN: 1573-7527. https://doi.org/10.1007/s10514-018-9746-1. https://doi.org/10.1007/s10514-018-9746-1 (visited on 12/30/2022)
32. Schulz, J., et al.: Estimation of collective maneuvers through cooperative multi-agent planning. In: 2017 IEEE Intelligent Vehicles Symposium (IV). Los Angeles, USA, June 2017, pp. 624–631 (2017). https://doi.org/10.1109/IVS.2017.7995788
33. Sun, L., et al.: Courteous autonomous cars. In: 2018 IEEE/RSJ International Conference on Intelligent Robots and Systems (IROS). Madrid, Spain, Oct. 2018, pp. 663–670 (2018). https://doi.org/10.1109/IROS.2018.8593969
34. Trautman, P., Krause, A.: Unfreezing the robot: navigation in dense, interacting crowds. In: 2010 IEEE/RSJ International Conference on Intelligent Robots and Systems. Taipei, Taiwan, Oct. 2010, pp. 797–803 (2010). https://doi.org/10.1109/IROS.2010.5654369
35. Treiber, M., Hennecke, A., Helbing, D.: Congested traffic states in empirical observations and microscopic simulations. Phys. Interaction-Aware Motion Plan. Game 27 Rev. E **62**(2), 1805–1824 (2000). ISSN: 1063-651X, 1095-3787. https://doi.org/10.1103/PhysRevE.62.1805
36. Wächter, A., Biegler, L.T.: On the implementation of an interior-point filter line-search algorithm for large-scale nonlinear programming. Math. Program. **106**(1), 25–57 (2006). ISSN: 1436-4646. https://doi.org/10.1007/s10107-004-0559-y
37. Wang, M., et al.: Game theoretic planning for self-driving cars in competitive scenarios. In: Robotics: Science and Systems XV. Freiburg, Germany: Robotics: Science and Systems Foundation, June 22 (2019). ISBN: 978-0-9923747-5-4. https://doi.org/10.15607/RSS.2019.XV.048
38. Wang, M., et al.: Game-theoretic plawang game theoretic planning risk aware 2020 inning for risk-aware interactive agents. In: 2020 IEEE/RSJ International Conference on Intelligent Robots and Systems (IROS). 2020 IEEE/RSJ International Conference on Intelligent Robots and Systems (IROS). IEEE, Las Vegas, pp. 6998–7005 (2020). ISBN: 978-1-72816-212-6. https://doi.org/10.1109/IROS45743.2020.9341137. https://ieeexplore.ieee.org/document/9341137/ (visited on 12/30/2022)
39. Wang, M., et al.: Game-theoretic planning for self-driving cars in multivehicle competitive scenarios. IEEE Trans. Robot. **37**(4), 1313–1325 (2021). ISSN: 1552-3098, 1941-0468. https://doi.org/10.1109/TRO.2020.3047521
40. Williams, G., et al.: Best response model predictive control for agile interactions between autonomous ground vehicles. In: 2018 IEEE International Conference on Robotics and Automation (ICRA). Brisbane, Australia, May 2018, pp. 2403–2410 (2018). https://doi.org/10.1109/ICRA.2018.8462831
41. Yoo, J.H., Langari, R.: A stackelberg game theoretic driver model for merging. In: ASME 2013 Dynamic Systems and Control Conference. Palo Alto, California, USA: American Society of Mechanical Engineers Digital Collection, Mar. 6 (2014). https://doi.org/10.1115/DSCC2013-3882
42. Yoo, J.H., Langari, R.: Stackelberg game based model of highway driving. In: ASME 2012 5th Annual Dynamic Systems and Control Conference Joint with the JSME 2012 11th Motion and Vibration Conference. Florida, USA: American Society of Mechanical Engineers Digital Collection, Sept. 17, 2013, pp. 499–508 (2013). https://doi.org/10.1115/DSCC2012-MOVIC2012-8703

Designing Maneuver Automata of Motion Primitives for Optimal Cooperative Trajectory Planning

Matheus V. A. Pedrosa, Patrick Scheffe, Bassam Alrifaee, and Kathrin Flaßkamp

Abstract Trajectory planning techniques form a central step to enable autonomous driving. The motion primitives method generates an automaton of precomputed maneuvers with structure-exploiting properties. Thereby, the trajectory planning problem can be reduced to finding an admissible/optimal sequence of motion primitives. In this chapter, we present ways to designing maneuver automata based on different system models and on either analytical or data-based approaches for automaton generation. Moreover, numerical methods for computing optimal maneuvers are listed and we discuss graph-based planning techniques. A subsequent chapter shows the evaluation of motion primitives automata in the Cyber-Physical Mobility Lab.

1 Introduction

The task of planning trajectories for multiple vehicles can be solved by many available techniques (see, for instance, [2, 3, 17]). However, there are still key challenges to be tackled for a multi-vehicle trajectories planner: (a) the admissibility of the planned trajectories, (b) the real-time capability of the optimization solvers on the respective vehicles and (c) the feasibility of the communication overhead between the vehicles.

M. V. A. Pedrosa (✉) · K. Flaßkamp
Chair of Systems Modeling and Simulation, Systems Engineering, Saarland University,
Saarbrücken, Germany
e-mail: matheus.pedrosa@uni-saarland.de

K. Flaßkamp
e-mail: kathrin.flasskamp@uni-saarland.de

P. Scheffe
Chair for Embedded Software, RWTH Aachen University, Aachen, Germany
e-mail: scheffe@embedded.rwth-aachen.de

B. Alrifaee
Department of Aerospace Engineering, University of the Bundeswehr Munich, Neubiberg,
Germany
e-mail: bassam.alrifaee@unibw.de

© The Author(s) 2024
C. Stiller et al. (eds.), *Cooperatively Interacting Vehicles*,
https://doi.org/10.1007/978-3-031-60494-2_8

231

Complemented into two chapters,[1] our work brings new methodologies for solving the cooperative trajectory planning problem for autonomous driving. We address the challenges mentioned above through graph-based optimal solutions. Before going into more detail about which methods we use and how we use them, let us first give an overview of how it is contextualized within vehicle automation.

Automated driving systems basically consist of three modules [31]:

1. Sensing or perception: capture the environment objects and conditions through sensors.
2. Planning: find a feasible trajectory.
3. Acting or control: track the trajectory by controlling the vehicle's actuators.

We place our focus on solving the second step. Motion planning aims to find a sequence of control inputs to move a vehicle from an original state to a set of possible goal states, while avoiding collisions during the trajectory [30]. At first, this task can be achieved by solving an optimal control problem (OCP). However, it could be computationally costly to get optimal control solutions when dealing with nonlinear vehicle models. Complex environments can also make it difficult to properly design all the obstacles into the optimization problem, which make the OCP unsuitable for many applications [14]. As an alternative, discrete planning techniques sample the state space, map it as a graph and perform a graph search for a minimum-cost path [22]. As disadvantages, we can cite the total neglect of the model in the case of the most famous graph search, the A*, or the numerically complex and non-time-critical solutions for the also well-known Hybrid A* search [10, 23, 26].

In order to get the best of both worlds, i.e., decreasing the motion planning problem complexity and avoiding a full discretization over the state space, we use the concept of motion primitives, originally proposed by [14]. Motion primitives are finite-time pieces of trajectories that can be concatenated. They are constructed from the dynamical system model. That is, the final path resulting from their interlocks is feasible with respect to the selected model. References [12, 14, 19] showed that, by using them, the highly complex problem of trajectory planning can be transformed into a graph search, in which solutions can be found with a suitable difficulty. However, for this to happen, it is of fundamental importance to have a library of primitives at hand that ensures appropriate routes for the desired road scenarios. At the same time, it should also have a size that makes the problem as computationally inexpensive as possible. Note that all this should also take into account the cooperative communication between agents, since it is desirable to have a trajectory planning with a sufficiently small communication effort.

The realization of motion primitives is only possible when the dynamic model has the symmetry property. To give an intuition, this property indicates that it is possible to perform rotations and translations in mechanical systems—without deformation of their path profile under the same sequence of control inputs. In the original work [14], the symmetry property of systems was exploited to develop two special kinds of

---

[1] Trajectory planning strategies for multiple vehicles are presented in the Chapter "Prioritized Trajectory Planning for Networked Vehicles Using Motion Primitives".

Fig. 1 Example of a
maneuver automaton with
four trim primitives:
$\{p_1, p_2, p_3, p_4\}$ and eight
maneuvers:
$\{m_{1,1}, m_{1,2}, m_{1,4},$
$m_{2,3}, m_{3,2}, m_{3,3}, m_{3,4}, m_{4,1}\}$
(figure from [24])

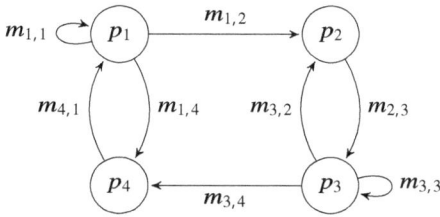

motion primitives: the trim primitives and the maneuvers. Trims are steady motions, where the control inputs are kept fixed, while the maneuvers are motions transitioning between the steady motions. Their rules for concatenation can be translated into a directed graph, which we call motion primitive automaton (MPA), also referred in [14] as maneuver automaton. Figure 1 illustrates an example of an MPA with four trims and eight maneuvers. Then, solving the motion planning problem consists of using a graph search method to find a sequence of primitives, which can be concatenated according to the MPA.

In this chapter, with the first part of our studies, we present the development of methods to architect the motion primitives selection and construction, as well as the relationship between them. The second part, written in the chapter "Prioritized Trajectory Planning for Networked Vehicles Using Motion Primitives", is devoted to detailing the cooperative trajectory planning algorithms that represent maneuvers primitives. The general workflow is given in Fig. 2.

This chapter is organized as follows. In Sect. 2, we evaluate, from a list of different vehicle models, the suitable dynamics for the planning problem and determine the symmetry group for a generic class of vehicle models. In Sect. 3, based on previous works, e.g., [12–14, 21, 23], we determine a method to analytically select trim primitives from a vehicle model and, alternatively, abstract typical trim primitives from traffic data. In Sect. 4, we model the computation of maneuvers as an OCP and solve the respective OCP to obtain the optimal maneuvers. Automata of different configurations with respect to their computational complexity and solution quality are analyzed in Sect. 5. It also investigates both time-optimal and maximum comfort motion graphs via the analysis of multi-objective maneuvers. In Sect. 6, we briefly present possible algorithms to solve the graph-based planning problem. Lastly, we give concluding remarks in Sect. 7.

2 Models and Symmetry

There are several ways to represent the dynamic system of vehicles, from the simplest cases, such as the point-mass model, the Dubins curves [11] and the Reeds–Shepp curves [25], to detailed, vehicle-specific models. Both Dubins and Reeds–Shepp curves take into account a kinematic car model consisting only of the pose, i.e., the position and orientation. Halfway through, the CommonRoad benchmarks present a

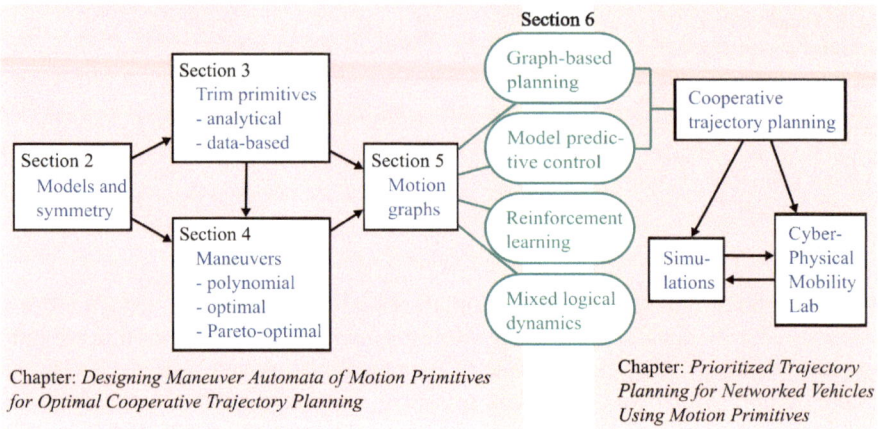

Fig. 2 Workflow of our methodology

hierarchical list of models that considers increasingly complex lateral vehicle dynamics and tire models [5]. This list includes, among others, the following models: kinematic single-track model, single-track model, and a multi-body model. It is assumed for these models the existence of controllers that can realize a commanded acceleration. The choice of the appropriate model depends on which detail you want to capture the physics of motion. In the Appendix, the reader can find the description of the equations for the kinematic single-track and the single-track models, as they will be used in this chapter.

All of the CommonRoad models have in common the following generic structure of ordinary differential equations:

$$\dot{x} = f(x, u) := \begin{bmatrix} f_1(r, u) \cos{(f_2(r, u) + \psi)} \\ f_1(r, u) \sin{(f_2(r, u) + \psi)} \\ f_\psi(r, u) \\ f_r(r, u) \end{bmatrix}, \tag{1}$$

with the vector of states $x = \begin{bmatrix} s_x & s_y & \psi & r \end{bmatrix}^{\mathsf{T}}$ belonging to a manifold X, where s_x and s_y are the positions of the center of gravity, ψ is the vehicle orientation, r is any vector of $n - 3$ states, $u \in \mathcal{U}$ is the vector of inputs and $f_1(r, u)$, $f_2(r, u)$, $f_\psi(r, u)$, and $f_r(r, u)$ are arbitrary nonlinear functions. For convenience, we omit the notation for dependence of $x(t)$ and $u(t)$ on time $t \in \mathbb{R}_{\geq 0}$. We assume the function $f(x, u)$ of Eq. (1) as being continuous and locally Lipschitz w.r.t. $x(t)$. Then, we guarantee the existence and uniqueness of solutions given by the flow

$$x(t) = \varphi_u(x(0), t) \tag{2}$$

for a given input function u on the time interval $t \in [0, T]$.

Many mechanical systems, including vehicles, exhibit the symmetry property, which acts as state transformations defined by Lie group representations [14]. They are necessary to build the primitives from the model. To describe them mathematically, we need to introduce some considerations.

Let the Lie group be denoted by G, its identity element by e, and its left action on X by $\Psi : G \times X \to X$ with Ψ smooth, $\Psi(e, x) = x$ for $x \in X$, and $\Psi(g, \Psi(h, x)) = \Psi(gh, x)$ for all $g, h \in G$ and $x \in X$.

Definition 1 (*Symmetry*) The tuple (G, Ψ) is a symmetry for $\dot{x} = f(x, u)$ on X, if for any fixed control $u \in \mathcal{L}_{\text{loc}}^{\infty}([0, \infty), \mathbb{R}^m)$,

$$\varphi_u(\Psi(g, x_0), t) = \Psi(g, \varphi_u(x_0, t)) \tag{3}$$

holds for all $g \in G$, $x_0 \in X$, and $t \geq 0$.

We can produce a symmetry group that fits the entire set of models described in [5]. It is given by combined rotations and translations on the pose, which we represent by $p = \begin{bmatrix} s_x & s_y & \psi \end{bmatrix}^{\mathrm{T}} \in \mathbb{R}^2 \times S^1$, in the following form [24]:

Theorem 1 *The symmetry group for Eq. (1) is given by*

$$G := \left\{ g \in \mathsf{SE}(n) : g := g(\Delta x) = \begin{bmatrix} R & \Delta x \\ 0 & 1 \end{bmatrix} \right\}, \tag{4}$$

where

$$R = \begin{bmatrix} R_{\mathsf{SO}(3)} & 0 \\ 0 & I \end{bmatrix} \in \mathsf{SO}(n), \tag{5}$$

$$\Delta x = \begin{bmatrix} \Delta s_x \\ \Delta s_y \\ \Delta \psi \\ 0 \end{bmatrix} \in \mathbb{R}^2 \times S^1 \times \{0\}^{n-3}, \tag{6}$$

$$R_{\mathsf{SO}(3)} = \begin{bmatrix} \cos(\Delta \psi) & -\sin(\Delta \psi) & 0 \\ \sin(\Delta \psi) & \cos(\Delta \psi) & 0 \\ 0 & 0 & 1 \end{bmatrix} \in \mathsf{SO}(3), \tag{7}$$

for I being the identity matrix with appropriate dimension, a vector Δx, and g given in homogeneous coordinates, such that the affine-linear group action can be represented by:

$$\Psi_g(x) = Rx + \Delta x. \tag{8}$$

To prove it, we show the equivariance of the system (1) w.r.t. the symmetry action (8). We will show the idea of the proof, while details can be found in [24].

Proof The vector field f is equivariant w.r.t. the symmetry action Ψ if

$$f(\Psi_g(x), u) = \frac{\mathrm{d}\Psi_g(x)}{\mathrm{d}x} \cdot f(x, u).$$
(9)

Let $\Delta p = \begin{bmatrix} \Delta s_x & \Delta s_y & \Delta \psi \end{bmatrix}^{\mathrm{T}}$. The group action (8) can be written as

$$\Psi_g(x) = \begin{bmatrix} R_{\mathrm{SO}(3)} p + \Delta p \\ r \end{bmatrix} = \begin{bmatrix} \cos(\Delta p)s_x - \sin(\Delta p)s_y + \Delta s_x \\ \sin(\Delta p)s_x + \cos(\Delta p)s_y + \Delta s_y \\ \psi + \Delta \psi \\ r \end{bmatrix}.$$
(10)

Then, from Eqs. (1), (10) and (5), we get that the left-hand side of Eq. (9) is:

$$f\left(\Psi_g(x), u\right) = Rf(x, u).$$
(11)

Considering $\Psi_g(x) = Rx + \Delta x$ as in Eq. (8),

$$\frac{\mathrm{d}\Psi_g(x)}{\mathrm{d}x} = R,$$
(12)

which we can replace in Eq. (11), proving the equivariance of the vector field by satisfying Eq. (9).

Given the proper considerations, we can now define motion primitives as equivalence classes of trajectories.

Definition 2 (*Motion primitive*) A motion primitive is the equivalence class of a representing pair (x, u) on $[t_i, t_f]$, if for any class member $(\bar{x}, \bar{u})$ on $[\bar{t}_i, \bar{t}_f]$, it holds that $t_f - t_i = \bar{t}_f - \bar{t}_i$ and there exists a group element $g \in G$ and a shift $\Delta t \in \mathbb{R}$, such that

$$(x(t), u(t)) = (\Psi(g, \bar{x}(t - \Delta t)), \bar{u}(t - \Delta t)) \quad \forall t \in [t_i, t_f].$$
(13)

In the next sections, we will introduce the two types of motion primitives: trim primitives and maneuvers.

3   Trim Primitives

These primitives are characterized by fixed, i.e., trimmed, controls and are symmetry-induced motions. They were introduced in [14], and the authors add that the trims are identified with steady-state motions, also known as relative equilibria of the system. Formally, they can be defined as follows.

Definition 3 (*Trim Primitive*) Following the Definition 1, let $\mathfrak{g}$ denote the Lie algebra of G with the exponential map $\exp : \mathfrak{g} \to G$, and $\bar{u} \in \mathcal{U}$ a fixed control input. The tuple (x, u) on $[0, T]$ with $x(0) = x_0$ is called a trim primitive if it is a solution to the system dynamics expressed, for all $t \in [0, T]$, by

$$\begin{cases} x(t) = \Psi(\exp(\xi t), x_0), \\ u(t) \equiv \bar{u}, \end{cases} \tag{14}$$

with $\xi \in \mathfrak{g}$ being a suitable chosen Lie algebra element.

The duration of a trim primitive is, in principle, not fixed and is called "coasting time". For the kinematic single-track or the single-track models, the trims are characterized by a fixed velocity and a constant curvature[2] (see [23, 24]).

A choice for a finite number of trim primitives has to be taken. The question of representation and well-spread trims arises. A "plain vanilla" approach is to uniformly grid the Lie algebra up until borders that seem physically plausible [19]. More sophisticated approaches choose representative trim primitives based on data, either the road-geometry of interest or from driving, as detailed in the following two subsections.

3.1 Choice Based on Road-Geometry

One way to select the trim primitives is to fit them to the geometry of the roads on which the vehicle is to drive. As an example, we take the map of the Cyber-Physical Mobility Lab (CPM Lab) [16] drawn using the CommonRoad interface [5], depicted in Fig. 3a. From information contained in the CommonRoad scenario file, we can decompose the roads into the discrete points taken from the center of each lane, as can be seen in the upper left of the Fig. 3b and generate trims by the sequence:

1. Calculate all the possible curvatures from the map.
2. Select the most frequent curvatures.
3. Choose an arbitrary set of speeds within the boundaries interval.
4. Combine curvatures and speeds in tuples that represent each trim.

The yaw angles at each point of the decomposed map could be calculated from the vectors tangent to the lane's center (see Fig. 3b). Then, we can get and store the set of different curvatures, which might be a large data set. To reduce it, we can cluster the data points, for instance via k-means, to get a smaller number of the representative curvatures [24]. However, we can directly steer the number of different curvatures by the considered rounding accuracy, as we can have many similar data points. For example, the set of curvatures $\{0.0507..., 0.0543..., 0.0539...\}$ could be reduced to the values $\{0.051, 0.054\}$ when considering three decimal places, or to just to the value of 0.05 if two decimal places are considered. See Fig. 4 for checking the number

---

[2] The curvature is calculated by dividing the yaw rate by the velocity. Then, the trims could, alternatively, be represented by a constant speed and constant yaw rate.

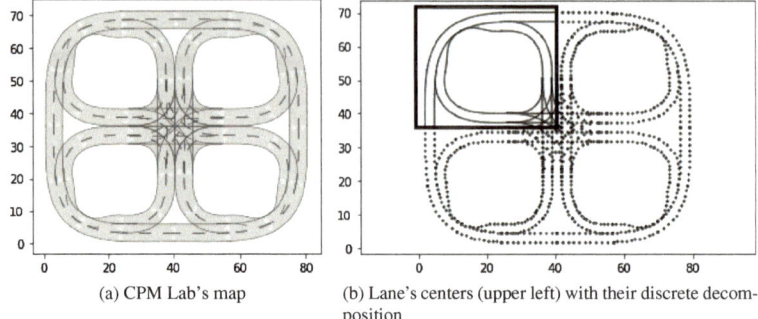

(a) CPM Lab's map   (b) Lane's centers (upper left) with their discrete decomposition

Fig. 3 Road geometry decomposition of the CPM Lab's map into 208 different points. The axes are the coordinates in meters

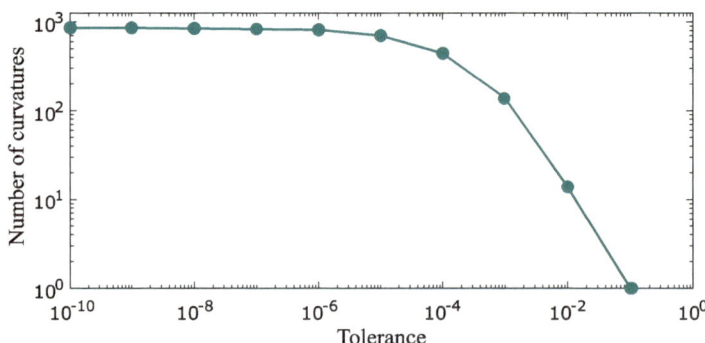

Fig. 4 The number of different curvatures computed in the example according to the tolerated decimal places

of different curvatures considered according to the accepted decimal places for this example. Having two decimal places, we get 14 classes of curvatures, that can be representative for this map. Lastly, a set of arbitrary speeds can be combined with these different classes and we get a set of trims to be used in the planning problem.

3.2 Choice Based on Driving Data

An automatic generation of data-based automata was proposed in [24]. The authors assumed that the data represents a dynamical model with symmetries. Also, this model is observable such that the full system state could be reconstructed from the available states on the data. Then, making an assumption on the model, the following sequence of steps was carried out:

1. Find invariances of trims in data.
2. Cluster trim primitives.

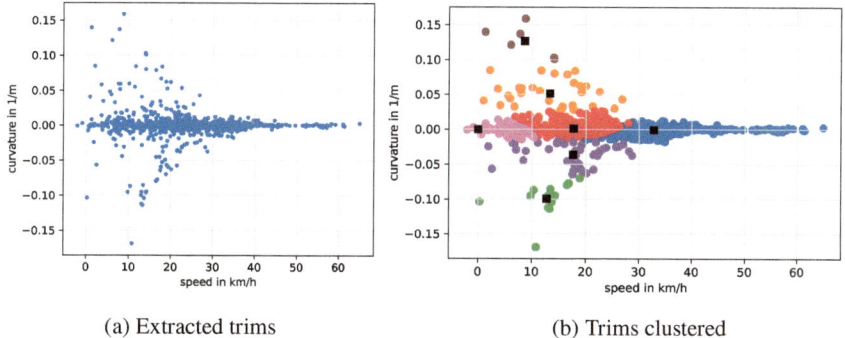

(a) Extracted trims   (b) Trims clustered

Fig. 5 Trims clustered using k-means in seven representative points, where the black squares are the centers of each cluster (figure from [24])

3. Evaluate a transition matrix.

The selected data in [24] was taken from the nuScenes data set [8], having multiple information about the vehicle's states, including the pose, the velocity, the acceleration, and the rotation rate recorded using an inertial measurement unit during urban driving in Singapore (Singapore) and Boston (United States). It is worth mentioning that this data set represents the interaction of the car in real traffic with other vehicles, including overtaking, braking, waiting on corners, etc. That is, several real interactions between vehicles are embedded in the selected primitives, ideal for a cooperative planning scenario.

Consider the data points being represented by the triples (t_i, x_i, u_i) for $i = 0, 1, 2, \ldots, D$, where $D \in \mathbb{R}$ is the number of elements in the data set. Consider that, for a suitable chosen symmetry $(\mathcal{G}, \Psi)$, there exist solutions (x, u) satisfying Definition 3 for a model $\dot{x} = f(x, u)$. Then, subsequent data points belong to the same trim if

$$
\begin{cases}
||u_{i+1} - u_i|| < \epsilon_u, \text{ and} \\
||\Psi(\exp(\xi(t_{i+1} - t_i)), x_i) - x_{i+1}|| < \epsilon_x
\end{cases}
\tag{15}
$$

for a sufficient small positive error margins ϵ_x and ϵ_u.

We can determine a minimum time length τ for the duration of a trim primitive, i.e., a minimal coasting time. Then, a trim will be considered only if, for a sequence of $N > 1$ points, the conditions (15) hold from i to $i + N - 1$ such that $t_{i+N} - t_i \geq \tau$.

However, the number of extracted trims from the data can be huge. Then, we need to look for a finite amount of clusters that define the most representative trims during a route in real traffic. In [24], they worked with the k-means algorithm, an unsupervised learning technique that finds clusters in a set of data points, where the amount of clusters is given [20]. The representative trims will be selected as the center points of each cluster. Figure 5 shows an example of trims being clustered for the kinematic single-track model (28) from [24]. The trims are represented by a constant speed (x-axis) and a constant curvature (y-axis).

At first, we could consider that a vehicle would be able to transit from any relative equilibrium to any other. For instance, in a kinematic robot model under nonholonomic constraints, given by

$$
\begin{bmatrix} \dot{s}_x(t) \\ \dot{s}_y(t) \\ \dot{\psi}(t) \end{bmatrix} = \begin{bmatrix} \cos(\psi(t)) \\ \sin(\psi(t)) \\ 0 \end{bmatrix} u_1(t) + \begin{bmatrix} 0 \\ 0 \\ 1 \end{bmatrix} u_2(t),
\tag{16}
$$

where the states are the pose and the controller manages the linear velocity $u_1(t) = v(t)$ and the angular velocity $u_2(t) = \dot{\psi}(t)$, trajectories can switch directly from one trim to another [13]. This is due to this model directly controlling the velocities and, thus, allowing discontinuities thereof. In this case, as well as in the (kinematic) single-track model, every constant control input defines a trim, either going straight with constant velocity or going in a circle with constant rotational velocity. However, in the (kinematic) single-track model, the control inputs correspond to the longitudinal acceleration and the steering angle velocity (see the Appendix for these models' equations). Thus, trims necessarily correspond to uncontrolled, i.e., constant-velocity motion. Smooth transitions between trims are then needed for, e.g., accelerating and decelerating to a new cruising speed, or for transitioning between straight and circular motions.

We can search and select these transitions according to their occurrence in the data. That is, only transitions with a high probability of occurrence will be considered. The probabilities are organized in a transition matrix, in which, for each trim cluster, the transitions from all points of this cluster to other clusters are counted in the data.

These transitions are another kind of primitive, called "maneuvers". The last step for the automatic generation of an automaton is the computation of the maneuvers. Their formal definition, as well as techniques to compute them, will be given in the next section.

4 Maneuvers

The second type of motion primitives is the maneuvers. They are responsible for smooth transitions in the system from one trim primitive to another. Formally, we can define them as follows.

Definition 4 (*Maneuver*) A maneuver is a finite-time trajectory that connects two trim primitives and is identified by:

- a time duration T;
- a sequence of control actions $u : [0, T] \to \mathbb{R}^m$;
- and an evolution in the form of (2) such that $(x(0), u(0))$ and $(x(T), u(T))$ belong to trim primitives characterized by (14).

In the class of vehicle models, the physics of maneuvers depends on the specific choice of the dynamical system model.

To derive maneuvers for the considered family of vehicle models, we present a geometric approach, in which polynomial equations define the transitions from the predecessor trim to the successor one. Alternatively, maneuvers can be computed as solutions of an OCP. In this case, we can also explore Pareto fronts in a multiobjective optimization problem.

4.1 Polynomial Approach

The paper [23] exemplifies a concrete case of formulating the geometric method using the single-track model (30) from [5]. In this case, a smooth transition needs to be made between the velocities v and steering angles δ from the predecessor trim to the successor one, both having these parameters fixed. Then, for a maneuver with the duration $T > 0$, we have the constraints $v(0) = v_0$ and $v(T) = v_T$. A jump in acceleration at the beginning or the end of the maneuver would theoretically result in infinite jerk, which can be avoided by setting $u(0) = u(T) = \begin{bmatrix} 0 & 0 \end{bmatrix}^{\mathrm{T}}$. Then, the control inputs are continuous, but we have additional constraints on the velocity $\dot{v}(0) = \dot{v}(T) = 0$. These constraints are met by the following cubic polynomial transitions for $0 \le t \le T$:

$$
\begin{cases}
v(t) = (v_T - v_0) \left(3 - 2\frac{t}{T} \right) \left(\frac{t}{T} \right)^2 + v_0, \\
\delta(t) = (\delta_T - \delta_0) \left(3 - 2\frac{t}{T} \right) \left(\frac{t}{T} \right)^2 + \delta_0.
\end{cases}
\tag{17}
$$

Then, the corresponding control signals $u_{\dot{v}}$ and $u_{\dot{\delta}}(t)$ are

$$
\begin{cases}
u_{\dot{v}}(t) = 6(v_T - v_0) \left(1 - \frac{t}{T} \right) \frac{t}{T^2}, \\
u_{\dot{\delta}}(t) = 6(\delta_T - \delta_0) \left(1 - \frac{t}{T} \right) \frac{t}{T^2}.
\end{cases}
\tag{18}
$$

In addition, to ensure the feasibility of the maneuver, constraints on the longitudinal acceleration and the derivative of the steering angle need to be considered. For the selected model, there exist the constraints

$$
|\dot{v}| \le \left| \frac{3}{2} \frac{v_T - v_0}{T} \right| \quad \text{and} \quad |\dot{\delta}| \le \left| \frac{3}{2} \frac{\delta_T - \delta_0}{T} \right|.
\tag{19}
$$

When the maneuvers have positive acceleration (i.e., $v_0 < v_T$), another constraint needs to be considered:

$$u_{\dot{v}} \leq a_{max} \frac{v_s}{v}, \tag{20}$$

with the switching velocity v_s, representing limited engine power, and a maximal longitudinal acceleration $a_{max} > 0$. Then, the duration of the maneuver can be chosen according to

$$\begin{cases} T = \max \left(\dfrac{3}{2} \dfrac{|v_T - v_0|}{a_{max}}, \dfrac{3}{2} \dfrac{|\delta_T - \delta_0|}{\dot{\delta}_{max}}, \dfrac{3}{2} \dfrac{(v_T - v_0)v_T}{a_{max} v_s}, T_{min} \right), \\ \quad \text{for } v_0 < v_T, \\ T = \max \left(\dfrac{3}{2} \dfrac{|v_T - v_0|}{a_{max}}, \dfrac{3}{2} \dfrac{|\delta_T - \delta_0|}{\dot{\delta}_{max}}, T_{min} \right), \quad \text{otherwise.} \end{cases} \tag{21}$$

where T_{min} is a defined shortest duration, set as a design choice.

4.2 Optimal and Pareto-Optimal Maneuvers

Alternatively, the maneuvers can be computed optimally with respect to a cost functional $J(T, x, u)$, for a duration T. Then, each maneuver is obtained by solving the following OCP:

$$\begin{array}{lrr} \underset{T,x,u}{\text{minimize}} & J(T, x, u) & (22a) \\ \text{subject to} & \dot{x}(t) = f(x(t), u(t)), \ 0 < t \leq T & (22b) \\ & 0 \geq g(x(t), u(t)), \ 0 < t \leq T & (22c) \\ & x(0) = x_0 & (22d) \\ & x(T) = x_T, & (22e) \end{array}$$

with x_0 and x_T as fixed states[3] evaluated at the predecessor and successor trims, respectively, and $g(\cdot)$ as the constraints for the states and inputs.

In the case of multiple cost functionals to be considered, the problem (22a) becomes a multiobjective optimal control problem. Then, we can select a Pareto-optimal maneuver by computing the so-called Pareto set of optimal compromises between the concurrent objectives [9] and choosing one of its points (see Fig. 6).

For instance, consider the kinematic single-track model (28), the costs $J_1 = T$ and $J_2 = \int_0^T \|u_{\dot{v}}\|_2^2$, for a trade-off between fast and comfortable trajectories. The maneuver goes from a trim described by $(v, \delta) = (0 \text{ km h}^{-1}, 0°)$ to a trim $(20 \text{ km h}^{-1}, 15°)$ and it is limited by 5 s. The Pareto front with 25 points is given in Fig. 6 together with their respective pose and inputs. Optimal control problems can be solved using numerical software tools, for instance CasADi [6] or TransWORHP [18]. We can

---

[3] Depending on the dynamical system, only part, not all, of the states x could be considered as fixed at the initial and final times of the maneuver.

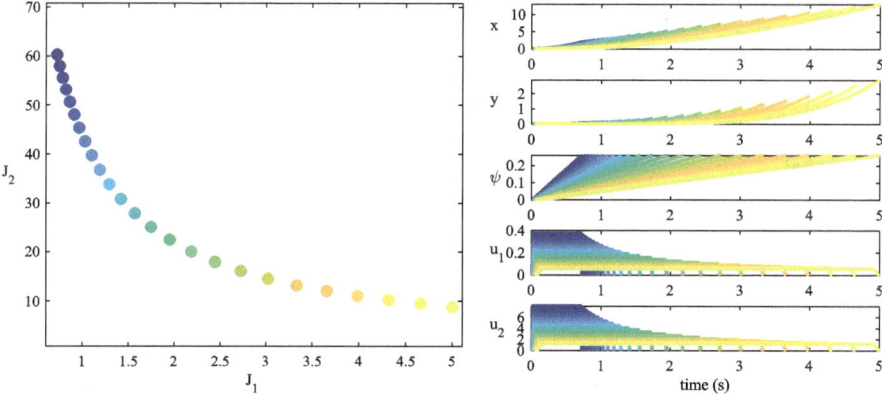

Fig. 6 Example of a Pareto front for a maneuver with $J_1 = T$ and $J_2 = \int_0^T ||u_{\dot{v}}||_2^2$

select a Pareto-optimal point based on a decision-making, to get the maneuver to be considered in the MPA.

5 Maneuver Automaton Selection

In [14], motion graphs are introduced as "maneuver automata", in which trims form the vertices and maneuvers the edges of the graph. This defines the concatenation rules, i.e., any path in the automaton defines a sequence of primitives. Together with a choice of coasting times, this sequence can be transformed into an admissible, controlled trajectory of the underlying dynamical system.

As presented in Sect. 3, maneuver automata can be constructed in an automatic way by extracting representative primitives from a data set. In [24], numerical examples were solved to compare handcrafted and extracted automata for the kinematic single-track model (Eq. (28)). The handcrafted automata consider a usual pragmatic way of designing it: a grid covering the entire space of allowed velocities and steering angles for the model [23]. For comparison, the handcrafted and extracted automata had the same quantity of trims and a similar number of maneuvers. A visual comparison of these two different ways of constructing an automaton is given in Fig. 7, considering the selection of 21 trim primitives. The difference in the trajectory planning when using each of these automata is replicated in Fig. 8. Note that the extracted primitives fit better to the road shape and the final goal position.

For the planning problem, a starting trim is assumed and an initial condition $x(0)$, i.e., a starting node in the MPA, is given. For a guarantee of the existence of a solution from an initial trim to a final trim (or node), it is shown in [14] that one of the requirements is the strong connectivity of the MPA. However, a priori, an MPA

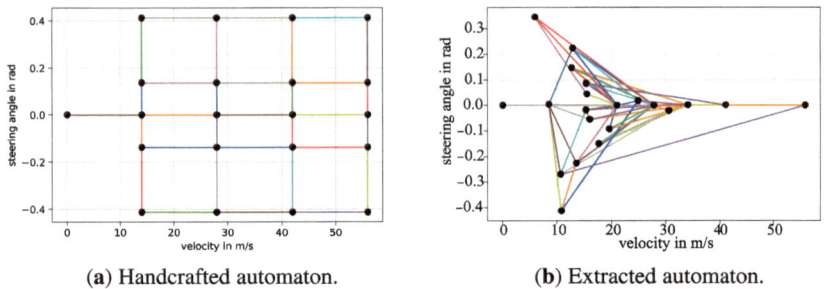

(a) Handcrafted automaton.   **(b)** Extracted automaton.

Fig. 7 Automata with 21 trim primitives. The dots correspond to trim primitives (axes: velocities versus steering angle) and the colored lines represent maneuvers connecting the trims (figures from [24])

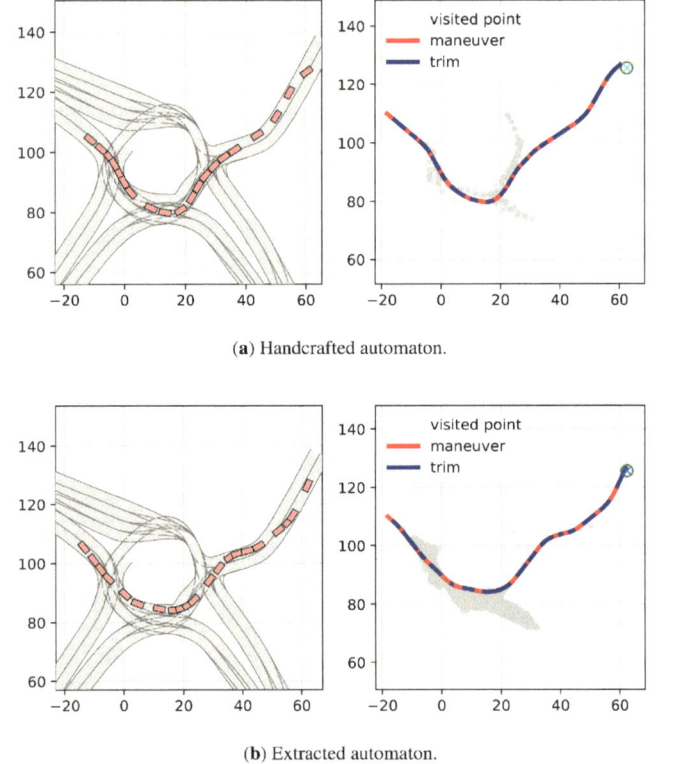

(a) Handcrafted automaton.

(b) Extracted automaton.

Fig. 8 Trajectories for the two different automata with 21 trim primitives (figures from [24])

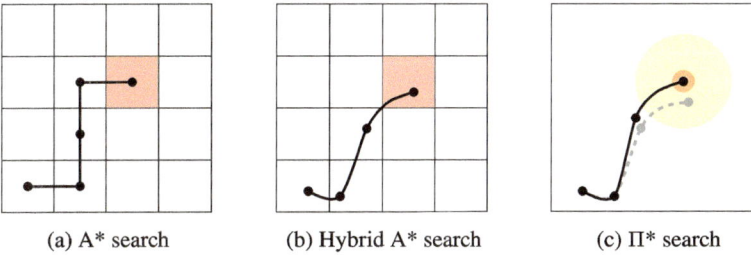

(a) A* search   (b) Hybrid A* search   (c) Π* search

Fig. 9 Comparison between different graph search methods: the goal regions are denoted in red and the yellow area is an optimization region, where the Π* will try to optimize the trim's coasting times to lead the vehicle to a goal point inside the goal region

does not need to be strongly connected. For the cases where there exists more than one admissible solution, an optimization problem can be posed.

6 Planning Algorithms

With a library of primitives condensed into a graph, path planning can be done using different techniques. In this section, we will mention some of the ideas developed. The complementary chapter will, however, delve into planning in a cooperative trajectory planning scenario.

6.1 *Optimized Primitives (Π*) Search*

The Π* search was developed in [23] and it is inspired by the Hybrid A* algorithm [10], an A*-based search. In the Hybrid A*, continuous states are associated with grid cells and the costs of the states, therefore, are the cost of their respective cell. However, in Π* search, each state is fully continuous, instead of being associated with discrete grid cells. The trims' coasting times can be adjusted by an online optimization problem of reduced complexity. The algorithm encapsulates the method of anytime search to deal with time deadlines [32]. The search, then, can lead the vehicle to an exact goal point in the state space while respecting computation time constraints. Figure 9 compares the different graph search methods.

Fig. 10 The interaction of
agent and environment in
reinforcement learning

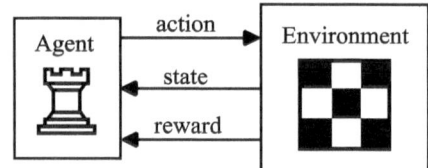

6.2 Reinforcement Learning

Reinforcement learning as a Markov decision process, as described in [29], is the
task of learning from the interaction between an agent and an environment to achieve
a goal. The agent is the decision-maker and learns which is the best action given the
current state. A numerical value evaluates an action and it is called "reward". Thus,
the action is selected to maximize the rewards. The environment, in turn, responds to
the agent with a new state and the reward for a given action. A schematic depiction
of this iterative process can be seen in Fig. 10.

It is possible to use primitives as the actions of a reinforcement learning agent,
as opposed to using a discrete or continuous set of control inputs as the action space
[15]. A work in this regard was developed in the Bachelor's thesis [28].

6.3 Graph-Based Receding Horizon Control

Introduced in [27], this method aims to transfer the receding horizon control approach
into graph-search problems, specially made for maneuver automata. Thus, nonlinear,
nonconvex optimization problems are solved in real-time, in opposite to traditional
graph-search approaches that keep the search until the goal vertex is found. This
approach was applied to cooperative planning of multiple networked and autonomous
vehicles on the CPM Lab [16]. Also, it was shown that the solutions are recursively
feasible by design of the finite state automaton. This method is explained in detail in
the chapter "Prioritized Trajectory Planning for Networked Vehicles Using Motion
Primitives".

6.4 Motion Graphs as Mixed Logical Dynamical System

We can model the motion graphs as a mixed logical dynamical (MLD) system to
transform the graph search into an OCP. MLD systems were introduced by [7] and
describe systems by a combination of continuous variables with Boolean ones. As an
example of application, an MLD system was modeled to solve collision avoidance
of collaborative vehicles in [4]. The authors did not use primitives, but linearized

the vehicle model over the operation points and solved mixed-integer linear and quadratic programming problems.

In short, the idea of our proposed MLD system is to formulate the execution of a primitive at a discrete-time k by "enabling" one primitive over all others available, given which node of the MPA is active for the vehicle. For that, we can define the Boolean variables, for $i = 1, 2, \ldots$, as

$$m_i(k) = \begin{cases} 1, & \text{if the primitive } \mathrm{p}_i \text{ is executed at } k, \\ 0, & \text{otherwise.} \end{cases} \tag{23}$$

where the set of available primitives at time k is $\{\mathrm{p}_i, i \in \mathbb{N}\}$. Then, given the current automaton state in the MPA and $x(k)$, the system dynamics can be written as:

$$\begin{cases} x(k+1) = \sum_i \Psi_{g_i}(x(k)) \cdot m_i(k), & \text{(8.24a)} \\ \sum_i m_i(k) = 1, & \text{(8.24b)} \end{cases}$$

for the continuous times given by

$$t_{k+1} = t_k + \sum_i \tau_i m_i \tag{25}$$

with τ_i representing the duration of the primitive p_i. This modeling approach leads to a mixed-integer nonlinear programming problem when searching for the optimal sequence for a given planning problem within an MPA.

Thus, it is possible to extend this modeling into a model predictive control (MPC) formulation and thus exploit the tools available for MPC, for example, stability, robustness, and inclusion of constraints, in the computation of trajectories with motion primitives.

7 Conclusion

We presented in this chapter a description of methods to design an automaton of motion primitives by properly selecting and constructing them. This automaton of primitives is implemented in trajectory planning for cooperative vehicles and its architecture is essential for efficient paths. We presented a list of vehicle models abstracted in a general formulation. Then, we showed how to abstract typical trim primitives from traffic data and derived maneuvers by the polynomial method and by an OCP. This last one is useful for finding Pareto-optimal maneuvers. We also compared different automata and presented possible algorithms to solve the graph-based planning problem.

Appendices

Here, we present two vehicle models from [5], the kinematic single-track and the single-track model.

A. The Kinematic Single-Track Model

The kinematic bicycle model has the state vector

$$x = \begin{bmatrix} s_x \ s_y \ \psi \ v \ \delta \end{bmatrix}^{\mathrm{T}} \in \mathbb{R}^5, \tag{26}$$

and the input vector:

$$u = \begin{bmatrix} u_{\dot{v}} \ u_{\dot{\delta}} \end{bmatrix}^{\mathrm{T}} \in \mathbb{R}^2, \tag{27}$$

where s_x and s_y are the positions of the rear axis, ψ is the vehicle orientation, v is the velocity, δ is the steering angle, $u_{\dot{v}}$ is the longitudinal acceleration, and $u_{\dot{\delta}}$ is the velocity of the steering angle. The state space equations are given by:

$$\begin{cases} \dot{s}_x(t) = v(t) \cdot \cos(\psi(t)), \\ \dot{s}_y(t) = v(t) \cdot \sin(\psi(t)), \\ \dot{\psi}(t) = \dfrac{v(t)}{L} \cdot \tan(\delta(t)), \\ \dot{v}(t) = u_{\dot{v}}(t), \\ \dot{\delta}(t) = u_{\dot{\delta}}(t), \end{cases} \tag{28}$$

for L being the wheelbase of the vehicle. In [24], it was used the wheelbase of the Renault Zoe, used in obtaining the nuScenes data, with value 2.588 m [1].

B. The Single-Track Model

The state vector

$$x = \begin{bmatrix} s_x \ s_y \ \psi \ \dot{\psi} \ v \ \delta \ \beta \end{bmatrix}^{\mathrm{T}} \in \mathbb{R}^7, \tag{29}$$

has the same variables described for Eq. (26) together with the slip angle at the center of gravity β (see Fig. 11). The inputs are the same as in Eq. (27). The state space equations are:

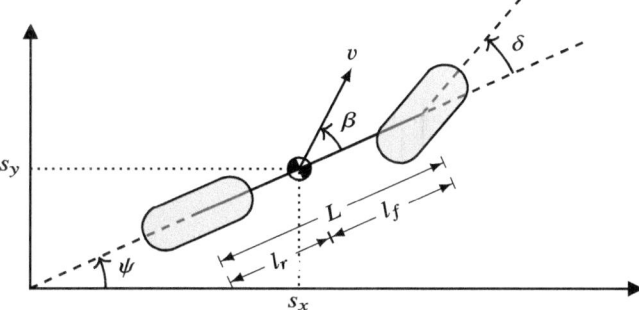

Fig. 11 Single-track model

$$
\begin{cases}
\dot{s}_x(t) = v(t) \cdot \cos(\psi(t) + \beta(t)), \\
\dot{s}_y(t) = v(t) \cdot \sin(\psi(t) + \beta(t)), \\
\dot{\psi}(t) = \dfrac{d}{dt}\psi(t), \\
\ddot{\psi}(t) = \dfrac{\mu M}{I_z L}\left(l_f \cdot \alpha_{f,r} \cdot \delta(t) + (l_r \cdot \alpha_{r,f} - l_f \cdot \alpha_{f,r})\beta(t) \right. \\
\qquad\qquad \left. - (l_f^2 \cdot \alpha_{f,r} + l_r^2 \cdot \alpha_{r,f})\dfrac{\dot{\psi}(t)}{v(t)} \right), \\
\dot{v}(t) = u_{\dot{v}}(t), \\
\dot{\delta}(t) = u_{\dot{\delta}}(t), \\
\dot{\beta}(t) = \dfrac{\mu}{L \cdot v(t)}\left(\alpha_{f,r} \cdot \delta(t) - (\alpha_{r,f} + \alpha_{f,r})\beta(t) \right. \\
\qquad\qquad \left. + (l_r \cdot \alpha_{r,f} - l_f \cdot \alpha_{f,r})\dfrac{\dot{\psi}(t)}{v(t)} \right) - \dot{\psi}(t),
\end{cases}
\tag{30}
$$

where $\alpha_{i,j} := \alpha_{i,j}(u_{\dot{\delta}}(t))$ is a function of the input $u_{\dot{\delta}}(t)$ defined as

$$
\alpha_{i,j} = C_i(g \cdot l_j - h \cdot u_{\dot{\delta}}(t))
\tag{31}
$$

for $i, j \in \{f, r\}$, L given by $L = l_f + l_r$ and the parameters described in Table 1 with the values used in [23].

Table 1 Single-track model's parameters

Parameter	Symbol	Unit	Value
Distance from the center of gravity to front axle	l_f	$[m]$	0.883
Distance from the center of gravity to rear axle	l_r	$[m]$	1.508
Total vehicle mass	M	$[kg]$	1.225
Moment of inertia about z axis	I_z	$[kg \cdot m^2]$	1.538
Center of gravity height of M	h	$[m]$	0.557
Cornering stiffness coeff. (front, rear)	C_f, C_r	$[1/rad]$	20.89
Friction coefficient	μ	$[-]$	1.048

Acknowledgements We thank Tristan Schneider and Matthias K. Hoffmann for their contributions on this chapter. This research is supported by the Deutsche Forschungsgemeinschaft (German Research Foundation) within the Priority Program SPP 1835 "Cooperative Interacting Automobiles" (grant number: KO 1430/17-1).

References

1. Renault Zoe dimensions & specifications. https://www.renault.co.uk/electric-vehicles/zoe/specifications.html. Accessed 25 Feb 2021
2. Coordinated non-cooperative distributed model predictive control for decoupled systems using graphs. IFAC-PapersOnLine **49**(22), 216–221 (2016). https://doi.org/10.1016/j.ifacol.2016.10.399. 6th IFAC Workshop on Distributed Estimation and Control in Networked Systems NECSYS 2016
3. Alrifaee, B.: Networked model predictive control for vehicle collision avoidance. Ph.D. thesis (2017). https://doi.org/10.18154/RWTH-2017-04199
4. Alrifaee, B., Mamaghani, M.G., Abel, D.: Centralized non-convex model predictive control for cooperative collision avoidance of networked vehicles. In: 2014 IEEE International Symposium on Intelligent Control (ISIC), pp. 1583–1588 (2014). https://doi.org/10.1109/ISIC.2014.6967623
5. Althoff, M., Koschi, M., Manzinger, S.: Commonroad: Composable benchmarks for motion planning on roads. In: Proceedings of the IEEE Intelligent Vehicles Symposium (2017). https://doi.org/10.1109/ivs.2017.7995802
6. Andersson, J.A.E., Gillis, J., Horn, G., Rawlings, J.B., Diehl, M.: CasADi—a software framework for nonlinear optimization and optimal control. Math. Program. Comput. **11**(1), 1–36 (2019). https://doi.org/10.1007/s12532-018-0139-4
7. Bemporad, A., Morari, M.: Control of systems integrating logic, dynamics, and constraints. Automatica **35**(3), 407–427 (1999). https://doi.org/10.1016/S0005-1098(98)00178-2
8. Caesar, H., Bankiti, V., Lang, A.H., Vora, S., Liong, V.E., Xu, Q., Krishnan, A., Pan, Y., Baldan, G., Beijbom, O.: Nuscenes: a multimodal dataset for autonomous driving. In: Proceedings of the IEEE/CVF Conference on Computer Vision and Pattern Recognition, pp. 11621–11631 (2020)
9. Dellnitz, M., Eckstein, J., Flaßkamp, K., Friedel, P., Horenkamp, C., Köhler, U., Ober-Blöbaum, S., Peitz, S., Tiemeyer, S.: Multiobjective optimal control methods for the development of an intelligent cruise control. In: Russo, G., Capasso, V., Nicosia, G., Romano, V. (eds.) Progress in

Industrial Mathematics at ECMI 2014, pp. 633–641. Springer International Publishing, Cham (2016)

10. Dolgov, D., Thrun, S., Montemerlo, M., Diebel, J.: Practical search techniques in path planning for autonomous driving. Ann. Arbor. **1001**(48105), 18–80 (2008)

11. Dubins, L.E.: On curves of minimal length with a constraint on average curvature, and with prescribed initial and terminal positions and tangents. Am. J. Math. **79**(3), 497–516 (1957)

12. Flaßkamp, K., Ober-Blöbaum, S., Kobilarov, M.: Solving optimal control problems by exploiting inherent dynamical systems structures. J. Nonlinear Sci. **22**, 599–629 (2012)

13. Flaßkamp, K., Ober-Blöbaum, S., Worthmann, K.: Symmetry and motion primitives in model predictive control. Math. Control Signals Syst. **31**, 455–485 (2019)

14. Frazzoli, E., Dahleh, M., Feron, E.: Maneuver-based motion planning for nonlinear systems with symmetries. IEEE Trans. Rob. **21**(6), 1077–1091 (2005). https://doi.org/10.1109/TRO.2005.852260

15. Kiran, B.R., Sobh, I., Talpaert, V., Mannion, P., Al Sallab, A.A., Yogamani, S., Pérez, P.: Deep reinforcement learning for autonomous driving: a survey. IEEE Trans. Intell. Transp. Syst. (2021)

16. Kloock, M., Scheffe, P., Maczijewski, J., Kampmann, A., Mokhtarian, A., Kowalewski, S., Alrifaee, B.: Cyber-physical mobility lab: an open-source platform for networked and autonomous vehicles. In: 2021 European Control Conference (ECC), pp. 1937–1944 (2021). https://doi.org/10.23919/ECC54610.2021.9654986

17. Kloock, M., Scheffe, P., Marquardt, S., Maczijewski, J., Alrifaee, B., Kowalewski, S.: Distributed model predictive intersection control of multiple vehicles. In: 2019 IEEE Intelligent Transportation Systems Conference (ITSC), pp. 1735–1740 (2019). https://doi.org/10.1109/ITSC.2019.8917117

18. Knauer, M., Büskens, C.: Real-time optimal control using TransWORHP and WORHP Zen, pp. 211–232. Springer International Publishing, Cham (2019). https://doi.org/10.1007/978-3-030-10501-3_9

19. Kobilarov, M.: Discrete geometric motion control of autonomous vehicles. University of Southern California (2008)

20. Lloyd, S.P.: Least squares quantization in PCM. IEEE Trans. Inf. Theory **28**, 129–137 (1982)

21. Lüttgens, L., Jurgelucks, B., Wernsing, H., Roy, S., Büskens, C., Flaßkamp, K.: Autonomous navigation of ships by combining optimal trajectory planning with informed graph search. Math. Comput. Model. Dyn. Syst. **28**(1), 1–27 (2022). https://doi.org/10.1080/13873954.2021.2007138

22. Paden, B., Čáp, M., Yong, S.Z., Yershov, D., Frazzoli, E.: A survey of motion planning and control techniques for self-driving urban vehicles. IEEE Trans. Intell. Veh. **1**(1), 33–55 (2016). https://doi.org/10.1109/TIV.2016.2578706

23. Pedrosa, M.V.A., Schneider, T., Flaßkamp, K.: Graph-based motion planning with primitives in a continuous state space search. In: 2021 6th International Conference on Mechanical Engineering and Robotics Research (ICMERR), pp. 30–39 (2021). https://doi.org/10.1109/ICMERR54363.2021.9680825

24. Pedrosa, M.V.A., Schneider, T., Flaßkamp, K.: Learning motion primitives automata for autonomous driving applications. Math. Comput. Appl. **27**(4) (2022). https://doi.org/10.3390/mca27040054. https://www.mdpi.com/2297-8747/27/4/54

25. Reeds, J., Shepp, L.: Optimal paths for a car that goes both forwards and backwards. Pac. J. Math. **145**(2), 367–393 (1990)

26. Russell, S., Norvig, P.: Artificial Intelligence: A Modern Approach, 3rd edn. Prentice Hall (2010)

27. Scheffe, P., Pedrosa, M.V.A., Flaßkamp, K., Alrifaee, B.: Receding horizon control using graph search for multi-agent trajectory planning. IEEE Trans. Control Syst. Technol. 1–14 (2022). https://doi.org/10.1109/TCST.2022.3214718

28. Schneider, T.: A comparison between reinforcement learning and graph search for motion planning in autonomous driving applications. Bachelor's thesis, Saarland University, Germany (2022)

29. Sutton, R.S., Barto, A.G.: Reinforcement Learning: An Introduction. MIT Press (2018)
30. Yang, Y., Pan, J., Wan, W.: Survey of optimal motion planning. IET Cyber-Syst. Robot. **1**, 13–19(6) (2019). https://doi.org/10.1049/iet-csr.2018.0003
31. Yurtsever, E., Lambert, J., Carballo, A., Takeda, K.: A survey of autonomous driving: common practices and emerging technologies. IEEE Access **8**, 58443–58469 (2020). https://doi.org/10.1109/ACCESS.2020.2983149
32. Zilberstein, S.: Using anytime algorithms in intelligent systems. AI Mag. **17**(3), 73 (1996). https://doi.org/10.1609/aimag.v17i3.1232

Prioritized Trajectory Planning for Networked Vehicles Using Motion Primitives

Patrick Scheffe, Matheus V. A. Pedrosa, Kathrin Flaßkamp, and Bassam Alrifaee

Abstract The computation time required to solve nonconvex, nonlinear optimization problems increases rapidly with their size. This poses a challenge in trajectory planning for multiple networked vehicles with collision avoidance. In the centralized formulation, the optimization problem size increases with the number of vehicles in the networked control system (NCS), rendering the formulation unusable for experiments. We investigate two methods to decrease the complexity of networked trajectory planning. First, we approximate the optimization problem by discretizing the vehicle dynamics with an automaton, which turns it into a graph-search problem. Our search-based trajectory planning algorithm has a limited horizon to further decrease computation complexity. We achieve recursive feasibility by design of the automaton which models the vehicle dynamics. Second, we distribute the optimization problem to the vehicles with prioritized distributed model predictive control (P-DMPC), which reduces the problem size. To counter the incompleteness of P-DMPC, we propose a framework for time-variant priority assignment. The framework expands recursive feasibility to every vehicle in the NCS. We present two time-variant priority assignment algorithms for road vehicles, one to improve vehicle progress and one to improve computation time of the NCS. We evaluate our approach for online trajectory planning of multiple networked vehicles in simulations and experiments.

Code https://github.com/embedded-software-laboratory/p-dmpc

P. Scheffe (✉)
Chair of Embedded Software, RWTH Aachen University, Aachen, Germany
e-mail: scheffe@embedded.rwth-aachen.de

M. V. A. Pedrosa · K. Flaßkamp
Chair of Systems Modeling and Simulation, Systems Engineering, Saarland University, Saarbrücken, Germany
e-mail: matheus.pedrosa@uni-saarland.de

K. Flaßkamp
e-mail: kathrin.flasskamp@uni-saarland.de

B. Alrifaee
Department of Aerospace Engineering, University of the Bundeswehr Munich, Neubiberg, Germany
e-mail: alrifaee@embedded.rwth-aachen.de; bassam.alrifaee@unibw.de

1 Introduction

Networked and autonomous vehicles (NAVs) have the potential to increase the safety and efficiency of traffic [42]. Realizing this potential requires advances in many fields of networked and autonomous vehicles (NAVs), among which is the field of decision making [56]. In decision making, we develop a plan and control the actuators of the vehicle to execute this plan. Planning can be decomposed into three hierarchical layers. The highest layer plans a route through the road network, the middle layer plans behaviors for the vehicle on the road, and the bottom layer plans motions to realize the behavioral plan [46]. The work in this article focuses on the middle and bottom layer of planning for a multi-agent system. We will refer to this area as trajectory planning for multiple NAVs. Section 1.1 motivates our work on networked trajectory planning, Sect. 1.2 presents the state of the art and Sect. 1.3 states our contribution to the state of the art. We introduce our notation in Sect. 1.4 and give an overview of this chapter in Sect. 1.5.

1.1 Motivation

Trajectory planning for multiple NAVs with collision avoidance can be modeled as a nonconvex, nonlinear optimal control problem (OCP). For trajectory planning in changing environments, this OCP must be solved within a duration of tenths of a second. With an increasing amount of controlled vehicles, the OCP grows large, and finding a solution quickly becomes intractable. This chapter investigates two approaches to decrease computation time of networked trajectory planning: simplifying and distributing the OCP.

When simplifying the OCP, a compromise between global optimality and computational efficiency must be found [12]. Trajectory planning approaches can be classified as optimization-based and graph-based [46]. Optimization-based algorithms are often based on convexification of the original nonconvex OCP [5, 6, 28, 52, 58]. The advantage of convexification is a short computation time, which comes at the cost of disregarding nonlinearities in the vehicle model and of disregarding parts of the solution space. Graph-based methods based on motion primitives (MPs) can retain the nonlinearities and the complete solution space. The coarseness of quantization of states and control inputs highly influences the computational complexity and the trajectory quality.

Distributing the centralized OCP, which plans trajectories for all vehicles at once, reduces the computational effort at the cost of global system knowledge. Prioritized trajectory planning for vehicles is first presented in [21]. In a prioritized approach, vehicles with lower priority adjust their objectives and constraints to respect coupled vehicles with higher priority. The core problem of prioritized planning algorithms is their incompleteness. That is, there might exist a priority assignment that leads to feasible optimization problems of all participating agents, but the algorithm can fail to find it.

1.2   Related Work

This section presents related work on trajectory planning with MPs and on prioritized trajectory planning.

1.2.1   Trajectory Planning with Motion Primitives

The goal of trajectory planning with MPs is to find an optimal sequence and duration of MPs that achieve a desired objective while satisfying constraints. MP consists of a control and state trajectory. Multiple MPs can be concatenated to form a vehicle trajectory plan. There are mainly two kinds of methods to plan trajectories using MPs: methods based on continuous optimization problem formulations, such as mixed integer programming (MIP), and methods with graph-based problem formulations, such as an A* algorithm or a rapidly-exploring random tree algorithm [39]. A literature review on both methods follows.

MIP formulates an OCP with both continuous and discrete variables. MIP can find the optimal sequence and duration of MPs for trajectory planning of a single vehicle [23, 26, 27]. When dealing with multiple NAVs, collision constraints can be modeled with binary decision variables [7]. The ability of MIP to find the optimal solution comes at the cost of high computation time, which rapidly increases with the size of the OCP. Centralized trajectory planning for multiple vehicles with MPs [2, 20, 22] encounters this problem.

A popular search algorithm for trajectory planning using MPs is A* and its variant hybrid A* [1, 19, 49]. When operating on a gridded environment representation, A* associates a cost value with a grid cell center and the cell center's state value, whereas hybrid A* associates a cost value alongside a continuous state value with a grid cell. A computationally demanding task in search algorithms for trajectory planning are edge evaluations, as they incorporate the collision constraints [39]. The number of edge evaluations can be reduced using a lazy approach, in which an edge is only evaluated when the connected vertex is chosen for expansion [17, 18, 43]. The computation time of graph-search algorithms increases with the length of its horizon. Limiting the horizon decreases computation time [9, 36, 45]. Algorithms for graph-based trajectory planning for multiple NAVs include a Monte Carlo tree search [37] and a traditional A* graph search [24, 25]. Graph searches with an infinite horizon suffer from high computation time [17–19, 43, 49]. This challenge can be overcome with a receding horizon at the cost of global optimality guarantees. Graph-based receding horizon approaches do not yet guarantee recursive feasibility [9, 40, 45].

1.2.2 Prioritized Distributed Control

The distributed control strategy for networked control system (NCS) examined in this work is prioritized distributed model predictive control (P-DMPC), in which each vehicle optimizes only its own decision variables. Lower prioritized vehicles consider a communicated optimized solution of coupled higher prioritized vehicles in both in their objective function and their constraints. The benefit of the greedy P-DMPC algorithm is its short computation time [3, 57]. One of the main challenges in P-DMPC is its incompleteness [40]. That means, a priority assignment might lead to an infeasible OCP of a vehicle although the problem is solvable with a different priority assignment. Additionally, the priority assignment influences the solution quality and the computation time.

The following works have designed priority assignments for robots and NAVs with the goal of feasibility and solution quality. In our work [32] the ordering is based on rules, i.e., we assign time-variant priorities to multiple vehicles competing on a racetrack based on their race position. Constraint-based heuristics increase the priority of a vehicle with the number of constraints it has [13, 16, 41, 48, 60]. The goal of these heuristics is to maintain feasibility of the control problems. In our work [35], we assign priorities to vehicles based on the time remaining before they enter an intersection. In our work [31], we assign priorities to vehicles based on the crowdedness of their goal location. Objective-based heuristics assign priorities to improve the solution quality of the NCS [15, 59]. A randomized priority assignment with hill-climbing is proposed in [10]. In [8], all priority assignments are considered to find the optimal one. Both approaches achieve higher solution quality with higher computation time. In [61], priorities are assigned based on machine learning and achieve results competitive to heuristics. The priority assignment can also influence computation time [4]. The number of simultaneous computations in prioritized planning is maximized in [38]. Despite the number of priority assignment strategies, the incompleteness of P-DMPC remains. Many works assign time-invariant priorities for a specific scenario [13, 16, 38, 41, 48, 59, 60]. Time-variant priority assignments improve feasibility in changing operating conditions over time-invariant priority assignments [10, 15]. In [38], time-invariant priorities are shown to produce recursively feasible solutions. Similarly, this property needs to be shown for time-variant priorities.

1.3   Contribution

The contribution of this chapter is twofold. First, we present our method of receding horizon graph search (RHGS), a search-based trajectory planning algorithm for road vehicles. We reduce the computation time by limiting the planning horizon. We prove that our method fulfills recursive feasibility by design of the motion primitive automaton (MPA) [55]. Second, we present a framework for distributed reprioritization of vehicles. We prove that it fulfills recursive NCS-feasibility for P-DMPC with any time-variant priority assignment algorithm [51].

We present two priority assignment algorithms, one for vehicle progression using a constraint-based heuristic, and a one for computation time reduction of the NCS using graph coloring. We demonstrate the effectiveness of the presented approach in a simulative case study of P-DMPC for trajectory planning.

1.4 Notation

A variable x is marked with a superscript $x^{(j)}$ if belonging to agent j, and with $x^{(-j)}$ if belonging to the neighbors of agent j. The actual value of a variable x at time k is written as $x(k)$, while values predicted for time $k + i$ at time k are written as $x_{k+i|k}$. A trajectory is denoted by substituting the time argument with $\cdot$ as in $x_{\cdot|k}$. An agent equals a vehicle in our application of prioritized trajectory plannning. In this chapter, we use the terms vehicle, road vehicle and NAV interchangeably.

1.5 Structure

The remainder of this chapter is structured as follows. Section 2 presents our vehicle model, our RHGS for trajectory planning, and our proof of recursive agent-feasibility. Section 3 presents the distribution of RHGS with P-DMPC for trajectory planning. We show recursive NCS-feasibility of our reprioritization framework before presenting two time-variant priority assignment algorithms, one for vehicle progression and one for computation time reduction. In Sect. 4, we evaluate both the RHGS and the P-DMPC in experiment, before combining both in a simulative case study.

2 Receding Horizon Graph Search for Trajectory Planning

This section presents how we transfer a receding horizon control (RHC) approach to graph-based trajectory planning. The content is based on our previous publication [55]. Section 2.1 states the RHC trajectory planning problem that we subsequently map to graph search based on an MPA. Section 2.2 presents our approximation of the vehicle dynamics as an MPA, Sect. 2.3 shows the graph-based optimization in our RHGS algorithm. In Sect. 2.4, we prove that our RHGS produces recursively agent-feasible trajectories by design of the MPA.

Fig. 1 Kinematic
single-track model of a
vehicle [55]

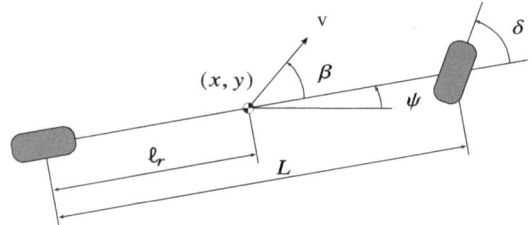

Fig. 1 Kinematic single-track model of a vehicle [55]

2.1 Trajectory Planning Problem

This section presents the ordinary differential equations describing the vehicle dynamics and our cost function before both are incorporated in a RHC problem for trajectory planning.

Figure 1 shows an overview of the variables for the nonlinear kinematic single-track model [47]. Assuming low velocities, we model no slip on the front and rear wheels, and no forces acting on the vehicle. The resulting equations are

$$\begin{cases} \dot{x}(t) = v(t) \cdot \cos(\psi(t) + \beta(t)), \\ \dot{y}(t) = v(t) \cdot \sin(\psi(t) + \beta(t)), \\ \dot{\psi}(t) = v(t) \cdot \dfrac{1}{L} \cdot \tan(\delta(t)) \cos(\beta(t)), \\ \dot{v}(t) = u_v(t), \\ \dot{\delta}(t) = u_\delta(t), \end{cases} \tag{1}$$

with

$$\beta(t) = \tan^{-1}\left(\frac{\ell_r}{L} \tan(\delta(t))\right), \tag{2}$$

where $x \in \mathbb{R}$ and $y \in \mathbb{R}$ describe the position of the center of gravity (CG), $\psi \in [0, 2\pi)$ is the orientation, $\beta \in [-\pi, \pi)$ is the side slip angle, $\delta \in [-\pi, \pi)$ and $u_v \in \mathbb{R}$ are the steering angle and its derivative respectively, $v \in \mathbb{R}$ and $u_v \in \mathbb{R}$ are the speed and acceleration of the CG respectively, L is the wheelbase length and ℓ_r is the length from the rear axle to the CG. The position of the CG and the orientation together form the pose p.

The system dynamics defined in (1) are compactly written as

$$\dot{x}(t) := \frac{d}{dt} x(t) = f\big(x(t), u(t)\big) \tag{3}$$

with the state vector

$$x = \big(x \ y \ \psi \ v \ \delta\big)^{\mathrm{T}} \in \mathbb{R}^5, \tag{4}$$

the control input

$$\boldsymbol{u} = \left(u_v \ u_v\right)^{\mathrm{T}} \in \mathbb{R}^2 \tag{5}$$

and the vector field f defined by (1). Transferring (3) to a discrete-time nonlinear system representation yields

$$\boldsymbol{x}_{k+1} = f_d\left(\boldsymbol{x}_k, \boldsymbol{u}_k\right) \tag{6}$$

with $k \in \mathbb{N}$, the vector field $f_d \colon \mathbb{R}^5 \times \mathbb{R}^2 \to \mathbb{R}^5$, the state vector $\boldsymbol{x} \in \mathbb{R}^5$ and the input vector $\boldsymbol{u} \in \mathbb{R}^2$.

We define the cost function to minimize in our trajectory planning problem as

$$J_{k \to k+N|k} = \sum_{i=1}^{N} \left(\boldsymbol{x}_{k+i|k} - \boldsymbol{x}_{\mathrm{ref},k+i|k}\right)^{\mathrm{T}} \boldsymbol{Q} \left(\boldsymbol{x}_{k+i|k} - \boldsymbol{x}_{\mathrm{ref},k+i|k}\right) \tag{7}$$

with the planning horizon length N, the positive semi-definite, block diagonal matrix

$$\boldsymbol{Q} = \begin{pmatrix} \boldsymbol{I}_2 & \boldsymbol{0}_{2\times 3} \\ \boldsymbol{0}_{3\times 2} & \boldsymbol{0}_3 \end{pmatrix} \in \mathbb{R}^{5\times 5} \tag{8}$$

and a reference trajectory $\boldsymbol{x}_{\mathrm{ref},\cdot|k} \in \mathbb{R}^5$.

We combine the system model (6) and the cost function (7) to an OCP

$$\underset{U_{k \to k+N|k}}{\text{minimize}} \quad J_{k \to k+N|k} \tag{9a}$$

subject to

$$\boldsymbol{x}_{k+i+1|k} = f_d\left(\boldsymbol{x}_{k+i|k}, \boldsymbol{u}_{k+i|k}\right), \quad i = 0, \ldots, N-1 \tag{9b}$$

$$\boldsymbol{u}_{k+i|k} \in \mathcal{U}, \quad i = 0, \ldots, N-1 \tag{9c}$$

$$\boldsymbol{x}_{k+i|k} \in \mathcal{X} \quad i = 1, \ldots, N-1 \tag{9d}$$

$$\boldsymbol{x}_{k+N|k} \in \mathcal{X}_f \tag{9e}$$

$$\boldsymbol{x}_{k|k} = \boldsymbol{x}(k) \tag{9f}$$

with the vector $U_{k \to k+N|k}$ of stacked control inputs $(\boldsymbol{u}_{k|k}, \boldsymbol{u}_{k+1|k}, \ldots, \boldsymbol{u}_{k+N-1|k})$, the input constraint set $\mathcal{U} \subseteq \mathbb{R}^2$, the state constraint set $\mathcal{X} \subseteq \mathbb{R}^5$ and the terminal set $\mathcal{X}_f \subseteq \mathbb{R}^5$. We assume a full measurement or estimate of the state $\boldsymbol{x}(k)$ is available at the current time k. The OCP (9) is solved repeatedly after a timestep duration T_s and with updated values for the states and constraints, which establishes the RHC.

2.2 Motion Primitive Automaton as System Model

This section presents how we model the state-continuous system (6) as an MPA, a type of maneuver automaton [23]. The MPA incorporates the constraints on system dynamics (9b), on control inputs (9c), and on both the steering angle and the speed (9d) and (9e).[1]

From the system dynamics (1), we derive a finite state automaton which we call MPA and define as follows.

Definition 1 (*Motion primitive automaton*) An MPA is a 5-tuple (Q, S, γ, q_0, Q_f) composed of:

- Q is a finite set of automaton states q;
- S is a finite set of transitions π, also called motion primitives;
- $\gamma : Q \times S \times \mathbb{N} \to Q$ is the update function defining the transition from one automaton state to another, dependent on the timestep in the horizon;
- $q_0 \in Q$ is the initial automaton state;
- $Q_f \subseteq Q$ is the set of final automaton states.

An automaton state is characterized by a specific speed v and steering angle δ. An MP is characterized by an input trajectory and a corresponding state trajectory which starts and ends with the speed and steering angle of an automaton state. It has a fixed duration which we choose equal to the timestep duration T_s. MPs can be concatenated to vehicle trajectories by rotation and translation. Our MPA discretizes both the state space with the update function γ and the time space with a fixed duration T_s for all MPs. This MPA replaces the system representation (6). Note that the dynamics model on which our MPA is based is exchangeable. Its complexity is irrelevant computation-wise for trajectory planning since MPs are computed offline.

2.3   Receding Horizon Graph Search Algorithm

This section demonstrates how our RHGS incorporates the constraints on the pose, which are included in (9d) and (9e), while minimizing the cost function (9a).

Our RHGS algorithm constructs a search tree $\mathcal{T}$ up to a limited depth N. A level i in the tree directly corresponds to the timestep $k + i$ in the OCP (9). The information contained in each vertex v of the tree is a tuple $\langle q, p, i, J \rangle$, whose elements are the automaton state, the vehicle pose, the distance to the root vertex, and the value of the cost function, respectively. When the algorithm finds the leaf vertex with the minimal cost value at the horizon $k + N$, it returns the path from the root vertex to this leaf vertex. The algorithm ensures optimality of the returned path with an admissible and underestimating cost estimation, similar to A*.

---

[1] A detailed explanation of modeling with MPAs is found in this book's chapter "Designing Maneuver Automata of Motion Primitives for Optimal Cooperative Trajectory Planning".

Algorithm 1 Receding Horizon Graph Search

Input: initial vertex v_0, MPA, goal set $\mathcal{X}_f$
Output: path from v_0 to best vertex v_p
 1: $L_{\text{open}} \leftarrow v_0$
 2: **while** $L_{\text{open}} \neq \emptyset$ **do**
 3:     Sort L_{open} ascending by $J = J_{\text{CTC}} + J_{\text{CTG}}$
 4:     $v_p \leftarrow L_{\text{open}}[0]$
 5:     $L_{\text{open}} \leftarrow L_{\text{open}} \setminus v_p$
 6:     **if not** IsVALID(v_p) **then**
 7:         **continue**
 8:     **if** $i_{v_p} = N$ and $\boldsymbol{x}_{v_p} \in \mathcal{X}_f$ **then**
 9:         **return** path from v_0 to v_p
10:     successors$\leftarrow$ EXPAND(v_p)
11:     **for all** $v_s \in$ successors **do**
12:         LAZYEVAL(v_s)
13:         $L_{\text{open}} \leftarrow L_{\text{open}} \cup v_s$
14: **return** failure

Algorithm 1 shows the main steps of our RHGS algorithm. At the beginning of the control loop at time k, the algorithm determines the search tree's root vertex v_0 from the state vector $\boldsymbol{x}(k)$ and initializes the open list with this root vertex (Line 1). Sorting the open list by the cost function value brings the vertex with the lowest cost v_p to the front (Line 3). It is removed from the open list (Line 5). We evaluate the edge to the selected vertex by checking inter-vehicle collisions and obstacle collisions (Line 6). If there is a collision, the algorithm continues to the next vertex in the open list. If the vertex is collision-free, satisfies the constraint (9e), and is at the planning horizon N, it is optimal (Line 8). The algorithm returns the path to the vertex (Line 9). Otherwise, the algorithm expands the vertex based on its automaton state q, the update function γ, and its state vector $\boldsymbol{x}$ (Line 10). The algorithm evaluates edges to successors lazily by computing only the cost function without collision checks to reduce computation time (Lines 11 to 12). In informed graph-search algorithms, the cost function consists of the cost-to-come (CTC) and the cost-to-go (CTG). Our algorithm minimizes (7) as the CTC is equal to (7) and the CTG is an underestimation of (7). We underestimate the cost from a vertex v at depth i_v by moving a vehicle towards its reference position at each subsequent timestep with maximum speed in a straight line

$$
J_{\text{CTG}}(i_v) = \sum_{i=i_v+1}^{N} \max \Bigg(0,
$$

$$
\left(\boldsymbol{x}_{k+i|k} - \boldsymbol{x}_{\text{ref},k+i|k}\right)^{\text{T}} \boldsymbol{Q} \left(\boldsymbol{x}_{k+i|k} - \boldsymbol{x}_{\text{ref},k+i|k}\right) - i \cdot v_{\max} \cdot T_s \Bigg) \tag{10}
$$

with the same $\boldsymbol{Q}$ as in (7). At the end of the loop, all successor vertices are added to the open list (Line 13).

2.4 Recursive Agent-Feasibility

This section proves recursive agent-feasibility of our RHGS. The property is commonly known as recursive feasibility or persistent feasibility. We design the time-variant update function γ of our MPA such that an equilibrium state can always be reached within the horizon N from expanded successors (Line 10).

A set $C_{\text{inv}} \subseteq X$ is a control invariant set for the system (6) subject to constraints (9b)–(9f) if

$$x(k) \in C_{\text{inv}} \implies \exists u(k) \in \mathcal{U} \text{ such that}$$
$$x(k+1) \in C_{\text{inv}}, \forall k \in \mathbb{N}. \tag{11}$$

Lemma 1 *If X_f is a control invariant set of the system (9) with $N > 1$, then (9e) ensures recursive agent-feasibility of the RHC.*

Proof The proof is given in [11].   □

We reformulate the condition of control invariant sets for MPAs as follows.

Definition 2 (*Control invariant set for an MPA*) A set $C_{\text{inv}} \subseteq X$ is a control invariant set for the system (6) given by an MPA if

$$x(k) \in C_{\text{inv}} \text{ with } q(k) \in Q_f \implies \exists \pi \in S \text{ such that}$$
$$x(k+1) \in C_{\text{inv}} \text{ with } q(k+1) \in Q_f \text{ and} \tag{12}$$
$$\gamma(q(k), \pi, k) = q(k+1), \forall k \in \mathbb{N}.$$

Note that the automaton state q follows from the state vector x.

Theorem 1 *RHGS achieves recursive agent-feasibility if the generated sequence of transitions ends in an automaton state and a state vector that together form a control invariant set.*

Proof Follows directly from Lemma 1 with Definition 2 of control invariant sets for MPAs.   □

In an equilibrium of the system, it holds that $f_d\big(x(k), u(k)\big) = x(k)$. If a sequence of transitions ends in an automaton state from where there exists a transition which keeps the system at an equilibrium, $x(k)$ represents a control invariant set. Such an automaton state in our MPA has a speed $v = 0\,\text{m s}^{-1}$. Figure 2 depicts a simple example of an MPA with a time-invariant update function. This MPA can generate sequences of transitions which are not recursively feasible. We design a time-variant update function which only generates recursively feasible sequences, as shown in an example MPA in Fig. 3.

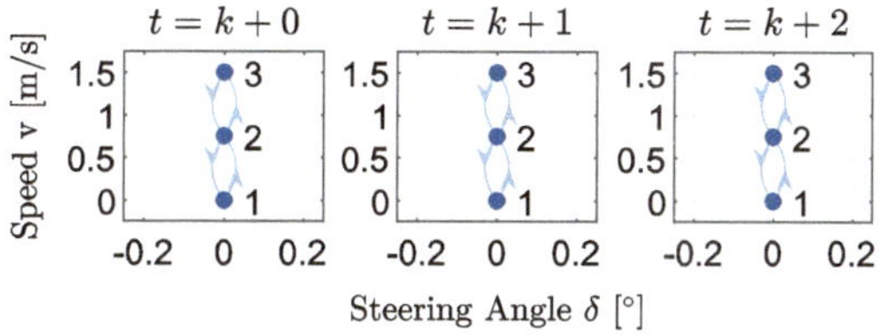

Fig. 2 MPA which does not guarantee recursive agent-feasibility, rolled out over a planning horizon with length $N = 3$

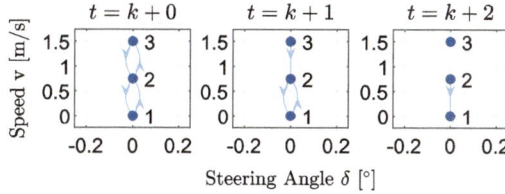

Fig. 3 MPA which guarantees recursive agent-feasibility by only allowing a speed of $0\,\mathrm{m\,s^{-1}}$ at the end of the horizon, rolled out over a planning horizon with length $N = 3$

3 Prioritized Trajectory Planning

This section presents our approach for distributed trajectory planning with distributed reprioritization while guaranteeing recursive NCS-feasibility. It is based on our publications [51, 53]. Our P-DMPC loop consists of the steps coupling, prioritization, trajectory planning, and communication of trajectories. We couple agents if they potentially interact during their planning horizon N. We represent couplings between agents with a coupling graph. Denote by $\mathcal{V} = \{1, \dots, N_A\}$ the set of agents and by $N_A = |\mathcal{V}| \in \mathbb{N}$ its cardinality.

Definition 3 (*Coupling graph*) A coupling graph $G = (\mathcal{V}, \mathcal{E})$ is a graph that represents the interaction between agents. Vertices represent agents and edges denote coupling objectives or constraints in the OCP associated with the vertex.

The agents connected to agent j are called its neighbors $\mathcal{V}^{(j)}$. Introducing priorities results in clear responsibilities to satisfy collision constraints. We direct edges in the coupling graph from a higher prioritized agent to a lower prioritized agent.

Definition 4 (*Directed coupling graph*) A directed coupling graph $G' = (\mathcal{V}, \mathcal{E}')$ results from a coupling graph $G = (\mathcal{V}, \mathcal{E})$ by keeping all vertices $\mathcal{V}$ and a subset of edges $\mathcal{E}' \subset \mathcal{E}$ of G. In a directed coupling graph, a directed edge denotes a coupling objective or constraint in the OCP associated with the ending vertex.

Vehicles determine their priorities using a priority assignment algorithm. A time-variant priority assignment algorithm yields an injective priority assignment function $p: \mathcal{V} \times \mathbb{N} \to \mathbb{N}$, which assigns a unique priority to each vehicle in the NCS at every timestep. If $p(l, k) < p(j, k)$, then vehicle l has a higher priority than vehicle j at timestep k. At each timestep k, every vehicle groups its current neighbors $\mathcal{V}^{(j)}(k)$ in a set of higher prioritized neighbors $\hat{\mathcal{V}}^{(j)}(k)$ and lower prioritized neighbors $\check{\mathcal{V}}^{(j)}(k)$. When a vehicle j has received the planned trajectories of all vehicles in $\hat{\mathcal{V}}^{(j)}(k)$, it plans its own trajectory while avoiding collisions with the received trajectories. It communicates its own trajectory to vehicles in $\check{\mathcal{V}}^{(j)}(k)$. Each vehicle j adds constraint functions $c^{(j,l)}$ to its OCP (9) to ensure collision-free trajectories with vehicles in $\hat{\mathcal{V}}^{(j)}(k)$

$$c^{(j,l)} \left(\boldsymbol{x}^{(j)}_{k+i|k}, \boldsymbol{x}^{(l)}_{k+i|k} \right) \leq 0, \quad \forall i = 1, \ldots, N, \quad \forall l \in \hat{\mathcal{V}}^{(j)}(k). \tag{13}$$

3.1 Reprioritization Framework for Recursive NCS-Feasibility

One of the main challenges for P-DMPC is its incompleteness: even though there exists a priority assignment that results in an NCS-feasible P-DMPC problem, a specific priority assignment might fail to produce a solution. Changing the priority assignment during runtime can prevent such a failure, but loses recursive NCS-feasibility of the P-DMPC problem.

Definition 5 (*NCS-feasible*) A P-DMPC problem is NCS-feasible if every agent in the NCS finds a feasible solution to its OCP.

A P-DMPC problem is recursively NCS-feasible if from NCS-feasibility at time k we can guarantee NCS-feasibility for all future times. Figure 4 illustrates our distributed reprioritization framework to maintain NCS-feasible P-DMPC trajectory planning problems while using a time-variant priority assignment function. At the beginning of every timestep k, each agent attempts to plan its trajectory given the priorities from time k. If any agent fails to find a feasible solution, it notifies all other agents. All agents stay on their recursively agent-feasible trajectory. At any point, if the P-DMPC problem is NCS-feasible, the corresponding input is applied.

A proof for recursive NCS-feasibility of time-invariant priorities is given in [38]. We need to prove recursive NCS-feasibility with time-variant priorities and our distributed reprioritization framework. We assume an initially NCS-feasible problem and bounded disturbances which an underlying controller can compensate.

Theorem 2 *A P-DMPC problem with our distributed reprioritization framework, the OCP (9) with coupling constraints (13), and any time-variant priority assignment function p is recursively NCS-feasible.*

Fig. 4 Distributed reprioritization framework which guarantees recursive NCS-feasibility, as seen from agent j. Figure adapted from [51]

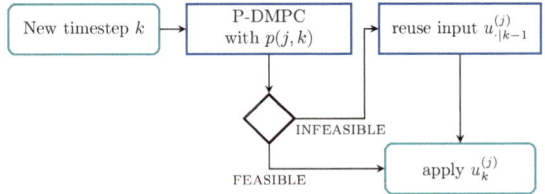

Proof Without loss of generality, assume the computation order resulting from the priority assignment function $p(j, k)$ to be $1, \ldots, N_A$. Assume an NCS-feasible solution $\left(u^{(j)}_{\cdot|k}, x^{(j)}_{\cdot|k}\right)$, $\forall j \in \mathcal{V}$ at timestep k. Because of bounded disturbances which an underlying controller can compensate, we have

$$x^{(j)}(k+1) \approx x^{(j)}_{k+1|k}, \quad \forall j \in \mathcal{V}. \tag{14}$$

Every agent shifts and extends the feasible solution of the previous timestep

$$\begin{aligned}
x^{(j)}_{k+1+i|k+1} &= x^{(j)}_{k+1+i|k}, \quad \forall j \in \mathcal{V}, \quad \forall i = 1, \ldots, N-1 \\
x^{(j)}_{k+1+N|k+1} &= x^{(j)}_{k+N|k}, \quad \forall j \in \mathcal{V}
\end{aligned} \tag{15}$$

For agent 1, who does not consider other agents, recursive feasibility is given by Theorem 1. For any agent $2 \le j \le N_A$, the coupling constraints (13) must also be considered. Substituting (15) in (13) yields

$$c^{(j,l)}\left(x^{(j)}_{k+1+i|k+1}, x^{(l)}_{k+1+i|k+1}\right) = c^{(j,l)}\left(x^{(j)}_{k+1+i|k}, x^{(l)}_{k+1+i|k}\right), \tag{16}$$

$\forall i = 1, \ldots, N-1$ and $\forall l \in \hat{\mathcal{V}}^{(j)}(k)$. Since the agents stand still at the horizon, we have for the last timestep $k + N + 1$

$$c^{(j,l)}\left(x^{(j)}_{k+N+1|k+1}, x^{(l)}_{k+N+1|k+1}\right) = c^{(j,l)}\left(x^{(j)}_{k+N|k}, x^{(l)}_{k+N|k}\right) \tag{17}$$

$\forall l \in \hat{\mathcal{V}}^{(j)}(k)$. This establishes recursive NCS-feasibility of the P-DMPC at time k. Because of a time-variant directed coupling graph, the set of higher prioritized agents $\hat{\mathcal{V}}^{(j)}(k+1)$ might differ from $\hat{\mathcal{V}}^{(j)}(k)$. Still, all coupling constraints are fulfilled. Our coupling constraints are symmetric, i.e., $c^{(j,l)} = c^{(l,j)}$. A new coupling constraint is guaranteed to be satisfied, as there was no collision possibility in timestep k. A vanished coupling constraint cannot interfere with feasibility. Since all constraints are satisfied at timestep $k + 1$, the P-DMPC problem with time-variant priorities is recursively NCS-feasible with our reprioritization framework. □

3.2 Priority Assignment Algorithms

This section introduces two priority assignment functions. Section 3.2.1 describes a constraint-based heuristic which aims at assigning priorities for NCS-feasibility. Section 3.2.2 presents a priority assignment function based on coloring of the coupling graph which reduces computation time.

3.2.1 Constraint-Based Heuristic

The goal of the priority assignment function presented in this subsection is to reduce the risk of standstill of the NCS due to infeasible OCPs of vehicles. We propose a distributed, time-variant priority assignment algorithm for road vehicles on road networks based on our previous work [51]. Each vehicle j first plans a trajectory without inter-vehicle collision constraints (13), which we call the free trajectory $y_{\text{free}}^{(j)}$. Then, each vehicle j counts the number of collisions N_c with other free trajectories $y_{\text{free}}^{(-j)}$ and possibly already planned, optimal trajectories $y^{(-j)^*}$. Vehicle w with most collisions receives the next priority and plans its trajectory considering already planned, optimal trajectories $y^{(-w)^*}$ by solving OCP (9) with coupling constraints (13). The loop repeats until all vehicles have planned their optimal trajectories. If a vehicle cannot find a feasible solution, all vehicles use the previous input as illustrated in Fig. 4. This algorithm results in a time-variant priority assignment function $p_{\text{fca}} : \mathcal{V} \times \mathbb{N} \to \mathbb{N}$. The index "future collision assessment (FCA)" reflects the inspiration of this approach from [41].

3.2.2 Graph Coloring

In P-DMPC, if there is no path between two vehicles in the coupling DAG, they can compute in parallel [4]. We call the number of necessary sequential computations the number of computation levels. This section presents a priority assignment function which minimizes the number of computation levels by vertex coloring based on our previous work [53]. Figure 5 illustrates the proposed problem solution with an example. From an example undirected graph, a baseline approach which assigns priorities equal to the vertex number results in four computation levels. Assigning

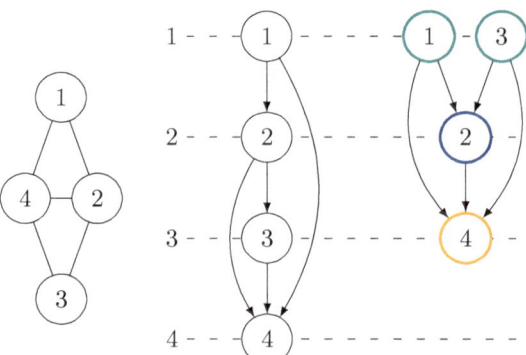

Fig. 5 Example of computation levels from graph coloring compared to baseline. Left: Undirected coupling graph. Middle: Coupling DAG with computation levels from baseline priorities equal to vertex number. Right: Coupling DAG with computation levels from priorities based on graph coloring. Figure adapted from [53]

priorities with our coloring approach reduces the number of computation levels to three, as each color corresponds to a computation level.

In vertex coloring, we map vertices $i \in \mathcal{V}(G)$ to colors $c \in C \subset \mathbb{N}_{>0}$ with the function $\varphi \colon \mathcal{V}(G) \to C$. In order to produce a valid coloring, φ has to satisfy

$$\varphi(i) \neq \varphi(j), \quad \forall i, j \in \mathcal{V}(G), \forall e_{ij} \in \mathcal{E}(G), i \neq j. \tag{18}$$

Our distributed graph coloring algorithm must produce the same coloring φ in every vehicle and must be fast enough for online execution. We propose a combination of saturation degree ordering, largest degree ordering and first-fit to achieve a deterministic coloring as detailed in [53]. We translate our graph coloring function φ to a priority assignment function p. Let $\mathcal{V}_c$ be all vertices of color c

$$\mathcal{V}_c = \{v \mid v \in \mathcal{V}, \varphi(v) = c\}. \tag{19}$$

We can generate a coupling DAG from an undirected coupling graph colored with φ with an injective priority assignment function p that fulfills the requirement

$$p(i) < p(j) \iff c_1 < c_2, \quad \forall i \in \mathcal{V}_{c_1}, \forall j \in \mathcal{V}_{c_2}. \tag{20}$$

4 Numerical and Experimental Results

This section describes the evaluation platform, our Cyber-Physical Mobility Lab (CPM Lab).[2] It presents the evaluation of our RHGS algorithm for recursive agent-feasibility and of our reprioritization framework for recursive NCS-feasibility. Our algorithms are implemented in MATLAB R2023a and openly available online.[3]

4.1 Cyber-Physical Mobility Lab

The evaluation hardware for this work is our 1:18 model-scale CPM Lab [34]. It is a remotely accessible open-source platform consisting of 20 networked and autonomous vehicles (μCars) [54]. Our trajectory planning algorithms run on a PC with an AMD Ryzen 5 5600X 6-core 3.7 GHz CPU and 32 GB of RAM. This PC communicates with the other components in the CPM Lab via the data distribution service standard over WLAN [33]. Figure 6 illustrates the road network in the CPM Lab. It replicates a wide variety of common traffic scenarios with a 16-lane urban intersection, a highway, highway on-ramps, and highway off-ramps.

---

[2] https://cpm.embedded.rwth-aachen.de.

[3] https://github.com/embedded-software-laboratory/p-dmpc.

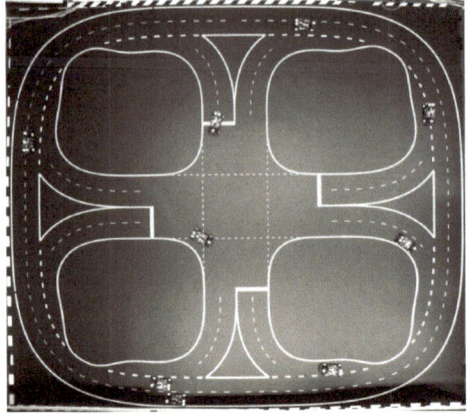

Fig. 6 1:18 model-scale road network in the CPM Lab with an intersection, a highway, highway on-ramps, and highway off-ramps

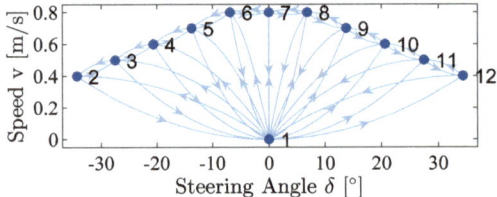

Fig. 7 The MPA for our experiments. The position of a state marks its speed v and its steering angle δ. For clarity of presentation, the figure omits the time dependency of transitions to ensure recursive feasibility

Our algorithm plans trajectories using the MPA shown in Fig. 7. It is based on a kinematic bicycle model (1) of our μCars with $\ell_r = 7.5$ cm and $L = 15$ cm. It is designed such that transitions between automata states respect input constraints of the μCars used in the experiments. The transitions change the control inputs linearly over the duration of the sampling time $T_s = 0.2$ s. The planning horizon is $N = 8$.

4.2 Evaluation of Receding Horizon Graph Search

In our RHGS algorithm, we achieve recursive agent-feasibility by design of the MPA, as illustrated in Fig. 3. The recursive agent-feasibility is verified in [55].

In [55], we compare our RHGS planner with a state-of-the-art graph search (SGS) planner. The SGS planner computes the trajectory once at the beginning of the experiment with a horizon spanning the whole experiment duration. The test scenario contains moving obstacles with known future trajectories. Both planners manage to avoid the obstacles. In the specific test scenario, the RHGS planner stops in front of the obstacles, while the SGS avoids the obstacles by steering early enough. Consequently, the cost function value is lower for the SGS than for the RHGS. However, in the worst case, the computation effort increases exponentially with the horizon length. A video of an experiment using RHGS with multiple vehicles in the CPM Lab is available online.[4]

4.3 Evaluation of Time-Variant Priority Assignment

This section presents P-DMPC trajectory planning with time-variant priority assignment using our reprioritization framework depicted in Fig. 4 to guarantee recursive NCS-feasibility. A time-invariant priority assignment algorithm and a time-variant random priority assignment algorithm represent state-of-the-art priority assignment algorithms for our evaluation. In the time-invariant priority assignment algorithm, each vehicle receives a unique priority corresponding to its unique number $j \in \mathcal{V}$ at the beginning of the experiment. The priority assignment function $p_{\text{const}} : \mathcal{V} \times \mathbb{N} \to \mathbb{N}$ is

$$p_{\text{const}}(j, k) = j. \tag{21}$$

In the random priority assignment algorithm, each vehicle receives a random priority in each timestep. The priority assignment function $p_{\text{rand}} : \mathcal{V} \times \mathbb{N} \to \mathbb{N}$ is

$$p_{\text{rand}}(j, k) = r(k). \tag{22}$$

The evaluation focuses on two criteria: (i) the ability to maintain progress of the vehicles, i.e., to avoid a standstill, and (ii) the ability to reduce computation time. We call the absence of progress a standstill, which we define as a situation where two or more vehicles stop for the rest of the experiment.

Our evaluation spans 720 numerical experiments with an individual duration of 180 s, a combination of the four priority assignment functions (p_{fca}, p_{color}, p_{rand}, and p_{const}) with vehicle amounts from 1 to 20 in 9 random scenarios. All scenarios are based on the map shown in Fig. 6. The vehicle starting positions and their reference paths in the map are determined randomly to replicate various traffic situations. We use the Mersenne Twister algorithm [44] with a manually set random seed for reproducible experiments.

Figure 8 depicts the performance on a scale of 0 to 1 of the four priority assignments in three aspects. The first aspect is the number of vehicles up to which all vehicles in all scenarios could maintain progress over the experiment duration. The functions p_{const} and p_{fca} are able to move up to 10 and 9 vehicles respectively, whereas p_{rand} and p_{color} produce a standstill with already 6 and 5 vehicles respectively. The second aspect is the percentage of scenarios from all scenarios with all numbers of vehicles, for which the corresponding priority assignment function successfully

---

[4] https://youtu.be/7LB7I5SOpQE.

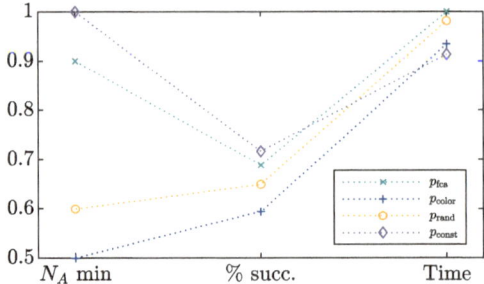

Fig. 8 Performance of priority assignment functions scaled from 0 to 1: N_A min: standstill-free up to number of agents ($\times 10$), % succ.: percentage of standstill-free scenarios ($\times 100$), Time: average time until standstill ($\times 145.1$ s)

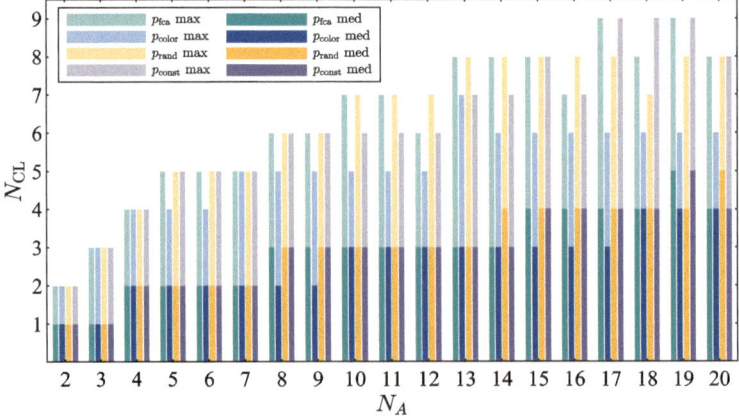

Fig. 9 Median and maximum number of computation levels N_{CL} in all timesteps of all standstill-free scenarios per priority assignment function over the number of vehicles N_A

maintained progress over the full experiment duration. The performance tendency is similar to the first aspect. Both aspects indicate that a change in the priority assignment can decrease NCS-feasibility. A constant priority might not be ideal in all situations, but can help maintaining NCS-feasibility and avoid standstills. The third aspect is the average time until standstill, in which p_{fca} performs best with an average time of 145.1 s. These results indicate that changing priorities might harm the systems performance. A better approach might be to change priorities only when the P-DMPC problem becomes NCS-infeasible.

The computation time in P-DMPC is mainly determined by number of computation levels, i.e., the minimum number of sequential computations of the NCS [4]. Figure 9 shows the median and maximum number of computation levels per priority assignment function in experiments without standstills. A scenario will develop differently for different priority assignment functions. To mitigate the effect of this difference, we consider each timestep from all experiments on its own. In every timestep, we assign priorities with all four priority assignment functions and analyze the resulting number of computation levels. The strength of the priority assignment function p_{color} lies in this criterion, as it produces the lowest amount of median and maximum computation levels for all experiments. In the scenarios with 17 to 19 vehicles, it reduces the number of computation levels by up to 33 %.

A video of an experiment in the CPM Lab is available online.[5] It presents the priority assignment function p_{fca} with our distributed reprioritization framework.

5   Conclusion

This chapter presented two approaches to deal with the complexity of a nonconvex trajectory planning problem: discretization of control inputs using motion primitives and distribution of the control problem using prioritization. We showed recursive agent-feasibility for our receding horizon graph search using motion primitives, making it a viable alternative to receding horizon approaches using optimization. The efficiency of the informed search algorithm is highly dependent on the quality of the cost-estimating heuristic. We showed recursive NCS-feasibility for time-variant priority assignment functions in prioritized planning. We presented and evaluated two priority assignment functions for road vehicles, one for maintaining progress of vehicles and one for reduced computation time. Changing the priorities during an experiment affects NCS-feasibility of the P-DMPC problem, as it alters the constraints of the vehicles' OCPs. Experiments with up to 17 vehicles in our CPM Lab showed efficient computation and effective results for networked trajectory planning problems.

The priority assignment function offers potential for improvement. A strategy that might be worth examining is the application of game theory to assign priorities [30]. Our framework for distributed reprioritization achieves recursive NCS-feasibility through standstill at the end of the prediction horizon. While ensuring safety, this counteracts the goal to maintain progress in traffic. Some of the scenarios we evaluated resulted in a standstill which could not be resolved through the priority assignment function. In these situations, the priority assignment function could be altered to explore different priority permutations. The trajectory planner could also switch to a cooperative or centralized trajectory planning algorithm, which is more flexible, but has higher computation time [29]. The minimum number of computation levels and thus the expected computation time in our P-DMPC is decided by the coupling graph. If the allowed computation time is fixed and the number vehicles increases, less computation time for each vehicle is available. This issue will be addressed in our future work. Another topic to explore is the cooperation of our distributed trajectory planning algorithm with others such as [14], and the cooperation with human-driven vehicles [50].

---

[5] https://youtu.be/RqwbHUwip10.

Acknowledgements We thank Julius Kahle and Georg Dorndorf for their contributions. This research is supported by the Deutsche Forschungsgemeinschaft (German Research Foundation) within the Priority Program SPP 1835 "Cooperative Interacting Automobiles" (grant number: KO 1430/17-1).

References

1. Ajanovic, Z., Lacevic, B., Shyrokau, B., Stolz, M., Horn, M.: Search-based optimal motion planning for automated driving. In: 2018 IEEE/RSJ International Conference on Intelligent Robots and Systems (IROS), pp. 4523–4530 (2018). https://doi.org/10.1109/IROS.2018.8593813
2. Alonso-Mora, J., Beardsley, P., Siegwart, R.: Cooperative collision avoidance for nonholonomic robots. IEEE Trans. Rob. **34**(2), 404–420 (2018). https://doi.org/10.1109/TRO.2018.2793890
3. Alrifaee, B.: Networked Model Predictive Control for Vehicle Collision Avoidance. Ph.D. Thesis, RWTH Aachen University (2017)
4. Alrifaee, B., Heßeler, F.J., Abel, D.: Coordinated non-cooperative distributed model predictive control for decoupled systems using graphs. IFAC-PapersOnLine **49**(22), 216–221 (2016). https://doi.org/10.1016/j.ifacol.2016.10.399
5. Alrifaee, B., Maczijewski, J.: Real-time trajectory optimization for autonomous vehicle racing using sequential linearization. In: 2018 IEEE Intelligent Vehicles Symposium (IV), pp. 476–483 (2018). https://doi.org/10.1109/IVS.2018.8500634
6. Alrifaee, B., Maczijewski, J., Abel, D.: Sequential convex programming MPC for dynamic vehicle collision avoidance. In: 2017 IEEE Conference on Control Technology and Applications (CCTA), pp. 2202–2207 (2017). https://doi.org/10.1109/CCTA.2017.8062778
7. Alrifaee, B., Mamaghani, M.G., Abel, D.: Centralized non-convex model predictive control for cooperative collision avoidance of networked vehicles. In: 2014 IEEE International Symposium on Intelligent Control (ISIC), pp. 1583–1588. IEEE, Juan Les Pins, France (2014). https://doi.org/10.1109/ISIC.2014.6967623
8. Altché, F., Qian, X., de La Fortelle, A.: Time-optimal coordination of mobile robots along specified paths. In: 2016 IEEE/RSJ International Conference on Intelligent Robots and Systems (IROS), pp. 5020–5026 (2016). https://doi.org/10.1109/IROS.2016.7759737
9. Andersson, O., Ljungqvist, O., Tiger, M., Axehill, D., Heintz, F.: Receding-horizon lattice-based motion planning with dynamic obstacle avoidance. In: 2018 IEEE Conference on Decision and Control (CDC), pp. 4467–4474. IEEE, Miami Beach, FL (2018). https://doi.org/10.1109/CDC.2018.8618964
10. Bennewitz, M., Burgard, W., Thrun, S.: Finding and optimizing solvable priority schemes for decoupled path planning techniques for teams of mobile robots. Robot. Auton. Syst. **41**(2), 89–99 (2002). https://doi.org/10.1016/S0921-8890(02)00256-7
11. Borrelli, F., Bemporad, A., Morari, M.: Predictive Control for Linear and Hybrid Systems. Cambridge University Press (2017)
12. Boyd, S.P., Vandenberghe, L.: Convex Optimization. Cambridge University Press, Cambridge, UK; New York (2004)
13. Buckley, S.: Fast motion planning for multiple moving robots. In: 1989 International Conference on Robotics and Automation Proceedings, pp. 322–326 vol.1 (1989). https://doi.org/10.1109/ROBOT.1989.100008
14. Burger, C., Fischer, J., Bieder, F., Taş, Ö.Ş., Stiller, C.: Interaction-aware game-theoretic motion planning for automated vehicles using bi-level optimization. In: 2022 IEEE 25th International Conference on Intelligent Transportation Systems (ITSC), pp. 3978–3985 (2022). https://doi.org/10.1109/ITSC55140.2022.9922600

15. Chalaki, B., Malikopoulos, A.A.: A priority-aware replanning and resequencing framework for coordination of connected and automated vehicles. IEEE Control Syst. Lett. **6**, 1772–1777 (2022). https://doi.org/10.1109/LCSYS.2021.3133416
16. Clark, C., Bretl, T., Rock, S.: Applying kinodynamic randomized motion planning with a dynamic priority system to multi-robot space systems. In: Proceedings, IEEE Aerospace Conference, vol. 7, pp. 7–7 (2002). https://doi.org/10.1109/AERO.2002.1035338
17. Cohen, B., Phillips, M., Likhachev, M.: Planning single-arm manipulations with n-arm robots. In: Robotics: Science and Systems X, pp. 226–. Robotics: Science and Systems Foundation (2014). https://doi.org/10.15607/RSS.2014.X.033
18. Dellin, C.M., Srinivasa, S.S.: A unifying formalism for shortest path problems with expensive edge evaluations via lazy best-first search over paths with edge selectors. In: Proceedings of the Twenty-Sixth International Conference on Automated Planning and Scheduling (ICAPS), p. 9 (2016)
19. Dolgov, D., Thrun, S., Montemerlo, M., Diebel, J.: Practical search techniques in path planning for autonomous driving. In: AAAI Conference on Artificial Intelligence, p. 6 (2008)
20. Eilbrecht, J., Stursberg, O.: Optimization-based maneuver automata for cooperative trajectory planning of autonomous vehicles. In: 2018 European Control Conference (ECC), pp. 82–88 (2018). https://doi.org/10.23919/ECC.2018.8550422
21. Erdmann, M., Lozano-Pérez, T.: On multiple moving objects. Algorithmica **2**(1), 477 (1987). https://doi.org/10.1007/BF01840371
22. Frazzoli, E.: Maneuver-based motion planning and coordination for multiple UAVs. In: Proceedings. The 21st Digital Avionics Systems Conference, vol. 2, pp. 8D3–8D3 (2002). https://doi.org/10.1109/DASC.2002.1052947
23. Frazzoli, E., Dahleh, M., Feron, E.: Maneuver-based motion planning for nonlinear systems with symmetries. IEEE Trans. Rob. **21**(6), 1077–1091 (2005). https://doi.org/10.1109/TRO.2005.852260
24. Frese, C.: Cooperative motion planning using branch and bound methods. In: Proceedings of the 2009 Joint Workshop of Fraunhofer IOSB and Institute for Anthropomatics, Vision and Fusion Laboratory, p. 15. KIT Scientific Publishing (2010)
25. Frese, C., Beyerer, J.: Planning cooperative motions of cognitive automobiles using tree search algorithms. In: Dillmann, R., Beyerer, J., Hanebeck, U.D., Schultz, T. (eds.) KI 2010: Advances in Artificial Intelligence, vol. 6359, pp. 91–98. Springer Berlin Heidelberg, Berlin, Heidelberg (2010). https://doi.org/10.1007/978-3-642-16111-7_10
26. Gray, A., Gao, Y., Lin, T., Hedrick, J.K., Tseng, H.E., Borrelli, F.: Predictive control for agile semi-autonomous ground vehicles using motion primitives. In: 2012 American Control Conference (ACC), pp. 4239–4244 (2012). https://doi.org/10.1109/ACC.2012.6315303
27. Ioan, D., Prodan, I., Olaru, S., Stoican, F., Niculescu, S.I.: Mixed-integer programming in motion planning. Annu. Rev. Control. **51**, 65–87 (2021). https://doi.org/10.1016/j.arcontrol.2020.10.008
28. Kapania, N.R., Subosits, J., Christian Gerdes, J.: A sequential two-step algorithm for fast generation of vehicle racing trajectories. J. Dyn. Syst. Meas. Control **138**(9) (2016). https://doi.org/10.1115/1.4033311
29. Kloock, M., Alrifaee, B.: Coordinated Cooperative Distributed Decision-Making using Synchronization of Local Plans. IEEE Transaction on Intelligent Vehicles **8**(2), 1292–1302 (2023). https://doi.org/10.1109/TIV.2023.3234189
30. Kloock, M., Dirksen, M., Kowalewski, S., Alrifaee, B.: Generation of coupling topologies for multi-agent systems using non-cooperative games. In: 2022 IEEE Intelligent Vehicles Symposium (IV), pp. 1–8 (2022). https://doi.org/10.1109/IV51971.2022.9827431
31. Kloock, M., Kragl, L., Maczijewski, J., Alrifaee, B., Kowalewski, S.: Distributed model predictive pose control of multiple nonholonomic vehicles. In: 2019 IEEE Intelligent Vehicles Symposium (IV), pp. 1620–1625 (2019). https://doi.org/10.1109/IVS.2019.8813980
32. Kloock, M., Scheffe, P., Botz, L., Maczijewski, J., Alrifaee, B., Kowalewski, S.: Networked model predictive vehicle race control. In: 2019 IEEE Intelligent Transportation Systems Conference (ITSC), pp. 1552–1557 (2019). https://doi.org/10.1109/ITSC.2019.8917222

33. Kloock, M., Scheffe, P., Greß, O., Alrifaee, B.: An Architecture for Experiments in Networked Control Systems. IEEE Open Journal of Intelligent Transportation Systems **4**, 175–186 (2023). https://doi.org/10.1109/OJITS.2023.3250951

34. Kloock, M., Scheffe, P., Maczijewski, J., Kampmann, A., Mokhtarian, A., Kowalewski, S., Alrifaee, B.: Cyber-physical mobility lab: An open-source platform for networked and autonomous vehicles. In: 2021 European Control Conference (ECC), pp. 1937–1944 (2021). https://doi.org/10.23919/ECC54610.2021.9654986

35. Kloock, M., Scheffe, P., Marquardt, S., Maczijewski, J., Alrifaee, B., Kowalewski, S.: Distributed model predictive intersection control of multiple vehicles. In: 2019 IEEE Intelligent Transportation Systems Conference (ITSC), pp. 1735–1740 (2019). https://doi.org/10.1109/ITSC.2019.8917117

36. Korf, R.E.: Real-time heuristic search. Artif. Intell. **42**(2–3), 189–211 (1990). https://doi.org/10.1016/0004-3702(90)90054-4

37. Kurzer, K., Zhou, C., Marius Zöllner, J.: Decentralized cooperative planning for automated vehicles with hierarchical monte carlo tree search. In: 2018 IEEE Intelligent Vehicles Symposium (IV), pp. 529–536 (2018). https://doi.org/10.1109/IVS.2018.8500712

38. Kuwata, Y., Richards, A., Schouwenaars, T., How, J.P.: Distributed robust receding horizon control for multivehicle guidance. IEEE Trans. Control Syst. Technol. **15**(4), 627–641 (2007). https://doi.org/10.1109/TCST.2007.899152

39. LaValle, S.M.: Planning Algorithms, 1stedn. Cambridge University Press (2006). https://doi.org/10.1017/CBO9780511546877

40. Li, J., Tinka, A., Kiesel, S., Durham, J.W., Kumar, T.K.S., Koenig, S.: Lifelong multi-agent path finding in large-scale warehouses. Proceed. AAAI Conf. Artif. Intell. **35**(13), 11272–11281 (2021)

41. Luo, W., Chakraborty, N., Sycara, K.: Distributed dynamic priority assignment and motion planning for multiple mobile robots with kinodynamic constraints. In: 2016 American Control Conference (ACC), pp. 148–154. IEEE, Boston, MA, USA (2016). https://doi.org/10.1109/ACC.2016.7524907

42. Malikopoulos, A.A.: Connected and Integrated Transportation Systems (2022). https://doi.org/10.48550/arXiv.2211.08600. ArXiv, Preprint

43. Mandalika, A., Salzman, O., Srinivasa, S.: Lazy Receding horizon A* for efficient path planning in graphs with expensive-to-evaluate edges. In: International Conference on Automated Planning and Scheduling, p. 9 (2018)

44. Matsumoto, M., Nishimura, T.: Mersenne twister: A 623-dimensionally equidistributed uniform pseudo-random number generator. ACM Trans. Model. Comput. Simul. **8**(1), 3–30 (1998). https://doi.org/10.1145/272991.272995

45. Mettler, B., Kong, Z.: Receding horizon trajectory optimization with a finite-state value function approximation. In: American Control Conference, p. 7. Seattle, WA, USA (2008)

46. Paden, B., Čáp, M., Yong, S.Z., Yershov, D., Frazzoli, E.: A survey of motion planning and control techniques for self-driving urban vehicles. IEEE Trans. Intell. Veh. **1**(1), 33–55 (2016). https://doi.org/10.1109/TIV.2016.2578706

47. Rajamani, R.: Vehicle Dynamics and Control. Mechanical Engineering Series. Springer Science, New York (2006)

48. Regele, R., Levi, P.: Cooperative multi-robot path planning by heuristic priority adjustment. In: 2006 IEEE/RSJ International Conference on Intelligent Robots and Systems, pp. 5954–5959 (2006). https://doi.org/10.1109/IROS.2006.282480

49. Richards, N., Sharma, M., Ward, D.: A hybrid A*/automaton approach to on-line path planning with obstacle avoidance. In: AIAA 1st Intelligent Systems Technical Conference, Infotech@Aerospace Conferences. American Institute of Aeronautics and Astronautics (2004). https://doi.org/10.2514/6.2004-6229

50. Scheffe, P., Alrifaee, B.: A scaled experiment platform to study interactions between humans and CAVs. In: 2023 IEEE Intelligent Vehicles Symposium (IV), pp. 1–6. IEEE, Anchorage, AK, USA (2023). https://doi.org/10.1109/IV55152.2023.10186623

51. Scheffe, P., Dorndorf, G., Alrifaee, B.: Increasing feasibility with dynamic priority assignment in distributed trajectory planning for road vehicles. In: 2022 IEEE International Conference on Intelligent Transportation Systems (ITSC), pp. 3873–3879 (2022). https://doi.org/10.1109/ITSC55140.2022.9922028
52. Scheffe, P., Henneken, T.M., Kloock, M., Alrifaee, B.: Sequential convex programming methods for real-time optimal trajectory planning in autonomous vehicle racing. IEEE Trans. Intell. Veh. 1–1 (2022). https://doi.org/10.1109/TIV.2022.3168130
53. Scheffe, P., Kahle, J., Alrifaee, B.: Reducing Computation Time with Priority Assignment in Distributed Control (2022). https://doi.org/10.36227/techrxiv.20304015.v1. Preprint
54. Scheffe, P., Maczijewski, J., Kloock, M., Kampmann, A., Derks, A., Kowalewski, S., Alrifaee, B.: Networked and autonomous model-scale vehicles for experiments in research and education. IFAC-PapersOnLine **53**(2), 17332–17337 (2020). https://doi.org/10.1016/j.ifacol.2020.12.1821
55. Scheffe, P., Pedrosa, M.V.A., Flaßkamp, K., Alrifaee, B.: Receding horizon control using graph search for multi-agent trajectory planning. IEEE Trans. Control Syst. Technol. 1–14 (2022). https://doi.org/10.1109/TCST.2022.3214718
56. Schwarting, W., Alonso-Mora, J., Rus, D.: Planning and decision-making for autonomous vehicles. Ann. Rev. Control Robot. Autonom. Syst. **1**(1), 187–210 (2018). https://doi.org/10.1146/annurev-control-060117-105157
57. Silver, D.: Cooperative pathfinding. Proceed. AAAI Conf. Artif. Intell. Inter. Digit. Entertain. **1**(1), 117–122 (2005). https://doi.org/10.1609/aiide.v1i1.18726
58. Tran, D.Q., Diehl, M.: An application of sequential convex programming to time optimal trajectory planning for a car motion. In: Proceedings of the 48h IEEE Conference on Decision and Control (CDC) Held Jointly with 2009 28th Chinese Control Conference, pp. 4366–4371 (2009). https://doi.org/10.1109/CDC.2009.5399823
59. van den Berg, J., Overmars, M.: Prioritized motion planning for multiple robots. In: 2005 IEEE/RSJ International Conference on Intelligent Robots and Systems, pp. 430–435 (2005). https://doi.org/10.1109/IROS.2005.1545306
60. Wu, W., Bhattacharya, S., Prorok, A.: Multi-robot path deconfliction through prioritization by path prospects. In: 2020 IEEE International Conference on Robotics and Automation (ICRA), pp. 9809–9815. IEEE, Paris, France (2020). https://doi.org/10.1109/ICRA40945.2020.9196813
61. Zhang, S., Li, J., Huang, T., Koenig, S.: Learning a priority ordering for prioritized planning in multi-agent path finding. Proceed. Symp. Combinat. Sear. **15**(1), 208–216 (2022)

Maneuver-Level Cooperation of Automated Vehicles

Matthias Nichting, Daniel Heß, and Frank Köster

Abstract Cooperative behavior of automated vehicles at the maneuver level is of utmost importance for the efficient and safe use of traffic space. This chapter discusses a vehicle-to-vehicle communication-based negotiation and cooperation method for maneuver cooperation. The method is based on the negotiation about explicitly defined reservation areas on the road for the exclusive use of a particular traffic participant. It covers all standard traffic situations occurring on regular streets and thus achieves universal applicability. The evaluation of simulations and driving tests shows the suitability of the method for effective maneuver cooperation in various traffic situations. Furthermore, based on this method, the planning and execution of cooperative maneuvers in emergency situations are investigated. Simulations show that collisions can be avoided in relevant cases by this method. Moreover, further simulations and driving tests show that joint maneuvers can avoid sharp braking maneuvers in many situations. In addition, research on a methodology for implicit maneuver cooperation is presented. Based on reinforcement learning methods, partially cooperative decision-making functions are studied in a setting that benefits from cooperative behavior. The evaluation shows that cooperative behaviors of road participants can be achieved using this technique.

M. Nichting (✉)
German Aerospace Center (DLR), Institute of Transportation Systems, Rutherfordstr. 2, 12489
Berlin, Germany
e-mail: matthias.nichting@dlr.de

D. Heß
German Aerospace Center (DLR), Institute of Transportation Systems, Lilienthalpl. 7, 38108
Braunschweig, Germany
e-mail: daniel.hess@dlr.de

F. Köster
German Aerospace Center (DLR), Institute for AI Safety and Security, Rathausallee 12, 53757
Sankt Augustin, Germany
e-mail: frank.koester@dlr.de

© The Author(s) 2024   277
C. Stiller et al. (eds.), *Cooperatively Interacting Vehicles*,
https://doi.org/10.1007/978-3-031-60494-2_10

1 Introduction

Road traffic rules should ensure efficient and, above all, safe traffic. In doing so, the participants in road traffic are taken into account with their capabilities. Automated vehicles have much greater potential than manually operated vehicles concerning the exchange and utilization of data so that more suitable solutions can replace static traffic rules on a situational basis. The cooperative behavior of automated cars at the maneuver level can contribute significantly to this. Joint maneuvers can, for example, increase efficiency and comfort in road traffic. For instance, explicitly coordinated cooperative behavior allows vehicles to keep shorter safety distances than human drivers or to give way in conflicting traffic situations, such as changing lanes, entering roundabouts, or at intersections. In summary, cooperative behavior at the maneuver level based on explicit communication enables the optimization of vehicle movements concerning shared objectives, whereas, without this cooperative behavior, vehicles act only based on their own goals.

Another possible use of cooperative maneuver execution addresses emergency situations. Unforeseen events may disrupt the planned movement of a vehicle and require a change in the preconditions for trajectory planning to achieve or maintain a safe state. This is often neither dangerous nor uncomfortable because other road users act considerately and do not force other participants to make last-minute changes in their motion planning, even if just out of self-interest. However, there are situations where prompt response is required to prevent or mitigate collisions. For example, the door of a car parked at the side of the road may suddenly be torn open and protrude into the planned path of the vehicle. Likewise, pedestrians or bicyclists may unexpectedly block the path of travel, for example, by suddenly changing the direction and speed of travel without correctly being predicted by the automated vehicle.

Figure 1 shows an exemplary traffic situation in which an immediate reaction of the automated vehicle is required. There, an automated vehicle approaches a suddenly occurring pedestrian on the right of two parallel lanes leading in the same direction. Depending on the time and location of the obstacle's occurrence and the speed of the approaching vehicle, a specific braking rate must be attained to avoid a collision with the obstacle without changing lanes. There may be constellations in which a lane change is more favorable in terms of an associated cost function than a pure

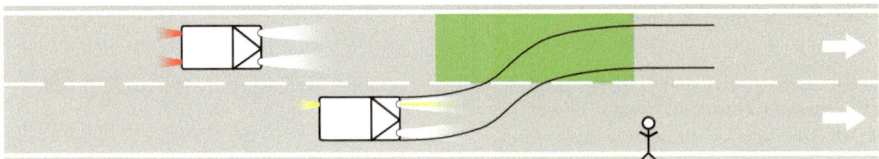

Fig. 1 Exemplary depiction of an emergency situation. The depicted pedestrian steps unexpectedly and irregularly into the lane, forcing the approaching automated vehicle to adapt its plan. The illustration indicates a cooperative lane change in response to the event

braking maneuver, e.g., a high braking rate is required without a lane change, or a collision cannot be avoided without changing the lane due to the physically limited braking rate. To execute a lane change, a suitable gap is required in the adjacent lane so that the lane change does not create a risk of collision. If an appropriate gap is available, it can be used by the swerving vehicle to resolve the situation. However, if this is not the case, cooperative behavior of vehicles in the target lane would be desirable so that the vehicle can tackle the emergency as smoothly as possible. Because of the dynamic nature of the situation, achieving this goal requires a quick agreement among the vehicles involved. Thus, a joint maneuver could increase road safety in safety-critical cases by allowing a coordinated, targeted response without the uncertainty and delay of inexplicit human communication.

Another critical point of cooperative maneuver-level behavior is the decision-making of an automated vehicle. The decisions that an automated vehicle has to make in road traffic range from very simple to complex. Examples include starting to move when a traffic light has just switched to green, selecting a cruising speed, choosing a distance to the vehicle in front, when to change lanes, and selecting a suitable gap for a lane change or crossing an intersection. Complexities are added by the traffic dynamics, differing or even unknown goals of road users, and their interactions. Reinforcement learning, a subcategory of machine learning, is particularly suitable for problems where it is relatively easy to evaluate the outcome of a decision, but engineering an algorithm to solve a given task is very complex or too time-consuming.

Up to this point, three essential aspects of Cooperative Automated Driving have been outlined. The research conducted in the CoInCiDE project on these three aspects is expounded in this chapter. Sect. 2 presents research on a foundational universal cooperation methodology based on explicit vehicle-to-vehicle communication (V2V). In the following Sect. 3, research on the further development of the method with regard to emergency situations is presented. Sect. 4 contains the research results on reinforcement learning methods for cooperative maneuver-level decision-making. Last, this chapter is concluded in Sect. 5.

2 Framework of Explicitly Negotiated Maneuver Cooperation via V2V

While human drivers on the road must rely mainly on implicit communication and communication methods that can rarely be interpreted beyond doubt, automated vehicles can easily exchange data via explicit communication. This enables explicit agreements between vehicles regarding joint maneuvers to be executed. This section presents a method, the Space-Time Reservation Procedure (STRP), based on the work already published on this topic [14, 15, 24, 25].

2.1 Related Work

Several approaches for the coordination and cooperation of automated vehicles based on explicit data exchange have already been documented in the literature. And there are already several message types that support cooperative driving functions defined or under development. Some of the messages have already been standardized or are in the process of being standardized. The Cooperative Awareness Message (CAM) is already standardized and contains basic information such as the position and velocity of the sender [6]. The likewise standardized Decentralized Environmental Notification Message (DENM) can be used for exchanging data on particular danger spots in the road network [7]. It contains information about the type and position of the area to be described. Collective Perception Messages (CPM) can be used to share information about obstacles and other road users detected by the sensors of the originating system [8]. This message is already standardized, too.

In many cases, specific, frequently occurring traffic situations are considered. One example is the change between parallel lanes. An approach is to equalize the speeds of the vehicles involved in the lane change to enable the maneuver [22]. This method is adapted from a technique for cooperation at intersections [28]. Another method for cooperative lane changes on highways achieves safe lane change maneuvers based on a minimum safety spacing model (MSS), even in complex situations [34]. The method performs trajectory planning based on the distances at different points in time between the involved vehicles calculated by the MSS.

The Maneuver Coordination Message (MCM) that is currently under standardization [5] allows the exchange of trajectories. Based on this, an approach in which vehicles continuously publish their currently planned trajectory is presented in [19]. In addition to the currently planned trajectory, a trajectory can be broadcast that is marked as desired and conflicts with the plans of other road users. Other vehicles can adjust their planned trajectory so that it no longer conflicts with other road users' desired trajectory. The desired trajectory can be executed once all trajectory conflicts are resolved. This method can be extended by a coordination protocol [35] which allows vehicles to form cooperative groups. A similar method that also relies on MCM and the continuous exchange of trajectories is presented in [21]. In this method, other trajectories in addition to the reference trajectory are sent that can be either more favorable for the sending vehicle or advantageous for other vehicles but to the disadvantage of the sending vehicle. Cooperation is achieved by evaluating the received trajectories and adjusting the reference trajectory.

A co-simulation framework for evaluating and testing cooperative driving functions is presented in [20]. The framework couples a vehicle dynamics simulation and a traffic flow simulation. It contains a machine learning module to generate and evaluate test scenarios. These three components together allow for extracting scenes from the traffic flow simulation, automatically testing them using the vehicle dynamics simulation, and evaluating the cooperative driving functions.

The space time reservation procedure (STRP) is another approach to achieve cooperative maneuver level behavior of automated vehicles [15]. The method is based on a structured negotiation about reservations of road space for agreeing on binding cooperative maneuvers. This approach has also been tested for more than two participating vehicles [24] and by test drives with two automated research vehicles [14]. Moreover, universality has been investigated to cover all traffic scenarios [25]. In the following, this method is presented in detail.

2.2 Definition of a Cooperative Maneuver

The foundation of the STRP is a set of rules for explicitly defining joint maneuvers. These rules avoid misunderstandings and allow the details of coordinated maneuvers to be described precisely. The reservation templates described in Sect. 2.3 are specifically adapted for different types of joint maneuvers. In this section, attributes that are used for all templates are explained. The method is based on reserving temporarily and spatially limited traffic space for the exclusive use of one automated vehicle. A data set represents the restriction of the traffic space reserved during a cooperative maneuver. First, this includes information for uniquely identifying the lane containing the reservation area. This is covered by two points P_0 and P_1 connected by the lane to be identified. Both points are described by their longitude, latitude, and elevation coordinates. The reservation area is longitudinally bounded by the length values s_0 and s_1. Both values refer to the point P_0 and determine the exact start and end of the reservation lengthwise. In the lateral direction, the reservation area is predetermined by the lane width. Therefore, the reservation area is spatially unambiguously defined. A time interval $[t_0, t_1]$ specifies the time limit within which the reserving vehicle must start to enter the reservation area. Otherwise, the cooperative maneuver becomes invalid. The reservation templates for different situations extend this basic definition as needed for specific traffic situations.

With this set of rules for explicitly defining reservation areas for cooperative maneuvers, a schematic negotiation process between road users can take place. A vehicle can use a definition of a reservation area to request cooperative behavior from other road users via vehicle-to-vehicle communication. To do this, a request message containing the reservation definition is broadcast. All receiving vehicles can then evaluate the request based on the requested reservation and ignore, reject, or accept it depending on their own goals. The evaluation of the responses is done solely by the requesting vehicle. It can execute the intended maneuver if the cars required for the coordinated maneuver have agreed to collaborate. Due to physical limits and incompatible objectives, a vehicle may not send an acceptance message. In this case, the requesting vehicle can cancel the reservation using an abort message so that other participants do not avoid the reservation area unnecessarily. If a vehicle has agreed to a reservation, the agreement is binding. The vehicle must then avoid the area according to the reservation definition, provided that the reserving car starts to enter the reservation area within the time interval $[t_0, t_1]$.

2.3   Reservation Templates

In order to make the method universally applicable for standard driving maneuvers occurring on regular streets, three patterns for reservation are defined. These differ, e.g., in terms of additional data that is transmitted and the end of a cooperative maneuver. The first template covers a vehicle's intention to change from a parallel lane to the lane containing the reservation area, e.g., a standard lane change. The second template covers cases in which vehicles leave the original lane, use another lane for a limited distance, and change back to the initial lane afterward. This can be used, for example, to drive around a traffic obstruction in the presence of oncoming traffic. The third template defines a reservation area located on the original lane of the vehicle. This template is suitable, for example, at intersection crossings for cooperation with cross-traffic. A more detailed presentation of the reservation templates can be found in [25].

2.3.1   Lane Change

To keep the length of the reservation area as short as possible and still allow the lane-changing vehicle a certain tolerance, an additional parameter v defines a speed at which the boundaries of the reservation area specified by s_0 and s_1 move along the direction of the road from time t_0. Furthermore, joint maneuvers agreed upon based on this reservation template end with their activation. That means the cooperative maneuver ends as soon as the reserving vehicle begins to enter the reservation area in the interval $[t_0, t_1]$. After that, the vehicles involved continue their journey individually.

The sequence of a cooperative maneuver with this reservation template is shown in Fig. 2. At the bottom of the figure, the two lanes are sketched, and the points P_0 and P_1, as well as the distances s_0 and s_1, are drawn in so that the reservation area marked in green is spatially clearly delimited. In the upper part of the figure, an s-t diagram is shown. In this, distances s_0 and s_1, as well as the time interval $[t_0, t_1]$, and an exemplary path on the target lane τ are drawn. Furthermore, the chart shows three different areas. The hatched area indicates the longitudinal positions and times where the vehicle must not be on the target lane. This is the case before t_0 and spatially before the lower limit of the reservation determined by s_0. The area in which the vehicle must begin to enter the reservation area is shown in dark green. Within the time interval, the spatial boundaries move with velocity v so that this area forms a parallelogram in the chart. The white space in the diagram marks time and space intervals on the target lane that may be used after the vehicle has activated the cooperative maneuver by entering the reservation area within the dark green area.

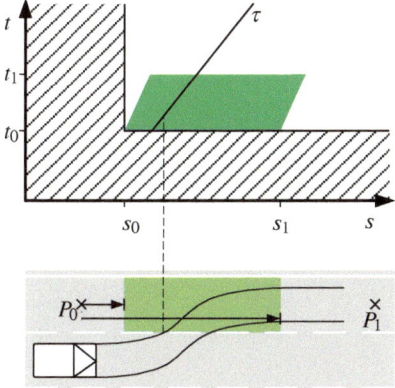

Fig. 2 Reservation shape for lane change: The s-t-diagram shows three different areas for the target lane. The hatched area must not be used by the reserving vehicle, the dark green area must be used for starting to enter the target lane, and the white area can be used after; τ is an exemplary path of the vehicle, adapted from [25]

2.3.2 Evasion With Oncoming Traffic

This reservation template allows the requesting vehicle to avoid an obstacle by using the lane of, e.g., the oncoming traffic. The vehicle must start entering the reservation area within the interval $[t_0, t_1]$. The defined reservation area is spatially static and must be left before reaching the upper longitudinal limit defined by s_1. The maneuver ends as soon as the vehicle has left the reservation area; there is no predefined time end. Figure 3a is analogous to Fig. 2. The s-t diagram refers to the target lane. The dark green room indicates when and in which area to enter the reservation. The hatched areas must not be entered at all within the target lane. The fading green color indicates the unlimited temporal validity of the reservation. The maneuver ends when the vehicle leaves the area. A possible path τ of the car on the target lane is drawn in black.

2.3.3 Lane Keeping

This reservation pattern is suitable, e.g., for a crossing passage. In this case, the reservation area is also spatially static and unrestricted in time. The start of the entry must lie in the interval $[t_0, t_1]$. Figure 3b shows this area in dark green in the s-t-diagram. After that, the reservation is valid for an unlimited time until the reservation area has been left.

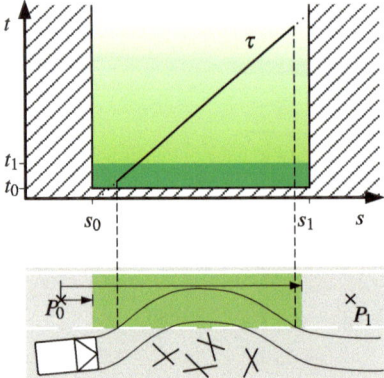 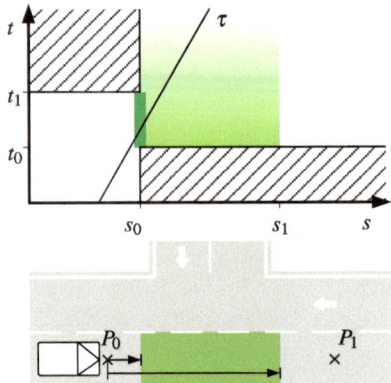

(a) Reservation shape for evasion with on-coming traffic: The s-t-diagram shows three different areas for the target lane. The hatched area must not be used by the re-serving vehicle, the dark green area must be used for starting to enter the target lane, and the shaded light green area can be used after without a time limit; τ is an exemplary path of the vehicle

(b) Reservation shape for lane keeping: The s-t-diagram shows four different areas for the target lane. The hatched area must not be used by the reserving vehicle, the dark green area represents the time interval within the reservation must be activated by being started to enter, and the light green area can be used after activation for an un-limited time. The white areas can be used before and after the reservation; τ is an ex-emplary path of the vehicle

Fig. 3 Reservation shapes for evasion and lane keeping, adapted from [25]

2.4 Simulations and Driving Experiments

Several experiments were conducted in simulation and using two automated research vehicles to analyze the method in more detail. Eclipse ADORe [13] is used to run the research vehicles and the simulation. For more information, please refer to this source. The research vehicles are equipped with hardware for vehicle-to-vehicle communication, special sensors, and other devices to operate the automation. An accurate map of an actual urban intersection in Braunschweig, Germany, is used for the experiments. The map was shifted accordingly to perform the driving experiments on a test site.

2.4.1 Simulation: Lane Change

In the simulation, two automated vehicles start about 200 m distant from a merging area. Coordination is required to drive through the area as efficiently as possible. The test drive results are shown in Fig. 4. Two vehicles are plotted at four consecutive time instants with $t_A < t_B < t_C < t_D$. At the earliest time, t_A, both vehicles approach the merging area in parallel lanes without coordination. Just before time

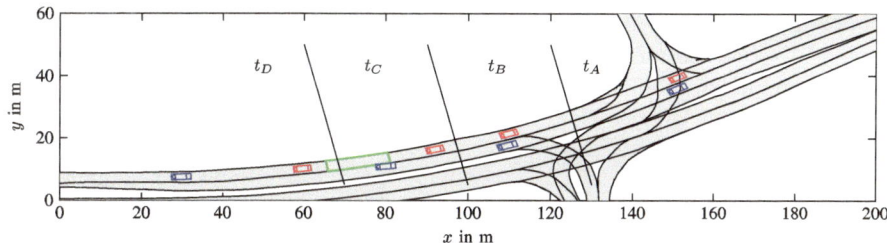

Fig. 4 Simulation of a cooperative lane change: The reservation area depicted in green is requested by the lane changing vehicle (blue), adapted from [25]

t_B, the reservation area marked in green is requested by the lane changing vehicle, depicted in blue. The lane following vehicle shown in red has evaluated this and agreed to the request. At the time t_C, the reservation area is just activated by the lane changing vehicle entering. Thus, the cooperative maneuver is finished, and both vehicles continue independently on the now single-lane road. As a result, the method is shown to coordinate the situation appropriately. The lane keeping vehicle brakes slightly, and the lane changing vehicle drives through the area without braking.

2.4.2 Driving Test: Three Vehicles at an Intersection

Since only two automated vehicles were available for the driving experiments, one of the three vehicles was simulated. Figure 5 shows the situation during the cooperative maneuver. The left-turning vehicle, shown in red, and the straight-out vehicle, shown in blue, are the two physical vehicles. The third car (green) is simulated. While approaching the intersection, the left-turning vehicle had requested the shown reservation area, and the other two conflicting cars had agreed to the maneuver. As a result, the left-turning vehicle can pass the intersection unimpeded. In contrast, the other two vehicles reduce speed to the required extent until the left-turning car has cleared the respective lane. Although the usefulness of this experiment in terms of traffic efficiency is not apparent at first glance, there are situations in which such

Fig. 5 Driving experiment with three vehicles at an intersection, adapted from [25]

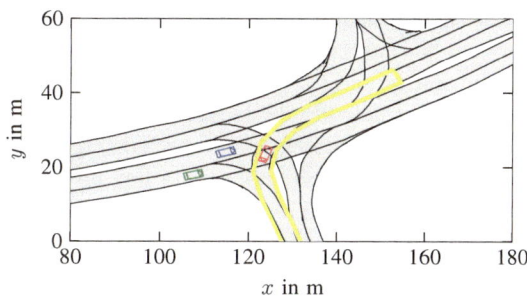

a cooperative maneuver is beneficial. For example, such cooperation can allow the automated vehicle to turn safely in heavy traffic, possibly including mixed traffic. Furthermore, it can enable fast and reliable priority for emergency vehicles.

2.5 Conclusion

Cooperation at the maneuver level between road users can contribute to the efficient use of road space. The presented approach uses vehicle-to-vehicle communication and a method designed for explicit negotiation and agreement of cooperative maneuvers. Various reservation templates establish the universal applicability of the technique. These templates are not limited to the traffic situations discussed in this section but may also be used for other conflicts between road users. The driving experiments and simulations conducted to research and improve the method show that it is suitable to ensure coordination in the studied situations. Furthermore, by design, the technique is inherently safe against message loss and suitable for mixed traffic scenarios. Its decentralized architecture allows flexible use at any place. The reader is referred to the publications [14, 15, 24, 25] for a deeper look at this method and more results of many simulations and driving experiments in various traffic situations.

3 Cooperation in Emergency Situations

This section discusses research on adapting the cooperation method presented in Sect. 2 to emergency situations. The effectiveness of the method to coordinate maneuvers of automated vehicles in emergency situations is evaluated by both simulations and driving tests.

3.1 Related Work

The related work regarding vehicle-to-vehicle communication-based cooperation of automated vehicles given in Sect. 2.1 is relevant here, too. In addition, a few publications concerning emergency maneuvers shall be presented here.

The authors of [16] propose a method for guaranteeing safety based on verifying the planned trajectory while the vehicle is in motion. The core of the approach is a two-step evasive strategy based on a discrete decision for an evasive maneuver and the computation of an appropriate low-level control to follow this maneuver. The method was validated in simulation.

An approach for lateral control in evasive maneuvers is proposed in [4]. The method, based on a sliding mode control, calculates a steering angle taking into account, among other factors, the tire slip saturation. Simulations show that lane changes are possible within 1.1 s at speeds of up to 130 km/h under certain circumstances. Another proposal involves taking into account the dynamics of the steering system during evasive maneuvers [27]. The model predictive control in this publication contains two models. Besides the vehicle model, also a steering model is included.

A parameterization of a geometric path for an evasive maneuver based on reinforcement learning is proposed in [9]. The path consisting of straight lines and clothoids is then executed by means of a model predictive control loop. Simulations of a common emergency situation show that the method significantly outperforms human drivers.

3.2 Approach

The basic framework of the cooperation and negotiation method has already been stated in Sect. 2. This approach is adapted to the particular requirements in emergencies. Negotiating the cooperative maneuver in the shortest possible time without avoidable delay is of the utmost importance in emergency situations. This is because these situations are highly dynamic, and any delay reduces the ability to respond to the situation. For example, evasive maneuvers may become impossible because of the intermediate progress of the surrounding traffic. Therefore, negotiation is started immediately after a cause for an emergency response is detected. Due to the safety-critical nature of emergencies, cooperative maneuver requests are of higher priority than other requests. The receiving vehicles can consider that during the evaluation of the request.

3.3 Simulations and Driving Experiments

To verify and investigate the method, simulations and driving experiments are conducted. The basis in each instance is the traffic situation shown in Fig. 1. The parameters, such as speeds and distances, vary in the different runs. A simulation run and a driving experiment are presented below with their evaluation and results.

3.3.1 Simulation

At the time $t = 0$, the obstacle occurs on the lane of the lane changing vehicle (lc-vehicle). At this point in time, the vehicle approaches the obstacle at a speed of 27.73 m/s, and the lane following vehicle (lf-vehicle) on the adjacent lane to the left is driving at a speed of 27.16 m/s. The distances measured along the road from the front bumpers of the vehicles to the position of the obstacle are 43.16 m for the lf-vehicle and 38.96 m for the lc-vehicle. In this scenario, the obstacle does not block the entire width of the right lane. The blocking is limited to the right side of the lane so that only the outer 50% of the width is blocked at the longitudinal position $d = 0$. The physically maximum possible braking rate of the lc-vehicle is assumed to be 9.81 m/s^2. Even with hypothetical constant deceleration at this rate, a collision would occur between the obstacle and the lc-vehicle since the braking distance exceeds the distance to the block. Therefore, the lc-vehicle immediately starts negotiating a cooperative maneuver and requests a reservation area just before the obstacle. The lf-vehicle in the target lane accepts the request and brakes to allow the requesting vehicle to change lanes.

Figure 6 shows the positions of the two vehicles in the distance-time diagram. Time $t = 0$ corresponds to the point in time of the obstacle occurrence. The obstacle is longitudinally located at $d = 0$. Two curves that are connected by a hatching are plotted for each vehicle. The two curves correspond to the longitudinal positions of the front and rear bumpers of both vehicles. The hatching patterns indicate which lateral zone the vehicles use at the respective time. The inclined single hatching corresponds to the left lane, which is unaffected by the obstacle at $d = 0$. The hori-

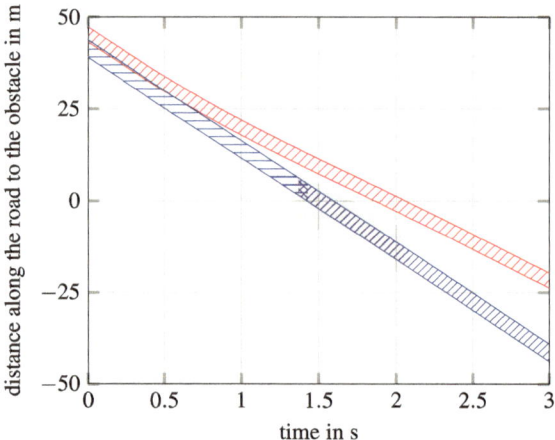

Fig. 6 Simulation: Distances between the front and rear bumpers of the lf-vehicle (red) and the lc-vehicle (blue) and the obstacle located at $d = 0$, with $t = 0$ being the point in time of the obstacle appearance; the distances are measured along the lane. The hatching patterns indicate the lateral area used by the vehicle: Horizontal single hatching indicates the use of the right half of the right lane, and inclined single hatching indicates the use of the left lane

Fig. 7 Simulation: Velocities of lc-vehicle (blue) and lf-vehicle (red) during the simulation. $t = 0$ is the point in time when the obstacle appears

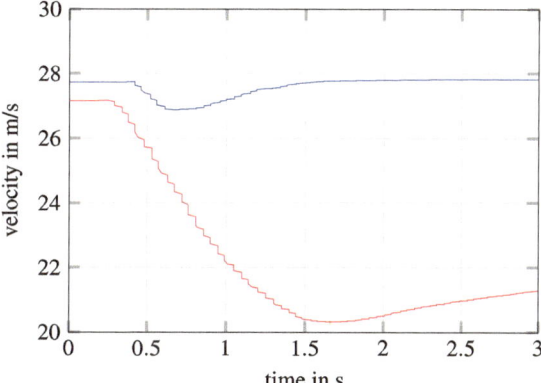

zontal parallel hatching indicates the usage of the lateral zone blocked from $d = 0$, i.e., the right 50% of the right lane. Directly after the occurrence of the obstacle, the lc-vehicle requests a reservation which is evaluated and accepted by the lf-vehicle. A gray box in the diagram depicts the reservation area. The lf-vehicle brakes sharply to respect the reserved area for the emergency lane change. Within the interval in time and longitudinal position, the lc-vehicle leaves the blocked part of the right lane and changes towards the left lane. The cross-hatching in the diagram indicates the short period in which the lc-vehicle uses both the left and the blocked part of the right lane. Figure 7 shows the development of the velocities of both vehicles during the scenario. While the lf-vehicle brakes and reduces its speed by approx. 7 m/s to assist the emergency evasion of the lc-vehicle, the latter reduces its velocity marginally.

Fig. 8 Automated research vehicles VIEWCar II (left) and FASCar E during the demonstration of a cooperative emergency lane change at the IEEE Intelligent Vehicles Symposium 2022

3.3.2 Driving Experiment

In addition to the simulations, physical tests were performed with automated research vehicles. These tests of the method were demonstrated at the IEEE Intelligent Vehicles Symposium 2022 in Aachen, Germany. Figure 8 shows the two automated research vehicles on the site during the driving demonstration. The results of the tests are presented in the following.

The two research vehicles, FASCar E and VIEWCar II, were used for the driving experiments and demonstrations. These vehicles are provided with the necessary hardware for communication via ITS-G5. The software framework for vehicle automation ADORe [13], further developed in the CoInCiDE project, is used in both cars for these tests. Currently, the maximum deceleration set by the automated research vehicles is limited to $3\,\text{m/s}^2$. This limitation is due to the vehicle interface and cannot be influenced by the automation software. To account for that limitation, the driving test distances are larger than those used in the simulation. In this way, meaningful results can be obtained despite the restriction.

The initial situation of the scenario is again two automated vehicles traveling in the same direction on adjacent lanes. The point in time of the virtual obstacle occurrence is defined as $t = 0$, and its longitudinal position is $d = 0$. In this scenario, the entire width of the right lane is blocked by the obstacle, so the vehicle must have left it entirely before passing this location. The lc-vehicle in the right, blocked lane approaches the obstacle at a speed of 13.65 m/s at a distance of 73.86 m at time $t = 0$. The lf-vehicle driving on the adjacent lane travels at this time with 13.45 m/s at a distance of 81.85 m measured along the lane in the same driving direction. Figure 9 shows the distances analogously to the evaluation in Sect. 3.3.1. The longitudinal distances from the front and rear bumpers to the obstacle are plotted for both vehicles. The hatching again gives information about the lateral position of the vehicles. Here, the inclined line hatching corresponds to the use of the unblocked left lane, and the horizontal line hatching indicates the use of the right lane, which is blocked from $d = 0$. The cross-hatching represents areas where both lanes are used at the same time.

Immediately after the virtual obstacle appears, the lc-vehicle requests a reservation area in the target lane so the obstacle can be passed without braking. After evaluating this emergency request, the lf-vehicle sends a confirmation message. Thus, the cooperative maneuver is bindingly agreed upon. The temporarily and spatially limited reservation area is indicated by a gray box in Fig. 9. Right at the beginning of this area, the lc-vehicle activates the reservation. The cross-hatching indicates the partial use of both lanes. Before reaching the obstacle at the longitudinal position $d = 0$, the lane change is completely finished, and both vehicles drive on the left lane one after the other. Figure 10 shows the speeds of the two vehicles during the experiment. The speed profile of the lc-vehicle is almost constant. The lf-vehicle, on the other hand, brakes and thus enables the cooperative maneuver.

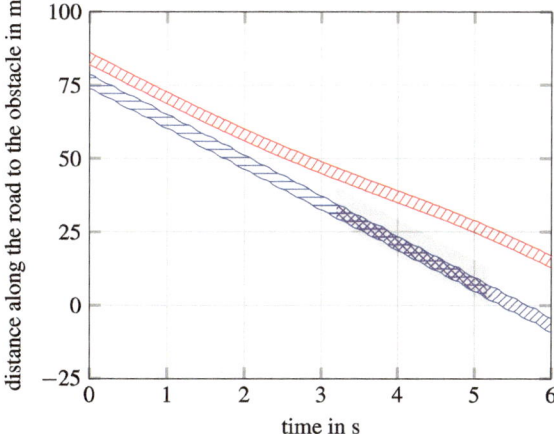

Fig. 9 Test drive: Distances between the front and rear bumpers of the lf-vehicle (red) and the lc-vehicle (blue) and the obstacle located at $d = 0$, with $t = 0$ being the point in time of the obstacle appearance; the distances are measured along the lane. The hatching patterns indicate the lateral area used by the vehicle: Horizontal single hatching indicates the use of the right lane, and inclined single hatching indicates the use of the left lane

Fig. 10 Test drive: Velocities of lc-vehicle (blue) and lf-vehicle (red) during the test drive. $t = 0$ is the point in time when the obstacle appears

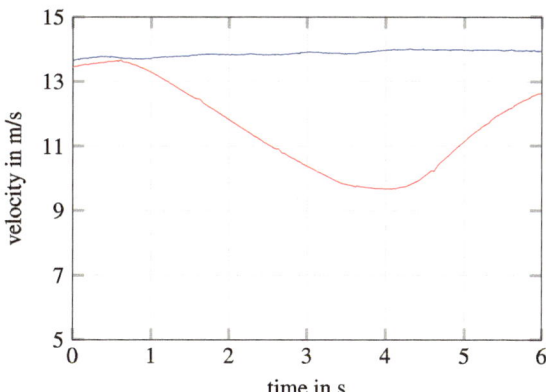

3.4 Conclusion

Emergencies in road traffic can hardly be avoided due to complexity and, not least, due to humans. Therefore, appropriate handling of such situations is of the utmost importance for developing automated vehicles. A basic example of such a hazardous situation is an obstacle's sudden and unforeseen occurrence within the planned path of movement. The most basic method of responding to such a situation is to brake the vehicle to avoid a collision or at least reduce the impact energy as much as possible. Evasive maneuvers can be used in some instances to avoid heavy braking or even to avoid collisions. The prerequisite is that the traffic space required for swerving is not in use by other road users. The method discussed in this section aims to use vehicle-to-vehicle communication to negotiate a reservation of the space required for an evasive maneuver with conflicting road users. Hence, evasive maneuvers should be possible in more situations than before, thus avoiding heavy braking maneuvers and collisions.

The evaluation of the performed simulation and test drive shows that the method is suitable for this purpose. It was shown in the test drive that the cooperative behavior reduced the impact of the obstacle. The simulation is parameterized so that a braking maneuver within the lane cannot avert a collision. Initially, a lane change is impossible because of the blocked adjacent lane. However, the cooperative behavior negotiated using the presented method can effectively resolve the emergency situation without causing a collision. Thus, the method can prevent collisions and reduce the impact of unforeseen obstacles. To further improve cooperative emergency behavior, future research can address, e.g., pre-negotiation of emergency responses and lane sharing in emergency situations. This could prevent collisions in a wider range of situations.

4   Implicitly Cooperative Decision-Making

The research presented here builds upon prior work on cooperation of automated vehicles [25] and the use of reinforcement learning for decision-making [26]. While the previous reference investigated deep Q-learning for the decision-making of an automated vehicle without considering interactions between road users, this section presents a method that does this based on the soft actor-critic approach [10] and proximal policy optimization algorithms [29]. For this purpose, a multi-agent system is built, and partly cooperative objective functions are designed. A common problem of road traffic is selected to show and research the methodology. Figure 11 shows a traffic situation similar to a highway entrance. Two lanes are in parallel for a limited stretch of way, with the right lane ending at the end of the segment and the left lane proceeding as part of a road with an arbitrary number of lanes. In such traffic situations, participants with different objectives interact, implicitly communicate and sometimes even cooperate.

4.1   Related Work

In literature, several methods have already been documented to implement parts of the decision-making of an automated vehicle using reinforcement learning methods. There are a few examples where end-to-end learning approaches are employed [3, 33] with the decision-making being part of the end-to-end architecture. But the task of automated driving is usually split into subtasks that are solved by different methods. Tram et al. [30] use deep Q-learning to adjust the speed of an automated vehicle as it passes through an intersection. The surrounding traffic, which consists of simulated manually driven vehicles, is used as input for an artificial neural network and a recurrent artificial neural network for comparison. As a result, the automated vehicle passes the intersection without collision in the majority of cases for both networks, with better results obtained from the recurrent network.

Wang et al. [31] consider lane changing and investigate a methodology to perform it in various situations. To do this, they model the problem with a state space consisting of road information such as curvature and width and vehicle dynamics information such as acceleration, speed, and position. Here, the reinforcement learning agent serves as the lateral controller, and the action space contains the yaw acceleration of the vehicle. The results show that the lane change controller manages the control task but lacks robustness and flexibility. The principal author later reformulated the task and published an approach for the lateral control during lane changing using deep deterministic policy gradient [32]. As a result, stable lane changing is achieved with the proposed architecture.

Kurzer et al. [18] propose a method to represent the environment in a generalized way with as few restrictions as possible. This is intended to improve the capability for generalization of the methods using this representation. To do this, the path in front of the vehicle is divided into segments and properties such as time to occupancy and time to vacancy are assigned to each segment. Together, these pieces of information form the state representation. Experiments presented in the paper show the successful abstraction of the environment representation from the concrete driving situation.

Bouton et al. [1] propose a decision-making algorithm for automated vehicles to navigate at intersections. In addition to a reinforcement learning algorithm, a model checker is used to make the decisions safe. Furthermore, perception errors are addressed with the help of a recurrent neural network. As a result, the algorithm proves to be robust and safe concerning the decisions.

For relevant examples of multi-agent reinforcement learning, reference is made to the survey by Hernandez-Leal et al. [12] and the article by Canese et al. [2]. Both references provide a literature review on multi-agent reinforcement learning.

4.2 Approach

The approach involves two independent agents that follow their goals defined by a reward function in a scenario. The lane change agent (lc-agent) has to change lanes in a limited time and on a limited road section while the lane following agent (lf-agent) follows the lane the lc-agent wants to change to. By adjusting the speed, the lf-agent

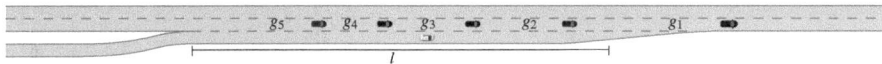

Fig. 11 Overview of the map and gap identifiers; gap g_5 has no longitudinal lower bound in this case, adapted from [26]

can let the lc-agent merge cooperatively. The agents differ in terms of the algorithm used, the state space, and the action space definitions. The specific components are described in Sects. 4.2.1 and 4.2.2.

The characteristics of the map are part of the state spaces of both agents. The map is depicted in Fig. 11, with l being the longitudinal distance between the first and last possibility to change lanes. In addition, a speed limit $v_{\text{speedlimit}}$ is defined for the area in which the examined traffic situation occurs. Other road characteristics, e.g., curvature, are not part of the state spaces to keep the definition general. Besides the map-specific part of the environment, the traffic participants themselves are part of the environment as they interact and limit the possibilities of the other participants in the scenario.

4.2.1 Lane Change Agent

The lc-agent controls the light-colored vehicle depicted in Fig. 11 and selects the gap on the target lane depending on the observed state of the environment. The proximal policy optimization algorithm [29] is used for this purpose. The state description consists of the longitudinal boundaries of the five gaps depicted in Fig. 11, the velocities of the vehicles on the target lane, the position and velocity of the ego-vehicle, and the longitudinal boundaries of the lane change area. The action space contains the discrete gap selection. For training the agents, a reward signal is used to induce the intended properties of the agents. For this purpose, a reward function is defined that rewards high values of v_{ego} and penalizes the use of the original lane in each time step. A systematic parameter study has been conducted to define the exact reward functions of both agents. The lc-agent's reward function R_{lc} is defined as follows:

$$R_{\text{lc}} = \begin{cases} -0.8 + \frac{1}{14} \times v_{\text{ego}}, & \text{if lane change is not finished} \\ \frac{1}{14} \times v_{\text{ego}}, & \text{if lane change is finished} \end{cases} \tag{1}$$

4.2.2 Lane Following Agent

For the lf-agent, the soft actor-critic algorithm [11] is used. This is an off-policy algorithm that seeks to maximize both expected reward and entropy. The state description of this agent consists of the longitudinal distances to the vehicle in front, the vehicle behind, the vehicle that attempts to perform a lane change, and the boundaries of the lane change area. Moreover, the ego velocity v_{ego} as well as the velocities of the lane changing vehicle and the vehicles in front and behind are part of the state representation. The continuous action space consists of a set-point velocity input to the trajectory planning. The reward function R_{lf} depends on the velocity v_{ego} and, to induce a partly-cooperative behavior, on the lane change state of the lc-agent:

$$R_{\text{lf}} = \begin{cases} 0.7 \times v_{\text{ego}}, & \text{if lane change is not finished} \\ 1 \times v_{\text{ego}}, & \text{if lane change is finished} \end{cases} \tag{2}$$

4.3   Experiment

The experiment involves training the agents in dense surrounding traffic. During the training, the policies of the agents are continuously evaluated. The scenario shown in Fig. 11 is used for the experiment.

4.3.1   Configuration

The length of the segment in which a lane change is possible is $l = 200\,\text{m}$. The speed limit in the scenario is set to $v_{\text{speedlimit}} = 13.89\,\text{m/s}$. Each episode consists of 40 training steps and takes 40 s of simulation time. A collision between traffic participants is impossible as the action space consists of inputs for the trajectory planning, which is inherently safe. Multiple simulations are conducted to identify proper hyperparameters. Table 1 gives the most important hyperparameters that are selected for the training process of the agents as they turned out to perform best after a limited parameter study. Besides, the standard parameters are chosen, as given in [11, 29].

The surrounding traffic on the middle and left lanes of the road shown in Fig. 11 is simulated by SUMO [23]. This traffic consists of differently parameterized vehicles, so random and busy traffic situations arise. The individual speed limit of each vehicle is taken from a normal distribution considering but not always obeying the global limit $v_{\text{speedlimit}}$. As a measure of the density, an emission probability of 43% for each of the middle and left lanes is specified. This value determines the probability of the emission of one vehicle each second.

Two agents are permanently trained during the simulation. The lc-agent starts from a standstill 150 m distant from the beginning of the merging lane and drives

Table 1 Hyperparameters

	lf-agent	lc-agent
Learning rate	Actor: 6e-5 critic: 1e-6 alpha: 1e-5	3e-4
Discount factor γ	0.999	
Activation function	Rectified linear unit	
Optimizing algorithm	Adam [17]	
Batch size	512	128
Reward scale	1.0	n.a.
Size of replay buffer	10,000	
Network topology	All networks: two fully connected layers with 256 nodes each	all networks: two fully connected Layers with 64 nodes each

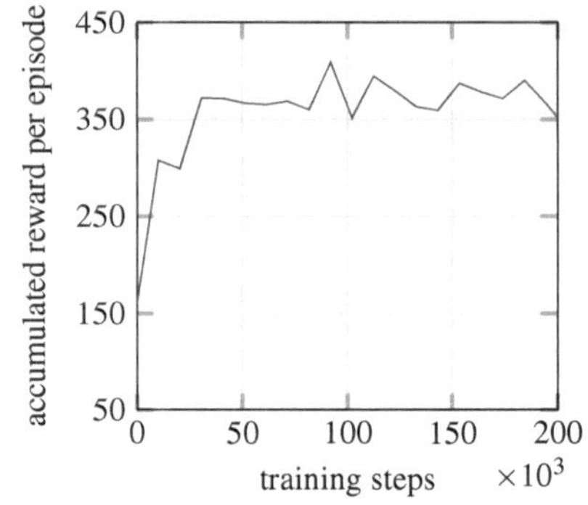

Fig. 12 Reward per episode during evaluation of the lane following agent

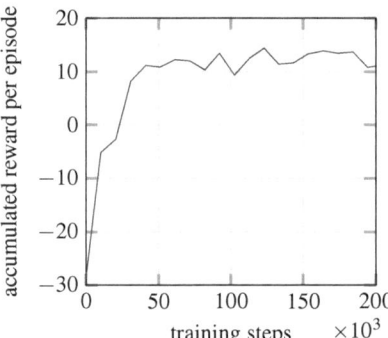

Fig. 13 Reward per episode during evaluation of the lane change agent

towards it. After the end of the episode, as defined above, the agent is reset to the starting position, and the next episode begins. The simulation control ensures that the lf-agent always controls a vehicle that is at a position suitable for potential cooperation during training. The vehicles of the lf-agent and the lc-agent are controlled by ADORe [13]. The interaction between the agents takes place solely implicitly through their behavior and the understanding of that behavior. Each agent executes decision-making at a frequency of 1 Hz.

4.3.2   Results

The training was conducted for 200,000 training steps. After every ten thousandth training step, twenty episodes were executed for evaluation. For each episode, the cumulative reward is logged. Figure 12 shows the accumulated reward per episode in relation to the training progress for the lf-agent, Fig. 13 shows that data analogously for the lc-agent. Initially, as the number of training episodes progresses, the rewards of both agents per episode increase continuously and reach their maxima. Then, the rewards remain approximately constant until the training is discontinued after 200,000 training steps. Figure 14 shows the number of evaluation episodes with

Fig. 14 Number of successful (—) and unsuccessful episodes (—)

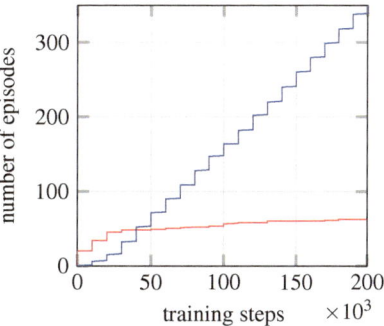

and without successful lane changes depending on the training progress. While the number of episodes without a successful lane change increases at the beginning of training, the slope flattens sharply with progressing training.

4.4 Conclusion

The experiment results show the suitability of reinforcement learning methods for the partially cooperative decision-making process. The agents use the soft actor-critic and the proximal policy optimization algorithms to cooperatively adapt their behavior to the other agent and maximize the reward per episode. Regardless of the presence of automated and manually driven vehicles, understanding the other vehicles' behavior is essential for efficiently accomplishing those situations. Extending the state spaces by a prediction of the vehicles in the scenario may further improve the performance.

The experiment does not consider direct communication via vehicle-to-vehicle communication. However, many cooperation methods work based on explicit communication. The combination of these two techniques can be addressed in the next steps. Furthermore, the variance of reward functions of the agents can be increased. More objectives can be considered, and thus more general applicability of the method can be reached.

5 Conclusion

The last three sections cover three important aspects of maneuver-level cooperation of automated vehicles. First, a fundamental method for defining, negotiating, and agreeing on cooperative maneuvers is presented. The STRP is based on reserving temporarily and spatially limited traffic space for exclusive use. The driving tests with automated research vehicles and simulations show that the method is suitable for effective cooperative resolutions of conflicts on the road. The different reservation templates allow universal applicability in conflict situations occurring in traffic. In the second part, research on cooperation in emergency situations is presented. The investigated approach is based on STRP and tested both in test drives and simulations. It is shown that the method allows to avoid collisions and to mitigate the impact of suddenly occurring obstacles by performing cooperative emergency maneuvers. The last part presents research on a cooperative high-level decision-making method. It is based on reinforcement learning algorithms and does not require explicit communication. Simulations show that cooperative behavior can be elicited by defining suitable objective functions for the vehicles present in a traffic scenario.

The results contribute to the achievement of safe and efficient behavior of automated vehicles in the three addressed aspects of cooperative automated driving. Based on the research presented in this chapter, the cooperative behavior of automated vehicles can be further researched. For example, an integration of STRP relying on explicit communication into the method for cooperative decision-making can be investigated. This could then be used to research an integrated approach for decision-making and explicitly negotiated cooperation of automated vehicles.

Acknowledgements The research presented in this chapter was funded within the priority program "Cooperatively Interacting Automobiles" (SPP 1835) of the German Research Foundation (DFG) in sub-project CoInCiDE under grant number KO 1990/3-2.

References

1. Bouton, M., Nakhaei, A., Fujimura, K., Kochenderfer, M.J.: Safe reinforcement learning with scene decomposition for navigating complex urban environments. In: 2019 IEEE Intelligent Vehicles Symposium (IV), pp. 1469–1476 (2019). https://doi.org/10.1109/IVS.2019.8813803
2. Canese, L., Cardarilli, G.C., Di Nunzio, L., Fazzolari, R., Giardino, D., Re, M., Spanó, S.: Multi-agent reinforcement learning: a review of challenges and applications. Appl. Sci. **11**(11) (2021). https://doi.org/10.3390/app11114948. URL https://www.mdpi.com/2076-3417/11/11/4948
3. Chen, L., Wang, Q., Lu, X., Cao, D., Wang, F.Y.: Learning driving models from parallel end-to-end driving data set. Proc. IEEE **108**(2), 262–273 (2020). https://doi.org/10.1109/JPROC.2019.2952735
4. Da Silva Junior, A., Birkner, C., Jazar, R.N., Marzbani, H.: Vehicle lateral dynamics with sliding mode control strategy for evasive maneuvering. In: 2021 9th International Conference on Systems and Control (ICSC), pp. 165–172 (2021). https://doi.org/10.1109/ICSC50472.2021.9666598

5. ETSI: Technical Specification ETSI TS 103 561 v0.0.1 draft (2018-01) (2018) Intelligent Transport Systems (ITS); Vehicular Communications; Basic Set of Applications; Maneuver Coordination Service

6. ETSI: European Standard ETSI EN 302 637-2 v1.4.1 (2019-04) (2019) Intelligent Transport Systems (ITS); Vehicular Communications; Basic Set of Applications; Part 2: Specification of Cooperative Awareness Basic Service

7. ETSI: European Standard ETSI EN 302 637-3 v1.3.1 (2019-04) (2019) Intelligent Transport Systems (ITS); Vehicular Communications; Basic Set of Applications; Part 3: Specifications of Decentralized Environmental Notification Basic Service

8. ETSI: Technical Report ETSI TR 103 562 v2.1.1 (2019-12) (2019) Intelligent Transport Systems (ITS); Vehicular Communications; Basic Set of Applications; Analysis of the Collective Perception Service (CPS); Release 2

9. Fehér, A., Aradi, S., Bécsi, T.: Hierarchical evasive path planning using reinforcement learning and model predictive control. IEEE Access **8**, 187470–187482 (2020). https://doi.org/10.1109/ACCESS.2020.3031037

10. Haarnoja, T., Zhou, A., Abbeel, P., Levine, S.: Soft actor-critic: off-policy maximum entropy deep reinforcement learning with a stochastic actor. CoRR **abs/1801.01290** (2018). http://arxiv.org/abs/1801.01290

11. Haarnoja, T., Zhou, A., Hartikainen, K., Tucker, G., Ha, S., Tan, J., Kumar, V., Zhu, H., Gupta, A., Abbeel, P., Levine, S.: Soft actor-critic algorithms and applications (2019)

12. Hernandez-Leal, P., Kartal, B., Taylor, M.E.: A survey and critique of multiagent deep reinforcement learning. Auton. Agent. Multi-Agent Syst. **33**(6), 750–797 (2019). https://doi.org/10.1007/s10458-019-09421-1

13. Heß, D., Lapoehn, S., Lobig, T., Nichting, M., Markowski, R., Lauermann, J., Dariani, R., Rieck, J.: ADORe: Automated driving open research. https://github.com/eclipse/adore (2022). https://github.com/eclipse/adore

14. Heß, D., Lattarulo, R., Pérez, J., Hesse, T., Köster, F.: Negotiation of cooperative maneuvers for automated vehicles: experimental results. In: 2019 IEEE Intelligent Transportation Systems Conference (ITSC), pp. 1545–1551. IEEE (2019)

15. Heß, D., Lattarulo, R., Pérez, J., Schindler, J., Hesse, T., Köster, F.: Fast maneuver planning for cooperative automated vehicles. In: 2018 21st International Conference on Intelligent Transportation Systems (ITSC), pp. 1625–1632 (2018). https://doi.org/10.1109/ITSC.2018.8569791

16. Iberraken, D., Adouane, L.: Safe navigation and evasive maneuvers based on probabilistic multi-controller architecture. IEEE Trans. Intell. Transp. Syst. **23**(9), 15558–15573 (2022). https://doi.org/10.1109/TITS.2022.3141893

17. Kingma, D.P., Ba, J.: Adam: A method for stochastic optimization. In: Bengio, Y., LeCun, Y. (eds.), 3rd International Conference on Learning Representations, ICLR 2015, San Diego, CA, USA, May 7–9, 2015, Conference Track Proceedings (2015). http://arxiv.org/abs/1412.6980

18. Kurzer, K., Schörner, P., Albers, A., Thomsen, H., Daaboul, K., Zöllner, J.M.: Generalizing decision making for automated driving with an invariant environment representation using deep reinforcement learning. In: 2021 IEEE Intelligent Vehicles Symposium (IV), pp. 994–1000 (2021). https://doi.org/10.1109/IV48863.2021.9575669

19. Lehmann, B., Günther, H.J., Wolf, L.: A generic approach towards maneuver coordination for automated vehicles. In: 2018 21st International Conference on Intelligent Transportation Systems (ITSC), pp. 3333–3339. IEEE (2018)

20. Lizenberg, V., Alkurdi, M.R., Eberle, U., Köster, F.: Intelligent co-simulation framework for cooperative driving functions. In: 2021 IEEE 17th International Conference on Intelligent Computer Communication and Processing (ICCP), pp. 109–115 (2021). https://doi.org/10.1109/ICCP53602.2021.9733618

21. Llatser, I., Michalke, T., Dolgov, M., Wildschütte, F., Fuchs, H.: Cooperative automated driving use cases for 5g v2x communication. In: 2019 IEEE 2nd 5G World Forum (5GWF), pp. 120–125 (2019). https://doi.org/10.1109/5GWF.2019.8911628

22. Lombard, A., Perronnet, F., Abbas-Turki, A., Moudni, A.E.: On the cooperative automatic lane change: speed synchronization and automatic "courtesy". In: Proceedings of the Conference on Design, Automation & Test in Europe, DATE '17, pp. 1659–1662. European Design and Automation Association, Leuven, BEL (2017)
23. Lopez, P.A., Behrisch, M., Bieker-Walz, L., Erdmann, J., Flötteröd, Y.P., Hilbrich, R., Lücken, L., Rummel, J., Wagner, P., Wießner, E.: Microscopic traffic simulation using sumo. In: The 21st IEEE International Conference on Intelligent Transportation Systems, pp. 2575–2582. IEEE (2018)
24. Nichting, M., Heß, D., Schindler, J., Hesse, T., Köster, F.: Explicit negotiation method for cooperative automated vehicles. In: IEEE International Conference on Vehicular Electronics and Safety (ICVES) (2019). https://doi.org/10.1109/ICVES.2019.8906401
25. Nichting, M., Heß, D., Schindler, J., Hesse, T., Köster, F.: Space time reservation procedure (strp) for v2x-based maneuver coordination of cooperative automated vehicles in diverse conflict scenarios. In: 2020 IEEE Intelligent Vehicles Symposium (IV), pp. 502–509 (2020). https://doi.org/10.1109/IV47402.2020.9304769
26. Nichting, M., Lobig, T., Köster, F.: Case study on gap selection for automated vehicles based on deep q-learning. In: 2021 International Conference on Artificial Intelligence and Computer Science Technology (ICAICST), pp. 252–257 (2021). https://doi.org/10.1109/ICAICST53116.2021.9497818
27. Park, Y., Nam, H.S., Ahn, C.: Evasive steering control using model predictive control. In: 2019 Australian and New Zealand Control Conference (ANZCC), pp. 204–204 (2019). https://doi.org/10.1109/ANZCC47194.2019.8945661
28. Perronnet, F., Abbas-Turki, A., El Moudni, A.: A sequenced-based protocol to manage autonomous vehicles at isolated intersections. In: 16th International IEEE Conference on Intelligent Transportation Systems (ITSC 2013), pp. 1811–1816 (2013). https://doi.org/10.1109/ITSC.2013.6728491
29. Schulman, J., Wolski, F., Dhariwal, P., Radford, A., Klimov, O.: Proximal policy optimization algorithms (2017). https://doi.org/10.48550/ARXIV.1707.06347. https://arxiv.org/abs/1707.06347
30. Tram, T., Jansson, A., Grönberg, R., Ali, M., Sjöberg, J.: Learning negotiating behavior between cars in intersections using deep q-learning. In: 2018 21st International Conference on Intelligent Transportation Systems (ITSC), pp. 3169–3174 (2018). https://doi.org/10.1109/ITSC.2018.8569316
31. Wang, P., Chan, C.Y., de La Fortelle, A.: A reinforcement learning based approach for automated lane change maneuvers. In: 2018 IEEE Intelligent Vehicles Symposium (IV), pp. 1379–1384 (2018). https://doi.org/10.1109/IVS.2018.8500556
32. Wang, P., Li, H., Chan, C.Y.: Continuous control for automated lane change behavior based on deep deterministic policy gradient algorithm. In: 2019 IEEE Intelligent Vehicles Symposium (IV), pp. 1454–1460 (2019). https://doi.org/10.1109/IVS.2019.8813903
33. Xu, H., Gao, Y., Yu, F., Darrell, T.: End-to-end learning of driving models from large-scale video datasets. In: 2017 IEEE Conference on Computer Vision and Pattern Recognition (CVPR), pp. 3530–3538 (2017). https://doi.org/10.1109/CVPR.2017.376
34. Xu, M., Luo, Y., Yang, G., Kong, W., Li, K.: Dynamic cooperative automated lane-change maneuver based on minimum safety spacing model. In: 2019 IEEE Intelligent Transportation Systems Conference (ITSC), pp. 1537–1544 (2019). https://doi.org/10.1109/ITSC.2019.8917095
35. Xu, W., Willecke, A., Wegner, M., Wolf, L., Kapitza, R.: Autonomous maneuver coordination via vehicular communication. In: 2019 49th Annual IEEE/IFIP International Conference on Dependable Systems and Networks Workshops (DSN-W), pp. 70–77 (2019). https://doi.org/10.1109/DSN-W.2019.00022

Hierarchical Motion Planning for Consistent and Safe Decisions in Cooperative Autonomous Driving

Jan Eilbrecht and Olaf Stursberg

Abstract The immersion of autonomous cars in continuously changing environments of on-road traffic requires procedures for decision-making with fast adaptation as well as guarantees on safe motion and collision-avoidance. This contribution proposes a three-layer hierarchic decomposition of the task of automatically steering the autonomous car along a designated route in cooperation with neighbored vehicles. The upper layer of the hierarchy identifies cooperative groups of those vehicles which are involved in a joint scenario for a phase of the planning horizon. The medium layer employs set-based computations of the free space for any vehicle of a joint scenario together with constrained optimal control to determine optimized motion plans. These plans are used on the lower layer as reference signals for tracking control in order to realize motion trajectories. The architecture ensures consistency of the vehicle motion with respect to safety for given assumptions, as well as relatively small computation times by combining offline with online computation.

1 Introduction

Autonomous driving of road vehicles promises to release passengers from paying attention to traffic, to enable car-sharing concepts relying on automated vehicles, and to enhance traffic flow by better coordination [34]. An anticipated additional advantage—and a required property at the same time—is the reduction of the number of accidents, injuries, and fatalities per driven distance. To see that this property indeed is achieved, the process of determining driving decisions for automated cars needs to continuously evaluate if an encountered scenario bears the risk of safety-critical evolutions, and to choose only driving options for which the motion remains safe as likely as possible. With respect to designing safe motion of single autonomous vehicles, intense research efforts in the past years have led to considerable insight into how to accomplish the main tasks of environment perception, vehicle localization,

J. Eilbrecht · O. Stursberg (✉)
Control and System Theory, Department of Electrical Engineering and Computer Science,
University of Kassel, Kassel, Germany
e-mail: stursberg@uni-kassel.de

C. Stiller et al. (eds.), *Cooperatively Interacting Vehicles*,
https://doi.org/10.1007/978-3-031-60494-2_11

303

and control, see [2, 36, 40] among many others. Recent tremendous progress in inter-vehicle communication [48] paves the path, however, to employ techniques of coordination and cooperation to further improve safe autonomous operation also for groups of vehicles. The exchange of driving plans among neighbored cars (or the distribution of jointly computed plans to these cars) can obviously reduce the uncertainty about the actions of other traffic participants, and thus can contribute to safety. This book chapter investigates how hierarchical concepts of cooperative motion planning for groups of autonomous vehicles can ensure driving decisions that are consistent with respect to safe interaction.

1.1 Relevant Work

Due to the complexity of the tasks to timely identify a current traffic situation, of computing a safe driving decision, and possibly to communicate with and align to the behavior of connected vehicles, the use of modular and hierarchic approaches has been investigated in various forms, see e.g. [5, 6, 10, 41, 50]. While such schemes are often expected to lead to quicker reactions, to more flexibility and suitability for maintenance (such as easier update of modules) [38], they also bear to the challenge of ensuring consistency between different decision units: For the information-flow from a top layer of a hierarchy, which typically determines a qualitative behavior (such as lane following, turning, emergeny braking, etc.), to the bottom-most layer, which takes care of the vehicle actuation, it must be ensured that decisions are not contradictive.

For the subtask of path planning of autonomous vehicles, a large set of different approaches has been proposed in the past, as reported in the survey papers [3, 21, 22]. One class of techniques is based on gridding of the state space of a vehicle and searching a path along a set of grid points, e.g., by path-velocity decomposition [27], by RRT*-algorithms [26, 30], or by Monte-Carlo trees [28]. The complexity of these approaches, however, grows exponentially with the dimension of the state space, as obtained for larger sets of vehicles. A second class of techniques is that of learning-based approaches comprising supervised learning based on data from human driving experience, e.g. [45, 47], and reinforcement learning [31]. While these methods do not require structural insight into traffic situations and the computational effort required online is relatively low, the data set required for learning offline is very large, and means to always guarantee safety are not known yet. A different class of methods, which is relevant for the approach to be proposed in this paper, uses elements for structuring driving behavior into maneuvers, motion primitives, or homotopies. The common idea is to group behaviors which satisfy a notion of similarity or symmetry (such as invariance to translation or rotation), see e.g. [18, 19]. Maneuvers together with principles of optimal control can be used for motion planning [19, 33, 44], as well as for verifying safety of vehicle motion [25]. To satisfy conditions under which maneuvers or motion primitives can be concatenated to longer driving plans, and the consideration of obstacles are challenging for these techniques. In addition, these concepts are so far not used for sets of cooperative autonomous vehicles. A fourth class of relevant techniques for the work to be proposed in this chapter are approaches based on mixed-integer programming. The underlying optimization problems combine logic conditions modeled by integer variables with continuous variables to represent vehicle motion, see e.g. [29, 35, 37, 39, 43]. While the subclasses of mixed-integer linear or quadratic programming ensure that globally optimal solutions can be found, the computational effort is typically an issue if used in online optimization.

For the subtask of tracking a reference on a lower layer of a control hierarchy, different variants of model-predictive control (MPC) have been considered in the past, since this class of techniques is suitable to consider constraints of the inputs and states (such as the adherence to the admissible regions). On the one hand, approaches of nonlinear MPC have been considered for this purpose [16, 17, 23], but it is questionable whether the nonlinear optimization problem can be solved with the frequency required for low-level tracking control. On the other hand, linearized-based variants of MPC [10, 24, 32, 50] need to account robustly for linearization errors, typically leading to conservativeness, thus implicitly reducing the space from which the reference signals can be chosen. Similar reasoning applies to methods relying on linear-parameter-varying models, see e.g. [1, 4]. For these reasons, the use of tracking techniques based on nonlinear dynamics, with consideration of constraints, but without embedding online optimization seems preferable. The techniques reported in [20, 49] for tracking by stabilizing feedback control motivate the concept used in this paper for the same purpose, but those techniques require extensions to account for the constraints imposed by the second layer.

1.2 Contribution

In order to timely adapt the driving behavior of sets of cooperative vehicles to changing situations, this book chapter proposes a hierarchic approach using three layers of decision: The first (and upper) layer structures the setting into cooperative groups, the second layer computes driving plans which are guaranteed to exclude collision while leading to the goals of the involved vehicles, and the third layer uses the plans as reference signals for the online control tasks. One particular focus with respect to the second layer is the division into an offline and an online part for computational efficiency. The offline part determines and stores admissible driving regions as well as selected optimized trajectories for the vehicles grouped together for a specific maneuver. The online part comprises only the relatively easy steps of interpolating between the pre-computed optimal trajectories. A second focus of this chapter is on the third layer, on which a feedback control task is solved for a more detailed representation of nonlinear vehicle dynamics such that following the plan obtained from the second layer (as reference signal) is achieved. At the same time, it is ensured that the admissible driving region is not left. Thus, a main benefit of the approach to be presented is that consistency of the decisions at the interfaces between the first and second layer, and between the second and third layer respectively, is guaranteed.

2   A Hierarchical Approach to Decision Making

This section first provides an overview of the proposed hierarchic procedure of decision making for cooperative autonomous driving, while details on the techniques assigned to the three layers will be described in the subsequent sections. According to Fig. 1, the upper-most layer is termed *group coordination* and aims at identifying the traffic scenario in a particular road area. The underlying (and required) information is the road topology in the respective area, the set of traffic participants and obstacles in this area, and the routes of the vehicles (leading essentially to the point at which each vehicle intends to leave the area). The sources of information are the route planners of the vehicles, the communication units of the vehicles (and possibly of road-side stations), and the on-board vision and perception systems. The mechanisms of the latter are not in the scope of this paper, but the assumption is that the onboard sensorics together with algorithms of object identification deliver the complete set of objects in the environment, together with predictions of the motion of dynamic objects. The available information is used to identify the current scenario and to select a maneuver from a maneuver library, which is computed a-priori in an offline phase. The selection of the maneuver comprises the formation of cooperative groups, i.e. the set of vehicles is partitioned into subsets which cooperate in executing a maneuver jointly, see Sect. 2. (Non-cooperative traffic participants are assigned to separate groups.)

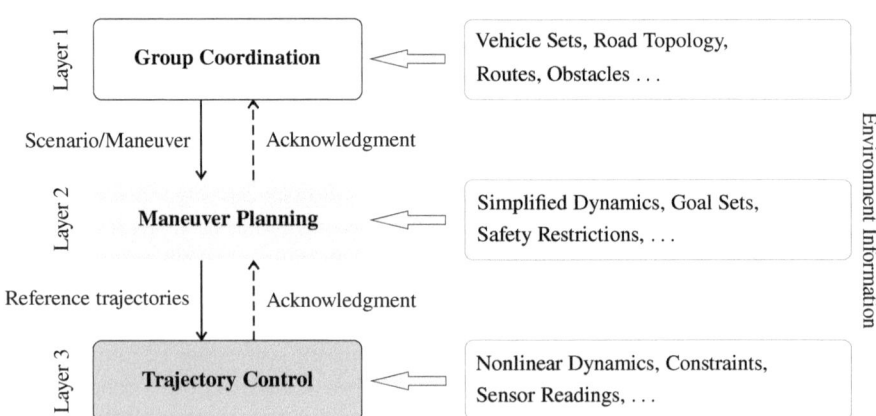

Fig. 1 Hierarchical structure of the decision-making procedure and additional information from the environment

For any of such cooperative groups, the maneuver selection is passed to layer 2, termed *maneuver planning*. For each maneuver, the set of possible behaviors for each car of the respective group is computed offline based on reachable sets of hybrid automata. These automata (to be defined in Sect. 4) combine the driving plans of the vehicles assigned to a cooperative group, and they ensure that any vehicle exclusively occupies a certain region of the road, thus excluding collision. The computation of reachable sets is based on simplified linear dynamics and considers the goal set for the chosen maneuver and constraints arising from obstacles and the road topology. Optimized sample trajectories within these reachable sets are computed offline, and are stored in a maneuver library for the respective situation. In the online execution, an interpolation between the optimized sample trajectories is computed for each vehicle and its exact position as observed by the sensors. These interpolated trajectories determine a reference forwarded to the third layer, and they remain contained in the set of admissible regions. An acknowledgement is sent back to the first layer to either report success of maneuver planning, or to initiate the selection of a different maneuver, if the one considered before was not found to be executable in the online procedure.

On the third layer, each vehicle controller locally aims at controlling the vehicle position and speed to the reference obtained from the second layer. As opposed to the linearized dynamics used on layer 2, this feedback control task is based on a higher-dimensional nonlinear model of the vehicle dynamics, which involves the quantities obtained from the on-board sensors, the actuated quantities, as well as possible disturbances (such as wind gusts). It is important for this layer that the local tracking error between the planned reference and the actual vehicle path is hold in suitable bounds. For the trajectory tracking, a tailored method is proposed which uses a nonlinear model of the tracking error represented in the Frenet frame. The tracking errors of the position, their speed of change, and of the yaw angle is controlled by state feedback control. Bounds on the maximum deviation arising for bounded uncertainties of wind and tire forces can be computed, and thus can be compared with the driving corridors used in maneuver planning on layer 2. If an inconsistency occurs, the acknowledgement signal from layer 3 to 2 needs to report this and triggers modification of the reference trajectory.

The computations on the three layers are iteratively updated with appropriate frequency: In the nominal case, formed groups on layer 1 would be expected to exist for a range of several seconds until a maneuver is terminated, but possible occurrence of new situations (including emergencies like suddenly occurring obstacles) requires to react within a few milliseconds. If layer 2 follows a selected maneuver over several seconds, incoming new measurements about the exact vehicle positions or communicated information allow to update the computation of updated reference trajectories—an update of every of 10–50 milliseconds can be deemed reasonable for implementations of the hierarchy, while emergency situations again require to switch much faster to, e.g., a mode of hard braking. On layer 3, the update of the feedback control action would be executed nominally in the range of milliseconds.

3 Group Coordination

The task of group coordination is explained based on the example scenario shown
in Fig. 2. Assume that a set of vehicles is present in a defined section of the road
network, here the area of a T-intersection. A decision unit performing the functions
of the upper two layers of the control hierarchy is assigned to this section. Physically,
this unit could either be embedded into the infrastructure of the intersection, or one
of the car controllers could temporarily assume this role. Let all vehicles be equipped
with devices for wireless communication, such that the intended routes, the current
positions and speed, and the driving plans (as outcome of the planning on layer 2)
can be exchanged among the vehicles and with the decision unit. If the five vehicles
follow the intended directions as indicated by the solid arrows, the assignment to
two cooperative groups is straightforward, namely the red-colored and the two gray-
colored vehicles form a group G_1, while the white and the green vehicle are assigned
to G_2. Obviously, a simple lane following maneuver needs to be planned for G_2,
whereas it has to be decided for G_1 at which position the car intending the right turn
merges into the convoy of the two others. Assuming that the green car intended a left
turn (indicated by the dashed arrow), the vehicle needed to be assigned to G_1.

The situation becomes more complicated if non-cooperative traffic participants
are present: If the red car were non-cooperative, the decision unit would assign it
to an additional group G_3. It had to be distinguished if this car communicates its
planned trajectory, or not. If it did, the trajectory would be considered as disturbance
for the maneuver planning of the gray cars in G_1. If it did not, the complete set of
possible behaviors of the red car would have to be considered (or estimated based on
the sensor data), leading to more conservative maneuver planning for the gray car
turning right. More details on considering non-cooperative traffic participants can be
found in [8, 42], where vulnerable road users (cyclists or pedestrians) are addressed.

Note that for some scenarios the assignment of vehicles to groups is not related
to a fixed section of a road, but to a moving section, as e.g. if a set of interacting
vehicles on a highway perform an over-taking maneuver.

While the approach presented in this paper determines cooperative maneuvers for
each group by the procedure described in the next section, the following alternative
approach from [9, 41] should be briefly mentioned: There, the different vehicles in
a road section compute a proposal for collision-free trajectories by MPC using lin-

Fig. 2 Example of forming
cooperative groups: The
green car is assigned to G_2
or G_1 depending on the
driving intention. In case of a
non-cooperative red car, it
forms another group G_3. All
vehicles communicate with a
decision unit

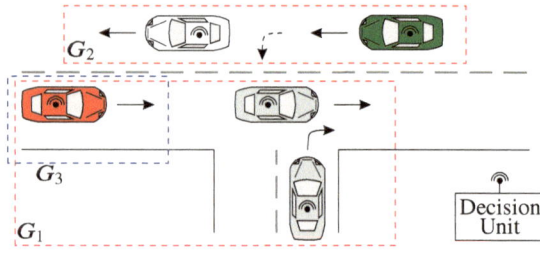

earized dynamics (as an alternative to layer 2). These trajectories are then negotiated by an auction-like bidding procedure, i.e. the vehicle controllers determine which proposed trajectories are agreeable for all vehicles of a group (as an alternative solution on layer 1).

4 Maneuver Planning

4.1 Planning Based on Hybrid Models and Controllable Sets

The maneuver planning for a cooperative group is based on joint modeling of the vehicle dynamics by hybrid automata. This choice is motivated by the observation that most maneuvers can be understood as a sequence of qualitatively different phases, as shown for the example of an over-taking scenario in Fig. 3, adopted from [11]. The process of the red car overtaking the gray one can be separated into the phase P_1 of accelerating and approaching the gray car, the phase P_2 of changing lane and passing the gray car, and the phase P_3 of changing back to the right lane and possibly decelerating. The maneuver is straightforwardly cast into a hybrid automaton, as shown in the bottom part of Fig. 3: Each phase P_i, $i \in \{1, 2, 3\}$ is modeled by a discrete state q_i, and transitions θ_{12}, θ_{23} represent the instantaneous change of the discrete state if an associated transition condition $g(\theta_{12}, x_k)$, or $g(\theta_{23}, x_k)$ holds true. These conditions depend on the continuous-valued state vector x_k at a discrete point of time indicated by k. The state vector comprises the position (in cartesian coordinates) and the speed of all vehicles in the cooperative group. Based on x_k, the condition $g(\theta_{12}, x_k)$ formulates, e.g., the conjunction of the facts that the red car has approached the gray one up to a lower distance threshold, that the speed of the red car is sufficiently

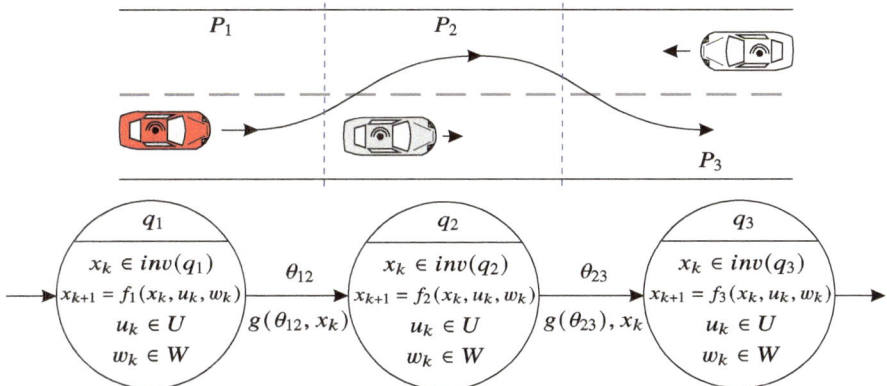

Fig. 3 Partitioning of the overtaking procedure of the red car into three phases and construction of a hybrid automaton modeling the procedures

larger that than the speed of the gray car, and that the distance to the white car is sufficiently large for a given speed of the white car. Conditions of similar type can be formulated to characterize the admissible values of x_k for each of the discrete states q_i, then referred to as *invariants* $x_k \in inv(q_i)$. They play an important role in separating safe from unsafe driving behavior, i.e., any x_k modeling a dangerously close distance between two vehicles is excluded from the invariants. Likewise, obstacles and regions outside of the drivable road space are not included in $inv(q_i)$. The evolution of the state is modeled by discrete-time difference equations $x_{k+1} = f_i(x_k, u_k, w_k)$, which depend on the vector of control inputs u_k (available to actively change x_k by accelerating, braking, and steering) and on a possible vector of disturbances w_k—for both vectors, static bounds $u_k \in U$ and $w_k \in W$ are assumed to be known. Formally, a hybrid automaton $HA = (T, Q, q_0, q_T, inv, U, W, X_0, X_T, \Theta, g, f)$ is introduced for a maneuver, containing the set $T \subset \mathbb{N}$ of discrete points of time, the set $Q \subset \mathbb{N}$ of discrete states, discrete initial and target states $q_0 \in Q$ and $q_T \in Q$, the assignment inv of invariants to discrete states, the bounded sets $U \subset \mathbb{R}^m$ and $W \subset \mathbb{R}^r$ of inputs and disturbances, the sets $X_0 \subset \mathbb{R}^n$ and $X_T \subset \mathbb{R}^n$ of possible continuous initial states and target states, the transition conditions g, and the discrete-time state transfer functions f_i (collected in f). See [7, 12] for more details on the semantics of the model.

Given this model, a *maneuver* is defined as a tuple $\mathcal{M} = (\mathcal{G}, \mathcal{N}, HA, h)$ of the group $\mathcal{G}$ of cooperating cars, the set $\mathcal{N}$ of non-cooperating vehicles, the hybrid automaton, and a planning horizon h (as a maximum number of time points to complete the maneuver). Such a tuple is modeled offline as a template for a class of scenarios of the same pattern, as for the example of overtaking procedures on a single lane road involving three vehicles. A tuple $\mathcal{M}$ is modeled offline for any class of scenarios the autonomous vehicles are expected to get in, and $\mathcal{M}$ is stored in a maneuver library.

The advantage of using maneuvers of this type for planning is that set-based offline computations allow (under some assumptions) to represent the set of feasible and collision-free maneuver instances. For this purpose, let all f_i be affine mappings of its arguments and all continuous sets in HA be chosen polytopic. Then, *j-step robust controllable sets* can be computed for HA as a sequence of polytopic subsets of the invariants:

$$\mathcal{K}_| = (K_0, \ldots, K_j)$$

for which an input trajectory $(u_0, \ldots, u_j)$ exists to definitely transfer the state $x_0 \in X_0$ in at most j steps into the target X_T, despite of the presence of the disturbances $w_k \in W$. These sets are instrumental for the guarantee of finding a winning control strategy for an arbitrary initialization $x_0 \in X_0$ measured in online operation, i.e. to determine $(u_0, \ldots, u_j)$ for:

$$(x_0 \in K_0, \ldots, x_j \in X_T \subseteq K_j) \tag{1}$$

(compare to [12, 46]). Such a trajectory ensures for $h \leq j$ that a maneuver $\mathcal{M}$ can be successfully completed, and it implies that the vehicles are coordinated in safe interaction.

Of course, the choice of $(u_0, \ldots, u_j)$ is not unique. In order to determine an optimal choice with respect to a cost functional, such as minimizing the control effort and the distance to a target point of the maneuver, a minimization problem:

$$\min_{(u_0, \ldots, u_h)} \sum_{k=1}^{h} \|C \cdot (x_k - x_T)\|_2^2 + \|D \cdot u_k\|_2^2 \tag{2}$$

(with weighting matrices C, D) could be solved (e.g. for worst-case values of the w_k). This minimization, however, is subject not only to (1), but also to the constraints arising from the dynamics of HA. Its encoding with respect to the assignment of f_i to $inv(q_i)$ and the transitions involves to use binary variables and several linear inequalities, leading to an optimization of type mixed-integer quadratic programming. The computation times found to be required to solve such problems are often too large compared to the update frequency targeted for layer 2. In addition, even smaller times are required to decide whether a selected maneuver leads to a feasible solution (in order to report back to layer 1 that an alternative maneuver is needed). Thus, the proposal is to use a combination of offline computation of optimal strategies and quick online interpolation [13]: First, polytopic inner approximations $\hat{\mathcal{K}}_k \subset \mathcal{K}_k$ of the control invariant sets are determined. The objective is to obtain good coverage of the $\mathcal{K}_k$, to consider the worst-case disturbances (if present), and to use only a relatively small number of facets of the polytopes $\hat{\mathcal{K}}_k$. A procedure for this step is described in [7], and it considers the transition dynamics and the invariants of HA. Note that, as the result of this procedure, still a control trajectory leading to X_T is guaranteed to exist for any initialization to a state in the $\hat{\mathcal{K}}_k$.

Secondly and still carried out offline, the vertices of the polytopes $\hat{\mathcal{K}}_k$ are optimally projected forward onto $\hat{\mathcal{K}}_{k+1}, k \in \{0, \ldots, h-1\}$ by solving an optimization problem similar to (2). This procedure is exemplary sketched in Fig. 4 by dashed gray-colored

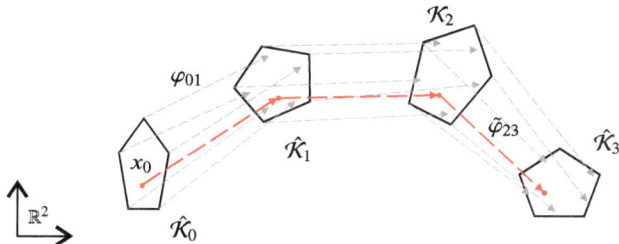

Fig. 4 Gray-colored arrows represent the optimal projections φ of the vertices of the polytopes $\hat{\mathcal{K}}_k$ (which are inscribed to the corresponding subset of $\mathcal{K}_k$). The path shown in red is obtained online from stepwise linear interpolation $\tilde{\varphi}$ between the optimal vertex proejctions

arrows, and is denoted by φ. The triples of vertex v_i, optimal input u_i^* and the optimal projected state $x_i^* = \varphi(v_i, u_i^*)$ are stored with the maneuver for online use.

In the online execution, the following is accomplished: First, it is checked whether the currently measured state, denoted by x_0, is contained in any of the sets $\hat{\mathcal{K}}_k$ of the selected maneuver $\mathcal{M}$. If not, the maneuver is not guaranteed to be executed safely, and an alternative maneuver has to be determined. If $x_0 \in \hat{\mathcal{K}}_k$ for any k, the barycentric coordinates of x_0 in $\hat{\mathcal{K}}_k$ (with respect to its vertices v_i) are computed. The input u_0 is then determined as the interpolation of the inputs u_i^* (as optimized offline for the vertices v_i) by use of the same barycentric coordinates. In Fig. 4, the outcome is shown by a red dashed line for the interpolated state path, and is denoted by $\tilde{\varphi}$. The corresponding sequence of interpolated inputs along this path determines the control strategy which transfers the state x_k eventually into the goal set X_T of the maneuver. The interpolation requires only relatively small computational effort.

4.2 Illustration for an Overtaking Maneuver

For illustration of the procedure, consider again the overtaking maneuver from Fig. 3, see also [7, 13]. Let the longitudinal position of the three vehicles be denoted by $p_x^{(i)}$ and the lateral position by $p_y^{(i)}$, where $i = 1$ refers to the red car, $i = 2$ to the gray car, and $i = 3$ to the white car approaching from opposite direction. To simplify the model, the relative longitudinal positions $p_r^{(2)} = p_x^{(2)} - p_x^{(1)}$ and $p_r^{(3)} = p_x^{(3)} - p_x^{(1)}$ are introduced, and the lateral positions of the gray and white car are assumed to be constant (thus their lateral speeds equal zero). The reduced state vector:

$$x = \begin{pmatrix} p_r^{(2)} & p_r^{(3)} & p_y^{(1)} & v_x^{(1)} & v_x^{(2)} & v_x^{(3)} & v_y^{(1)} \end{pmatrix}$$

is thus defined on a 7-dimensional space. The inputs are chosen identical to the accelerations in longitudinal and lateral direction ($u_x^{(i)} := \dot{v}_x^{(i)}$, $u_y^{(i)} := \dot{v}_y^{(i)}$), leading to a 4-dimensional input vector:

$$u = \begin{pmatrix} u_x^{(1)} & u_x^{(2)} & u_x^{(3)} & u_y^{(1)} \end{pmatrix}.$$

A simple double integrator dynamics is used for longitudinal and lateral motion, leading to a linear model $\dot{x}(t) = A \cdot x(t) + B \cdot u(t)$ with only entries 0 and 1 in the matrices A and B. While very simple, such a model (with subsequent discretization of time) is very frequently used in path planning—and well justifiable for the hierarchic approach, since it only serves the purpose of computing a reference trajectory for a control problem with more detailed dynamics on layer 3.

The model is complemented by constraints for the state and input vector, by a target set formulating the ranges of the vehicle positions for completing the maneuver, by a nominal longitudinal speed (as x_T), and by weighting matrices C, D of the cost functional (here chosen to identity matrices). The hybrid model HA is obtained

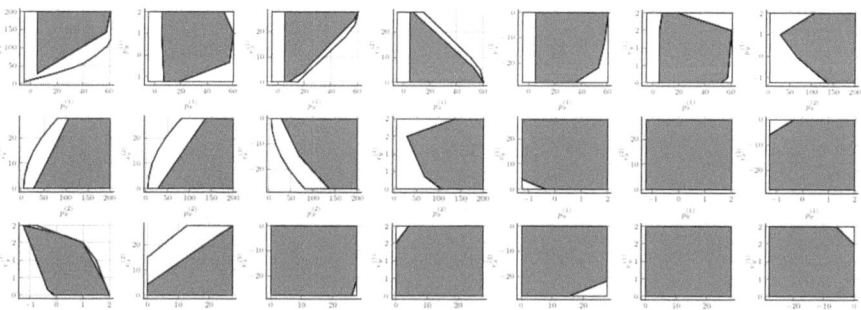

Fig. 5 Two-dimensional projections of the controllable sets obtained for the overtaking maneuver; the white sets represent the exact controllable sets, while the gray sets establish polytopic inner approximations; from [7]

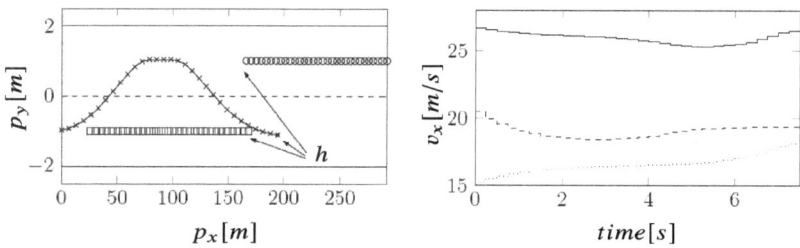

Fig. 6 Results for the overtaking procedure: left part—lateral over longitudinal position (crossed: red vehicle, boxed—gray vehicle, circled—white vehicle), h denotes the end of the planning horizon; right part—longitudinal speed over time (solid—red car, dashed—gray car, dotted—white car)

(according to Fig 3) by adding the discrete states, the invariants (which consider a safe minimal distance between the vehicles in both coordinates), and the transitions including the conditions $g(\theta)$ (corresponding to reaching the boundaries of the invariants); see [7] for a full parametrization of these components.

Based on this hybrid model (which does not include disturbances), the controllable sets are computed over $j = 5$ steps, and are shown in Fig. 5. After optimizing the trajectories originating from the vertices of the controllable sets, the interpolation procedure is applied. Figure 6 shows example trajectories of the positions and longitudinal speeds for a chosen initialization. The trajectories demonstrate that the maneuver is successfully completed without collision, and they indicate cooperation in the sense that the gray vehicle lowers its speed and the white vehicle keeps a low speed until the red vehicle has passed. The references [7, 13] provide insights into the computation times for the proposed scheme of online interpolation between offline optimized trajectories compared to online trajectory optimization within the controllable sets: For a large number of test instances for this example (with varying initialization) it was found that the interpolation approach required an average time of 0.25 milliseconds, while the results for the online optimization were obtained in average after 12.1 milliseconds. The first value can be deemed sufficiently small for execution within the hierarchic approach.

For a second example describing the cooperative maneuver of vehicles merging into highway traffic at an on-ramp, the interested reader is referred to [15].

5 Trajectory Control

The technique described in the preceding section provides a feasible path for any vehicle involved in a scenario with respect to the simplified (piecewise) linear dynamics used on layer 2. The control strategies obtained there for realizing the paths are, however, not immediately useful for controlling the autonomous vehicles for the following reasons: (a) vehicle motion comprises nonlinear effects which need to be considered for low-level vehicle control, (b) the state and input vectors used for layer 2 do not contain the full set of controlled and actuated variables typically employed for vehicle control. In consequence, it is not ensured that the vehicle would indeed follow the plan computed on layer 2—thus, the control hierarchy includes the additional layer 3 for local low-level vehicle control. This layer employs more accurate models, and the paths computed on layer 2 serve as reference signals for the local vehicle controllers. Two questions immediately arise from this choice: First, a more accurate and nonlinear model lets one expect higher computational effort. Reference tracking by nonlinear MPC, as in [10], may not be feasible if the time required for the online solution of the optimization problem is not compatible to high execution frequencies for low-level control. Hence, the approach proposed below uses a state-feedback controller with very low computational demands. Secondly, the use of different models for plan generation and tracking control may compromise consistency between the two decision levels—this point is addressed below. (See also [14] for a discussion on the relation between linear integrator models for planning and the nonlinear motion of vehicles on curved roads.)

To prepare the control design, it is convenient to transform the setting for a given reference trajectory into so-called Frenet-coordinates, which specify positions relative to their projection onto the reference path. This means that a point on the path is described by a path coordinate s, and an offset that is measured in normal direction to the reference.

The standard bicycle model is used here as starting point for vehicle modeling, with a state vector χ containing the positions p_x and p_y (in cartesian coordinates), the vehicle orientation ψ, and the longitudinal and lateral speeds v_X, v_Y, and the yaw rate ω. The input vector μ is defined to contain longitudinal tire slip s_X and the steering angle δ. With a rotation matrix $R(\psi)$ and external accelerations $a := (a_X, a_Y, a_\psi)^T$ to account for effects like wind and tire forces, the model is formulated to:

$$
\begin{pmatrix} \dot{p}_x \\ \dot{p}_y \\ \dot{\psi} \\ \dot{v}_X \\ \dot{v}_Y \\ \dot{\omega} \end{pmatrix} = \begin{pmatrix} R(\psi) \begin{pmatrix} v_X \\ v_Y \end{pmatrix} \\ \omega \\ \begin{pmatrix} a_X \\ a_Y \\ a_\psi \end{pmatrix} + \begin{pmatrix} v_X \omega \\ -v_Y \omega \\ 0 \end{pmatrix} \end{pmatrix}. \tag{3}
$$

It is assumed that the accelerations $a := \bar{a} + \Delta a$ can be written as the sum of a nominal part $\bar{a}$ and a bounded offset Δa, $||\Delta a|| \leq \Delta a_{max}$. As detailed in [7], the dependency of (3) on s_X is obtained from an appropriate tire model leading to the function $\bar{a}(\delta, s_X, s_Y, v_Y, \omega)$.

By defining the error state vector (with tangential and normal part to the reference indicated by indices t and n):

$$
e := \begin{pmatrix} e_t & e_n & \dot{e}_t & \dot{e}_n & e_\psi & e_\omega \end{pmatrix}^T = \begin{pmatrix} e_{pos}^T & \dot{e}_{pos}^T & e_{yaw}^T \end{pmatrix}^T,
$$

the error dynamics can be derived to:

$$
\frac{d}{dt} \begin{pmatrix} e_{pos} \\ \dot{e}_{pos} \end{pmatrix} = \begin{pmatrix} 0 & 0 & 1 & 0 \\ 0 & 0 & 0 & 1 \\ \dot{\theta}^2 & \ddot{\theta} & 0 & 2\dot{\theta} \\ -\ddot{\theta} & \dot{\theta}^2 & -2\dot{\theta} & 0 \end{pmatrix} \begin{pmatrix} e_{pos} \\ \dot{e}_{pos} \end{pmatrix} - \begin{pmatrix} 0 \\ 0 \\ \ddot{s} \\ \dot{\theta}\dot{s} \end{pmatrix} + \begin{pmatrix} 0 & 0 \\ 0 & 0 \\ R(e_\psi) \end{pmatrix} \begin{pmatrix} \bar{a}_X \\ \bar{a}_Y \end{pmatrix} + \begin{pmatrix} 0 & 0 \\ 0 & 0 \\ R(e_\psi) \end{pmatrix} \begin{pmatrix} \Delta a_X \\ \Delta a_Y \end{pmatrix} \tag{4}
$$

and:

$$
\frac{d}{dt} \begin{pmatrix} e_\psi \\ \dot{e}_\omega \end{pmatrix} = \begin{pmatrix} 0 & 1 \\ 0 & 0 \end{pmatrix} \begin{pmatrix} e_\psi \\ \dot{e}_\omega \end{pmatrix} + \begin{pmatrix} 0 \\ 1 \end{pmatrix} (\bar{a}_\psi + \Delta a_\psi - \ddot{\theta}). \tag{5}
$$

In here, θ describes the orientation angle between the longitudinal direction (indicated by X, tangential to the reference signal) and the cartesian coordinate x.

For given v_X, v_Y, and ω, the nominal acceleration $\bar{a}$ is controlled by appropriate choice of δ and s_X (while Δa is a disturbance). By defining a virtual input vector $\tilde{\mu}$, $\bar{a}$ can be computed:

$$
\tilde{\mu} := \begin{pmatrix} \dot{\theta}^2 & \ddot{\theta} & 0 & 2\dot{\theta} \\ -\ddot{\theta} & \dot{\theta}^2 & -2\dot{\theta} & 0 \end{pmatrix} \begin{pmatrix} e_{pos} \\ \dot{e}_{pos} \end{pmatrix} - \begin{pmatrix} \ddot{s} \\ \dot{\theta}\dot{s} \end{pmatrix} + R(e_\psi) \begin{pmatrix} \bar{a}_X \\ \bar{a}_Y \end{pmatrix} \tag{6}
$$

$$
\Leftrightarrow \begin{pmatrix} \bar{a}_X \\ \bar{a}_Y \end{pmatrix} = R(e_\psi)^T \left(\tilde{\mu} + \begin{pmatrix} \ddot{s} \\ \dot{\theta}\dot{s} \end{pmatrix} - \begin{pmatrix} \dot{\theta}^2 & \ddot{\theta} & 0 & 2\dot{\theta} \\ -\ddot{\theta} & \dot{\theta}^2 & -2\dot{\theta} & 0 \end{pmatrix} \begin{pmatrix} e_{pos} \\ \dot{e}_{pos} \end{pmatrix} \right) \tag{7}
$$

To reduce the tracking error by feedback compensation, define a feedback law:

$$
\tilde{\mu} = -K \begin{pmatrix} e_{pos} \\ \dot{e}_{pos} \end{pmatrix} \tag{8}
$$

with controller matrix K, by which the closed-loop position and speed error are obtained to:

$$\frac{d}{dt}\begin{pmatrix} e_{pos} \\ \dot{e}_{pos} \end{pmatrix} = f_{pos}(e, \Delta a) := \begin{pmatrix} 0_{2\times 2} & I_{2\times 2} \\ & -K \end{pmatrix}\begin{pmatrix} e_{pos} \\ \dot{e}_{pos} \end{pmatrix} + \begin{pmatrix} 0 & 0 \\ 0 & 0 \\ & R(e_\psi) \end{pmatrix}\begin{pmatrix} \Delta a_X \\ \Delta a_Y \end{pmatrix}. \quad (9)$$

The transformation of a given $\bar{a}$ into δ and s_X is detailed in [7], as well is the derivation of an expression for the yaw error dynamics:

$$\frac{d}{dt}\begin{pmatrix} e_\psi \\ e_\omega \end{pmatrix} = f_{yaw}(e, \bar{x}, \Delta a).$$

For several variables contained in the nonlinear model, physical constraints need to be satisfied in order to establish safe driving. Only those reference trajectories from layer 2 which satisfy these constraints, can be characterized as realizable. Furthermore, deviations from the reference signal (in terms of the introduced tracking errors) must be located inside of those constraints. Thus, for given bounds Δa_{max} of the uncertainties, admissible ranges for the tracking errors can be computed in order to satisfy the constraints. While out of scope of this book chapter, the reader is referred to the approach described in [7] for synthesizing the matrix K in the feedback control law (8). This law keeps the tracking error (for e_{pos} and $\dot{e}_{pos}$) within the admissible ranges (for single reference values). The synthesis is based on solving a semi-definite program constrained by linear matrix inequalities. The analysis of the tracking error of the yaw dynamics is more intricate due to its nonlinear and time-varying nature, but boundedness of this error can be shown, too, under appropriate assumptions [7].

6 Conclusions

This book chapter has proposed the concept of a hierarchic decision architecture to enable cooperative driving of a set of autonomous vehicles (even in presence of non-cooperative traffic participants). While the implementation and testing of this proposal in practice (i.e. on an autonomous vehicle) is still matter of future investigations, the following principal advantages are highlighted:

- The decomposition of the overall problem into three layers for separating the tasks of group determination, planning of joint maneuvers, and local vehicle control makes the problem tractable, even for the challenging timing of real-world traffic scenarios. The additional partitioning of the maneuver planning on layer 2 into an offline and an online part further increases the computational efficiency.
- With the objective to provide guarantees on safety of the cooperative driving strategies, particular emphasis has been set on embedding constraints for excluding collision, and on reasoning about consistency of the decisions across the layers.

Along this line, mechanisms for checking admissibility of the information received from the super-ordinated layers have been proposed.

- The feedback control scheme on layer three is not only computationally very efficient, but it allows to derive conditions for which constraint compliance is obtained. Upper bounds on the tracking errors (for bounded disturbances) allow to construct reference trajectories on layer 2 which have sufficient safety margins for excluding collision.

Aspects of future research include the construction of an as complete as possible maneuver library for layer 2. If a scenario is encountered for which no compliant maneuver has been defined, stopping one or more vehicles is the only and undesired choice. (Of course, this lack of completeness applies for all existing approaches, including those relying on learning from massive data sets.) Systematic classification of scenarios and structured modeling of atoms of maneuvers (and concatenation hereof) may help to run into this situation very rarely. With respect to the third layer, the inclusion of measurement uncertainties constitutes a valuable future extension.

Acknowledgements The research leading to the results described in this chapter was financially supported by the German Research Foundation (DFG) within the priority program *SPP 1835— Kooperativ interagierende Automobile* under grant number STU-262/6-1/2. This support is gratefully acknowledged.

References

1. Alcalá, E., Puig, V., Quevedo, J., Rosolia, U.: Autonomous racing using linear parameter varying-model predictive control (LPV-MPC). Control. Eng. Pract. **95**, 104270 (2020)
2. Bresson, G., Alsayed, Z., Yu, L., Glaser, S.: Simultaneous localization and mapping: a survey of current trends in autonomous driving. IEEE Trans. Intell. Veh. **2**(3), 194–220 (2017)
3. Claussmann, L., Revilloud, M., Gruyer, D., Glaser, S.: A review of motion planning for highway autonomous driving. IEEE Trans. Intell. Transp. Syst. **21**(5), 1826–1848 (2019)
4. Corno, M., Panzani, G., Roselli, F., Giorelli, M., Azzolini, D., Savaresi, S.M.: An LPV approach to autonomous vehicle path tracking in the presence of steering actuation nonlinearities. IEEE Trans. Control Syst. Technol. 1–9 (2020)
5. Donges, E.: A conceptual framework for active safety in road traffic. Veh. Syst. Dyn. **32**(2–3), 113–128 (1999)
6. Du, J., Masters, J., Barth, M.: Lane-level positioning for in-vehicle navigation and automated vehicle location (AVL) systems. In: Proceedings of the IEEE International Conference on Intelligent Transportation Systems, pp. 35–40 (2004)
7. Eilbrecht, J.: Consistent Hierarchical Control in Cooperative Autonomous Driving. Kobra (2022). https://doi.org/10.17170/kobra-202205116162
8. Eilbrecht, J., Bieshaar, M., Zernetsch, S., Doll, K., Sick, B., Stursberg, O.: Model-predictive planning for autonomous vehicles anticipating intentions of vulnerable road users by artificial neural networks. In: Proceedings of the IEEE Symposium Series on Computational Intelligence, pp. 2869–2876 (2017)
9. Eilbrecht, J., Stursberg, O.: Auction-based cooperation of autonomous vehicles using mixed-integer planning. In: Proceedings of the AAET, pp. 266–286. ITS Automotive Nord (2017)

10. Eilbrecht, J., Stursberg, O.: Cooperative driving using a hierarchy of mixed-integer programming and tracking control. In: Proceedings of the IEEE Intelligent Vehicles Symposium, pp. 673–678 (2017)
11. Eilbrecht, J., Stursberg, O.: Optimization-based maneuver automata for cooperative trajectory planning of autonomous vehicles. In: Proceedings of the European Control Conference, pp. 82–88 (2018)
12. Eilbrecht, J., Stursberg, O.: Hierarchical solution of non-convex optimal control problems with application to autonomous driving. Eur. J. Control. **50**, 188–197 (2019)
13. Eilbrecht, J., Stursberg, O.: Reducing computation times for planning of reference trajectories in cooperative autonomous driving. In: Proceedings of the IEEE Intelligent Vehicles Symposium, pp. 114–120 (2019)
14. Eilbrecht, J., Stursberg, O.: Challenges of trajectory planning with integrator models on curved roads. In: Proceedings of the 21st IFAC World Congress, pp. 15588–15595 (2020)
15. Eilbrecht, J., Stursberg, O.: Set-based scheduling for highway entry of autonomous vehicles. In: Proceedings of the 21st IFAC World Congress, pp. 15396–15403 (2020)
16. Falcone, P., Borrelli, F., Asgari, J., Tseng, H., Hrovat, D.: Low complexity MPC schemes for integrated vehicle dynamics control problems. In: Proceedings of the International Symposium on Advanced Vehicle Control, pp. 875–880 (2008)
17. Frasch, J.V., Gray, A., Zanon, M., Ferreau, H.J., Sager, S., Borrelli, F., Diehl, M.: An auto-generated nonlinear MPC algorithm for real-time obstacle avoidance of ground vehicles. In: Proceedings of the European Control Conference, pp. 4136–4141 (2013)
18. Frazzoli, E.: Robust hybrid control for autonomous vehicle motion planning. Ph.D. thesis, Massachusetts Inst. of Technology (2001)
19. Frazzoli, E., Dahleh, M.A., Feron, E.: Maneuver-based motion planning for nonlinear systems with symmetries. IEEE Trans. Robot. **21**(6), 1077–1091 (2005)
20. Fuchshumer, S., Schlacher, K., Rittenschober, T.: Nonlinear vehicle dynamics control-a flatness based approach. In: Proceedings of the IEEE Conference on Decision and Control, pp. 6492–6497 (2005)
21. Goerzen, C., Kong, Z., Mettler, B.: A survey of motion planning algorithms from the perspective of autonomous UAV guidance. J. Intell. Rob. Syst. **57**(1–4), 65 (2010)
22. González, D., Pérez, J., Milanés, V., Nashashibi, F.: A review of motion planning techniques for automated vehicles. IEEE Trans. Intell. Transp. Syst. **17**(4), 1135–1145 (2016)
23. Gray, A., Gao, Y., Lin, T., Hedrick, J.K., Tseng, H.E., Borrelli, F.: Predictive control for agile semi-autonomous ground vehicles using motion primitives. In: Proceedings of the American Control Conference, pp. 4239–4244 (2012)
24. Gutjahr, B., Gröll, L., Werling, M.: Lateral vehicle trajectory optimization using constrained linear time-varying MPC. IEEE Trans. Intell. Transp. Syst. **18**(6), 1586–1595 (2016)
25. Heß, D., Althoff, M., Sattel, T.: Formal verification of maneuver automata for parameterized motion primitives. In: Proceedings of the IEEE/RSJ International Conference on Intelligent Robots and Systems, pp. 1474–1481 (2014)
26. Hwan Jeon, J., Cowlagi, R.V., Peters, S.C., Karaman, S., Frazzoli, E., Tsiotras, P., Iagnemma, K.: Optimal motion planning with the half-car dynamical model for autonomous high-speed driving. In: Proceedings of the American Control Conference, pp. 188–193 (2013)
27. Kant, K., Zucker, S.W.: Toward efficient trajectory planning: the path-velocity decomposition. Int. J. Robot. Res. **5**(3), 72–89 (1986)
28. Kurzer, K., Zhou, C., Zöllner, J.M.: Decentralized cooperative planning for automated vehicles with hierarchical Monte Carlo tree search. In: Proceedings of the IEEE Intelligent Vehicles Symposium, pp. 529–536 (2018)
29. Kuwata, Y.: Trajectory planning for unmanned vehicles using robust receding horizon control. Ph.D. thesis, Massachusetts Inst. of Technology (2007)
30. Kuwata, Y., Fiore, G.A., Teo, J., Frazzoli, E., How, J.P.: Motion planning for urban driving using RRT. In: Proceedings of the IEEE/RSJ International Conference on Intelligent Robots and Systems, pp. 1681–1686 (2008)

31. Li, X., Xu, X., Zuo, L.: Reinforcement learning based overtaking decision-making for highway autonomous driving. In: Proceedings of the International Conference on Intelligent Control and Information Processing, pp. 336–342 (2015)
32. Li, Y., Chen, X., Mårtensson, J.: Linear time-varying model predictive control for automated vehicles: Feasibility and stability under emergency lane change. In: Proceedings of the 21st IFAC World Congress, pp. 15928–15933 (2020)
33. Majumdar, A., Tedrake, R.: Funnel libraries for real-time robust feedback motion planning. Int. J. Robot. Res. **36**(8), 947–982 (2017)
34. Maurer, M., Christian Gerdes, J., Lenz, B., Winner, H.: Autonomous Driving: Technical, Legal and Social Aspects. Springer Nature (2016)
35. Nilsson, J., Sjöberg, J.: Strategic decision making for automated driving on two-lane, one way roads using model predictive control. In: Proceedings of the IEEE Intelligent Vehicles Symposium, pp. 1253–1258 (2013)
36. Paden, B., Čáp, M., Yong, S.Z., Yershov, D., Frazzoli, E.: A survey of motion planning and control techniques for self-driving urban vehicles. IEEE Trans. Intell. Veh. **1**(1), 33–55 (2016)
37. Park, J., Karumanchi, S., Iagnemma, K.: Homotopy-based divide-and-conquer strategy for optimal trajectory planning via mixed-integer programming. IEEE Trans. Rob. **31**(5), 1101–1115 (2015)
38. Payton, D.: An architecture for reflexive autonomous vehicle control. In: Proceedings of the IEEE International Conference on Robotics and Automation, vol. 3, pp. 1838–1845 (1986)
39. Qian, X., Altché, F., Bender, P., Stiller, C., de La Fortelle, A.: Optimal trajectory planning for autonomous driving integrating logical constraints: an MIQP perspective. In: Proceedings of the IEEE International Conference on Intelligent Transportation Systems, pp. 205–210 (2016)
40. Ranft, B., Stiller, C.: The role of machine vision for intelligent vehicles. IEEE Trans. Intell. Veh. **1**(1), 8–19 (2016)
41. Rewald, H., Stursberg, O.: Cooperation of autonomous vehicles using a hierarchy of auction-based and model-predictive control. In: Proceedings of the IEEE Intelligent Vehicles Symposium, pp. 1078–1084 (2016)
42. Schneegans, J., Eilbrecht, J., Zernetsch, S., Bieshaar, M., Doll, K., Stursberg, O., Sick, B.: Probabilistic vru trajectory forecasting for model-predictive planning - a case study: overtaking cyclists. In: Proceedings of the IEEE Intelligent Vehicles Symposium, pp. 272–279 (2021)
43. Schouwenaars, T.: Safe trajectory planning of autonomous vehicles. Ph.D. thesis, Massachusetts Inst. of Technology (2006)
44. Schürmann, B., Heß, D., Eilbrecht, J., Stursberg, O., Köster, F., Althoff, M.: Ensuring drivability of planned motions using formal methods. In: Proceedings of the IEEE International Conference on Intelligent Transportation Systems, pp. 1–8 (2017)
45. Silver, D., Bagnell, J.A., Stentz, A.: Learning autonomous driving styles and maneuvers from expert demonstration. In: Proceedings of the 13th International Symposium on Experimental Robotics, pp. 371–386 (2013)
46. Stursberg, O., Lohmann, S.: Synthesizing safe supervisory controllers for hybrid nonlinear systems. In: Proceedings of the 17th IMACS World Congress (2005)
47. Vallon, C., Ercan, Z., Carvalho, A., Borrelli, F.: A machine learning approach for personalized autonomous lane change initiation and control. In: Proceedings of the IEEE Intelligent Vehicles Symposium, pp. 1590–1595 (2017)
48. Wang, J., Liu, J., Kato, N.: Networking and communications in autonomous driving: a survey. IEEE Commun. Surv. Tutor. **21**(2), 1243–1274 (2018)
49. Werling, M., Groll, L., Bretthauer, G.: Invariant trajectory tracking with a full-size autonomous road vehicle. IEEE Trans. Robot. **26**(4), 758–765 (2010)
50. Xu, Y., Zheng, H., Wu, W., Wu, J.: Robust hierarchical model predictive control for trajectory tracking with obstacle avoidance. In: Proceedings of the 21st IFAC World Congress, pp. 15954–15959 (2020)

Specification-Compliant Motion Planning of Cooperative Vehicles Using Reachable Sets

Edmond Irani Liu and Matthias Althoff

Abstract Automated vehicles must comply explicitly with specifications, including traffic-based and handcrafted rules, in order for them to safely and effectively participate in mixed traffic. In addition to driving individually, there are many traffic situations in which cooperation between vehicles maximizes their collective benefits, including preventing collisions. To realize these benefits, we compute specification-compliant reachable sets for vehicles, i.e., sets of states which can be reached by vehicles over time that are constrained by a set of considered specifications. We summarize and combine our previous works on computing specification-compliant reachable sets and negotiating conflicting reachable sets within a group of cooperating vehicles. As a result, conflicts between specification-compliant reachable sets of vehicles are resolved, and specification-compliant trajectories can be individually planned for each vehicle within the negotiated reachable sets using arbitrary motion planners.

1 Introduction

When compared with human-driven vehicles, automated vehicles are expected to deliver enhanced road safety, passenger comfort, and traffic efficiency compared with human-driven vehicles. To safely and effectively participate in mixed traffic, in which both automated and human-driven vehicles share the road, automated vehicles must comply explicitly with specifications, including traffic regulations and handcrafted rules. Compliance with the former is essential in order to exempt manufacturers from liability claims in the event of an accident, while compliance with the latter allows motion plans to be generated that satisfy additional requirements. An example of a handcrafted rule is: *Follow vehicle 1 up to step k_1, then completely overtake it from the left before step k_2*. Generating a drivable trajectory that satisfies a set of specifications

E. I. Liu (✉) · M. Althoff
Technical University of Munich, Boltzmannstr. 3, 85748 Garching, Germany
e-mail: edmond.irani@tum.de

M. Althoff
e-mail: althoff@tum.de

© The Author(s) 2024
C. Stiller et al. (eds.), *Cooperatively Interacting Vehicles*,
https://doi.org/10.1007/978-3-031-60494-2_12

for an automated vehicle involves reasoning not only with continuous states (which may reflect the physical motion of the vehicle) but also discrete states (possibly due to discretization of the continuous state space or action space) of the vehicle. This poses computational challenges from a variety of aspects, including vehicle dynamics, the specifications under consideration (including collision avoidance), and dependencies between planned trajectories and constraints originating from the specifications. On the one hand, planning solely in the discrete state space may produce plans that meet specifications but violate vehicle dynamic constraints or lead to collisions. On the other hand, motion planners may generate dynamically drivable trajectories that do not comply with the specifications.

One solution to this problem is to guide the motion planning of an automated vehicle using its specification-compliant reachable set, which is defined as the set of states reachable by the vehicle over time that is constrained by a set of considered specifications. Computing the reachable sets in an over-approximative fashion will enclose all drivable trajectories of the automated vehicle [34]. The smaller the solution space is, the faster reachable sets can be computed, as demonstrated in [22]. In addition, the search space for the motion planner is greatly reduced particularly in critical situations. In contrast to conventional approaches, both effects result in quick computations even in critical situations. Low-level trajectory planning constraints can be extracted from the computed reachable sets and passed on to motion planners to generate specification-compliant trajectories.

In addition to driving individually, there are many traffic situations that demand cooperation between vehicles in order to maximize their collective benefits and to prevent collision in a potential emergency. Human drivers typically interact with each other through implicit communication and by anticipating the most likely behaviors of others. In comparison, automated vehicles can communicate and collaborate explicitly to jointly offer and suggest more sophisticated and efficient solutions in an ongoing traffic situation. One of the challenges of such cooperation lies in developing a computationally efficient scheme that does not compromise the optimality of the output solutions.

Reachable sets can be employed to tackle this challenge. The reachable sets of a group of cooperating vehicles can be computed and negotiated where conflicts in the position domain arise. This negotiation can be systematically organized such that each vehicle unambiguously receives its own negotiated reachable set, within which trajectories can be planned. This prevents exponential complexity of the collaborative motion planning.

In this chapter, we summarize and combine our previous works on computing specification-compliant reachable sets for an ego vehicle [13] as well as on negotiating conflicting reachable sets between a group of cooperating vehicles [21]. As a result, conflicts between specification-compliant reachable sets of vehicles are resolved, and each vehicle plans its own specification-compliant trajectories within its negotiated reachable set, for example, using the planners described in [22, 36].

The remainder of this article is organized as follows: Sect. 2 reviews related work on specification-compliant motion planning and cooperative motion planning. Section 3 presents the necessary preliminaries and definitions. The computation of specification-compliant reachable sets is summarized in Sect. 4 and the negotiation of reachable sets in Sect. 5. Example results are presented in Sect. 6, and we conclude in Sect. 7.

2 Related Work

In this section, review related works on specification-compliant motion planning and cooperative motion planning of vehicles.

2.1 Specification-Compliant Motion Planning

The efforts to obtain a specification-compliant trajectory can be categorized on the basis of whether compliance with specifications is examined *after*, *during*, or *before* motion planning.

2.1.1 Considering Compliance After Motion Planning

The most straightforward approach to obtain a specification-compliant trajectory is to examine the compliance with specifications after the trajectories have been generated. The process of checking whether an execution of a system satisfies the expected behaviors is often referred to as *runtime verification* or *monitoring*. For example, article [29] presents a monitor for formally examining the compliance of automated vehicles with traffic rules (safe distances and overtaking); a monitor for so-called responsibility-sensitive safety rules [31] is described in [10]. While monitoring can be performed efficiently, monitors typically only provide a verdict, i.e., a *true* or *false* appraisal, on whether the specifications have been satisfied. If the trajectory under examination is rejected, no alternative trajectory is returned. This often necessitates the (re)planning of multiple trajectories in order to locate a valid solution for more complex specifications.

2.1.2 Considering Compliance During Motion Planning

Works in this category often adopt a mechanism that simultaneously handles planning in both the continuous and discrete state spaces of a system, with the generated discrete plans guiding the trajectory planning process. For example, a satisfiability modulo convex programming framework for cyber-physical systems was introduced in [32] that handles both convex constraints on a continuous model and Boolean constraints on a discrete model; article [16] puts forth a multilayered synergistic framework for motion planning of robots considering linear temporal logic (LTL);

timed automata are used in [37] to synthesize timed paths for indoor robots that comply with specifications expressed in metric temporal logic. In these works, discrete plans are generated in the discrete state space based on abstractions of the considered systems, and trajectories are planned in the continuous state space by motion planners, with the discrete plans taken into consideration. In most cases, the dynamic constraints of the system are not reflected in the discrete plans. Thus, the drivability of these plans is often not ensured, requiring frequent replanning of both the discrete plans and the trajectories.

2.1.3 Considering Compliance Before Motion Planning

The final category of works considers the specifications prior to trajectory planning, e.g., in high-level maneuver planners, from which trajectory planning constraints can be extracted. The work in [15] generates maneuvers that respect simple traffic rules by traversing a graph defined in a discretized state space of the ego vehicle; article [8] embraces a similar concept and produces maneuvers satisfying specifications expressed in LTL; in [33], so-called driving corridors are extracted from reachable sets of an ego vehicle that reflect different position relations to other vehicles over time. Our approach to computing specification-compliant reachable sets [13] falls into this category. It can handle propositional logic with predicates related to positions, velocities, accelerations, and certain traffic regulations introduced in [18, 19].

2.2 Cooperative Motion Planning

Survey articles [9, 24, 28] reviewed recent advances in cooperative driving of automated vehicles with varied focuses on architecture, maneuver planning, and motion planning use cases. Optimization-based and reservation-based approaches are common paradigms for cooperative motion planning [9, 28]. In optimization-based approaches, one or more optimization problems are formulated based on the motion planning constraints and cost functions of cooperating vehicles. The optimization problems are solved with a (centralized) optimizer, which corresponds to trajectories to be followed by the cooperating vehicles. The complexity of the optimization problem increases dramatically with the number of vehicles considered, which requires either a high computation power or a limit to the number of vehicles in a group.

Our approach to cooperative motion planning falls into the reservation-based category and employs auction algorithms for resolving conflicts in reachable sets of vehicles. Reservation-based methods assign free space to vehicles for trajectory planning. Earlier works with a focus on intersection management were introduced by Dresner [4]: Tiles are created from the intersection region, which can be requested by vehicles approaching the intersection. A centralized intersection manager proceeds to assign tiles with multiple requests to vehicles, using a first-come-first-served protocol, ensuring that no tile is occupied by more than one vehicle at any one time. Its extensions and variations are presented in [5, 6]. As the first-come-first-served policy for reservation assignment may be inefficient in situations with higher traffic density, it was replaced in [3, 27, 35] by auction-based methods. In auction-based methods, each bidder (cooperating vehicle) bids for offered packages (e.g., combinations of tiles representing road areas) in a way that reflects its interests or utilities. An auction algorithm is then executed to maximize the total revenue of the packages. Instead of tiles, some works identify possible conflicting points, regions, or moving space-time corridors and allocate them to vehicles in the event of a conflict [17, 20, 23, 38]. The corridors correspond to predefined behaviors, such as following a lane or performing a lane change; vehicles receiving such corridors must act accordingly. In [11, 25], an efficient and explicit space-time reservation protocol was devised for cooperative maneuver planning, through which a vehicle broadcasts requested space envelopes over time and drives within the envelopes once the request has been accepted by surrounding vehicles of interest.

3 Preliminaries

This section introduces the necessary preliminaries, including the general setup, coordinate systems, definitions of reachable sets, and propositional logic.

3.1 Setup and Coordinate System

In this work, the considered scenarios are described in the CommonRoad[1] [1] format, which consists of (1) a road network constructed of lanelets [2], whose left and right bounds are represented by polylines, (2) dynamic and static obstacles, and (3) traffic rule elements (such as road markings, traffic signs, and traffic lights). Figure 1 depicts an exemplary traffic scenario. We denote by $\mathcal{V}^c = \{V_1^c, \ldots, V_N^c\}$ the set of cooperative vehicles V_n^c with IDs $\mathcal{N} = \{1, \ldots, N\}$ for which trajectories are planned. Each V_n^c is associated with a planning problem with a planning horizon of up to $k_h \in \mathbb{N}_0$, which includes the initial state of V_n^c and a set of goal states. A reference path Γ_n is constructed for a planning problem with a given route planner, which is then used to establish a local curvilinear coordinate system F_n^L of V_n^c as described in [2]. Within F_n^L, (s_n, d_n) describes the longitudinal coordinate s_n and the lateral coordinate d_n. Adopting this coordinate system facilitates the formulation of maneuvers from the perspective of V_n^c, examples of which include lane-following and preventing driving backwards. We use $\mathcal{LL}$ to denote the set of lanelets in the road network of a considered scenario. Without loss of generality, we assume obstacles present in the scenarios to be non-cooperating vehicles, denoted by $\mathcal{V}^o = \{V_1^o, \ldots, V_M^o\}$ with IDs

---

[1] https://commonroad.in.tum.de/.

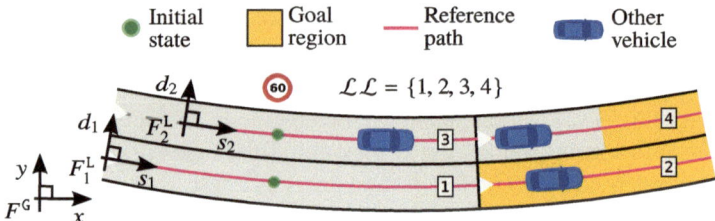

Fig. 1 A scenario containing planning problems with two cooperating ego vehicles V_1^c and V_2^c, and four lanelets with IDs 1–4. The triangles at the beginning of each lanelet indicate the driving directions

$\mathcal{M} = \{1, \ldots, M\}$. In addition, we assume that the most likely predictions of trajectories of other vehicles V_m^o are given as input. The conflicts between the reachable sets of vehicles in $\mathcal{V}^c$ are detected and resolved in the global Cartesian coordinate system F^G.

3.2 System Dynamics

The dynamics of an ego vehicle V_n^c is abstracted by a point-mass model with the center of the vehicle as the reference point. Notably, the reachable sets of the point-mass model over-approximate those of high-fidelity vehicle models; thus, this abstraction does not exclude possible behaviors of V_n^c. This model is represented with two double integrators in its longitudinal s_n and lateral d_n directions. Let $\square_n$ be a variable of V_n^c, with minimum and maximum values denoted by $\underline{\square}_n$ and $\overline{\square}_n$, respectively. The system dynamics of V_n^c is

$$x_{n,k+1} = f(x_{n,k}, u_{n,k}) = \begin{pmatrix} 1 & \Delta_t & 0 & 0 \\ 0 & 1 & 0 & 0 \\ 0 & 0 & 1 & \Delta_t \\ 0 & 0 & 0 & 1 \end{pmatrix} x_{n,k} + \begin{pmatrix} \frac{1}{2}\Delta_t^2 & 0 \\ \Delta_t & 0 \\ 0 & \frac{1}{2}\Delta_t^2 \\ 0 & \Delta_t \end{pmatrix} u_{n,k}, \quad (1)$$

where $k \in \mathbb{N}_0$ is a step corresponding to time $t_k = k\Delta_t$, with $\Delta_t \in \mathbb{R}_+$ being a predefined time increment. The variable $x_{n,k} \in \mathcal{X}_{n,k} \subset \mathbb{R}^4$ represents the state of V_n^c in the state space $\mathcal{X}_{n,k}$, and $u_{n,k} \in \mathcal{U}_{n,k} \subset \mathbb{R}^2$ represents an input in the input space $\mathcal{U}_{n,k}$ of V_n^c, each at step k. The states and inputs are modeled as $x_{n,k} = (s_{n,k}, \dot{s}_{n,k}, d_{n,k}, \dot{d}_{n,k})^\mathsf{T}$ and $u_{n,k} = (\ddot{s}_{n,k}, \ddot{d}_{n,k})^\mathsf{T}$, respectively. The velocities and accelerations at a position $(s_{n,k}, d_{n,k})$ are bounded by

$$\underline{\dot{s}}(\Gamma_n) \leq \dot{s}_{n,k} \leq \overline{\dot{s}}(\Gamma_n), \quad \underline{\dot{d}}(\Gamma_n) \leq \dot{d}_{n,k} \leq \overline{\dot{d}}(\Gamma_n), \tag{2a}$$

$$\underline{\ddot{s}}(\Gamma_n) \leq \ddot{s}_{n,k} \leq \overline{\ddot{s}}(\Gamma_n), \quad \underline{\ddot{d}}(\Gamma_n) \leq \ddot{d}_{n,k} \leq \overline{\ddot{d}}(\Gamma_n). \tag{2b}$$

The bounds are chosen conservatively to consider the kinematic limitations and effects of representing the system dynamics using the point-mass model within a curvilinear coordinate system, see, for example, article [7]. We define an operator $\mathrm{proj}_\Diamond(\cdot)$ for subsequent computations, which maps the input to its elements $\Diamond$. An example is: $\mathrm{proj}_{(s,\dot{s})}(\tilde{\mathbf{x}}_{n,k}) = (s_{n,k}, \dot{s}_{n,k})^\mathsf{T}$ for $\tilde{\mathbf{x}}_{n,k} = (s_{n,k}, \dot{s}_{n,k}, \ddot{s}_{n,k})^\mathsf{T}$. A set $\tilde{X}_{n,k}$ can be projected using the same operator:

$$\mathrm{proj}_\Diamond(\tilde{X}_{n,k}) = \left\{ \mathrm{proj}_\Diamond(\tilde{\mathbf{x}}_{n,k}) \middle| \tilde{\mathbf{x}}_{n,k} \in \tilde{X}_{n,k} \right\}.$$

3.3   Reachable Set

We denote the occupancy of V_n^c by $Q_n(\mathbf{x}_{n,k}) \subset \mathbb{R}^2$ and the occupancies of all vehicles in $\mathcal{V}^\circ$ as well as the regions outside the road surface by $O_{n,k} \subset \mathbb{R}^2$, both within F_n^L. The set of forbidden states $X_{n,k}^\mathrm{F}$ of V_n^c at k is defined as

$$X_{n,k}^\mathrm{F} := \left\{ \mathbf{x}_{n,k} \in X_{n,k} \middle| Q_n(\mathbf{x}_{n,k}) \cap O_{n,k} \neq \emptyset \right\}.$$

Let $\mathcal{R}_{n,0}^* = X_{n,0}$ be the initial reachable set of V_n^c, with $X_{n,0}$ being the initial set of states. The reachable set $\mathcal{R}_{n,k+1}^*$ of the next step is defined as the set of states reachable from the current reachable set $\mathcal{R}_{n,k}^*$ while avoiding the forbidden states:

$$\mathcal{R}_{n,k+1}^* := \left\{ \mathbf{x}_{n,k+1} \in X_{n,k+1} \middle| \exists \mathbf{x}_{n,k} \in \mathcal{R}_{n,k}^*, \exists \mathbf{u}_{n,k} \in \mathcal{U}_{n,k} : \right.$$
$$\left. \mathbf{x}_{n,k+1} = f(\mathbf{x}_{n,k}, \mathbf{u}_{n,k}) \wedge \mathbf{x}_{n,k+1} \notin X_{n,k+1}^\mathrm{F} \right\}.$$

Efficient computation of $\mathcal{R}_{n,k}^*$ is generally difficult; hence, we compute its over-approximation $\mathcal{R}_{n,k} \approx \mathcal{R}_{n,k}^*$, which encloses all trajectories of V_n^c. We adopt the union of so-called *base sets* $\mathcal{R}_{n,k}^{(i)}$, $i \in \mathbb{N}$ as a set representation for $\mathcal{R}_{n,k}$ [34]. Each base set $\mathcal{R}_{n,k}^{(i)} = \hat{\mathcal{P}}_{s,n,k}^{(i)} \times \hat{\mathcal{P}}_{d,n,k}^{(i)}$ is chosen to be a Cartesian product of two convex polytopes that enclose the reachable positions and velocities of V_n^c in the $(s_n, \dot{s}_n)$ and $(d_n, \dot{d}_n)$ planes, respectively (see Fig. 2a, b). To simplify the notation, we also denote the collection (set of sets) of $\mathcal{R}_{n,k}^{(i)}$ by $\mathcal{R}_{n,k} = \left\{ \mathcal{R}_{n,k}^{(1)}, \ldots, \mathcal{R}_{n,k}^{(i)}, \ldots \right\}$. The projection of $\mathcal{R}_{n,k}^{(i)}$ onto the position domain yields axis-aligned rectangles $\mathcal{D}_{n,k}^{(i)}$ (see Fig. 2c), whose union is referred to as the drivable area $\mathcal{D}_{n,k}$. Similarly, we use $\mathcal{D}_{n,k}$ to denote the collection of $\mathcal{D}_{n,k}^{(i)}$.

In this study, each base set $\mathcal{R}_{n,k}^{(i)}$ carries a set of semantic labels $\mathcal{L}_{n,k}^{(i)}$, whose collection is denoted by $\mathcal{L}_{n,k}$. The generation of $\mathcal{L}_{n,k}^{(i)}$ will be explained in Sect. 4.6.3. To store the relationships of $\mathcal{R}_{n,k}^{(i)}$ in terms of reachability and time, we create a directed and acyclic graph G_n, which is referred to as a *reachability graph*, see Fig. 3. Each

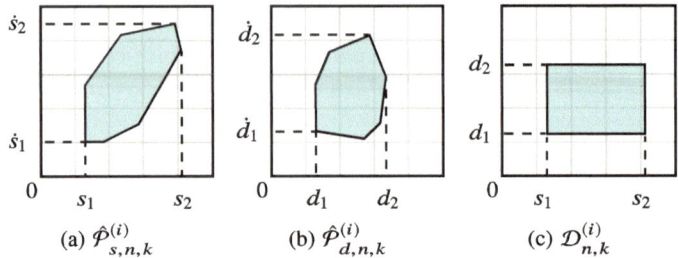

Fig. 2 Polytopes and drivable area of a base set $\mathcal{R}_{n,k}^{(i)}$ (adapted from [13])

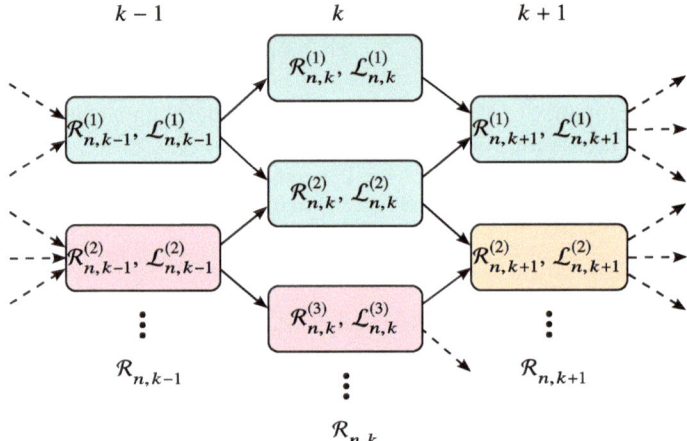

Fig. 3 Reachability graph G_n connecting nodes of different steps. Nodes of the same color have the same labels (adapted from [13])

node in G_n corresponds to one base set with its labels. An edge connecting $\mathcal{R}_{n,k}^{(i)}$ and $\mathcal{R}_{n,k+1}^{(j)}$ indicates that $\mathcal{R}_{n,k+1}^{(j)}$ is reachable from $\mathcal{R}_{n,k}^{(i)}$ after one step.

3.4 Propositional Logic

We consider specifications expressed in propositional logic [12] for V_n^c, denoted by $\mathcal{F}_n$, which are directly integrated during the computation of the reachable sets (see Sect. 4.6.4). Let $\varphi_n \in \mathcal{F}_n$ be a propositional logic formula, we introduce an additional syntax $\mathbf{G}_I(\varphi_n)$, $I = [a, b]$, $0 \le a \le b \le k_h$, where I is an integer interval specifying steps for which φ_n should hold. If I is not specified, we assume it to be the entire planning horizon $[0, k_h]$. For example, the following specification requires V_n^c to follow V_1° between steps 0 and 10, and never to be on the right of V_1°:

$$\mathbf{G}_{[0,10]}\left(\text{behind}(V_1^\circ) \wedge \text{aligned\_with}(V_1^\circ)\right) \wedge \mathbf{G}\left(\neg \text{right\_of}(V_1^\circ)\right).$$

Table 1 Selection of considered predicates inspired by [19] (adapted from [13])

Category	Type	Predicate
Position	VI	in _ lanelet, on _ main _ carriageway, on _ access _ ramp, ...
	VD	behind, beside, in _ front _ of, left _ of, aligned _ with, right _ of, ...
Velocity	VI	below _ fov _ velocity _ limit, below _ type _ velocity _ limit, ...
	VD	safe _ following _ velocity _ speed _ limit, safe _ leading _ velocity _ speed _ limit, ...
Acceleration	VD	admissible _ braking, ...
General	VI	change _ lanelet, preserve _ traffic _ flow, standing _ still, ...
	VD	in _ congestion, exists _ slow _ leading _ vehicle, ...

4 Computing Specification-Compliant Reachable Sets

To obtain specification-compliant and negotiated reachable sets for V_n^c, we (1) semantically label reachable sets considering relevant predicates, (2) constrain reachable sets to subsets satisfying specifications $\mathcal{F}_n$, and (3) negotiate conflicting reachable sets with other cooperating vehicles in $\mathcal{V}^c$. This section summarizes our previous work [13] covering steps 1 and 2; step 3 will be covered in the next section. A selection of considered predicates is listed in Table 1: The evaluation of a *vehicle-dependent (VD)* predicate is dependent on other vehicles $\mathcal{V}^\circ$, whereas that of a *vehicle-independent (VI)* predicate is not.

4.1 State Space Partitioning

To expedite the labeling of reachable sets, we partition the state space of V_n^c based on considered position predicates. Velocity predicates are not considered in the partitioning since they require computationally demanding splitting of the state space of V_n^c with (non)linear curves (see Fig. 5c, d). For efficiency, we instead directly evaluate them on individual reachable sets (see Sect. 4.6.2). Set operations such as intersection and difference are required to compute the partitions of the state space. We model the partitions for V_n^c with a set of hyperrectangles $R_{n,q}$ to avoid gross approximations while keeping computational complexity at a reasonable level. This choice is not mandatory; any other set representation that captures the partitions will also suffice. $R_{n,q}$ is defined as the Cartesian product of intervals over the position and velocity domains within F_n^L:

$$R_{n,q} := \left([\underline{s}_{n,q}, \overline{s}_{n,q}] \times [\underline{\dot{s}}_{n,q}, \overline{\dot{s}}_{n,q}] \right) \times \left([\underline{d}_{n,q}, \overline{d}_{n,q}] \times [\underline{\dot{d}}_{n,q}, \overline{\dot{d}}_{n,q}] \right), \qquad (3)$$

where $s_{n,q}$ and $\dot{s}_{n,q}$ denote the position and velocity of the q-th hyperrectangle in the s_n direction, respectively. The same applies to $d_{n,q}$ and $\dot{d}_{n,q}$ in the d_n direction. A regular grid of axis-aligned cells is formed along Γ_n and the q-th cell in the grid occupies $[\underline{s}_{n,q}, \overline{s}_{n,q}] \times [\underline{d}_{n,q}, \overline{d}_{n,q}] \subset \mathbb{R}^2$. The default values of the velocity intervals $[\underline{\dot{s}}_{n,q}, \overline{\dot{s}}_{n,q}]$ and $[\underline{\dot{d}}_{n,q}, \overline{\dot{d}}_{n,q}]$ are set according to (2a).

The set of considered position predicates as well as its power set are denoted by $\mathcal{P}^{\mathrm{pos}} = \{\sigma_1, \sigma_2, \dots\}$ and $2^{\mathcal{P}^{\mathrm{pos}}}$, respectively. We also denote by $\mathrm{part}_n(k; \mathcal{Z}_{n,j})$ the set of hyperrectangles of V_n^{c} for which the predicates in $\mathcal{Z}_{n,j} \in \mathcal{Z}_n \subseteq 2^{\mathcal{P}^{\mathrm{pos}}}$ evaluate to true at step k. Figures 4 and 5b illustrate example partitions projected onto the (s_n, d_n) and $(s_n, \dot{s}_n)$ planes, respectively.

4.2 Position Predicates

only a few example evaluations position predicates. Vehicle-independent position predicates do not depend on other vehicles; examples are:

- in _ lanelet$(R_{n,q}; L_{\mathrm{id}}) \Leftrightarrow \mathrm{proj}_{(s,d)}(R_{n,q}) \cap \mathrm{occ}_n(L_{\mathrm{id}}) \neq \varnothing$, where $L_{\mathrm{id}} \in \mathcal{LL}$ denotes the lanelet with ID id, and $\mathrm{occ}_n(L_{\mathrm{id}})$ returns its occupancy within F_n^{L}.
- drives _ rightmost$(R_{n,q}; X^{\mathrm{RM}}) \Leftrightarrow \mathrm{proj}_{(s,d)}(R_{n,q}) \cap X^{\mathrm{RM}} \neq \varnothing$, where $X^{\mathrm{RM}} \subset \mathbb{R}^2$ denotes the rightmost region of lanelets. Within this region, the distance between any point to the right bound of a lanelet does not exceed a predefined distance [19].

For the sake of brevity, we omit $R_{n,q}$ in the arguments of the predicates in the rest of this work.

Vehicle-dependent position predicates describe position relationships between an ego vehicle V_n^{c} and non-cooperating vehicles in $\mathcal{V}^{\circ}$. Following [19], we define necessary helper functions to assist the evaluation of predicates. The functions $\mathrm{front}(k; n; m)$ and $\mathrm{rear}(k; n; m)$ return the maximum and mini-

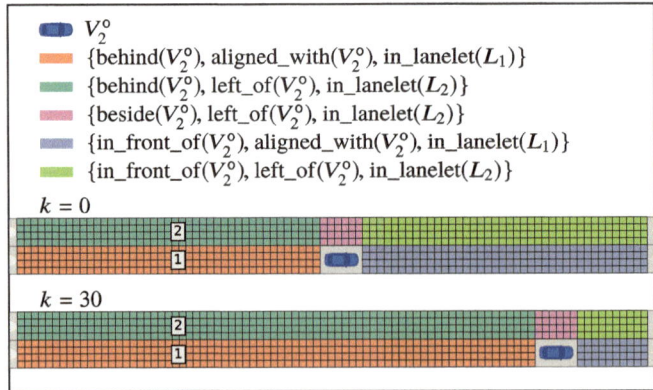

Fig. 4 Projection of the partitions of realizable sets of position predicates onto the position domain. Lanelet IDs are shown with numbered boxes. In this example we only consider position predicates related to L_1, L_2, and V_2° (adapted from [13])

mum longitudinal coordinates of V_m° within F_n^L, respectively, each at step k. Along the longitudinal direction, the mutually exclusive predicates $\mathcal{P}_{n,m,s}^{\text{pos}} = \big\{\{\text{in\_front\_of}(V_m^\circ)\}, \{\text{behind}(V_m^\circ)\},$ $\{\text{beside}(V_m^\circ)\}\big\}$ can be evaluated as follows:

- $\text{in\_front\_of}(V_m^\circ) \Leftrightarrow \underline{s}_{n,q} - l_n/2 > \text{front}(k; n; m)$,
- $\text{behind}(V_m^\circ) \Leftrightarrow \overline{s}_{n,q} + l_n/2 < \text{rear}(k; n; m)$,
- $\text{beside}(V_m^\circ) \Leftrightarrow \qquad \neg \text{in\_front\_of}(V_m^\circ) \wedge \neg \text{behind}(V_m^\circ) \wedge (\text{left\_of}(V_m^\circ) \vee \text{right\_of}(V_m^\circ))$.

We define the mutually exclusive set of predicates $\mathcal{P}_{n,m,d}^{\text{pos}} = \big\{\{\text{left\_of}(V_m^\circ)\},$ $\{\text{right\_of}(V_m^\circ)\}, \{\text{aligned\_with}(V_m^\circ)\}\big\}$ similarly along the lateral direction.

4.3 Realizable Sets of Position Predicates

The partitions of the collection $\mathcal{Z}_n$ of realizable sets of position predicates of V_n^c are used for splitting the reachable sets (see Sect. 4.6.2). Sets $\mathcal{Z}_{n,j} \in \mathcal{Z}_n$ are said to be realizable for V_n^c if $\exists k \in 0, \ldots, k_h : \text{part}_n(k; \mathcal{Z}_{n,j}) \neq \varnothing$, with k_h being the planning horizon. We refer the readers to [13, Sect. III.C] for the computation of $\mathcal{Z}_n$. Figure 4 shows an example of the partitions of $\mathcal{Z}_{n,j}$ projected onto the position domain for a scenario containing two lanelets and one non-cooperating vehicle. It follows from our formulation of the predicates that the aforementioned projection is collision-free with respect to other vehicles.

4.4 Velocity Predicates

We briefly present examples of the evaluation of velocity predicates required for the subsequent computation of reachable sets. Vehicle-independent velocity predicates often relate to extremum requirements on velocities. For example, rule R-G3 [19] specifies maximum velocity limits originating from different sources, which should be respected. These include velocity limits introduced by the type of lane(let), the type of vehicle, and the limited field of view of the ego vehicle.

The evaluation of vehicle-dependent velocity predicates depends on other vehicles $\mathcal{V}^\circ$. Examples are predicates indicating whether the ego vehicle V_n^c is driving at a safe velocity with respect to a leading or a following vehicle V_m° [19, cf. Sect. IV.C]. See [19] for further examples.

Algorithm 1 One-Step Computation of Specification-Compliant Reachable Sets

---

Inputs: Specifications $\mathcal{F}_n$, base sets $\mathcal{R}_{n,k-1}$, realizable sets of predicates $\mathcal{Z}_n$.
Output: Updated reachability graph G_n.

1: $\mathcal{R}_{n,k}^{\mathrm{P}} \leftarrow \mathrm{PROPAGATE}(\mathcal{R}_{n,k-1})$ ▷ Sect. 4.6.1
2: $\mathcal{R}_{n,k}^{\mathrm{S}} \leftarrow \mathrm{SPLIT}(\mathcal{R}_{n,k}^{\mathrm{P}}, \mathcal{Z}_n)$ ▷ Sect. 4.6.2
3: $\mathcal{L}_{n,k} \leftarrow \mathrm{LABEL}(\mathcal{R}_{n,k}^{\mathrm{S}}, \mathcal{F}_n)$ ▷ Sect. 4.6.3
4: $\mathrm{CHECKCOMPLIANCE}(\mathcal{R}_{n,k}^{\mathrm{S}}, \mathcal{L}_{n,k}, \mathcal{F}_n)$ ▷ Sect. 4.6.4
5: $\mathcal{R}_{n,k} \leftarrow \mathrm{CREATENEWBASESETS}(\mathcal{R}_{n,k}^{\mathrm{S}})$ ▷ Sect. 4.6.5
6: **for** $\mathcal{R}_{n,k}^{(i)} \in \mathcal{R}_{n,k}$ **do**
7:      $G_n.\mathrm{ADDNODE}(\mathcal{R}_{n,k}^{(i)}, \mathcal{L}_{n,k}^{(i)})$
8: **end for**

4.5 General Traffic Situation Predicates

General traffic situation predicates may reveal the states of a cooperating or non-cooperating vehicle. These include whether V_n^{c} or V_m^{o} has conducted a lane change maneuver, whether a slow leading vehicle exists for V_n^{c}, and whether V_n^{c} is stuck in traffic congestion. See [19] for further examples.

4.6 Computation of Reachable Sets

Algorithm 1 details one step of the computation of specification-compliant reachable sets for an ego vehicle. The reachable sets of subsequent steps are computed analogously.

4.6.1 Forward Propagation

Each base set $\mathcal{R}_{n,k-1}^{(i)} \in \mathcal{R}_{n,k-1}$ from the previous step is forward-propagated based on the discrete-time system model (1), resulting in the propagated sets $\mathcal{R}_{n,k}^{\mathrm{P},(i)}$ (see Fig. 5a). We perform the forward propagation as described in [34], except that additional acceleration constraints originating from the specifications can be imposed (for example, unnecessary braking rule R_G2 in [19]).

4.6.2 Splitting

The propagated sets $\mathcal{R}_{n,k}^{\mathrm{P},(i)}$ are split into new sets $\mathcal{R}_{n,k}^{\mathrm{S},(i)}$ with respect to position and velocity predicates:

1. $\mathcal{R}_{n,k}^{\mathrm{P},(i)}$ are split such that the new sets only intersect with a single partition (see Fig. 5b).

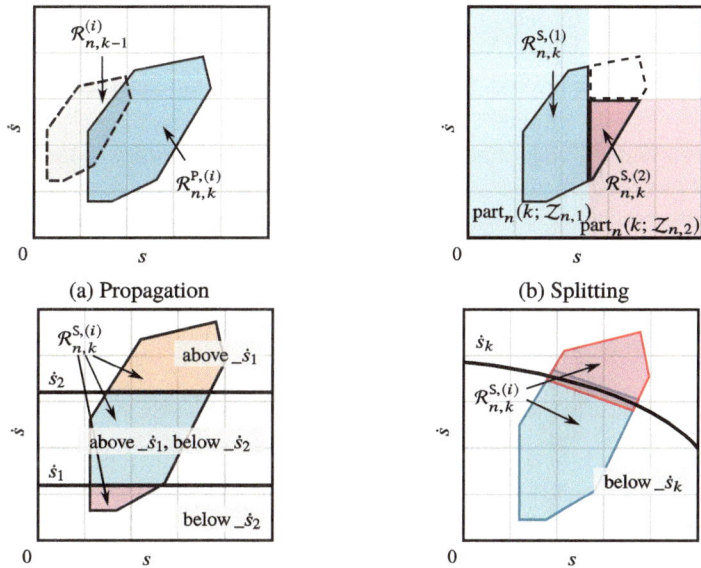

(a) Propagation

(b) Splitting

(c) Splitting with respect to vehicle-independent velocity predicates above_$\dot{s}_1$ and below_$\dot{s}_2$

(d) Splitting with respect to vehicle-dependent velocity predicate below_$\dot{s}_k$

Fig. 5 Propagation, splitting, and labeling of base sets. We only show the operations in the s-direction. Labels of polytopes are shown in gray boxes. Notably, in **d**, the two newly split polytopes are slightly over-approximated and convexified due to the nonlinearity introduced by the velocity predicate (adapted from [13])

2. The split sets are further split, over-approximated, and convexified with respect to velocity predicates (see Fig. 5c, d).

4.6.3 Semantic Labeling

The semantic labels $\mathcal{L}_{n,k}^{(i)}$ of reachable sets $\mathcal{R}_{n,k}^{S,(i)}$ are updated as follows:

1. $\mathcal{R}_{n,k}^{S,(i)}$ propagated with acceleration-specific specifications include atomic propositions $\sigma \in \mathcal{AP}$ corresponding to acceleration predicates in their set of labels.
2. $\mathcal{R}_{n,k}^{S,(i)}$ include atomic propositions $\sigma \in \mathcal{AP}$ corresponding to the position predicates associated with the partition with which it intersects, velocity predicates, and traffic situation predicates that hold in $\mathcal{R}_{n,k}^{S,(i)}$ in their set of labels.

4.6.4 Compliance Check

In this step, we iterate through $\mathcal{R}_{n,k}^{\mathrm{S},(i)}$ and examine the compliance of the labels $\mathcal{L}_{n,k}^{(i)}$ with the given specifications $\mathcal{F}_n$. We discard $\mathcal{R}_{n,k}^{\mathrm{S},(i)}$ if $\exists \varphi_n \in \mathcal{F}_n : \mathcal{L}_{n,k}^{(i)} \not\models \varphi_n$. If all sets are discarded, $\mathcal{F}_n$ cannot be complied with by any trajectory of the ego vehicle (recall that our reachable sets are over-approximative). In this case, one can either recompute the reachable sets with respect to a different set of specifications or execute a previously computed fail-safe trajectory [26].

4.6.5 Creation of New Base Sets

Finally, the new base sets are created by computing the drivable areas $\mathcal{D}_{n,k}^{\mathrm{S},(i)}$ of $\mathcal{R}_{n,k}^{\mathrm{S},(i)}$, repartitioning $\mathcal{D}_{n,k}^{\mathrm{S},(i)}$, and producing $\mathcal{R}_{n,k}^{(i)}$. We refer the reader to [34] for a detailed explanation of these steps. The reachability graph G_n is updated by adding $\mathcal{R}_{n,k}^{(i)}$ along $\mathcal{L}_{n,k}^{(i)}$ as new nodes.

5 Negotiation of Reachable Sets

This section summarizes our previous work on the negotiation of conflicting reachable sets $\mathcal{R}_{n,k}$ among a group of cooperating vehicles [21]. We use the notation $[\square_n]_1^N = [\square_1, \ldots, \square_N]$ to denote a list of elements $\square_n$ of vehicles V_n^c. Algorithm 2 details the steps for resolving conflicts between cooperating vehicles at each step k:

1. Compute specification-compliant reachable sets for each cooperating vehicle.
2. Identify conflicting cells based on reachable sets of cooperating vehicles (see Sect. 5.1).
3. Determine the optimal allocation of packages of cells among cooperating vehicles (see Sect. 5.2).
4. Compute negotiated reachable sets for each cooperating vehicle (see Sect. 5.2).

Step 1 is computed as described in Sect. 4; we will now elaborate on steps 2–4.

5.1 Problem Statement

We denote by $C = \{C_0, C_1, \ldots, C_{\bar{j}}, \ldots\}$ a grid with cells $C_{\bar{j}}$ of rectangular shape, created by tessellation of the position domain within the global Cartesian coordinate system F^{G}. Each cell is an individual asset representing an area of the road surface and can be combined into unions of assets, which we refer to as packages $C_{\bar{j}}$. We specify the mapping $\mathrm{cell}_n : 2^{X_{n,k}} \to 2^C$ that returns the cells $C_{\bar{j}} \in C$ occupied by vehicle V_n^c due to its set of states $X_{n,k}$ at step k and its shape. The cooperating vehicles in $\mathcal{V}^c$ act

---

Algorithm 2 Computation of Negotiated Reachable Sets

1: **function** COMPUTENEGOTIATEDREACHABLESET($[\mathcal{R}_{n,0}]_1^N$,C)
2:     $[\mathcal{R}_{n,0}^N]_1^N \leftarrow [\mathcal{R}_{n,0}]_1^N$ ▷ Initialization
3:     **for** $k = 1$ to k_h **do**
4:         **for** $n = 1$ to N **do**
5:             $\mathcal{R}_{n,k} \leftarrow$ COMPUTEREACHABLESET($\mathcal{R}_{n,k-1}^N$)   ▷ Sect. 4.6
6:         **end for**
7:         $C_k^C \leftarrow$ IDENTIFYCONFLICTINGCELLS($[\mathcal{R}_{n,k}]_1^N$,C)   ▷ Sect. 5.1
8:         $\mathcal{W}^* \leftarrow$ DETERMINEOPTIMALALLOCATION($[\mathcal{R}_{n,k}]_1^N$,C_k^C)   ▷ Sect. 5.2
9:         **for** $n = 1$ to N **do**
10:             $\mathcal{R}_{n,k}^N \leftarrow$ COMPUTENEGOTIATEDREACHABLESET($\mathcal{R}_{n,k}$,$\mathcal{W}^*$)   ▷ Sect. 5.2
11:         **end for**
12:     **end for**
13:     **return** $[\cup_k \mathcal{R}_{n,k}^N]_1^N$
14: **end function**

---

as bidders and propose bids to packages $C_{\bar{j}}$ for which $C_{\bar{j}} \cap \mathrm{cell}_n(\mathcal{R}_{n,k}) \neq \varnothing$ holds. Let us introduce $2_{\geq=2}^N$ to denote all subsets of the power set of N with a cardinality greater than one, $C_k^C \subseteq C$ denotes the set of conflicting cells requested by at least two vehicles at step k:

$$C_k^C := \bigcup_{I \in 2_{\geq=2}^N} \bigcap_{n \in I} \mathrm{cell}_n(\mathcal{R}_{n,k}). \tag{4}$$

We restrict the packages to those containing at least one conflicting cell, denoted by $C_k^p \subseteq 2^{C_k^C}$ (see Fig. 6). We assume that every cooperating vehicle V_n^c bids its true value, with $\overline{b}_k(C_{\bar{j}})$ being the maximum bid of the package $C_{\bar{j}}$ proposed by $\mathcal{V}^c$. The overall revenue is maximized, while no single cell is assigned to multiple bidders:

$$\max_{\delta_k(C_{\bar{j}})} \sum_{C_{\bar{j}}} \delta_k(C_{\bar{j}}) \overline{b}_k(C_{\bar{j}}), \tag{5}$$

where $\delta_k(C_{\bar{j}}) = 1$ if package $C_{\bar{j}}$ is assigned to the bidder with the highest bid at step k. Problem (5) is known as the *winner determination problem*, and its solution is NP-hard [30]. Furthermore, accepting every package $C_{\bar{j}}$ demands that each bidder V_n^c bids for $2^{|C_k^C|} - 1$ packages at step k, which becomes more computationally demanding as $|C_k^C|$ grows. Using a hierarchical tree structure for the packages allows us to attain computational tractability and ensures that the optimal allocation of packages will be found in the time $O(|C_k^C|^2)$ [30].

5.2 Conflict Resolution

We employ an auction-based mechanism to resolve conflicts with occupied road cells between cooperating vehicles. At every step k, the conflicts are resolved as follows:

Fig. 6 Visualization of the road grid C, the set of conflicting cells C_k^C, the set of packages C_k^P, and the individual packages $C_{\bar{j}}$ (adapted from [21])

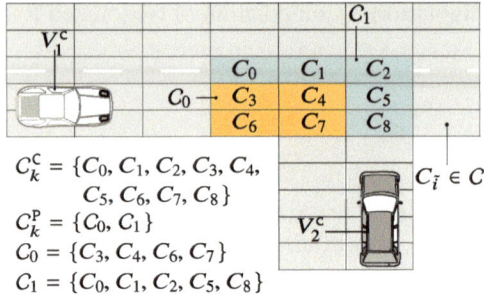

$$C_k^C = \{C_0, C_1, C_2, C_3, C_4, \\ C_5, C_6, C_7, C_8\}$$
$$C_k^P = \{C_0, C_1\}$$
$$C_0 = \{C_3, C_4, C_6, C_7\}$$
$$C_1 = \{C_0, C_1, C_2, C_5, C_8\}$$

1. Determine packages $C_{\bar{j}}$ based on C_k^C and their position within the hierarchical tree (see Sect. 5.2.1).
2. Evaluate individual bids of packages $C_{\bar{j}}$ and determine the maximum bid $\overline{b}_k(C_{\bar{j}})$ (see Sect. 5.2.2).
3. Determine the optimal allocation $\mathcal{W}^*$ of packages to cooperating vehicles (see Sect. 5.2.3).

5.2.1   Hierarchical Tree of Packages

All conflicting cells C_k^C at k are included in the root node of a hierarchical tree T. At each level of the tree, the cells in a parent node are decomposed into disjoint sets of cells, each of which is a package associated with a child node (see Fig. 7). To decompose the cells into more granular packages, we consider the following levels:

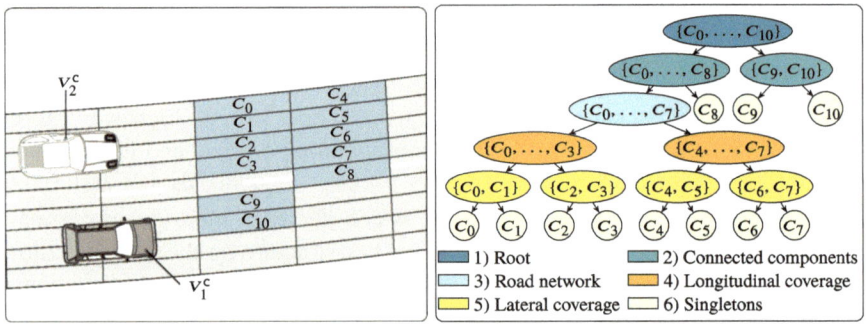

(a) V_1^c intends to perform a lane change maneuver; both V_1^c and V_2^c request cells $\{C_0, \ldots, C_{10}\}$

(b) A possible hierarchical tree constructed using the decomposition strategy outlined in Sect. 5.2.1

Fig. 7 Example grouping of conflicting cells (adapted from [21])

1. Connected components: Connected regions on the road surface prevents ego vehicles having disjointed drivable areas, which would complicate subsequent motion planning. We aggregate connected cells into packages.
2. Road network: Vehicles have to obey the traffic rules imposed by the road network; therefore, we encourage the creation of packages based on lanelets. A cell is assigned to the lanelet with which it has the largest intersecting area.
3. Longitudinal position coverage: The packages of the parent nodes are decomposed in the longitudinal direction such that the longitudinal coverage of each new package does not exceed a predefined threshold.
4. Lateral position coverage: The packages of the parent nodes are decomposed in the lateral direction such that the lateral coverage of each new package does not exceed a predefined threshold.
5. Singletons: The packages comprise only a single cell.

5.2.2 Bids on Packages

We adopt a common utility function for cooperating vehicles to avoid a situation in which a vehicle could continuously outbid others due to differences in the scales and weights used to calculate the bids on packages. We use the following sets as the basis for computing the utility of V_n^c for $C_{\bar{j}}$ to determine $b_{n,k}(C_{\bar{j}})$:

1. the conflict-free reachable set: $\mathcal{R}_{n,k}^{CF} := \left\{ x_{n,k} \in \mathcal{R}_{n,k} \middle| \mathrm{cell}_n(\{x_{n,k}\}) \cap C_k^C = \varnothing \right\}$.
2. the conflicting reachable set depending on package $C_{\bar{j}}$ that would be lost if $C_{\bar{j}}$ was not assigned to V_n^c: $\mathcal{R}_{n,k}^{CP}(C_{\bar{j}}) := \left\{ x_{n,k} \in \mathcal{R}_{n,k} \middle| \mathrm{cell}_n(\{x_{n,k}\}) \cap C_{\bar{j}} \neq \varnothing \right\}$.
3. the assigned reachable set that V_n^c possesses given that $C_{\bar{j}}$ is assigned to V_n^c: $\mathcal{R}_{n,k}^{AS}(C_{\bar{j}}) := \mathcal{R}_{n,k}^{CF} \cup \mathcal{R}_{n,k}^{CP}(C_{\bar{j}})$.

For computational reasons, the sets $\mathcal{R}_{n,k}^{CF}, \mathcal{R}_{n,k}^{CP}(C_{\bar{j}})$, and $\mathcal{R}_{n,k}^{AS}(C_{\bar{j}})$ are approximated by the union of base sets (see Sect. 3.3) and are denoted by $\cup_i \mathcal{R}_{n,k}^{CF,(i)}$, $\cup_i \mathcal{R}_{n,k}^{CP,(i)}$, and $\cup_i \mathcal{R}_{n,k}^{AS,(i)}$, respectively. To take the objectives of the vehicles into account while preventing the complete loss of the reachable set of a vehicle (so that a trajectory can still be found), the utilities of vehicles are computed differently for regular mode and survival mode:

$$b_{n,k}(C_{\bar{j}}) := \begin{cases} U_{n,k}^R(C_{\bar{j}}), & \mathrm{area}(\mathcal{R}_{n,k}^{CF}) > \underline{A}, \text{ (regular mode)} \\ U_{n,k}^S(C_{\bar{j}}), & \text{otherwise, (survival mode)} \end{cases}$$

where area($\cdot$) returns the size of the drivable area of the input (see Sect. 3.3) and $\underline{A}$ is a threshold. We now proceed with explaining the regular mode and survival mode. (1) *Regular Mode*: The utility of $\mathcal{R}_{n,k}^{AS}$ (or $\mathcal{R}_{n,k}^{CF}$, as the case may be) is defined as the sum of the utilities of $\mathcal{R}_{n,k}^{AS,(i)}$ (or $\mathcal{R}_{n,k}^{CF,(i)}$), weighted by their areas. The function $U_{n,k}^R(C_{\bar{j}})$ reflects the utility of $C_{\bar{j}}$ for V_n^c by computing the ratio of the utility of $\mathcal{R}_{n,k}^{AS}$ to that of $\mathcal{R}_{n,k}^{CF}$:

$$U_{n,k}^{\text{R}}(C_{\tilde{j}}) = \frac{\sum_i \left(u^{\text{pos}}(\mathcal{R}_{n,k}^{\text{AS},(i)}) + u^{\text{ref}}(\mathcal{R}_{n,k}^{\text{AS},(i)}) \right) \times \text{area}\left(\mathcal{R}_{n,k}^{\text{AS},(i)}\right)}{\sum_i \left(u^{\text{pos}}(\mathcal{R}_{n,k}^{\text{CF},(i)}) + u^{\text{ref}}(\mathcal{R}_{n,k}^{\text{CF},(i)}) \right) \times \text{area}\left(\mathcal{R}_{n,k}^{\text{CF},(i)}\right)},$$

with partial utility functions u^{pos} and u^{ref}. To encourage advances in traffic flow, we reward progression in the longitudinal direction with

$$u^{\text{pos}}(\Box_{n,k}) = y \left(\frac{\max(\text{proj}_{(s)}(\Box_{n,k})) - \max(\text{proj}_{(s)}(\mathcal{R}_{n,k-1}^{\text{N}}))}{\frac{1}{2}\,\overline{\overline{s}}_{n,k}\,\Delta_t^2 + \overline{s}_{n,k}\,\Delta_t} \right),$$

where $\overline{\overline{s}}_{n,k}$ and $\overline{s}_{n,k}$ are determined according to (2a), and y is a generalized logistic function that maps the utility to $(0, 1)$; in addition to [21], we also consider the deviation of V_n^{c} from its reference path:

$$u^{\text{ref}}(\Box_{n,k}) = e^{-w\,d'},\, d' = \min(\{|d''|\,|\,d'' \in \text{proj}_{(d)}(\Box_{n,k})\}),$$

where $w \in \mathbb{R}_+$ is a tunable weight that dictates how fast $u^{\text{ref}}(\Box_{n,k})$ approaches 0 as the deviation increases.

(2) *Survival Mode*: Two countermeasures are introduced to prevent reachable sets of V_n^{c} from vanishing: (1) if any V_n^{c} is in survival mode, no other vehicle in regular mode can bid on the package $C_{\tilde{j}}$; (2) the utility function is switched to

$$U_{n,k}^{\text{S}}(C_{\tilde{j}}) = \frac{\text{area}\left(\mathcal{R}_{n,k}^{\text{CP}}(C_{\tilde{j}})\right)}{\text{area}\left(\mathcal{R}_{n,k}\right)},$$

which reflects how close the reachable set of V_n^{c} is to vanishing given that $C_{\tilde{j}}$ is not assigned to V_n^{c}.

5.2.3 Optimal Allocation of Packages

The algorithm for finding the optimal allocation $\mathcal{W}^*$ of packages $C_{\tilde{j}}$ is based on [30]. In each iteration, we retrieve the deepest node N^{deep} in the hierarchical tree T (see Sect. 5.2.1), its parent node N^{parent}, and the set of child nodes $\mathcal{N}^{\text{child}} = \left\{ \ldots, N_{\tilde{q}}^{\text{child}}, \ldots \right\}$ of N^{parent}. Next, we compare the summed maximum bids (revenue) of all child nodes $\text{rev}(\mathcal{N}^{\text{child}}) := \sum_{\tilde{q}} \overline{b}_k(N_{\tilde{q}}^{\text{child}})$ with the maximum bid of the parent node $\overline{b}_k(N^{\text{parent}})$:

- If $\overline{b}_k(N^{\text{parent}}) > \text{rev}(\mathcal{N}^{\text{child}})$, $\mathcal{N}^{\text{child}}$ is excluded from $\mathcal{W}^*$.
- If $\overline{b}_k(N^{\text{parent}}) \leq \text{rev}(\mathcal{N}^{\text{child}})$, N^{parent} is excluded from $\mathcal{W}^*$.

Following this comparison, N^{child} is removed from the tree. The process is repeated until N^{parent} becomes the root node. After obtaining $\mathcal{W}^*$, each ego vehicle V_n^c proceeds to determine its negotiated reachable sets:

$$\mathcal{R}_{n,k}^N := \left\{ x_{n,k} \in \mathcal{R}_{n,k} \,\big|\, \text{cell}_n(\{x_{n,k}\}) \cap C_{n,k}^{\text{UA}} = \varnothing \right\},$$

where $C_{n,k}^{\text{UA}} \subseteq C_k^C$ denotes the set of unassigned cells of V_n^c based on $\mathcal{W}^*$.

6 Evaluation

This section provides example results for specification-compliant reachable sets for a single ego vehicle and its extension to cooperative vehicles. The implementation is based on the CommonRoad-Reach toolbox [14] for computing the reachable sets of vehicles.

6.1 Scenario I: Precise Overtaking

This scenario depicts a situation in which the vehicle V_1^c should overtake a leading vehicle V_1^o in the presence of another vehicle V_2^o. Let the following specification be issued by a high-level maneuver planner of V_1^c:

$$\mathbf{G}_{[0,15]} \left(\text{behind}(V_1^o) \wedge \text{aligned\_with}(V_1^o) \right) \wedge$$
$$\mathbf{G}_{[16,38]} \left(\text{in\_lanelet}(L_2) \vee \text{in\_lanelet}(L_4) \right) \wedge$$
$$\mathbf{G}_{[39,45]} \left(\text{in\_front\_of}(V_1^o) \wedge \text{behind}(V_2^o) \wedge \text{in\_lanelet}(L_3) \right).$$

The specification-compliant reachable sets are computed as described in Sect. 4. The non-empty result implies that it is possible to find a trajectory that meets the specifications. Figure 8 visualizes the drivable areas of V_1^c along with a trajectory planned within the reachable sets using the motion planner described in [22]. For a more detailed evaluation of computing specification-compliant reachable sets for a single ego vehicle, we refer the reader to [13].

6.2 Scenario II: Highway

In this scenario, we negotiate the reachable sets of four cooperating vehicles driving on a highway. Figure 9 shows the computation results at different steps. As can be seen, our method can allocate road areas to cooperating vehicles even in such complex traffic situations with many non-cooperating traffic participants. For a more detailed

Fig. 8 Overtaking scenario.
a Drivable area at different
steps. **b** A trajectory planned
within the reachable set
(adapted from [13])

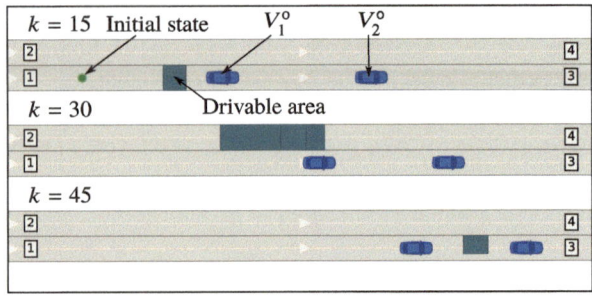

(a)

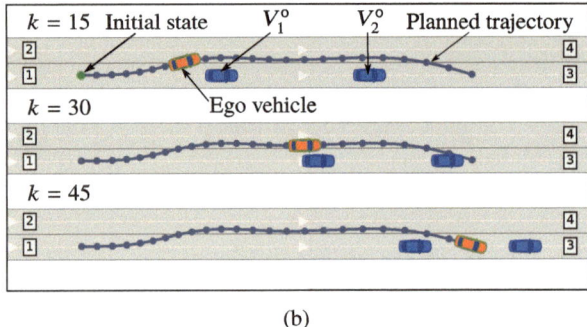

(b)

evaluation of negotiating reachable sets among a group of cooperating vehicles, we
refer the reader to [21].

6.3 Scenario III: Roundabout

This scenario illustrates a situation in which two vehicles V_1^c and V_2^c should cooperate
to go around a roundabout. We show the computation results under different settings:
(1) no specification is considered; (2) V_1^c yields to V_2^c; (3) V_2^c yields to V_1^c. The
latter two settings are relevant when a *yield* traffic sign is present at the junction and
specifies which vehicle has to yield to other vehicles entering with a higher passing
priority. The specification can be expressed as follows:

$$\mathbf{G}(exists\_yield\_sign \wedge exists\_other\_entering\_vehicle \Rightarrow brake\_to\_stop),$$

which can be regarded as a simplified version of the intersection rules described
in [18] but without temporal logic connectives. Figure 10 illustrates the computation
results under these settings. In Fig. 10b, V_2^c can either accelerate and enter the roundabout ahead of V_1^c or decelerate to enable V_1^c to enter first. In Fig. 10c, d, the yielding
vehicles have to brake in order to stop and yield to the other entering vehicle.

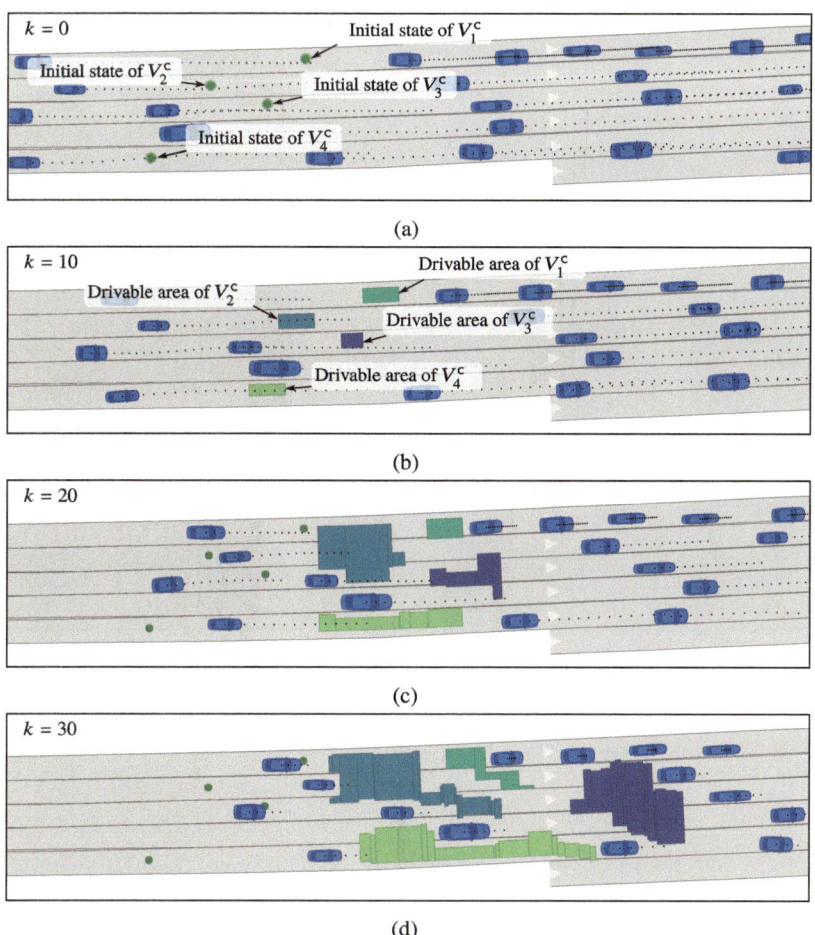

Fig. 9 Highway scenario. Subfigures **b–d** show the drivable areas of the negotiated reachable sets of vehicles at different steps

7 Conclusions

In this chapter, we summarized our previous works on computing specification-compliant reachable sets for an ego vehicle and negotiating conflicting reachable sets between a group of cooperating vehicles. The specification-compliant and negotiated reachable set is used to guide subsequent motion planners to find specification-compliant trajectories. As a result, the cooperative vehicles can consider traffic rules and handcrafted rules expressed in propositional logic that involve position, velocity, acceleration, and general traffic situation predicates. A limitation of the method is that it does not yet handle specifications formulated in temporal logic, which reflects

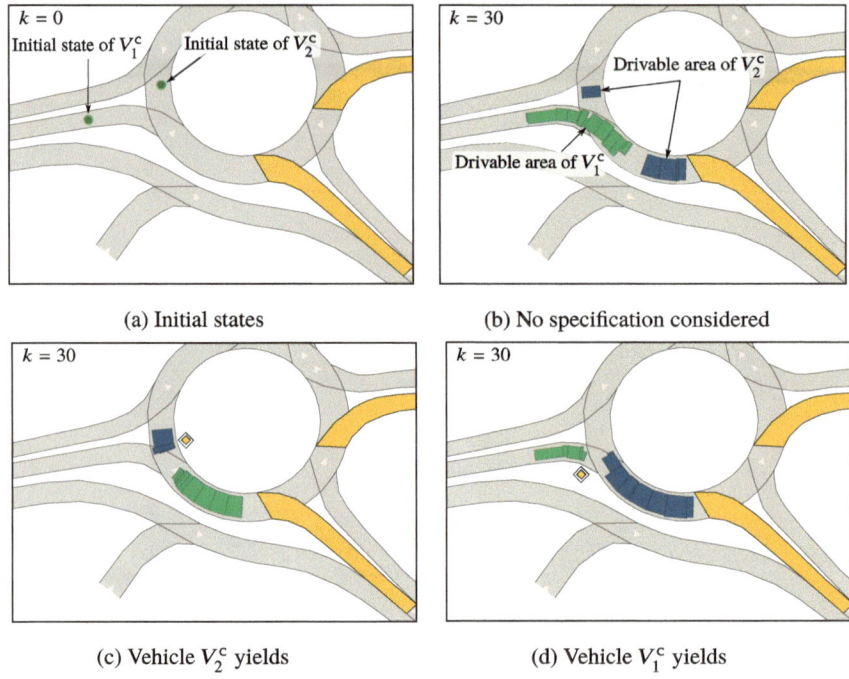

(a) Initial states

(b) No specification considered

(c) Vehicle V_2^c yields

(d) Vehicle V_1^c yields

Fig. 10 Roundabout scenario. V_2^c intends to reach the first exit, and V_1^c intends to reach the second exit. Subfigures b–d show the drivable areas of the negotiated reachable sets of vehicles at step $k = 30$

temporal requirements on vehicles, both in the computation and negotiation of the reachable sets. This will be a subject of future research.

Acknowledgements This work was funded by the Deutsche Forschungsgemeinschaft (German Research Foundation) within the Priority Program SPP 1835 *Cooperative Interacting Automobiles* under grant No. AL 1185/4-2, Huawei-TUM collaboration project *Research on Key Technologies of Safety Assurance for Autonomous Vehicles*, and the Central Innovation Program of the German Federal Government under grant No. KK5116401KG0. The authors appreciate the fruitful collaborations with the project partners.

References

1. Althoff, M., Koschi, M., Manzinger, S.: CommonRoad: composable benchmarks for motion planning on roads. In: Proceedings of the IEEE Intelligent Vehicles Symposium, pp. 719–726 (2017)
2. Bender, P., Ziegler, J., Stiller, C.: Lanelets: efficient map representation for autonomous driving. In: Proceedings of the IEEE Intelligent Vehicles Symposium, pp. 420–425 (2014)

3. Carlino, D., Boyles, S.D., Stone, P.: Auction-based autonomous intersection management. In: Proceedings of the IEEE International Conference on Intelligent Transportation Systems, pp. 529–534 (2013)
4. Dresner, K., Stone, P.: Multiagent traffic management: a reservation-based intersection control mechanism. In: Proceedings of the International Conference on Autonomous Agents and Multiagent Systems, pp. 530–537 (2004)
5. Dresner, K., Stone, P.: Turning the corner: improved intersection control for autonomous vehicles. In: Proceedings of the IEEE Intelligent Vehicles Symposium, pp. 423–428 (2005)
6. Dresner, K., Stone, P.: Human-usable and emergency vehicle-aware control policies for autonomous intersection management. In: Workshop on Agents in Traffic and Transportation, pp. 17–25 (2006)
7. Eilbrecht, J., Stursberg, O.: Challenges of trajectory planning with integrator models on curved roads. In: Proceedings of the IFAC World Congress, pp. 15588–15595 (2020)
8. Esterle, K., Aravantinos, V., Knoll, A.: From specifications to behavior: maneuver verification in a semantic state space. In: Proceedings of the IEEE Intelligent Vehicles Symposium, pp. 2140–2147 (2019)
9. Häfner, B., Bajpai, V., Ott, J., Schmitt, G.A.: A survey on cooperative architectures and maneuvers for connected and automated vehicles. IEEE Commun. Surv. Tutorials 24(1), 380–403 (2021)
10. Hekmatnejad, M., Yaghoubi, S., Dokhanchi, A., Amor, H.B., Shrivastava, A., Karam, L., Fainekos, G.: Encoding and monitoring responsibility sensitive safety rules for automated vehicles in signal temporal logic. In: Proceedings of the ACM/IEEE International Conference on Formal Methods and Models for System Design, pp. 1–11 (2019)
11. Heß, D., Lattarulo, R., Pérez, J., Schindler, J., Hesse, T., Köster, F.: Fast maneuver planning for cooperative automated vehicles. In: Proceedings of the International Conference on Intelligent Transportation Systems, pp. 1625–1632 (2018)
12. Huth, M., Ryan, M.: Logic in Computer Science: Modelling and Reasoning About Systems. Cambridge University Press (2004)
13. Irani Liu, E., Althoff, M.: Computing specification-compliant reachable sets for motion planning of automated vehicles. In: Proceedings of the IEEE Intelligent Vehicles Symposium, pp. 1037–1044 (2021)
14. Irani Liu, E., Würsching, G., Moritz, K., Althoff, M.: CommonRoad-Reach: a toolbox for reachability analysis of automated vehicles. In: Proceedings of the IEEE International Conference on Intelligent Transportation Systems, pp. 1–8 (2022)
15. Kohlhaas, R., Bittner, T., Schamm, T., Zöllner, J.M.: Semantic state space for high-level maneuver planning in structured traffic scenes. In: Proceedings of the IEEE International Conference on Intelligent Transportation Systems, pp. 1060–1065 (2014)
16. Lahijanian, M., Maly, M.R., Fried, D., Kavraki, L.E., Kress-Gazit, H., Vardi, M.Y.: Iterative temporal planning in uncertain environments with partial satisfaction guarantees. IEEE Trans. Rob. 32(3), 583–599 (2016)
17. Levin, M.W., Fritz, H., Boyles, S.D.: On optimizing reservation-based intersection controls. IEEE Trans. Intell. Transp. Syst. 18(3), 505–515 (2017)
18. Maierhofer, S., Moosbrugger, P., Althoff, M.: Formalization of intersection traffic rules in temporal logic. In: Proceedings of the IEEE Intelligent Vehicles Symposium, pp. 1135–1144 (2022)
19. Maierhofer, S., Rettinger, A.K., Mayer, E.C., Althoff, M.: Formalization of interstate traffic rules in temporal logic. In: Proceedings of the IEEE Intelligent Vehicles Symposium, pp. 752–759 (2020)
20. Manzinger, S., Althoff, M.: Negotiation of drivable areas of cooperative vehicles for conflict resolution. In: Proceedings of the IEEE International Conference on Intelligent Transportation Systems, pp. 1–8 (2018)
21. Manzinger, S., Althoff, M.: Tactical decision making for cooperative vehicles using reachable sets. In: Proceedings of the IEEE International Conference on Intelligent Transportation Systems, pp. 444–451 (2018)

22. Manzinger, S., Pek, C., Althoff, M.: Using reachable sets for trajectory planning of automated vehicles. IEEE Trans. Intell. Veh. **6**(2), 232–248 (2020)
23. Marinescu, D., Čurn, J., Bouroche, M., Cahill, V.: On-ramp traffic merging using cooperative intelligent vehicles: a slot-based approach. In: Proceedings of the IEEE International Conference on Intelligent Transportation Systems, pp. 900–906 (2012)
24. Montanaro, U., Dixit, S., Fallah, S., Dianati, M., Stevens, A., Oxtoby, D., Mouzakitis, A.: Towards connected autonomous driving: review of use-cases. Veh. Syst. Dyn. **57**(6), 779–814 (2019)
25. Nichting, M., Heß, D., Schindler, J., Hesse, T., Köster, F.: Space time reservation procedure (STRP) for V2X-based maneuver coordination of cooperative automated vehicles in diverse conflict scenarios. In: Proceedings of the IEEE Intelligent Vehicles Symposium, pp. 502–509 (2020)
26. Pek, C., Althoff, M.: Fail-safe motion planning for online verification of autonomous vehicles using convex optimization. IEEE Trans. Rob. **37**(3), 798–814 (2020)
27. Rewald, H., Stursberg, O.: Cooperation of autonomous vehicles using a hierarchy of auction-based and model-predictive control. In: Proceedings of the IEEE Intelligent Vehicles Symposium, pp. 1078–1084 (2016)
28. Rios-Torres, J., Malikopoulos, A.A.: A survey on the coordination of connected and automated vehicles at intersections and merging at highway on-ramps. IEEE Trans. Intell. Transp. Syst. **18**(5), 1066–1077 (2017)
29. Rizaldi, A., Keinholz, J., Huber, M., Feldle, J., Immler, F., Althoff, M., Hilgendorf, E., Nipkow, T.: Formalising and monitoring traffic rules for autonomous vehicles in Isabelle/HOL. In: International Conference on integrated Formal Methods, pp. 50–66 (2017)
30. Rothkopf, M.H., Pekeč, A., Harstad, R.M.: Computationally manageable combinational auctions. Manag. Sci. **44**(8), 1131–1147 (1998)
31. Shalev-Shwartz, S., Shammah, S., Shashua, A.: On a formal model of safe and scalable self-driving cars (2017). arXiv:1708.06374
32. Shoukry, Y., Nuzzo, P., Sangiovanni-Vincentelli, A.L., Seshia, S.A., Pappas, G.J., Tabuada, P.: SMC: satisfiability modulo convex programming. Proc. IEEE **106**(9), 1655–1679 (2018)
33. Söntges, S., Althoff, M.: Computing possible driving corridors for automated vehicles. In: Proceedings of the IEEE Intelligent Vehicles Symposium, pp. 160–166 (2017)
34. Söntges, S., Althoff, M.: Computing the drivable area of autonomous road vehicles in dynamic road scenes. IEEE Trans. Intell. Transp. Syst. **19**(6), 1855–1866 (2018)
35. Vasirani, M., Ossowski, S.: A market-inspired approach for intersection management in urban road traffic networks. J. Artif. Intell. Res. **43**, 621–659 (2012)
36. Würsching, G., Althoff, M.: Sampling-based optimal trajectory generation for autonomous vehicles using reachable sets. In: Proceedings of the IEEE International Conference on Intelligent Transportation Systems, pp. 828–835 (2021)
37. Zhou, Y., Maity, D., Baras, J.S.: Timed automata approach for motion planning using metric interval temporal logic. In: Proceedings of the European Control Conference, pp. 690–695 (2016)
38. Zhu, F., Ukkusuri, S.V.: A linear programming formulation for autonomous intersection control within a dynamic traffic assignment and connected vehicle environment. Transp. Res. Part C: Emerg. Technol. **55**, 363–378 (2015)

AutoKnigge—Modeling, Evaluation and Verification of Cooperative Interacting Automobiles

Christian Kehl, Maximilian Kloock, Evgeny Kusmenko, Lutz Eckstein, Bassam Alrifaee, Stefan Kowalewski, and Bernhard Rumpe

Abstract The development of cooperative driving functions to optimize traffic systems shows high potential to improve individual autonomous driving systems with respect to topics like traffic flow, vehicle safety and user comfort. The core concept of the presented solutions is the Local Traffic System (LTS). Following the messages defined in European Telecommunications Standards Institute (ETSI) Intelligent Transport Systems (ITS) G5 for Vehicle-to-everything (V2X) cooperation we introduce concepts and implementations to intelligently group vehicles based on the exchanged V2X data with respect to the individual vehicle capability for cooperation. Based on the determined grouping, we present algorithms for cooperative trajectory planning. We develop a verification method for the cooperatively planned trajectories

C. Kehl, M. Kloock, E. Kusmenko—These authors contributed equally.

C. Kehl (✉) · L. Eckstein
Institute for Automotive Engineering (ika), RWTH Aachen University, Aachen, Germany
e-mail: christian.kehl@ika.rwth-aachen.de

L. Eckstein
e-mail: lutz.eckstein@ika.rwth-aachen.de

M. Kloock · S. Kowalewski
Chair of Embedded Software, RWTH Aachen University, Aachen, Germany
e-mail: kloock@embedded.rwth-aachen.de

S. Kowalewski
e-mail: kowalewski@embedded.rwth-aachen.de

E. Kusmenko · B. Rumpe
Chair of Software Engineering, RWTH Aachen University, Aachen, Germany
e-mail: kusmenko@se-rwth.de

B. Rumpe
e-mail: rumpe@se-rwth.de

B. Alrifaee
Department of Aerospace Engineering, University of the Bundeswehr Munich, Neubiberg, Germany
e-mail: bassam.alrifaee@unibw.de

© The Author(s) 2024
C. Stiller et al. (eds.), *Cooperatively Interacting Vehicles*,
https://doi.org/10.1007/978-3-031-60494-2_13

347

within a LTS. The verification guarantees collision avoidance and deadlock-freeness in real-time. Finally we introduce a model language based on MontiArc to enable a systematic representation and description of the presented concepts for grouping, cooperation and interaction.

1 Introduction

Rapid technological advancements in the area of automated driving functions in recent years make large-scale deployment of SAE Level 4 and 5 [45] automated vehicles likely in the next few years. While technological progress is mainly limited to the development of vehicle-specific automation functions, the development of cooperative automation functions for the optimization of traffic systems already shows high potential to significantly improve current topics of concern such as traffic flow, vehicle safety and user comfort.

Current Vehicle-to-everything (V2X) systems show a beacon-like behavior without a direct sender or receiver and are rather designed to transmit one-time events to alert other traffic participants. The next logical step towards the development of cooperatively interacting vehicles requires a significant extension of existing V2X systems at all levels. The extension of these systems from a one-time event-based communication to a continuous data exchange for the execution of cooperatively interacting algorithms [4, 28, 39], raises questions regarding the grouping of the involved road users [29], reliability vehicle communication [32], the type of information exchanged, the underlying algorithms as well as the basic model description of these systems. Methods that present cooperative trajectory planning of vehicles in different scenarios are, e.g., the works in [31, 33, 34]. These works focus on the applicability of cooperative trajectory planning in intersections, pose control, and vehicle racing.

The core concept of the solutions presented in the following is the Local Traffic System [7]. Local Traffic Systems can be understood as cooperating C-ITS subsystems as defined in European Telecommunications Standards Institute (ETSI) Intelligent Transport Systems (ITS) G5. Based on this concept, different approaches for the detection of the corresponding traffic scenarios, the formation of Local Traffic Systems as well as their evaluation are presented. Within these systems the cooperation takes place. In the context of this work, the cooperative trajectory planning, as well as a real-time verification of the cooperatively planned trajectories are presented. The verification guarantees the absence of collisions and deadlocks for the trajectories of all vehicles in one or multiple LTS. Finally, a model language based on MontiArc is presented for the systematic representation and description of the presented concepts for grouping, cooperation and interaction.

2 Learning-Based and Vehicle Capability-Aware Architecture for Clustering of Cooperative Interacting Automobiles

One of the central aspects within the overall process for cooperation and interaction of vehicles is the clustering of traffic participants relevant for cooperation. The formation of these clusters for the purpose of cooperation inevitably leads to the following questions: When is cooperation and interaction between traffic participants useful? What kind of vehicle data must be exchanged before and during cooperation? How can relevant traffic participants be identified?

In order to group the corresponding traffic participants, this work takes up the concept of Local Traffic Systems (LTS) [7] and develops it further. Local Traffic Systems are defined as a grouping of road users for the purpose of information exchange as well as cooperation. The cooperation take place exclusively within the LTS.

Previous work [8] in the area of Local Traffic Systems has been based on a single evaluation function. This evaluation function consists of various normalized distance metrics such as the distance between individual vehicles, the derivative of the distance function, the direction of travel, etc. The position information is based on a predefined road graph that must be known to all road users. The nodes of the road graph represent different points within the traffic network and have a distance of a few meters. The edges of the road graph represent the roads themselves. Road properties such as the maximum permitted speed are assigned to the edges. The individual distance metrics are then normalized and multiplied by a developer defined weighting factor. The now normalized and weighted metrics are finally added to an overall evaluation function. The objective is to minimize the evaluation function. The LTS configuration with the most minimized evaluation function is considered as an optimal solution. The information exchanged here to determine the individual metrics is already based on current standardizations such as the Cooperative Awareness Message (CAM) [13] and are extended when necessary. The vehicle data is exchanged cyclically. After the LTS formation, the cooperation takes place through data exchange within the system.

However, this approach has several disadvantages. The recurring calculation of the entire LTS configuration leads to an enormously high computing load, which makes a calculation in real time almost impossible. In [8] therefore a greedy algorithm is recommended, which makes only small changes at the past configuration in each time step without recalculating the total configuration. Additionally, the number of permitted LTS participants is limited to a maximum of 5-10 participants. This serves on the one hand to reduce the total computation time, and on the other hand to reduce the amount of information exchanged within the LTS in order to prevent an overload of the available bandwidth.

The necessity for a common road graph model shared by all vehicles represents a considerable limitation. The formation based on a road graph is here not limited by the mere necessity of the graph itself, but by the required correspondence of the graph between all road users. It is already apparent today that predefined map data will play a decisive role in the implementation of autonomous driving functions

[47]. However, they often take on a supporting role for localization [41]. Due to the frequent changes in the road network and the resulting inaccurate data, possible cooperation approaches should be map-independent. Furthermore, planning based on the road graph limits the accuracy of LTS formation to the accuracy of the existing map material because all positions are defined relative to the underlying graph. This poses a problem especially for cooperative maneuvers when the required vehicle distance is below the minimal accuracy level defined by the road graph.

Another disadvantage is the decoupling of exchanged vehicle data, LTS generation, the underlying cooperation algorithms, and the current driving situation. The permanently high amount of exchanged vehicle data leads to an unnecessarily high utilization of the available V2X data rate in the vehicle. IEEE 802.11p and LTE V2X can support data rate of up to 27 Mbit s^{-1} and 28.8 Mbit s^{-1} [44]. The lack of a link to the current vehicle situation and the underlying algorithms not only makes it difficult to prioritize individual LTS systems, but also ignores the influence of the current driving situation on LTS generation. The following section is intended to give a better impression of the resulting problems and derive additional requirements for improvements.

2.1 Requirements for an Extended LTS Architecture

Using selected examples, this section attempts to provide insight into the motivation for extending the previous approach and to derive possible requirements for an extended architecture. The goal is to preserve the general concept while identifying concept limitations and avoiding the disadvantages identified in the course of previous work.

The question of when a Local Traffic System should be formed at all and which road users should participate in it is closely linked to the respective traffic scenario. Possible traffic scenarios are shown in Fig. 4. In the following an exemplary traffic scenario of a roundabout with five vehicles is displayed in Fig. 1. The planned routes of the vehicles are marked in color along the center of the lane. The drawn rectangles indicate the possible LTS groupings for the scenario at hand. Vehicles can be grouped based on their respective vehicle state relative to other road users as well as relative to their surroundings. This can be based on various vehicle data such as the spatial proximity to the next vehicle, the overlap of the planned trajectories, the general overlap of the planned routes, or the spatial proximity of the vehicle under consideration relative to a relevant traffic node such as an intersection or roundabout.

One of the central problems here is the influence of cooperation, or cooperation capability, as well as the driving situation on LTS generation itself. The scenario shown in Fig. 2 illustrates the problem. A fast moving vehicle is approaching a slow moving vehicle on the right lane. In order to avoid heavy braking of the right vehicle behind, a lane change to the left lane is attempted. In general, two possible LTS are conceivable in this situation. One LTS consisting of the rear two vehicles to coordinate the lane change and one overall LTS consisting of all vehicles. However, if the rear

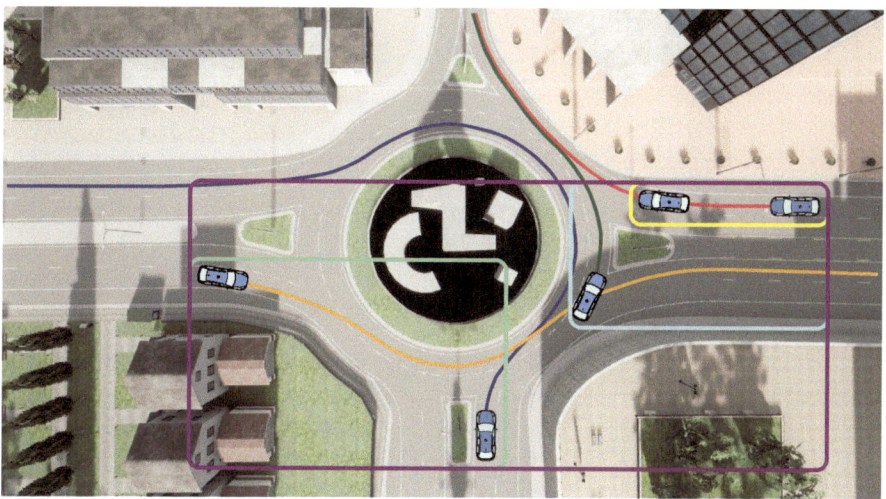

Fig. 1 Possible local traffic systems—roundabout scenario

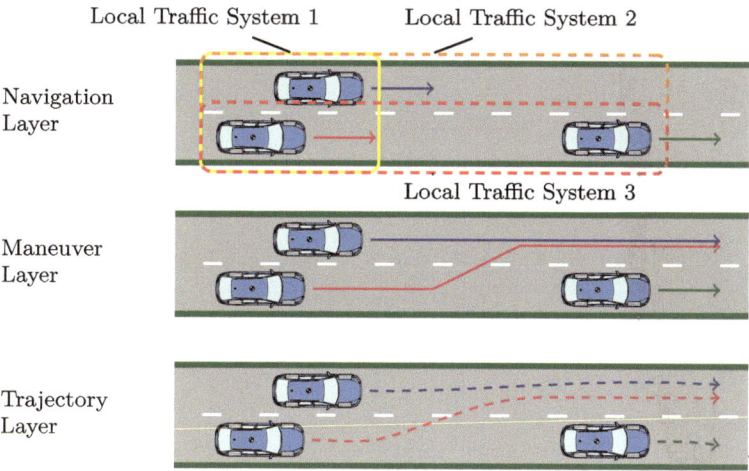

Fig. 2 Motion planning LTS layers

vehicle does not have the ability to change lanes cooperatively, the formation of a third LTS from the two right vehicles for the purpose of speed adaptation is necessary. A downstream cooperation without consideration of the vehicle capabilities leads to an incorrect LTS formation.

If we now extend the given scenario as shown in Fig. 3, assuming the ability to change lanes, another problem becomes apparent. In order to enable a lane change of the right vehicle, in principle three vehicles would have to slow down their speed, which is unfavorable from the point of view of a global optimization. However, a

Local Traffic System 1
Local Traffic System 2
Navigation Layer
Maneuver Layer
Trajectory Layer

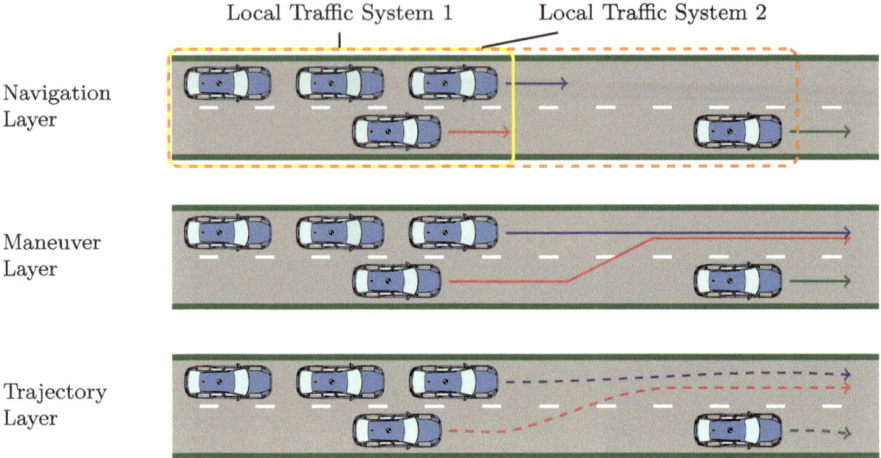

Fig. 3 Motion planning LTS layers conflict

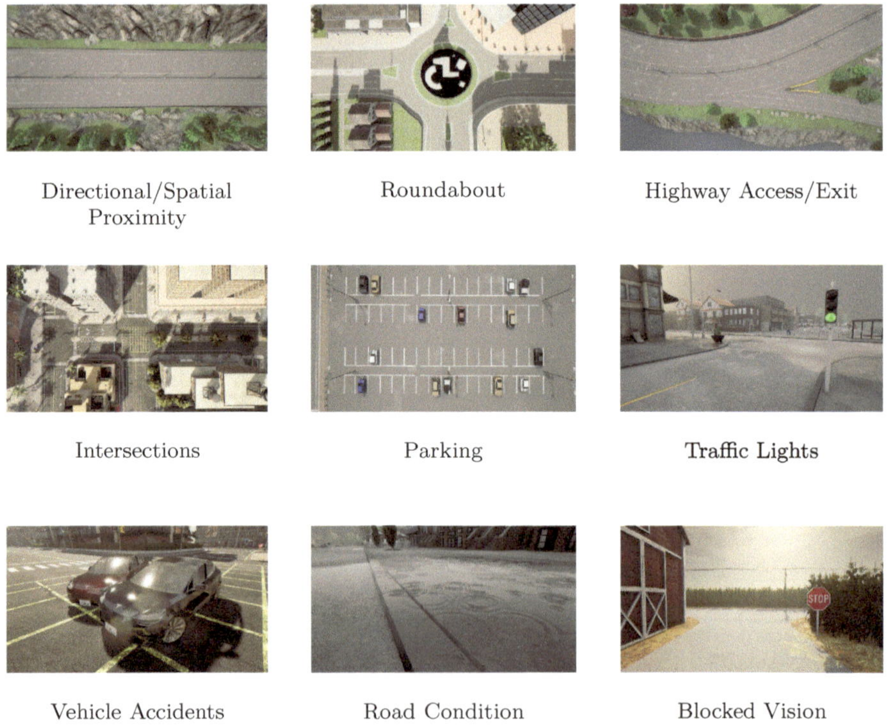

Directional/Spatial Proximity
Roundabout
Highway Access/Exit
Intersections
Parking
Traffic Lights
Vehicle Accidents
Road Condition
Blocked Vision
STOP

Fig. 4 Possible LTS scenarios

human actor would possibly prefer the light braking of several vehicles to the strong braking maneuver of a single vehicle. The respective driving situation therefore also has a decisive influence on the vehicle grouping here. The simple static weighting of different distance metrics is highly unlikely to meet this requirement.

The examples described above show that the selection of LTS participants is complex and a wide variety of conflict situations can occur within a single scenario. These are strongly dependent on the direct driver environment and require a high level of algorithmic understanding of the driving situation which, as shown in Fig. 3, cannot be solved solely within the cooperation algorithm but has direct repercussions on the LTS formation. The incorrect too narrow selection of the LTS makes an optimal solution impossible.

In addition to the wrong selection of the LTS participants, there is also the possibility of potential target conflicts between different LTSs. If a vehicle is simultaneously a member of several competing LTSs, it is necessary to establish a prioritization between the individual systems.

2.2 Extended LTS Architecture

The disadvantages and problems of the previous concept described in the previous sections are to be solved by an extension of the architecture. The basic principles and advantages of the previous approach are to be preserved.

Figure 5 describes the novel approach to the clustering of vehicles. The most obvious difference is the division into different LTS levels between level 0 and level 4. The individual levels represent an increasing urgency in the need for cooperation between the road users and allow prioritization between individual LTS. Systems with a higher level are always given priority. In case of identical levels, no cooperation is performed. The system waits for escalation to higher levels. If several systems reach the highest level at the same time, cooperation between all traffic participants is required. The traffic systems are merged into a larger system. Each LTS level is associated with a specific set of exchanged vehicle data, boundary conditions, and available cooperation algorithms. At the beginning, every vehicle that has not yet been assigned to a specific group is at level 0. No active cooperation takes place here. Only simple awareness based information, like the current vehicle position or additional road information, like emergency warnings, are exchanged. This also provides a way to integrate passive road users unable to participate in a cooperative effort such as pedestrians, cyclists or infrastructure components like traffic lights. Each level is assigned a cooperation algorithm in addition to the vehicle data and associated boundary conditions. The LTS level is increased if the exchanged vehicle data exceeds the level specific boundary conditions. The type of cooperation algorithm increases according to the intensity of the intervention in the longitudinal and lateral control of the vehicle. The vehicle data required for the cooperation must not exceed the scope of the data exchange planned for the level. The amount of data exchanged increases here because more complex cooperation maneuvers usually require a larger pool of

Need for cooperation and interaction

LTS Level	LTS 0	LTS 1	LTS 2	LTS 3	LTS 4
Exchanged Vehicle Data	Position, Additional Road Information	Position, Velocity, Route	Position, Velocity, Route, Acceleration, Planned Trajectory	Position, Velocity, Route, Acceleration, Planned Trajectory	Position, Velocity, Route, Acceleration, Planned Trajectory
		Data Volume and required V2X Bandwidth			
LTS Boundary Condition	-	Distance	Route Intersection	Route Intersection, Time Threshold, Trajectory Intersection	Route Intersection, Time Threshold, Trajectory Intersection, TTC
		Severity of the interference with the individual planning behavior			
Cooperation and Interaction	-	Velocity Adaption	Velocity Adaption, Lane Change	Velocity Adaption, Lane Change, Cooperative Trajectory Planning	Velocity Adaption, Lange Change, Cooperative Trajectory Planning, Hazard Braking

Fig. 5 LTS structure

data. The data exchanged is roughly based on the data specified by the ETSI ITS G5 standard. Large parts of the described ETSI ITS G5 functionalities are still in an early stage of development at the time of this work and are therefore susceptible to possible changes. ETSI ITS G5 defines Cooperative Intelligent Transport Systems (C-ITS) as ITS subsystems such as people, vehicles, roadside units that exchange information or cooperate with each other to improve driving safety, traffic guidance or driving experience. The cooperation capabilities to be realized are referred to as services.

A general distinction is made between three categories of services. Cooperative Awareness Services [12–14] define the lowest level and describe the exchange of simple status information such as position and speed for the purpose of simple warnings. Cooperative Perception Services [17, 18] describe the second type of information exchange on top of status data. Within this service, other traffic participants are not only warned but also enabled to perform more complex functions such as Cooperative Adaptive Cruise Control. Cooperative Maneuver Coordination Services [15, 16] describe the highest level of cooperation. In addition to status and observation data road users can share their intention in order to allow cooperation in complex driving situations. These include scenarios like platooning or cooperative lane changes. The approach to LTS education presented here is oriented along the escalating nature of these services in terms of user interaction and user collaboration. In addition to minimizing intervention in the longitudinal and lateral control of the vehicle to increase user comfort, this also reduces the required bandwidth. Instead of exchanging all driving information periodically, only the information required for cooperation at the current level is exchanged. Further development and replacement of individual cooperation algorithms is possible without adaptation to the overall system.

For further development of the described architecture, a stimulative implementation approach is used. The developed framework is structured according to the diagram in Fig. 6. The CARLA simulator [10] serves as the basic foundation. The CARLA simulator is an open source driving simulator providing a virtual environment to simulate different driving scenarios and test autonomous driving functions in a virtual environment. The simulator is using a server/client concept. While the server is responsible for the simulation itself, the client controls the simulator by reading and writing data from and to the simulation using a TCP/IP. The client exposes the provided functions using an API to control traffic generation, pedestrian behavior, weather, sensors, maps and much more. During this project the provided Python API is used by the simulator interface to expose relevant functionality to the other components of the Autoknigge Framework. All components are connected using a ROS2 Communication Layer. This applies to messages controlling the simulator itself as well as messages exchanged between simulated vehicles. ROS2 uses the Data Distribution Service (DDS) which is also part of the Automotive Open System Architecture (AUTOSAR) Adaptive Platform. DDS is a middleware specified by the Object Management Group for data-centric communication in distributed systems. Based on the ROS Communication Layer there are higher level components like the World component. The World acts as a central repository for all data relevant to the simulation, such as vehicle positions, velocities, planned routes and trajectories, LTS

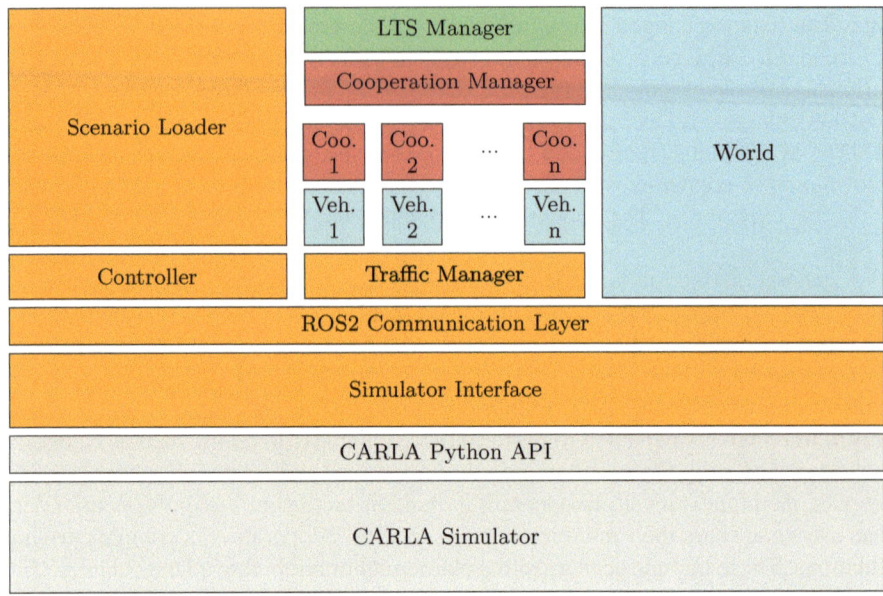

Fig. 6 Autoknigge architecture (Cooperation (Coo.), Vehicle (Veh.))

allocations, etc. The World component is also the central repository for all data relevant to the simulation. Unlike the data stored in the individual vehicle components, all data is available here. The controller is responsible for controlling the simulation itself and provides functions for selecting the map, adding vehicles and people. The controller is in turn used by the Scenario Loader to load various traffic scenarios. The Traffic Manager is used to abstract the management and communication of individual vehicles in the simulation. Here, for example, a distance-based forwarding of V2X messages takes place in order to simulate a range limitation of the vehicle messages. The number of managed vehicles is determined by the vehicles currently present in the simulation. Each vehicle component can theoretically contain its own Cooperation Manager to start and stop cooperation maneuvers. For simplicity in the context of the simulation, all vehicles currently share a Cooperation Manager. This also applies to the LTS Manager which assigns vehicles to the individual LTS.

The architecture described so far still lacks a concrete cooperation algorithm. Therefore Sect. 2.3 presents the implementation of a method for cooperative velocity adaptation for LTS level 1 systems.

2.3 Cooperative Velocity Adaption Algorithm

Cooperative adjustment of vehicle velocity represents one of the most minimally invasive forms of active cooperation, as it only interferes with the longitudinal control of the vehicle. By intervening in the vehicle's longitudinal control at an early stage, it is often possible to resolve conflict situations without impairing the vehicle occupants' sense of comfort due to strong longitudinal or lateral acceleration. The cooperative speed adaptation algorithm presented in the following is designed as a constraint optimization problem. The relevant boundary conditions are formulated as hard constraints and soft constraints. Hard constraints are unbreakable rules that must be fulfilled in any case, otherwise the result is not considered as a valid solution of the problem. Soft constraints represent less strict constraints and are understood as a kind of optimization parameter to distinguish several valid solutions in their quality. The problem is formulated in the form of a model and then passed to a solver. The concrete algorithm uses the Google OrTools CP-Sat Solver [42]. Due to the limitations of the solver, all variables and parameters of the model are formulated as integer values. Input and output values that are represented as floating point numbers are appropriately scaled by the algorithm beforehand.

The algorithm expects for each vehicle j two position arrays specifying the planned x,y-trajectories for each timestep i as well as additional parameters like the allowed minimum speed $v_{j,min}$, the maximum speed $v_{j,max}$. In addition, limits for the permitted longitudinal acceleration a_{max} as well as a minimum time gap $t_{gap,min}$ to ensure collision avoidance need to be defined. The variables are defined for each vehicle j involved.

The algorithm defines a vehicle velocity variable $v_{j,i}$ as well as a resulting timestamp $t_{j,i}$ for each vehicle position $p_{j,i} = (x_{j,i}, y_{j,i})$. The maximum acceleration is used to determine the permitted velocity change $v_{j,\Delta,max,i}$ for each distance step $d_{j,i}$. The timestamp $t_{j,i+1}$ is automatically calculated in the solver model using the distance $d_{j,i}$ between the position $p_{j,i}$ and $p_{j,i+1}$ as well as the velocity $v_{j,i}$ determined by the solver. To avoid a collision the model requires the time gap between two timestamps of two vehicles to be greater than the predefined time gap $t_{gap,min}$ threshold if the positions are closer than d_{thres}. The algorithm currently does not take into account the actual vehicle geometry. Therefore, the position distance value d_{thres} must be chosen sufficiently large. At every position the calculated velocity $v_{j,i}$ must be between $v_{j,min}$ and $v_{j,max}$ to be considered as valid result. As an optimizable soft constraint, the algorithm determines the maximum total duration of the maneuver as the time at which the last vehicle reaches the last target position in the planned trajectory.

The goal of the optimization is to minimize the total maneuvering time while taking into account the constraints described above.

As an example the algorithm is applied to the intersection scenario presented in Fig. 7. The vehicles are each positioned 10 m from the center of the intersection. The given speed is chosen for both vehicles so that the trajectories intersect at the same time and place. The usage of the presented algorithm with a spatial resolution of 1 m results in an optimal solution shown in Fig. 8. Compared to the individual use of the

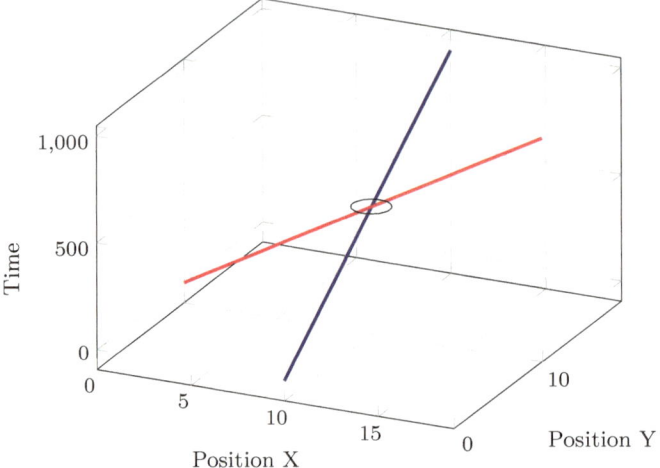

Fig. 7 Intersection scenario—conflict

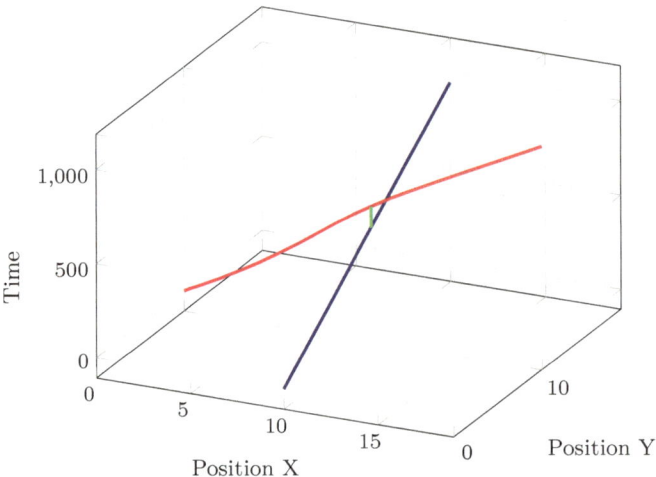

Fig. 8 Intersection scenario—conflict resolution by velocity adaption

intersection by each vehicle, the travel time for the braking vehicle is increased by approximately 13.5 %. The specified time gap is marked in green. The labeling of the displayed time axis does not correspond to the actual time in seconds but represents the direct integer solution value of the solver.

The formulation as a constraint optimization problem offers several advantages. On the one hand, the solver is able to capture the problem completely and detect conflicting constraints or prove the unsolvability of the problem at an early stage. The LTS system can thus detect the unsolvability of the cooperation task at an early

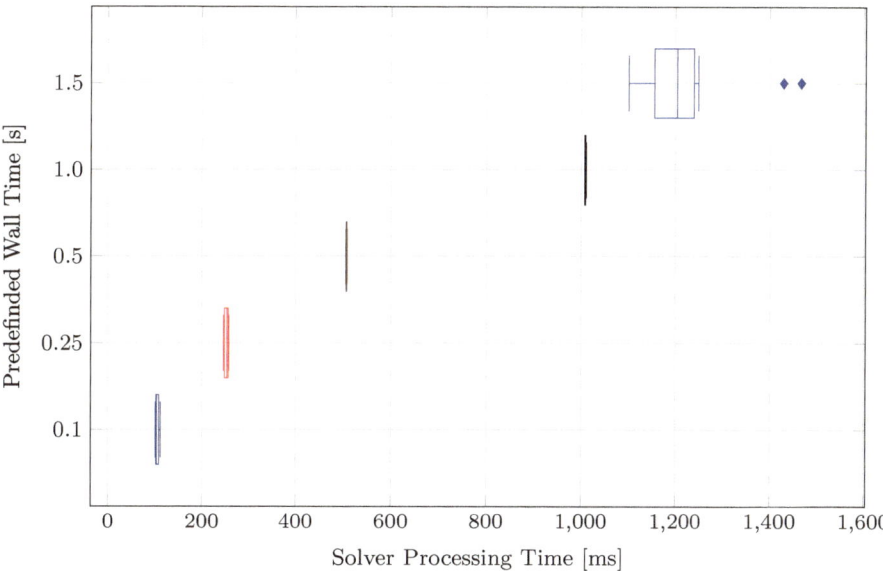

Fig. 9 Intersection scenario—solver time

stage with the algorithms available at this LTS level and increase the level. On the other hand, the solver is able to recognize optimal solutions as such and abort further optimization at an early stage. The term optimal here refers only to the given solution space based on the discretization used. For example, a finer discretization of the vehicle position would lead to an improved optimal solution.

A disadvantage of the used approach is the generally slower solution of complex constraint optimization problems with many variables. If in the given example the accuracy of the trajectory is increased from 1 m to 10 cm, the calculation time of the solution increases by a factor of 10-12 to around 1.2 s. The solver allows to set a wall time to reduce the calculation time. This represents the allowed calculation time. The best available solution at this time is used. Figure 9 shows the computation time of the algorithm for different given maximum computation times. For each calculation time, 10 runs were performed. It can be seen that the algorithm respects the specified maximal calculation time with a deviation of a few milliseconds. The lower value at a maximal calculation time of 1.5 s shows the automatic termination process, since on average an optimal solution is already found at 1.2 s.

The limitation of the calculation time has a significant influence on the reduction of the solution quality. Figure 10 shows that below 1.0 s in most cases no optimal solution can be found. Between 0.1 s and 1.0 s the algorithm finds sufficient solutions with constantly decreasing quality. The percentage increase in the duration of the cooperation maneuver relative to an optimal solution is shown in Fig. 11. With a limited calculation time of 0.1 s, the algorithm only finds a valid solution in 50 % of the cases. The measurements also show that the initial abandonment of an opti-

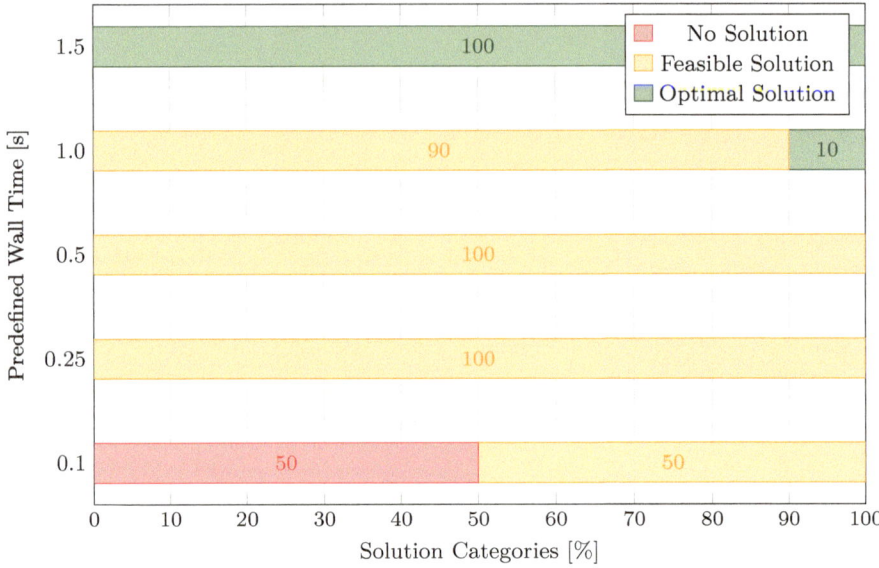

Fig. 10 Intersection scenario—solution categories

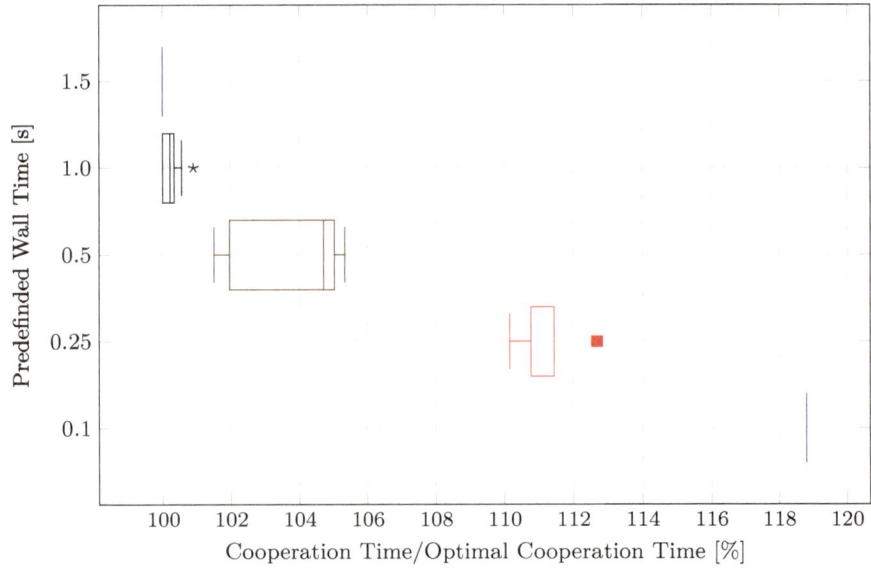

Fig. 11 Intersection scenario—cooperation time increase relative to optimal solution

mal solution brings significant speed advantages without a dramatic loss in solution quality and increase in maneuver duration.

2.4 Learning-Based Clustering

The architecture described in Fig. 6 successfully decouples the formation of Local Traffic Systems from the actual cooperation between the participating vehicles and the associated algorithm.

However, the limits for determining the LTS level are still statically defined by the developer. This static specification of the LTS parameters has several disadvantages. On the one hand, static optimization of the corresponding parameters is often suboptimal. The system is only adapted to a small set of possible conflict situations and traffic scenarios and is likely biased towards these scenarios used as test cases during the development. The administration and maintenance of a corresponding traffic scenario collection is time-consuming and often does not meet the requirement of completeness. On the other hand, there is no direct connection between the exchanged vehicle information shown in Fig. 5, the respective LTS Boundary Conditions and the capabilities of the underlying cooperation algorithm. However, changes to the underlying algorithms should logically also have an impact on the transmitted data as well as the LTS formation. Due to the disadvantages presented, a static parameter definition should be considered unsuitable for fully meeting the requirements of an LTS generation architecture described before.

A deep learning based approach offers a possible solution to the aforementioned problems. Here, the formation algorithm based on static parameters is replaced by a deep learning model. The model decides whether the LTS level should be increased or decreased, based on the exchanged vehicle data. The internal decision process is learned by the model during the training process based on a stimulative approach. The system can be trained in a simulation environment without managing a complex data set of conflict scenarios.

In this way, the model learns the link between the exchanged vehicle data and the underlying algorithms. The system learns not only the influence of a single parameter on the formation of the respective LTS level, but also the implicit relationships and similarities of individual traffic scenarios represented by the exchanged vehicle data. The detection of the traffic scenario takes place implicitly. If the underlying cooperation algorithms or the training scenarios are changed, the model can be trained again without additional changes.

However, a training process as described in Fig. 12 is unfortunately not applicable to the current problem. On the one hand, it is not possible to provide a static data set for training the deep learning model, since already after the first time step the LTS formation has an impact on the training environment surrounding the vehicle. Another central problem is caused by the time-delayed verifiability of the LTS formation for correctness. Common supervised/unsupervised deep learning approaches are based on presenting the model with input data based on a training data set. From this input data, the model generates the output data, which is then compared with a label. Labels are part of the training data set in the case of supervised learning and are generated from it in the case of unsupervised learning. The label is considered as the correct output of the model on the existing input data. The deviation from the

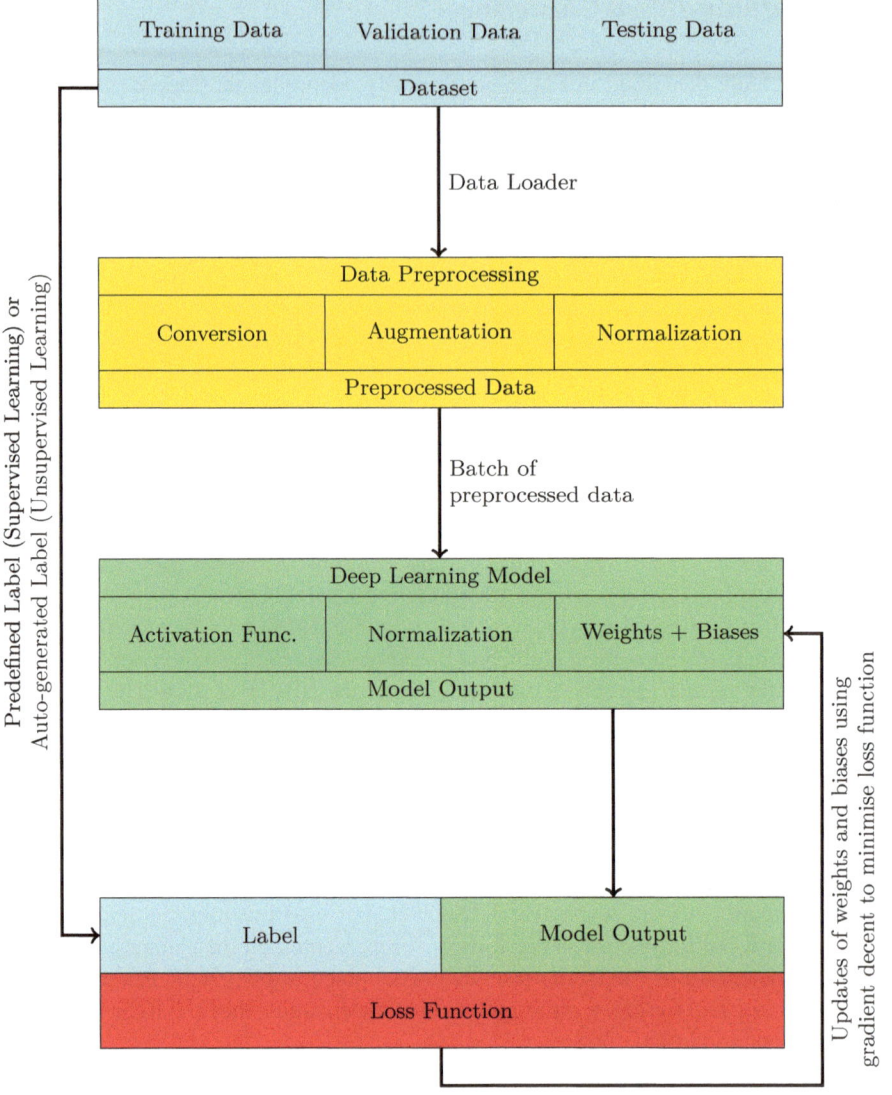

Fig. 12 Basic training architecture of supervised/unsupervised deep learning models

output of the model is represented by a loss function. The underlying parameters of the model are adjusted with the goal of minimizing the loss function. However, in the present case, such a label does not exist for a given set of input data. Whether an LTS formation was goal-directed becomes apparent only in the course of the executed cooperation maneuver several time steps after the actual LTS formation. Therefore, a reinforcement-based approach is used in the following.

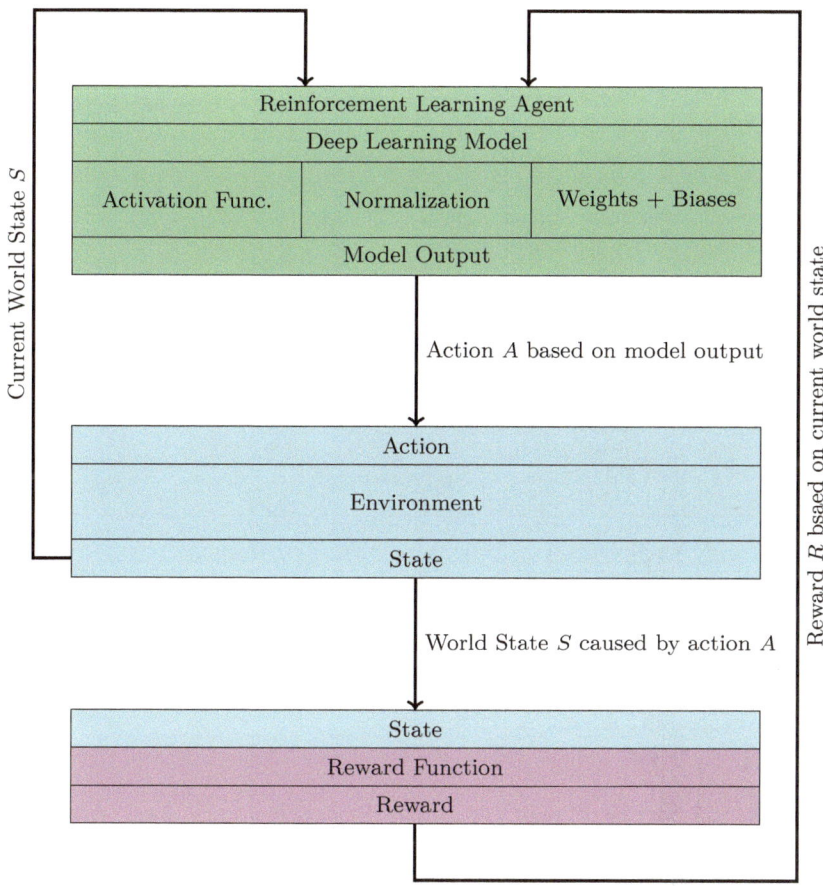

Fig. 13 Basic training architecture of reinforcement learning models

The diagram in Fig. 13 shows the general structure of a reinforcement learning algorithm. The algorithm consists of three main components. The reinforcement learning agent, the surrounding environment and a reward function. The reinforcement learning agent has the task to make optimal decisions based on the surrounding environment. The decision made by the agent at a time t is called action A_t. The action A_t is determined on the basis of the current environment. This is represented by the current state S_t. To evaluate the quality of a decision, the reward R_t is calculated by a reward function. Thus, the agent's goal is to maximize the total reward.

To transform the previous concept into a reinforcement-based approach, modifications to the architecture described in Fig. 6 are necessary. The changes are shown in Fig. 14.

A higher-level component RL-Agent is introduced. The previous algorithm based on static thresholds for determining whether a local traffic system is formed is removed from the LTS Manager. The LTS formation is made in the LTS Manager

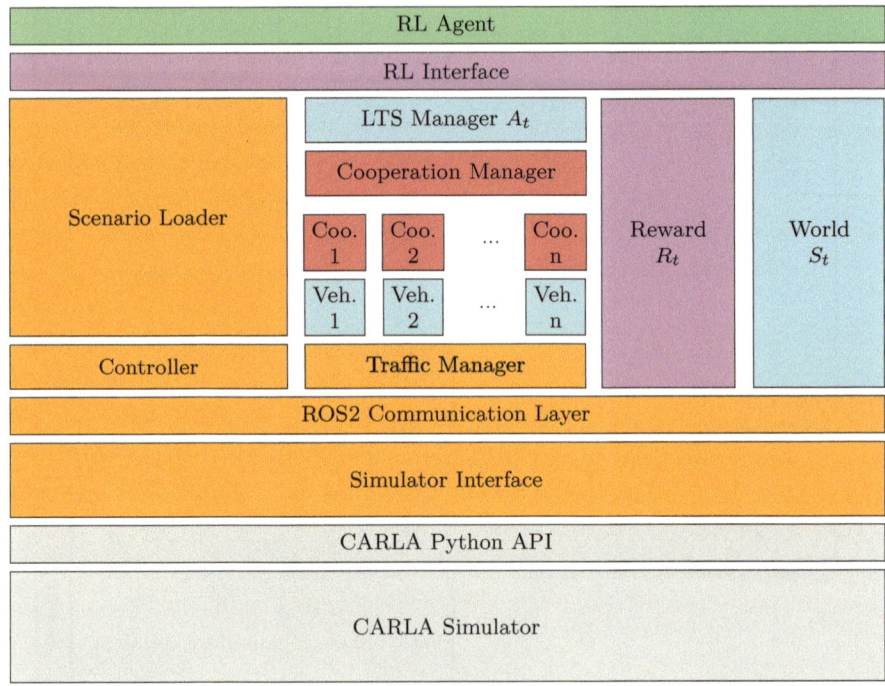

Fig. 14 Learning enabled Autoknigge architecture

on the basis of the action A_t by the RL agent. These actions are based on the current state S_t which is determined by the already existing World component. Furthermore, the previous pure data collection of the World component has been extended by a reward component to determine R_t. The required data S_t, R_t for the computation of A_t are provided by a reinforcement learning interface to the agent during training. The access to the Scenario Loader allows switching between different conflict scenarios during the training process.

2.5 Example Cooperation Intersection and Highway Access

The following section shows two example applications of the described concepts described and gives a visual impression about the cooperation result. The first scenario describes the conflict situation between two vehicles crossing an intersection. The second scenario describes the conflict situation at a highway access. The environment perception of the vehicles involved was completely deactivated. The only information exchanged is the data specified for LTS formation. Cooperative driving maneuvers are visualized within the simulator by a green line between the involved participants.

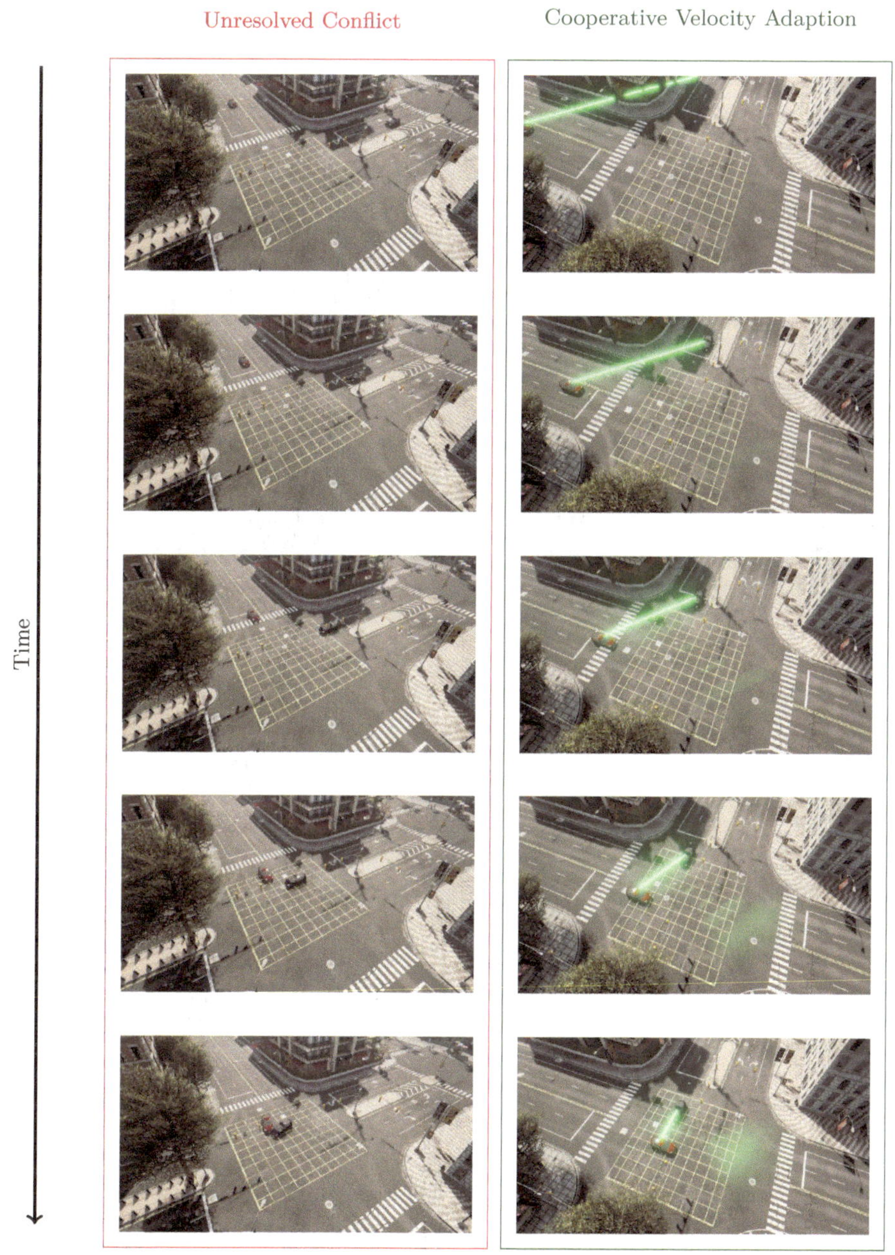
Unresolved Conflict
Cooperative Velocity Adaption
Time

Fig. 15 Example cooperation intersection

Unresolved Conflict   Cooperative Velocity Adaption

Time

Fig. 16 Example cooperation highway access

As visible in Figs. 15 and 16, a short time gap was deliberately chosen in order to test the system at its limits. Both conflict situations are solved successfully on the LTS Level 1 by early cooperative adjustment of the vehicle speed. The fact that the present different conflict scenarios can be successfully solved with the same cooperation maneuver supports the chosen approach of using simple cooperation maneuvers, similar to human behavior, to solve different conflict situations.

2.6 Conclusion and Outlook

The presented architecture fulfills the requirements placed on the system. The transmitted vehicle data, the LTS formation as well as the underlying cooperation algorithms are successfully separated without neglecting the retroactive influences of the driving situation and cooperation algorithm on the LTS formation. Successful separation avoids the black box behavior of an end-to-end trained machine learning architecture. The cooperation algorithms are exchangeable. The introduction of LTS levels allows for easy prioritization in case of conflicting goals. The vehicle data assigned to the individual levels and the quantity of transmitted data, which increases proportionally to the urgency, as well as the constantly increasing interference in the longitudinal and lateral guidance of the vehicle, both reduce the necessary quantity of data for simple cooperation maneuvers and increase driving comfort.

Although the current approach is promising, there is still a need for research in the field of LTS education. This can be found in three main areas. First—The cost function. In addition to simplified basics such as a traffic flow optimization function, this should take into account other factors such as the CO_2 emissions of vehicles. Second—The underlying cooperation algorithms and the exchanged data. Since the focus of this work is on the optimization of approaches to LTS formation, there is still a high need for research in this area. In particular, as standardization continues, changes in V2X message definitions are to be expected. In the long run, V2X communication should be realized by frameworks like Veins, Artery [20] instead of ROS2 messages. Third—Further consideration of single-agent and multi-agent concepts of the reinforcement learning approach. The current system uses a single agent that learns the LTS formation. A multi-agent system where each vehicle uses an individual agent could offer significant advantages as there is no need to ensure that all vehicles have the same agent. This offers advantages in simplifying the learning process or realizing vehicle individual optimizations relevant to specific user preferences. Whether such a system contributes at all to the minimization of a global cost function if each agent follows an individual optimization remains to be researched.

3 Verification of Cooperative Interacting Automobiles

3.1 Introduction

This section proposes an approach to use formal methods for verifying trajectories in our LTS framework. Our algorithm generates behavior patterns that guarantee collision-free and deadlock-free trajectories. In order to generate the behavior patterns, we use the model checker nuXmv [6], which is specialised in synchronous finite-state systems.

This section is structured as follows. We introduce our verification architecture in Sect. 3.2. Section 3.3 presents the offline part of our approach, i.e., our modeling and verification of traffic scenarios and the generation of rule sets. Section 3.4 introduces the implementation of our rule checker and Sect. 3.5 evaluates the verification and rule checker. Finally, Sect. 3.6 concludes this section.

3.2 Verification Architecture

Figure 17 shows our verification architecture from [30]. The verification works in an offline and an online part. The offline part consists of modeling and verification of traffic scenarios. The verification classifies the traffic scenarios as collision-free and deadlock-free or provides a counter example in case of possible collisions or deadlocks. We generalize the counter examples to traffic rules for networked and autonomous vehicles. The traffic rules are stored in a rule set. The online part is a rule checker, which uses the map and planned trajectories of the current driving situation and the rule set generated by the offline part as input. The rule checker checks if the trajectories comply with the traffic rules of the rule set. If no rule is violated, the trajectories are considered safe.

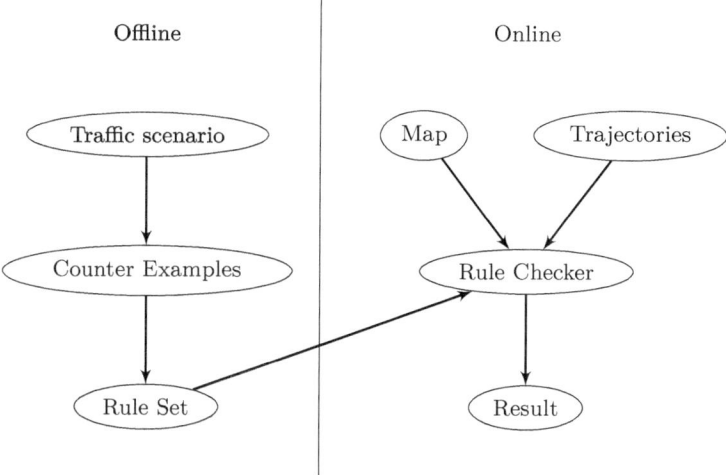

Fig. 17 Verification architecture of [30], consisting of an offline and an online part. Before deployment, counter examples of safety verification are generalized into rules that guarantee the absence of collisions and deadlocks. At run-time, a rule checker classifies trajectories of vehicles into safe and unsafe trajectories, depending if they follow the rules for the scenario

3.3 Rule Set Generation

We decompose the traffic scenario model into two parts: the map and vehicles' trajectories. Through this modular approach, it becomes easier to develop general purpose encodings for vehicles and maps independently of each other. We call the map the *static model* and we call the trajectory model *dynamic model*. We model both components time and spatial discrete. Section 3.3.1 and 3.3.2 summarize our modeling of [30, 46]. Section 3.3.3 introduces our extension to combined models of connected LTS. Section 3.3.4 presents our NuXmv encoding and Sect. 3.3.5 introduces the rule generation.

3.3.1 Roadway Model

The map consists of blocks and transitions. Each block represents a part of the physical road. Blocks are non-overlapping and identified by unique Identities (IDs). A discretization takes care of the vehicles' dynamics and safety distances. Each vehicle can occupy only one block at each time step. If a vehicle holds a block, the block is *occupied*, otherwise the block is *free*. To model valid transitions between blocks, each block has a list of successor-tuples.

Definition 1 (*Successor-Tuple* [30]) Successor-tuples are defined as

$$t_{suc} = (ID_{suc}, Cost, Watchlist, I), \tag{1}$$

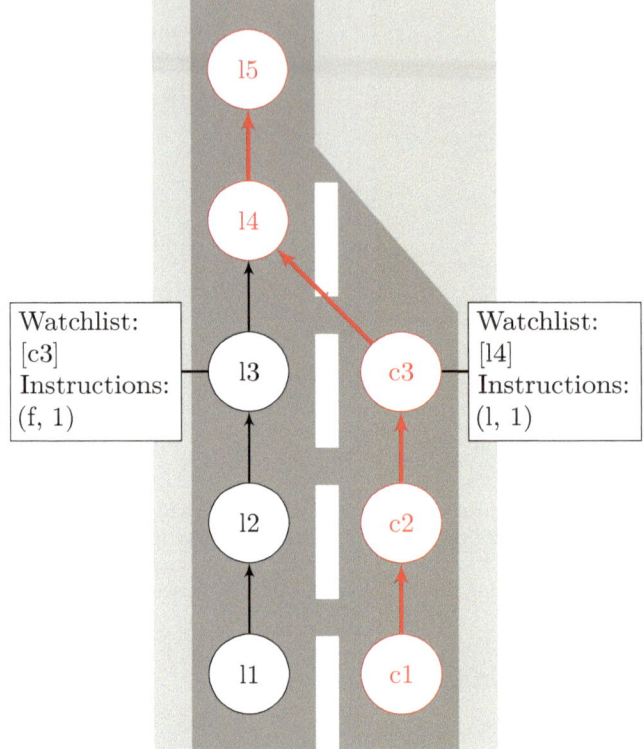

Fig. 18 Example model of a narrowing road, adapted from [30]

where:

- ID_{suc} denotes the ID of the successor block,
- $Cost$ stores the costs for the transition,
- $Watchlist$ is a list of block IDs. A vehicle can only use the transition if all blocks in its watchlist are free, and
- $I = (Type, Velocity) \in (String \times \mathbb{Z})$ is a scenario-dependent instruction. $Type$ describes which behavior is expected by the vehicle, e.g., "move forward", "turn right", and "switch to left lane".

Figure 18 shows an example model of a narrowing scenario. Only physically possible transitions respective to road boundaries and vehicle dynamics are included.

3.3.2 Trajectory Model

Trajectories consist of a sequence of adjacent blocks. The first block of a trajectory is the vehicle's current position. The last block represents the vehicle's destination.

Each time step, the vehicles transit to the next block in their trajectories. The same block may be used multiple times in a trajectory. Trajectories can have different lengths. After leaving the LTS, the vehicle moves into a final block with the ID n with no further process. One of the following two statements hold for any consecutive blocks in each trajectory:

1. The blocks have the same ID, i.e., the vehicle does not move.
2. There is a valid transition between the blocks in the direction of movement.

In traffic scenarios, some vehicles may be important for more than one LTS. We propose a method to create connected traffic scenarios. A connected traffic scenario combines two traffic scenarios with transitions from one traffic scenario to the other traffic scenario. Using these transitions, vehicles can travel between both traffic scenarios. In connected traffic scenarios, different rules may apply in comparison to separate traffic scenarios. Our approach extends the methods from our previous works done in [30, 46] by connecting traffic scenarios and generating rules for connected traffic scenarios. We classify pairs of traffic scenarios into overlapping traffic scenarios and non-overlapping traffic scenarios. Overlapping Traffic scenarios are scenarios where both single scenarios have entrance blocks, which have the same ID. In Fig. 19 two single traffic scenarios are sketched. Both have blocks with the same IDs.

3.3.3 Connected LTS

In order to verify connected LTS, this subsection extends the modeling of Sects. 3.3.1 and 3.3.2. We start with an example of collision-free and deadlock-free single LTS, while the combination of both LTS is collision-free but not deadlock-free.

Motivating Example

In the following, we give an example of rule sets generated for single traffic scenarios that do not provide deadlock-freeness in overlapping LTS. The rules were generated by our method in [30]. In this example, we use two overlapping intersections, both with 4 entrances. Each single intersection has only one rule. This rule does not allow vehicles in the center to drive in 4 different directions. In Fig. 20 we can see an initial configuration. Each center of the model is filled with vehicles and all cars try to reach the end of the opposite center. Using this configuration, we were able to show that this rule is not sufficient to avoid deadlocks. As seen in Fig. 21, this configuration leads to a deadlock in both centers of the intersections. In the upper part, both vehicles on position $c0$ and $d3$ try to move to $c1$. Since only one vehicle may occupy block $c1$, a deadlock is caused. The same holds for position $c2$ in the bottom part, which is blocked by vehicles at position $c3$ and $u0$. These two situations cause a deadlock, since every vehicle tries to take the entrance to get to the opposite center and block one another. The entrances are blocked by the vehicles on block $d2$ in the upper part and $u1$ in the bottom intersection. Both vehicles cannot make any progress. This example shows that rules that apply for a single scenario must

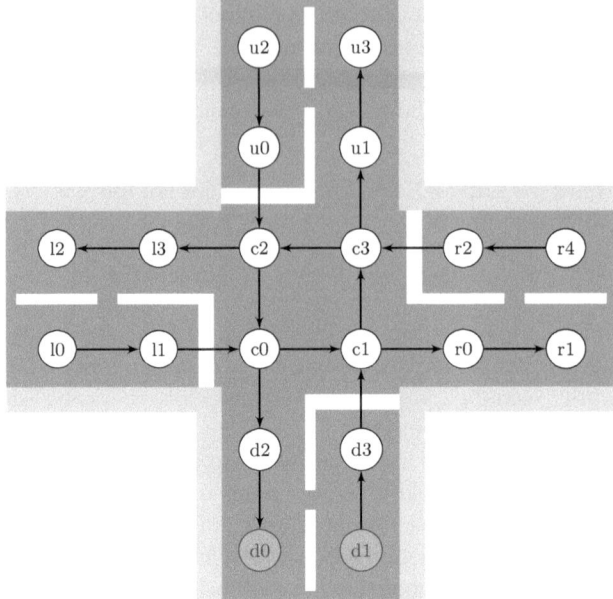

(a) Model 1, where the bottom entrance is overlapping with
model 2

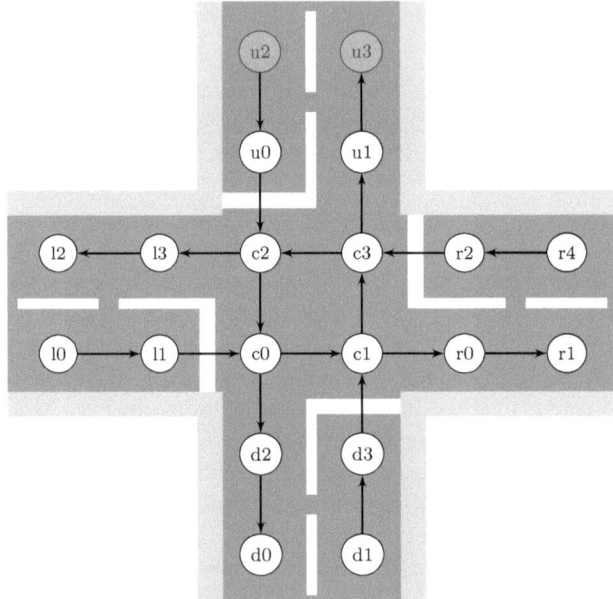

(b) Model 2, where the upper entrance is overlapping with
model 1

Fig. 19 Two single traffic scenarios, here crossroads, with overlapping borders in blue

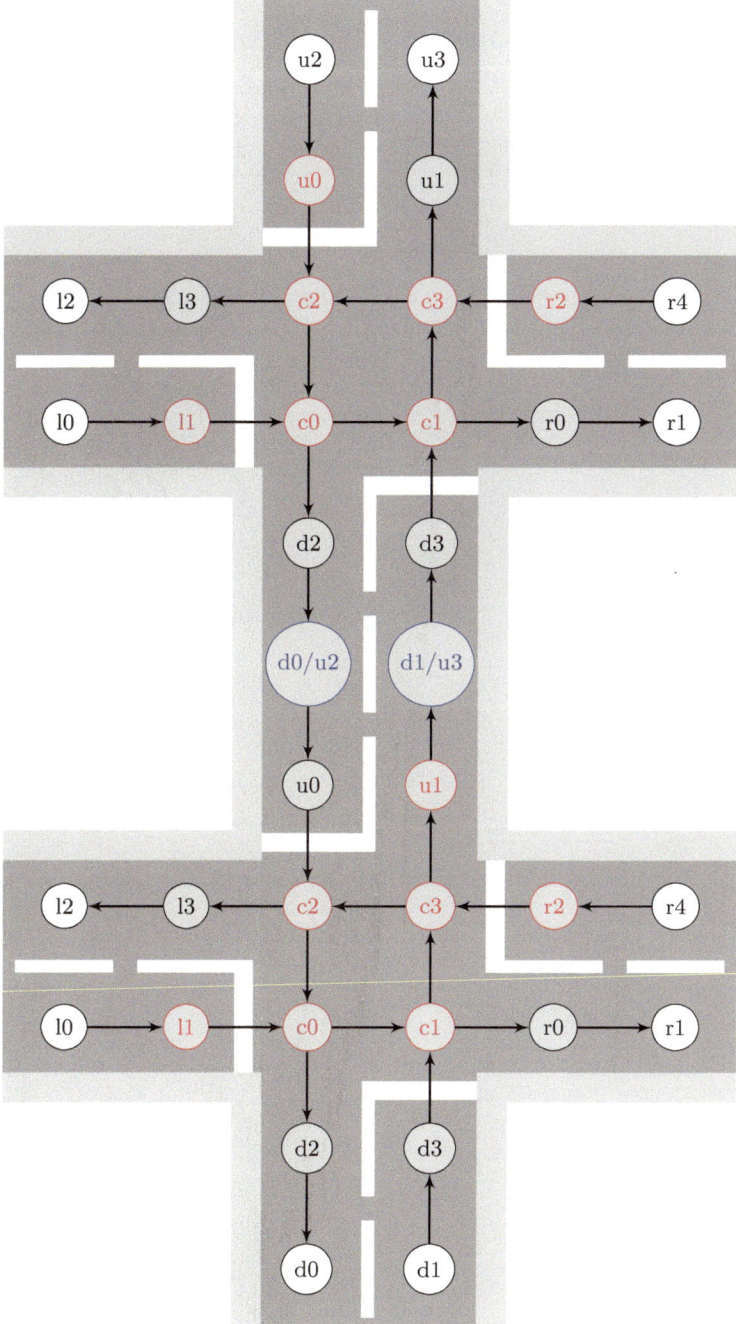

Fig. 20 Starting positions of vehicles for deadlock scenario in nuXmv

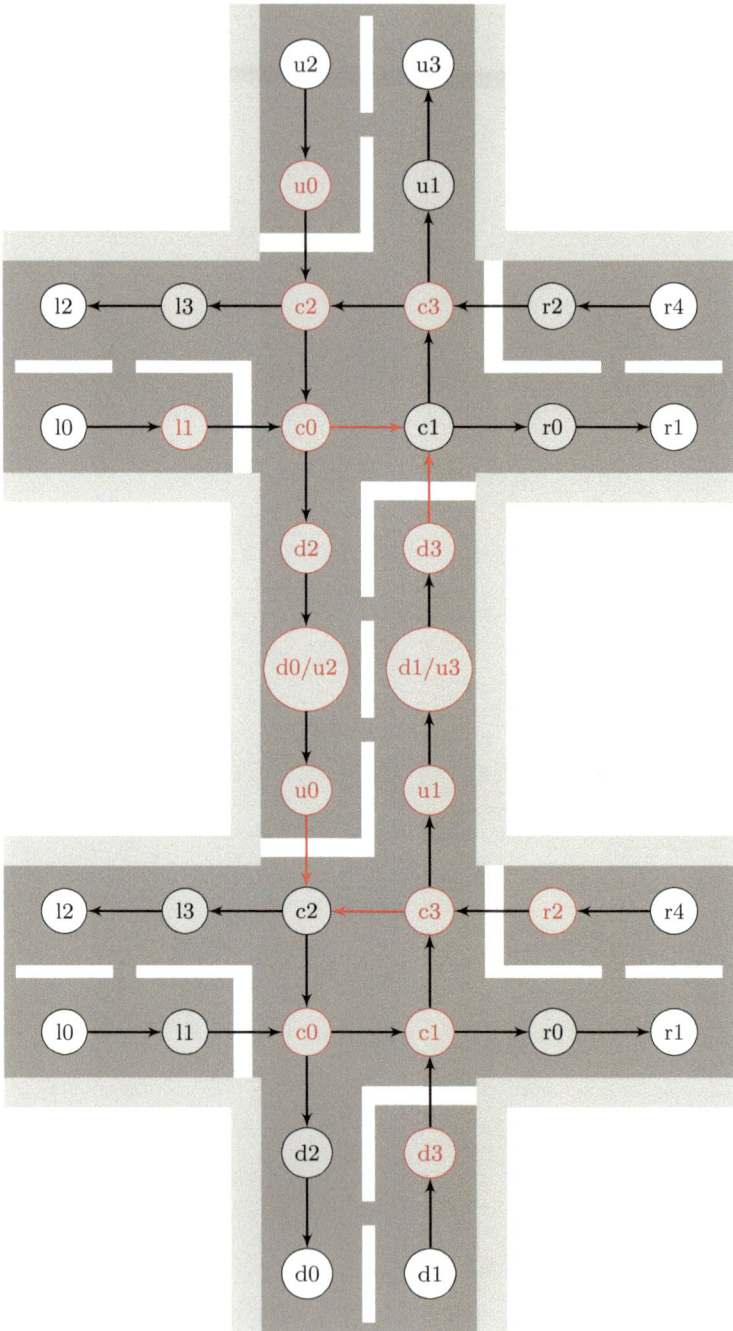

Fig. 21 Deadlock situation for a connected traffic scenario

not be enough to guarantee safety properties also in the connected traffic scenario. Therefore we generated rule sets for overlapping LTS.

The rest of this subsection introduces our model of connected LTS in order to guarantee collision-freeness and deadlock-freeness in connected LTS.

Border Blocks

Each entrance consists of multiple blocks. We divide these blocks in three categories:

1. blocks which border to a center block c1, c2, c3 or c4, e.g., the blocks l1 and l3 in Fig. 19a,
2. blocks which leave or enter the traffic scenarios, e.g., l2 and in Fig. 19a, and
3. all other blocks that are forming the middle of each lane.

In the following, we will call blocks of category 2 border blocks. Border blocks are critical states, because taking only single traffic scenarios into consideration when verifying for safety properties, e.g. collisions, everything that happens outside the current traffic scenario is not considered. For example, it could be that a vehicle's position is currently a border block which is the exit of the traffic scenario. Not considering the next traffic scenario after the border block could lead to deadlocks or collisions. Verifying only separate traffic scenarios could lead to false negative classification.

Transition States

There may be traffic scenarios that are not overlapping. In this case, no border blocks lead from one traffic scenario into another traffic scenario. We model connection points between traffic scenarios. Possible connection points are two border blocks, each belonging to the other traffic scenario. These two border blocks form a *connection pair*.

We use lane information to identify connection pairs of multiple traffic scenarios. In order to be connectable, the traffic scenarios require the same number of lanes. If two traffic scenarios have the same number of lanes, we use the lane positions to identify the border blocks of both traffic scenarios that form a connection pair. We introduce an argument *TransitionStates* to model connection pairs in nuXmv. The argument TransitionStates contains the following information:

- the number of border blocks contained in the input traffic scenario,
- the successor blocks of each border block,
- the corresponding lane to which a block belongs,
- the number of lanes existing in the input traffic scenario, and
- each lane's position compared to the other ones in the same entrance.

The argument TransitionStates is a list that contains all possible blocks for a connection pair. Each element of the list represents an entrance or exit of the traffic system.

Definition 2 (*Entrance*) An entrance consists of multiple blocks that form a group of pairing border blocks. In Fig. 19a, the pairs (l0, l2), (d0, d1), (r1, r4), (u2, u3) form four entrances.

Our method checks for opposing border blocks and create connection pairs. Then, a transition is created between the corresponding border blocks. The transition starts at the border block which is an exit block of its traffic scenario and is connected to the corresponding input block from the connection pair.

An example combined model is the model in Fig. 20. The combination of multiple LTS increases the size of the scenarios to be verified. Since large models cause performance issues during verification, we reduce the combined model, while maintaining the correctness of verification.

Model Reduction

We reduce the models of combined LTS to keep computational efficiency of the offline verification. To this end, we reduce the number of blocks in the resulting model. We include the center blocks of both single LTS models and all blocks connecting the center blocks. Each center block that lead to an exit state becomes an exit state, while each center block connected to an entrance becomes an entry block. The gray states in Fig. 20 are the states included in the reduced model of connected LTS.

3.3.4 NuXmv Encoding

Based on our work in [46], we translate our models into the nuXmv input language. NuXmv distinguishes four input types: variables, transitions, dictionaries, and specifications.

Variables

We model vehicles as variables in nuXmv. The possible states of each variable are the blocks the corresponding vehicle will occupy in the scenario. In the example shown in Fig. 18, Vehicle v_0 has the following path: c0 - c1 - c2 - c3 - l4 - l5.

In nuXmv, the path is represented as follows:

VAR
    v_0 : {c0, c1, c2, c3, l4, l5, n};

The initial state of each vehicle is the first block of its trajectory, e.g., vehicle v_0 in Fig. 18 starts at block c0. The corresponding nuXmv code is

INIT
    v_0 = c0.

Transitions To encode transitions, we use the *case* statement for every pair of consecutive blocks in a vehicle's trajectory (B_i, B_{i+1}). We use the following two statements:

$$(Pos = B_i)\&(\phi) : B_{i+1} \tag{2}$$

$$(Pos = B_i)\&(\neg\phi) : B_i, \tag{3}$$

where Pos denotes the current block and ϕ is a Boolean expression. ϕ evaluates to true, if the transition is safe to use, i.e., if all blocks in the watchlist of this transition are free. Equation (2) allows the vehicle to move to its next block B_{i+1}, if the transition is safe. Equation (3) forces the vehicle to remain in its current block, if the transition is not safe. The block c0 in Fig. 18 is given as

c0: [(c1, 1, [c1], (f,1))],

where $ID_{suc} = c1$, $cost = 1$, $Watchlist = [c1]$, and $I = (f, 1)$. The instruction I states that the vehicle moves forward one block.

Suppose an example vehicle v_1 that moves from c0 to c1 in the scenario in Fig. 18. If there is another vehicle in the LTS with ID v_0, the nuXmv statements for this transition are as follows:

$((v_1) = (c0))$ & $((v_0)$!= $(c1))$
                                            : c1;
$((v_1) = (c0))$ & $(!((v_0)$!= $(c1)))$
                                            : c0;

The first statement states that if vehicle v_1 is on block c0 and vehicle v_0 is not on block c1, v_1 moves to c1. The second statement states that if vehicle v_0 is on block c1, vehicle v_1 remains on block c0.

This is done for every pair of consecutive blocks in the vehicle's trajectory. Once a vehicle reaches the last block of its trajectory, it moves to block n and stays there through the following equations:

$$Pos = B_{end} : n \tag{4}$$

$$Pos = n : n \tag{5}$$

Equation 4 causes all vehicles to move to block n after their trajectory ended. Equation 5 states that vehicles that reached block n will remain there.

Dictionaries

We use dictionaries to represent LTS entrances. Each entrance is represented by two dictionaries, because there are always at least two lanes per entrance, one exit and one entrance into the traffic scenario. Each dictionary can have multiple border blocks. The number of elements in this dictionary represents the number of lanes of the corresponding entrance's exit or entry and the position of a block in this dictionary represents its corresponding lane, to which the block belongs. An example TransitionStates argument for the intersection of Fig. 19a looks like the following:

```
[
# border blocks of upper entrance
  {'u2': [('u0', 1, ['u0'], ('f', 1))]}
, {'u3': [('u3', 1, ['u3'], ('s', 0))]}
# border blocks of right entrance
, {'r4': [('r2', 1, ['r2'], ('f', 1))]}
, {'r1': [('r1', 1, ['r1'], ('s', 0))]}
# border blocks of bottom entrace
, {'d1': [('d3', 1, ['d3'], ('f', 1))]}
, {'d0': [('d0', 1, ['d0'], ('s', 0))]}
# border blocks of left entrance
, {'l0': [('l1', 1, ['l1'], ('f', 1))]}
, {'l2': [('l2', 1, ['l2'], ('s', 0))]}
].
```

The first two elements represent the border blocks of the upper entrance, the next two the border blocks of the right entrance, the next two the border blocks of the bottom entrance, and the last two for the border blocks of the left entrance. Each block that is an element of an entrance lane stores the information about its successor block.

Specifications

We verify the safety of our traffic system. To this end, we formulate *specifications* by *invariants* and *temporal logic*. We give more details on the specifications in [30].

We use invariants to check for collision-freeness. A collision occurs, if multiple vehicles occupy the same block at the same time. The invariants to check collision-avoidance are

$$(pos_1 \neq n) \Rightarrow (pos_1 \neq pos_2), \tag{6}$$

where pos_i is the position of vehicle i. Equation (6) models that two vehicles 1 and 2 do not occupy the same block, unless vehicle 1 finished its trajectory and moved to the end block n. We check this invariant for each pair of vehicles at each time step.

We use temporal logic to check deadlock-freeness. We use Linear Temporal Logic (LTL) [43]. In LTL, we model deadlock-freeness as

$$F(pos_1 = n \wedge pos_2 = n \wedge \ldots), \tag{7}$$

where $F(\cdot)$ denotes the *eventually* operator of LTL. Equation (7) models that each vehicle eventually reaches block n, i.e., finished its trajectory.

3.3.5   Summarize Rules

We alter the static and dynamic models to create different verification scenarios. NuXmv provides counter examples if a verification scenario is not collision-free and

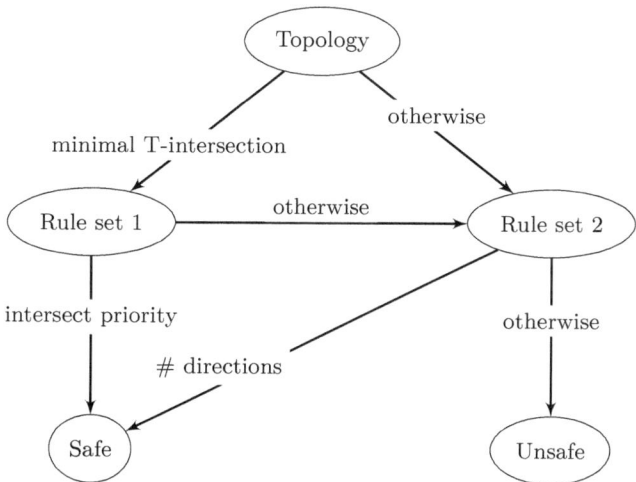

Fig. 22 Evaluations of rule sets in an intersection scenario. Different rule sets apply, depending on the intersection topology

deadlock-free. We generalize counter examples derived from the same group of static models to generate rules for this group. The rule sets formulated for each scenario group prevent any collisions or deadlocks found during verification. Depending on the static and dynamic model, different rule sets have to be applied.

We demonstrate the generalized rule sets for intersection models. In the following, "inner lanes" refer to the leftmost lane of each direction and "center" refers to the area of the intersection, where the lanes intersect. In the intersection model, the rule set depends on the intersection topology and the priority rule, i.e., the right of way, applied in the trajectories. Figure 22 gives an overview of the rule set selection process. If the map represents a T-intersection with only one lane in each direction, it is called a *minimal* T-intersection. For minimal T-intersections, we need to check the priority rule for vehicles, denoted as Rule set 1. If the vehicles in the intersection consistently have priority over vehicles outside, the rules of Rule set 1 are met and the trajectories are always safe to execute. In all other cases, Rule set 2 is applied. In Rule set 2, trajectories are considered safe if the center never has vehicles traveling in four different directions, denoted by the red arrows in Fig. 23.

3.4 Rule Checker

The rule checker takes the static and dynamic model, i.e., the map and trajectory data, as input. The output of the rule checker is the classification of the trajectories according to the rule sets generated in Sect. 3.3. The rule checker classifies the trajectories into *safe* and *unsafe* trajectories. The rule checker detects the vehicle's

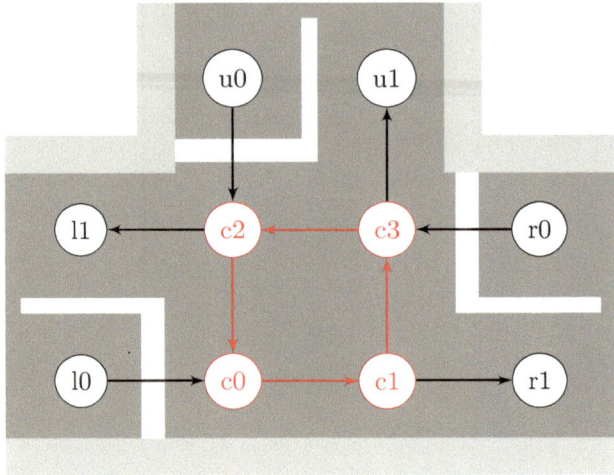

Fig. 23 Minimal T-intersection deadlock example

behavior, e.g., the applied priority rules. Different scenarios have different rule sets. We form groups of scenarios with similar rule sets. We demonstrate the idea using the intersection example in Fig. 19a.

One behavior that need to be identified is if the vehicles inside or outside the intersection have priority at the entrances. In the intersection model shown in Fig. 19a, the blocks c1 and r2 are a pair of interest, since vehicles on both blocks are able to move into c3. The rule checker checks for all pairs of interest if the following two conditions are satisfied at each time-step:

- both blocks in the pair of interest are occupied, and
- the vehicle in the intersection does not leave the intersection.

If both conditions are satisfied, the rule checker checks the next instructions of the vehicles. If the vehicle in the intersection is the only one instructed to move forward, then we have a case where vehicles in the intersection have priority. If the vehicle that tries to enter the intersection is the only one instructed to move forward, then we have a case where vehicles entering the intersection have priority. If none of the mentioned possibilities happened, then we cannot decide what has happened and conclude that there is no consistent priority rule between them. There are three possible cases after the rule checker iterated over each pair of interest:

- vehicles in the intersection consistently have priority over vehicles outside of the intersection,
- vehicles that enter the intersection consistently have priority over vehicles in the intersection, and
- there are no consistent priority rules.

We formulate rules for all three cases.

3.5 Evaluation

We evaluate the feasibility of our verification process for single LTS in Sect. 3.5.1 and for connected LTS in Sect. 3.5.2. Moreover, we evaluate the computation time of the offline verification and online rule checker in Sect. 3.5.3.

3.5.1 Feasibility in Single LTS

This section presents evaluation results of the verification process for single LTS. We evaluate the generation of rule sets and the performance of the rule checker. We divide combinations of roadways and trajectories into different classes. We evaluate scenarios of multiple classes.

We define classes of roadways according to their generated rule sets. Within the same model class, different rules need to be applied depending on the road topology and vehicles' trajectories. For example of an intersection model, the rules to apply depend on the number of entries of the intersection and the trajectories' priority rules. As such, a T-intersection has 3 potential classes:

1. The model is a T-intersection with one lane in each direction, vehicles in the intersection area always have priority over vehicles that are outside of the intersection.
2. The model is a T-intersection with one lane in each direction, all vehicles give priority to vehicles on the right.
3. The model is not a T-intersection with one lane in each direction.

We evaluate our rule checker on a four-way intersection. Table 1 presents the input trajectories for vehicles v_1, v_2, v_3, and v_4 and compares the expected and actual rule checker results. Figure 24 visualizes the first example of Table 1. The roadway is a minimal T-intersection and the vehicles' trajectories give priority to vehicles in the intersection. Please note that the lower entrance (the blocks d0 and d1) are not included in the T-intersection model. The rule checker gives the expected results in all cases. It classifies collision-free and deadlock-free scenarios as safe and unsafe otherwise. Nevertheless, the rule checker may classify collision-free and deadlock-free scenarios as unsafe. Figure 25 shows such a false positive result. The rule checker rejects these trajectories since vehicles in the center are traveling in all four directions on a non-minimal T-intersection. However, the rule checker will not classify unsafe scenarios as safe.

3.5.2 Feasibility in Overlapping LTS

We extend the rule sets for single LTS to guarantee collision-free and deadlock-free trajectories also in connected LTS. As an example we present two new rules for the intersection scenario:

Table 1 Evaluation scenarios for the intersection model. We show the trajectories of four vehicles in each scenario, the expected result and the actual result of the rule checker. The trajectories of the first evaluation are visualized in Fig. 24, represented by the corresponding color

Minimal T-intersection with priority of vehicles in the intersection

Input trajectory	Expected	Result
v_1: c2 - c0 - d1	Safe	Safe
v_2: l0 - l0 - c0 - c1 - r1		
v_3: d0 - c1 - c3 - c2		
v_4: r0 - c3 - c2 - l1		
v_1: c2 - c0 - d1	Safe	Safe
v_2: c0 - c1 - r1		
v_3: c1 - c3 - c2 - l1		
v_4: c3 - c2 - l2		

Any other intersection model

Input trajectory	Expected	Result
v_1: c2 - c2 - c0 - d1	Unsafe	Unsafe
v_2: l0 - c0 - c1 - r1		
v_3: d0 - c1 - c3 - c2		
v_4: r0 - c3 - c2 - l1		
v_1: c2 - c2 - c0 - d1	Safe	Safe
v_2: l0 - c0 - d11		
v_3: d0 - c1 - c3 - c2		
v_4: r0 - c3 - c2 - l1		
v_1: r0 - c3 - u1	Safe	Safe
v_2: u0 - c2 - c0 - d1		
v_3: l0 - c0 - c1 - r1		
v_4: d0 - c1 - c3 - u1		
v_1: r0 - c3 - c2 - l1	Unsafe	Unsafe
v_2: u0 - c2 - c0 - d1		
v_3: l0 - c0 - c1 - r1		
v_4: d0 - c1 - c3 - u1		

- Vehicles entering the center must be able to exit the center. We refer to this rule as *exit free*.
- Vehicles may not leave the center in the same entrance, which was used to enter the center. We will call this rule *entry and exit differ*.

All developed models have been checked for correctness. To verify the correctness of the scenario 2-intersection, we generate trajectories. Since the 2-intersection model is a connection of two single crossroad scenarios, we only generated trajectories that are valid for single intersection models. Tables 2 and 3 show the test cases for the newly generated rules *exit free* and *entry and exit differ*. For each rule, we show an expected positive classification and an expected negative classification. The results

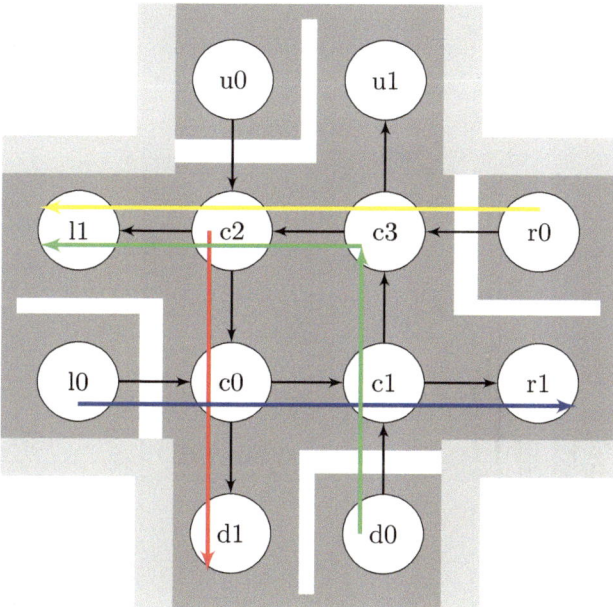

Fig. 24 Visualization of Table 1

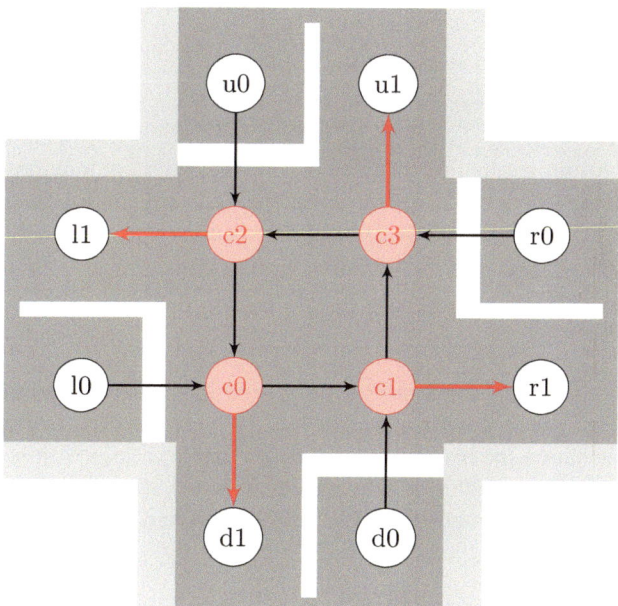

Fig. 25 False-positive result

Table 2 Rule *entry and exit differ* test cases for traffic scenario 2-intersection

Input trajectories						Expected	Results
u2	u0	c2	c0	d2	d0	Safe	Safe
l1	c0	c1	c3	u1	u3		
d3	c1	c3	c2	l3	l2		
r4	r4	r3	c3	u1	u3		
u0	c2	c0	c1	c3	u1	Unsafe	Unsafe
l1	c0	c1	c3	u1	u3		
d3	c1	c3	c2	l3	l2		
r4	r4	r3	c3	u1	u3		

Table 3 Rule *exit free* test cases for connected traffic scenario 2-intersection

Input trajectories								Expected	Results
l0	l1	c0	c1	c3	u1	u3	*n*	Safe	Safe
d1	d3	c1	c3	c2	l2	l4	*n*		
r4	r2	c3	u1	u3	*n*	*n*	*n*		
l0	l1	l1	c0	c1	c3	u1	u3	Unsafe	Unsafe
d1	d3	c1	c3	c2	l2	l4	*n*		
r2	c3	u1	u3	*n*	*n*	*n*	*n*		

of the rule checker were as expected. Both rules are also valid in single intersection scenarios.

3.5.3 Computation Time

As extension to our evaluation in [30], we evaluate the computation time of the rule generation for connected LTS. We measured the computation times on a laptop running nuXmv 2.0.0 on Windows 10 using a processor with 2x 3.20GHz and 8 GB RAM.

Offline Computations

We present the results for the 2-intersection scenario. Each intersection model consists of 4 entries, each consisting of one lane. We execute 100 runs per measurement. Figure 26 shows the results. The execution time increases exponentially for an increasing number of vehicles. For 6 vehicles, the execution time is less than 15 s. For more vehicles, the execution time increases to around 1.7 min for 12 vehicles.

Online Performance

Figure 27 shows the execution time of one run of the rule checker. The execution times are average values of 100 runs to reduce measurement inaccuracies. For up to

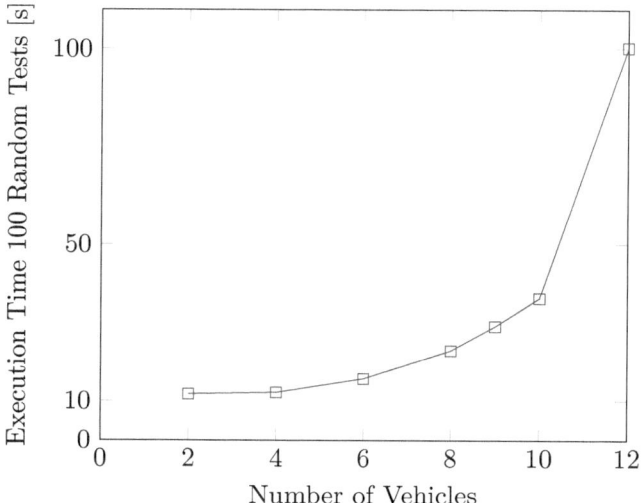

Fig. 26 Execution time of counter example generator for model of length 1

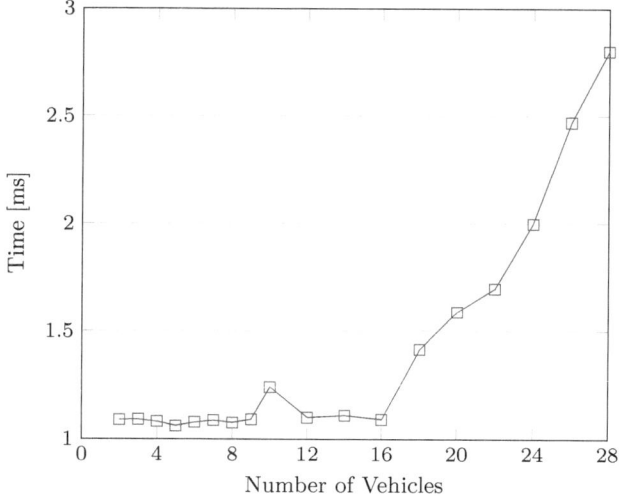

Fig. 27 Execution time of rule checker

16 vehicles, the execution time is almost constant and below 1.5 ms. For more than 17 vehicles, the execution time increases to around 2.8 ms for 28 vehicles.

3.6   Conclusion

This verification method is able to verify collision-freeness and deadlock-freeness of trajectories in one or more LTS. We summarized the space and time discrete model of [46] and the verification architecture of [30]. Our verification architecture generalizes counterexamples of the offline verification to generate rule sets. The rule sets restrict the solution space of valid trajectories so that all trajectories that fulfill the corresponding rule set are collision-free and deadlock-free. This work considered the verification of connected LTS. Evaluation results show that this method is real-time capable even for scenarios with a high number of vehicles. Further research may include evaluations on more complex LTS with more than one overlap.

4   Modeling Dynamic Systems

4.1   Why Modeling?

Vehicles of all SAE levels are safety critical systems and hence, their development needs to comply with legal regulations and safety standards. For instance, the functional safety standard ISO26262 highly recommends the usage of formal and semi-formal notations, hierarchical components of restricted size, the usage of strong type systems, range and plausibility checks, as well as the avoidance of hidden data-flows. In this section we are going to discuss the EmbeddedMontiArc (EMA) language family, a model-driven design approach for dynamic cyber-physical systems such as cooperative vehicles based on the component-and-connector (C&C) principle [37, 38]. The C&C paradigm views a software system as a composition of hierarchically organized components communicating with each other over connectors. The approach can help the development team to enforce the design principles required by ISO26262 by providing a domain-oriented syntax, a strong type system, verification mechanisms and a code generation toolchain.

4.2   The EMA Data Type System

Type systems are an important error avoidance mechanism of many programming languages. Strong typing is highly recommended by the ISO26262 for the development of automotive software. While most type systems are based on technical types such as integers, floats, and doubles, we are going to show how more abstract type systems can support modeling of cyber-physical systems on a more domain-oriented level. The type system of EMA is based on primitive types, which can be refined or grouped together, enabling the developer to create new types tailored to the application. The primitive types are abstract in the sense that they are not bound to a specific realization or standard such as IEEE754 [25]. Instead, EMA types resemble mathematical sets they aim to represent. EMA supports the following basic types: N represents the set of positive integers including 0, i.e. $\mathbb{N}$, N1 represents the set of positive integers not including 0, i.e. $\mathbb{N} \setminus \{0\}$, Z represents the set of signed integers $\mathbb{Z}$, Q represents the set of signed rational numbers $\mathbb{Q}$, C represents the set of Gaussian rationals $\mathbb{Q}[j] = \{z \in \mathbb{C} : z = a + jb : a, b \in \mathbb{Q}\} \subset \mathbb{C}$, B represents the set of Booleans (true and false). For the sake of convenience the alias Boolean can be used interchangeably.

The types N1, N, Z, Q, and C form a directed compatibility relation, where a type is compatible with another type if the latter can represent all the elements of the former. For instance, N is compatible with Z, Q, and C, but not with N1, since the latter does not include zero. A variable of type N can hence be assigned to variables of types Z, Q, and C, but not to variables of type N1. Note that these types represent infinite sets of numbers. Since no technical system can represent arbitrarily large numbers, using primitive EMA types leads to a model that can only be implemented partially by definition. Obviously, this does not hold for Booleans (B). The decision how to implement such types is delegated to the compiler and can depend on the application.

Technical systems are generally bounded, e.g. a vehicle has a maximum velocity, a minimum turning radius, etc. To model such bounds explicitly, EMA types can be refined by ranges consisting of a lower and an upper bound. A bounded type is defined as T (minValue : maxValue), where T can be any primitive type except B. The bounded type covers a subset of the primitive type T bounded by minValue and maxValue. minValue and maxValue must be of type T themselves and their values are included in the bounded type. For instance, the bounded type N(5:7) represents the set $\{5, 6, 7\}$. A type can be defined as half-open using the infinity operator oo as one of the bounds. For instance, N(5:oo) is a type covering all integers in $\{n \in \mathbb{N} | n \geq 5\}$.

Bounded types are not completely implementable if the base type is Q or C, as a technical system cannot handle arbitrarily high resolutions. To obtain a completely realizable type, a bounded type needs to be refined by a resolution or step size. This parameter is written between the minimum and maximum value of a bounded type, i.e. T (minValue : resolution : maxValue). The refined type only contains values of the form minValue+k×resolution satisfying minValue $\leq$ minValue+k×resolution $\leq$ maxValue, where $k \in \mathbb{N}$. For instance, the type Q(5:0.5:6.5) represents the set $\{5.0, 5.5, 6.0, 6.5\}$ Similarly to the lower and the upper bounds, the step size needs to be of the basic type it is restricting.

Different levels of type refinements can be employed in different phases of a systems engineering process such as the specification method for requirements, design, and test (SMArDT) [11, 22] during the development of a cyber-physical system (CPS).

In complex technical systems, data is often multidimensional. For this reason, primitive types of EMA can be organized as one-, two- or multidimensional arrays. The syntax to do so is based on the LATEX syntax for raising a base to a power. To specify the dimensionality of an array type, we need to append a circumflex followed by a list of comma-separated integer-valued dimension sizes in curly brackets to the primitive type's name: `T^{a,b,...}`. Each argument initializes the size of the respective array dimension. For instance, `Q^{5}` represents the set of all five-dimensional rational vectors $\mathbb{Q}^5$, `Z^{2,3}` represents the set of all integer-valued 2×3 matrices, and so on. We refer to one-dimensional arrays as vectors, to two-dimensional arrays as matrices, to three-dimensional arrays as cubes, and to multidimensional arrays as (*n*-dimensional) hypercubes. The base type of an array can also be a bounded type. For instance, the type `N(0:255)^{3,w,h}`, is often used to represent images with three channels, a size of w×h, and a color depth of 8 bit. In contrast to dynamic types systems as used by MATLAB or Python, dimensions are set at compile-time and cannot be changed at runtime. Variables of the aforementioned matrix type `Z^{2,3}` can only be assigned 2×3 matrices.

In EMA, a data type can be refined by the SI unit of the physical quantity it represents. For instance, `Q(0m:1dm:1km)` is a rational variable representing a length between 0 m and 1 km with a resolution of 1 dm. If the type has no range, only the unit is given in brackets. For instance, `Q(m)` denotes the rangeless rational number type to be interpreted as meters. Two EMA variables are only compatible if they represent the same physical quantity. Conversions are carried out automatically in assignments featuring compatible but different units. This way, the developer does not need to keep track of the physical quantities of the variables used in a program, nor does she have to carry out the conversions of units manually. EMA supports all SI units as well as common prefixes.

4.3   Components, Ports, and Connectors

In EMA components are first-level citizens. A component type is defined using the keyword `component` followed by a name which can later be used to create instances of this component type.[1] For instance, we declare the component type `Main` in L.1 of Fig. 28. Optionally, a component type declaration can include a list of generic parameters in angle brackets and another list of component parameters in round brackets. While generic parameters are allowed to change a component's interface, component parameters can only be used to parameterize a component's implementation. Depending on the use case, a generic parameter can be set to a component type, a data type, or a concrete value.

The syntax for declaring a generic component or data type in a component header definition is just the parameter name, cf. parameter `T` in L.1. If the generic parameter is a concrete value, its name needs to be preceded by its data type, cf. generic parameter `n`, which is of type `N(2:10)` in this example. Component parameters, in contrast to generic parameters, can only be of a data type. The syntax resembles the definition of

---

[1] The component type system is not to be confused with the data type system introduced in Sect. 4.2.

```
 1  component Main<T, N(0:10) n> (Q param1, N param2,…) {        EMA
 2    ports in T A,
 3          in T B,
 4          out T C;
 5
 6    instance Add<T, n> adder(0);
 7    instance Mult<T, n> multiplier(1);
 8
 9    connect A -> adder.A;
10    connect B -> adder.B;
11    connect adder.C -> multiplier.A;
12    connect B -> multiplier.B;
13    connect multiplier.C -> C;
14  }
```

Fig. 28 A basic example of an EMA architecture. The component `Main` contains two subcomponents `Add` adder and `Mult` multiplier

function parameters in many languages, where a type is followed by a unique name, cf. parameters `Q param1` and `N param2` in L.1.

The body of a component definition is enclosed in curly brackets and contains an interface and a structure definition. The interface definition is initiated with the keyword `ports` and is followed by a port list. A port definition consists of the port kind, which can be either `in` or `out` (EMA ports are strictly unidirectional), a data type, and a unique port name, cf. L.2-4 in Fig. 28. A component must have at least one input and one output port, since a major assumption of EMA is the absolute absence of side effects. Clean side effect-free models are crucial for testability, maintainability, and extensibility. An exception are components outputting a constant or a (possibly parameterizable) constant sequence. Such components obviously do not need an input port, but can require a component parameter, which alone defines the output behavior in every execution step.

Subcomponents are created using the keyword `instance` followed by the component type to instantiate and a component instance name, which is unique in the scope. If the component type to be instantiated has generic and/or component parameters, these have to be set by providing appropriate arguments in angle and/or round brackets, respectively. In L.6-7 of Fig. 28 two components are instantiated with their generic parameters being set to the type `T` and the value `n`. Furthermore, both subcomponents receive a component parameter in round brackets, which is 0 in L.6 and 1 in L.7.

To interconnect the subcomponents and to connect them to the parent component in the first place, we need to create connectors. The source of a connector must be either an output port of a sibling or subcomponent or an input port of the enclosing component. Similarly, the target of a connector must be either an input port of a sibling or subcomponent or an output port of the enclosing component. A connector is created using the `connect` keyword followed by the source port, the arrow operator `->`, and a target port. Ports of subcomponents can be referenced by using the subcomponent's name and the dot access operator. Connector examples are given

```
1    component Main<N(2:10) n> {                          EMA
2      ports in Q A[n],
3            in Q B[n],
4            out Q C[n];
5            out Q D;
6
7      instance Add2 adder[n];
8      instance Mult2n<2*n> multiplier;
9
10     connect A[1:n] -> adder[1:n].firstSummand;
11     connect B[:] -> adder[:].secondSummand;
12     connect adder[:].sum -> C[:];
13
14     connect A[:] -> multiplier.factors[1:n];
15     connect B[:] -> multiplier.factors[n+1:2*n];
16     connect multiplier.product -> D;
17   }
```

Fig. 29 An EMA architecture example featuring port and component arrays. The component Main contains n Add2 components, each operating on one of n operand pairs coming from the port arrays A and B. The Mult2n component computes the product of 2n operands passed through the port arrays A and B of the Main component to the port array factors of Mult2n

in L.9-13. Connectors define explicit dataflows. At execution time, data is exchanged only between ports connected by connectors.

Once a component cannot be subdivided into smaller subcomponents, it can be linked to a concrete behavior as will be discussed later. In standard EMA, the structure, i.e. the subcomponents as well as the connectors between them, is fixed at design-time.

Modeling cooperative systems and agent networks often requires the replication of large numbers of similar components and the interconnection thereof. EMA enables the designer to create multiple similar components and/or ports by means of arrays. Based on the array syntax of many languages, an array is created by appending the array size to the port or component name in brackets. For instance, in Fig. 29 we define the input ports A and B as well as the output port C as port arrays of length n. Since parameter n affects the interface of Main by changing the length of the port arrays A, B, and C, it cannot be defined as a component parameter, but must be a generic parameter instead.

In this example we demonstrate two interconnection patterns which are commonly used when dealing with port and component arrays. In the first one, we instantiate an array of components to deal with an array of incoming streams. Therefore, we create n adders of the component type Add2 in L.7, each instance to operate on two scalar inputs. Now, we need to connect the ports of the two arrays A and B of the parent component to the respective subcomponents, i.e. A[1] and B[1] should be connected to adder[1] and so on. This can be done in just one line, cf. L.10, by selecting the elements 1 to n from the port array A and, similarly, the components 1 to n from the adder component array. The connect operator connects each source element to the respective target element based on the index. Since this connection

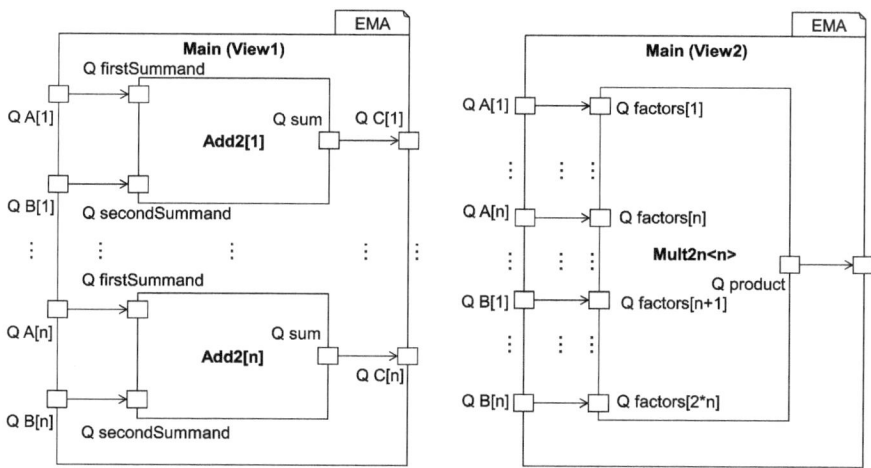

Fig. 30 Graphical views of the component defined in Fig. 29. On the lhs, the elements of two port arrays are connected to target ports of a component array. On the rhs, a port array is connected to another port array

pattern is often applied to *all* elements of an array, EMA offers syntactic sugar allowing the developer to leave out the indices of the first and last elements as is done in L.11. Similarly, in L.12 the output of each component in the `adder` array is connected to a corresponding port in the target port array C. This structural pattern is depicted graphically in the view on the left side of Fig. 30.

Furthermore, we can connect a port array to the port array of a target component, let this component aggregate the data and output a single result or a constant number of values. In our example, the port array A is connected element-wise to the first n elements of the input port array of the `multiplier` component of type `Mult2n` in L.14, while the port array B is connected to the remaining n input ports of `multiplier` in L.15. The output of the `multiplier` component is forwarded to the output port D of the enclosing component in L.16. This connection pattern is depicted graphically in the view on the rhs of Fig. 30.

4.4 Execution Semantics

Standard EMA has a synchronous and weakly causal execution semantics, which is based on the FOCUS theory [3] and inspired by Simulink [40]. In each cycle, every component is executed exactly once. Once a component has finished its execution, the computation results are immediately available at its output ports. We assume that data transmission over connectors is lossless and has no delay. Connectors transmit data instantly, i.e. when a source port of a connector is updated, the data is replicated immediately to the target port. A component is only allowed to be executed, once

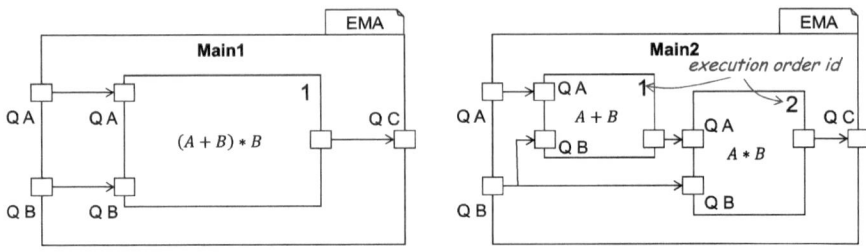

Fig. 31 This example shows two C&C architectures `Main1` and `Main2`, which are semantically equivalent in EMA due to its synchronized and weakly causal execution model, but which might have different interpretations in a language with strongly causal semantics

all of its predecessors, i.e. components connected to its input ports via a connector, have finished execution. Therefore, the identification of a dataflow-based execution order is crucial for a correct realization of the model semantics. A fixed execution order is established at compile-time and no re-scheduling needs to be performed at runtime. This is similar to Simulink's sorted execution order list.[2] In EMA, the C&C model is flattened at compile-time before the execution order is computed. Hence, only atomic components receive an execution order id. In EMA, multiple component instances can share a single execution order id if the execution order of these respective component instances can be exchanged without affecting the computation results. For instance, the adders of the `adder` component array of Fig. 28 can be executed independently.

At runtime all the components are executed sequentially based on the execution order list in each cycle. A cycle is finished when all components have been executed. The next cycle can be started, once the preceding cycle is finished. In EMA the input until time t completely determines the output until time t rendering the semantics weakly causal [3], which is convenient for modeling algorithms and physical processes. As an example consider the two architectures in Fig. 31. Both systems have the same semantics in EMA and can be described mathematically using the equation $C_k = (A_k + B_k) B_k$, where k is a sequential index. In contrast, if the system were strongly causal under the assumption that each subcomponent required n timesteps to compute and communicate the output, the equations describing `Main1` and `Main2` would become $C_k = (A_{k-n} + B_{k-n}) B_{k-n}$ for the left and $C_k = (A_{k-2n} + B_{k-2n}) B_{k-n}$ for the right model, respectively.

Finding an execution order for linear models, i.e. models without cyclic port dependencies, is straightforward: each component instance is put on the execution list after all component instances its input ports depend on. When structural loops are present in the model, i.e. when there is a path from a subcomponent's output to its own input without a delay, the compiler checks if the loop is algebraic. If yes, the compiler tries to transform the algebraic loop to a loop-free equivalent model.

_____

[2] https://de.mathworks.com/help/simulink/ug/controlling-and-displaying-the-sorted-order.html, accessed November 25, 2022.

If an explicit solution cannot be found, i.e. if the loop does not correspond to a known (solvable) pattern, it can be solved at runtime using an algebraic solver. Since this must be done in each timestep and there is no guarantee that a solution exists, a runtime solver would not only affect the runtime performance heavily, but might also lead to unpredictable behavior, which must be avoided in safety-critical systems. For this reason we only allow loops, which can be transformed into loop-free architectures at compile-time. If no such transformation can be found, the model is considered invalid.

To resolve algebraic loops, knowledge of the component behavior is required. A means to integrate behavior models into EMA components will be discussed in Sect. 4.5.

4.5 MontiMath

MontiMath is an imperative language developed for the design and implementation of math-heavy algorithms and to describe physical processes. It has been inspired by MATLAB's matrix-oriented paradigm. However, in contrast to MATLAB, MontiMath uses the EMA type system, which makes it a statically and strictly typed language similar to EMA itself. An example showing the basic language constructs is given in Fig. 32. The declaration of a MontiMath variable requires a type definition, which is expressed by preceding the newly declared variable by an EMA type, e.g. Q(0 Ohm : 1 nOhm : 1 MOhm)^{2,2} impedance. The syntax to define a matrix constant is the same as in MATLAB, but the literals inside the

```
                                                                    MontiMath

1   N nrows = 2;                        variables are statically and strongly
2   N ncols = 3;                        typed using the EMA type system
3   N (0 m: 10 km) x = 1 m;
4   Q(-oo m: 0.1mm : 10km)^{nrows, ncols} A = [-1, x, 1; x, 2*x, 0];
5
6   for c = 1:ncols                     indices are 1-based in
7     for r = 1:nrows                   EMA and MontiMath   2x3 matrix literal
8       if r == c
9         A(r,c) = 2;
10        elseif abs(r-c) == 1
11          A((r+2)%r, (c*3)%c) = -1;   for loop header defining a counter variable r
12        else                          and letting it run from 1 to nrows
13          A(r, c) = 0;
14        end                           if clause with a conditional and an
15      end                             unconditional alternative
16  end
```

similar to MATLAB, MontiMath uses   assigning a value to the entry at r-th
the end keyword to delimit blocks   row and c-th column of the matrix A

Fig. 32 This listing shows a simple MontiMath example exhibiting the main language constructs including variable declarations, matrix literal definitions, loops and conditions

matrix can be enriched by système international d'unités (SI) units if needed. As in MATLAB, a matrix constant is defined in square brackets. Thereby, columns and rows are separated by commas and semicolons, respectively. The initialization of the impedance matrix `impedance` modeling a two-port network can hence be written as `impedance = [10 Ohm, 5 Ohm; 6 Ohm, 8 Ohm];`.

To maintain compatibility to MATLAB, MontiMath indices start with 1 as opposed to most general purpose programming languages (GPLs), where arrays are zero-based. Scalars are treated as 1×1 matrices, but the square brackets can be dropped when defining a scalar literal. Other than in MATLAB, statements, except conditional statements and loops, need to be terminated with a semicolon.

MontiMath supports the typical operators needed in many computations including addition (+), subtraction (-), multiplication (*), division (-), and power (^). If applied to matrices, these operators perform the corresponding algebraic matrix operation, e.g. a matrix multiplication. Division by a matrix, e.g. `A/X`, is semantically equivalent to multiplying the dividend with the inverse of X, i.e. `A/X` is equivalent to `A*X^-1` or `A*inv(X)`.

Furthermore, MontiMath supports the Hadamard product or element-wise multiplication (.*), inverse Hadamard product (./), and element-wise power (.^). The transpose operation for real and the Hermitian transpose operation for complex-valued matrices can be expressed by appending the apostrophe operator (') to a matrix name, e.g. `A'`. Furthermore, the entries are conjugated in the complex case. Since matrix dimensions are statically typed, incompatibilities are detected at compile-time.

MontiMath supports the standard control flow constructs including `for` loops and `if` clauses, enabling us to write arbitrarily complex algorithms. Many tasks in CPS engineering can be expressed as optimization problems, e.g. model-predictive controllers. For this reason, we introduce optimization statements in MontiMath. The syntax provides dedicated keywords for optimization problems to come as close as possible to the original mathematical formulation enabling the developer to write down the objective function, to define the optimization variable, as well as a set of constraints.

A MontiMath program can be embedded into an EMA component by means of an implementation block as is shown in Fig. 33. This way the MontiMath script is executed in every execution cycle of the EMA component. It can read the input ports

```
1  component NormalizedLaplacian<N1 n> {
2    ports in Q^{n,n} A,
3          out Q^{n,n} L;
4                              Reading the input port
5    implementation Math {
6      Q^{n,n} D = diag(A * ones(n,1));        EMA with embedded MontiMath
7      L = D^-0.5 * A * D^-0.5;                behavior specification
8    }
9  }            Writing to the output port
```

Fig. 33 An EMAM model embeds a MontiMath script into an EMA component

of the EMA component and write the computation results to the output ports. To let a MontiMath script pass variable values from one execution cycle to another, we introduce the `static` keyword. A variable declared with this modifier, e.g. `static Q cumulativeError`, is saved in a cycle-independent scope. Its value does not get lost when an execution cycle is finished and can be reused in the next cycle. Alternatively, variables can be passed between cycles by feeding the output of a component back to one of its input ports and putting a delay block in between.

The modular structure of the EMA language family enables an easy composition with other modeling languages to be used in the *implementation* block of an EMA component for the definition of the component behavior. The language used can be another domain-specific language (DSL) or a GPL such as C++ or Java. For the composition to work, the embedded language must have a MontiCore implementation [23]. A particularly important DSL for component behavior definition is the deep learning modeling language MontiAnna [27, 35, 36]. It enables a concise modeling of deep neural networks as directed acyclic graphs (DAGs) of neuron layers. The MontiAnna generator produces code for data loading, training, and execution of the neural network. Furthermore, it controls the machine learning lifecycle of the deep learning component, e.g. supporting data management [2] and deciding whether a training phase is needed or can be skipped if a trained model is already available, based on a machine learning artifact model [1]. MontiAnna has been applied to model deep neural networks for various domains, including image processing convolutional neural networks (CNNs) [35], language processing networks [35], reinforcement learning applications [19], generative adversarial networks (GANs), variational autoencoders (VAEs), etc. A CNN for the recognition of handwritten digits embedded into an EMA component is depicted in Fig. 34. The neural network is assembled from predefined layers and the custom layer `conv` in L.13-21. While the example is a linear graph of layers, arbitrary DAGs can be constructed using MontiAnna.

4.6   Cooperative Agents and EmbeddedMontiArc Dynamics

Until now the focus was on static architecture modeling of closed, isolated systems such as autonomous vehicles using EMA. The elements of a static architecture are fixed at design time and cannot be altered, removed, or added at runtime. With this approach we can cover the majority of closed systems such as embedded devices and control software. However, cooperative driving systems which are highly dynamic by nature require the ability to restructure or reconfigure parts of their architecture according to changing circumstances and requirements at runtime. For this reason, we are going to discuss an extension for EMA introducing dynamics to architectural elements such as ports, connectors, and components based on [26].

Different forms of dynamic architecture description languages (ADLs) are known in the literature tackling different concerns of architectural dynamics [5]. In particular, the choice of appropriate means of architectural runtime reconfiguration depends on the kind of system under development and the application domain. The concepts

```
                                                                    EMADL
1    component Detector<Z(2:oo) classes = 10>{
2      ports in Z(0:255)^{1, 28, 28} data,
3      out Q(0:1)^{classes} softmax;
4
5      implementation CNN
6      {
7        def conv(channels, kernel=1, stride=1){
8          Convolution(kernel=(kernel,kernel),channels=channels) ->
9          Relu() ->
10         Pooling(pool_type="max", kernel=(2,2), stride=(stride,stride))
11       }
12
13       data ->
14       conv(kernel=5, channels=20, stride=2) ->
15       conv(kernel=5, channels=50, stride=2) ->
16       FullyConnected(units=500) ->
17       Relu() ->
18       Dropout() ->
19       FullyConnected(units=classes) ->
20       Softmax() ->
21       softmax;
22   } }
```

Fig. 34 A CNN for handwritten digit recognition embedded into an EMA component, also referred to as an EMADL component

discussed in this chapter are intended for the Local Traffic System (LTS) domain discussed in the previous sections. Our design decisions will hence be based on the following list of assumptions:

- The agents are instances of compatible types or share a common interface. In the automotive domain, for instance, agents are equal or similar vehicles or roadside units (RSUs). The agents are independent processes with proprietary goals. They are not part of and do not contribute to the functioning of a bigger system (in contrast to an aircraft architecture designed using a language like Architecture Analysis & Design Language (AADL), where architectural dynamics is used to model functional variations of a single but complex system).
- The agents do not know each other by default and there is no communication between them at the beginning. Furthermore, the total number of agents living in the system is not known to an agent. Each agent's knowledge about its peers is limited to what it perceives through its sensors and communication.
- The number of agents in the system can vary throughout time. Agents can be spawned without existing agents to be notified explicitly. In the cooperative vehicles domain, new vehicle instances can come into existence by being manufactured or by entering the area of interest from outside.
- There is a communication channel which can be used by the agents to send and receive messages to and from other agents, respectively. This channel can be used for both directed and broadcast communication. However, since we are dealing with the application layer, we will not care about lower network protocols in this

work, assuming an end-to-end channel connecting the logical interfaces, e.g. EMA ports, of two different agents directly.

To be able to model interactions between participants of a dynamically changing traffic system, the C&C language used needs to support changes in the component structure and variations of the dataflows at runtime. Such changes can be induced by specific events, such as the occurrence of a new traffic participant, which the developer should be able to model with the same language, as well.

The aim of this section is to introduce the main concepts of an EMA language extension for dynamic reconfiguration, which we are going to refer to as Embedded-MontiArc Dynamics (EMAD). The extension is conservative [24], meaning that standard, non-dynamic models can be parsed and generated by EMAD without changes.

4.7 EMAD Execution Semantics

In Sect. 4.4 we have discussed the synchronous execution semantics of EMA. The system is executed stepwise. In each step all the subcomponents are executed according to an execution order determined at compile-time. To enable reconfiguration and to support dynamically evolving architectures, we extend the execution semantics of EMA by a reconfiguration phase which takes place in each execution cycle.

In the reconfiguration phase, reconfiguration triggers are checked and, if present, the corresponding reconfigurations are performed. This possibly activates further reconfiguration triggers which are then handled as well, until the reconfiguration queue is empty. We introduce two main concepts for runtime reconfiguration in EMAD: 1. Data-triggered and 2. Service-based reconfiguration.

4.7.1 Data-Triggered Internal Reconfiguration

The simplest way to trigger and model reconfiguration is the data-triggered approach. Thereby, a reconfiguration is initiated when a signal fulfills a given condition, e.g. a port value exceeds a predefined threshold. The reconfiguration is executed and maintained as long as the condition is satisfied. The approach can be easily motivated and illustrated by non-linear components used in electronics. For instance, a diode is conductive only if the applied voltage is higher than the threshold voltage; a multiplexer passes the data signal chosen by a control signal; when a battery electric vehicle (BEV) is connected to a charging station, the connection is signaled to the charging electronics which reacts by enabling the charging process as long as the connection signal is active.

To enable modeling data-triggered reconfiguration, we extend the body of an EMA component definition by a list of reconfiguration blocks. The header of such a reconfiguration block contains a condition formulated as a Boolean expression over port values and architectural properties, which needs to be fulfilled in order to trigger

```
1    component BMux4<T>                                      EMAD
2      ports in T inSig[4],
3             in B ctrSig[2],
4             out T outSig;
5
6      instance BMux2<T> mux2;
7
8      connect ctrSig[1] -> mux2.ctrSig;
9      connect mux2.outSig -> outSig;
10         reconfiguration condition
11     @ ctrSig[2]::value() == true {
12       connect inSig[3]  ->  mux2.inSig[1];  ⎤ Value-triggered
13       connect inSig[4]  ->  mux2.inSig[2];  ⎦ reconfiguration
14     }
15
16     @ ctrSig[2]::value() == false {
17       connect inSig[1]  ->  mux2.inSig[1];
18       connect inSig[2]  ->  mux2.inSig[2];
19     }
20   }
```

Fig. 35 A multiplexer component choosing two of its inputs to be passed to the inner multiplexer dependent on a control signal

the reconfiguration. The body of the reconfiguration block follows for the most part the same syntax as the body of a standard non-dynamic component and contains a declarative definition of the architectural changes to be performed as a response to the triggering event. These changes are rolled back as soon as the reconfiguration condition in the reconfiguration block header ceases to hold.

To illustrate the syntax and the mechanics behind data-triggered reconfiguration, we introduce a simple multiplexer example in Fig. 35. The BMux4 component has four data inputs of a generic type T and two Boolean control inputs. The purpose of the component is to choose one of the four input signals of the inSig port array based on the values of the control signals (ctrSig port array) and to forward it to the output port. The idea is to realize this behavior by altering the connectors corresponding to the control signal. Therefore, we first choose two of the four data signals (the first two *or* the second two ports of the inSig array) based on the value of inSig[1] and then forward them as well as a further control signal inSig[2] to a subcomponent of type BMux2, which in turn uses the received control signal inSig[2] to choose one of the remaining two data signals. Its choice is then output through the parent component's output port.

The static connectors of the component are defined in L.8-9 to connect the first control signal with the inner multiplexer and its output to the output of the parent BMux4. There are two reconfiguration definitions given in L.11-14 and L.16-19. In L.11 and L.16 the @ symbol denotes the beginning of a reconfiguration condition. The actual reconfiguration code is a block enclosed in curly brackets following the condition. As can be seen in L.12-13 and in L.17-18, the configuration code is composed of ordinary connect statements as we know them from the static EMA syntax. The connections defined in these two blocks are established and released in the reconfiguration phase at the beginning of an execution cycle as discussed earlier. In this example, this is used to choose two of the four incoming inputs to be forwarded to the child component `mux2`.

A reconfiguration is executed once the condition becomes true and remains active as long as the condition remains true, i.e. as long as the value at the port `ctrSig[1]` is true in L.11 and as long as it is false for L.16. When the condition of an active reconfiguration goes back to false, the reconfiguration is rolled back, i.e. all the architectural elements defined in the reconfiguration block are removed (irrespective of whether or not another reconfiguration becomes active instead). In our example, the two reconfiguration conditions are mutually exclusive, but their disjunction is always true. Consequently, exactly one of the two reconfigurations is active at any given point in time. In general, arbitrarily many reconfigurations (including zero) can be active in parallel. However, each combination must result in a valid architecture. That is, an input port must not be the target of more than one connector. Furthermore, under no circumstances an input port may be floating. This is verified by context conditions at compile-time. Consequently, none of the two reconfigurations can be removed from the component in the multiplexer example: when no dynamic reconfiguration is active, only the static part of the architecture is present. In this case, the `inSig` ports of `mux2` would be floating.

Note that in order to access the value of a port in an EMAD reconfiguration, we use the port function `value()` accessible for each port of the component using the `::` operator. The syntax highlights that we are not trying to use a model element in a conventional manner (which would require a dot), but want to perform a runtime query related to a model element instead. The function is available in reconfiguration conditions and bodies only. If the port we are referring to belongs to a subcomponent, we can access it by specifying the port's name preceded by the (subcomponents') instance name, e.g. `mux2.outSig::value()`. Note that a component can only query the values visible in its scope, i.e. values of its own or of its immediate subcomponents', but not of its subsubcomponents' or the parent component's ports.

A reconfiguration condition can be an arbitrary Boolean expression. Similarly to other languages the Boolean OR and the Boolean AND operators are denoted by `||` and `&&`, respectively. For equalities and inequalities we use the following operators: `==, <=, >=, <, >`.

Reconfiguration conditions can be formulated in terms of an expression sequence in order to identify sequence patterns. A value sequence can be notated similarly to an EMA row vector with the oldest value coming leftmost. To avoid confusions with vector-valued variables, the `tick` keyword is used as a separator instead of a comma. For instance, the condition `ctrSig[1]::value()` == `[true tick false tick false]` is evaluated to true at execution cycle n if the following sequence of values was observed: `true` at $n - 2$, `false` at $n - 1$, `false` at n. The type of each expression in the sequence must be compatible with the corresponding port type. The sequence notation implies that past values of the underlying port need to be stored at runtime. In this particular example, in addition to the current value at

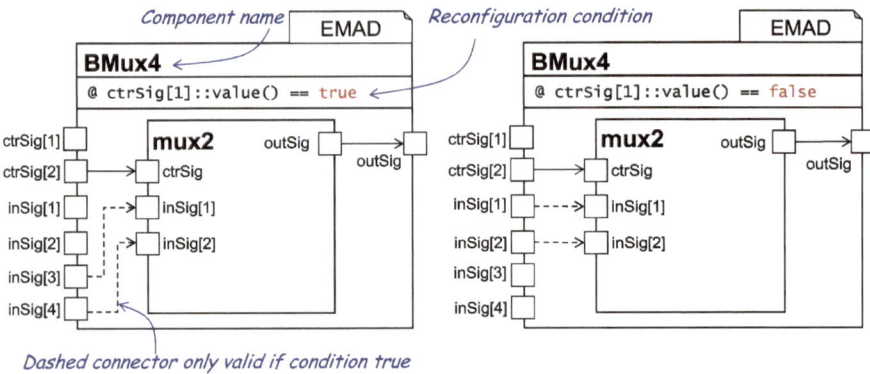

Fig. 36 The two architectural states of the BMux4 component

the ctrSig[1] port, the component needs to store two of this port's past values in order to be able to evaluate the reconfiguration condition in each execution step.

Until now, we have been using a graphical representation of EMA models to facilitate the understanding of the architecture. Given the fact that there is no single representation of an EMAD model, we need an appropriate extension of the graphical syntax. Diagrams representing the two reconfigurations of the BMux4 model are depicted in Fig. 36. Thereby, we introduce two syntactic elements: first, the reconfiguration condition triggering the reconfiguration is specified in a box under the component's name. Second, model elements, which are added in this reconfiguration, are denoted by dashed figures instead of solid ones. In this example, only connectors are created dynamically at runtime. Components and ports can be added in a similar way by the means of dynamic arrays, which will be discussed in Sect. 4.7.2.

The aim of the example in Figs. 35 and 36 was to introduce the main ideas behind data-triggered reconfiguration. The exactly same behavior can be achieved with a mode model with two states [21]. Using a mode finite state machine (FSM) for a system with a small number of states and state transitions can be favorable as it facilitates a state-centric model analysis. In cases with many, possibly partially overlapping reconfiguration conditions and state transitions between all possible states, however, the data-triggered reconfiguration concept presented in this chapter can lead to much more concise models, since we don't need to define all possible states explicitly and no transitions need to be modeled at all. On the other hand, modes are more powerful since reconfigurations can depend on the current *architectural* state, which is not possible with our concept. We recommend using modes and data-triggered reconfiguration interchangeably depending on the requirements and the nature of the modeled system.

4.7.2 Service-Based External Reconfiguration

To enable the creation of more complex, propagating reconfigurations, we introduce a second way of triggering architectural changes at runtime, the service-based reconfiguration. The idea behind it is to trigger reconfigurations by external architectural change requests and to propagate such requests from component to component.

We are going to present the concepts of service-based reconfiguration by the example of a cooperative collision prediction component given in Fig. 37. The `CollisionSystem` component receives the planned trajectories from other vehicles of an LTS and checks each of these trajectories for a collision with its own one. Each trajectory is input into the component through a dedicated port. Furthermore, each pairwise collision check is executed by a dedicated subcomponent of type `CollisionCalculator`.

Before we proceed with the discussion of the service-based trigger mechanism, we need to introduce the concept of dynamic component and port arrays. In Sect. 4.3, static component and port arrays were introduced, allowing us to model an arbitrary but fixed number of similar components and ports in a single line of code. In the collision detection example described here we don't know at design time, how many traffic participants will be present in the LTS. Furthermore, the number of peers can change over time. The concept of dynamic arrays enables us to cope with this modeling challenge by allowing us to specify a range instead of a fixed number of elements in the array. At runtime the concrete number of elements in the array can change.

indicates component with
dynamic interface and behavior

```
1   dynamic component CollisionSystem {                           EMAD
2     ports in Trajectory ownTrajectory,   dynamic number range
3         keyword
4   dynamic  dynamic in StatusMsg otherStatus [0:32],
5   ports   dynamic in TrajectoryMsg otherTrajectory [0:32],
6           out CollisionMsg msgOut;
7
8     instance CollisionCalculator cc[0:32];
9     instance CollisionMessageBuilder cmb;
10
11    connect cmb.msgOut -> msgOut;
12                    port connection event
13    @ otherStatus::connect () && otherTrajectory::connect () {
14      connect ownTrajectory -> cc[?].ownTraj;
15      connect otherStatus[?] -> cc[?].otherStatus;
16      connect otherTrajectory[?] -> cc[?].otherTraj;
17      connect cc[?].collisionOut -> cmb.collisionIn[?];
18    }
19  /* other modes & connections */ }
```

Fig. 37 Collision system of an autopilot calculating potential collisions with up to 32 other vehicles

The syntax is based on the range syntax of EMA types: the modeler needs to specify the minimum and the maximum number of elements inside the square brackets of an array declaration separated by a colon instead of a single length value. This is done in L.4 and L.5 of Fig. 37 to define a dynamic port array and in L.8 to define a dynamic component array. In the case of port arrays it is obligatory to use the `dynamic` keyword. If the component interface contains dynamic port arrays, it is also necessary to mark the component with the `dynamic` keyword in the header, cf. L.1 of Fig. 37.

In case the lower bound of the element count is greater than zero, the minimum number of elements will be created at instantiation of the component. Once the upper bound of the elements in an array has been reached, events leading to an instantiation of further elements cannot be handled. The availability of free port and/or component slots in an array can hence be regarded as a further implicit condition of a reconfiguration. Upper bounds on elements in an array have been introduced with embedded systems in mind often having very limited resources and strict performance requirements. The upper bound can be set to infinity by putting `oo`, similarly to EMA type bounds. However, since this can have a negative impact on the performance of an overloaded system, this is not an advisable modeling pattern and results in a warning. A system knowing its limits can react to an overly high demand in a controlled manner.

In our collision system example, the port arrays `otherStatus` and `otherTrajectory` are supposed to receive status and trajectory messages from other cooperative vehicles in the LTS. The maximum number of connections is limited to 32. On the other hand, if there are no other vehicles in the network, the port arrays can be empty.

For each connected vehicle, the `CollisionSystem` component provides an individual `CollisionCalculator` component instance. Accordingly, the number of these instances varies between 0 and 32, as well. At system start up, the minimum number of components and ports is instantiated, i.e. zero.

The question arises how the free slots in the component and port arrays can be used and released at runtime. We realize this by introducing a *reconfiguration service interface*. This interface allows external components or even external software to request reconfigurations. More precisely, it allows external clients to request a port from a dynamic array.

The reconfiguration interface is defined not just by declaring a dynamic port array, but by the reconfiguration conditions using it, cf. L.13 in Fig. 37. To query reconfiguration requests in a reconfiguration condition, we introduce the new port property `connect`, which is basically a Boolean flag indicating whether a connect request for this port has been issued, bundled with an id to avoid confusions with other requests sent to the same port. Similarly to the value at a port, the `connect` property can be queried using the `::` operator, i.e. as `portName::connect()`. A reconfiguration condition can be composed as a conjunction of arbitrarily many connect atoms, i.e. `portName1::connect() &&,...,&& portNameN::connect()`, where the port names used must be dynamic port arrays declared in the component's interface. Disjunctions and negations of connect atoms are forbidden by a context condition to prevent inconsisten-cies (in a disjunction we do not know at design-time which port(s) will be actually requested and hence, cannot define meaningful reconfigurations using these ports).

The resulting reconfiguration interface can be used by issuing connect request for all the ports required by the reconfiguration condition simultaneously. In our example this means that, due to the reconfiguration condition in L.13 of Fig. 37, connections to the otherStatus *and* the otherTrajectory port must be requested at once. Such a request is created in an EMAD model in the reconfiguration body of a parent component as connect statements targeting the corresponding dynamic port arrays. This is shown in Fig. 38, where a component holding an instance of CollisionSystem connects to the aforementioned port arrays otherStatus and otherTrajectory of the latter in L.4-5 of its own reconfiguration body.

Note that the reconfiguration bodies of Figs. 37 and 38 are chained: the reconfiguration of the latter triggers the one of the former. If ReconfigurationCondition in L.3 of Fig. 38 is a data-driven reconfiguration as discussed in Sect. 4.7.1, the chain starts in Fig. 38. If ReconfigurationCondition defines a reconfiguration interface similar to L.13 in Fig. 37, it must be triggered from another reconfiguration body itself. Hence, arbitrarily long service-based reconfiguration chains can be initiated by a data-driven reconfiguration.

Note that the reconfiguration request issued by the parent component of the CollisionSystem component in L.4-5 of Fig. 38 matches the reconfigura-tion interface defined in L.13 of Fig. 37 exactly. This is verified at compile-time by a context condition. An invalid usage of the reconfiguration interface of the CollisionService component is shown in Fig. 39. Here we are trying to connect to the otherStatus port only. However, this is not supported and results in a compile-time error as there is no such reconfiguration condition in the CollisionSystem component.

To be able to deal with dynamic port and component arrays in reconfiguration descriptions, we need a syntax allowing us to access the newly created elements. To do so, we introduce the ?-operator. It is used instead of the element number in square brackets to request and access new elements in a dynamic port or component array, e.g. myArray[?]. Usage of the operator is restricted to reconfiguration bodies.

triggers a reconfiguration condition in the CollisionSystem component by requesting the two ports otherStatus and otherTrajectory simultaneously

```
1   instance CollisionSystem cs;
2
3   @ ReconfigurationCondition {
4       connect somePort1 -> cs.otherStatus[?];
5       connect somePort2 -> cs.otherTrajectory[?];
6   }
```

EMAD

Fig. 38 The listing shows a valid usage of the reconfiguration service interface of the Collision System component of Fig. 37 by a parent component

invalid port request results in a compile-time error: the CollisionSystem
component requires otherStatus and otherTrajectory to be requested together

```
1   instance CollisionSystem cs;                        ┌─────────┐
2                                                       │  EMAD   │
3   @ ReconfigurationCondition {                        └─────────┘
4       connect somePort1 -> cs.otherStatus[?];
5   }
```

Fig. 39 The listing leads to a compile-time error since CollisionSystem does not have a reconfiguration triggered by requesting only the otherStatus port

An example is given in L.14-17 of the `CollisionSystem` model in Fig. 37. In L.14 the `?`-operator is used to connect the `ownTrajectory` port to a new component `cc[?]`. Since this is the first access to `cc[?]` in this reconfiguration body, it implicitly triggers the creation of a new component instance. In contrast, further accesses to `cc[?]` in L.15-17 are pure access operations, no implicit instantiation is involved. If the component type of the component array requires component parameters, the parameter list can be passed in parenthesis right after the array brackets and before the dot operator, e.g. `cc[?](param1, param2,...).ownTraj`.

Since the `cc` array has a maximum capacity which cannot be exceeded, a further implicit reconfiguration condition is that the maximum capacity of this array has not yet been reached. If, however, the array is maxed out, the reconfiguration condition will evaluate to false and the reconfiguration will thus not be activated.

The reconfiguration service interface is available not only at modeling level allowing other components to use it, but also in the generated code. The latter can be used by any client. For instance, C++ code can be generated for the `CollisionSystem` component. Then it can be compiled to a library to be deployed as a building block of the vehicle run-time environment (RTE). The RTE receives a stream of vehicle to vehicle (V2V) messages and redirects them to the right ports of the `CollisionSystem` library (each sender is assigned to one port). If a new LTS participant starts sending, the RTE can request a new port from the `CollisionSystem` library by calling a generated request function. The library in turn checks whether the request is satisfiable. If yes, it provides a new port instance the RTE can forward messages of the new vehicle to. Otherwise no reconfiguration is carried out and the library call returns with an error. The client can then withdraw the request or wait until the dynamic component satisfies the request in a future reconfiguration cycle.

To facilitate the usage of the generated reconfiguration interface, we generate request methods allowing the client to require all necessary ports to activate a reconfiguration with a single function call, e.g. `requestOther StatusAndOtherTrajectory(Port<T1> *otherStatus, Port<T2> *otherTrajectory)`, where `Port <T>` is a generic class representing an EMA port of type `T` at C++ level. This way, it is not possible to create invalid request, e.g. requiring only an `otherStatus`, but no `otherTrajectory` port, when using the generated code as a library.

```
1   dynamic component DynamicSum {                              EMAD
2     port dynamic in Q summands[0:32],
3                   out Q sum;
4
5     implementation Math {
6       Q tmp = 0;
7       for i = 1:size(summands)        iterates over all ports in the
8           tmp = tmp + summands(i);     input port array summands
9       end
10      sum = tmp;
11  } }
```

Fig. 40 Adder with 0 to 32 inputs

Figure 40 shows an example combining a dynamic interface with a MontiMath implementation. The purpose of the component is to compute a sum of all inputs and to output the result. This is a typical data aggregation example working on a varying number of inputs. The dynamic input port array summands can contain 0 to 32 elements, i.e. at instantiation the component has no inputs and outputs zero due to the initial assignment tmp = 0 in L.6. The loop in L.7-9 iterates over all ports in the summands array and adds each port's value to the overall sum, which is accumulated in the tmp variable. In this example, we treat the dynamic port array in a stateless anonymous way. We iterate over the port array and are only interested in the value present at each available port without caring about its history. This is the natural way to deal with dynamic port arrays in MontiMath. Tracking states related to dynamic ports using MontiMath is possible but should be avoided. Instead, to track a concrete dynamic port's history, we need to replicate a dynamic subcomponent for each dynamic port instance, as was done in Fig. 37. This way, each communication partner requiring a port in a dynamic port array is assigned a dedicated processing subcomponent maintaining the corresponding state. Each of these dedicated processing subcomponents only sees a single input port of the dynamic port array it is assigned to instead of the whole port array. This pattern enforces the separation of concerns and high cohesion principle as the processing related to each communication partner is clearly encapsulated and limited to the actual logic (no explicit iterating over the port array is needed in the behavior implementation).

Based on the reconfiguration mechanism described in this section, we can model whole *reconfiguration chains* to realize deep or flat reconfigurations. A deep reconfiguration means that reconfiguration of a parent component triggers reconfigurations in child components. A connect to a subcomponent's port activates this port's connect flag which can in turn be used to trigger a reconfiguration in the subcomponent. In the same way, the subcomponent can trigger reconfigurations in its subcomponents and so on. When a parent component instantiates a static subcomponent in an EMAD model, it can connect its output ports immediately, e.g. as is done in L.11 of Fig. 37. However, the subcomponent might be dynamic and new output ports might be added throughout the subcomponent's reconfiguration procedures. In this case, the parent component can react to newly created ports of the subcomponent by observing the dynamic ports' `connect` flags in the same way as it would observe connect request to its own input ports. This enables us to create reconfiguration chains propagating downwards into the hierarchy as well as those coming from the bottom and propagating upwards.

A reconfiguration chain is always performed in one single reconfiguration phase as an atomic transaction, i.e. if the chain breaks at some point, the whole reconfiguration is considered infeasible. If a failure occurs after some reconfiguration steps of the chain have already been carried out, these steps will be rolled back.

As in data-triggered reconfiguration, a reconfiguration remains active as long as the respective condition is fulfilled. Whenever a new port request is issued, the port is created and a connector connected to it, the `port::connect()` property is activated for this port. This flag and hence, the configuration remain active until the requesting client removes its connector to the dynamic port. If the client created the connector as part of an EMAD reconfiguration, it would remove it, when the condition of this original reconfiguration ceased to hold. If the client is an external software, it can use the reconfiguration service interface to roll back a reconfiguration available in the generated code. Such a rollback would remove all architectural elements created in the reconfiguration and trigger the rollback of reconfigurations of subcomponents. This way, a reconfiguration chain is rolled back completely. The rollback interface is not usable explicitly in an EMAD model to prevent arbitrary removals of ports leading to inconsistencies in an architecture.

The service-based reconfiguration procedure of EMAD models boils down to the following steps:

1. Request: an external component sends a set of connect requests.
2. Reservation: the receiving component checks if the requested ports are available, i.e. if the corresponding dynamic port arrays do not violate their respective upper limit constraint. If yes, the component returns references for the new ports, i.e. the newly allocated array indices, to the requester so that explicit access is possible in the future. Otherwise, the requester is informed that its request has been rejected.
3. Reconfiguration: in the reconfiguration phase of the component, the reconfiguration bodies of all valid reconfiguration requests, i.e. those fulfilling a reconfiguration condition, are realized (L.14-17 in the `CollisionSystem` example). Consequently, the component reacts to the external reconfiguration request by internal self-modifications.
4. Follow-up requests: possibly, the reconfiguration instructions of the previous step contain the creation of new ports and/or subcomponents, as well. In this case, the component becomes a requester itself initiating a follow-up reconfiguration in its subcomponents or external components.

In our target domain of interconnected vehicles we mostly need the combination of both data-driven and service-based reconfiguration, which, when used together, can result in a powerful symbiosis. Reconfigurations which emerge as reactions to environmental changes measured by sensors or to incoming messages can be modeled using the following pattern: a data-driven event stands at the beginning of an

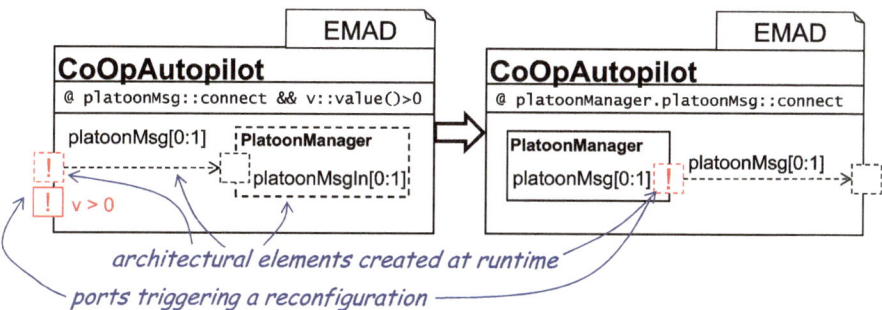

Fig. 41 A reconfiguration chain involving input and output ports of the `PlatoonManager` component. An arriving platoon message causes the creation of new input ports in the diagram on the left. Follow-up reconfigurations inside the `PlatoonManager` result in a new output port and a new outgoing connector as depicted in the diagram on the right

event chain. The reconfiguration caused by this event requests new components and ports triggering service-based reconfigurations, which in turn trigger further service-based reconfigurations. As soon as the original trigger vanishes, the reconfiguration chain is rolled back completely and the architecture returns to its initial state. A data-driven source event can be based on a sensor measurement (including the vehicle's antenna receiving messages from other cooperating traffic participants). A particular measurement value or the reception of a specific message would trigger a reconfiguration of the controller architecture, the internal reconfigurations of which are mostly service-based.

An important aspect of EMAD is that there is no explicit way to *remove* architectural elements. Instead, elements are removed implicitly, whenever the triggering reconfiguration condition switches back to false. This guarantees that an architecture can always be put back into its original state.

A further important property is that all possible reconfigurations are fixed by the design time model. Component and port replication is limited by an upper dynamic array size. Consequently, there is only a finite number of possible architectural states at runtime. This is an important design decision preventing a system to reach unexpected states and behaviors and facilitating verification.

Often reconfigurations trigger each other resulting in reconfiguration chains. We can visualize such chains using reconfiguration views, each view only showing the part of the model which is being changed in the current reconfiguration step. One such reconfiguration chain is depicted using views in Fig. 41. In the first reconfiguration view, depicted on the left, the `CoOpAutopilot` component, a controller of a cooperative vehicle, instantiates a platoon manager when a platoon port is requested and the velocity is greater than 0. In a second reconfiguration step, an inner component of the platoon manager requests a new output port and the `CoOpAutopilot` component reacts by creating a new connector. The ports triggering the reconfigurations are emphasized with an exclamation mark. Additionally, the data condition (v>0) is set next to the corresponding v port. Note that the `PlatoonManager` component is depicted using a dashed line in the left view, while it is solid in the view on the rhs. This is because the component is already there, when the second reconfiguration event is triggered. A big arrow between the two views stresses the order of the reconfigurations. Obviously, a reconfiguration must have taken place inside the `PlatoonManager` component to request the creation of its new output port `PlatoonManager.platoonMsg`. This reconfiguration (chain) is not part of the depicted sequence as it is not in the scope of the `CoOpAutopilot` component and should be visualized in a separate view chain.

4.8   Conclusion

In this chapter we discussed EMA, an architecture description language based on the component-and-connector paradigm. The language facilitates the component-based design of technical systems such as cooperative vehicles thereby enforcing a compliance with functional safety standards. While core EMA only allows the description of static architectures, its conservative extension EMAD enables the developer to model architectural changes such as the creation, removal, and (re)connection of components which are performed at runtime. Due to the conservative extension property, each valid EMA model is also a valid EMAD model [24].

EMAD introduces an event-based reconfiguration system which can react to data-driven as well as architectural events. An EMAD component can instantiate ports, subcomponents, and connectors at runtime as a reaction to a trigger event. Thereby, it can trigger further events of its subcomponents, enabling the modeler to define complex reconfiguration chains.

In EMAD, all possible configuration states are implicitly defined at design time, maintaining the possibility to analyze, predict, and verify the behavior of dynamic components at design and compile-time. A set of context conditions ensures that reconfigurations never clash, making the language applicable to safety-critical systems.

In particular, EMAD can be used to model cooperative systems and their dynamically changing communication channels and processing chains, e.g. in the context of local traffic systems.

To embed behavior into EMA and EMAD components, two behavior description languages are presented: first, MontiMath is a strongly typed matrix-based scripting language offering common constructs such as loops and conditions; second, the MontiAnna language can be used to describe deep neural networks as DAGs of neuron layers, enabling the integration of AI components into larger software architectures.

5 Conclusion

This work demonstrates basic concepts for cooperation and interaction of autonomous vehicles. Basic approaches to the architecture and formation of local traffic systems are shown which are subsequently verified in a real-time verification method for cooperative vehicles. The verified trajectories are collision free and dead-lock free. The presented modeling language allows a formal description of vehicle software architectures as well as cooperation and interaction of distributed systems.

Although the presented concepts represent and implement the feasibility and basic approaches, there is a need for further research. The focus of future work should be on further investigation of the reciprocal influence of local traffic systems as well as the cooperation algorithms used and the resulting requirements for necessary modeling languages for distributed systems. Further research is needed in the standardization of necessary V2X messages as well as algorithms used. In the area of V2X messages, it is still not clear whether WLAN-based standards such as ITS-G5 or cellular network-based standards such as C-V2X will prevail. While WLAN-based standards are already used by some manufacturers, C-V2X offers significantly greater potential. With regard to the algorithms used for grouping and actual cooperation, there is a need for more research when considering possible failure cases such as a spontaneous communication interruption with regard to functional safety and achieving required safety standards like ISO 26262 ASIL D [9].

Acknowledgements This research is supported by the Deutsche Forschungsgemeinschaft (DFG, German Research Foundation) within the Priority Program SPP 1835 "Cooperative Interacting Automobiles".

References

1. Atouani, A., Kirchhof, J.C., Kusmenko, E., Rumpe, B.: Artifact and reference models for generative machine learning frameworks and build systems. In: Tilevich, E., De Roover, C., (eds.) Proceedings of the 20th ACM SIGPLAN International Conference on Generative Programming: Concepts and Experiences (GPCE 21), pp. 55–68. ACM SIGPLAN (2021). http://www.se-rwth.de/publications/Artifact-and-Reference-Models-for-Generative-Machine-Learning-Frameworks-and-Build-Systems.pdf
2. Baumann, N., Kusmenko, E., Ritz, J., Rumpe, B., Weber, M.B.: Dynamic data management for continuous retraining. In: Burgueño, L., Bork, D., Nguyen, P., Zschaler, S. (eds.) Proceedings of MODELS 2022. Workshop MDE Intelligence (2022)
3. Broy, M., Stølen, K.: Specification and Development of Interactive Systems: Focus On Streams, Interfaces, and Refinement. Springer Science & Business Media (2012)
4. Burger, C., Lauer, M.: Cooperative multiple vehicle trajectory planning using miqp. In: 2018 21st International Conference on Intelligent Transportation Systems (ITSC), pp. 602–607. IEEE (2018)
5. Butting, A., Heim, R., Kautz, O., Ringert, J.O., Rumpe, B., Wortmann, A.: A classification of dynamic reconfiguration in component and connector architecture description languages. In: Proceedings of MODELS 2017. Workshop ModComp, CEUR 2019

(2017). http://www.se-rwth.de/publications/A-Classification-of-Dynamic-Reconfiguration-in-Component-and-Connector-Architecture-Description-Languages.pdf

6. Cavada, R., Cimatti, A., Dorigatti, M., Griggio, A., Mariotti, A., Micheli, A., Mover, S., Roveri, M., Tonetta, S.: The nuXmv symbolic model checker. In: International Conference on Computer Aided Verification, pp. 334–342. Springer (2014)

7. Dankert, J., Dernehl, C., Eckstein, L., Kowalewski, S., Kusmenko, E., Rumpe, B.: Rapidcoop-robuste architektur durch geeignete paradigmen für kooperativ interagierende automobile. Automatisiertes und Vernetztes Fahren (AAET'17) **7**, 1–6 (2017)

8. Dankert, J., Kowalewski, S., Eckstein, L.: Architekturen und algorithmen für kooperative automobile. Technical report, Lehrstuhl und Institut für Kraftfahrzeuge (ika) (2021)

9. Debouk, R., et al.: Overview of the 2nd Edition of iso 26262: Functional Safety-Road Vehicles. General Motors Company, Warren, MI, USA (2018)

10. Dosovitskiy, A., Ros, G., Codevilla, F., Lopez, A., Koltun, V.: Carla: An open urban driving simulator. In: Conference on Robot Learning, pp. 1–16. PMLR (2017)

11. Drave, I., Greifenberg, T., Hillemacher, S., Kriebel, S., Kusmenko, E., Markthaler, M., Orth, P., Salman, K.S., Richenhagen, J., Rumpe, B., Schulze, C., Wenckstern, M., Wortmann, A.: SMArDT modeling for automotive software testing. Softw.: Pract. Exper. **49**(2), 301–328 (2019)

12. ETSI, T.: 102 637-1 v1.1.1 intelligent transport systems (its) vehicle communications basic set of applications part 1: Functional requirements. Intelligent transport systems (ITS) (2010)

13. ETSI, T.: 102 637-2 v1.3.1 intelligent transport systems (its) vehicle communications basic set of applications part 2: Awareness basic service. Intelligent transport systems (ITS) (2014)

14. ETSI, T.: 102 637-3 v1.2.1 intelligent transport systems (its) vehicle communications basic set of applications part 3: Specifications of decentralized environmental notification basic service. Intelligent transport systems (ITS) (2014)

15. ETSI, T.: Etsi tr 103 578 "intelligent transport systems (its); vehicular communication; informative report for the maneuver coordination service. Intelligent transport systems (ITS) (2018)

16. ETSI, T.: Etsi ts 103 561 vehicular communications basic set of applications maneuver coordination service (2018). Draft

17. ETSI, T.: 103 562 v2.1.1 intelligent transport systems (its) vehicle communications basic set of applications analysis of the collective perception service (cps). Intelligent transport systems (ITS) (2019)

18. ETSI, T.: Etsi tr 103 324 intelligent transport systems (its) cooperative perception services (2022). Draft

19. Gatto, N., Kusmenko, E., Rumpe, B.: Modeling deep reinforcement learning based architectures for cyber-physical systems. In: Proceedings of MODELS 2019. Workshop MDE Intelligence, pp. 196–202 (2019). http://www.se-rwth.de/publications/Modeling-Deep-Reinforcement-Learning-based-Architectures-for-Cyber-Physical-Systems.pdf

20. Hegde, A., Festag, A.: Artery-c: An omnet++ based discrete event simulation framework for cellular v2x. In: Proceedings of the 23rd International ACM Conference on Modeling, Analysis and Simulation of Wireless and Mobile Systems, pp. 47–51 (2020)

21. Heim, R., Kautz, O., Ringert, J.O., Rumpe, B., Wortmann, A.: Retrofitting controlled dynamic reconfiguration into the architecture description language MontiArcAutomaton. In: Software Architecture - 10th European Conference (ECSA'16). LNCS, vol. 9839, pp. 175–182. Springer (2016). http://www.se-rwth.de/publications/Retrofitting-Controlled-Dynamic-Reconfiguration-into-the-Architecture-Description-Language-MontiArcAutomaton.pdf

22. Hillemacher, S., Kriebel, S., Kusmenko, E., Lorang, M., Rumpe, B., Sema, A., Strobl, G., von Wenckstern, M.: Model-based development of self-adaptive autonomous vehicles using the SMARDT methodology. In: Proceedings of the 6th International Conference on Model-Driven Engineering and Software Development (MODELSWARD'18), pp. 163 – 178. SciTePress (2018)

23. Hölldobler, K., Kautz, O., Rumpe, B.: MontiCore Language Workbench and Library Handbook: Edition 2021. Aachener Informatik-Berichte, Software Engineering, Band 48. Shaker Verlag (2021). http://www.monticore.de/handbook.pdf

24. Hölldobler, K., Rumpe, B.: MontiCore 5 Language Workbench Edition 2017. Aachener Informatik-Berichte, Software Engineering, Band 32. Shaker Verlag (2017). http://www.se-rwth.de/phdtheses/MontiCore-5-Language-Workbench-Edition-2017.pdf

25. IEEE: IEEE-754, Standard for Floating-Point Arithmetic. IEEE Std 754-2008 pp. 1–58 (2008)

26. Kaminski, N., Kusmenko, E., Rumpe, B.: Modeling dynamic architectures of self-adaptive cooperative systems. J. Object Technol. 18(2), 1–20 (2019). https://doi.org/10.5381/jot.2019.18.2.a2. http://www.se-rwth.de/publications/Modeling-Dynamic-Architectures-of-Self-Adaptive-Cooperative-Systems.pdf. The 15th European Conference on Modelling Foundations and Applications

27. Kirchhof, J.C., Kusmenko, E., Ritz, J., Rumpe, B., Moin, A., Badii, A., Günnemann, S., Challenger, M.: MDE for machine learning-enabled software systems: a case study and comparison of MontiAnna & ML-Quadrat. In: Burgueño, L., Bork, D., Nguyen, P., Zschaler, S. (eds.) Proceedings of MODELS 2022. Workshop MDE Intelligence (2022)

28. Kloock, M., Alrifaee, B.: Coordinated cooperative distributed decision-making using synchronization of local plans (2021). https://doi.org/10.36227/techrxiv.16622017.v2. Submitted to IEEE Transactions on Intelligent Vehicles (T-IV)

29. Kloock, M., Dirksen, M., Kowalewski, S., Alrifaee, B.: Generation of coupling topologies for multi-agent systems using non-cooperative games. In: 2022 IEEE Intelligent Vehicles Symposium (IV). IEEE (2022)

30. Kloock, M., He, Q., Kowalewski, S., Alrifaee, B.: Trajectory verification for networked and autonomous vehicles using temporal logic and model checking. In: 2021 IEEE International Intelligent Transportation Systems Conference (ITSC), pp. 244–250. IEEE (2021)

31. Kloock, M., Kragl, L., Maczijewski, J., Alrifaee, B., Kowalewski, S.: Distributed model predictive pose control of multiple nonholonomic vehicles. In: 2019 IEEE Intelligent Vehicles Symposium (IV), pp. 1620–1625. IEEE (2019)

32. Kloock, M., Muehleisen, M., Calvo, J.A.L., Mathar, R.: Adaptive modulation and coding for reliable vehicular real-time communication. In: Mobile Communication-Technologies and Applications; 25th ITG-Symposium, pp. 1–9. VDE (2021)

33. Kloock, M., Scheffe, P., Botz, L., Maczijewski, J., Alrifaee, B., Kowalewski, S.: Networked model predictive vehicle race control. In: 2019 IEEE Intelligent Transportation Systems Conference (ITSC), pp. 1552–1557. IEEE (2019)

34. Kloock, M., Scheffe, P., Marquardt, S., Maczijewski, J., Alrifaee, B., Kowalewski, S.: Distributed model predictive intersection control of multiple vehicles. In: 2019 IEEE Intelligent Transportation Systems Conference (ITSC), pp. 1735–1740. IEEE (2019)

35. Kusmenko, E., Nickels, S., Pavlitskaya, S., Rumpe, B., Timmermanns, T.: Modeling and training of neural processing systems. In: Conference on Model Driven Engineering Languages and Systems (MODELS'19), pp. 283–293. IEEE (2019). http://www.se-rwth.de/publications/Modeling-and-Training-of-Neural-Processing-Systems.pdf

36. Kusmenko, E., Pavlitskaya, S., Rumpe, B., Stüber, S.: On the engineering of AI-driven systems. In: ASE'19. Software Engineering Intelligence Workshop (SEI'19), pp. 126–133. IEEE (2019). http://www.se-rwth.de/publications/On-the-Engineering-of-AI-Powered-Systems.pdf

37. Kusmenko, E., Roth, A., Rumpe, B., von Wenckstern, M.: Modeling architectures of cyber-physical systems. In: European Conference on Modelling Foundations and Applications (ECMFA'17), LNCS 10376, pp. 34–50. Springer (2017). http://www.se-rwth.de/publications/Modeling-Architectures-of-Cyber-Physical-Systems.pdf

38. Kusmenko, E., Rumpe, B., Schneiders, S., von Wenckstern, M.: Highly-optimizing and multi-target compiler for embedded system models: C++ compiler toolchain for the component and connector language EmbeddedMontiArc. In: Conference on Model Driven Engineering Languages and Systems (MODELS'18), pp. 447 – 457. ACM (2018). http://www.se-rwth.de/publications/Highly-Optimizing-and-Multi-Target-Compiler-for-Embedded-System-Models.pdf

39. Malikopoulos, A.A., Beaver, L., Chremos, I.V.: Optimal time trajectory and coordination for connected and automated vehicles. Automatica 125, 109469 (2021)

40. Mathworks Inc.: Simulink User's Guide. Technical Report. R2016b, MATLAB & SIMULINK (2016)
41. de Paula Veronese, L., Auat-Cheein, F., Mutz, F., Oliveira-Santos, T., Guivant, J.E., de Aguiar, E., Badue, C., De Souza, A.F.: Evaluating the limits of a lidar for an autonomous driving localization. IEEE Trans. Intell. Transp. Syst. **22**(3), 1449–1458 (2020)
42. Perron, L., Furnon, V.: Or-tools. https://developers.google.com/optimization/
43. Pnueli, A.: The temporal logic of programs. In: 18th Annual Symposium on Foundations of Computer Science (sfcs 1977), pp. 46–57. IEEE (1977)
44. Sakaguchi, K., Fukatsu, R., Yu, T., Fukuda, E., Mahler, K., Heath, R., Fujii, T., Takahashi, K., Khoryaev, A., Nagata, S., et al.: Towards mmwave v2x in 5g and beyond to support automated driving. IEICE Trans. Commun. **104**(6), 587–603 (2021)
45. Shuttleworth, J.: Levels of Driving Automation are Defined in New Sae International Standard j3016. SAE International, Warrendale, PA, USA (2014)
46. Völker, M., Kloock, M., Rabanus, L., Alrifaee, B., Kowalewski, S.: Verification of cooperative vehicle behavior using temporal logic. IFAC-PapersOnLine **52**(8), 99–104 (2019)
47. Wong, K., Gu, Y., Kamijo, S.: Mapping for autonomous driving: opportunities and challenges. IEEE Intell. Transp. Syst. Mag. **13**(1), 91–106 (2020)

Implicit Cooperative Trajectory Planning with Learned Rewards Under Uncertainty

Karl Kurzer, Philipp Stegmaier, Nikolai Polley, and J. Marius Zöllner

Abstract Urban traffic scenarios often require high interaction between traffic participants to ensure safety and efficiency. While the capabilities of automated driving systems have made remarkable progress in the past decade, they lack two critical abilities: anticipation and provision of cooperation between traffic participants without communication, i.e., implicit cooperation. Observing the behavior of other traffic participants, humans infer the need to cooperate and act accordingly. Our work presents a system that utilizes a sampling-based cooperative trajectory planner that accounts for all possible actions of other traffic participants, enabling cooperation. Further, we extend the planner employing learned reward models based on expert trajectories to demonstrate its ability to adapt to a desired human driving style for smooth integration into today's traffic. Lastly, we address the issue of measurement uncertainties to robustify the decision-making process in real-world environments utilizing return distributions over start states according to a belief. We exemplify the effectiveness of our solutions on 15 challenging multi-agent scenarios.

1 Introduction

While the capabilities of Automated Vehicles (AVs) are evolving rapidly, they still lack an essential component that distinguishes them from their human counterparts: the ability to cooperate (implicitly, i.e., without explicit communication, e.g., through eye contact or indicators) with others. Many of the remaining challenges in Auto-

K. Kurzer · P. Stegmaier (✉) · N. Polley · J. M. Zöllner
Karlsruhe Institute of Technology, Kaiserstraße 12, 76131 Karlsruhe, Germany
e-mail: philipp.stegmaier@kit.edu

K. Kurzer
e-mail: karl.kurzer@kit.edu

N. Polley
e-mail: nikolai.polley@kit.edu

J. M. Zöllner
e-mail: marius.zoellner@kit.edu

© The Author(s) 2024
C. Stiller et al. (eds.), *Cooperatively Interacting Vehicles*,
https://doi.org/10.1007/978-3-031-60494-2_14

413

mated Driving (AD) are due to the complex interactions between AVs and humans. Thus, it is paramount to equip future AVs with the ability to implicitly demand and provide cooperation where necessary to integrate smoothly into today's heterogeneous traffic.

Our work addresses this challenge with three key contributions:

1. A cooperative multi-agent trajectory planner in continuous space, see Sect. 2.
2. A method to infer reward functions from human expert demonstrations to ensure a human-like behavior, see Sect. 3.
3. A method to integrate measurement uncertainties into the algorithm to robustify the planning, see Sect. 4.

The code of our work is publicly available on GitHub[1] under the BSD 3-Clause License.

2   Implicit Cooperative Trajectory Planning

Cooperative planning considers all traffic participants' possible actions, intentions, and interdependencies, and seeks to maximize the total utility by following the best combination of actions. It is important to note that cooperation does not need to be beneficial for an individual agent (rational cooperation) but that it is sufficient if the combined utility increases compared to a reference point (i.e., a possible action sequence that fulfills the goals of individuals less efficiently) [15, 39].

2.1   Related Work

While the reward models vary depending on the desired behavior of the respective method [37], all approaches model the problem of implicit cooperative trajectory planning as a Markov decision process (MDP) or a partially observable Markov decision process (POMDP). And all approaches aim to find an optimal trajectory or policy by planning or learning.

The following methods were developed for cooperative planning. The majority of methods take the interdependence of each traffic participant's choices into account; nevertheless, some call for communication, and only one can plan cooperative maneuvers for continuous action spaces required by obstructed road designs and congested urban areas.

---

[1] https://github.com/ProSeCo-Planning.

2.1.1 Planning

Schwarting et al. [46] propose to determine the best individual plan using an egoistic driver model. If this strategy causes a conflict, a pairwise recursive conflict resolution procedure is started assuming that the traffic ahead influences the decision. The algorithm chooses the best non-conflicting solution. Similarly, Düring et al. [15] also conduct an exhaustive search over a communicated set of discrete actions between two vehicles and choose the joint action with minimum cost. Using this as a foundation, extensions have been created that incorporate fairness enhancements to prevent cooperation from becoming one-sided [39] and precomputed maneuver templates that match a given traffic scene with particular initial constraints to expedite the search [35]. Others propose a game-theoretic, interactive lane change method based on an exhaustive search [4].

Conducting an exhaustive search has the drawback of quickly becoming intractable for a larger number of traffic participants or longer time horizons. Thus, the problem of cooperative decision-making has been addressed by employing Monte Carlo Tree Search (MCTS) to estimate the best maneuver combination over multiple time steps. While some approaches rely on higher-level abstractions, such as the probability of participants yielding and driver models such as the Intelligent Driver Model (IDM) [22, 33], other approaches model the complete interaction and reasoning [29, 55].

Hubmann et al. [20, 21] propose interactive intersection-navigation and lane-change systems, modeled using POMDPs. They are solved via the Adaptive Belief Tree algorithm [23, 48].

Based on game-theoretical modeling, Schwarting et al. present an interactive lane-changing and intersection-navigation method [47]. Each of the traffic participants is considered with its entire action space. The cooperative component is included by assessing the agent's social value orientation. A state-of-the-art nonlinear optimizer is employed to solve the nonlinear program.

Other approaches are not explicitly cooperative. However, they capture the interdependencies of actions as they evaluate the risk resulting from different maneuver combinations. They predict the motions of vehicles [32] and can generate proactive and cooperative driving actions [3].

2.1.2 Learning

Model-free Reinforcement Learning (RL) is used by Saxena et al. to propose an interactive lane change method in dense traffic [42]. Similar to the Social Generative Adversarial Network (SGAN) [2], the interactions of traffic participants do not need to be explicitly modeled, as they are learned implicitly. The data for training is generated with driver models based on IDM and Minimizing Overall Braking Induced by Lane Changes (MOBIL) in a simulator.

Bouton et al. propose an intersection navigation method capable of handling occlusions using model-free RL and a probabilistic model checker to ensure safety [8]. Two other traffic participants are modeled using rule-based models. The resulting interactions are learned implicitly from data during training. Further, the extension to scenarios with additional traffic participants is realized through scene decomposition, possibly leading to extremely conservative behavior.

Similarly, they propose an interactive single-lane lane change method in dense traffic using model-free RL [7]. The interactions of traffic participants do not need to be explicitly modeled, as they are learned implicitly from data during training generated in a simulator based on IDM driver models. The cooperativeness of other drivers is estimated using a recursive Bayesian filter. It can be seen that an MCTS approach with full observability performs best.

2.1.3   Combined Learning and Planning

Bae et al. propose an interactive lane change method in dense traffic using the predictions of an SGAN [17] as a basis for planning controls using Model Predictive Control (MPC) [2]. The SGAN is trained on simulations of dense traffic using the IDM and MOBIL driver models. The controller uses Monte Carlo rollouts to create a set of trajectories during planning that is assessed alongside the SGAN predictions. The trajectory with the lowest cumulative cost, which does not violate the constraints, is chosen. This approach is limited to lane changes and uses simple driver models for other traffic participants. Further, the amount of interaction is reduced due to the planning horizon of 2 s.

Hoel et al. propose an interactive lane change method using a combination of MCTS and RL [18]. The driver models used for other traffic participants are also based on IDM and MOBIL. The driver state is only partially observable, and its belief state is inferred using a particle filter. The action space does not include actions that result in collisions. An artificial neural network (ANN) that guides the sampling within the MCTS also generates state-value targets for the MCTS.

2.2   Problem Formulation

We model the cooperative trajectory planning problem with a variable number of non-communicating agents interacting in a Markov Game (MG). In a multi-agent system, the state transition and reward depend on the behaviors of all actors. But for every timestep, each agent individually picks an action without being aware of the choices made by others. All agents receive an immediate reward, and the system is transferred to the next state.

Therefore, it is the aim of each agent to maximize its expected cumulative reward,

$$G(\tau) = \sum_{(s_t, a_t) \in \tau} \gamma^t \mathcal{R}_{s_t}^{a_t}, \tag{1}$$

in the MG, by choosing a trajectory τ consisting of state-action tuples (s_t, a_t) that maximize the sum of rewards $\mathcal{R}_{s_t}^{a_t}$ over the time horizon discounted by the discount factor γ.

While there are multiple ways to maximize the expected cumulative reward, one of them is value-based RL. Here, the agent attempts to estimate the state value function $V^\pi(s)$ for each state s of the MDP to later determine the optimal policy π^* by maximizing the state value for each state,

$$\pi^*(a \mid s) = \arg \max_\pi V^\pi(s). \tag{2}$$

Value-based RL can find the optimal policy; however, it is very time intensive. Therefore, in contrast to learning the optimal policy (global optimization for all states), we frame the problem as searching for an optimal action (local optimization for a single state, i.e., the current state).

Hence, the problem is formulated as the search for the optimal action, given a state in an MG.

2.3 Approach

We employ Monte Carlo Tree Search (MCTS) to search for the optimal action. MCTS, a reinforcement learning method [56], has shown great potential when facing problems with huge branching factors. The most prominent example, AlphaGo, reached super-human performance in the game Go [49]. The following introduces the basics of MCTS.

2.3.1 Monte Carlo Tree Search

An exhaustive tree search can find the optimal trajectory through any MDP with a finite set of states and actions [9]. However, as the action space grows, it quickly becomes intractable to search for the optimal trajectory through the entire tree.

Tree Search combined with Monte Carlo sampling addresses this issue by approximating the optimal solution asymptotically. Monte Carlo Tree Search (MCTS) is a computationally efficient, highly selective best-first search [13, 24], that explores different trajectories through the MDP to discover the trajectory that maximizes the return G from the root state.

Given an initial root state of the MDP, MCTS approximates the state-action value in four sequential steps during each iteration until a terminal condition is met (e.g., until a time budget or computational budget is exceeded). Since MCTS is an anytime algorithm [24], it returns an estimate after the first iteration.

Selection   Expansion

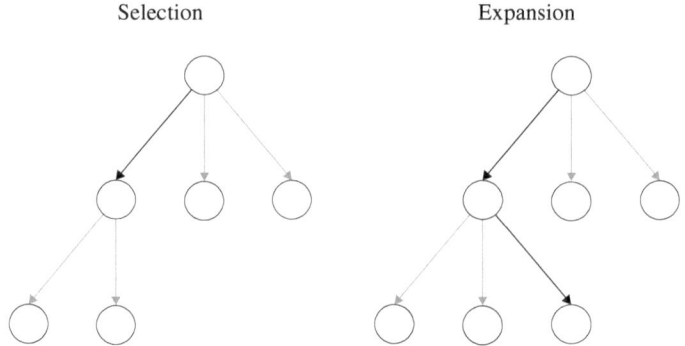

Fig. 1 Selection and expansion in MCTS: circles denote states and edges denote actions

Selection

The most popular form of MCTS uses an Upper Confidence Bound for Trees (UCT), to control the selection of successor states. The UCT value [24] of all explored actions from the current state is calculated during the selection phase, see (3), and the state action tuple with the maximum UCT value is selected. This process repeats until a state is selected that has not been fully explored (i.e., not all available actions in the state have been expanded), see Fig. 1.

Using UCT, MCTS solves the exploration—exploitation dilemma [24], an upper confidence bound for the estimation of the true state-action value. The first term in (3), the estimated state-action value function $\widehat{Q}^\pi(s, a)$, fosters exploitation of previously explored actions with high state-action values. The second term guarantees that all actions for a given state are being explored at least once, with $N(s)$ being the visit count for state s and $N(s, a)$ the number of times action a has been chosen in that state. To balance the exploration-exploitation trade-off, a constant factor c is used [24].

$$\mathrm{UCT}(s, a) = \widehat{Q}^\pi(s, a) + c\sqrt{\frac{2 \log N(s)}{N(s, a)}} \tag{3}$$

Expansion

Once the selection policy encounters a state with untried actions left, it expands that state by randomly sampling an action from a uniform distribution over the action space, see (4), and executing the action reaching a successor state, see Fig. 1.

$$a \sim U[\min(\mathcal{A}), \max(\mathcal{A})] \tag{4}$$

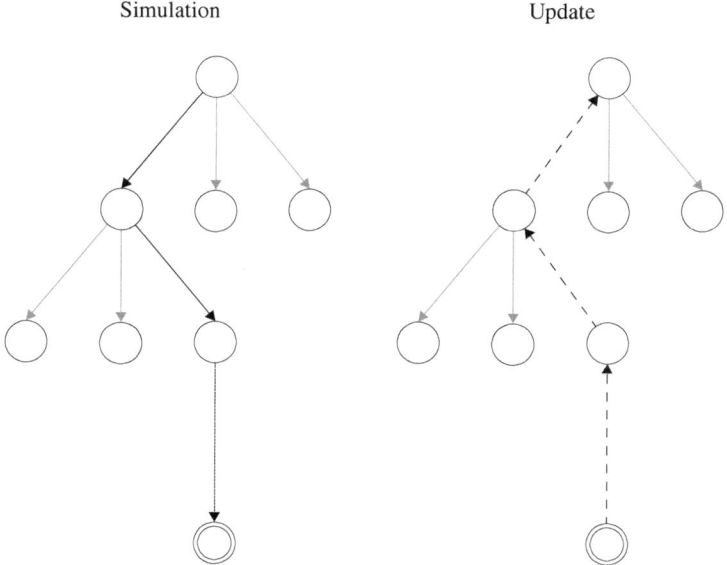

Fig. 2 Simulation and update in MCTS: circles denote states and edges denote actions

Simulation

After the expansion of an action completes, a simulation (i.e., a roll-out) of subsequent random actions is conducted until a terminal condition is met (i.e., the planning horizon is reached or an action resulting in a terminal state is sampled). This generates an estimate of the state-action value for the previous expansion, see Fig. 2.

Update

Lastly, the return G of the trajectory τ generated by the iteration is backpropagated to all states along the trajectory, see Fig. 2, and the state-action values and visit counts for all actions of the trajectory are updated, see (5) and (6), respectively.

$$N(s, a) \leftarrow N(s, a) + 1 \tag{5}$$

$$Q^\pi(s, a) \leftarrow Q^\pi(s, a) + \frac{1}{N(s, a)}(G(\tau) - Q^\pi(s, a)) \tag{6}$$

2.3.2 Multi-agent Driving Simulator

To simulate an agent's experience, RL-based methods need access to an environment, such as a simulator or the real world. However, as RL is founded on the notion of trial and error, it is challenging to guarantee safety while learning in the actual world. Consequently, we use simulation; the multi-agent driving simulator created for and utilized throughout this work is described in the following.

State Space

The state s of a traffic participant is defined with its

- longitudinal position x_{lon},
- lateral position x_{lat},
- longitudinal velocity $\dot{x}_{\text{lon}}$,
- lateral velocity $\dot{x}_{\text{lat}}$,
- longitudinal acceleration $\ddot{x}_{\text{lon}}$,
- lateral acceleration $\ddot{x}_{\text{lat}}$, and
- heading ϕ.

Further, each traffic participant is a vehicle denoted by its

- width w,
- length l,
- wheelbase l_{wb},
- maximum acceleration a_{max} and
- maximum steering angle δ_{max}.

An agent υ represents a traffic participant. Generally, this could be any participant, such as a car, truck, trailer, motorcycle, bicycle, or even a human. This work uses the kinematics of a single-track model. Thus, only car-like vehicles are considered.

Action Space

The action space of an agent is two-dimensional. The two dimensions are the longitudinal velocity change $\Delta \dot{x}_{\text{lon}}$ and the lateral change in position Δx_{lat}. The tuple describes the desired state change over the action duration $\Delta T = t_1 - t_0$, with t_0 denoting the start and t_1 the end of an action. Based on the current state and the chosen action, a jerk-optimal trajectory is calculated using quintic polynomials [54], one for the longitudinal and lateral direction, respectively, see (7),

$$
\begin{aligned}
x(t) &= a_0 + a_1 t + a_2 t^2 + a_3 t^3 + a_4 t^4 + a_5 t^5 \\
\dot{x}(t) &= a_1 + 2a_2 t + 3a_3 t^2 + 4a_4 t^3 + 5a_5 t^4 \\
\ddot{x}(t) &= 2a_2 + 6a_3 t + 12a_4 t^2 + 20a_5 t^3
\end{aligned}
\tag{7}
$$

as well as in matrix form (8).

$$
Ma = \begin{bmatrix} 1 & t_0 & t_0^2 & t_0^3 & t_0^4 & t_0^5 \\ 0 & 1 & 2t_0 & 3t_0^2 & 4t_0^3 & 5t_0^4 \\ 0 & 0 & 2 & 6t_0 & 12t_0^2 & 20t_0^3 \\ 1 & t_1 & t_1^2 & t_1^3 & t_1^4 & t_1^5 \\ 0 & 1 & 2t_1 & 3t_1^2 & 4t_1^3 & 5t_1^4 \\ 0 & 0 & 2 & 6t_1 & 12t_1^2 & 20t_1^3 \end{bmatrix} \begin{bmatrix} a_0 \\ a_1 \\ a_2 \\ a_3 \\ a_4 \\ a_5 \end{bmatrix} \tag{8}
$$

With each polynomial requiring six coefficients, six constraints need to be defined. The constraints are denoted by c, see (9).

$$
c = \begin{bmatrix} x(t_0) \\ \dot{x}(t_0) \\ \ddot{x}(t_0) \\ x(t_1) \\ \dot{x}(t_1) \\ \ddot{x}(t_1) \end{bmatrix} \tag{9}
$$

The start constraints are determined by the current state of the vehicle. The end constraints for $\dot{x}_{lon}$ and x_{lat} are based on the selected action. The end position is given by the mean of the start and end velocity and the action duration ΔT. Further, the acceleration in longitudinal and lateral direction, as well as the velocity in lateral direction are set to zero (10).

$$
\begin{aligned}
x_{lon}(t_1) &= x_{lon}(t_0) + \frac{\dot{x}_{lon}(t_0) + \dot{x}_{lon}(t_1)}{2} \Delta T \\
\dot{x}_{lon}(t_1) &= \dot{x}_{lon}(t_0) + \Delta \dot{x}_{lon} \\
\ddot{x}_{lon}(t_1) &= 0 \\
x_{lat}(t_1) &= x_{lat}(t_1) + \Delta x_{lat} \\
\dot{x}_{lat}(t_1) &= 0 \\
\ddot{x}_{lat}(t_1) &= 0
\end{aligned} \tag{10}
$$

Using the constraints, (7) can be solved for its coefficients a, see (11) (assuming M is invertable).

$$
\begin{aligned}
Ma &= c \\
M^{-1}Ma &= M^{-1}c \\
a &= M^{-1}c
\end{aligned} \tag{11}
$$

We generate trajectories utilizing the Frenet frame, a dynamic reference frame that aligns with the road's centerline rather than Cartesian coordinates [59]. This transformation enables the separation of longitudinal and lateral trajectory planning.

Transition Function

Since this work does not focus on trajectory planning close to physical limits (e.g., required by evasive maneuvers), trajectories are evaluated using a single track model [45] as it has been shown to perform sufficiently well for trajectory planning tasks [25]. The execution of trajectories derived from the selected action is deterministic.

Physical Constraints

To ensure that a chosen trajectory is drivable for a single-track model, the differential and kinematic constraints, i.e., the maximal acceleration and the minimum curve radius, must be accounted for [31].

Based on the polynomials that describe the trajectories in longitudinal and lateral directions, the heading

$$\phi = \arctan\left(\frac{\dot{x}_{\text{lat}}}{\dot{x}_{\text{lon}}}\right), \tag{12}$$

curvature

$$\kappa = \frac{\dot{x}_{\text{lon}}\ddot{x}_{\text{lat}} - \dot{x}_{\text{lat}}\ddot{x}_{\text{lon}}}{\left(\dot{x}_{\text{lon}}^2 + \dot{x}_{\text{lat}}^2\right)^{\frac{3}{2}}}, \tag{13}$$

steering angle

$$\delta = \arctan\left(l_{\text{wb}}\kappa\right), \tag{14}$$

acceleration

$$\ddot{x} = \sqrt{\ddot{x}_{\text{lon}}^2 + \ddot{x}_{\text{lat}}^2}, \tag{15}$$

and velocity

$$\dot{x} = \sqrt{\dot{x}_{\text{lon}}^2 + \dot{x}_{\text{lat}}^2} \tag{16}$$

are calculated.

A new action is sampled if a selected action violates either the maximum steering angle or the maximum acceleration. Resampling also occurs if an action would lead a traffic participant off the road, see Fig. 3. However, resampling is limited to a maximum number of retries.

The physical constraints are considered in the validation reward of the reward function, see Sect. 2.3.2.

Reward Function

The reward function $\mathcal{R}_s^a$ is the basis for the agent's behavior. It considers the state s and the action a of an agent.

Fig. 3 Invalid states: invalid successor states resulting from different actions

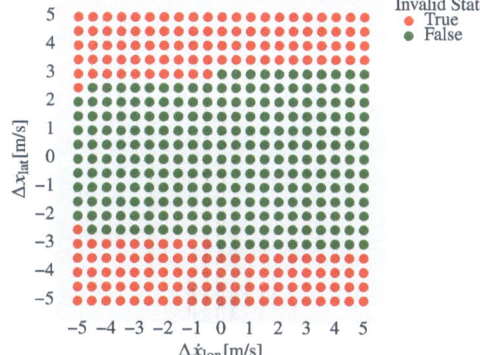

$$r = r_s + r_a + r_{\text{validation}} \tag{17}$$

The importance of each of the features mentioned below is adjusted with a corresponding weight.

State Reward

The state reward r_s is based on the divergence of the current state to the desired state. The desired state is defined by a longitudinal velocity v_{des} and a lane index l_{des}. The desired values must be estimated for all agents by a separate module. We assume that the desired longitudinal velocity and the lane index are equal to the values before the interaction in each scenario.

Additionally, the agent is encouraged to drive close to the center of a lane by reducing the reward for deviations from the center.

Action Reward

Actions are selected to minimize the deviation from the desired state, ideally leading to an equilibrium state. Thus, all actions are penalized, and the action reward r_a is always negative, i.e., a cost. In this work, r_a considers longitudinal and lateral acceleration as well as lane changes, with w being the weights and Δl the number of lane changes an action results in. If desired, both the state and action reward can easily be extended to capture additional safety, efficiency, and comfort-related aspects of the generated trajectories.

Validation Reward

The last term is the action validation reward, see (18). It evaluates whether a state and action are valid, i.e., being inside the drivable environment and adhering to the physical constraints and whether a state action combination is collision-free.

$$
\begin{aligned}
r_{\text{validation}} = {}& w_{\text{invalid state}}\, \mathbb{1}_{\text{invalid state}} \\
& + w_{\text{invalid action}}\, \mathbb{1}_{\text{invalid action}} \\
& + w_{\text{collision}}\, \mathbb{1}_{\text{collision}}
\end{aligned}
\tag{18}
$$

Cooperative Reward

To achieve cooperative behavior, a cooperative reward r_{coop}^i is defined. The cooperative reward of an agent i is the sum of its own rewards, see (17), as well as the sum of all other rewards of all other agents multiplied by a cooperation factor λ^i, see (19), [29, 33]. The cooperation factor determines the agent's willingness to cooperate with other agents with $\lambda^i = 0$ being purely interactive and $\lambda^i = 1$ being fully cooperative.

$$
r_{\text{coop}}^i = r + \lambda^i \sum_{j=0, j \neq i}^{n} r^j
\tag{19}
$$

The cooperation factor can be used to represent different driver types. For example, an offensive driver weighs his own goals more than the goals of other road users. The joint reward function (19) is agent-individual and does not represent a global cost function. Therefore, the cooperative rewards of the individual agents cannot be compared since they have different values depending on the respective cooperation factor.

2.3.3 Decentralized Continuous MCTS

We depict an exemplary application of MCTS to a traffic scenario requiring cooperation in Fig. 4. It is important to note that while it is possible to apply MCTS directly, specific extensions are required to model the interdependence of actions between traffic participants and allow for continuous action spaces.

The original MCTS algorithm uses UCT [24], designed for sequential decision-making games with a finite set of states and actions. However, if traffic participants interact without communication, the actions of other traffic participants are not known until they are observed. Thus the basic MCTS used in turn-based games is not applicable, and we need to extend it to simultaneous move games. In addition, the requirement of trajectory planning in a continuous state and action space

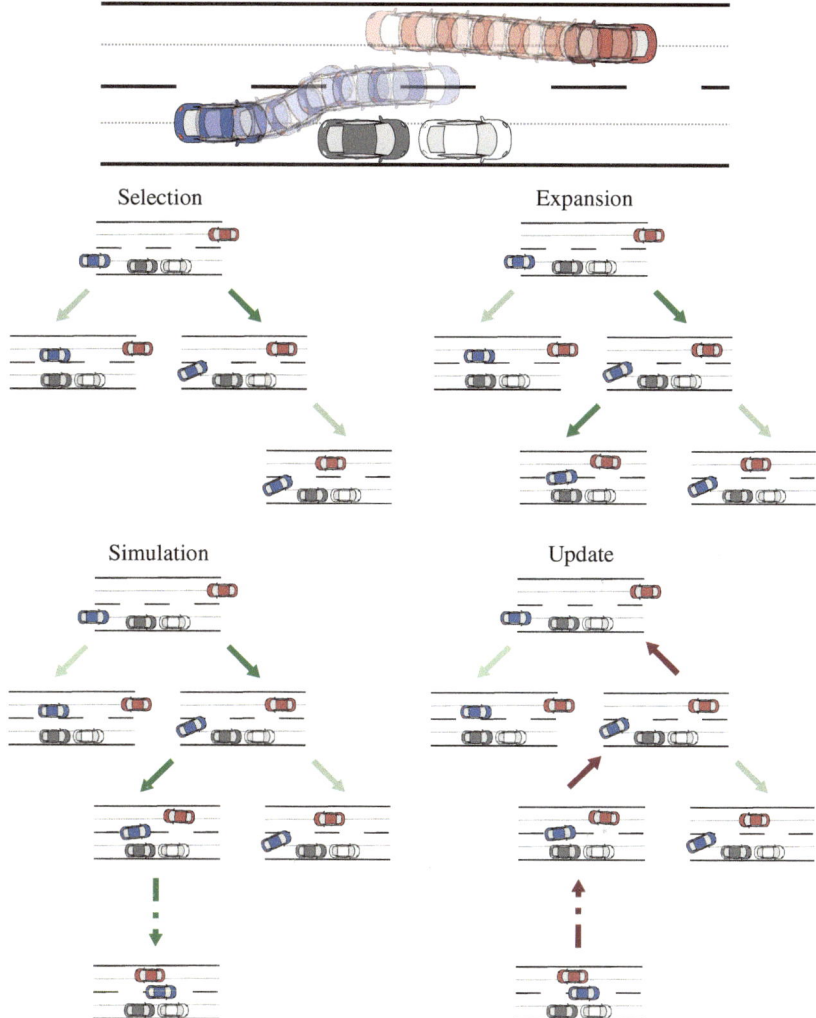

Fig. 4 Phases of MCTS for a scenario with a narrow passage; The gray vehicles are parked, and the red and the blue vehicle can pass the narrowing simultaneously if the red vehicle deviates from its optimal trajectory. During the selection, MCTS traverses the tree by selecting auspicious future states until a state is encountered that has untried actions left. After the expansion of the state, a simulation of subsequent actions is run until the planning horizon is reached. Next, the result is backpropagated to all states along the selected path. Reference [29] © 2018 IEEE

requires alterations to the standard selection procedure of actions. These alterations are necessary because UCT would degenerate MCTS to Monte Carlo Search (MCS), as each action in each state needs to be selected at least once; see (3).

Decoupled UCT

We employ Decoupled UCT (DUCT) to address the problem of decentralized, simultaneous decision-making [53]. While the decoupled version of UCT does not guarantee to converge to an optimal policy [43], it has been shown to perform best compared
to other variants [53].

The complexity of simultaneous decision-making results from incomplete information regarding the decision of other agents. Thus, the state-action value for an
action a_i from agent i can only be approximated by averaging over all possible actions
of the other agents. Based on the description of MCTS in Sect. 2.3.1, DUCT hence
tracks the state-action value and visit count on a per-agent basis, and the dependency
between different agents i is not considered when calculating the DUCT value,

$$\text{DUCT}(s, a_i) = \widehat{Q}^\pi(s, a_i) + c\sqrt{\frac{2 \log N(s)}{N(s, a_i)}}, \quad a_i \in \mathcal{A}^i. \tag{20}$$

Each agent selects the action that maximizes its DUCT value during the selection
step. The resulting joint action $\boldsymbol{a}$ leads to the successor node if it exists or expands a
new node.

For two agents with identical action spaces with three actions, Fig. 5 depicts
all possible successor states. Since DUCT tracks the state-action value and visit
count separately for each agent, s_0 is considered fully expanded once each agent
has executed each of its available actions (a_0, a_1, a_2) at least once (s_2, s_6, s_7), rather
than all possible combinations resulting from the joint action space. Thus, an action
can only be executed again if all of an agent's actions have already been explored at
least once. Only by randomly selecting a joint action $\boldsymbol{a}$ that has not previously been
selected the remaining combinations $(s_1, s_3, s_4, s_5, s_8, s_9)$ are added to the search
tree.

Similarly, the final selection is conducted independently of other agents.

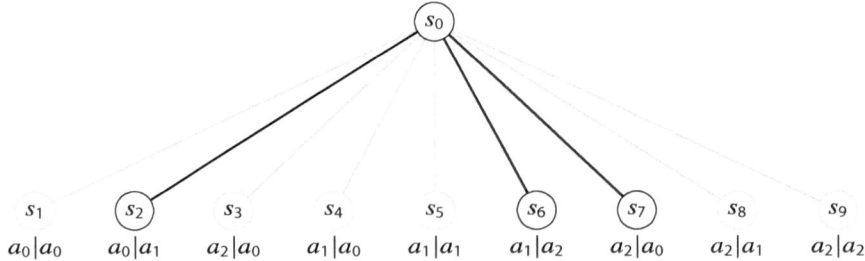

Fig. 5 Action combinations in DUCT: the resulting successor states are based on an identical action
space $[a_0, a_1, a_2] \in \mathcal{A}(s_0)$ for two agents, with $\cdot|\cdot$ denoting the joint action $\boldsymbol{a}$ that led to a state

Progressive Widening

The use of UCT, see (3), in MCTS requires exploring all possible actions from a given state [24]. Actions of the successor states are only explored [9, 24] once all actions of the predecessor have been explored. Thus, if the action space is continuous, the application of UCT within MCTS degenerates to MCS.

Progressive widening, sometimes called progressive unpruning, aims to address the issue of large action spaces [10, 13], with additional research considering infinitely many actions within UCT [57].

Progressive widening gradually expands the existing action space of a node by adding additional actions. The number of actions for a state follows a sublinear function of the visit count N, see (21), with $c \in \mathbb{R}^+$ and $\alpha \in (0, 1)$ being determined empirically. The expansion of the action space can be random or follow a heuristic.

$$|\mathcal{A}(s)| = \lfloor cN(s)^\alpha \rfloor \tag{21}$$

Picking a random action from the theoretical action space is the simplest technique to add new actions. Another more advanced method is to utilize a heuristic, such as blind value, as described below.

Since the visit count for individual actions decreases with increasing search depth due to branching, the application of progressive widening is restricted to a specific depth within the tree that we determined empirically. Therefore, the restriction ensures that the available actions at larger depths are still sufficiently explored.

Expansion Strategies

The expansion strategy is either random or guided. The random expansion strategy samples uniformly from the entire action space. In order to be able to use a continuous action space, we employ progressive widening to decide whether the action space should be expanded. The guided expansion strategy uses a heuristic such as blind value to find a promising node for expansion.

Final Selection Strategies

After the computational budget of the planning phase is exhausted, an action must be selected to be executed. While UCT defines a clear selection criterium within the search tree of MCTS, different strategies for the final selection exist. Two of the most common are

$$\text{MaxVisitCount}(s) = \arg\max_{a \in \mathcal{A}(s)} N(s, a) \tag{22}$$

and

$$\text{MaxActionValue}(s) = \arg\max_{a \in \mathcal{A}(s)} \widehat{Q}^{\pi}(s, a). \qquad (23)$$

Blind Value

Whenever a node is to be expanded, MCTS needs to add an action to the action space of the node. The standard strategy for discrete as well as small continuous action spaces is to employ uniform sampling over the entire action space [9, 11]. However, as the action space grows, it becomes less likely that promising regions are sampled.

A heuristic that aims to increase the likelihood of sampling promising regions is called blind value (BV) [11]. Blind value uses the previously explored actions of a node to guide the next expansion. It first focuses on regions away from previously explored actions and then shifts towards regions with many highly valued actions.

The blind value for an action a_i of a set of randomly sampled actions $\mathcal{A}_{\text{rnd}}$ is calculated using the set of explored actions $\mathcal{A}_{\text{exp}}$ as well as an adaptation coefficient ρ, see (24) and (26), respectively.

$$BV(a_i, \mathcal{A}_{\text{exp}}, \rho) = \min_{a_j \in \mathcal{A}_{\text{exp}}} UCT(s, a_j) + \rho d(a_i, a_j) \qquad (24)$$

$$d(a_i, a_j) = \sqrt{\left(a_i^{\Delta \dot{x}_{\text{lon}}} - a_j^{\Delta \dot{x}_{\text{lon}}}\right)^2 + \left(a_i^{\Delta x_{\text{lat}}} - a_j^{\Delta x_{\text{lat}}}\right)^2} \qquad (25)$$

$$\rho(\mathcal{A}_{\text{exp}}, \mathcal{A}_{\text{rnd}}) = \frac{\sigma\left(\{UCT(s, a_j) \mid \forall a_j \in \mathcal{A}_{\text{exp}}\}\right)}{\sigma\left(\{d(0, a_i) \mid \forall a_i \in \mathcal{A}_{\text{rnd}}\}\right)} \qquad (26)$$

The action with the highest blind value is finally selected, see (27) (Fig. 6).

$$a^* = \arg\max_{a_i \in \mathcal{A}_{\text{rnd}}} BV(a_i, \mathcal{A}_{\text{exp}}, \rho) \qquad (27)$$

Fig. 6 Blind values of actions in the continuous space: assuming that the previously explored actions have identical UCT values, $a_i \in \mathcal{A}_{\text{rnd}}$ has the highest blind value, since its distance to other actions is largest, cf. (24) and (25)

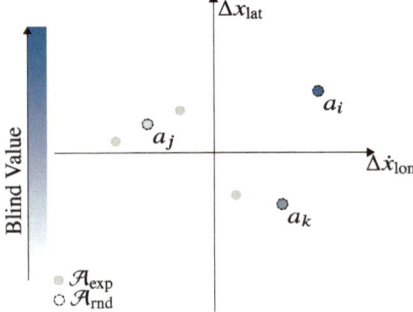

2.4 Experiments

We evaluate the algorithm's performance on 15 challenging multi-agent scenarios, which can be found online.[2] Each scenario consists of at least two agents with conflicting goals. Further, the start state of each agent is determined randomly within predefined nonoverlapping areas.

Each scenario has a defined terminal condition; we deem it solved once reached. The resulting average scenario length is 16.5 s.

Due to the inherent random nature of the algorithm, we evaluated each configuration using 250 random seeds to generate statistically relevant results.[3]

We assess the results with the success rate indicating the fraction of collision-free and valid trajectories. An agent's trajectory is valid if the agent stays within the road boundaries and if the trajectory is physically drivable. We compare the results for different numbers of iterations.

First, we demonstrate that resampling invalid actions (i.e., randomly sampling a new action to sample a valid action), either due to violations of physical constraints, collisions with static obstacles, or driving off the road, improves the algorithm's performance substantially, see Fig. 7a. With this, scenarios one to six reach a mean success rate close to 98.07 % for 1280 iterations, Fig. 7b. For scenarios seven to twelve, the mean success rate decreases to 95.13 % for 1280 iterations. This demonstrates the general applicability of our approach in cooperative driving scenarios. The more challenging scenarios, thirteen to fifteen, reach a mean success rate of 42.13 % for 1280 iterations and thus require a larger computational budget due to their complexity.

Second, we show that applying the blind value heuristic improves the baseline slightly, Fig. 7c. However, while most deviations are significant, no selected number of samples increases the performance consistently in a significant manner. This observation is surprising since the insignificant change only occurs for 160 iterations. Further, increasing the number of samples does not necessarily increase performance either. Overall, 10 and 20 samples perform best. The results for ten samples are depicted in Fig. 7d. The changes are negligible for scenarios one to six, already reaching high overall performance. The scenarios that profit the most from the blind value heuristic to guide the exploration of the action space are scenarios seven to nine and scenario fifteen. The overall improvement for scenarios seven to fifteen is 3.21 % points over all iterations.

---

[3] If the deviation from the baseline for any configuration is significant, we annotate it by an asterisk. In addition, if a configuration is consistently deviating from the baseline, i.e., the deviation is significant for all numbers of iterations, we mark that trace with an asterisk as a suffix in the legend.

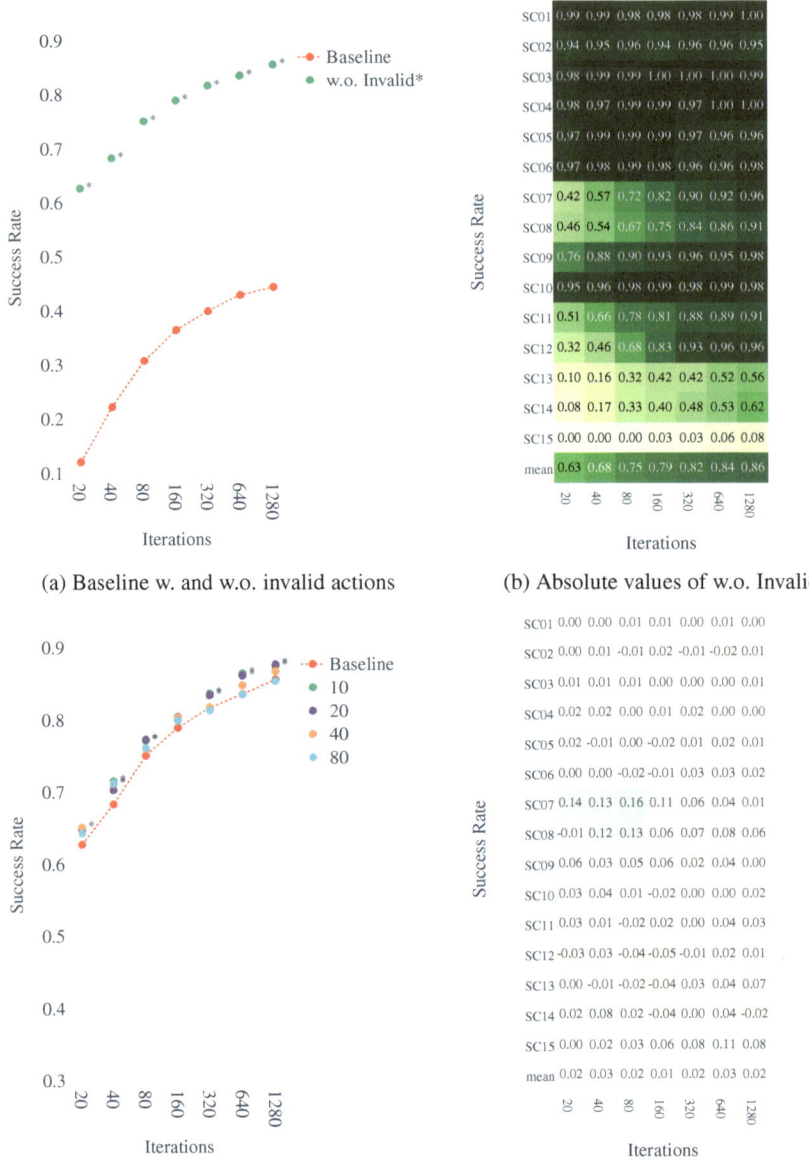

(a) Baseline w. and w.o. invalid actions

(b) Absolute values of w.o. Invalid

(c) BV for differing number of samples and baseline

(d) Absolute difference of BV for 10 samples
compared to the baseline

Fig. 7 Comparison of the mean success rates of the baselines if invalid actions are included or resampled (**a**), with a detailed view of the absolute performance of resampling invalid actions (**b**). In (**c**) the performance is assessed in comparison with the blind value (BV) heuristic. The absolute difference of BV compared to the baseline is shown in (**d**)

3 Learning Reward Functions

Reinforcement Learning (RL) based approaches frequently use manually specified reward models [30, 60]. In environments where systems must interact with humans, their decisions must be comprehensible and predictable. However, as the complexity of the reward model rises its manual parametrization to generate a desired behavior becomes infeasible.

In the case of driving, it is clear that various features influence the reward of any given trajectory [37]. While tuning the weighting of features to create the desired behavior in a diverse set of scenarios is tedious and error-prone, Inverse Reinforcement Learning (IRL) has proven to be able to recover the underlying reward model from recorded trajectories that demonstrate expert behavior in areas such as robotics and automated driving [1, 26, 38, 63, 65].

Our work builds upon the previously introduced cooperative trajectory planning algorithm to generate expert trajectories. Our contribution is twofold. The first is a system that conducts Guided Cost Learning (GCL) [16], a sampling-based Maximum Entropy Inverse Reinforcement Learning method with Monte Carlo Tree Search (MCTS) to efficiently solve the forward RL problem in a cooperative multi-agent setting. The second is an evaluation that compares a linear and nonlinear reward model to a manually designed one. We demonstrate that the performance of the learned models is similar to or better than the tediously tuned baseline. An overview of the system is depicted in Fig. 8.

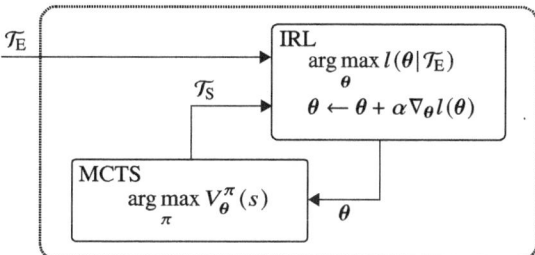

Fig. 8 Overview of the system: at first, an initial set of expert trajectories $\mathcal{T}_E$ is generated. Then the cooperative trajectory planning algorithm computes a set of sample trajectories $\mathcal{T}_S$ using the randomly initialized reward model. Next, using the $\mathcal{T}_E$ and $\mathcal{T}_S$, the likelihood of the parameters θ given the expert trajectories is increased using gradient ascent. Finally, the process repeats with the cooperative trajectory planning algorithm, sampling new trajectory samples until convergence. Reference [28] © 2022 IEEE

3.1 Related Work

While RL's task is deducing an optimal policy from an agent's interactions with the environment based on a reward model [52], the opposite is the case for IRL. Here, the task is to infer the underlying reward model that the optimal policy aims to maximize [38]. Since the reward model is the most succinct and transferable description of an agent's behavior [1], a close approximation of the underlying reward model will yield a behavior similar to the behavior that results from the optimal policy, i.e., the expert behavior.

Early work in IRL performed feature matching rather than estimating the true underlying reward function [1] to learn driving styles in a discrete driving simulator. More recently, driving styles have been learned using continuous trajectories and action spaces [26], including additional features that impact driver preference [37].

Wulfmeier et al. demonstrate the effectiveness of learning nonlinear reward models building on Maximum Entropy IRL [65] using Deep Neural Networks [62], which they extended to learning cost maps for path planning from raw sensor measurements [63].

Further improvements in the approximation of the partition function and the efficiency of IRL in combination with RL have been proposed by Guided Cost Learning [16]. By adapting the IRL procedure, the method yields both a cost function and policy given expert demonstrations using sampling-based methods. In addition, more efficient one-shot sampling methods have been proposed [61].

3.2 Problem Formulation

Using the definitions of a policy

$$\pi(a \mid s) \quad a \in \mathcal{A}(s), \tag{28}$$

and a trajectory

$$\tau = (s_0, a_0, s_1, a_1, \ldots, s_T), \tag{29}$$

a policy over trajectories can be defined as

$$\rho(\tau) = \rho(s_0, a_0, s_1, a_1, \ldots, s_T) = \prod_{t=0}^{T} \pi(a_t \mid s_t), \tag{30}$$

assuming a deterministic start state distribution and transition model. The return of a trajectory τ equals its accumulated discounted reward at time step t, taking action a_t in state s_t [52], see (1). The value function of a policy π for an MDP with a reward model parameterized by $\boldsymbol{\theta}$ is the expectation of the return of trajectories sampled from that policy,

$$V_\theta^\pi(s) = \mathbb{E}_{\tau \sim \rho}[G_\theta(\tau)].\qquad(31)$$

While the forward RL problem is solved by finding the optimal policy, see (2), the inverse RL problem is solved by finding the parameters θ so that,

$$V_\theta^{\pi_E}(s) \geq V_\theta^\pi(s) \quad \forall \pi \in \Pi,\qquad(32)$$

with π_E being the expert policy as part of the policy space Π.

This work aims to learn the parameters θ of a reward model for cooperative trajectory planning so that the optimal trajectories of the planning algorithm are similar to the demonstrated expert trajectories, i.e., that the expert policy yields the highest state value of all policies given the parametrization of the reward model (32).

3.3 Approach

IRL is used to learn a reward model from expert demonstrations so that the behavior sampled from the optimal policy based on this reward model resembles the expert demonstrations. Concisely, this work makes use of the previously introduced cooperative trajectory planning algorithm based on MCTS [29], and Maximum Entropy IRL [65], yielding a system that is similar to Guided Cost Learning [16].

The MCTS is used to solve the (forward) RL problem, i.e. finding the optimal policy/action given a reward model and generating near optimal trajectory samples $\mathcal{T}_S$ for that policy. Using these trajectories in combination with the expert trajectories $\mathcal{T}_E$, Maximum Entropy IRL is used to conduct a gradient ascent step increasing the likelihood of the parameters θ given the expert trajectories, (50), (51), see Fig. 8.

3.3.1 Reward Model

The reward model is a central part of an RL system, as the goal of RL is to maximize the cumulative discounted reward by finding the optimal policy [52].

Initially, IRL applied solely linear reward models that are represented as a linear combination of features $\phi(s, a)$ and parameters θ (42) [1]. However, especially for larger RL problems, linear reward models have been outperformed by nonlinear reward models such as ANNs [16, 62]. This work uses both a linear reward model and a nonlinear reward model in the form of an ANN.

Features

Similar to many other planning methods, the cooperative trajectory planner assumes the desired lane l_{des} as well as the desired velocity v_{des} for each agent [37]. State and action dependent features $\phi(s, a)$ are scalar values that consider specific characteristics of a state and action. Each feature is evaluated for each time step t of the trajectory,

$$\phi(\tau) = \frac{1}{T} \sum_{(s_t, a_t) \in \tau}^{T} \phi(s_t, a_t). \tag{33}$$

The parameters θ are identical for all agents, however features are not. All features are normalized to lie between $[-1, 1]$ for the length T of a trajectory τ. The feature for the desired lane is defined as

$$\phi_{\text{desLane}}(\tau) = \frac{1}{T} \sum_{(s_t, a_t) \in \tau}^{T} \max\left(1 - |l_t - l_{\text{des}}|, -1\right), \tag{34}$$

encouraging the agent to drive on the desired lane. A deviation from the desired velocity v_{des} larger than 10 % results in a negative feature value.

$$\phi_{\text{desVelocity}}(\tau) = \frac{1}{T} \sum_{(s_t, a_t) \in \tau}^{T} \max\left(1 - 10\left|\frac{v_t}{v_{\text{des}}} - 1\right|, -1\right) \tag{35}$$

Similarly, deviating more than a quarter of the lane width l_{width} from the lane center l_{center} yields a negative feature value.

$$\phi_{\text{laneCenter}}(\tau) = \frac{1}{T} \sum_{(s_t, a_t) \in \tau}^{T} \max\left(1 - \frac{|l_{\text{center}} - y_t|}{l_{\text{width}}/4}, -1\right) \tag{36}$$

To avoid excessive accelerations, a proxy value for the acceleration a of an action is determined,

$$C_{\text{acceleration}} = \frac{1}{g}\sqrt{\frac{\int_t^{t+\Delta T} (a(t))^2 \, dt}{\Delta T}}. \tag{37}$$

If this value is larger than an eighth of the gravity g, the feature turns negative,

$$\phi_{\text{acceleration}}(\tau) = \frac{1}{T} \sum_{(s_t, a_t) \in \tau}^{T} \max\left(1 - \frac{C_{\text{acceleration}}}{g/8}, -1\right). \tag{38}$$

In addition, the following binary features are defined for trajectories that either result in collisions (39), invalid states (i.e. an agent drives off the road) (40) or invalid actions (41) (i.e. an agent executes a physically impossible action). Each of these binary features mark a terminal state.

$$\phi_{\text{collision}}(\tau) = \begin{cases} 1 & \text{if } \tau \text{ contains a collision} \\ 0 & \text{if } \tau \text{ does not contain a collision} \end{cases} \tag{39}$$

$$\phi_{\text{invalid state}}(\tau) = \begin{cases} 1 & \text{if } \tau \text{ contains an invalid state} \\ 0 & \text{if } \tau \text{ does not contain an invalid state} \end{cases} \tag{40}$$

$$\phi_{\text{invalid action}}(\tau) = \begin{cases} 1 & \text{if } \tau \text{ contains an invalid action} \\ 0 & \text{if } \tau \text{ does not contain an invalid action} \end{cases} \tag{41}$$

Linear Reward Model

The linear reward model $\mathcal{R}_\theta$ is a linear combination of the parameters θ and their respective features $\phi(s, a)$,

$$\mathcal{R}_\theta(s_t, a_t) = \theta^\top \phi(s_t, a_t). \tag{42}$$

The feature count is normalized using the length of the trajectory. Since each feature is bounded between $[-1, 1]$, the return of a trajectory is bounded between $[-||\theta||, ||\theta||]$.

Nonlinear Reward Model

To allow for a more complex reward structure, a nonlinear reward model in the form of an ANN is introduced,

$$\mathcal{R}_\theta(s_t, a_t, s_{t-1}) = W_2 \Gamma(W_1 \phi(s_t, a_t, s_{t-1})). \tag{43}$$

It consists of two fully connected layers, with parameters W_1 and W_2 respectively. The first layer is followed by a ReLU activation function Γ. The inputs to the ANN are the features for the linear model in addition to the values of ϕ_{desLane}, $\phi_{\text{desVelocity}}$ and $\phi_{\text{laneCenter}}$ at the previous time step.

3.3.2 Maximum Entropy Inverse Reinforcement Learning

IRL learns the parameters θ of a parameterized reward model $\mathcal{R}_\theta$ so that the expert policy π_E becomes the optimal policy given the reward model [38].

Instead of requiring access to the expert policy π_E itself, it is sufficient to observe trajectories $\mathcal{T}_E$ that originate from that policy [1].

$$\tau_E = (s_0, a_0, s_1, a_1, \ldots, s_T) \quad a_t \sim \pi_E(a_t|s_t, \theta) \tag{44}$$

Similarly to the policy π as a distribution over actions (28), a policy ρ as a distribution over trajectories can be defined (30).

A prominent method for IRL is Maximum Entropy Inverse Reinforcement Learning [65], which assumes a probabilistic model for expert behavior. Using the definition of a policy over trajectories (30), Maximum Entropy IRL specifies the distribution over expert trajectories conditioned on the parameters of the reward model

$$\rho_{\mathrm{E}}(\tau) = \frac{e^{G_\theta(\tau)}}{Z_\theta}. \tag{45}$$

This model implies that the probability of an expert trajectory increases exponentially with its return. The numerator is the exponentiated return of a trajectory (1) and the denominator the partition function (46), the integral of the exponentiated return of all trajectories.

$$Z_\theta = \int e^{G_\theta(\tau)}\, \mathrm{d}\tau \tag{46}$$

The likelihood of the parameters θ given the expert trajectories $\mathcal{T}_{\mathrm{E}}$ is defined with

$$L(\theta|\mathcal{T}_{\mathrm{E}}) = \prod_{\tau \in \mathcal{T}_{\mathrm{E}}} \frac{e^{G_\theta(\tau)}}{Z_\theta}. \tag{47}$$

Applying the logarithm to (47) yields the log-likelihood

$$l(\theta|\mathcal{T}_{\mathrm{E}}) = \sum_{\tau \in \mathcal{T}_{\mathrm{E}}} \left(G_\theta(\tau) - \log Z_\theta\right), \tag{48}$$

which is proportional to

$$\frac{1}{|\mathcal{T}_{\mathrm{E}}|} \sum_{\tau \in \mathcal{T}_{\mathrm{E}}} G_\theta(\tau) - \log Z_\theta. \tag{49}$$

The maximization of the log-likelihood[4] (49) through the parameters θ will result in the parameters that best explain the expert trajectories.

$$\max_{\theta \in \Theta} \frac{1}{|\mathcal{T}_{\mathrm{E}}|} \sum_{\tau \in \mathcal{T}_{\mathrm{E}}} G_\theta(\tau) - \log Z_\theta \tag{50}$$

Using the gradient of the log-likelihood in a gradient ascent step, locally optimal parameters can be found (51).

$$\theta \leftarrow \theta + \alpha \nabla_\theta l(\theta) \tag{51}$$

---

[4] In the following, the log-likelihood refers to the proportional log-likelihood.

Using the logarithmic and exponential derivatives, the gradient of the log-likelihood can be formulated as an expectation (52) [28].

$$\nabla_\theta l(\boldsymbol{\theta}) = \frac{1}{|\mathcal{T}_E|} \sum_{\tau \in \mathcal{T}_E} \nabla_\theta G_\theta(\tau) - \mathbb{E}_{\tau \sim \rho_E(\tau)} [\nabla_\theta G_\theta(\tau)] \tag{52}$$

3.3.3 Guided Cost Learning

GCL is an algorithm that combines sampling-based Maximum Entropy IRL with RL [16].

Since the partition function (46) can only be calculated for small and discrete MDPs, it cannot be computed for the cooperative trajectory planning problem. Instead, GCL circumvents this problem by sampling to approximate it.

It estimates the partition function (46) using the distribution over trajectories generated by a sampling-based method (in this work, the MCTS-based cooperative trajectory planner [29]) (55). The optimal proposal density for importance sampling

$$\rho_S^*(\tau) \propto e^{G_\theta(\tau)} \tag{53}$$

is the distribution that yields the lowest variance [16]. The key concept of GCL is the adjustment of this sampling distribution to the distribution that follows from the current reward model (45). In order to achieve this within the MCTS, this work introduces a probabilistic final selection policy named *Softmax Q-Proposal*,

$$\pi_{\mathrm{MCTS}}(a|s_0) = \frac{e^{cQ^\pi(s_0,a)}}{\sum_{a \in \mathcal{A}(s_0)} e^{cQ^\pi(s_0,a)}}. \tag{54}$$

The numerator is the exponentiated state-action value $Q^\pi(s_0, a)$ (i.e. the expected return (1)) of taking action a in root state s_0 over the sum of the state-action values of all explored actions $\mathcal{A}$ in the root state s_0. The coefficient c can be used to scale the variance of the distribution, its value is determined empirically. Based on (30), this results in the following distribution over trajectories

$$\rho(\tau) = \rho_{\mathrm{MCTS}}(s_0, a_0, s_1, a_1, \ldots, s_T) = \prod_{t=0}^{T-1} \pi_{\mathrm{MCTS}}(a_t|s_t). \tag{55}$$

Applying importance sampling, the expectation in (52) can be calculated using the policy $\rho_S(\tau)$ [28],

$$\mathbb{E}_{\tau \sim \rho_E(\tau)} [\nabla_\theta G_\theta(\tau)] = \mathbb{E}_{\tau \sim \rho_S(\tau)} \left[\frac{e^{G_\theta(\tau)}}{\rho_S(\tau)Z_\theta} \nabla_\theta G_\theta(\tau) \right] \tag{56}$$

Further, the partition function can be approximated using importance sampling as well,

$$\widehat{Z}_\theta := \frac{1}{|\mathcal{T}_S|} \sum_{\tau \in \mathcal{T}_S} \frac{e^{G_\theta(\tau)}}{\rho_S(\tau)}. \tag{57}$$

Substituting the expectation in (52) with (56) as well as the partition function (46) with (57), the final approximation of the gradient can be obtained (58).

$$\nabla_\theta l(\theta) = \frac{1}{|\mathcal{T}_E|} \sum_{\tau \in \mathcal{T}_E} \nabla_\theta G_\theta(\tau) - \mathbb{E}_{\tau \sim \rho_E(\tau)} \left[\nabla_\theta G_\theta(\tau) \right]$$

$$= \frac{1}{|\mathcal{T}_E|} \sum_{\tau \in \mathcal{T}_E} \nabla_\theta G_\theta(\tau) - \mathbb{E}_{\tau \sim \rho_S(\tau)} \left[\frac{e^{G_\theta(\tau)}}{\rho_S(\tau) Z_\theta} \nabla_\theta G_\theta(\tau) \right] \tag{58}$$

$$\approx \frac{1}{|\mathcal{T}_E|} \sum_{\tau \in \mathcal{T}_E} \nabla_\theta G_\theta(\tau) - \frac{1}{|\mathcal{T}_S|} \sum_{\tau \in \mathcal{T}_S} \frac{e^{G_\theta(\tau)}}{\rho_S(\tau) \widehat{Z}_\theta} \nabla_\theta G_\theta(\tau)$$

Given this form of the gradient, the proposed Softmax Q-IRL algorithm (Algorithm 1) performs gradient ascent, converging towards the expert behavior.

Algorithm 1: Softmax Q-IRL [28] © 2022 IEEE.

Input: $\mathcal{T}_E$
Output: θ
1 $\theta_0 \sim U[-1, 1]$;
2 **for** $i \leftarrow 0$ **to** M **do**
3 $\quad \mathcal{T}_S \leftarrow \varnothing$;
4 $\quad$ **for** $j \leftarrow 0$ **to** N **do**
5 $\quad \quad \mathcal{T}_S, \Pi_S \leftarrow (\mathcal{T}_S, \Pi_S) \cup \texttt{generateSamples}(\theta)$;
6 $\quad \widehat{\nabla_\theta l(\theta)} = \frac{1}{|\mathcal{T}_E|} \sum_{\tau \in \mathcal{T}_E} \nabla_\theta G_\theta(\tau) - \frac{1}{|\mathcal{T}_S|} \sum_{\tau \in \mathcal{T}_S} \frac{e^{G_\theta(\tau)}}{\rho_S(\tau) \widehat{Z}_\theta} \nabla_\theta G_\theta(\tau)$;
7 $\quad \theta_{i+1} \leftarrow \theta_i + \alpha \widehat{\nabla_\theta l(\theta_i)}$;
8 **return** θ

The necessary data sampling routine (Algorithm 1 Line 5) is depicted in Algorithm 2. It generates the sample trajectories $\mathcal{T}_S$ as well as their policies Π. Here, Υ denotes the number of agents in the respective scenario.

3.4 Experiments

Using our cooperative trajectory planner, we create a set $\mathcal{T}_E$ of 50 expert trajectories for each scenario that resembles expert behavior. Each trajectory has a length of 10.4 s. The number of agents and obstacles in a scenario is set, but we sample its start

---

Algorithm 2: Sampling of Trajectories and Policies [28] © 2022 IEEE.

---

1 **Function** generateSamples($\boldsymbol{\theta}$)
2 | $\mathcal{T} \leftarrow \varnothing; \Pi \leftarrow \varnothing, \mu(\cdot) \leftarrow 1,$
| // sample from the start state distribution
3 | $s_0 \sim d;$
4 | **for** $t \leftarrow 0$ **to** $T - 1$ **do**
| | // estimate for each action explored at the root state
5 | | $\widehat{Q}(s_t, a_0), \ldots, \widehat{Q}(s_t, a_m) \leftarrow$ MCTSQEstimate($\boldsymbol{\theta}, s_t$);
6 | | $\pi_{\text{MCTS}}(a|s_t) \leftarrow \dfrac{e^{c\widehat{Q}(s_t,a)}}{\sum_{a \in \mathcal{A}(s_t)} e^{c\widehat{Q}(s_t,a)}};$
| | // for each agent in the scenario
7 | | **for** $i \leftarrow 0$ **to** $|\Upsilon|$ **do**
8 | | | $a_i \sim \pi_{\text{MCTS}}(a|s_t);$
9 | | | $\rho(\tau_i) \leftarrow \rho(\tau_i)\pi_{\text{MCTS}}(a_i|s_t);$
10 | | | $\tau_i \leftarrow \tau_i \cup (s_t, a_i);$
11 | | | **if** $t = T - 1$ **then**
12 | | | | $\mathcal{T} \leftarrow \mathcal{T} \cup \tau_i;$
13 | | | | $\Pi \leftarrow \Pi \cup \rho(\tau_i);$
14 | | $s_t \leftarrow$ EnvironmentStep($s_t, a_0, \ldots, a_m$);
15 | **return** $\mathcal{T}, \Pi$

---

state from a distribution. The agents' lateral and longitudinal positions are drawn from a normal distribution. Further, different random seeds are used to initialize the sampling-based trajectory planning algorithm.

We trained the linear and nonlinear reward models for 2000 gradient steps with a learning rate of 5×10^{-4}.

We compare the performance of the linear and nonlinear models with the manually tuned baseline in Table 1. The reward model of the baseline has been hand-tuned over numerous days through an iterative process of parameter modification and quantitative and qualitative analysis of the resulting trajectories. The results were shown in Sect. 2.4. It can be seen that both models perform well in all scenarios. The nonlinear model outperforms the manually tuned baseline in SC02 and SC06, as it does not generate any collisions or invalid trajectories. Further, the learned models manage to reach the desired velocity v_{des} in an additional 46 % of the cases, while the desired lane is reached less frequently -5 %. Finally, both models yield a lower mean distance to the expert trajectories than the manually tuned baseline, while only the nonlinear model is consistently better. A Euclidean distance metric is depicted in Fig. 9. As expected, the linear and the nonlinear model converge toward the expert trajectories. However, neither model converges completely but stall at a distance of 1.98 and 1.30 m for the linear and nonlinear, respectively. A possible remedy could be a reward model with a higher capacity.

Table 1 Absolute change in performance: the performances of the linear and nonlinear reward models are compared with the manually tuned baseline on the scenarios. Columns Δcollision, Δinvalid, Δl_{des}, and Δv_{des} denote the fraction of trajectories compared to the baseline that reach that feature. Similarly, $\Delta\mu(d)$ and $\Delta\sigma(d)$ represent the mean and standard deviation of the Euclidean distance d to the k-nearest neighboring ($k = 3$) expert trajectories compared to the baseline.

Scenario	Model	Δ collision	Δinvalid	Δl_{des}	Δv_{des}	$\Delta\mu(d)$	$\Delta\sigma(d)$
SC01	Linear	0.00	0.00	0.06	0.03	−0.02	0.06
	Nonlinear	0.00	0.00	0.07	0.01	−0.04	0.03
SC02	Linear	0.01	0.00	0.31	0.50	−1.56	0.15
	Nonlinear	0.00	−0.01	0.09	0.79	−3.58	−0.80
SC03	Linear	0.00	0.00	0.02	0.61	−1.14	−0.25
	Nonlinear	0.00	0.00	0.03	0.60	−1.22	−0.35
SC04	Linear	0.00	0.00	−0.20	0.61	−0.49	0.34
	Nonlinear	0.00	0.00	−0.11	0.48	−0.40	0.35
SC05	Linear	0.05	0.01	−0.25	0.51	1.54	0.27
	Nonlinear	0.00	0.00	−0.17	0.41	−0.38	0.27
SC06	Linear	−0.01	0.00	−0.27	0.58	−0.67	0.42
	Nonlinear	−0.01	0.00	−0.15	0.41	−0.76	0.28
Mean	Linear	0.01	0.00	−0.06	0.47	−0.39	0.16
	Nonlinear	0.00	0.00	−0.04	0.45	−1.06	−0.04

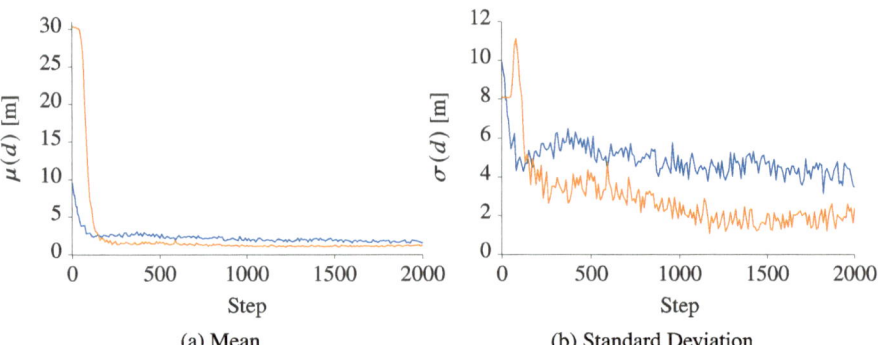

(a) Mean   (b) Standard Deviation

Fig. 9 Distance mean and standard deviation between $\mathcal{T}_E$ and $\mathcal{T}_S$: the Euclidean distance to the k-nearest neighboring ($k = 3$) expert trajectories throughout the training for the linear (blue) and nonlinear reward model (orange). Reference [28] © 2022 IEEE

The visual resemblance of the generated samples to the expert trajectories by the nonlinear reward model can be assessed in Fig. 10. Some of the trajectories that deviate significantly from the majority of the optimal trajectories could be the result of the inherently stochastic sampling-based trajectory planning algorithm.

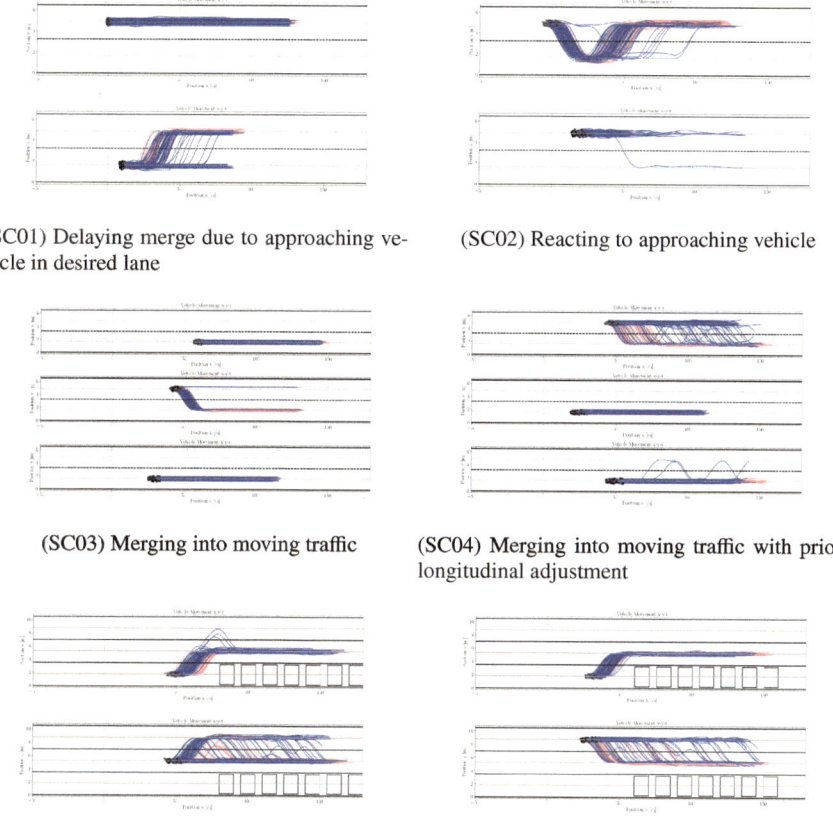

(SC01) Delaying merge due to approaching vehicle in desired lane

(SC02) Reacting to approaching vehicle

(SC03) Merging into moving traffic

(SC04) Merging into moving traffic with prior longitudinal adjustment

(SC05) Changing lane as other vehicle needs to merge onto lane

(SC06) Delaying lane change as other vehicle needs to merge first

Fig. 10 Sample trajectories from the nonlinear reward model: expert trajectories (red) that are used to learn the parameters of the reward model and the optimal trajectories (blue, after 2000 training steps of the nonlinear reward model). Reference [28] © 2022 IEEE

4   Planning Under Uncertainties

In reality, the true state of the surrounding environment is never exactly known. Autonomous vehicles use perception and localization sensors to observe their environment and try to deduce the current state from these observations. The real state often differs from the observations due to noisy measurements and imperfections in the perception algorithms. Hence, the true state of the environment is uncertain. If this uncertainty is not considered in the planning process, the chosen actions may lead to unwanted outcomes such as collisions or undesirable driving behavior. Therefore, we also present an approach that considers uncertainty by constructing search trees from different start states with MCTS and applying kernel regression and risk metrics afterward to find robust actions.

4.1 Related Work

Uncertainties have been addressed with various MCTS approaches in the literature. For example, Kernel Regression UCT [64] tackles the execution uncertainty in continuous action spaces by selecting actions according to a modified upper confidence bound value that incorporates all action assessments by kernel regression. Furthermore, the estimated value of a selected action is refined by applying progressive widening to add similar actions to the search tree. Another concept is *determinization* which describes sampling several deterministic problems with perfect information from a stochastic problem with imperfect information, solving these problems, and fusing their results to get a solution for the original problem [9]. For instance, Couëtoux et al. [12] employ UCT with Double Progressive Widening in a setting with stochastic state transitions and continuous action and state spaces. Their method expands the set of available actions and the set of sampled outcomes iteratively. Another example is HOP-UCT [6], which uses Hindsight Optimization (HOP). Several deterministic UCT search trees are constructed, and their results are averaged to determine the action assessments. Ensemble-Sparse-UCT [6] creates multiple trees and restricts the number of outcomes for an action to a sampling width parameter. Afterward, the results of the search trees are combined. A different approach is Information Set MCTS [14], which uses nodes representing information sets. An information set comprises all indistinguishable states for an agent. This approach restricts the search tree to regions consistent with a determinization. The real-state uncertainty can be modeled by a POMDP approach [50] that uses a search tree with history nodes and an unweighted particle filter representing the belief state. A black box simulator samples successor states and observations, and actions are selected according to the UCT criterion. Generally, POMDPs can be solved online with search trees of belief states with Branch-and-Bound Pruning, Monte Carlo Sampling, or Heuristic Search [41].

In the domain of automated driving the Toolkit for Approximating and Adapting POMDP solutions In Realtime (TAPIR) [23] based on Adaptive Belief Trees [27] is successfully applied if the action space is sufficiently small [19, 20, 44]. Another approach to account for uncertainty is to use Distributional Reinforcement Learning in combination with risk metrics [5]. Instead of learning a return for each state and action, Distributional Reinforcement Learning learns a return distribution. This can be done offline, exposing the agent to various uncertainties during training. Online, during inference, a risk metric is applied to the (risk-neutral) return distribution to select the best action.

4.2   Problem Formulation

Besides the execution uncertainty, which is already modeled by the transition function of an MDP/MG, a POMDP also models the uncertainty emerging from the fact that the real state of the environment can only be partially observed in reality.

A POMDP introduces an observation space $\mathcal{D}$ as a set of possible observations and an observation function $O_a^{s'}$ that defines the probability of an observation after reaching state s' by executing action a. With the help of past observations, an agent can determine a belief b about the real state in the form of a probability distribution over the state space $\mathcal{S}$. The objective is to find a policy $\pi(a|b)$ that specifies which action shall be executed given the belief b.

In our multi-agent setting [51], the ego-agent only receives an observation with stochastic features of the environment. The position, orientation, and velocity of other agents and the length and width of the agents' vehicles are modeled as Gaussian distributions. Furthermore, obstacles' position, orientation, length, and width are normally distributed. Lastly, the observation of the lane width of the road is also regarded as stochastic. All used Gaussian distributions use mean vectors corresponding to the true state since the sensor system is assumed to be unbiased. The uncertainty is hence specified by the covariance matrixes, which are derived from the accuracy statistics of standard perception systems using lidar or radar sensor systems. For simplicity reasons, we assume that the stochastic observation features are mutually independent, and thus each feature follows an individual Gaussian distribution.

In the next section, we present an approach that considers the uncertainties mentioned above explicitly to find more robust actions.

4.3   Approach

In this section, we modify our previous approach so that the uncertainty about the real state due to noisy measurements is explicitly considered during planning. The objective is to find robust actions given an initial belief state b_0 that represents a probability distribution over the state space $\mathcal{S}$. Our approach [51] does not propose a procedure to update the belief state after each new observation in a new planning cycle. Instead, it uses the information of a belief state to select robust actions in the current planning cycle.

Similar to the modeling of the stochastic observations, we also use Gaussian distributions for the belief about the real state features. However, since we assume for simplicity reasons that the features are mutually independent, each belief feature is individually normally distributed with a mean corresponding to the observed value and the variance representing the accuracy of the sensor system.

Our approach samples multiple start states $\mathcal{S}_0$ according to a modified procedure given the initial belief state b_0. It constructs a search tree with MCTS from each start state $s_0 \in \mathcal{S}_0$ to obtain action values $Q(s_0, \cdot)$. Action assessments from all search

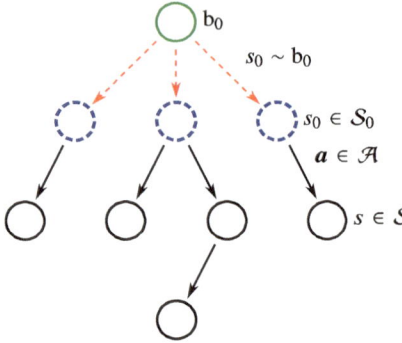

Fig. 11 Using the distribution of the belief state b_0, possible start states are sampled. Then, for each start state s_0, an MCTS is run to determine the action values $Q(s_0, \cdot)$, yielding a return distribution over start states $\mathcal{S}_0$. Reference [51] © 2022 IEEE

trees are combined with kernel regression to yield a return distribution over the start states $\mathcal{S}_0$ (see Fig. 11). After the computational budget for the planning phase is exhausted, a risk metric is applied to the return distribution to select the best action vector $\boldsymbol{a}^*$. This concept is sensible since robust actions should perform well from multiple likely start states according to the belief about the environment so that the risk of failure is reduced.

Algorithm 3 summarizes the uncertainty handling concept. The sampling of start states and the construction of search trees are conducted iteratively until the computational budget is exhausted. Afterward, the best action vector is selected, given the search trees. Before we present further details about the construction of the search trees and the final selection of actions, the next sections briefly summarize the concepts of kernel regression and risk metrics.

4.3.1 Kernel Regression

The general objective of regression is to find the regression function

$$m(x) = \mathbb{E}[Y|X = x] = \frac{\int_{-\infty}^{\infty} y f(x, y) dy}{\int_{-\infty}^{\infty} f(x, y) dy} \tag{59}$$

specifying the conditional expected value of a random variable Y given the realization of a random variable X [58]. Since the joint probability density function f is unknown, *kernel regression* [36, 58] estimates f by

$$\hat{f}(x, y) = \frac{1}{n} \sum_{i=1}^{n} \tilde{K}(x - x_i, y - y_i) \tag{60}$$

Algorithm 3: Uncertainty Handling Concept [51] © 2022 IEEE

Input: belief state b_0
Output: best action vector a^*
1 create initial start states $\mathcal{S}_0$
2 **for** *iteration* $i \leftarrow 1, \ldots, I$ **do**
3   **if** *start states $\mathcal{S}_0$ shall be expanded* **then**
4       $s_0 \leftarrow$ create a new start state
5       $\mathcal{S}_0 \leftarrow \mathcal{S}_0 \cup \{s_0\}$
6   **else**
7       $s_0 \leftarrow$ select an existing start state $\in \mathcal{S}_0$
8    $v \leftarrow$ select a node in the tree of s_0 that shall be expanded
9    $v' \leftarrow$ expand v by creating a new node according to the selected action vector
10   add v' as a child node of v
11   run a simulation from v'
12   update the tree of s_0 according to the simulation results
13   update the start state s_0
14 $a^* \leftarrow$ `FinalSelectionPolicy` selects the best action vector given the search trees

---

with a kernel $\widetilde{K}$ (i.e., a non-negative smoothing function whose integral over both dimensions equals one) from a finite set of data samples $\{(x_1, y_1), \ldots, (x_n, y_n)\}$. Under mild requirements, the estimated regression function can then be formulated as

$$\hat{m}(x) = \frac{\sum_{i=1}^n y_i \widetilde{K}(x - x_i)}{\sum_{i=1}^n \widetilde{K}(x - x_i)} = \frac{\sum_{i=1}^n y_i K(x, x_i)}{\sum_{i=1}^n K(x, x_i)}, \tag{61}$$

where $\widetilde{K}(\cdot)$ is the "marginal" kernel and the kernel value $K(x, x_i)$ is just a simplified notation for $\widetilde{K}(x - x_i)$ [58].

In the context of an MDP/MG, kernel regression can be used to assess actions. The kernel regression value

$$\mathrm{KR}(s, a) = \frac{\sum_{a' \in \mathcal{A}(s)} K(a, a') Q(s, a') N(s, a')}{\mathrm{W}(s, a)} \tag{62}$$

for a state s and action a combines the state-action value estimates $Q(s, a')$ of all actions a' of a finite action set $\mathcal{A}(s)$ weighted by the visit counts $N(s, a')$ and a kernel K that specifies the similarity between the actions [64]. The denominator in (62) is the kernel density

$$\mathrm{W}(s, a) = \sum_{a' \in \mathcal{A}(s)} K(a, a') N(s, a') \tag{63}$$

and denotes the exploration of action a and similar actions. The kernel regression lower confidence bound

$$\text{KRLCB}(s, a) = \text{KR}(s, a) - c\sqrt{\frac{\log \sum_{a' \in \mathcal{A}(s)} W(s, a')}{W(s, a)}} \tag{64}$$

penalizes poorly explored actions by subtracting a normalized exploration term scaled by a constant $c \in \mathbb{R}_{\geq 0}$ from the kernel regression value [64].

4.3.2 Risk Metrics

A risk metric is a measure of risk, but it is not a metric in a mathematical sense since it does not represent a distance function. Let $Z : \Omega \to \mathbb{R}$ be a random variable assigning costs (in monetary units) that are caused by an action to each outcome ω of the sample space Ω. Furthermore, let $\mathcal{Z}$ denote the set of all cost random variables, then a risk metric $\rho : \mathcal{Z} \to \mathbb{R}$ maps each cost random variable Z to a real number that represents the amount of risk [34]. Sensible risk metrics should fulfill the following axioms [34]: *Monotonicity, Translation Invariance, Positive Homogeneity, Subadditivity, Comonotone Additivity*, and *Law Invariance*.

One example for a risk metric that fulfills all six axioms is the Conditional Value at Risk (CVaR) [34]. For the definition of CVaR, the Value at Risk (VaR)

$$\text{VaR}_\alpha(Z) := \min \{z \in \mathbb{R} \mid \mathbb{P}(Z > z) \leq \alpha\} \tag{65}$$
$$= \min \{z \in \mathbb{R} \mid \mathbb{P}(Z \leq z) \geq 1 - \alpha\} \tag{66}$$

for a cost random variable Z is necessary, which specifies the smallest $(1 - \alpha)$-quantile of Z for a given probability α [34, 40]. CVaR is then defined as the conditional expected value

$$\text{CVaR}_\alpha(Z) := \mathbb{E}\left[Z \mid Z \geq \text{VaR}_\alpha(Z)\right]. \tag{67}$$

under the condition that all costs are less than or equal to $\text{VaR}_\alpha(Z)$. In addition, the Upper Value at Risk

$$\text{VaR}_\alpha^+(Z) := \inf \{z \in \mathbb{R} \mid \mathbb{P}(Z > z) < \alpha\} \tag{68}$$
$$= \inf \{z \in \mathbb{R} \mid \mathbb{P}(Z \leq z) > 1 - \alpha\} \tag{69}$$

specifies the largest $(1 - \alpha)$-quantile of Z for a given probability α [40]. VaR_α^+ and VaR_α only differ if the cumulative distribution function is constant around the probability level $(1 - \alpha)$.

4.3.3   Construction of Search Trees

Search trees are constructed iteratively from the start states sampled according to the belief state. We apply the idea of progressive widening (cf. Sect. 2.3.3) to balance the number and the depth of the trees. In each iteration, it is checked whether the condition

$$|S_0| \geq c_{pw} \, N^{\alpha_{pw}} \tag{70}$$

with start states S_0, constant parameters $c_{pw} \in \mathbb{R}_{\geq 0}$ and $\alpha_{pw} \in [0, 1)$, and the total visit count N holds true. If this condition is fulfilled, an already existing start state is selected uniformly from S_0, and its subtree is expanded according to MCTS. Otherwise, a new start state associated with a new search tree is generated.

The generation of a new start state is a modified sampling process. Given the belief that is modeled as a random vector following a Gaussian distribution $\mathcal{N}(\boldsymbol{\mu}, \boldsymbol{\Sigma})$ with mean vector $\boldsymbol{\mu}$ and covariance matrix $\boldsymbol{\Sigma}$, a start state is sampled from the scaled Gaussian distribution $\mathcal{N}(\boldsymbol{\mu}, c\boldsymbol{\Sigma})$. If the sampled start state represents a collision or is invalid, the scaling factor c is increased, and a new sampling attempt is conducted. The scaling factor c is determined by

$$c = \min \left\{ \left(c_{step}\right)^{l_{step}}, c_{\max} \right\}, \tag{71}$$

where $c_{step} \in \mathbb{R}_{\geq 0}$ is a constant, l_{step} is a step counter and $c_{\max} \in \mathbb{R}_{\geq 0}$ is the maximally allowed value. The step counter

$$l_{step} = \left\lfloor \frac{l_{attempt}}{l_{stepSize}} \right\rfloor \tag{72}$$

is dependent on the number of total attempts conducted so far $l_{attempt} \in \mathbb{R}_{\geq 0}$, and the step size $l_{stepSize} \in \mathbb{R}_{\geq 0}$ specifies the number of attempts for one step.

Since this process is executed independently for sampling each start state, the i-th start state is a realization of the random vector $\mathbf{X}_i \sim \mathcal{N}(\boldsymbol{\mu}, c_i \boldsymbol{\Sigma})$ with an individual factor c_i.

Increasing the scaling factor c_i increases the variance of the sampled start states. Hence the probability of obtaining a collision-free and valid start state from which planning is sensible rises. Furthermore, actions must perform well from a greater variety of start states, increasing robustness. On the other hand, we restrict the variance to a realistic level by limiting the scaling factor to a maximum scaling factor $c_{\max}$.

4.3.4   Final Selection

After the computational budget is exhausted for constructing the search trees, a robust action is finally selected. We present two variants using the Kernel Regression

---

Algorithm 4: Basic Final Selection Policy [51] © 2022 IEEE

Input: search trees of the planning phase
Output: best action vector a^*
1 $a^* \leftarrow \langle\rangle$ // empty best action vector
2 **foreach** *agent* $i \in \Upsilon$ **do**
3 | $C^i \leftarrow$ getActionCandidates (i)
4 | **if** $C^i = \emptyset$ **then**
5 | | store default action "Maintain (0)" in a^*
6 | **else**
7 | | $\mathcal{W}^i, \mathcal{KR}^i \leftarrow$ kernelRegression (C^i)
8 | | $Q^i \leftarrow$ getACValues $(C^i, \mathcal{W}^i, \mathcal{KR}^i)$
9 | | $a^* \leftarrow \arg\max_{a \in C^i} Q^i(a)$
10 | | store a^* in a^*

11 **return** a^*

Lower Confidence Bound (KRLCB) and the Conditional Value at Risk (CVaR). Both variants follow the general steps of Algorithm 4. For each agent $i \in \Upsilon$, action candidates C^i are determined at first. This set comprises all actions from collision-free and valid start states that have been visited more often than a threshold. This condition ensures that unreliable action value estimates with large variances are not considered. If the set C^i is empty, the default action "Maintain (0)" is selected for the agent i. Otherwise, kernel regression is applied to calculate the densities $\mathcal{W}^i$ and kernel regression values $\mathcal{KR}^i$. Given these values, each final policy variant computes action candidate values Q^i that assess the performance and robustness of the action candidates. Finally, the action with the largest value Q^i is selected.

Similar to (63), the density for an action candidate $a \in C^i$ is calculated as

$$W[a|C^i] = \sum_{a' \in C^i} K(a, a') N(s_{a'}, a') \tag{73}$$

with the kernel

$$K(a, a') = \exp\left(-\gamma \left\| a - a' \right\|^2\right) \tag{74}$$

and the visit count $N(s_{a'}, a')$ of action candidate a' from the corresponding start state $s_{a'}$. Furthermore, the kernel regression value (c.f., 62) is then defined as

$$KR[a|C^i] = \frac{\sum_{a' \in C^i} K(a, a') Q(s_{a'}, a') N(s_{a'}, a')}{\sum_{a' \in C^i} K(a, a') N(s_{a'}, a')} \tag{75}$$

with $Q(s_{a'}, a')$ being the estimated action value of action a' from the start state $s_{a'}$.

The first final policy variant calculates the Kernel Regression Lower Confidence Bound (KRLCB) by

$$\text{KRLCB}[a|C^i] = \text{KR}[a|C^i] - c\sqrt{\frac{\log \sum_{a' \in C^i} \text{W}[a'|C^i]}{\text{W}[a|C^i]}} \tag{76}$$

with a constant $c \in \mathbb{R}_{\geq 0}$. Actions with high KRLCB values are well explored from many start states due to the exploration term (subtrahend) and provide large returns from many start states due to the exploitation term (minuend). Hence, the KRLCB selects robust actions.

The second policy variant employs the Conditional Value at Risk (CVaR). Since CVaR has been defined for a cost random variable Z in Sect. 4.3.2, we derive a consistent definition for a reward random variable $R := -Z$. The Value at Risk (VaR) of Z can be transformed with

$$\begin{aligned} \text{VaR}_\alpha(Z) &= \min\{z \in \mathbb{R} \mid \mathbb{P}(-R > z) \leq \alpha\} \\ &= -\max\{r \in \mathbb{R} \mid \mathbb{P}(R < r) \leq \alpha\} \\ &= -\inf\{r \in \mathbb{R} \mid \mathbb{P}(R \leq r) > \alpha\} \\ &= -\inf\{r \in \mathbb{R} \mid \mathbb{P}(R > r) < 1 - \alpha\} \\ &\overset{(68)}{=} -\text{VaR}^+_{1-\alpha}(R) \end{aligned} \tag{77}$$

where $\text{VaR}^+_{1-\alpha}$ is the Upper Value at Risk and α is a probability.

The objective is to select the best action a^* from a set of available actions $\tilde{\mathcal{A}}^i$ that minimizes the risk of costs according to risk metric CVaR (see (78)). The following derivations

$$\begin{aligned} a^* &= \arg\min_{a \in \tilde{\mathcal{A}}^i} \text{CVaR}_\alpha(Z) \\ &= \arg\min_{a \in \tilde{\mathcal{A}}^i} \mathbb{E}[-R \mid -R \geq \text{VaR}_\alpha(Z)] \\ &\overset{(77)}{=} \arg\min_{a \in \tilde{\mathcal{A}}^i} \mathbb{E}[-R \mid R \leq \text{VaR}^+_{1-\alpha}(R)] \\ &= \arg\max_{a \in \tilde{\mathcal{A}}^i} \mathbb{E}[R \mid R \leq \text{VaR}^+_{1-\alpha}(R)] \end{aligned} \tag{78, 79}$$

show that this idea is equivalent to choosing the action that maximizes a conditional expected value based on the reward random variable R. We denote this expected value as *Complementary Conditional Value at Risk (CCVaR)* and define it as

$$\overline{\text{CVaR}}_{1-\alpha}(R) := \mathbb{E}[R \mid R \leq \text{VaR}^+_{1-\alpha}(R)]. \tag{80}$$

In the context of MDPs and POMDPs, an agent selects the action with the largest $\overline{\text{CVaR}}_{1-\alpha}(G)$ value, where G denotes the return (1).

Before the CVaR idea can be applied, a return distribution must be determined. We use kernel regression to calculate the density $\mathcal{W}^i_{a,s_0} := \text{W}[a|C^i_{s_0}]$ (see (73)) and the kernel regression value $\mathcal{KR}^i_{a,s_0} := \text{KR}[a|C^i_{s_0}]$ (see (75)) of each action a with respect to the action candidates $C^i_{s_0} \subseteq C^i$ from each start state s_0. Each kernel regression value $\mathcal{KR}^i_{a,s_0}$ for a specific start state s_0 serves as an unweighted particle in the set

---

Algorithm 5: CVaR Final Selection Policy [51] © 2022 IEEE

Input: action candidates C^i for agent i
Output: density distributions $\mathcal{W}^i$ and kernel regression value distributions $\mathcal{KR}^i$ for agent i

1 **Function** kernelRegression(C^i)

2     $\mathcal{W}^i \leftarrow \emptyset$

3     $\mathcal{KR}^i \leftarrow \emptyset$

4   **foreach** *action* $a \in C^i$ **do**

5   **foreach** *start state* $s_0 \in \mathcal{S}_0$ **do**

6           $C^i_{s_0} \leftarrow \{$action candidates from $s_0\} \subseteq C^i$

7           $\mathcal{W}^i_{a,s_0} \leftarrow \mathrm{W}[a|C^i_{s_0}]$ // see (73)

8   **if** $\mathcal{W}^i_{a,s_0} \geq w_{\min}$ **then**

9              $\mathcal{KR}^i_{a,s_0} \leftarrow \mathrm{KR}[a|C^i_{s_0}]$ // see (75)

10   append $\mathcal{W}^i_{a,s_0}$ to $\mathcal{W}^i$ and $\mathcal{KR}^i_{a,s_0}$ to $\mathcal{KR}^i$

11   **return** $\mathcal{W}^i, \mathcal{KR}^i$

Input: action candidates C^i, densities $\mathcal{W}^i$, kernel regression values $\mathcal{KR}^i$ for agent i
Output: CCVaR values $\mathcal{Q}^i$ for agent i

12 **Function** getACValues(C^i, $\mathcal{W}^i$, $\mathcal{KR}^i$)

13     $\mathcal{Q}^i \leftarrow \emptyset$

       /* number of start states from which actions have been
          appended to the action candidates   */

14     $N_{s_0} \leftarrow$ see (82)

15   **foreach** *action* $a \in C^i$ **do**

16        $\mathcal{KR}^i_a \leftarrow \left\{ \mathcal{KR}^i_{a,s_0} \mid s_0 \in \mathcal{S}_0 \wedge \mathcal{KR}^i_{a,s_0} \in \mathcal{KR}^i \right\}$

17   **if** $|\mathcal{KR}^i_a| \geq c_m N_{s_0}$ with $c_m \in [0, 1]$ **then**

18           $\mathcal{Q}^i_a \leftarrow \overline{\mathrm{CVaR}}_{1-\alpha}(\mathcal{KR}^i_a)$ // see (80)

19   **else**

20           $\mathcal{Q}^i_a \leftarrow -\infty$

21   append $\mathcal{Q}^i_a$ to $\mathcal{Q}^i$

22   **return** $\mathcal{Q}^i$

$$\mathcal{KR}^i_a = \left\{ \mathcal{KR}^i_{a,s_0} \mid s_0 \in \mathcal{S}_0 \wedge \mathcal{KR}^i_{a,s_0} \in \mathcal{KR}^i \right\} \subseteq \mathcal{KR}^i \tag{81}$$

representing the return distribution of an action a from different start states. A particle is not added to the set if the density $\mathcal{W}^i_{a,s_0}$ is smaller than a threshold $w_{\min} \in \mathbb{R}_{\geq 0}$. This condition ensures that unreliable return estimates due to a low visit count of similar actions do not distort the distribution.

After the particle sets have been constructed, the action candidate values $\mathcal{Q}^i$ can be determined. Let

$$N_{s_0} = |\{C^i_{s_0} \mid s_0 \in \mathcal{S}_0 \wedge C^i_{s_0} = \{\text{action candidates from } s_0\} \subseteq C^i \wedge C^i_{s_0} \neq \emptyset\}| \tag{82}$$

be the number of start states from which actions have been appended to the action candidates C^i. If the final selection policy follows the process described at the beginning of Sect. 4.3.4 for the initialization of C^i, N_{s_0} corresponds to the number of collision-free and valid start states visited more often than a threshold. Then, the condition

$$|\mathcal{KR}_a^i| \geq c_m N_{s_0} \tag{83}$$

with a constant $c_m \in [0, 1]$ indicates whether $\mathcal{KR}_a^i$ has enough elements to be a meaningful representation for the return distribution of action a. If this is the case, the action candidate value Q_a^i is set to $\overline{\text{CVaR}}_{1-\alpha}(\mathcal{KR}_a^i)$ (see (80)). Otherwise, Q_a^i is set to negative infinity, which prevents the selection of this action.

The CVaR final selection policy, summarized in Algorithm 5, favors robust actions since it combines the performances from several start states and limits the influence of high-return outliers from a few specific start states.

4.4 Experiments

We evaluate the presented approach in the same manner as described in Sect. 2.4. In addition, we compare our approach with a baseline corresponding to our basic approach of Sect. 2.3.3, which neglects uncertainties. The results are summarized in Fig. 12.

In an uncertain environment with noisy measurements, the performance of the baseline decreases significantly in comparison to a deterministic environment (see Fig. 12a and c). For instance, the mean success rate drops from 85 % to 24 % for 1280 iterations. If the number of iterations increases, the success rate of the baseline also rises due to better action assessment and exploration. Our approach that uses the KRLCB final selection policy outperforms the baseline significantly. For 1280 iterations, the success rate reaches 62 %, for instance. As indicated in Fig. 12b, scenarios one to ten are solved nearly completely, with a success rate approaching 100 %. Furthermore, a greater number of iterations improves the performance in general. However, more complex scenarios with multiple agents, such as scenarios 11–15, need more iterations than 1280 for satisfactory performance.

Our approach that employs the CVaR final selection policy also outperforms the baseline with 320 iterations or more but consistently performs worse than KRLCB over all evaluated iteration counts (see Fig. 12a). The experiments show that CVaR needs more iterations to unfold its potential. For instance, more than 160 iterations are necessary to improve the success rate upon the baseline. This is plausible since a sufficient amount of particles is necessary for a reasonable representation of the return distribution.

If our approach is applied in a deterministic environment (see Fig. 12c), the success rates are lower on average compared to the baseline. This behavior is expected since the baseline can plan from the environment's real state without needing robustification. On the other hand, the KRLCB and CVaR improve their performances with increasing iteration counts. Fig. 12d shows that KRLCB even solves scenarios two

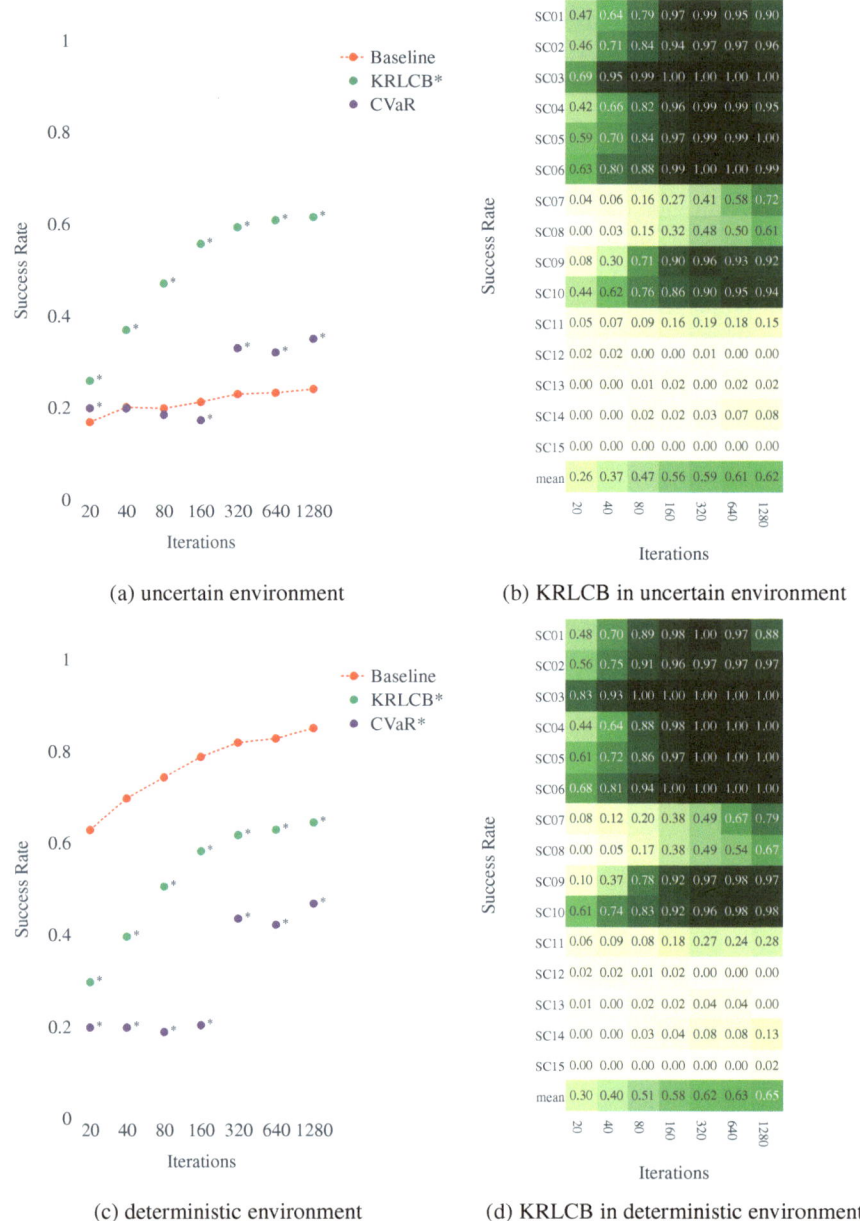

(a) uncertain environment

(b) KRLCB in uncertain environment

(c) deterministic environment

(d) KRLCB in deterministic environment

Fig. 12 Comparison of the mean success rates between the baseline, KRLCB, and CVAR final selection policies in an uncertain (**a**) and a deterministic (**c**) environment. The performance of KRLCB is depicted in more detail for uncertain (**b**) and deterministic (**d**) scenarios (SC)

to six better than the baseline in a deterministic environment. However, the success rates are significantly lower for the other scenarios.

5 Conclusion

Our work proposes a method to plan cooperative trajectories in challenging interactive urban scenarios and tight spaces. At its core, MCTS powers the search for optimal actions. We employ Decoupled-UCT to model the simultaneous action selection process of multiple interacting agents. In addition, we augment the planning algorithm with progressive widening to enable a continuous action space that allows for arbitrary trajectories.

Further, we combine Maximum Entropy IRL with MCTS to learn reward models for the cooperative trajectory planning problem. The efficacy of the MCTS to quickly generate (approximately) optimal samples for arbitrary reward models, in combination with adjusting the sampling distribution after gradient updates, yields reward models that quickly converge towards the experts. Furthermore, with this approach, it is no longer necessary to manually tune reward functions to mimic human drivers.

Last, we extend the resulting cooperative trajectory planning algorithm to explicitly model uncertainties and thus account for noisy perception and partially unknown environments. By constructing search trees from multiple start states, combining their action assessments by kernel regression to yield a return distribution, and applying risk metrics to this distribution, our approach significantly improves the performance in uncertain environments.

Acknowledgements We wish to thank the German Research Foundation (DFG) for funding the project Cooperatively Interacting Automobiles (CoInCar) within which the research leading to this contribution was conducted. The information as well as views presented in this publication are solely the ones expressed by the authors.

References

1. Abbeel, P., Ng, A.Y.: Apprenticeship learning via inverse reinforcement learning. In: Proceedings of the Twenty-First International Conference on Machine Learning. Association for Computing Machinery (2004). https://doi.org/10.1145/1015330.1015430
2. Bae, S., Saxena, D., Nakhaei, A., Choi, C., Fujimura, K., Moura, S.: Cooperation-aware lane change maneuver in dense traffic based on model predictive control with recurrent neural network. In: 2020 American Control Conference (ACC) (2020). https://doi.org/10.23919/ACC45564.2020.9147837
3. Bahram, M., Hubmann, C., Lawitzky, A., Aeberhard, M., Wollherr, D.: A combined model- and learning-based framework for interaction-aware maneuver prediction. IEEE Trans. Intell. Transp. Syst. (2016). https://doi.org/10.1109/TITS.2015.2506642

4. Bahram, M., Lawitzky, A., Friedrichs, J., Aeberhard, M., Wollherr, D.: A game-theoretic approach to replanning-aware interactive scene prediction and planning. IEEE Trans. Veh. Technol. (2016). https://doi.org/10.1109/TVT.2015.2508009

5. Bernhard, J., Pollok, S., et al.: Addressing inherent uncertainty: risk-sensitive behavior generation for automated driving using distributional reinforcement learning. In: IEEE Intelligent Vehicles Symposium (2019). https://doi.org/10.1109/IVS.2019.8813791

6. Bjarnason, R., Fern, A., et al.: Lower bounding Klondike solitaire with Monte-Carlo planning. In: Proceedings of the International Conference on Automated Planning and Scheduling (ICAPS) (2009)

7. Bouton, M., Nakhaei, A., Fujimura, K., Kochenderfer, M.J.: Cooperation-aware reinforcement learning for merging in dense traffic. In: 2019 IEEE Intelligent Transportation Systems Conference (ITSC) (2019). https://doi.org/10.1109/ITSC.2019.8916924

8. Bouton, M., Nakhaei, A., Fujimura, K., Kochenderfer, M.J.: Safe reinforcement learning with scene decomposition for navigating complex urban environments. In: 2019 IEEE Intelligent Vehicles Symposium (IV) (2019). https://doi.org/10.1109/IVS.2019.8813803

9. Browne, C.B., Powley, E., Whitehouse, D., Lucas, S.M., Cowling, P.I., Rohlfshagen, P., Tavener, S., Perez, D., Samothrakis, S., Colton, S.: A survey of Monte Carlo tree search methods. IEEE Trans. Comput. Intell. AI Games (2012). https://doi.org/10.1109/TCIAIG.2012.2186810

10. Chaslot, G.M.J.B., Winands, M.H.M., Herik, H.J.V.D., Uiterwijk, J.W.H.M., Bouzy, B.: Progressive strategies for Monte-Carlo tree search. New Math. Nat. Comput. (2008). https://doi.org/10.1142/S1793005708001094

11. Couëtoux, A., Doghmen, H., Teytaud, O.: Improving the exploration in upper confidence trees. In: Learning and Intelligent Optimization. Springer Berlin Heidelberg (2012)

12. Couëtoux, A., Hoock, J.B., Sokolovska, N., Teytaud, O., Bonnard, N.: Continuous upper confidence trees. In: Learning and Intelligent Optimization. Springer Berlin Heidelberg (2011)

13. Coulom, R.: Efficient selectivity and backup operators in Monte-Carlo tree search. In: Computers and Games. Springer Berlin Heidelberg (2007)

14. Cowling, P.I., Powley, E.J., et al.: Information set Monte Carlo tree search. IEEE Trans. Comput. Intell. AI Games (2012). https://doi.org/10.1109/TCIAIG.2012.2200894

15. Düring, M., Pascheka, P.: Cooperative decentralized decision making for conflict resolution among autonomous agents. In: INISTA 2014 - IEEE International Symposium on Innovations in Intelligent Systems and Applications, Proceedings. IEEE (2014). https://doi.org/10.1109/INISTA.2014.6873612

16. Finn, C., Levine, S., Abbeel, P.: Guided cost learning: deep inverse optimal control via policy optimization. In: Proceedings of the 33rd International Conference on International Conference on Machine Learning, vol. 48. JMLR.org (2016)

17. Gupta, A., Johnson, J., Fei-Fei, L., Savarese, S., Alahi, A.: Social GAN: socially acceptable trajectories with generative adversarial networks. In: 2018 IEEE/CVF Conference on Computer Vision and Pattern Recognition (2018). https://doi.org/10.1109/CVPR.2018.00240

18. Hoel, C.J., Driggs-Campbell, K., Wolff, K., Laine, L., Kochenderfer, M.J.: Combining planning and deep reinforcement learning in tactical decision making for autonomous driving. IEEE Trans. Intell. Veh. (2020). https://doi.org/10.1109/TIV.2019.2955905

19. Hubmann, C., Becker, M., Althoff, D., Lenz, D., Stiller, C.: Decision making for autonomous driving considering interaction and uncertain prediction of surrounding vehicles. In: 2017 IEEE Intelligent Vehicles Symposium (IV) (2017). https://doi.org/10.1109/IVS.2017.7995949

20. Hubmann, C., Schulz, J., Becker, M., Althoff, D., Stiller, C.: Automated driving in uncertain environments: planning with interaction and uncertain maneuver prediction. IEEE Trans. Intell. Veh. (2018). https://doi.org/10.1109/TIV.2017.2788208

21. Hubmann, C., Schulz, J., Xu, G., Althoff, D., Stiller, C.: A belief state planner for interactive merge maneuvers in congested traffic. In: 2018 21st International Conference on Intelligent Transportation Systems (ITSC) (2018). https://doi.org/10.1109/ITSC.2018.8569729

22. Isele, D.: Interactive decision making for autonomous vehicles in dense traffic. In: 2019 IEEE Intelligent Transportation Systems Conference (ITSC) (2019). https://doi.org/10.1109/ITSC.2019.8916982
23. Klimenko, D., Song, J., Kurniawati, H.: Tapir: A software toolkit for approximating and adapting pomdp solutions online. In: Proceedings of the Australasian Conference on Robotics and Automation, Melbourne, Australia (2014)
24. Kocsis, L., Szepesvári, C.: Bandit based Monte-Carlo planning. In: Machine Learning: ECML 2006. Springer Berlin Heidelberg (2006)
25. Kong, J., Pfeiffer, M., Schildbach, G., Borrelli, F.: Kinematic and dynamic vehicle models for autonomous driving control design. In: 2015 IEEE Intelligent Vehicles Symposium (IV) (2015). https://doi.org/10.1109/IVS.2015.7225830
26. Kuderer, M., Gulati, S., Burgard, W.: Learning driving styles for autonomous vehicles from demonstration. In: 2015 IEEE International Conference on Robotics and Automation (ICRA) (2015). https://doi.org/10.1109/ICRA.2015.7139555
27. Kurniawati, H., Yadav, V.: An Online POMDP Solver for Uncertainty Planning in Dynamic Environment. Springer International Publishing (2016). https://doi.org/10.1007/978-3-319-28872-7_35
28. Kurzer, K., Bitzer, M., Zöllner, J.M.: Learning reward models for cooperative trajectory planning with inverse reinforcement learning and Monte Carlo tree search. In: 2022 IEEE Intelligent Vehicles Symposium (IV) (2022). https://doi.org/10.1109/IV51971.2022.9827031
29. Kurzer, K., Engelhorn, F., Zöllner, J.M.: Decentralized cooperative planning for automated vehicles with continuous Monte Carlo tree search. In: International Conference on Intelligent Transportation Systems (ITSC). IEEE (2018). https://doi.org/10.1109/ITSC.2018.8569988
30. Kurzer, K., Schörner, P., Albers, A., Thomsen, H., Daaboul, K., Zöllner, J.M.: Generalizing decision making for automated driving with an invariant environment representation using deep reinforcement learning. In: Intelligent Vehicles Symposium (IV). IEEE (2021). https://doi.org/10.1109/IV48863.2021.9575669
31. LaValle, S.M.: Planning Algorithms. Cambridge University Press (2006)
32. Lawitzky, A., Althoff, D., Passenberg, C.F., Tanzmeister, G., Wollherr, D., Buss, M.: Interactive scene prediction for automotive applications. In: 2013 IEEE Intelligent Vehicles Symposium (IV) (2013). https://doi.org/10.1109/IVS.2013.6629601
33. Lenz, D., Kessler, T., Knoll, A.: Tactical cooperative planning for autonomous highway driving using Monte-Carlo tree search. In: 2016 IEEE Intelligent Vehicles Symposium (IV) (2016). https://doi.org/10.1109/IVS.2016.7535424
34. Majumdar, A., Pavone, M.: How Should a Robot Assess Risk? Towards an Axiomatic Theory of Risk in Robotics (2017)
35. Manzinger, S., Leibold, M., Althoff, M.: Driving strategy selection for cooperative vehicles using maneuver templates. In: 2017 IEEE Intelligent Vehicles Symposium (IV) (2017). https://doi.org/10.1109/IVS.2017.7995791
36. Nadaraya, E.A.: On estimating regression. Theory Probab. Its Appl. (1964). https://doi.org/10.1137/1109020
37. Naumann, M., Sun, L., Zhan, W., Tomizuka, M.: Analyzing the suitability of cost functions for explaining and imitating human driving behavior based on inverse reinforcement learning. In: 2020 IEEE International Conference on Robotics and Automation (ICRA) (2020). https://doi.org/10.1109/ICRA40945.2020.9196795
38. Ng, A.Y., Russell, S.J.: Algorithms for inverse reinforcement learning. In: Proceedings of the Seventeenth International Conference on Machine Learning. Morgan Kaufmann Publishers Inc. (2000)
39. Pascheka, P., Düring, M.: Advanced cooperative decentralized decision making using a cooperative reward system. In: 2015 International Symposium on Innovations in Intelligent Systems and Applications (INISTA) (2015). https://doi.org/10.1109/INISTA.2015.7276779
40. Rockafellar, R.T., Uryasev, S.: Conditional value-at-risk for general loss distributions. J. Bank. Finance (2002). https://doi.org/10.1016/S0378-4266(02)00271-6

41. Ross, S., Pineau, J., et al.: Online planning algorithms for POMDPs. J. Artif. Intell. Res. (2008). https://doi.org/10.1613/jair.2567

42. Saxena, D.M., Bae, S., Nakhaei, A., Fujimura, K., Likhachev, M.: Driving in dense traffic with model-free reinforcement learning. In: 2020 IEEE International Conference on Robotics and Automation (ICRA) (2020). https://doi.org/10.1109/ICRA40945.2020.9197132

43. Schaeffer, M.S.N.S.J., Shafiei, N., et al.: Comparing UCT versus CFR in simultaneous games. In: IJCAI Workshop on General Game Playing (2009)

44. Schörner, P., Töttel, L., et al.: Predictive trajectory planning in situations with hidden road users using partially observable Markov decision processes. In: IEEE Intelligent Vehicles Symposium (IV) (2019). https://doi.org/10.1109/IVS.2019.8814022

45. Schramm, D., Hiller, M., Bardini, R.: Vehicle Dynamics. Springer Berlin, Heidelberg (2018). https://doi.org/10.1007/978-3-662-54483-9

46. Schwarting, W., Pascheka, P.: Recursive conflict resolution for cooperative motion planning in dynamic highway traffic. In: 17th International IEEE Conference on Intelligent Transportation Systems (ITSC) (2014). https://doi.org/10.1109/ITSC.2014.6957825

47. Schwarting, W., Pierson, A., Alonso-Mora, J., Karaman, S., Rus, D.: Social behavior for autonomous vehicles. Proc. Natl. Acad. Sci. (2019). https://doi.org/10.1073/pnas.1820676116

48. Seiler, K.M., Kurniawati, H., Singh, S.P.N.: An online and approximate solver for POMDPs with continuous action space. In: 2015 IEEE International Conference on Robotics and Automation (ICRA) (2015). https://doi.org/10.1109/ICRA.2015.7139503

49. Silver, D., Huang, A., Maddison, C.J., Guez, A., Sifre, L., van den Driessche, G., Schrittwieser, J., Antonoglou, I., Panneershelvam, V., Lanctot, M., Dieleman, S., Grewe, D., Nham, J., Kalchbrenner, N., Sutskever, I., Lillicrap, T., Leach, M., Kavukcuoglu, K., Graepel, T., Hassabis, D.: Mastering the game of go with deep neural networks and tree search. Nature (2016). https://doi.org/10.1038/nature16961

50. Silver, D., Veness, J.: Monte-Carlo planning in large POMDPs. In: Advances in Neural Information Processing Systems (2010)

51. Stegmaier, P., Kurzer, K., Zöllner, M.J.: Cooperative trajectory planning in uncertain environments with Monte Carlo tree search and risk metrics. In: 2022 IEEE 25th International Conference on Intelligent Transportation Systems (ITSC) (2022). https://doi.org/10.1109/ITSC55140.2022.9921861

52. Sutton, R.S., Barto, A.G.: Reinforcement learning: an introduction. MIT Press (2018)

53. Tak, M.J.W., Lanctot, M., et al.: Monte Carlo tree search variants for simultaneous move games. In: 2014 IEEE Conference on Computational Intelligence and Games (2014). https://doi.org/10.1109/CIG.2014.6932889

54. Takahashi, A., Hongo, T., Ninomiya, Y., Sugimoto, G.: Local path planning and motion control for AGV in positioning. In: Proceedings. IEEE/RSJ International Workshop on Intelligent Robots and Systems'. (IROS '89) 'The Autonomous Mobile Robots and Its Applications (1989). https://doi.org/10.1109/IROS.1989.637936

55. Tian, R., Sun, L., Tomizuka, M., Isele, D.: Anytime game-theoretic planning with active reasoning about humans' latent states for human-centered robots. In: 2021 IEEE International Conference on Robotics and Automation (ICRA) (2021). https://doi.org/10.1109/ICRA48506.2021.9561463

56. Vodopivec, T., Samothrakis, S., Šter, B.: On Monte Carlo tree search and reinforcement learning. J. Artif. Int. Res. (2017)

57. Wang, Y., Audibert, J.Y., Munos, R.: Algorithms for infinitely many-armed bandits. In: Proceedings of the 21st International Conference on Neural Information Processing Systems. Curran Associates Inc. (2008)

58. Watson, G.S.: Smooth regression analysis. Indian J. Stat. (1964)

59. Werling, M., Ziegler, J., Kammel, S., Thrun, S.: Optimal trajectory generation for dynamic street scenarios in a frenét frame. In: 2010 IEEE International Conference on Robotics and Automation (2010). https://doi.org/10.1109/ROBOT.2010.5509799

60. Wolf, P., Kurzer, K., Wingert, T., Kuhnt, F., Zöllner, J.M.: Adaptive behavior generation for autonomous driving using deep reinforcement learning with compact semantic states. In: Intelligent Vehicles Symposium (IV). IEEE (2018). https://doi.org/10.1109/IVS.2018.8500427
61. Wu, Z., Sun, L., Zhan, W., Yang, C., Tomizuka, M.: Efficient sampling-based maximum entropy inverse reinforcement learning with application to autonomous driving. IEEE Robot. Autom. Lett. (2020). https://doi.org/10.1109/LRA.2020.3005126
62. Wulfmeier, M., Ondruska, P., Posner, I.: Maximum entropy deep inverse reinforcement learning (2016)
63. Wulfmeier, M., Wang, D.Z., Posner, I.: Watch this: Scalable cost-function learning for path planning in urban environments. In: 2016 IEEE/RSJ International Conference on Intelligent Robots and Systems (IROS) (2016). https://doi.org/10.1109/IROS.2016.7759328
64. Yee, T., Lisy, V., Bowling, M.: Monte carlo tree search in continuous action spaces with execution uncertainty. In: Proceedings of the Twenty-Fifth International Joint Conference on Artificial Intelligence. AAAI Press (2016)
65. Ziebart, B.D., Maas, A., Bagnell, J.A., Dey, A.K.: Maximum entropy inverse reinforcement learning. In: Proceedings of the 23rd National Conference on Artificial Intelligence, vol. 3. AAAI Press (2008)

Learning Cooperative Trajectories at Intersections in Mixed Traffic

Shengchao Yan, Tim Welschehold, Daniel Büscher, Christoph Burger, Christoph Stiller, and Wolfram Burgard

Abstract Intersections are a significant bottleneck in traffic and have been a topic of much research. Optimization approaches incorporating traffic models are often limited by the intractable complexity resulting from the combinatorial explosion associated with increasing numbers of vehicles. Learning cooperative maneuver policies with deep neural networks from traffic data is a promising approach to address this issue. This chapter presents two approaches for managing traffic at intersections using deep reinforcement learning. The first approach learns an adaptive traffic signal controller, serving as a trajectory planner for all vehicles at the intersection. For smaller intersections with less traffic and fewer lanes, traffic signs are preferred over traffic lights due to their lower cost and higher efficiency. The second approach uses a centralized control unit to optimize efficiency and equity by ordering automated vehicles to yield to vehicles on conflicting routes with lower priorities. Self-driving cars have the potential to improve traffic flow in mixed environments with human-driven vehicles. The chapter evaluates the approaches using a traffic simulator with simulated and real-world traffic data. The approaches achieve state-of-the-art performance in terms of traffic efficiency and equity compared to non-learning and other

S. Yan · T. Welschehold · D. Büscher
Department of Computer Science, University of Freiburg, Freiburg, Germany
e-mail: yan@cs.uni-freiburg.de

T. Welschehold
e-mail: twelsche@cs.uni-freiburg.de

D. Büscher
e-mail: buescher@cs.uni-freiburg.de

C. Burger · C. Stiller
Institute of Measurement and Control Systems, KIT, Karlsruhe, Germany
e-mail: christoph.burger@kit.edu

C. Stiller
e-mail: stiller@kit.edu

W. Burgard (✉)
Department of Engineering, University of Technology Nuremberg, Nuremberg, Germany
e-mail: wolfram.burgard@utn.de

459

learning-based methods. The chapter concludes with a discussion of possible future work on learning cooperative trajectories in mixed traffic.

1 Traffic Signal Controller with Deep Reinforcement Learning

Traffic congestion is a tremendous cost factor in terms of fuel and time and many cities all over the world suffer from it [19]. Moreover, the emissions of road transport have been considered as the main cause for air pollution [6, 39]. To alleviate traffic congestion and the associated problems, smarter and cleaner vehicles have been investigated [23, 28]. In addition to that, the effectiveness of road traffic can also be improved by optimizing the scheduling of traffic lights.

In this section, we focus on reducing congestion by improving automated traffic light controllers. More specifically, we focus on traffic signal controllers (TSCs) for isolated intersections [31], i.e., signalized intersections whose traffic is unaffected by any other controllers or supervisory devices.

The performance of conventional fixed-time or actuated TSCs is limited by the restricted setup and the relative primitive sensor information available. Recently, adaptive TSCs [20] attracted more attention due to their high degree of flexibility. Advances in perception and vehicle-to-everything (V2X) communication [18] could make such controllers even better by providing additional (real-time) information, such as locations and velocities of the vehicles. With more detailed information available, adaptive TSCs have the potential to provide optimal control according to current traffic situations. One approach to achieve this is to consider traffic signal optimization as a scheduling problem [20, 47], in which a junction is considered as a production line and the input vehicles as different products to be processed. However, this type of methods suffers from the curse of dimensionality which limits their applicability to small numbers of vehicles [1]. As a result, these methods in general only satisfy real-time requirements for either oversimplified intersections or under small traffic flow rates.

A recent line of research proposes to design adaptive TSCs based on deep reinforcement learning (DRL). DRL has been shown to reach state-of-the-art performance in various domains [30, 37]. However, we believe that the performance of DRL approaches in the traffic domain can be pushed further, in particular with regards to the following limitations:

- Most previous approaches have focused on improving efficiency, which is calculated according to the throughput of intersections. However, we argue that the equity of the travel time of individual vehicles is also of vital importance. Previous works have been mostly evaluating in scenarios with relatively low traffic flow, in which case the trade-off between efficiency and equity might not have a great influence on the performance of the controller. However, in dense traffic

with nearly- or even over-saturated intersections and unbalanced traffic density on incoming lanes, the efficiency-equity trade-off can be an important factor.
- The flexibility of adaptive TSCs has not been sufficiently explored. Instead, most approaches employ fixed green traffic light duration or fixed traffic light cycles.
- Previously proposed DRL agents are trained and evaluated in relatively simplified traffic scenarios: very few traffic demand episodes with limited variation or evenly distributed flow for each incoming lane [16]. Thus, their experimental results might not be sufficient indicators of their performance in real traffic scenarios.
- Current DRL-based approaches have shown performance improvement mainly against fixed-time or actuated TSCs. They either have not compared with state-of-the-art adaptive TSCs, such as the Max-pressure controller [43], or do not surpass state-of-the-art performance [15, 16].

To overcome these limitations, we present a novel method that introduces the following innovations:

- An *equity factor* to trade off efficiency (average travel time) against equity (variance of individual travel times) as well as a solution to calculate a rough bound for it.
- An *adaptive discounting* method to account for the issues brought by transitional phases of traffic signals, which is shown to substantially stabilize learning.
- A learning strategy that surpasses state-of-the-art baselines. It is generic with regards to different traffic flow rates, traffic distributions among incoming lanes and intersection topologies.

In line with the aforementioned DRL approaches, we conduct experimental studies in the traffic simulation environment SUMO [25]. We show that our method achieves state-of-the-art performance, which had been held by traditional non-learning methods, on a wide range of traffic flow rates with varying traffic distributions on the incoming lanes. The content of Sect. 1 is based on [49].

1.1 Related Works

In traditional fixed-time TSC designs [31], the traffic flow rates at intersections are treated as constants, and the green-red phases for each route are scheduled in a cyclic manner. Then the duration for each green phase is optimized using history flow rates. The Uniform TSC with the same fixed duration for all green phases and the Webster's method [44] with pre-timed duration according to latest traffic history are usually used as baselines in TSC works [16]. As the real traffic flow rates generally vary across lanes and across time, the performance of such TSCs is restricted.

Actuated TSCs [31] make use of loop detectors, which are electromagnetic sensors mounted within the road pavement. Such sensors can detect the incoming vehicles and estimate their velocity when they pass by, so that actuated TSCs can dynamically react to the vehicles driving into the intersection. Yet, their performance are still restricted due to the limited information provided by the sensor.

Since decades researchers have investigated on developing adaptive TSCs, which can schedule traffic lights acyclic and with flexible green phase duration according to the real-time traffic situation. Some early works like [14, 26] have been largely applied in real traffic designs. Yet it is still believed that the performance of TSC can be further improved. In recent years, analytical [18, 50], heuristic [17, 43] and learning-based [11, 15, 16, 46] approaches have been proposed. Among these, the heuristic Max-pressure method [43] is reported to be holding state-of-the-art performance [16]. DRL-based methods hold great promise with the possibility to learn generalized and flexible controller policies by interacting with traffic simulators, and that they could provide scheduling decisions in real-time, as opposed to some non-learning methods that need optimization iterations before giving out each decision.

A few works have deployed DRL for isolated intersection TSCs [15, 16, 21, 33]. However, none of them were able to surpass state-of-the-art performance achieved by the Max-pressure method. Each of these method proposes its own reward functions for training the agent, but the connection between them has not been clear. In this work we attempt to give such an analysis of those different reward functions that have been proposed (Sect. 1.2.3).

While efficiency has been the main objective for most of these works, some previous algorithms actually had considered equity implicitly. They [33, 45, 46] design the reward as a weighted sum of several different quantities about the intersection. However, finding the optimal weighting is non-trivial. In this work we instead propose an *equity factor* along with a method to calculate its rough bound.

1.2 Methods

We consider the task of TSC in standard reinforcement learning settings. At each step, from its state $s \in S$ the agent selects an action $a \in \mathcal{A}$ according to the policy $\pi(\cdot|s)$. It then transits to the next state $s' \in S$ and receives a scalar reward $r \in \mathbb{R}$. The state and action spaces and the reward function in our work are discussed in the next subsections.

For learning the optimal policy that maximizes the discounted (by γ) cumulative expected rewards, we use proximal policy optimization (PPO) [37] as the backbone DRL algorithm. For a policy π_θ parameterized by θ, PPO maximizes the following objective:

$$\mathcal{J}_\theta = \mathbb{E}_t \left[\min \left(\rho_t(\theta) A_t, \text{clip} \left(\rho_t(\theta), 1 - \epsilon, 1 + \epsilon \right) A_t \right) + \beta_{\text{entropy}} \cdot H \left(\pi_\theta(s_t) \right) \right],$$
(1)

where the expectation is taken over samples collected by following $\pi_{\theta_{\text{old}}}$, $\rho_t(\theta) = \pi_\theta(a_t|s_t)/\pi_{\theta_{\text{old}}}(a_t|s_t)$ is the importance sampling ratio, and ϵ is a hyperparameter for clipping the probability ratio. H represents the entropy of the current policy, β_{entropy} adjusts the strength of entropy regularization. A_t is a truncated version (on trajectory segments of length up to K) of the generalized advantage estimator [36], which is an exponentially-weighted average (controlled by λ):

$$A_t = \delta_t + (\gamma\lambda)\delta_{t+1} + \cdots + (\gamma\lambda)^{K-1-t}\delta_{K-1}, \tag{2}$$

where $\delta_t = r_t + \gamma V_{\phi_{\text{old}}}(s_{t+1}) - V_{\phi_{\text{old}}}(s_t)$. The value function V_ϕ, parameterized by ϕ, is learned by minimizing the following loss (with coefficient β_{value}):

$$\mathcal{L}_\phi = \beta_{\text{value}} \cdot \mathbb{E}_t \left[\left\| V_\phi(s_t) - \left(V_{\phi_{\text{old}}}(s_t) + A_t \right) \right\|_2^2 \right]. \tag{3}$$

1.2.1 Action Space

We carry out our method on a four-road intersection where each road contains three incoming lanes (one forward-only, one forward+right-turning, one left-turning, Fig. 1a). We note that our approach can easily generalize to other intersections by adjusting the state and action representations accordingly.

The agent has an action space of size 4: while one of the two sets of facing directions (north and south, east and west) has only red light, the other set can schedule either of the following two traffic light signal combinations (Fig. 1b):

• Green for the forward-only and forward+right-turning lanes and red for the rest;
• Green for the left-turning lanes and red for the rest.

In order to give the agent more flexibility, we set the duration for each of the 4 actions as 1 second.

We note that choosing one action means scheduling a distinct green phase. During the transition between different green phases, yellow or all-red phases must be scheduled. In our work, a 3s-yellow and 2s-all-red phase is scheduled before activating a new green phase. We denote the constant $T_{yr} = 5\,s$ as the duration for the yellow-red phase.

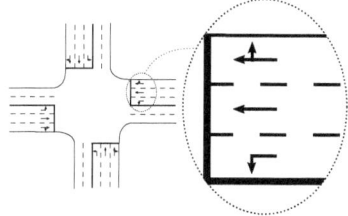

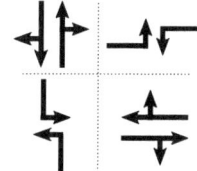

(a) Four-road intersection with three incoming lanes for each road.

(b) Four green phases (actions) for the intersection.

Fig. 1 The intersection and its corresponding action space

Due to this setting, if two different actions (green phases) are scheduled consecutively, the effective duration of the second action is 6 s instead of 1 s; while if the same action (green phase) is scheduled twice in a row, then the effective duration for the second action is still 1 s. During the learning process, the aforementioned two scenarios should not be treated equally. To cope with this we propose the method of *adaptive discounting* which will be presented when discussing the reward function (Sect. 1.2.3).

1.2.2 State Space

At each process step, the state s_t the agent receives is comprised of the following components:

- The distance along the lane to the traffic light and the velocity of each vehicle that has not passed the light and is within 150 m range (each lane has a maximum capacity of 19 vehicles) to the center of the intersection. A block of 19×2 scalars in the state vector is reserved for the vehicles in each incoming lane. The vehicles' states in each block are sorted according to their distance values. The order of lanes in the state vector has to be kept unchanged. All the values are normalized to be within $[-1, 1]$. If any lane does not reach its maximum capacity, the corresponding position and velocity values will be set to 1 and -1.
- The action of the last step a_{t-1} (in one hot encoding so a 4-dimensional vector).
- A counter that contains for each action the time in seconds since its last execution. The 4-dimensional vector is normalized by 500 s. This component along with the last action a_{t-1} helps to avoid state-aliasing.

1.2.3 Reward Function

Several different reward functions have been proposed in previous works to train DRL agents for controlling traffic signals. However, the reasoning behind different designs have not been clearly presented, also the connections between those different choices and the different effects they are causing have not been thoroughly analyzed. We attempt for such an analysis below, which indicates that the vanilla versions of those rewards tend to result in policies that only consider time efficiency (average travel time in an intersection). We then propose solutions that also take equity (variance of individual travel time) into consideration.

Definitions
We first give the definitions of several important concepts in traffic intersection systems. We visualize the important ones in Fig. 2.

- Total number of vehicles in the intersection (N): At t, the number of vehicles in the intersection system N_t is the total number of vehicles that are within a certain

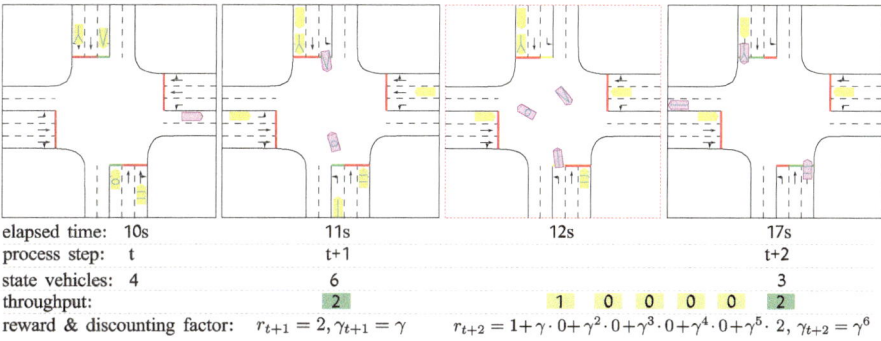

elapsed time:	10s	11s	12s	17s
process step:	t	t+1		t+2
state vehicles:	4	6		3
throughput:		2	1 0 0 0 0 2	
reward & discounting factor:		$r_{t+1} = 2, \gamma_{t+1} = \gamma$	$r_{t+2} = 1 + \gamma \cdot 0 + \gamma^2 \cdot 0 + \gamma^3 \cdot 0 + \gamma^4 \cdot 0 + \gamma^5 \cdot 2, \gamma_{t+2} = \gamma^6$	

Fig. 2 Illustration on the proposed *adaptive discounting*, as well as several important concepts in the traffic intersection domain. In the figure we show each released car with a distinguish symbol on top. As for the colors, the cars in yellow are those that have not yet passed through the traffic light, and they would be depicted in purple immediately after they pass through their traffic lights (judged by the head of the car). The cars in yellow are considered in the state representation, while the cars that turned from yellow to purple are calculated into the throughput. The 1st, 2nd and 4nd sub-figures correspond respectively to system elapsed time: {10, 11, 17}s, and to learning process step: {t, t + 1, t + 2}. Since the action a_{t+1} chooses to schedule a different green phase than that of a_t, a 3s-yellow and 2s-red phase will be scheduled before the new green phase. The 3rd sub-figure in red dashed bounding box shows the 1st second in the yellow phase. In previous works, the discounting has been conducted with respect to the process step. While we propose to discount according to system elapsed time which is shown in our experiments to be of vital importance for stable learning

range to the intersection center (e.g., 150 m) but have not yet passed through the corresponding traffic lights.

- Throughput (N^{TP}): The number of vehicles that pass through the traffic lights of their corresponding incoming lanes within $(t - 1, t]$ is denoted N_t^{TP}.
- Travel time (T_{travel}): For a single vehicle, its travel time is counted as the time period starting from when it enters the intersection and ending when it passes through the traffic light. The total travel time of the intersection is the summation of the individual T_{travel} of each vehicle in the intersection. We note an equivalent way of calculating the total travel time is to count N_t at every second and sum it over a given time period.
- Delay time (T_{delay}): Similar to travel time, except that a constant is subtracted from each individual travel time: $T_{delay} = T_{travel} - T_{free}$, where T_{free} is the constant time length for a vehicle to pass through the intersection system with no cars ahead and green lights always on.
- Traffic flow rate (F): The number of vehicles that pass through an intersection in unit time. A commonly used unit is the number of vehicles per hour v/h.
- Saturation flow rate (F_s): This is a constant representing the traffic flow rate for one lane under the condition that the traffic light stays green during unit time and that the flow of traffic is as dense as it could be [5].

Reward Function Categories

Given the above definitions, the majority of the reward functions proposed in the TSC domain can be categorized into the following two types:

- Throughput-based reward functions $\mathcal{R}^{TP}$ [46]. The vanilla form of this type uses the throughput N_t^{TP} as the reward for step t. Learning on this reward function means maximizing the cumulative throughput of the intersection. The change in throughput $N_t^{TP} - N_{t-1}^{TP}$ has also been used as a reward function [35].
- Travel-time-based reward functions $\mathcal{R}^{TT}$ [11, 15, 16, 33, 46]. As mentioned before, the total travel time of an intersection for a given period of time $[\tau_{start}, \tau_{end}]$ can be calculated as the summation of N_t during that time: $\sum_{\tau_{start}}^{\tau_{end}} N_t$. The vanilla reward function of this type thus uses $-N_t$ as the reward for step t. We note that $N_t = N_{t-1} - N_t^{TP} + N_t^{in}$ where N_t^{in} denotes the number of new vehicles input into the system from $t-1$ to t, which is commonly assumed to be determined solely by the traffic flow distribution thus is out of the control of TSC. Learning on this reward function would result in policies that minimize cumulative travel time. Reward functions utilizing the change of cumulative delay time between actions and the total delay time of the intersection have also been investigated.

The above description indicates that maximizing cumulative throughput and minimizing total travel time could both result in policies that puts efficiency in the top priority. During research we observed that throughput-based reward generally leads to more stable learning with smaller variance across different runs. Therefore, we focus on throughput-based reward in the following content.

Adaptive Discounting

Treating the two scenarios discussed in Sect. 1.2.1 equally when discounting rewards in either of the two reward categories can result in suboptimal policy performance. We propose the method of *adaptive discounting* that properly discount for those scenarios and is shown to be critical for convergence in our experiments.

We illustrate this method under the throughput based reward $R_t^{TP} = N_t^{TP}$ in Fig. 2: At system elapsed time 10 s the reinforcement learning process is at step t. The action a_t is chosen that schedules green lights for the left-turning lanes for the north-south roads. Transitioning from t to $t+1$, the throughput reward obtained is $r_{t+1} = 2$. This is a normal RL iteration and no special adjustments need to be done. But at step $t+1$ when the system elapsed time is at 11 s, the action a_{t+1} is chosen to schedule green lights for the forward+right turning directions of the north-south roads, which is a different green phase than that of a_t. This means a 3s-yellow and a 2s-all-red phase will be automatically scheduled before the new green phase. The 5 s intermediate phase and the chosen 1 s green phase are both within step $t+2$ of the learning process. During this step the throughput obtained at elapsed times $\{12, 13, 14, 15, 16, 17\}$s are $\{1, 0, 0, 0, 0, 2\}$. With no special treatment when calculating the reward for step $t+2$ it would be $r_{t+2} = 3$. But this could lead to undesired properties since the agent gets the intermediate phase "for free" for collecting extra rewards whenever it chooses to schedule a different green phase, and that the subsequent states are not sufficiently discounted. Furthermore, given that the throughput of two episodes matches at every system elapsed second, the agent should obtain exactly the same return, even with different traffic light schedules. However, with the transitional phases it is not anymore a one-to-one mapping between the system time and the process step. So when discounting according to process steps, those two episodes of interest could lead to different returns. This issue has been overlooked in the current literature of DRL based TSC designs [11, 16]. Thus we propose the method of *adaptive discounting* to account for the mismatch between the two timing paradigms, in which we discount the reward according to system elapsed time instead of learning process steps. As a result, the reward for $t + 2$ is calculated as:

$$r_{t+2} = 1 + \gamma \cdot 0 + \gamma^2 \cdot 0 + \gamma^3 \cdot 0 + \gamma^4 \cdot 0 + \gamma^5 \cdot 2, \qquad (4)$$

and a discount factor of γ^6 instead of γ will be used for the subsequent reward or value.

The Equity Factor

Having presented the *adaptive discounting* technique, now we present the *equity factor* for reward functions for training TSC. The aforementioned two types of reward functions (throughput-based and travel-time based) both treat efficiency, i.e., average travel time of the intersection as the major concern. Equity, the variance of individual travel times, is not explicitly considered. Take the following scenario as an example: Assuming that the north-south roads are saturated, while the east-west roads have lighter traffic, the policy to maximize the cumulative throughput should always keep the north-south traffic lights green, while keeping the east-west lights red. Consequently, the vehicles on the east-west roads might have to wait for an intolerable long time to pass through the intersection. This is due to that in the vanilla reward definitions, every vehicle contributes equally to the throughput or to the travel time, regardless of how long it has been waiting.

Following the above analysis, we propose to use the vehicle's travel time together with an equity factor η in the reward function. The basic idea is to adapt the contribution of each vehicle to the throughput-based reward according to its travel time in the intersection while passing the traffic light. Instead of just counting value 1 when a vehicle passes through, we consider three ways to incorporate η into the reward calculation: linear ($\eta \cdot T_{travel}$), power (T_{travel}^{η}) and base ($\eta^{T_{travel}}$). Since simply scaling the rewards does not change the value function landscape, we mainly considered the power and base forms. During research our experiment results show that the power form equity factor leads to convergence to better policies than the base form. Therefore, we focus on the analysis of the T_{travel}^{η} in the following.

To define the proper range of η, two special scenarios are considered.

- Scenario 1: Only one vehicle is before the traffic light, and its travel time at step t is τ. With the equity factor η and the discount factor γ, the return contributed by this vehicle would be τ^{η} if it passes through the traffic light at t, and $\gamma \cdot (\tau + 1)^{\eta}$ if one second later. We require $\tau^{\eta} > \gamma \cdot (\tau + 1)^{\eta}$ so that releasing this vehicle sooner is more desired. With this we get $\eta < \ln(\gamma)/\ln\frac{\tau}{\tau+1}$,

- Scenario 2: One lane with green light is over-saturated, while a single car is waiting at red light in another lane. In the case where the over-saturated lane always has green light on and the single vehicle is never released, the highest return for any state is:

$$G^e = \mathsf{T}_{free}{}^\eta \left(1 + \gamma^{\frac{1}{\mathsf{F}_s}} + (\gamma^{\frac{1}{\mathsf{F}_s}})^2 + \cdots \right) = \mathsf{T}_{free}{}^\eta / 1 - \gamma^{\frac{1}{\mathsf{F}_s}}$$

(denoted as G^e as in this case efficiency is the top priority). If the waiting vehicle is released at step t when its travel time is τ, the upper limit of the return the system can obtain at state s_t is:

$$\sup(G^{e+e}) = \mathsf{T}_{free}^\eta + \tau^\eta \cdot \gamma^{\mathsf{T}_{yr}} + \left(\mathsf{T}_{free} + 2 \cdot \mathsf{T}_{yr} + 1\right)^\eta \cdot \gamma^{2 \cdot \mathsf{T}_{yr}+1} / 1 - \gamma^{\frac{1}{\mathsf{F}_s}}$$

(we use G^{e+e} since this strategy cares about efficiency and equity). The three terms in the summation are all calculated out of the best case scenario (the traffic light on the saturated lane turns yellow then red for a total of T_{yr} elapsed time, then the light on the single vehicle lane turns green for one second then turns yellow) to get the upper limit: the first term is the reward obtained from the vehicle on the saturated lane that manages to pass through at the beginning of the yellow phase; the second term is contributed by the single vehicle passing through the traffic light in its 1 s green phase; the last term is the summation of the reward obtained by the vehicles on the saturated lane after the green phase switches back to this lane. We require $G^e < \sup(G^{e+e})$ to release the single vehicle after certain travel time τ.

With these analysis a range of η can be found. We note that this is a rough calculation under our system settings as for example the traffic flow in the saturated lane does not recover instantaneously to F_s after the green light switches back. Nevertheless the analysis gives a general solution to calculate a rough bound for η. The experimental results show that the desirable TSC policies could be learned in this bound.

1.3 Experiments

1.3.1 Experimental Setup

We conduct experiments using the urban traffic simulator SUMO [25] and evaluate the trained agents in both simulated one-hour traffic demand episodes (with the intersection type described above) and a real-world whole-day traffic demand (with a different type of intersection in Freiburg, Germany). Both intersections have a speed limit of 50 km/h. We compare with the following common baselines in the TSC domain:

- Uniform: This controller circulates ordered green phases in the intersection. Each green phase is scheduled for a same fixed period, the duration of which is a hyper-parameter of this algorithm.
- Webster's [44]: Same as the Uniform controller, it schedules traffic phases in a cyclic manner. But each phase duration is adjusted in accordance with the latest traffic flow history. It has three hyperparameters: the length $T_{history}$ of how long the traffic flow history to take into account for deciding the phase duration for the next $T_{history}$ period, and the minimum and maximum duration for one complete cycle.
- Max-pressure [43]: Regarding vehicles in lanes as substances in pipes, this algorithm favors control schedules that maximizes the release of pressure between incoming and outgoing lanes. More specifically, with incoming lanes containing all lanes with green traffic light in a certain phase, and outgoing lanes being those lanes where the traffic from the incoming lanes exit the intersection system, this controller tends to minimize the difference in the number of vehicles between the incoming and outgoing lanes. The minimum green phase duration is a hyper-parameter.

We note that previous learning methods were not able to surpass the state-of-the-art performance held by the non-learning method Max-pressure TSC [16].

Regarding our network architecture for the intersection in Fig. 1a, the input size for both the policy network θ and the value network ϕ is $4 + 4 + 2 \cdot 19 \cdot 12 = 464$. Then θ consists of fully connected layers of sizes 2 048 (ReLU), 1 024 (ReLU) and 4 (SoftMax), where 4 is the size of the action space. For ϕ the fully connected layers are of sizes 2 048 (ReLU), 1 024 (ReLU) and 1. We perform a grid search to find the hyperparameters. We use 2.5e−5 as the learning rate for the Adam optimizer, 1e−3 as the coefficient for weight decay. For PPO, we use 32 actors, 0.2 for the clipping ϵ. In each learning step a total number of around 20 mini-batches of size 1 000 is learned for 8 epochs.

1.3.2   Training

Previous methods focused on relatively limited traffic situations, for example a single one-hour demand episode [11] and traffic input less than 3 000ᵛ/h [15, 16]. In this paper we challenge our method to experience a wider range of traffic demand. For the four-way junction we consider, the upper bound of the traffic flow can be calculated as $4 \cdot F_s$, where F_s is the saturation flow rate for one incoming lane. This maximum flow is reached when all 4 forward-going lanes of either the north-south or the east-west roads have green lights and are in full capacity. However, this extreme scenario rarely happens in real traffic. In our experiments we found that the intersection already starts to saturate with around 3 000ᵛ/h of total traffic input. In our training we set the range of traffic flow rate to be $[F_{min}, F_{max}] = [0, 6\,000ᵛ/h]$ which is much wider than that used in previous works.

With this flow rate range, we sample traffic demand episodes for training. Each episode is 1 200 s long and defined by these randomly sampled parameters: the total traffic flow at the beginning and end F_{begin} and F_{end}, and for each incoming lane its traffic flow ratio of the total input at the beginning and end. F_{begin} is randomly sampled from $[F_{min}, F_{max}]$. Then F_{end} is sampled uniformly within $[\max(F_{min}, F_{begin} - 1\,500), \min(F_{max}, F_{begin} + 1\,500)]$. The flow ratios are decided by sampling 12 uniform random numbers then normalized by their sum. The traffic flow during the episode is then linearly interpolated. The sampled episodes with possibly big change of traffic flow and unbalanced distribution should be enough to cover real traffic scenarios.

1.3.3   Evaluation During Training

During training, we conduct evaluation to monitor the learning progress every 20 learning steps, which corresponds to 640 episodes experienced by 32 actors. For each evaluation phase 5 evaluators are deployed, corresponding to traffic flow ranges of $[500, 1\,500]$, $[1\,500, 2\,500]$, $[2\,500, 3\,500]$, $[3\,500, 4\,500]$ and $[4\,500, 5\,500]$ respectively. Each evaluator samples traffic demand for evaluation in the corresponding range similar to how training episodes are sampled except that the flow rates at the beginning and end are independently sampled from the same corresponding range.

An ablation study is conducted to analyze the individual contributions of different components in our proposed algorithm. The plots are shown in Fig. 3, where the following agent configurations are compared: $[\times] + [\eta = 0]$, $[\times] + [\eta = 0.25]$, $[ad] + [\eta = 0]$, $[ad] + [\eta = 1]$, $[ad] + [\eta = 0.25]$. $[ad]$ means the agent utilizes *adaptive discounting* while $[\times]$ means not; $[\eta = \cdot]$ denotes the value of the equity factor used by the agent, where the $[\eta = 0]$ agents, which use exactly the vanilla throughput-based reward, care only about efficiency while the $[\eta = 1]$ ones favor equity.

Interestingly, from Fig. 3 we can observe that the two agents without the technique of *adaptive discounting* struggle to learn successful policies in both low and high flow rates. We can also observe the influence of the equity factor η: the $[ad][\eta = 0]$ agent who does not care about equity converges to a better policy than the $[ad][\eta = 1]$ agent in lower traffic density, while the latter agent outperforms the former one in denser traffic. This makes sense, since with little traffic input the equity problem is not critical, while with higher traffic flow the intersection could be saturated with continuously growing queues even under optimal policies. The efficiency-first policies favor releasing more vehicles in saturated traffic, thus vehicles in other lanes could have long waiting time.

We observe that the $[ad] + [\eta = 0.25]$ configuration obtains the best performance across different traffic flow rates, thus this is used for the agent *Ours* in the following experiments.

Having compared the plots of travel time (for released vehicles) and waiting time (for not released vehicles), we notice that the average waiting time always decreases during training when the policy gets better, while the average travel time may vary in different ways. This is because the travel time only considers the released vehicles. Some initial poor policies may choose the same action all the time, which leads to fast

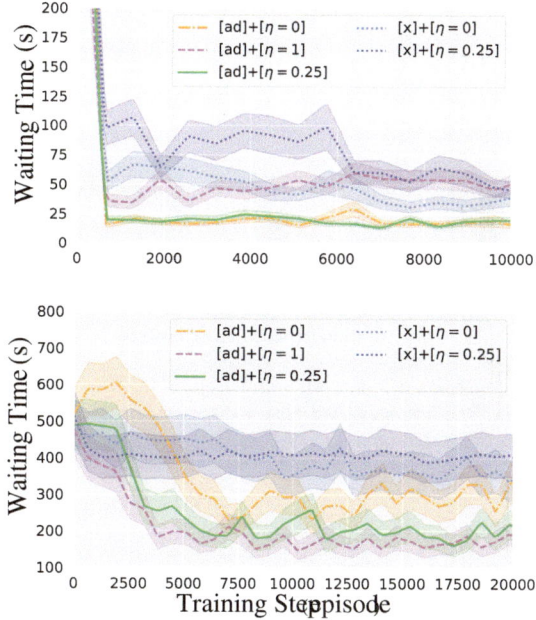

Fig. 3 Waiting time obtained in evaluation during training for all agent configurations under ablation study. Each plot shows the mean with $\pm 1/5$ standard deviation over 3 non-tuned random seeds (we show $1/5$ of the standard deviation for clearer visualization). The left figure shows the logs of the evaluators of traffic flow range [500, 1 500], while the right one shows that of [4 500, 5 500]. The vehicles passed the traffic light are not considered for the waiting time. The waiting time for a vehicle is calculated with $T_{\mathrm{episode}} - T_{\mathrm{in}}$, where T_{episode} is the episode duration and T_{in} is the time when it enters the intersection

throughput for vehicles on the lanes with green light while extremely long waiting time for other vehicles. The waiting time, however, considers only the vehicles not passed the intersection during the episode. As the policy gets better, the number of vehicles staying in the intersection at the end becomes smaller. In order to show the training process clearly, we choose to use the plot of waiting time.

1.3.4 Evaluation on Simulated Traffic Demand

To test the performance of our agent we first evaluate on simulated traffic demand episodes that each lasts one hour. For each of the 5 traffic flow rate ranges as used for the evaluators during training, we randomly sample 30 episodes; this exact set of $5 \cdot 30$ episodes are used to test all compared algorithms. These demand episodes are sampled following the similar procedure to that for evaluation during training.

To ensure a fair comparison, in each demand episode, we use the exactly same vehicles generation time for different methods. Via the sampling process described

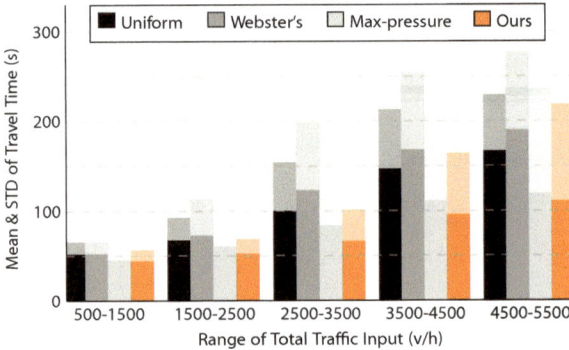

Fig. 4 Performance comparison of our work with baselines on 150 one-hour simulated demand episodes (30 from each of the 5 ranges). We note that the baselines are optimized for each of the test episodes before they are tested on it

above, our test set covers a very wide range of traffic scenarios and could in turn provide a more thorough evaluation.

The evaluation results are shown in Fig. 4. We observe that our method reaches state-of-the-art performance on all traffic flow ranges. It is worth noting that for each baseline that we compare with, we find its optimized hyperparameters for each of the 150 test episodes; while our agent is trained only once and a single agent is used to evaluate on all 150 test episodes. This means that the overall performance of our one trained model outperforms that of the 150 individually optimized models. The performance improvement at about 1 000 and 5 000 v/h is not very obvious, because in light traffic many vehicles do not have to wait in queue and in over-saturated traffic, where there is a queue in every incoming lane, the best policy is similar to scheduling the green phases cyclically. The capability of our agent to react to real time traffic situation can be fully utilized for the traffic flow ranges in the middle, where the improvement against the Max-pressure controller and the fixed-time controllers could be over 20 and 40%. The Webster's method performs worse than the Uniform controller due to the quick change and short duration of the test episodes, which is most of the time not the case in real traffic (Fig. 5).

As mentioned, the travel time only indicates how fast the released vehicles drive through. In order to show that our agent can also benefit more drivers than baselines, we present the testing statistics for throughput in Table 1. The percentage values are the ratio of the released vehicles in the total vehicle number generated. With traffic flow lower than about 3 000 v/h, all TSCs can properly release traffic input. Not 100 percent of the generated vehicles can be released, because the test is stopped directly after one hour. Some vehicles generated at the end do not have enough time to travel through. From about 3 000 v/h the throughput of the baselines start to drop, which means the TSC can not fully release the input traffic flow and traffic jam starts to form, while our agent can avoid traffic jams in much denser traffic. With the increased

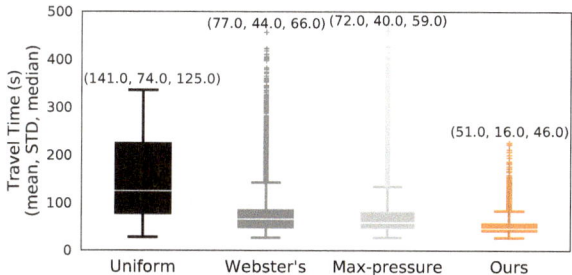

Fig. 5 Performance comparison of our work with baseline models on a whole-day real-world traffic demand

Table 1 Throughput (%) of considered methods in Fig. 4

Traffic flow input (v/h)	Throughput (%)			
	Uniform	Webster's	Max-pressure	Ours
500–1 500	97.38	97.47	97.59	**97.76**
1 500–2 500	97.09	97.52	97.54	**97.95**
2 500–3 500	93.77	95.34	95.91	**97.76**
3 500–4 500	87.16	86.65	88.84	**92.88**
4 500–5 500	77.62	74.90	81.75	**86.27**

efficiency, our agent can still guarantee equity, which is shown by the low standard deviation of vehicles travel time and the high throughput.

1.3.5 Evaluation on Real-World Traffic Demand

To further measure the performance of our agent in more realistic traffic scenarios, we conduct additional tests with a whole-day traffic demand of a real-world intersection of Loerracherstrasse and Wiesentalstrasse located in Freiburg, Germany. This intersection has different layout than the one in Fig. 1a. Here each road has one forward+right turning lane with one additional short lane for protected left turn. So the size of the state changes to 224. We regard the short left-turning lane, the forward+right-turning lane and the lane segment before the branching as separated when we construct the state.

Since the size of the input is different from the experiments above, we need to train another agent. As we want to test the generalization capabilities of our method, the training traffic demand is sampled in the same way as before (only the maximum limit of the traffic flow is reduced to $1/2$ to reflect the change in the intersection layout) The trained agent is tested on the real-world traffic demand of February 4, 2020, with typical traffic flow peaks at rush hours. The input traffic flow is in the range

[0, 1 740]ᵛ/ₕ. We sincerely appreciate the support of city Freiburg (www.freiburg.de/verkehr), which provides us with the traffic flow data measured with inductive-loop detectors.

The results of this real-world experiment is shown in Fig. 5. All the TSCs can properly release all vehicles, because the traffic flow is nearly zero in the night when the demand episode ends. We can observe that our method is again outperforming all baseline methods, even though the baselines are firstly optimized with exactly this whole-day demand and our model is only trained on the simulated episodes with 1 200 s duration. The substantial improvement of nearly 30% on average travel time is even greater than the performance gain in the simulated evaluations. This validates that out proposed method has great generalization capabilities and can adapt to a wide range of traffic scenarios.

2   Courteous Behavior of Automated Vehicles at Unsignalized Intersections

Traffic control signals are not panacea for intersection problems [42]. For example, they may reduce traffic efficiency for low or unbalanced traffic demand. Although recent works [43, 49] developed more intelligent adaptive traffic signal control methods, for the majority of all intersections, which often have only one lane per road and mostly small traffic volume [13], the use of static road signs assigning priority has proven to be more efficient [42]. Ulbrich et al. [41] summarized how humans cooperate with other traffic participants to improve the whole traffic utility. Consider, as an example, the situation shown in Fig. 6. Even though vehicle 1 has higher priority and can proceed through the intersection before vehicle 2, its driver might prefer to yield to vehicle 2 so that the traffic behind vehicle 2 can be released sooner.

Based on the expectation that future traffic will consist of connected autonomous vehicles (CAVs), a large majority of current research excludes human-driven vehicles (HVs) in their development of traffic management approaches. However, it might

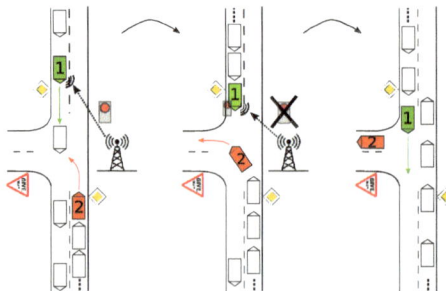

Fig. 6 Our intersection management agent optimizes traffic flow by assigning virtual red traffic lights to connected autonomous vehicles (vehicle number 1). Once vehicle 2 is released, the vehicles following it can also proceed through the intersection

take decades for the technology, the infrastructure and the users to be ready for traffic with only connected autonomous vehicles [24]. We therefore believe that, for the near future, applicable traffic management solutions must *(i)* consider various degrees of mixed traffic, *(ii)* pose no complications or major adjustment requests for human-driven vehicles, and *(iii)* not present a traffic disturbance or danger when the communication between the connected autonomous vehicles fails.

In this section, we propose a novel centralized method to improve intersection management in mixed traffic. Our approach learns a policy for CAVs that maximizes the overall utility while at the same time showing courteous behavior [27]. We make the following contributions:

- We present a centralized intersection management method based on deep reinforcement learning that improves traffic performance at unsignalized intersections in mixed traffic scenarios.
- We introduce *return scaling* for training in environments with a large imbalance of cumulative rewards at different states. In our case, this helps to balance policy updating of states with different traffic densities, in particular to counteract the large cumulative reward collected in heavy traffic, which would otherwise dominate the stochastic gradient descent process and make the policy unstable for states in sparse traffic.
- We present a comprehensive performance comparison for various traffic densities and changing rates of CAVs to demonstrate the potential of our approach.

We conduct experimental studies in the traffic simulation environment SUMO [25] and show that our method outperforms the state-of-the-art intersection management method on a wide range of traffic densities with varying traffic distributions. The content of Sect. 2 is based on [48].

2.1 Related Work

Among the first ones to propose an intelligent intersection management system were Dresner and Stone whose reservation-based approach [7, 8] divides the junction with intersecting trajectories into a grid of tiles. Their autonomous intersection management approach, realized as a centralized controller, applies a first-come-first-served (FCFS) strategy to deal with the requests by CAVs for time slots of the tiles along their trajectories. To accommodate HVs they employ traffic lights and the so-called FCFS-light policy [9, 10]. Later, this framework was extended to allow for the centralized intersection management to set the speed profiles of vehicles with cruise control [3]. To improve the performance of FCFS-light, Sharon and Stone introduced hybrid autonomous intersection management [38]. With this extension, requests of CAVs can be approved regardless of the traffic lights if there are no HVs in the intersecting routes.

In general, the methods based on autonomous intersection management [7] provide a relative advantage to CAVs over HVs, which, in our opinion, should be avoided as it might cause the public to repel automated vehicles. Furthermore, human drivers will be more sensitive to stopping and waiting than the passengers in CAVs. We therefore suggest that the benefit brought by intersection management and CAVs in general should be evenly shared with human drivers.

Lin et al. developed a method similar to the FCFS-light policy [22]. It reserves conflicting sections among different routes instead of the grid of tiles. Another first-come-first-served reservation based method has been proposed by Bento et al. [4]. They suggest to control both CAVs and HVs via speed profiles sent by the intersection management unit. This places an undesirable burden on human drivers to follow a given speed profile and additionally even requires all HVs to be connected.

The described approaches make the vehicles roughly follow a first-come-first-serve strategy to traverse intersections. However, as shown by Meng et al. [29], the performance of an intersection management strategy mainly depends on the passing order of vehicles and not so much on the individual trajectory planning algorithms. As the computation time grows exponentially with the number of considered vehicles [29], often simplifying assumptions are made including linear constraints, no overtaking, no lane changing, constant speed, and constant traffic input. The coordination of the passing order can mitigate control uncertainties, which makes it more suitable for mixed traffic. Based on this idea, our work is aimed at finding better passing orders, while having vehicles drive based on their own trajectory planning model.

Qian et al. [34] assign priorities representing the passing order to vehicles. While CAVs receive the priority from a central control unit and plan trajectories accordingly, the passing order of HVs is regulated by traffic lights. With high rates of HVs, this potentially results in an inefficient, mostly first-come-first-served control. Fayazi et al. [12] propose to formulate the intersection management problem as a mixed-integer linear program. Their controller assigns times of arrivals to a virtual access area around the junction to CAVs, while HVs are regulated by traffic lights.

The approaches of these related works are already outperformed by Webster's method or fixed-time traffic signal controllers when over 10–20% of the vehicles are driven by humans [7, 9, 12, 22]. The exception is our previous state-of-the-art learning-based adaptive traffic signal controller, which further outperforms these two controllers in any traffic flow range and reduces the average travel time by up to 30–60% in the experiment with real-world traffic input. Therefore, we evaluate our proposed method mainly against Sect. 1 in a wide range of dynamic traffic demands and show that the performance gain is available even with a small portion of CAVs in the traffic system.

2.2 Methods

Like Sect. 1.2, we model the intersection management task at unsignalized intersections as a Markov Decision Process, and use proximal policy optimization [37] for training due to its stability, good performance and ease of implementation. Our work

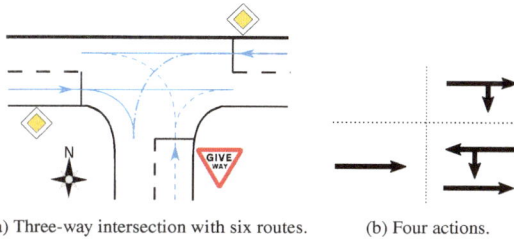

(a) Three-way intersection with six routes.   (b) Four actions.

Fig. 7 Common regulation of a right-hand traffic three-way intersection (**a**). The high-priority-routes are W-E, W-S and E-W. The low-priority-routes are S-W and S-E. Route E-S has intersecting routes with higher and lower priority. The proposed set of actions (**b**) stops CAVs on routes along the indicated directions

is aimed at training a centralized agent for an intersection that timely stops the CAVs on the routes with higher priority to let the vehicles on conflicting routes with lower priority pass, so that the performance of the whole system is optimized. Since this is similar to red traffic lights for CAVs on the routes with higher priority, we denote our method as *Courteous Virtual Traffic Signal Control* (CVTSC). We evaluate our proposed approach on the most common type of three-way intersections as illustrated in Fig. 7. By adjusting the state and action representations, our approach can easily be generalized to other intersection layouts, as we show for the real-world intersection in Sect. 2.3.5.

As we focus on an isolated intersection, we assume that the vehicles can drive freely after they passed the junction and entered the outgoing lanes. Thus the vehicles on the outgoing lanes do not influence the intersection management. However, unlike in Sect. 1, in which we only considered vehicles in front of the stop lines, we here also take the vehicles into account, which already passed the stop line but not yet entered the outgoing lanes. This is necessary as at unsignalized intersections vehicles very often choose to wait after stop lines and coordination may happen there inside the junction.

2.2.1   Action Space

For the intersection in Fig. 7a we assume that vehicles drive according to the priorities predefined by the road signs, where the diamond indicates priority roads and the triangle indicates yield. Vehicles on the routes with lower priority have to wait until there is enough gap on the conflicting routes with higher priority before passing the junction. Note that in Fig. 7a the route E-S has intersecting routes with higher and lower priority.

To obtain courteous behavior for CAVs on routes with higher priority, without loss of generality, we define a discrete set of four actions {(), (W-E), (W-E, W-S), (W-E, E-W, E-S)} as the action space $\mathcal{A}$ in relation to Fig. 7b. The indicated directions show the corresponding routes on which the intersection management unit commands CAVs to halt before the respective stop lines to give priority to vehicles waiting on intersecting routes with lower priority. The action restricting no routes uses the default priorities to manage the intersection. We set the duration of each action to 1 s. A categorical policy is learned: during training the actions are sampled according to the output distribution, while during testing the action with the highest probability is always chosen. When a new action a_t is chosen, CAVs on the routes indicated in a_t will receive stopping commands, while the instruction for the routes restricted by a_{t-1} is canceled, if they are not regulated by a_t. If a CAV receives a stopping command while being too close to the stop line, it will continue through the intersection thus ignoring the command. Acceleration, collision avoidance and safe distance are managed by the low-level controllers of the individual vehicles (both CAVs and HVs).

2.2.2 State Space

Due to the restriction of sensors and wireless communication, we assume that the intersection management unit can collect information of vehicles that are within a distance of 150 m along the road measured from the center of the intersection. We assume that every vehicle's state, composed of continuous values (its position along the road, velocity and time since entering intersection) and discrete values (a binary value for CAV or HV and optionally a route index indicating the driving direction if the lane contains more than one route), is available to the control unit. Similar to Sect. 1.2.2, the state s_t of the intersection at time t is given by a vector that contains the structured information of vehicles in it.

As described in Sect. 2.2.1, only CAVs are controlled by the agent. Every 1 s a new action should be chosen according to the new state. However, at certain points in time there are no CAVs in the intersection and including these states in training hinders the learning process. We therefore remove states without CAVs from the training data. As a result, the influence of actions is not limited to a fixed interval and the duration of one step in the learning process can be any positive integer in seconds. To deal with this variable step length, we employ the method of *adaptive discounting* as proposed in Sect. 1.2.3.

2.2.3 Reward Function

The common objective of intersection management methods is to improve the efficiency while keeping a certain level of fairness for all vehicles. Here, we extend the idea of a reward function with equity factor. Instead of using $T_{travel}{}^{\eta}$, we propose to use $\eta_a \cdot T_{travel} + \eta_b$ as the reward for each released vehicle, where η, η_a and η_b are equity factors. Due to the flexible step lengths discussed above, the reward of each step r_t is calculated by accumulating discounted rewards generated during step t which might contain up to k environment steps (each one second). I.e., we accumulate the contribution of N_t^{TP} released vehicles by

$$r_t = \sum_{i=0}^{k-1} \gamma^i \sum_{j=1}^{N_{t\_i}^{TP}} (\eta_a \cdot \tau_j + \eta_b), \tag{5}$$

where $N_{t\_i}^{TP}$ is the throughput of the ith second in step t and τ_j is the travel time of the jth released vehicle in the ith second.

The values of η_a and η_b are selected as in Sect. 1.2.3 based on two heuristics. First, we favor releasing each vehicle as soon as possible for the purpose of efficiency. The second heuristic aims at equity by considering a traffic situation, where one vehicle waits for saturated traffic flow on an intersecting route. Since efficient traffic flow on the high priority route should not be achieved on the expense of accumulating too large waiting time on the single vehicle, we increase the reward contributed by each released vehicle according to its travel time. This linear relation between reward and travel time is more intuitive than the previous exponential formulation. Moreover, the additional free variable in this formulation can be used to scale the rewards of single released vehicles to keep them around unity, which is beneficial for hyperparameter tuning in common deep reinforcement learning setups.

2.2.4 Return Scaling

According to the reward definition, the return $G(s_t)$ is mainly influenced by the throughput and the travel time of released vehicles. Since both of them increase with the traffic input, the scale of $G(s_t)$ could vary from less than 5 to over 100 if the state of the intersection changes from nearly empty s_{low} to saturated s_{high}. Consequently, s_{high} would have a much larger impact on π_θ and V_ϕ during the update phase, making the learning process of a policy for light traffic less stable.

We introduce *return scaling* to resolve the issues caused by imbalanced return of states, which has shown to be critical for convergence with low traffic volumes in our experiments. In order to reduce the difference between $G(s_{low})$ and $G(s_{high})$, we scale the cumulative rewards before the update phase with

$$G(s_t) = \rho(s_t) \cdot \sum_{i \geq t} (\gamma^{\sum_{j=t}^{i-1} k_j}) r_i, \tag{6}$$

where k is the number of environment steps (each one second) in one step of learning process. The scaling factor ρ is defined as

$$\rho(s_t) = (N_c^V / n^V)^\xi, \tag{7}$$

where n^V and N_c^V are the current number of vehicles in the intersection and its capacity, respectively, and ξ is a hyperparameter.

2.3   Experiments

2.3.1   Experimental Setup

We use the open-source traffic simulator SUMO [25] to train and evaluate various intersection management agents. Besides simulated traffic episodes we also evaluate our approach on real-world rush hour traffic demand. For all roads we set a speed limit of 50 km/h. We compare our approach CVTSC to baselines managing the intersection with *road signs (RS)* defining static priorities for routes and with *traffic lights (TL)* controlled by a deep reinforcement learning agent according to Sect. 1. A possible set of green phases for the three-way intersection is shown in Fig. 9.

Two fully connected networks θ and ϕ are used as the policy and value function estimators. They have an input layer of size 343 and the same structures as in Sect. 1.3.1. A grid search was used to select the hyperparameters. We use $5e-6$ as the

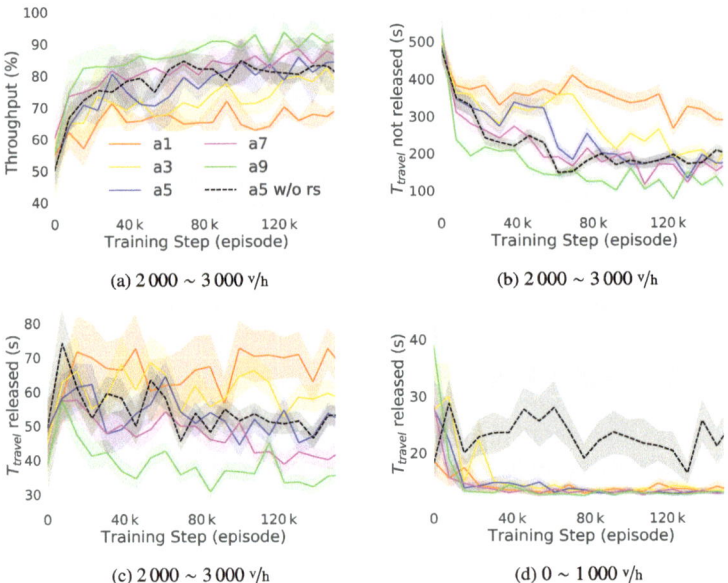

Fig. 8 Results obtained in evaluation during training for all agents with varying CAV rates in traffic (solid lines) and an ablation study for the usage of the return scaling (dashed black line). The plots show the mean with standard deviation, where the latter is scaled by $\pm 1/10$ for the travel times (for clearer visualization), over three non-tuned random seeds. By the end of each episode there are still some vehicles, which have not passed the junction. The travel time for such a vehicle is calculated with $T_{\text{episode}} - T_{\text{spawn}}$, where T_{episode} is the episode duration and T_{spawn} is its scheduled spawning time in the simulator

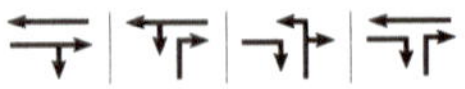

Fig. 9 Traffic light green phases for the intersection in Fig. 7a

learning rate for the Adam optimizer and 0.001 as the coefficient for weight decay. For proximal policy optimization algorithm, we use 32 actors, the clipping threshold of $c = 0.001$ and the discount factor of $\gamma = 0.98$. For the return scaling factor, we use $\xi = 0.2$, which is found to be the optimal value in the range of $(0, 1]$. In each learning step mini-batches of size 100 are used to update the agents in 8 epochs. The number of mini-batches in each learning step is, however, variable due to the varying step lengths. The equity factors η_a and η_b for reward calculation are set to 0.0027 and 0.946. The training process of 150 k steps takes about 40–60 h (depending on the corresponding CAV rate) running on four NVIDIA TITAN X GPUs, while CPU computation is not a limiting factor.

2.3.2   Training Setup

Most current related work has been developed and tested with simplified traffic demand, such as constant traffic input to the intersection. We challenge our approach and train it with more dynamic traffic input ranges to cover as many real traffic scenarios as possible. For the three-way junction in Fig. 7a the saturation flow rate F_s of each incoming lane is $1\,670$ v/h and as it is very rare that two non-conflicting routes are simultaneously saturated, we set the traffic demand range to $[F_{min}, F_{max}] = [0, 3\,000]$ v/h. The simulated traffic episodes are sampled in the same way as Sect. 1.3.2. We train five agents ($a1$, $a3$, $a5$, $a7$, $a9$), each corresponding to a fixed CAV rate of $[10, 30, 50, 70, 90]$ %, corresponding to the expected increasing CAV rates in the future traffic.

2.3.3   Evaluation During Training

To monitor the learning process the performance is evaluated for traffic input of different ranges: $[0, 1\,000]$, $[500, 1\,500]$, $[1\,000, 2\,000]$, $[1\,500, 2\,500]$, $[2\,000, 3\,000]$. The generation of traffic demand is analogous to that of training episodes except that the total traffic inputs at the beginning F_{begin} and end F_{end} are sampled independently in the five given ranges.

The plots in Fig. 8 show the performance of agents trained with different CAV rates and present an ablation study for the usage of the *return scaling*. The agent *a5 w/o rs* is trained with a CAV rate of 50% without using return scaling. We analyze the throughput in percentage of released vehicles among all spawned vehicles, the travel time of released and not released vehicles at the highest traffic density level and the travel time of released vehicles at the lowest level. The calculated travel time is the mean among all released or not released vehicles during three evaluation episodes. We analyze the throughput and travel times instead of the accumulated reward as they give us a better estimate of the overall performance. The variance of the travel times is of particular interest as it is a good indicator for the equity. Large variances correspond to some vehicles with long waiting times at the intersection.

As illustrated in Fig. 8a, b and c, CVTSC with higher CAV rate leads to more throughput, more efficient clearance (lower average T_{travel}) of the intersection and more fairness (shown by lower standard deviation of T_{travel}) to all the vehicles. As expected, from Fig. 8c and d, we observe that the agent without return scaling fails to learn an efficient policy for light traffic, although its performance is similar to that of $a5$ in heavy traffic. We plan to conduct further investigation on return scaling, in particular whether it is applicable to a broader class of problems or can be replaced with other methods like γ-tuning.

2.3.4 Evaluation on Simulated Traffic Demand

We first test our agents with simulated traffic episodes, each with a duration of one hour. For each of the five traffic demand levels described above, we first create 50 traffic episodes with spawning time of each vehicle following the procedure to that for evaluation during training. Then we generate five sets of mixed traffic episodes with different CAV rates by randomly setting each vehicle as CAV or HV according to the penetration rate. Note that the baseline methods *road sign (RS)* and *traffic light (TL)* do not distinguish between CAV and HV. Following this setup, we test both baselines and our trained agents with identical number of vehicles and same spawning times. In the following, the five agents are first tested with their corresponding CAV rates to evaluate their performance against the baseline methods. Then we cross-evaluate them on settings corresponding to different CAV penetration rates.

Performance of Intersection

The performance is shown in Fig. 10 and Table 2. For all the tested traffic density levels, our CVTSC agents can improve the performance of the unsignalized intersec-

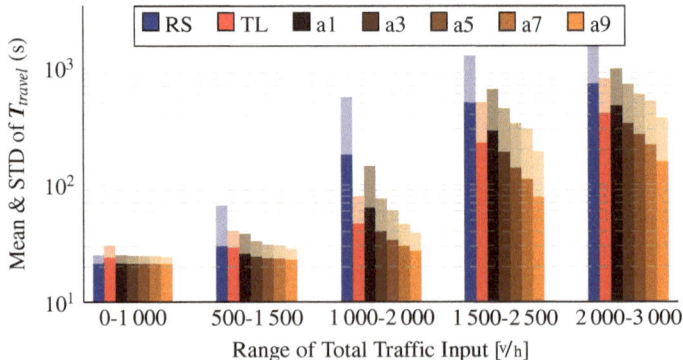

Fig. 10 Performance comparison of our CVTSC with baselines RS and TL in traffics with different CAV rates. For each controller with each traffic density, the mean (opaque bars) and positive standard deviation (translucent bars) of T_{travel} are calculated over all vehicles (including released and not released) of 50 simulated traffic episodes. Each CVTSC agent is trained and evaluated in traffics with its corresponding CAV rate

Table 2 Throughput (%) of considered methods in Fig. 10

Traffic input (v/h)	Throughput (%)						
	RS	a1	a3	a5	a7	a9	TL
0–1 000	99.4	99.4	99.4	99.4	99.4	99.4	99.4
500–1 500	99.2	99.3	99.3	**99.4**	99.3	**99.4**	99.2
1 000–2 000	91.1	97.7	98.6	99.0	99.1	**99.2**	98.5
1 500–2 500	72.2	85.3	90.6	93.5	94.7	**96.8**	88.5
2 000–3 000	59.8	74.6	82.1	85.8	88.5	**91.9**	77.9

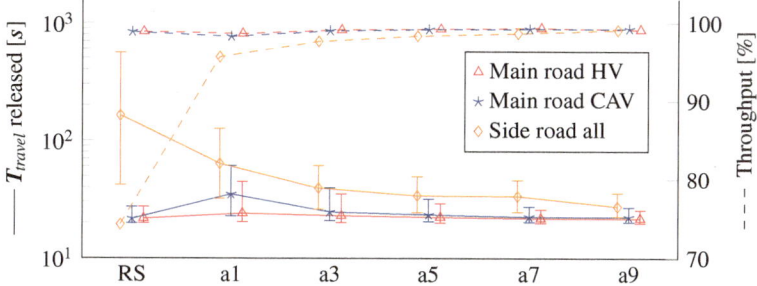

Fig. 11 Performance comparison of different vehicle groups at traffic demand 1 000–2 000 v/h. The plotted travel times show the median, lower quartile and higher quartile over all released vehicles among all evaluated episodes. The plotted throughput is the percentage of released vehicles among all spawned vehicles throughout all episodes

tion. Not only more vehicles are released during the same period, but also the mean and standard deviation of their travel times are reduced. The higher the CAV rate is, the better our approach performs. The performance gain of CVTSC on the lowest traffic density is not obvious, because nearly no vehicles have to stop at the junction. When there is little traffic, employing *TL* can cause unnecessary stopping due to the transition phase (amber or red lights). In heavier traffic over 1 500 v/h *TL* outperforms *a1* by a little margin. However, it is outperformed by CVTSC when 30% or more vehicles are CAVs.

Performance of Vehicle Groups

In contrast to the relative advantage of CAVs over HVs suggested by the methods based on autonomous intersection management [9], our CVTSC tends to share the performance gain evenly between the two types of vehicles. Figure 11 shows how CVTSC can increase the intersection management performance while keeping the balance between different vehicle categories. Since the actions are executed only for CAVs on the main road, we divide vehicles on the main road into *Main road CAV* and *Main road HV* and assign all vehicles on the side road to a third group *Side road all*. As illustrated, the performance gain against *RS* is mainly caused by the improvement of the traffic on the side road. With only 10% CAVs the throughput of the side road traffic is increased from 74.3% to 95.6% and the median travel time is decreased by

Table 3 Performance comparison of different agents with different traffic input settings. For each agent with each traffic setting, the average T_{travel} is calculated over all vehicles (including released and not released) of 50 simulated traffic episodes

Traffic input		Average T_{travel} [s]					Throughput [%]				
CAV rate (%)	Input flow [v/h]	a1	a3	a5	a7	a9	a1	a3	a5	a7	a9
10	500–1 500	25.8	**25.7**	25.8	26.3	27.5	99.3	99.3	99.3	99.3	99.3
	1 000–2 000	**63.2**	69.9	80.8	102.1	131.4	**97.7**	97.4	96.9	95.8	94.2
	1 500–2 500	**287.8**	299.3	339.0	364.2	432.3	**85.3**	84.7	82.1	80.6	77.1
	2 000–3 000	**471.0**	482.5	517.8	554.4	610.1	**74.6**	73.9	72.0	69.3	65.9
30	500–1 500	24.7	**24.3**	24.4	24.9	24.8	99.3	99.3	99.3	99.3	99.3
	1 000–2 000	42.1	**40.0**	43.4	49.4	58.7	98.5	**98.6**	98.6	98.2	98.0
	1 500–2 500	213.4	**190.0**	204.2	237.9	274.1	89.5	**90.6**	90.0	88.1	85.9
	2 000–3 000	367.3	**334.2**	347.0	411.7	430.6	80.3	**82.1**	81.4	77.5	76.6
50	500–1 500	24.1	**23.6**	**23.6**	23.9	23.9	99.3	99.3	**99.4**	99.3	99.3
	1 000–2 000	36.0	33.7	**33.5**	35.6	38.9	98.9	**99.0**	**99.0**	98.9	98.7
	1 500–2 500	191.0	145.5	**138.8**	159.7	174.7	90.6	93.2	**93.5**	92.3	91.8
	2 000–3 000	346.9	269.0	**267.0**	308.6	313.4	81.4	**85.9**	85.8	83.5	83.3
70	500–1 500	23.6	23.2	**23.1**	23.4	23.3	**99.4**	99.4	**99.4**	99.3	99.3
	1 000–2 000	32.6	29.8	**28.6**	29.9	30.0	99.0	**99.1**	99.1	99.1	99.1
	1 500–2 500	176.3	120.7	**101.1**	111.2	112.1	91.2	94.4	**95.6**	94.7	95.0
	2 000–3 000	323.2	234.2	**203.5**	219.0	217.0	82.6	87.3	**89.3**	88.5	88.5
90	500–1 500	23.1	22.8	**22.6**	23.1	22.9	**99.4**	99.4	99.4	99.3	99.4
	1 000–2 000	30.3	27.9	**26.7**	27.4	27.3	99.0	**99.2**	99.2	99.2	99.2
	1 500–2 500	164.8	105.5	77.0	**76.5**	77.9	91.8	95.2	**96.8**	96.7	96.8
	2 000–3 000	311.5	192.0	161.6	**154.5**	157.8	83.2	90.0	91.9	**92.2**	91.9

61%. As a necessary side effect, the courteous behavior adds about 13 s to the median travel time of CAVs on the main road and slows down some HVs following the CAVs consequently. However, the median travel time of Main road HV and the throughput of both vehicle groups on the main road are nearly not influenced. With growing rate of CAVs in traffic, the performance of the traffic on the side road continues to be improved while the initial disadvantage for the main road is compensated.

Comparison of Agents

To cross-evaluate their performance on other traffic settings than their natives, we further test each agent (a1 to a9) on the five different CAV rates on 50 simulated episodes on each of the five traffic densities. Since CVTSC brings nearly no measurable difference for the lowest traffic density, only the results for the other four traffic densities are listed in Table 3.

We observe that all trained CVTSC agents outperform *RS* in any mixed traffic setting. Furthermore, two significant patterns can be observed in the results. First, for each CAV rate the agents trained with similar rate values are among the best, as expected. Second, as the CAV rate increases the performance of all agents is continuously improved. Interestingly, *a5*, the one trained with CAV rate of 50%, outperforms or performs equally well as *a7* and *a9* even in settings where CAVs are the majority. We suppose this is because *a5* during training is exposed to more diverse traffic situations, especially ones with fewer CAVs in the intersection. As shown in Figs. 10 and 11, the margin of the performance gain decreases with increased CAV rate. Even though *a7* and *a9* can handle highly automated traffic better than *a5*, the performance gain is so small that it can not compensate the performance loss when occasionally more HVs drive in the intersection.

2.3.5   Evaluation on Real-World Traffic Demand

To further evaluate CVTSC in more realistic traffic situations, we conduct additional tests with real-world traffic demand recorded at an intersection in Freiburg, Germany, which is sketched in Fig. 12. Unlike the intersection in Fig. 7a, one part of the main road (Tullastrasse) forks before the stop line. After adjusting the state representation and the intersection structure in the simulator we trained two new agents *a3* and *a5* and employ them in the test. The traffic demand, listed in Table 4, was manually recorded on October 19, 2017 by the traffic department of Freiburg. The total traffic input was about $1\,000-1\,500$ v/h with roughly 20% on the side road.

Figure 13 shows box plots of the travel times of released vehicles controlled by *RS* and CVTSC agents in traffic scenarios with different CAV rates. The agent *a3* is employed for 10% and 30% automated traffic, while *a5* is employed for the other three. In all scenarios over 99.7% of all vehicles traverse the intersection. Our method continuously improves the traffic flow with increasing rate of CAVs in traffic. We notice that the median of travel times in all scenarios stay similar, which means the performance gain comes mainly from the vehicles with long travel times on the side road. CVTSC agents manage to release them faster without delaying the traffic on the main road.

Fig. 12 Intersection of Tullastrasse and Hans-Bunte-Strasse in Freiburg, Germany

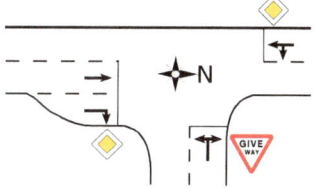

Table 4 Traffic in rush hours on the morning and afternoon of October 19, 2017 at the intersection of Fig. 12

Direction	Traffic input (Number of vehicles every 15 min)															
	7:15	7:30	7:45	8:00	8:15	8:30	8:45	9:00	16:15	16:30	16:45	17:00	17:15	17:30	17:45	18:00
N-S	55	63	101	80	98	85	60	111	102	104	79	97	148	122	104	67
N-E	44	29	38	44	32	44	28	31	32	44	26	28	32	37	38	19
S-N	71	76	96	111	78	86	80	65	105	88	119	116	112	86	100	108
S-E	35	41	32	53	68	42	52	43	29	32	29	36	33	30	29	27
E-N	11	26	29	20	40	29	20	22	58	48	56	35	55	50	47	35
E-S	16	25	51	26	31	21	32	22	53	32	43	23	32	19	31	25

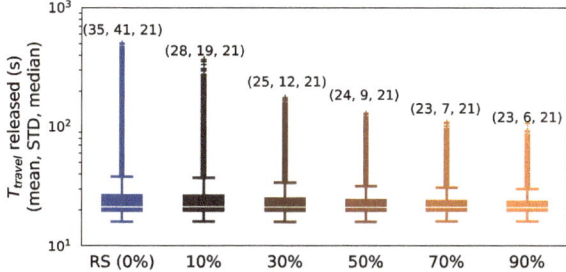

Fig. 13 Box plot of travel times with different CAV rates over all released vehicles in the simulation based on the real-world intersection of Fig. 12. The whiskers extend $1.5 \cdot$ IQR (interquartile range) from the upper and lower quartiles

3 Conclusion and Future Work

3.1 Conclusion

In this chapter we first present an approach to learning traffic signal controllers using deep reinforcement learning. Our approach extends existing reward functions by a dedicated equity factor. We furthermore proposed a method that utilizes adaptive discounting to comply with the learning principles of deep reinforcement learning agents and to stabilize training. We validated the effectiveness of our approach using simulated and real-world data.

Then we present an approach to improving mixed traffic management at unsignalized intersections using deep reinforcement learning. Our proposed method CVTSC creates courteous behavior for automated vehicles in order to optimize the overall traffic flow at intersections. Furthermore, we introduce return scaling to counteract the imbalance of cumulative rewards at different states and to stabilize training. We validate the effectiveness of CVTSC using simulated and real-world traffic data and show that CVTSC improves the traffic performance continuously with increasing percentage of automated vehicles. With more than 10% of automated vehicles it also outperforms the state-of-the-art adaptive traffic signal controller. Besides the performance gain, our method does not require a change of the current driving habits of humans. Moreover it is fault-tolerant, since the method is an add-on to the existing traffic rules and thus the intersection will still be fully functional even if the intersection management unit fails. Besides outperforming state-of-the-art methods, both of our approaches can be easily adopted to different intersection topologies.

3.2 Future Work

Given the current status, there are still some interesting problems to be solved for learning cooperative driving behaviors at intersections in mixed traffic. In this section, we explain two of them and try to propose some possible approaches to the solutions.

Simulation and Real World
Due to the difficulty of real-world experiments in traffic, most of the research in this area are developed in simulators. However, discrepancies between simulators and the real world traffic system can make it challenging to transfer the learned behaviors from simulation. Some important discrepancies include human driver models, behavior models of other traffic participants and physical models of vehicles. There are three possible solutions for this problem. First we can utilize Sim2Real methods like domain randomization to have enough variability in the simulator, the real world may appear to the model as just another variation [40]. The second approach is training the agent via offline reinforcement learning with history real-world traffic data instead of in simulators. Although lacking online interactions with the environment, it is still optimistic that the agent can learn high quality policies [2]. The third solution is to use more naturalistic data-driven behavior models of traffic participants including human drivers [32].

Decentralized Control
Both of the presented approaches use centralized controller, which requires road infrastructures for collecting traffic data and communicating with the CAVs. To enable cooperative trajectory planning without additional infrastructures, decentralized version of the presented controllers can be an interesting future work. Decentralized control in this task has some main challenges. First, perception systems of CAVs are limited due to sensors property or physical occlusion, leading to a partial observable environment for each agent. Secondly, the number of agents in the environment is constantly changing, which makes it impossible to assign a policy to each agent. The recent advance of multi-agent reinforcement learning [52] and neural networks operating on sets [51] show us very promising approaches.

Acknowledgements The authors thank the German Research Foundation (DFG) for being funded within the German collaborative research center "SPP 1835—Cooperative Interacting Automobiles" (CoInCar).

References

1. Abdulhai, B., Kattan, L.: Reinforcement learning: introduction to theory and potential for transport applications. Can. J. Civ. Eng. **30**(6), 981–991 (2003)
2. Agarwal, R., Schuurmans, D., Norouzi, M.: An optimistic perspective on offline reinforcement learning. In: International Conference on Machine Learning, pp. 104–114. PMLR (2020)
3. Au, T.C., Zhang, S., Stone, P.: Autonomous intersection management for semi-autonomous vehicles. Handbook of transportation pp. 88–104 (2015)

4. Bento, L.C., Parafita, R., Santos, S., Nunes, U.: Intelligent traffic management at intersections: Legacy mode for vehicles not equipped with v2v and v2i communications. In: Proceedings of the IEEE International Conference on Intelligent Transportation Systems (ITSC), pp. 726–731. IEEE (2013)
5. Bester, C.J., Meyers, W.L.: Saturation flow rates. In: Proceedings of the South African Transport Conference, pp. 560–568 (2007)
6. Denney, R.W., Jr., Curtis, E., Olson, P.: The national traffic signal report card. ITE J. **82**(6), 22–26 (2012)
7. Dresner, K., Stone, P.: Multiagent traffic management: a reservation-based intersection control mechanism. In: Autonomous Agents and Multiagent Systems, International Joint Conference on, vol. 3, pp. 530–537. IEEE Computer Society (2004)
8. Dresner, K., Stone, P.: Multiagent traffic management: An improved intersection control mechanism. In: Proceedings of the Fourth International Joint Conference on Autonomous Agents and Multiagent Systems, pp. 471–477 (2005)
9. Dresner, K., Stone, P.: Sharing the road: autonomous vehicles meet human drivers. In: The 20th International Joint Conference on Artificial Intelligence, pp. 1263–68 (2007)
10. Dresner, K., Stone, P.: A multiagent approach to autonomous intersection management. J. Artif. Intell. Res. **31**, 591–656 (2008)
11. El-Tantawy, S., Abdulhai, B., Abdelgawad, H.: Design of reinforcement learning parameters for seamless application of adaptive traffic signal control. J. Intell. Transp. Syst. **18**(3), 227–245 (2014)
12. Fayazi, S.A., Vahidi, A.: Mixed-integer linear programming for optimal scheduling of autonomous vehicle intersection crossing. IEEE Trans. Intell. Veh. **3**(3), 287–299 (2018)
13. Ferreira, M., Fernandes, R., Conceição, H., Viriyasitavat, W., Tonguz, O.K.: Self-organized traffic control. In: Proceedings of the Seventh ACM International Workshop on VehiculAr InterNETworking, VANET '10, pp. 85–90. New York, NY, USA (2010). https://doi.org/10.1145/1860058.1860077
14. Gartner, N.: OPAC: a demand-responsive strategy for traffic signal control. Transp. Res. Rec. **906**, 75–81 (1983)
15. Genders, W., Razavi, S.: Using a deep reinforcement learning agent for traffic signal control (2016). arXiv:1611.01142
16. Genders, W., Razavi, S.: An open-source framework for adaptive traffic signal control (2019). arXiv:1909.00395
17. Gregoire, J., Qian, X., Frazzoli, E., De La Fortelle, A., Wongpiromsarn, T.: Capacity-aware backpressure traffic signal control. IEEE Trans. Control Netw. Syst. **2**(2), 164–173 (2014)
18. Guler, S.I., Menendez, M., Meier, L.: Using connected vehicle technology to improve the efficiency of intersections. Transp. Res. Part C: Emerg. Technol. **46**, 121–131 (2014)
19. INRIX: INRIX 2018 Global Traffic Scorecard (2005). https://inrix.com/scorecard/. Accessed 16 Jan 2020
20. Li, L., Wen, D., Yao, D.: A survey of traffic control with vehicular communications. IEEE Trans. Intell. Transp. Syst. **15**(1), 425–432 (2013)
21. Liang, X., Du, X., Wang, G., Han, Z.: A deep reinforcement learning network for traffic light cycle control. IEEE Trans. Veh. Technol. **68**(2), 1243–1253 (2019)
22. Lin, P., Liu, J., Jin, P.J., Ran, B.: Autonomous vehicle-intersection coordination method in a connected vehicle environment. IEEE Intell. Transp. Syst. Mag. **9**(4), 37–47 (2017)
23. Lipson, H., Kurman, M.: Driverless: Intelligent Cars and the Road Ahead. The MIT Press (2017)
24. Litman, T.: Autonomous vehicle implementation predictions: implications for transport planning (2021). https://www.vtpi.org/avip.pdf. Accessed 06 Mar 2021
25. Lopez, P.A., Behrisch, M., Bieker-Walz, L., Erdmann, J., Flötteröd, Y., Hilbrich, R., Lücken, L., Rummel, J., Wagner, P., Wiessner, E.: Microscopic traffic simulation using sumo. In: Proceedings of the IEEE International Conference on Intelligent Transportation Systems (ITSC), pp. 2575–2582 (2018)

26. Lowrie, P.R.: SCATS, Sydney co-ordinated adaptive traffic system: a traffic responsive method of controlling urban traffic. Roads and Traffic Authority NSW, Darlinghurst, NSW Australia (1990)
27. Menéndez-Romero, C., Sezer, M., Winkler, F., Dornhege, C., Burgard, W.: Courtesy behavior for highly automated vehicles on highway interchanges. In: IEEE Intelligent Vehicles Symposium (IV), pp. 943–948 (2018). https://doi.org/10.1109/IVS.2018.8500407. http://ais.informatik.uni-freiburg.de/publications/papers/menendez18iv.pdf
28. Menéndez-Romero, C., Winkler, F., Dornhege, C., Burgard, W.: Maneuver planning for highly automated vehicles. In: Proceedings of the IEEE Intelligent Vehicles Symposium (IV), pp. 1458–1464 (2017). https://doi.org/10.1109/IVS.2017.7995915
29. Meng, Y., Li, L., Wang, F.Y., Li, K., Li, Z.: Analysis of cooperative driving strategies for nonsignalized intersections. IEEE Trans. Veh. Technol. **67**(4), 2900–2911 (2017)
30. Mnih, V., Kavukcuoglu, K., Silver, D., Rusu, A.A., Veness, J., Bellemare, M.G., Graves, A., Riedmiller, M., Fidjeland, A.K., Ostrovski, G., et al.: Human-level control through deep reinforcement learning. Nature **518**(7540), 529–533 (2015)
31. Papageorgiou, M., Diakaki, C., Dinopoulou, V., Kotsialos, A., Wang, Y.: Review of road traffic control strategies. Proc. IEEE **91**(12), 2043–2067 (2003)
32. Papathanasopoulou, V., Antoniou, C.: Towards data-driven car-following models. Transp. Res. Part C: Emerg. Technol. **55**, 496–509 (2015). https://doi.org/10.1016/j.trc.2015.02.016. https://www.sciencedirect.com/science/article/pii/S0968090X15000716. Engineering and Applied Sciences Optimization (OPT-i) - Professor Matthew G. Karlaftis Memorial Issue
33. Van der Pol, E., Oliehoek, F.A.: Coordinated deep reinforcement learners for traffic light control. In: NIPS'16 Workshop on Learning, Inference and Control of Multi-Agent Systems (2016)
34. Qian, X., Gregoire, J., Moutarde, F., De La Fortelle, A.: Priority-based coordination of autonomous and legacy vehicles at intersection. In: Proceedings of the IEEE International Conference on Intelligent Transportation Systems (ITSC), pp. 1166–1171. IEEE (2014)
35. Salkham, A., Cunningham, R., Garg, A., Cahill, V.: A collaborative reinforcement learning approach to urban traffic control optimization. In: Proceedings of the IEEE/WIC/ACM International Conference on Web Intelligence and Intelligent Agent Technology, vol. 2, pp. 560–566 (2008)
36. Schulman, J., Moritz, P., Levine, S., Jordan, M., Abbeel, P.: High-dimensional continuous control using generalized advantage estimation. In: Proceedings of the International Conference on Learning Representations (ICLR) (2016)
37. Schulman, J., Wolski, F., Dhariwal, P., Radford, A., Klimov, O.: Proximal policy optimization algorithms (2017). arXiv:1707.06347
38. Sharon, G., Stone, P.: A protocol for mixed autonomous and human-operated vehicles at intersections. In: International Conference on Autonomous Agents and Multiagent Systems, pp. 151–167. Springer (2017)
39. Silva, R.A., Adelman, Z., Fry, M.M., West, J.J.: The impact of individual anthropogenic emissions sectors on the global burden of human mortality due to ambient air pollution. Environ. Health Perspect. **124**(11), 1776–1784 (2016)
40. Tobin, J., Fong, R., Ray, A., Schneider, J., Zaremba, W., Abbeel, P.: Domain randomization for transferring deep neural networks from simulation to the real world. In: Proceedings of the IEEE/RSJ International Conference on Intelligent Robots and Systems (IROS), pp. 23–30. IEEE (2017)
41. Ulbrich, S., Grossjohann, S., Appelt, C., Homeier, K., Rieken, J., Maurer, M.: Structuring cooperative behavior planning implementations for automated driving. In: 2015 IEEE 18th International Conference on Intelligent Transportation Systems, pp. 2159–2165. IEEE (2015)
42. U.S. Federal Highway Administration: Manual on Uniform Traffic Control Devices. https://mutcd.fhwa.dot.gov/pdfs/2009r1r2/pdf_index.htm (2009). [Online; accessed 06-Mar-2021]
43. Varaiya, P.: The max-pressure controller for arbitrary networks of signalized intersections. In: Advances in Dynamic Network Modeling in Complex Transportation Systems, pp. 27–66. Springer (2013)
44. Webster, F.V.: Traffic Signal Settings. H.M.S.O, London (1958)

45. Wei, H., Zheng, G., Gayah, V., Li, Z.: A survey on traffic signal control methods (2019). arXiv:1904.08117
46. Wei, H., Zheng, G., Yao, H., Li, Z.: Intellilight: a reinforcement learning approach for intelligent traffic light control. In: Proceedings of the ACM SIGKDD International Conference on Knowledge Discovery & Data Mining, pp. 2496–2505 (2018)
47. Xie, X.F., Smith, S.F., Lu, L., Barlow, G.J.: Schedule-driven intersection control. Transp. Res. Part C: Emerg. Technol. **24**, 168–189 (2012)
48. Yan, S., Welschehold, T., Büscher, D., Burgard, W.: Courteous behavior of automated vehicles at unsignalized intersections via reinforcement learning. IEEE Robot. Autom. Lett. **7**(1), 191–198 (2022). https://doi.org/10.1109/LRA.2021.3121807. http://ais.informatik.uni-freiburg.de/publications/papers/yan21ral.pdf
49. Yan, S., Zhang, J., Büscher, D., Burgard, W.: Efficiency and equity are both essential: a generalized traffic signal controller with deep reinforcement learning. In: Proceedings of the IEEE/RSJ International Conference on Intelligent Robots and Systems (IROS), pp. 5526–5533 (2020). https://doi.org/10.1109/IROS45743.2020.9340784. http://ais.informatik.uni-freiburg.de/publications/papers/yan20iros.pdf
50. Yang, K., Guler, S.I., Menendez, M.: Isolated intersection control for various levels of vehicle technology: conventional, connected, and automated vehicles. Transp. Res. Part C: Emerg. Technol. **72**, 109–129 (2016)
51. Zaheer, M., Kottur, S., Ravanbakhsh, S., Poczos, B., Salakhutdinov, R.R., Smola, A.J.: Deep sets. Advances in Neural Information Processing Systems 30 (2017)
52. Zhang, K., Yang, Z., Başar, T.: Multi-agent reinforcement learning: a selective overview of theories and algorithms. Handbook of Reinforcement Learning and Control, pp. 321–384 (2021)

Human Factors

Cooperative Hub for Cooperative Research on Cooperatively Interacting Vehicles: Use Cases, Design and Interaction Patterns

Frank Flemisch, Nicolas Herzberger, Marcel Usai, Marcel Baltzer, Maximilian Schwalm, Gudrun Voß, Josef Krems, Laura Quante, Daniel Trommler, Nadine Strelau, Christoph Burger, and Christoph Stiller

Abstract This chapter describes central cooperative activities in the research priority program Cooperatively Interacting Vehicles (CoInCar). If the whole research program CoInCar can be seen as a wheel, which is turning research questions into answers, knowledge and hopefully progress for society, the individual research projects described in the other chapters could be seen as spokes of the wheel, and the aspects described in this chapter as an informal cooperative hub of the wheel. Starting with common essential definitions, a use case catalogue was derived and documented. Based on that, cooperation and interaction pattern were sketched and documented into a pattern database. While the details of the research hub described here are specific for this DFG priority program, the general principles of a research hub could be transferred to any other research and development activity.

F. Flemisch (✉) · N. Herzberger · M. Usai
Institute of Industrial Engineering and Ergonomics, RWTH Aachen University, Eilfschornsteinstr. 18, 52062 Aachen, Germany
e-mail: f.flemisch@iaw.rwth-aachen.de; frank.flemisch@fkie.fraunhofer.de

F. Flemisch · N. Herzberger · M. Usai · M. Baltzer
Fraunhofer FKIE, Fraunhoferstr. 20, 53343 Wachtberg, Germany

M. Schwalm
Chair and Institute of Highway Engineering, RWTH Aachen University, Mies-van-der-Rohe Str. 1, 52074 Aachen, Germany

G. Voß
Projektträger Jülich, Wilhelm-Johnen-Str., 52428 Jülich, Germany

J. Krems · D. Trommler
TU Chemnitz, Str. der Nationen 62, 09111 Chemnitz, Germany

L. Quante
German Aerospace Center (DLR), Lilienthalpl. 7, 38108 Braunschweig, Germany

N. Strelau
ifab, KIT Karlsruhe, Engler-Bunte-Ring 4, 76131 Karlsruhe, Germany

C. Burger · C. Stiller
Institute of Measurement and Control Systems (MRT), KIT Karlsruhe, Engler-Bunte-Ring 21, 76131 Karlsruhe, Germany

© The Author(s) 2024
C. Stiller et al. (eds.), *Cooperatively Interacting Vehicles*,
https://doi.org/10.1007/978-3-031-60494-2_16

Fig. 1 One of the spears of Schoeningen, used for cooperative hunting as an example for cooperative movement, cooperative reasoning and acting

1 Introduction: The Big Picture—From Cooperative Homo Heidelbergensis to Cooperative Human Machine Systems

In general, movement through space and time is a vital feature of life. Movement in form of mobility, by foot, bicycle, car, train or airplane is an important part of our life as individuals, organizations and societies. Cooperation in contrast to competition is a central aspect of mobility already for a long time, and was already quite important for our development as homo sapiens, as the following example shows.

Figure 1 shows one of ten wooden spears which were excavated in 1994ff at Schoeningen near Braunschweig in the north of Germany. These throwing spears, dated between 380,000 and 400,000 years old, represent the oldest preserved hunting weapons of prehistoric Europe yet discovered [34]. These spears are not only an early example of weapon technology, but also for an art which is much later called Human Factors (Engineering), for which Homo Heidelbergensis, a prerunner of Homo Sapiens, was already able to combine different techniques like cutting to carve, and fire to harden an effective tool and adapt it to the individual bearer. These spears are also an early example of Human Systems Integration, which in the 21st millennium is understood as integration of humans, technology, organization, and environment, and which was already an important factor for Homo Heidelbergensis: Close to the location of the spears, many horse bones were found. Anthropologists reconstructed that a tribe of Homo Heidelbergensis hunted, rounded up, speared, and ate these horses. Especially the production of the spears, which can be seen as a clever use of or integration with the environment, and the cooperative hunting took a degree of organization, which was not available to other rival species.

It is obvious that movement and mobility in combination with these tools was one of the key factors for success of these homo tribes. But how could these relatively slow species round up and eat other species which were physically much faster and stronger? The key issue can be found not so much on the physical layer, but on the cognitive layer of evolution: Tomasello [35] describes how human cognition evolved together with the ability to create and handle such tools, and especially how the cooperation and shared intentionality e.g. of hunting AND tool manufacturing fostered the evolution of homo towards homo sapiens as one of the most dominant species on this planet. Cooperation and teaming were obviously essential for early hunting. Cooperation and teaming might also be essential for cooperatively interacting vehicles, and for researchers and developers addressing these complex cooperative systems. It is not by accident that communities of research institutions applying for a research grant nickname themselves "hunting communities" or "hunting tribes", but by insight that the ability to cooperate might be similarly important with hunting and with research.

Fig. 2 Shared mental models to create shared intentionality, example Homo Heidelbergiensis hunting

What Tomasello [35] describes as shared intentionality, other researchers like Norman [28] or Gentner [16] describe as shared mental models (c.f. Fig. 2), where mismatches between the mental model of system designer and system users might lead to dangerous errors in design or use of sociotechnical systems. Mental models are also at the very core of cooperatively interacting vehicles, and of the cooperative research on cooperative interacting vehicles.

Applied to cooperatively interacting vehicles, the setup of cooperatively interacting vehicles might use similar cognitive capacities which already helped homo heidelbergensis to move cooperatively, but might include a new complexity: Here, not only individuals and groups of homo sapiens are involved, but also a new player on the cognitive evolution: The computer. In less than a century from its first invention by Konrad Zuse in 1941, the computer and later Artificial Intelligence (AI) has become a central player in sociotechnical systems. The teaming of computer and humans is already hinted in Wieners famous book about Cybernetics in 1950, where he describes feedback loops as the central mechanism of intelligence both in the animal and the machine. Later Licklider [24] describes symbiotic human—computer systems. Rasmussen [31] proposed the term cooperation, Hollnagel and Woods [22] and Sheridan [33] defined initial principles, Hoc and Lemoine [20] and Hoc [19] described the common ground and know-how-to-cooperate as important parts of developing human computer cooperation.

A major breakthrough was to think of cognition not only as something separated/assigned to individual agents, but also as something which is combined or joined between the different players, i.e. Joint Cognition or Joint Cognitive Systems. Hollnagel sketches how these Joint Cognitive Systems can be nested, from the small to the big, and already prepared the ground for a system of systems approach. System of systems can be understood as the joining of individual systems which "deliver important emergent properties, which have an evolving nature that stakeholders must recognize, analyze and understand" (e.g. Maier [27]). Flemisch et al. [9] described how humans and machine cognition in system of systems can cooperate on levels with different time frequencies, yet still work together like the blunt end and the sharp end of a spear. Flemisch et al. [13] extended this view also to conflicts that can happen between agents in cognitive systems, and how these can be mitigated. Examples for conflicts are different intentions of humans and machines, e.g. vehicle automation, of where to go and how fast. Flemisch et al. [11] describe a holistic bowtie model of meaningful and effective control, which brings together the individual human-machine system with a system-of-systems, organizational, societal and environmental perspective. Cooperation between these layers are—once again—a central key for failure or success of these systems.

As already hinted by Hoc [19], the key of any cooperative activity is to have sufficient common ground and common work space between humans and computers in the form of shared mental models. This proved to apply even more so for cooperatively interacting vehicles. Related to common ground is the concept of inner and outer compatibility (e.g. Flemisch et al. [8]), which describes the ability of interfaces on the outside system border between humans and machines to play together, and the ability of inner mental models to interact in a cooperative way.

In general, the development of shared mental models does not start from scratch but is always a development and migration. Starting point could be basic image schemes which we inherited from our ancestors (e.g. Lakoff [23], Baltzer [3]), patterns we learned during our life to deliberate discussions in our research and development community on how sociotechnical systems, here cooperatively interacting vehicles, should work together amongst themselves and with the humans involved. In this ongoing effort to shape the mental models, it is important that mental models evolve cooperatively. They are never all up to date at the same time, as Fig. 3 shows.

Applied to cooperatively interaction vehicles, Fig. 3 shows an example of inconsistent mental models, where the human on board of an automated vehicle assumes that the vehicle automation has the control, while the other user assumed that the human was in control. Such a misunderstanding already led to a deadly accident in 2018 with a highly automated vehicle operated by Uber [29].

Figure 4 depicts a very simplified model of how shared mental models in the research community of automated and interconnected driving might have evolved, starting with a simple "black and white" model of manual or fully automated driving, then the intense discussions on different levels of automation, sparked by the theoretic work of Parasuraman et al. [30], boosted with insights like the H(orse)-Metaphor [7], the first formulation of highly automated driving, and its practical solutions like H-Mode [2] or Conduct by Wire [37], leading to the BASt and SAE levels of driving

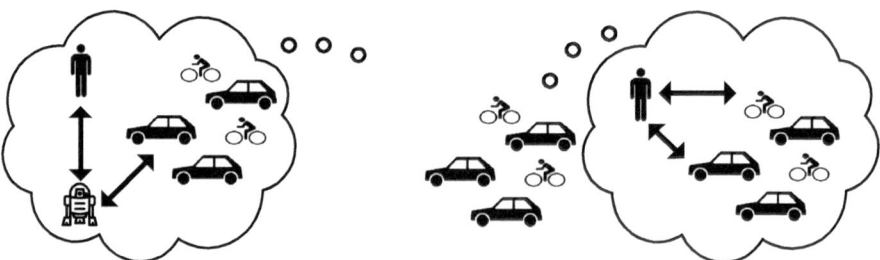

Fig. 3 Shared mental models between humans and computers on cooperatively interacting vehicles (Flemisch et al. [12] based on Flemisch et al. [9])

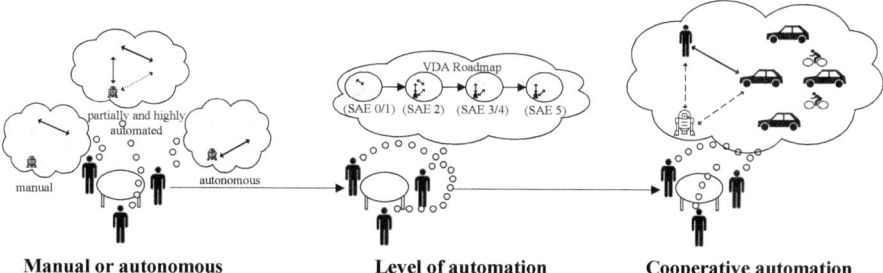

Manual or autonomous   **Level of automation**   **Cooperative automation**

Fig. 4 Vehicle and traffic automation from "on/off-automation" of the 1990ties via "Levels of Automation" of 2000ff to Cooperative Automation 2010ff, as a cooperative development of shared mental models in the research and development community [12]

automation [15, 32]. With CoInCar, we entered a new stage of automation, which still uses levels of driving automation, but connects the differently automated users and automations with cooperative driving patterns.

2 Bringing Researchers Together: Concepts and Definitions Wiki, Ph.D. Workshops

In general, common ground and common mental models for researchers usually do not start with definitions, but with common inspiration, ideas and visions, as vague or fuzzy as they might initially be. Only if this inspirational common ground is assured first, the more tedious work on common concepts and definitions has a chance to succeed. Even then, with complex systems and interdisciplinary teams, it is often impossible to achieve a similar crispness of definitions, as scientist were able to achieve in physical sciences. Especially in the integration of humans, technology, organization and the environment, so many disciplines are involved, that crisp definitions like in physics are highly unpractical, but a higher plasticity of concepts and definitions has to be tolerated from the very beginning, if the definitions should really open the chance to converge between disciplines.

Applied to CoInCar, in a series of workshops in mixed subgroups, concepts and definitions were worked out, documented in a Wiki and deconflicted over the duration of the project (see Fig. 5). It is important to note that the approach was not to deconflict all differences in the definitions—this alone would have consumed most of the research budgets—but to find a common ground just big enough to start cooperations, and to further consolidate the Wiki "on the job".

For further networking within the priority program, a series of structured activities took place for the Ph.D. researchers. For example, every two years there was a cross-project Ph.D. workshop in which the researchers presented their research topics, discussed, and identified links between the subprojects. Furthermore, individual

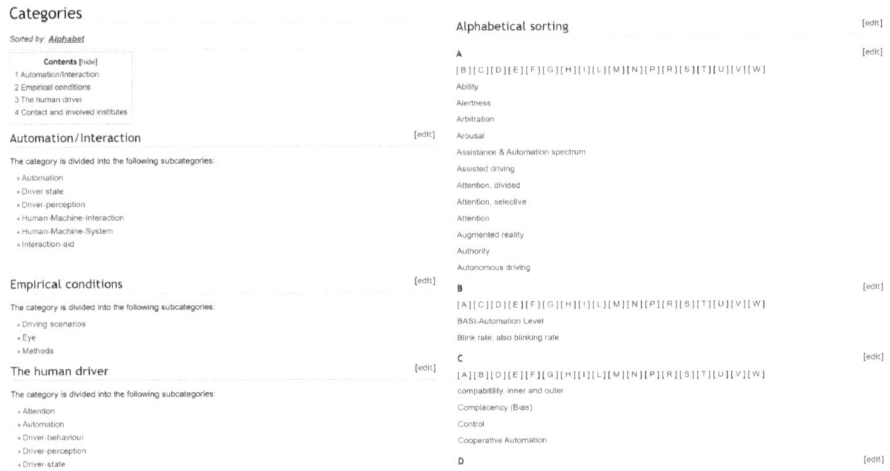

Fig. 5 Concept and definition Wiki of the CoInCar project. (Weßel and Herzberger [36], exemplary screenshots taken form the CoInCar intranet)

disciplines had regular Ph.D. workshops. One example is the regular meeting for human factors researchers, which met once a month for one hour in a hybrid format. Here, short presentations were given in a rotating process and the researchers own progresses and difficulties were exchanged and discussed within the group in order to benefit from the experiences of the other research groups.

3   Bringing Researchers and Developers Together: Use Case Catalogues

In general, to really understand and master complex sociotechnical systems with all their possible combinations, a system analysis can lay the structure for inter-disciplinary teams. Over the years of Systems Engineering and Human Systems Integration, a structuring of systems in problem and solution space (e.g. Haberfell-ner et al. [18]), and in design space, use space and value space has shown good results of mastering the complexity (for an overview and history of these concepts and their application to the exploration of human-machine systems, see e.g. Flemisch et al. 2022a).

Applied to CoInCar, the alignment of mental models of researchers started with the use space, i.e. the dimensions of use and their combination into use cases and use situations. Based on the positive experience in EU-projects on highly automated driving of working with use cases (e.g. Hoeger et al. [21]), and research efforts to find define a unified ontology for test and use case catalogues in DFG-projects before

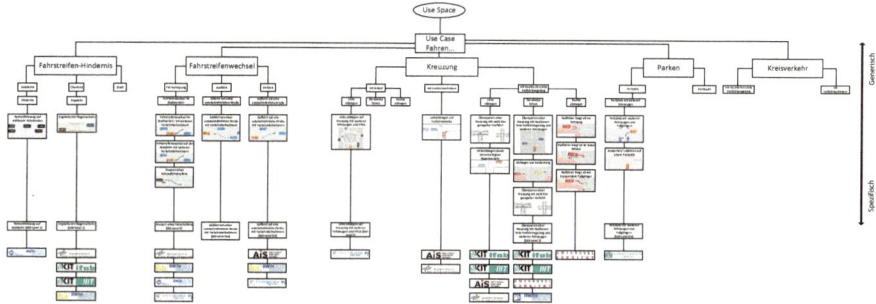

Fig. 6 Use case catalogue representing the research topics of the subprojects within the CoInCar project (Canpolat et al. [5], published in the CoInCar intranet SSELab)

CoInCar (e.g. Geyer et al. [17]), CoInCar started with the discussion, selection and definition of initial use cases of cooperatively interacting vehicles.

Figure 6 shows the use case tree of CoInCar as an overview of use cases addressed in the consortium. Starting with the use case family of obstacle avoidance, the use case families of lane change, intersection, parking and roundabout are identified, and individual use cases documented. In deconflicting sessions the use cases were discussed and if possible aligned. The use case catalogue also served as a map to explain the priority program, and to onboard new researchers.

4   Bringing Researchers, Developers and Users Together: Pattern Catalogue of Cooperatively Interacting Vehicles

In general, complex systems can be decomposed into system models, use space, design space and value space. This helps with the understanding of the individual components of the system, but not yet with the understanding of the relations and only partially with the recomposing and designing of system variants. Seeing design, use and value space as systems themselves, and taking Luhmann's argument "Contact happens at the border" [26] of these systems seriously, it is crucial to find a way to describe the interplay of these dimensions as a combination which has a combined effect: How a certain system design is used by the user, and what effects this has on users and the surrounding system. The challenge for this to find a representation that really grasps the essence of a specific design, use and value in a way that it is general enough to be reused, and understandable enough that it can connect researchers, developers and users.

An essential concept to achieve this are patterns, here as design and interaction patterns. Patterns can be traced back to the philosophical theory of Forms (e.g. Plato 427 B.C.). Just neglecting the long philosophical dispute whether forms are something outside of the physical world or just mental models in the brain of the analysts,

Christopher Alexander described architecture as a language of design patterns [1]. This concept was transferred to software design patterns by Gamma et al. [14], to human computer interaction by Borchers [4] and to human-machine systems e.g. by Flemisch [6], Baltzer [3]. Based on Alexander's initial definition, Flemisch et al. [10] understand patters as follows:

> A pattern describes something that occurs over and over again. An example for this is a problem and/or its solutions. If this can be observed, and its core can be mapped and modelled, you can either observe and match the pattern over and over again, without ever making the identical observation twice. And/or you can instantiate and design with this pattern over and over again, not necessarily doing it the same way twice. Examples for this are designing, engineering and using of artefacts like human-machine systems. (Flemisch et al. [10] based on Alexander et al. [1])

Patterns bridge the more concrete world of applications with the more abstract world of concepts, and can provide a common mental model of the sociotechnical system and its principal understanding, design and usage. With that, patterns can be a crucial technique to bring designers, engineers, users and other stakeholder together (see Fig. 7).

Patterns can be based on use cases or use situations, and then describe how the use is usually happening with which results. This can be described on different levels of detail, e.g. very general usage (e.g. Baltzer [3]) up to a more detailed description of the interaction happening in a certain use situation (e.g. Flemisch [6], López Hernández et al. [25]).

Patterns can be freely formed, or transferred between domains, e.g. from the biosphere to the technosphere. A striking example, shown in Fig. 8, for the potential of transferring design and interaction patters is flying, where Otto Lilienthal systematically evaluated the flight of birds, and transferred the most important principles to design patterns e.g. of wings, foils etc., which still form the basis of flying today.

Applied to CoInCar, patterns influenced the scientific undertaking from the very beginning, e.g. in form of the H-Metaphor as an inspiration for the cooperation between automation and the driver, but also the cooperation between vehicles as a herd or flock. More inspiration came from other domains like dancing, where a common understanding of figures and movements, i.e. patterns, allow dancers to move together and enjoy it (see Fig. 9).

More concretely, the pattern concept was introduced in the second half of the CoInCar focus program, discussed and refined in a series of workshops.

Based on this conceptual work between the individual research groups, a first database was built up and filled for a first test. Figure 10 shows a fundamental pattern "inform, warn, intervene" as an example for a cooperation pattern, which was used in a couple of use cases of CoInCar. Figure 11 shows another fundamental family of patterns "Transition of control", which were investigated in a couple of explorations and experiments in CoInCar.

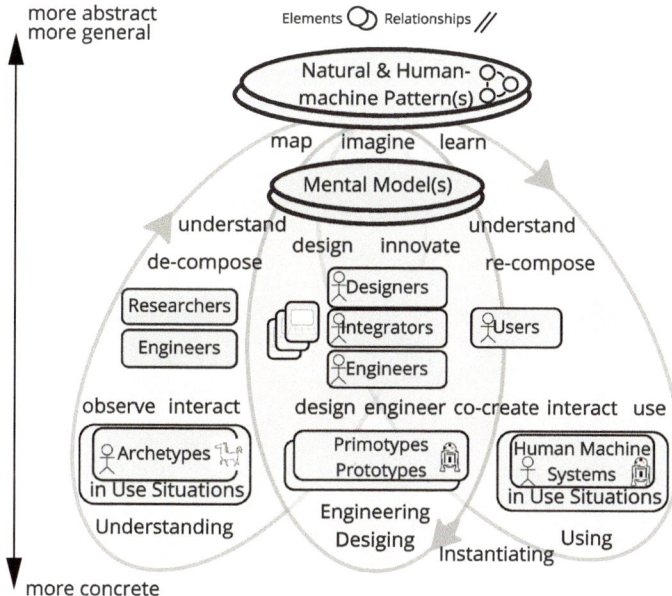

Fig. 7 Human-machine pattern as an interplay of researchers, designers, engineers and users in understanding, designing, engineering and using human-machine systems (adapted from Flemisch et al. [10])

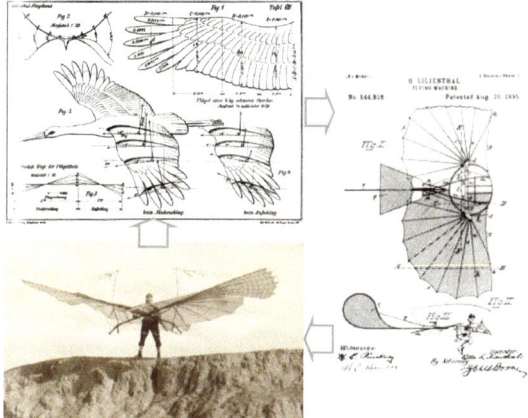

Fig. 8 Design and interaction patters as a connection of decomposing/understanding, recomposing/engineering and using, example flying (Otto Lilienthal, 1874ff)

5   Conclusion and Outlook

It is important to note that focus programs are usually not as rigidly organized as e.g. excellence clusters or even industrial research and development projects. Organizing

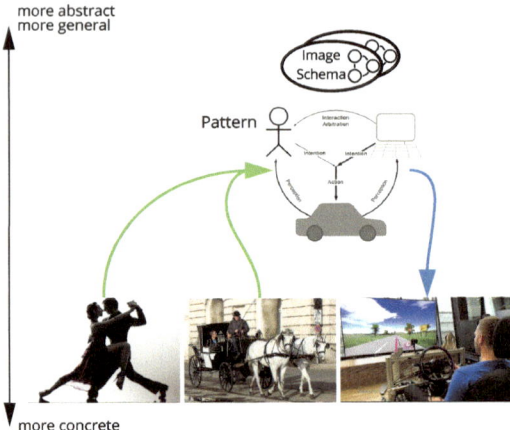

Fig. 9 Example for cooperation patterns and image schemes, derived in other domains and translated to cooperatively interacting vehicles [10]

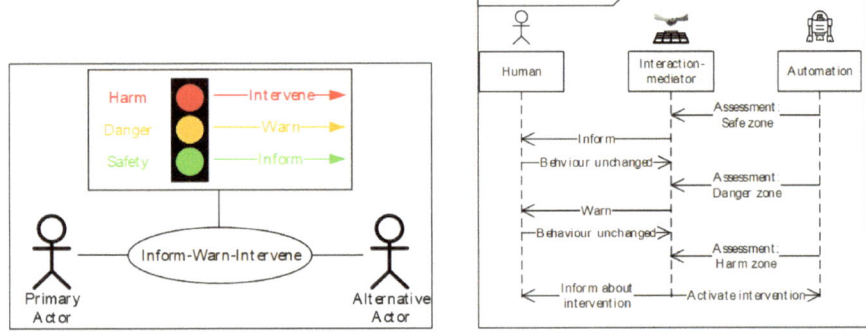

Fig. 10 Example for cooperation pattern "inform, warn, intervene", which can be applied between humans, humans and machines, or between machines [3]

a research hub like in CoInCar, based on a concept Wiki, a use case catalogue and a first pattern database was an exploration of ideas, with promising first results, but far from providing complete catalogues or databases which are now ready to use in industry. Nevertheless, these results can provide an inspiration, or a concrete first core for more rigid research and development projects in the realm of cooperatively interacting vehicles, or beyond in the realm of cooperating human-machine systems including human—AI systems.

We see a huge potential in the combination of use cases and design/interaction patterns, which can clearly help to manage the complexity of future cooperative systems. Our vision is that the know-how about human machine patterns and their usage is increasingly collected in easy to access and easy to use data bases, ideally globally (c.f. Fig. 12). The key will be to provide the human machine patterns in a way

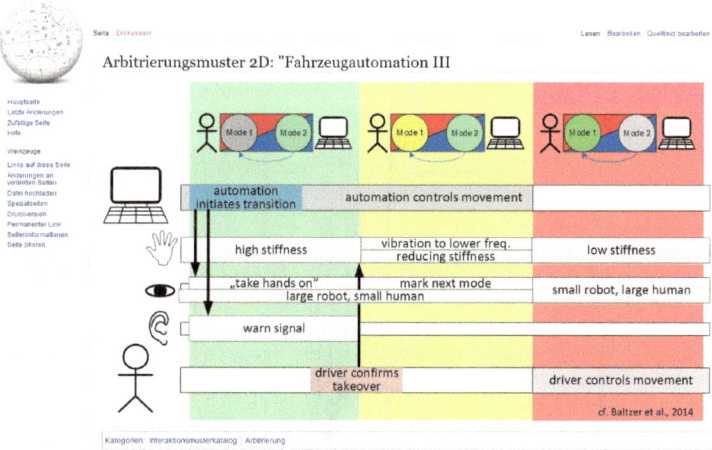

Fig. 11 Cooperation pattern "Transition of control" as an example of pattern in the online pattern database

Fig. 12 Our vision are Global Databases for Design and Interaction patterns to collect knowledge and know-how about the design, engineering and usage of sociotechnical systems [10]

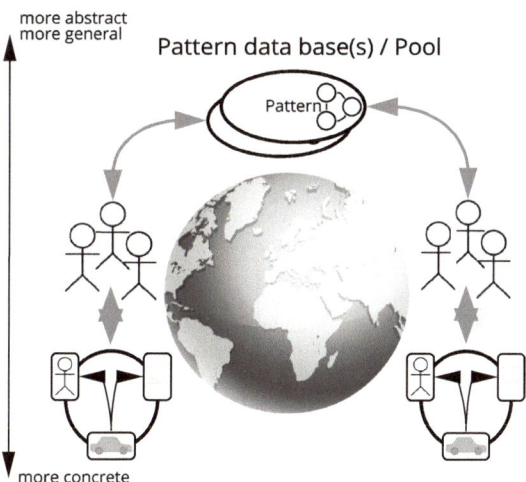

that it can be easily used in design, engineering, and research activities, so that know-how can flow freely back and forth between researchers, designers, engineers, users and policy makers. This could also be the core for incident and accident databases that, along with cooperative research and development, could make our world safer, more sustainable, and more fun and joy to live in.

Acknowledgements This publication was funded within the Priority Programme 1835 "Cooperative Interacting Automobiles (CoInCar)" of the German Science Foundation (DFG).

References

1. Alexander, C., Ishikawa, S., Silverstein, M.: A Pattern Language: Towns, Buildings, Construction, Center for Environmental Structure Series, vol. 2. Oxford University Press, New York, NY (1977)
2. Altendorf, E., Baltzer, M., Kienle, M., Meier, S., Weißgerber, T., Heesen, M., Flemisch, F.: H-Mode 2D. In: Handbuch Fahrerassistenzsysteme, pp. 1123–1138. Springer Vieweg, Wiesbaden (2015). https://doi.org/10.1007/978-3-658-05734-3_60
3. Baltzer, M.C.A.: Interaktionsmuster der kooperativen Bewegungsführung von Fahrzeugen. Dissertation, Shaker Verlag and Dissertation, RWTH Aachen University, 2020, Aache DOI 40345 (2021). https://publications.rwth-aachen.de/record/818952
4. Borchers, J.O.: A pattern approach to interaction design. In: Boyarski, D., Kellogg, W.A. (eds.) Proceedings of the Conference on Designing Interactive Systems Processes, Practices, Methods, and Techniques—DIS '00, pp. 369–378. ACM Press, New York, New York, USA (2000). https://doi.org/10.1145/347642.347795
5. Canpolat, Y., Voß, G.M.I., Herzberger, N.D.: Use Case Catalogue (2016)
6. Flemisch, F.: Pointillistische Analyse der visuellen und nicht-visuellen Interaktionsressourcen am Beispiel Pilot-Assistentensystem. Ph.D. thesis, Universität der Bundeswehr München (2001)
7. Flemisch, F., Adams, C.A., Conway, S.R., Goodrich, K.H., Palmer, M.T., Schutte, P.C.: The H-Metaphor as a guideline for vehicle automation and interaction (2003). https://ntrs.nasa.gov/citations/20040031835
8. Flemisch, F., Schieben, A., Kelsch, J., Löper, C.: Automation spectrum, inner/outer compatibility and other potentially useful human factors concepts for assistance and automation. In: de Waard, D., Flemisch, F., Lorenz, B., Oberheid, H., Brookhuis, K.A. (eds.) Human Factors for Assistance and Automation. Shaker Publishing (2008). https://elib.dlr.de/57625
9. Flemisch, F., Abbink, D.A., Itoh, M., Pacaux-Lemoine, M.P., Weßel, G.: Joining the blunt and the pointy end of the spear: towards a common framework of joint action, human-machine cooperation, cooperative guidance and control, shared, traded and supervisory control. Cogn. Technol. Work 21(4), 555–556 (2019)
10. Flemisch, F., Usai, M., Herzberger, N.D., Baltzer, M.C.A., Hernandez, D.L., Pacaux-Lemoine, M.P.: Human-machine patterns for system design, cooperation and interaction in socio-cyber-physical systems: introduction and general overview. In: 2022 IEEE International Conference on Systems, Man, and Cybernetics (SMC), pp 1278–128. IEEE (2022). https://doi.org/10.1109/SMC53654.2022.9945181
11. Flemisch, F., Baltzer, M.C.A., Abbink, D.A., Siebert, L.C., van Diggelen, J., Herzberger, N.D., Draper, M., Boardman, M., Pacaux-Lemoine, M.P., Wasser, J.: Towards a dynamic balance of humans and AI-based systems within our global society and environment—holistic bowtie model of meaningful human control over effective systems. In: van den Hoven, J., Abbink, D.A., Santoni de Sio, F., Amoroso, D., Mecacci, G., Siebert, L. (eds.) Meaningful Human Control of Artificial Intelligence Systems (in Press)
12. Flemisch, F., Abendroth, B., Bengler, K., Peters, S., Vortisch, P.: Migration of Road Vehicle Automation (Submitted)
13. Flemisch, F.O., Pacaux-Lemoine, M.P., Vanderhaegen, F., Itoh, M., Saito, Y., Herzberger, N., Wasser, J., Grislin, E., Baltzer, M. (2020) Conflicts in human-machine systems as an intersection of bio- and technosphere: cooperation and interaction patterns for human and machine interference and conflict resolution. In: 2020 IEEE International Conference on Human-Machine Systems (ICHMS), pp. 1–6. https://doi.org/10.1109/ICHMS49158.2020.9209517
14. Gamma, E., Helm, R., Johnson, R., Vlissides, J.: Design patterns: abstraction and reuse of object-oriented design. In: European Conference on Object-Oriented Programming, pp. 406–431. Springer, Berlin, Heidelberg (1993). https://doi.org/10.1007/3-540-47910-4_21
15. Gasser, T.M., Arzt, C., AYoubi, M., Bartels, A., Buerkle, L., Eier, J., Flemisch, F., Haecker, D., Hesse, T., Huber, W., Lutz, C., Maurer, M., Ruth-Schumacher, S., Schwarz, J., Vogt, W.: Rechts-

folgen zunehmender Fahrzeugautomatisierung. 83, Wirtschaftsverlag NW (2012). http://bast. opus.hbz-nrw.de/volltexte/2012/587/pdf/F83.pdf

16. Gentner, D.: Mental models, psychology of. In: Sills, D.L. (ed.) International Encyclopedia of the Social and Behavioral Sciences/edited by Neil J. Smelser and Paul B. Baltes, pp. 9683–9696. Elsevier Science, New York (2001). https://doi.org/10.1016/B0-08-043076-7/01487-X
17. Geyer, S., Baltzer, M., Franz, B., Hakuli, S., Kauer, M., Kienle, M., Meier, S., Weißgerber, T., Bengler, K., Bruder, R., Flemisch, F., Winner, H.: Concept and development of a unified ontology for generating test and use-case catalogues for assisted and automated vehicle guidance. IET Intell. Transp. Syst. **8**(3), 183–189 (2014). https://doi.org/10.1049/iet-its.2012.0188
18. Haberfellner, R., de Weck, O., Fricke, E., Vössner, S.: Systems Engineering: Fundamentals and Applications. Birkhäuser, Cham, Switzerland (2019)
19. Hoc, J.M.: From human-machine interaction to human-machine cooperation. Ergonomics **43**(7), 833–843 (2000). https://doi.org/10.1080/001401300409044
20. Hoc, J.M., Lemoine, M.P.: Cognitive evaluation of human-human and human-machine cooperation modes in air traffic control. Int. J. Aviat. Psychol. **8**(1), 1–32 (1998). https://doi.org/10.1207/s15327108ijap0801_1
21. Hoeger, R., Zeng, H., Hoess, A., Kranz, T., Boverie, S., Strauss, M., Jakobsson, E., Beutner, A., Bartels, A., To, T.B., Stratil, H., Fürstenberg, K., Ahlers, F., Frey, E., Schieben, A., Mosebach, H., Flemisch, F., Dufaux, A., Manetti, D., Amditis, A., Mantzouranis, I., Lepke, H., Szalay, Z., Szabo, B., Luithardt, P., Gutknecht, M., Schoemig, N., Kaussner, A., Nashahibi, F., Resende, P., Vanholme, B., Glaser, S., Allemann, P., Seglö, F., Nilsson, A.: The future of driving–HAVEit (Final Report, Deliverable D61. 1) (2011)
22. Hollnagel, E., Woods, D.D.: Cognitive systems engineering: new wine in new bottles. Int. J. Man Mach. Stud. **18**(6), 583–600 (1983). https://doi.org/10.1016/s0020-7373(83)80034-0
23. Lakoff, G.: Women, Fire, and Dangerous Things: What Categories Reveal about the Mind. University of Chicago Press, Chicago (1990)
24. Licklider, J.C.R.: Man-computer symbiosis. IRE Trans. Hum. Factors Electron. HFE-1(1), 4–11 (1960). https://doi.org/10.1109/thfe2.1960.4503259
25. López Hernández, D., Vorst, D., Baltzer, M.C.A., Bielecki, K., Flemisch, F.: Parts of a whole: First Sketch of a block approach for interaction pattern elements in cooperative systems. In: Mařík, V. (ed.) International Conference on Systems, Man, and Cybernetics. IEEE (2022)
26. Luhmann, N.: Soziale systeme: Grundriss einer allgemeinen Theorie. Suhrkamp (1984). https://ixtheo.de/record/040204065
27. Maier, M.W.: Architecting principles for systems-of-systems. Syst. Eng. **1**(4), 267–284 (1998). https://doi.org/10.1002/(SICI)1520-6858(1998)1:4<267::AID-SYS3>3.0.CO;2-D
28. Norman, D.A.: The Psychology of Everyday Things. Basic Books (1988)
29. NTSB, National Transportation Safety Board: Highway Accident Report NTSB/HAR-19/03: Collision Between Vehicle Controlled by Developmental Automated Driving System and Pedestrian (2019). https://www.ntsb.gov/investigations/AccidentReports/Reports/HAR1903.pdf
30. Parasuraman, R., Sheridan, T.B., Wickens, C.D.: A model for types and levels of human interaction with automation. IEEE Trans. Syst. Man Cybern. Part A Syst. Hum. **30**(3), 286–29 (2000). https://doi.org/10.1109/3468.844354
31. Rasmussen, J.: Skills, rules, and knowledge; signals, signs, and symbols, and other distinctions in human performance models. IEEE Trans. Syst. Man Cybern. SMC-13 (3), 257–266 (1983). https://doi.org/10.1109/tsmc.1983.6313160
32. SAE: SAE International Standard J3016: Taxonomy and Definitions for Terms related to Driving Automation Systems for On-Road Motor Vehicles (2021)
33. Sheridan, T.B.: Humans and automation: system design and research issues (2002). https://www.cambridge.org/core/services/aop-cambridge-core/content/view/s0263574702274858
34. Thieme, H.: Lower Palaeolithic hunting spears from Germany. Nature **385**(6619), 807–81 (1997)
35. Tomasello, M.: A Natural History of Human Thinking. Harvard University Press (2014)

36. Weßel, G., Herzberger, N.D.: Concept and definition Wiki of the CoInCar project (2018)
37. Winner, H., Hakuli, S.: Conduct-by-wire–following a new paradigm for driving into the future (2006)

Cooperation Between Vehicle and Driver: Predicting the Driver's Takeover Capability in Cooperative Automated Driving Based on Orientation Patterns

Nicolas Herzberger, Marcel Usai, Maximilian Schwalm, and Frank Flemisch

Abstract This chapter first describes central development steps of cooperative vehicle control before focusing on the cooperation within the vehicle, between driver and co-system. To enable smooth transitions within this internal cooperation, both agents (driver and co-system) need a mutual understanding of the current capabilities for safely executing the driving task. For this purpose, first the model of confidence horizons is briefly introduced, which represents these mutual capability assessments. In the following, the focus of this chapter is on the assessment of the driver's ability to take over. First, the state of the art of Driver State Monitoring Systems (DSMS) as well as current challenges are presented. Here it is shown that a prediction based purely on driver observation is not yet possible. Therefore, an alternative approach, the diagnostic takeover request (TOR), is presented, which predicts the takeover capability based on the driver's initial orientation reaction. In the following, two driving simulator studies are presented in which the diagnostic TOR was used for the first time and thereafter the results are presented and discussed. Finally, a brief outlook is given on how both the diagnostic TOR and the concept of confidence horizons will be further developed.

1 Cooperative Driving

The intensive research and development of the last decades on driver assistance systems and automated vehicles already enables a synthesis of assistance and automation systems. And we are seeing the first automated vehicles [28] in real road traffic (in Germany), although still in limited operational design domains (ODD), driving certain sections of the highway in a highly automated mode (SAE level 3, SAE International Standard J3016 2021). During active automation, the driver may turn

N. Herzberger (✉) · M. Usai · F. Flemisch
RWTH-IAW, Eilfschorsteinstr. 18, 52062 Aachen, Germany
e-mail: nicolas.herzberger@fkie.fraunhofer.de

M. Schwalm
RWTH-isac, Mies-Van-Der-Rohe-Str. 1, 52074 Aachen, Germany

to non-driving related tasks, but must still be able to take over the driving task at any time. However, increasing vehicle automation also raises the question of how the cooperation and interaction between driver and vehicle should be designed in the future.

An initial idea in the early days of assistance systems and automation was to take over either longitudinal or lateral driving. By switching a function on or off, the driver decides which task is to be taken over, i.e. either lateral (LKAS) or longitudinal control (ACC). Afterwards, the driver cannot intervene further in the execution of the task, except to deactivate the automation. However, this form of interaction was already identified as complicated and disadvantageous by Schieben et al. [35] and Hoeger et al. [27].

An approach that goes beyond this black and white view of automation is shared and cooperative control. This can be explained with the H(orse)-Metaphor. The H-Metaphor describes how a driver and a highly automated vehicle cooperate similar to a rider and a horse, sharing and trading control in an assistance and automation scale of assisted, partially and highly automated driving. This insight was the starting point for the invention of highly automated driving (e.g. [13], Hoeger et al. [27]), for the German Federal Highway Research Institute (BASt) [16] and later on for the SAE levels of automation [34]. In a more general way, this parallels the thought of Christoffersen and Woods [2], who proposed based on ideas like assistant systems, e.g. Flemisch and Onken [11], to design an automation as a *team player*.

The concept of cooperative highly automated driving describes that the cooperation increases with a mutual understanding between the human and the co-system e.g. regarding the abilities of the partner and the distribution of control (e.g. Hoeger et al. [27, 15]).

A basic requirement for the co-system is to detect and understand the status of the driver and to use this information in order to balance between the driver and the co-system, e.g. by trading control towards the partner who still has the ability to control the vehicle. This allows harmonizing the driving strategies of the two agents (co-system and driver) into a common strategy [29]. Griffiths and Gillespie [18] use the term *shared control* to describe that the driver as well as the automation can have control over the vehicle at the same time. Flemish et al. [5] describe more precisely a design space of cooperative control, which combines shared and traded control.

A concept that includes shared control but has a wider scope can be referred to as *cooperative automation* and *cooperative guidance and control* (e.g. [4, 6]). Cooperation in this context implies working together towards the same goal. Cooperative automation is mainly understood as the cooperation between vehicles, e.g. Stiller et al. [39], Völker et al. [44]. However, cooperation can also be applied to the cooperation between the driver/operator and the automated driving system, as hinted already by Onken [32], Schulte [37] and Flemisch [9]. This driver-vehicle-cooperation requires a common mental model about the capabilities and limits of the automation and the driver [14]. In highly automated driving, the ability of the automation or co-system to observe and assess the abilities of the human partner has been an integral aspect of the concept of cooperative driving from the very beginning, e.g. in the EU-HAVEit project (Hoeger et al. 27).

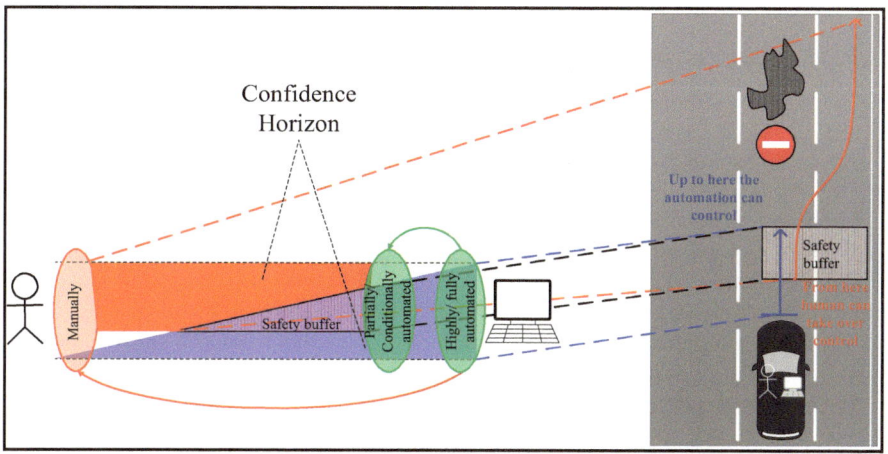

Fig. 1 Confidence horizon concept with an example of a potential safety buffer (Flemisch et al. [5], based on Flemisch et al. [10] Herzberger et al. [26], Usai et al. [43])

One possible implementation of such an embedded mental model of the capabilities of cooperation partners is the concept of *confidence horizons* (Flemisch et al. [5], based on Flemisch et al. [10], Herzberger et al. [26], Usai et al. [43]). In this concept, the capabilities of the driver are continuously compared with those of the automated subsystem, resulting in two horizons: First, the confidence of the technical subsystem in its own ability to safely control the vehicle, and second, the confidence of the technical subsystem in the driver's ability to take over the vehicle control. With that, it is possible to quickly identify whether transitions between different levels of automation are safe, whether there is a balanced distribution of control, and whether, when, and how a maneuver with minimal risk might be required. Figure 1 depicts the Confidence Horizon concept.

A fundamental cornerstone of a dynamic balance between driver and co-system, here within the concept of confidence horizon, is the assessment of the driver's (takeover) capabilities, which is addressed in the following section. A more detailed description of driver assessment can be found in the dissertation by Herzberger [23], which was written as part of the Priority Program *Cooperative Interacting Automobiles* (CoInCar) of the German Science Foundation DFG.

2 Driver Monitoring—State of the Art

In their meta-study on driver state monitoring systems (DSMS), Hecht et al. [20] point out that there is currently no commonly used definition of the term driver state. However, not only the basic definition, but also the possible states differ greatly: For example, Rauch's [33] model focuses on vigilance and drowsiness, Marberger et al. [30] model focuses on drivers' understanding of presented information during transitions, and Herzberger's [23] model focuses on drivers' assessment of driving performance. The majority of research on the classification of possible states focuses on the following constructs, or excerpts thereof: situational awareness, attention, stress, fatigue, strain, and confidence in automation (e.g., Heikoop et al. [21], or Guettas et al. [19]).

In this context, involvement in the driving task as well as the associated awareness for relevant information (ARI) plays a role in the assessment of takeover quality [17]. This ARI concept is also followed by the definition of Herzberger et al. [22]. Regardless of which definition for potential driver states is followed, it will be essential to have a reliable detection of those operator states by the technical system in order to avoid handing over the driving task in critical situations to drivers who are not ready for takeovers. In the following, therefore, an overview of the state of the art of current DSMS as well as current research approaches will be given.

In the past, vehicle manufacturers have mainly focused on drowsiness detection and the suitable warning. Most systems have focused on monitoring the driver's steering behavior and concluded that a change in steering behavior, such as jerkiness, indicates a change in vigilance, e.g., the "Drowsiness Detection System" (Volkswagen AG). The detection of steering behavior is often additionally coupled with lane departure detection systems that register deviations from the zero line, such as "Attention Assist" (Daimler AG). In the case of newer, SAE level 2-capable vehicles with traffic jam assistant system, the vehicles sometimes also drive independently for several minutes. Here, however, the attention checks by the systems differ greatly: Some systems allow longer subsequent periods of driving in traffic jams without deactivation if hands are always detected to be on the steering wheel, e.g. "Traffic Jam Assist" (Audi AG). Other systems, such as "Driving Assistant Professional" (BMW AG) or "Blue Cruise" (Ford Motor Company), enable several-minutes periods of driving without hand contact with the steering wheel provided that the driver's gaze is always directed on the road. For this purpose, the gaze is monitored by a camera system, e.g., above the instrument cluster. This system enables a warning and deactivation if the driver turns away from the driving task, since in SAE level 2 the driving task must be permanently monitored despite activated assistance systems and the responsibility lies with the driver ([23, 34]).

In addition to systems that detect the direction of the driver's gaze, a great number of research projects are also focusing on different systems for measuring physiological parameters, such as electrocardiography (ECG), photoplethysmography (PPG), electroencephalography (EEG) or the measurement of electrodermal activity (EDA, also known as skin conductivity measurement). Sensors are usually attached to the driver's body to record the signals most accurately. In addition, non-invasive methods are also being researched, where the sensors are built into the steering wheel rim, or the seat, for example Guettas et al. [19], but these require continuous contact with the body. Most common are studies that use ECG to record cardiac parameters such as heart rate and heart rate variability, e.g., Minhad et al. [31] and Taherisadr et al. [40]. This is used to draw conclusions about fatigue, stress, emotional responses, and general health of the driver. Much more complex are studies on EEG, which records electrical potentials of cerebral cortical neurons at the scalp. These studies attempt to detect cognitive states, such as activity or boredom, or even to transmit individual driving commands to the automation, such as hazard braking, e.g. Teng et al. [41]. Another approach are EDA measurements, where changes in sweat gland activity are recorded and analyzed. This methodology is used to assess emotional state, emotional arousal, or sleepiness [42]. However, this cannot validly capture which emotion is being measured because, for example, stress and anger elicit similar responses [45].

A disadvantage of some physiological measurement systems, e.g., EEG is that they are expensive, which would significantly increase the total vehicle cost, and that their use is hardly practical. In addition, other systems, such as EDA, are sensitive to surrounding temperature, and, clearly more serious, however, is that some of the signals detected may have different causes and vary greatly from person to person, making automated interpretation of these signals virtually impossible to date [1].

An alternative could be DSMS that focus on measuring the direction of the driver's gaze. These have the distinct advantage that they can be permanently installed in the vehicle and do not need to be attached to the driver in any way. Further advantages are that vehicle manufacturers as well as suppliers already have experience with the series use of camera systems for observation, these systems can now be procured at low cost and require little installation space. For these reasons, many development approaches currently focus on camera-based state estimation. Hecht et al. [20] therefore describe DSMS based on eye tracking as the technology with the greatest potential.

However, for the use of such DSMS, it is necessary to identify measurable criteria that correlate with possible driver states or the future takeover quality. Despite various efforts to compile such a set of criteria (e.g. [17, 25]), the authors are not aware of any valid set of criteria to date. And even if the prediction of the driver's future takeover capability should remain the major goal in driver assessment, the question arises whether today's DSMS already make takeovers safer. The concept of the *Diagnostic Takeover Request* (Diagnostic TOR), which is described below, pursues this idea.

3 Theoretical Concept of the Diagnostic Takeover Request

Since the desired criteria that would enable the prediction of a future takeover quality are not (yet) available, an alternative approach, first published by Herzberger et al. [24] and Schwalm and Herzberger [38], and now patented [36], was developed.

This method, hereafter referred to as Diagnostic TOR, no longer focuses on inferring a state from the operator's behavior during automated driving, but on predicting risky takeovers based on a driver's response to a takeover request (TOR). The general idea is to detect missing or reduced takeover capability based on classifying drivers' orientation reactions after a TOR. For that, orientation reactions will first be recorded and evaluated for a large number of drivers, together with the subsequent takeover quality. Based on this data set, post-hoc safe and thus good takeovers can be separated from riskier, poor takeovers. After this classification is done, the previously shown orientation responses to the TOR can be analyzed. The hypothesis here is that the

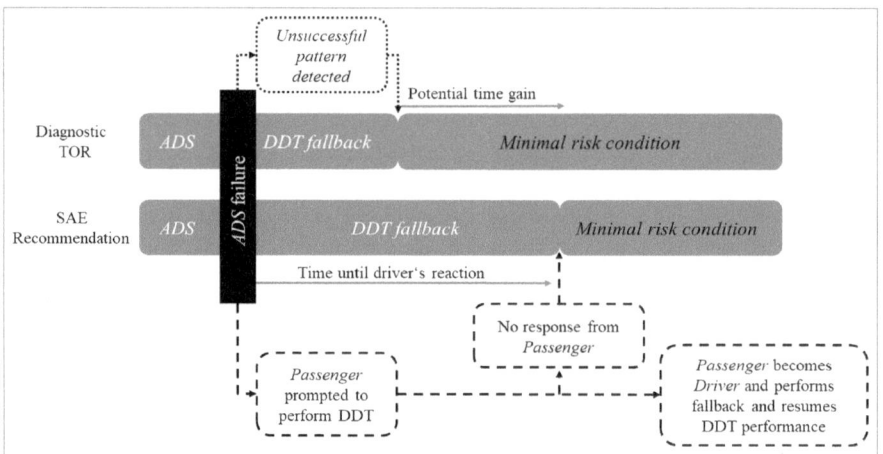

Fig. 2 Pattern for a diagnostic takeover: potential time gain for a minimum risk maneuver (MRM) using the Diagnostic TOR in comparison to the recommendation for level 4 by the SAE [24]

orientation responses of drivers before safe and unsafe takeovers differ significantly [23].

If the orientation reactions do indeed differ significantly, this would allow to predict a risky takeover already after the orientation phase of the driver, which would enable multiple early intervention options such as the initiation of a minimum risk maneuver (MRM), or the transition to a higher level of automation. This is illustrated in Fig. 2 using the example of a SAE level 4 automated driving system (ADS) with a system failure. Following the SAE recommendation, the vehicle has to wait (dynamic driving task (DDT) fallback) for the driver's response to take over control. The Diagnostic TOR concept aims to significantly reduce this waiting time, since it is not the driver's actual intervention that has to be waited for, but only his or her orientation reaction.

4 Review of the Concept and First Application

In this study, first published by Herzberger [23], the concept of the Diagnostic TOR was applied for the first time to find out whether it is indeed possible to detect distinguishable orientation reactions before safe and unsafe takeovers.

The study was conducted in the static driving simulator of the Institute of Automotive Engineering (ika) at RWTH Aachen University (see Fig. 3). A 5-series BMW (F10) served as mock-up and Virtual Test Drive (VTD) was used as driving simulation software. By using a curved projection surface, a visual range of up to 210° horizontally and 40° vertically was achieved. A three-lane highway with hard shoulder (RQ 31; FGSV [3]), with both straight and curved passages, was chosen as the driving

Fig. 3 In-vehicle study setup in the static driving simulator of the Institute of Automotive Engineering (ika) at RWTH Aachen University

situation. Since the environment was to be designed with as little stimulus as possible, there was no flowing or oncoming traffic.

The study design included an SAE level 2 driving function, which was activated by pressing an orange button on the steering wheel. The automation was designed so it could be activated at speeds above 120 km/h, whereupon it displayed a confirmatory feedback in the HMI and accelerated or decelerated to the maximum permissible speed of 130 km/h. After activation, longitudinal and lateral guidance was fully executed by the automation, provided the system limit (unforeseen situation ahead) was not reached. Furthermore, it was not necessary to have the hands on the steering wheel continuously or to re-engage with the steering wheel after a certain period of time. This system design was chosen because the goal was to have as many as 50% of the subjects (not) manage a safe takeover in order to have a data set as equal as possible for each of the takeover qualities (safe/unsafe). The system could be deactivated by both braking and steering interventions and by pressing the orange steering wheel button again.

After 15 min of SAE level 2 driving at 130 km/h in the center lane, a critical situation occurred, in the form of a broken-down vehicle (white Audi Q5) in the center lane ahead. The vehicle was parked with its hazard warning lights activated, but without any other markings e.g., a warning triangle. In addition, starting two minutes before the TOR, there was heavy traffic on the left lane at a higher speed, so that it was not possible to change to the left lane.

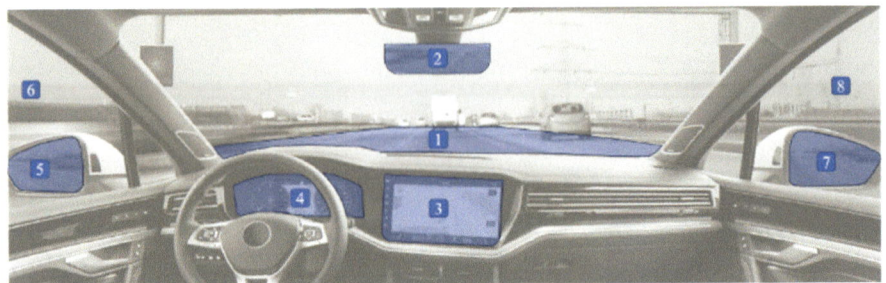

Fig. 4 AOIs according to the numbering of ISO 15007:2020 (adapted from Herzberger [23])

During automated driving, half of the participants were offered a non-driving related task (NDRT). This, visual-haptic NDRT, was a Tetris game running on a tablet (Samsung Galaxy Tab S3) mounted in the center stack. The NDRT was chosen to provide a relevant variance in takeover performance shown in the experiment. Under real-world conditions, Tetris, during a SAE level 2 drive, would not be an acceptable NDRT and thus represents a miss-use.

The subjects' gaze direction was measured both during the automated drive and during the takeover. The head-mounted eyetracker Dikablis Professional by Ergoneers was used for this purpose. The defined areas of interest (AOIs) were based on the recommendations of ISO 15007:2020, see Fig. 4, which include the road ahead (1), the rearview mirror (2), the TICS (Transport information and control system) display (3), the instrument cluster (IC, 4), the driver-side rearview mirror (5), the driver-side side window (6), the passenger-side rearview mirror (7), and the passenger-side side window (8). For this study, the AOIs highlighted in blue (1, 2, 3, 4, 5, 7) were selected because additional hardware would have been required for the side window detection, which was not available at the time the experiment was conducted.

$N = 50$ subjects participated in the study (52% female). The age of the participants ranged from 20 to 69 years ($M = 32.18$ years, $SD = 11.03$ years) and the average annual mileage was $M = 16{,}076$ ($SD = 18{,}221$ km). The results of the Karolinska Sleepiness Scale (KSS) as well as the Sofi scale, which measure the fatigue of test subjects, did not differ significantly between the groups *safe takeover* (ST) and *unsafe takeover* (UST)—the participants of both groups thus assessed themselves comparably awake. ST were defined as emergency braking in front of the broken-down vehicle, or swerving into the clear right lane, UST as swerving into the right lane occupied by faster moving vehicles, hitting a broken-down vehicle, or hitting the guardrail.

The subjects' takeovers of the driving task were analyzed after the study. The evaluation of the objective driving simulator data revealed 34% ST by the driver without NDRT and 14% in the Tetris group. The UST break down as follows: 16% without NDRT and 36% in the Tetris group—in each case as a percentage of the total takeovers. Thus, the goal of obtaining data from both successful and unsuccessful takeovers in each group was achieved. The gaze sequences were then analyzed. No distinction was made as to whether the gaze sequences prior to a UST were from individuals with or without NDRT, since future algorithms would have to be able to handle both drivers with and without NDRT. Several gaze sequences were identified that were unique to UST (results see Table 1). A discussion of the results, together with those from the second study, follows after Table 1.

However, the most obvious limitation is that the number of participants per gaze sequence is very small, which is due to the fact that there are six possible AOIs that can be stringed together in any combinatorial order. Accordingly, there are a large number of possible combinations, which minimizes the probabilities of each occurring. In order to obtain a representative sample for each of the cases, driving simulation studies are simply not suitable, as they are too time-consuming and too expensive. However, the aim of the study was not to generate an exhaustive data set, but to perform a first analysis based on the question whether the Diagnostic TOR could be implemented in principle. This analysis showed that the orientation responses differ before ST and UST (at least for this first sample), which strongly supports the usefulness of this approach. But, given the small number of participants, a replication

Table 1 Gaze patterns from the first and second study (S1 and S2) and the likelihood (L) of an unsuccessful takeover (UST) after a gaze pattern (adapted from Herzberger [23])

Gaze Pattern	ST S1	UST S1	ST S2	UST S2	UST L S1	UST L S2	UST L total
T	0	0	0	1		100%	100%
T R	2	4	0	1	67%	100%	71%
T R T	0	1	0	0	100%		100%
T R IC	0	3	2	3	100%	60%	75%
T R ML	1	1	0	0	50%		50%
T R MM	0	0	1	1		50%	50%
T R MR	1	0	0	0	0%		0%
T IC R	1	8	4	4	89%	50%	71%
T IC T	0	1	0	0	100%		100%
T IC ML	0	0	0	1		100%	100%
R	4	0	5	1	0%	17%	10%
R IC R	4	5	3	2	56%	40%	50%
R ML R	6	0	1	0	0%	0%	0%
R MM R	1	0	1	0	0%	0%	0%
R MR R	2	0	0	0	0%		0%
IC	0	1	0	0	100%		100%
IC R	0	1	0	0	100%		100%
IC R IC	0	0	2	0		0%	0%
IC R ML	1	0	0	0	0%		0%
IC T R	0	0	1	0		0%	0%
ML R	1	0	0	0	0%		0%
ML R IC	0	1	0	0	100%		100%
MM R IC	0	0	3	0		0%	0%
MR R IC	0	0	0	1		100%	100%
N = 88	24	26	23	15			

study (see next section) was needed to find out whether another sample shows similar gaze orientations and whether, despite further data, distinctive sequences are still preserved.

5   Generalizability of Orientation Reactions

The purpose of this replication study, first published by Herzberger [23] was to find out whether comparable orientation reactions could be found in a different sample under conditions that were as comparable as possible. Furthermore, the recorded orientation patterns should be merged with those from the previous study to check whether distinguishable sequences can still be identified in a larger sample.

This replication study was conducted in the static driving simulator of the IAW of the RWTH Aachen University. Since this is not based on a real vehicle mock-up but on a Bosch-Rexroth setup, it was possible to replicate the dimensions of the vehicle exactly. For this purpose, the mock-up at the ika was precisely measured to record both the distances between the mirrors and the exact positions of the IC and TICS. This ensured that as few confounding variables as possible were introduced into the study. A four-camera remote eye tracking system (Smart Eye Pro 6) was used. The system uses four cameras of which two were placed at the bottom of the A-pillars, one above the dashboard, and one at the bottom right of the TICS to reliably detect the relevant AOIs. This system was chosen because the head-based system had difficulties with sudden head movements and to best meet the requirement for a method that could be used in the real world. The study design was replicated as closely as possible.

$N = 38$ subjects participated in the study (55% male). Participants' ages ranged from 18 to 65 years ($M = 33.26$ years, $SD = 15.01$ years), and mean annual mileage was $M = 8,860$ ($SD = 11,289$ km). After the study, the takeovers were classified. This resulted in the following takeover qualities by group: Successful were 36.8% of participants without NDRT and 23.7% with NDRT. Unsuccessful were 7.9% of the subjects without NDRT and 31.6% with NDRT. Thus, the replication study also achieved its goal of collecting data from both successful and unsuccessful takeovers in each group.

Table 1 provides the participants' gaze patterns from both studies as well as the likelihood for a subsequent UST. The AOIs were labeled according to the following classification: Road ahead (1) is R, TICS display (3) is T, instrument cluster (4) is IC, driver-side rearview mirror (5) is Mirror left ML, passenger-side rearview mirror (7) is Mirror right MR, and rearview mirror (2) is MM (Mirror middle, since RM was too similar to MR). It becomes apparent that even after merging the data sets from the first and the second study (S1 and S2), it is still possible to identify distinct sequences. The gaze patterns that have a probability >0.5 and a sample $n > 1$ are highlighted in gray.

It is noticeable that takeovers are most often unsuccessful when they start at the TICS display (T). This is not surprising since drivers who have not lowered their gaze to T may at least still be peripherally aware of their surroundings or may even be able to recognize the critical situation at an early stage. These results should be taken as an opportunity to reconsider the warning strategy in the course of takeover requests, since warnings in the instrument cluster (IC) do not seem to be very helpful for safe takeovers, to say the least. An alternative could be to display the warning (additionally) in the TICS or the head-up display, so that the driver does not have to look into the instrument cluster to grasp the content or trigger of the warning. Since even with an enlarged sample separating orientation reactions were identified, the Diagnostic TOR seems to be a promising approach to detect a UST at an early stage in case of a necessary handover of the driving task to the driver by comparing the detected orientation reaction with previously recorded gaze patterns and thus to be able to initiate safeguarding measures.

6   Conclusion and Outlook

The presented concept of the Diagnostic TOR shows that a meaningful use of DSMS is already possible today, which could enable early detections of unsuccessful takeovers. Nevertheless, the major goal remains to identify criteria in the future that can be used to predict the driver's capability to take over from the driver's behavior during the automated drive. Until this is possible, however, approaches such as the Diagnostic TOR could be helpful in gaining more reaction time during critical transitions. Importantly, the limit in terms of reaction time as well as accuracy is far from being reached: In the presented studies, only the gaze pattern was used to estimate the human's capability to take over. For more advanced approaches, such as confidence horizons, the human horizon can be determined in much greater detail. For example, there are ongoing studies that also include the driver's body posture, weight shift and grip strength. The gaze direction, therefore, is only a small part of a takeover pattern [12, 8], and a large number of variables forms of takeover responses. More detailed takeover patterns are being investigated in the Exploroscope [7] of the IAW at the RWTH Aachen University and will also be included in the design of the confidence horizon in the future. This concept, which allows the two agents (driver and vehicle) to have a shared mental model, will enable dynamic in-vehicle cooperation of the driving task. The information about the mutual capabilities or limits of driver and automated system could also be made available to other automated vehicles in a next development step. If, for example, a transition to the driver fails and an evasive maneuver has to be performed at the last moment, surrounding automated vehicles could react adaptively based on this data, taking vehicle-vehicle-cooperation to a next level.

Acknowledgements This project was funded within the Priority Program 1835 *Cooperative Interacting Automobiles (CoInCar)* of the German Science Foundation DFG.

References

1. Barua, S., Ahmed, M.U., Ahlström, C., Begum, S.: Automatic driver sleepiness detection using EEG, EOG and contextual information. Expert Syst. Appl. **115**, 121–135 (2019). https://doi.org/10.1016/j.eswa.2018.07.054
2. Christoffersen, K., Woods, D.D.: How to make automated systems team players. In: Advances in Human Performance and Cognitive Engineering Research (2002)
3. FGSV: Richtlinien für die Anlage von Autobahnen: RAA; R1, 2008th edn. Forschungsgesellschaft für Straßen- und Verkehrswesen: FGSV, 202: R1. FGSV-Verl., Köln (2009)
4. Flemisch, F., Winner, H., Bruder, R., Bengler, K.: Cooperative guidance, control and automation. In: Winner, H., Hakuli, S., Wolf, G. (eds.) Handbuch Fahrerassistenzsysteme: Grundlagen, Komponenten und Systeme für aktive Sicherheit und Komfort. Vieweg + Teubner, Wiesbaden (2015)
5. Flemisch, F., Abbink, D., Itoh, M., Pacaux-Lemoine, M.-P., Weßel, G.: Shared control is the sharp end of cooperation: towards a common framework of joint action, shared control and human machine cooperation. IFAC-PapersOnLine **49**(19), 72–77 (2016). https://doi.org/10.1016/j.ifacol.2016.10.464
6. Flemisch, F., Abbink, D.A., Itoh, M., Pacaux-Lemoine, M.-P., Weßel, G.: Joining the blunt and the pointy end of the spear: towards a common framework of joint action, human–machine cooperation, cooperative guidance and control, shared, traded and supervisory control. Cogn. Tech. Work **21**(4), 555–568 (2019). https://doi.org/10.1007/s10111-019-00576-1
7. Flemisch, F., Preutenborbeck, M., Baltzer, M.C.A., Wasser, J., Kehl, C., Grünwald, R., Pastuszka, H.-M., Dahlmann, A.: Human systems exploration for ideation and innovation in potentially disruptive defense and security systems. In: Advanced Sciences and Technologies for Security Applications, pp. 79–117. Springer International Publishing, Cham (2022a)
8. Flemisch, F., Usai, M., Herzberger, N.D., Baltzer, M.C.A., Hernandez, D.L., Pacaux-Lemoine, M.-P.: Human-machine patterns for system design, cooperation and interaction in socio-cyber-physical systems: introduction and general overview. In: 2022 IEEE International Conference on Systems, Man, and Cybernetics (SMC). 2022 IEEE International Conference on Systems, Man, and Cybernetics (SMC), Prague, Czech Republic, pp. 1278–1283. IEEE (2022b). https://doi.org/10.1109/SMC53654.2022.9945181
9. Flemisch, F.: Pointillistische Analyse der visuellen und nicht-visuellen Interaktionsressourcen am Beispiel Pilot-Assistentensystem (2001)
10. Flemisch, F., Schieben, A., Schoemig, N., Strauss, M., Lueke, S., Heyden, A.: Design of human computer interfaces for highly automated vehicles in the EU-project HAVEit. In: Universal Access in Human-Computer Interaction. Context Diversity: 6th International Conference, UAHCI 2011, Held as Part of HCI International 2011, Orlando, FL, USA, July 9–14, 2011, Proceedings, Part III 6, pp. 270–279. Springer, Berlin Heidelberg (2011)
11. Flemisch, F., Onken, R.: The cognitive assistant system and its contribution to effective man/machine interaction. In: RTO SCI Symposium on "The Application of Information Technologies (Computer Science) to Mission Systems". Monterey, California, USA (1998)
12. Flemisch, F., Herzberger, N.D., Usai, M., Baltzer, M.C.A., Schwalm, M., Voß, G., Krems, J., Quante, L., Burger, C., Stiller, C.: Cooperative hub for cooperative research on cooperatively interacting vehicles: use-cases, design and interaction patterns. In: Stiller, C. (ed.) Cooperative interacting vehicles. Springer (in press)
13. Flemisch, F., Kelsch, J., Schieben, A., Schindler, J.: Stücke des Puzzles hochautomatisiertes Fahren: H-Metapher und H-Mode: Zwischenbericht 2006. Workshop Fahrerassistenz (2006)
14. Flemisch, F., Schieben, A., Kelsch, J., Löper, C.: Automation spectrum, inner/outer compatibility and other potentially useful human factors concepts for assistance and automation. In: de Waard, D., Flemisch, F., Lorenz, B., Oberheid, H., Brookhuis, K.A. (eds.) Human Factors for assistance and automation. Annual Meeting Human Factors & Ergonomics Society, European Chapter, Braunschweig, 2007. Shaker Publishing (2008)

15. Flemisch, F., Heesen, M., Hesse, T., Kelsch, J., Schieben, A., Beller, J.: Towards a dynamic balance between humans and automation: authority, ability, responsibility and control in shared and cooperative control situations. Cogn. Tech. Work **14**(1), 3–18 (2012). https://doi.org/10.1007/s10111-011-0191-6
16. Gasser, T.M., Arzt, C., AYoubi, M., Bartels, A., Buerkle, L., Eier, J., Flemisch, F., Haecker, D., Hesse, T., Huber, W., LOTZ, C., Maurer, M., Ruth-Schumacher, S., Schwarz, J., Vogt, W.: Rechtsfolgen zunehmender Fahrzeugautomatisierung 83. Wirtschaftsverlag NW (2012)
17. Grazioli, F.: Design of a driver state detector based on eye-tracking, unpublished Master Thesis. RWTH Aachen University (2017)
18. Griffiths, P.G., Gillespie, R.B.: Sharing control between humans and automation using haptic interface: primary and secondary task performance benefits. Hum. Factors **47**(3), 574–590 (2005). https://doi.org/10.1518/0018720057748 59944
19. Guettas, A., Ayad, S., Kazar, O.: Driver state monitoring system: a review. In: Proceedings of the 4th International Conference on Big Data and Internet of Things, New York, NY, USA. ACM, New York, NY, USA (2019). https://doi.org/10.1145/3372938.3372966
20. Hecht, T., Feldhütter, A., Radlmayr, J., Nakano, Y., Miki, Y., Henle, C., Bengler, K.: A review of driver state monitoring systems in the context of automated driving. In: Congress of the International Ergonomics Association, pp. 398–408. Springer, Cham (2019). https://doi.org/10.1007/978-3-319-96074-6_43
21. Heikoop, D.D., de Winter, J.C., van Arem, B., Stanton, N.A.: Psychological constructs in driving automation: a consensus model and critical comment on construct proliferation. Theor. Issues Ergon. Sci. **17**(3), 284–303 (2016). https://doi.org/10.1080/1463922X.2015.1101507
22. Herzberger, N.D., Voß, G.M.I., Schwalm, M.: Identification of criteria for drivers' state detection. In: International Conference on Applied Human Factors and Ergonomics, pp. 798–806. Springer, Cham (2017). https://doi.org/10.1007/978-3-319-60441-1_76
23. Herzberger, N.D.: Erfassung der Übernahmefähigkeit von Fahrpersonen im Kontext des automatisierten Fahrens. Shaker Verlag. (2023)
24. Herzberger, N.D., Eckstein, L., Schwalm, M.: Detection of missing takeover capability by the orientation reaction to a takeover request. In: 27th Aachen Colloquium Automobile and Engine Technology 2018, pp. 1231–1240 (2018)
25. Herzberger, N.D., Schwalm, M., Voß, G.M.I., Flemisch, F., Schmidt, E., Sitter, A.: Erfassung der Fahrerübernahmefähigkeit im automatisierten Fahren anhand von Fahrerbeobachtungen. In: Mensch-Maschine-Mobilität 2019, pp. 53–66. VDI Verlag (2019)
26. Herzberger, N.D., Usai, M., Flemisch, F.: Confidence horizon for a dynamic balance between drivers and vehicle automation: first sketch and application. In: Human Factors in Transportation. 13th International Conference on Applied Human Factors and Ergonomics (AHFE 2022), July 24–28, 2022. AHFE International (2022). https://doi.org/10.54941/ahfe1002431
27. Hoeger, R., Zeng, H., Hoess, A., Kranz, T., Boverie, S., Strauss, M. et al.: Final report, Deliverable D61.1. Highly automated vehicles for intelligent transport (HAVEit). 7th Framework programme (2011)
28. Kraftfahrt-Bundesamt: Kba erteilt erste genehmigung zum automatisierten Fahren, Flensburg, Germany (2021)
29. Löper, C., Kelsch, J., Flemisch, F.: Kooperative, manöverbasierte Automation und Arbitrierung als Bausteine für hochautomatisiertes Fahren. In: AAET—Automatisierungs-, Assistenzsysteme und eingebettete Systeme für Transportmittel: Beiträge zum gleichnamigen 9. Braunschweiger Symposium vom 13. und 14. Februar 2008, Deutsches Zentrum für Luft- und Raumfahrt e.V. am Forschungsflughafen, Braunschweig. Net_372work. GZVB, Braunschweig (2008)
30. Marberger, C., Mielenz, H., Naujoks, F., Radlmayr, J., Bengler, K., Wandtner, B.: Understanding and applying the concept of "Driver Availability" in automated driving. In: International Conference on Applied Human Factors and Ergonomics, pp. 595–605. Springer, Cham (2018). https://doi.org/10.1007/978-3-319-60441-1_58
31. Minhad, K.N., Ali, S.H.M., Reaz, M.B.I.: Happy-anger emotions classifications from electrocardiogram signal for automobile driving safety and awareness. J. Transp. Health **7**, 75–89 (2017). https://doi.org/10.1016/j.jth.2017.11.001

32. Onken, R.: Human process control and automation—still compatible concepts? In: Burys, B.-
    B., Wittenberg, C. (eds.) From Muscles to Music: a Festschrift to celebrate the 60th Birthday
    of Gunnar Johannsen, pp. 75–87. Kassel University Press (2002)
33. Rauch, N.: Ein verhaltensbasiertes messmodell zur Erfassung von situationsbewusstsein im
    fahrkontext (2009)
34. SAE International: Taxonomy and definitions for terms related to driving Automation Systems
    for On-Road Motor Vehicles (No. J3016_202104). 400 Commonwealth Drive, Warrendale,
    PA, United States. SAE International (2021)
35. Schieben, A., Temme, G., Köster, F., Flemisch, F.: Intermediate results of the Human Factors
    work in the EU-Project HAVEit: Interaction design and simulator testing of the Joint System
    for highly automated vehicles. In: HFES Europe Conference, Berlin, 13–15 Oct 2010 (2010)
36. Schories, L., Erggelet, M., Schwalm, M., Herzberger, N.D.: Verfahren zum Feststellen einer
    Übernahmefähigkeit eines Fahrzeugnutzer eines Fahrzeugs. DE 102018007508 A8. Accessed
    26 Mar 2020
37. Schulte, A.: Co-operating Cognitive Machines—An Automation Approach to Improve Situ-
    ation Awareness in Distributed Work Systems. NATO RTO System Concepts and Integration
    (SCI) Panel Workshop on Tactical Decision Making and Situational Awareness for Defense
    against Terrorism, Turin, Italy (2006)
38. Schwalm, M., Herzberger, N.D.: Die Erfassung des Fahrerzustands als Voraussetzung für höher
    automatisierte Fahrfunktionen–Eine kritische Diskussion und ein Lösungsvorschlag. In: 12.
    Workshop Fahrerassistenzsysteme, Walting (Albmühltal), Germany, 28.09. (2018)
39. Stiller, C., Farber, G., Kammel, S.: Cooperative cognitive automobiles. In: 2007 IEEE
    Intelligent Vehicles Symposium. IEEE (2007). https://doi.org/10.1109/ivs.2007.4290117
40. Taherisadr, M., Asnani, P., Galster, S., Dehzangi, O.: ECG-based driver inattention identifica-
    tion during naturalistic driving using Mel-frequency cepstrum 2-D transform and convolutional
    neural networks. Smart Health **9–10**, 50–61 (2018). https://doi.org/10.1016/j.smhl.2018.07.022
41. Teng, T., Bi, L., Liu, Y.: EEG-based detection of driver emergency braking intention for brain-
    controlled vehicles. IEEE Trans. Intell. Transport. Syst. **19**(6), 1766–1773 (2018). https://doi.
    org/10.1109/tits.2017.2740427
42. Urbano, M., Alam, M., Ferreira, J., Fonseca, J., Simioes, P.: Cooperative driver stress sensing
    integration with eCall system for improved road safety. In: IEEE EUROCON 2017-17th Inter-
    national Conference on Smart Technologies. IEEE (2017). https://doi.org/10.1109/eurocon.
    2017.8011238
43. Usai, M., Herzberger, N.D., Yu, Y., Flemisch, F.: Confidence horizons: dynamic balance of
    human and automation control ability in cooperative automated driving. In: Stiller, C. (ed.)
    Cooperative Interacting Vehicles. Springer (in press)
44. Völker, M., Kloock, M., Rabanus, L., Alrifaee, B., Kowalewski, S.: Verification of cooperative
    vehicle behavior using temporal logic. IFAC-PapersOnLine **52**(8), 99–104 (2019). https://doi.
    org/10.1016/j.ifacol.2019.08.055
45. Wu, G., Liu, G., Hao, M.: The analysis of emotion recognition from GSR based on PSO. In:
    2010 International Symposium on Intelligence Information Processing and Trusted Computing.
    IEEE (2010). https://doi.org/10.1109/iptc.2010.60

Confidence Horizons: Dynamic Balance of Human and Automation Control Ability in Cooperative Automated Driving

Marcel Usai, Nicolas Herzberger, Yang Yu, and Frank Flemisch

Abstract This chapter presents the concept of confidence horizon for cooperative vehicles. The confidence horizon is designed to let the automation predict its own and the human's abilities to control the vehicle in the near future. Based on the pattern approach originating from Alexander et al. [1], the confidence horizon concept is instantiated with a pattern framework. In case of a necessary takeover of the driving task by the human, a mode transition pattern is initiated. In order to determine when the takeover is required, which pattern to start and when to omit the takeover attempt and directly start a minimum risk maneuver, the confidence horizon for both human and co-system is an important parameter. A visual representation of the confidence horizon for the driver in different scenarios prior to a takeover request was explored. Intermediate results of a simulator study are presented, which assess the confidence horizon in automation safety-critical takeover scenarios involving an intersection and a broken-down vehicle on a highway.

1 Cooperation Between Human, Co-system and Environment

Cooperation in automated driving is a bridging paradigm connecting many facets, e.g., cooperation between machines and machines, between humans and humans as well as between humans and machines. Cooperation does not necessarily need simi-

M. Usai (✉) · N. Herzberger · F. Flemisch
Institute of Industrial Engineering and Ergonomics, RWTH Aachen University, Eilfschornsteinstr. 18, 52062 Aachen, Germany
e-mail: m.usai@iaw.rwth-aachen.de

N. Herzberger
e-mail: n.herzberger@iaw.rwth-aachen.de

F. Flemisch
e-mail: frank.flemisch@fkie.fraunhofer.de

Y. Yu
RWTH Aachen University, Aachen, Germany
e-mail: yang.yu2@rwth-aachen.de

larity among cooperation partners. Compatibility, however, is a crucial requirement. It needs to be sufficiently developed between the outer borders of the cooperating sub-systems (outer compatibility) and between the inner, often cognitive, aspects of the cooperating sub-systems (inner compatibility) [9], leading to outer and inner cooperation. In these complex systems, not only the humans and machines in the directly acting human-machine system should cooperate, but also the people and machines in the meta-system, e.g., in research and development.

The cooperation between multiple vehicles reflects the outer cooperation from the viewpoint of a single automated vehicle and is examined in various details in many other chapters of this book. The following chapter focuses on the cooperation of a single human with a single automation within a highly automated vehicle. Any cooperation with other vehicles, between these vehicles and with the ego vehicle itself are considered as part of the environment.

In general, there are three main entities within the system of the ego automated vehicle: The human, the co-system (including the automation and other technical subsystems), both of which are considered agents within the system, and the environment. As shown in Fig. 1, the human and co-system influence the environment through joint actions. To enable a joint action [25], the human and the co-system have to cooperate either through direct communication or through a mediator, which is represented by the center element of the diagram.

In this system model, it is assumed that both the human and the co-system may share the vehicle control and transition control between one another. The direct communication between the two agents is crucial for the co-system to communicate decisions made by a network of cooperating vehicles as well as possible actions needed by the human if the co-system reaches its limitations.

In order to successfully design human-machine cooperation, it is necessary to align the "mental model" of the co-system with the mental model of the human

Fig. 1 Simple model for the cooperation between human and co-system. Human and co-system both act on and control the vehicle as a part of the environment. (Based on Flemisch et al. [8] and Löper [21])

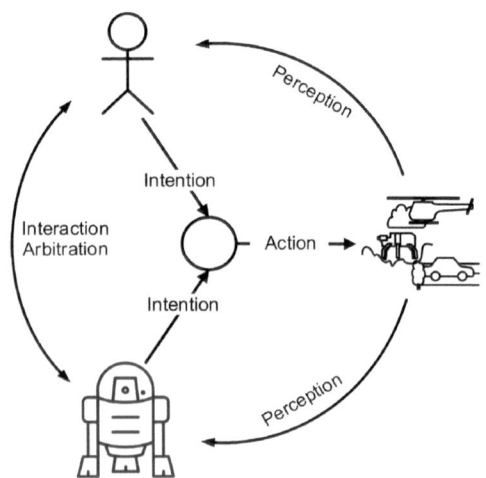

[9] to include the environment, and to keep it transparent and repeatable. One tool to achieve this is a design metaphor, which has been successfully applied e.g. in the form of the desktop-metaphor (as established by Alan Kay from Xerox PARC in 1970) or the H(orse)-metaphor [8], transferring the mental model of a rider and horse to the domain of highly automated vehicles. A more generalized approach is the pattern approach based on Alexander et al. [1], applied to music by Borchers [5], to software by Gamma et al. [16] and applied to human-machine systems by Baltzer [2], Herzberger et al. [20], López Hernández [22], and others. For more details on patterns see Flemisch et al. [15] and the chapter of Flemisch et al. [7].

2 The Concept of Confidence Horizons

The idea behind the confidence horizon concept is to bring together the prediction of the time points of when and until when the human and the automation are able to control the joint system, in this case an automated vehicle.

In this sense, the confidence horizon is coupled to the prediction of the ability to execute control over the joint system. Combining the predictions for human and automation makes clear when a safe transition of control between human and automation can be expected and how automation and human need to communicate, depending on the severity of the situation. Figure 2 depicts the confidence horizon concept.

As shown on the left, human and automation are more or less involved in the current driving task, depending on the current automation mode (e.g. manually, partially or highly automated), and the resulting distribution of control [10]. As stated in the SAE [24], starting from SAE Level 4 automation, the driver is explicitly allowed to disengage completely from the driving task, which results in a potential loss of

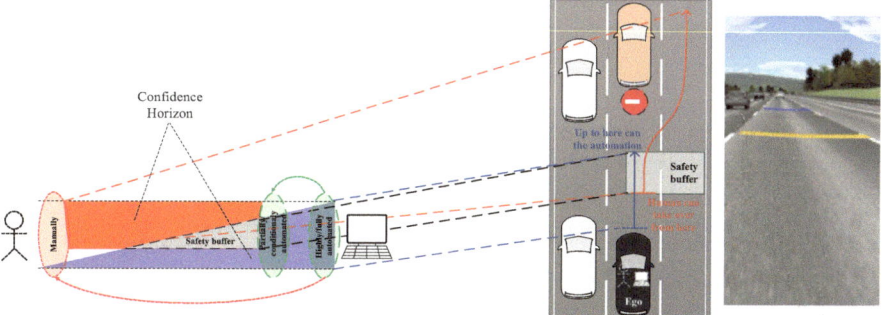

Fig. 2 The confidence horizon as a product of the distribution of the ability to control for human and automation. Displayed in the control distribution according to (left, based on Herzberger et al. [20]) and projected onto a use case of highly automated driving (center). Application of the confidence horizon start (human) and end (automation) to the driving simulator (right)

situation awareness for the driving task, especially when engaging in a non-driving related task (NDRT) [28]. Even in lower automation levels (automation according to SAE Level 2), despite the driver's obligation to be ready to intervene and ongoing liability for the vehicle's actions, the driver may tend to lose awareness, a mechanism described as the unsafe valley of automation [11]. With the confidence horizon concept, we propose to make this unsafe valley visible at least to the automation and its developers, as an option also for the driver, so that she can act accordingly. The control distribution in Fig. 2 (left) shows, on the one hand, who has to control the vehicle in a given automation mode and, on the other hand, the ability of the human (in orange) and automation (in blue) to actually execute the vehicle control. Projecting the ability distribution for the human and the automation into a real situation (see Fig. 2, right) directly shows the need for a control transition due to a lack of ability of the automation to handle an obstacle in this situation. Furthermore, it shows the available time frame for a transition to the human, in which this transition has to take place (shown as safety buffer).

In a critical situation (system boundary or system failure), the confidence horizons clearly show a safety gap, i.e., a time frame in which neither automation nor human are able to control the driving related task. The confidence horizon concept enables the automation to detect such cases as early as possible and act accordingly. Depending on the time remaining until the system failure is reached and the current ability of the driver, the automation either triggers a diagnostic take-over request (TOR), in case that there is a safety buffer present before the system would fail, or a minimum risk maneuver (MRM).

We propose to use the confidence horizon concept for the design of highly automated human-machine systems to identify the proper transition strategy in case of an upcoming control gap and to predict the future ability of the human and the automation to control the joint system. However, based on our exploration results, we recommend using the confidence horizon as a basis for designing HMI designs in situations of varying criticality, including the communication strategy of the automation, rather than a simple visual representation of the confidence horizon as in Fig. 2 (right).

3 Application of the Pattern Approach to Cooperative Automated Driving

To achieve good cooperation between two agents, both need to understand each other. When designing human-machine cooperation, the challenge is to find a common language. A promising solution is the approach of interaction patterns to find common ground at large scale. Based on Alexander et al. [1], Flemisch et al. [14] describe a pattern as follows:

> A pattern describes something that occurs over and over again. An example for this is a problem and/or its solutions. If this can be observed, and its core can be mapped and

modelled, you can either observe and match the pattern over and over again, without ever making the identical observation twice. And/or you can instantiate and design with this pattern over and over again, not necessarily doing it the same way twice. Examples for this are designing, engineering and using of artefacts like human-machine systems. Flemisch et al. [14]

Alexander et al. [1], Borchers [5] and Baltzer [3] use patterns to describe a solution to a given problem and propose a pattern language for the design of patterns. Another focus is set by Flemisch [6] and López Hernández [22] on the structure of the solution by describing in detail the sequence of interaction within a pattern. This focus, however, further tailored to matching a given pattern instance for the case of cooperatively interacting vehicles, is also applied in the proposal of the authors.

When using the pattern approach for active cooperation, the pattern structure is extended by a set of properties to detect which cooperation partner should, wants and is currently performing a given pattern, resulting in a new subset of patterns: Cooperation patterns. In the proposed setup, all properties are predicted by the co-system. Each property can be described by a sub-pattern, so that if the sub-pattern matches the co-system, the activation value and confidence for the respecting property increases as well.

The fundamental properties of a cooperation pattern are utility, ability, intention and execution. Utility describes how useful the activation of the current pattern would be for the respecting agent. Ability represents the agent's ability to execute the pattern now and in the near future. Intention describes the agent's inner determination to execute the pattern, while the execution property describes the matching of the agent's actual current action with the actions required to execute the pattern at hand.

Derived from the cooperation pattern, the relevant patterns are activity patterns and transition patterns. Applied to cooperative vehicles, there are driving related and non-driving related activities (see Fig. 3).

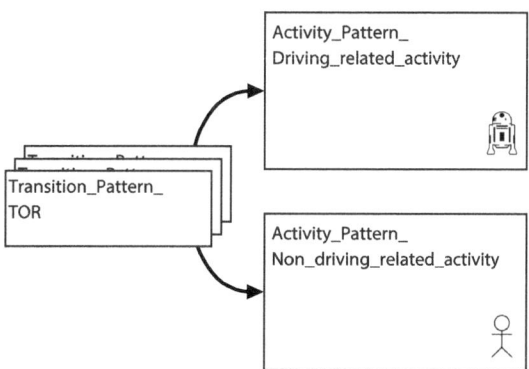

Fig. 3 Simple illustration of the change of focus on an activity using transitions. The symbol in the right corner of the activity shows which agent is currently active the most in the respecting activity

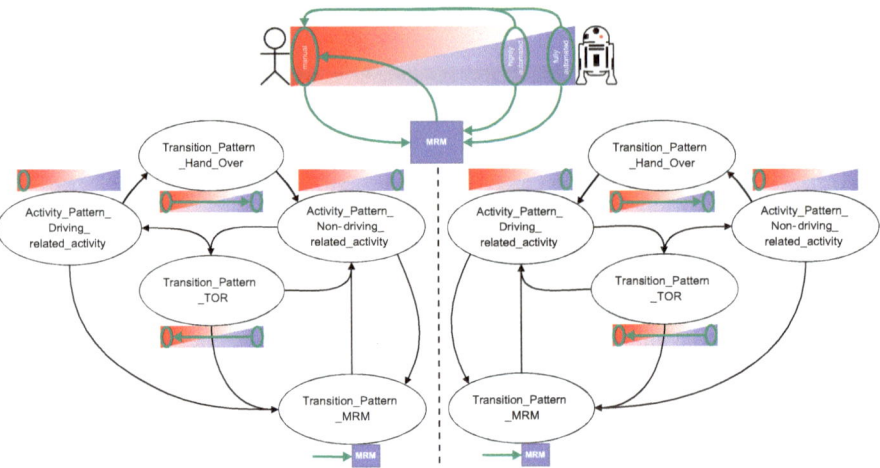

Fig. 4 Possible transitions between activity and transition patterns for human (left) and automation (right) in case of a take-over initiated by the co-system

Both agents, the human as well as the automation, can focus on one of these activities. They can change their own focus and try to change the other's focus by starting a transition pattern, e.g., a takeover request (TOR).

Figure 4 depicts the pattern network for the application in transition control for highly automated driving. It displays the same process as in Fig. 4 with the patterns as states and for each agent individually. On the most basic level, the activity of human and automation can be considered as driving related or non-driving related. Since activity patterns are derived from cooperation patterns, they contain their properties for the utility, ability, intention and execution of the activity by both agents according to the co-system's prediction. The same applies to transition patterns.

The detection of the ability of both human and automation to execute the driving related ability directly reflects the current state of the confidence horizon. Transitions are used to switch from one activity to the other. Various transitions are available based on the initiator of the transition, the current size of the safety buffer in the confidence horizon and the predicted ability to execute the target activity for human and automation. It should be noted that, in the case of a transition, both human and automation have to change their activity. As part of the co-system, a mediator arbitrates conflicts between human and machine [4] and provides transparency of the automation's behavior to maximize the overall utility of the human-machine system. This mediator makes all joint decisions. It is the mediator's responsibility to let the co-system initiate a certain transition or prevent the human from using a transition that is not feasible for the system. Figure 4 illustrates the possible transitions between activity and transition patterns for both human and automation, assuming that each agent is focused on a single task at any given time. In this application, the automation can initiate a take-over request (TOR) that, if successful, leads to a change in activity for both agents, or be pushed into a minimum risk maneuver (MRM).

A combined representation of both diagrams of Fig. 4 is shown in Fig. 3, highlighting that all activities are considered states with the properties of utility, ability, intention and execution for each agent. Additionally, an agent is not limited to focus on a single activity, but rather uses transition patterns to change focus from one activity to another.

Applied to human-automation cooperation in cooperatively interacting vehicles, this could be implemented as follows (Fig. 5): The co-system detects a safety gap ahead and needs to transition the human activity from a non-driving related to the driving related task. This has to be done before the safety gap comes to close. Otherwise, the co-system has to initiate a minimum risk maneuver, which, however, might involve a higher risk than a successful take-over of the human. Figure 5 depicts this situation at time $t_{1.1}$. If there is enough time to hand over control to the human, the co-system starts a two-stage take-over pattern (as based on e.g. Rhede et al. [23], Winkler et al. [27] or Guo et al. [17]) to let the driver gain situational awareness and enable the driver to take back control safely. Depending on the predicted ability of the driver, the first warning might be sufficient, or, the second warning stage has to be triggered, starting at $t_{1.2}$. If the transition fails because the human is either unwilling or unable to take over in time, according to Herzberger et al. [19], the co-system starts another transition to the MRM and aborts the take-over transition, leading to $t_{2.1}$. Only if the transition is successful, control is transferred to the human and the automation accordingly loses control over the driving related activity ($t_{2.2}$).

4 Exploration of the Confidence Horizon Cooperation Design

To explore the design options for the cooperation between human and co-system and in particular the HMI used in the use case of a breakdown vehicle, a Human Systems Exploration (as described by Flemisch et al. [13]) was conducted at the IAW Exploroscope.

The chosen use case was the appearance of a stopped vehicle on a three-lane highway in the center lane with traffic to the left lane. To avoid a collision with the vehicle in front, there are two possibilities: Either one breaks and stops in front of the vehicle, staying vulnerable to traffic from behind, or one changes to the right lane to avoid a collision. It is assumed that the automation is unable or not allowed to perform[1] the evasive maneuver.

The setup consisted of two scenarios representing the safety buffer and safety gap cases in two different severity levels of time to collision (TTC) with $TTC = 10$ s and $TTC = 3$ s, indicated by the distance between the ego and the breakdown vehicle.

---

[1] In Germany and other European countries, traffic rules do not allow the overtaking of other road users from the right when driving out of town, e.g. on highways.

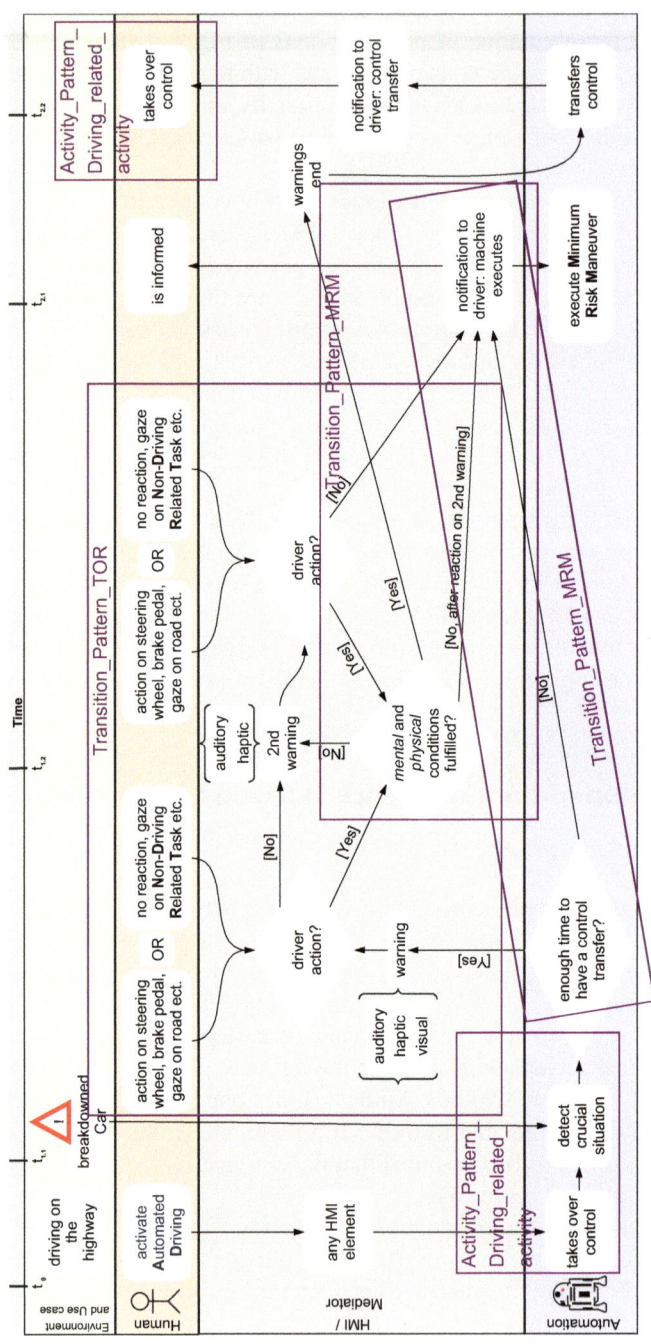

Fig. 5 Swinlane diagram for the take-over process. The red frames show the connection of the actions to the respecting pattern. (Herzberger et al. [20], modified)

Fig. 6 Snapshot from the on-site Exploration (left) and situation as it was displayed in the online Exploration (right) to show the view of one situation as in the simulator

A total of $N = 12$ persons (41, 67% female, 58, 33% male) with an average age of $30 (\sigma^2 = 7, 98)$ years participated in the exploration. Due to the Covid-19 restrictions in 2020, the exploration was conducted partly on-site ($n = 5$ participants) and online ($n = 7$ participants). A digital whiteboard tool was used for documentation in both cases.

Participants were shown all four resulting situations on a digital whiteboard with the confidence horizon markings (cf. Fig. 6, right) displayed for reference and asked to share their thoughts on how the co-system should communicate a take-over request to the human. They were given the task of drawing a sketch of their proposed head up display (HUD) concept.

As a first finding, it should be noted, that only one in 12 (8%) would display the confidence horizon (as in Fig. 6, right) directly to the driver. 42% of participants would display the confidence horizon only for the ability of the co-system and under certain conditions. And 50% would never display it to the driver, especially because predicting human capability is perceived as confusing or uncanny, and displaying information in an area where the co-system cannot control the vehicle is considered plausible. From these results, it is concluded that the confidence horizon might be a useful tool for cooperation design and to initiate transitions in foresight but should only be used with caution as a too detailed HMI element.

Participants also noted that the information displayed in the visual HMI should be limited to focus attention and that they prefer not to read text in a critical situation. 33% indicated that a general warning message in the corners of the visible area would be useful. 42% commented positively on the visualization of a lane change trajectory as well as the display of the center lane trajectory with changing colors indicating the criticality of the distance to the obstacle ahead. Figure 7 shows the proposal for the safety buffer scenarios combined from all the results collected. The participants wanted to be shown how much way they still have before the situation becomes too critical if they do not react. The left lane is shown as blocked and an arrow indicates the possible lane change to the right lane. An icon in the center of the field of view

Fig. 7 Examples for combined hand-drawn HMI concepts from the exploration workshops on safety buffer scenarios. Left: scenario for $TTC = 10$ s; right: $TTC = 3$ s

indicates necessary action. The broken-down vehicle is highlighted with a frame in warning color (red), annotated with the remaining distance in meters. In the corners of the field of view (might be realized as part of the HUD or ambient lighting), light flashing colors emphasize the possible and impossible directions.

The safety gap scenarios were not fully understood by most of the participants. The main reason for that was that it is difficult to understand why the co-system would provide information on the situation despite it itself failing in the very moment. This shows that it was unclear to the participants that situation awareness and ability to execute the driving task are separated in the case of the co-system. Most importantly, participants wanted transparency of the automation's actions for both cases. For example, the co-system should inform the driver, that a minimum risk maneuver is being executed and that the driver may only take over control after the maneuver is completed.

5 Simulator Study of the Confidence Horizon Cooperation Design

To evaluate the proposed application of the confidence horizons, a study with $N = 20$ participants was conducted in the static driving simulator at the IAW Exploroscope of the RWTH Aachen University. The study produced much more results than can be shown in the last part of this chapter, so that only an overview can be given, with more detailed publications to follow. The study tested three different designs in two different use cases. The use cases were:

Use case 1 "Avoidance of broken-down vehicle", starting on the highway in SAE level 3/4, where drivers engaged in a non-driving related task had to take over control and avoid to the obstacle by changing from the center to the right lane, as the left lane is blocked by fast dense traffic.

Use case 2 "Avoidance of collision at X-intersection", starting on a rural road in SAE level 3/4, where drivers engaged in a non-driving related task had to take over control and avoid a collision with a vehicle coming from the right.

Since the use cases are already very detailed here, they could be considered as use situations. In order to maintain the conceptual connection to the other chapters, we will nevertheless continue to refer to use cases here.

Each participant experienced both use cases and one of the three cooperation designs:

Design 1 is the baseline: Here, the driver only receives an acoustic takeover request from the automation combined with an immediate dropout/ deactivation of the automated system.

Design 2 is a combination of the first design with an MRM (Minimum Risk Maneuver). If the driver does not intervene after the drop out, emergency braking is automatically initiated.

Design 3 is a more complex attention sensitive design that combines the ideas of the confidence horizon: On the one hand, the driver's ability to take over is determined by his orientation reaction, as proposed in the diagnostic TOR approach [19]. On the other hand, the capabilities of the automation are derived from the tested use cases. If the driver is classified as not ready to take over, a second warning stage is initiated. Here, depending on the human's reaction to the TOR, her or his ability to execute the driving task, and the time remaining before the accident, the interaction mediator decided to either immediately return control to the automation, wait until the human was ready to take over, or immediately transfer control to the human. Thus, the time advantage resulting from the detection of the readiness to take over (see chapter by Herzberger et al. [18]) is used to either trigger a second warning, with a still possible strong MRM, or an early and comfortable MRM. As in design 1 and 2, the driver in design 3 receives a TOR that is combined with visual warnings, based on the results from the exploration (see Fig. 8), in the HUD.

The photo at the bottom of Fig. 8 shows the HMI from design 3 in the highway use case with the broken-down vehicle. Here, the left lane, which is occupied by fast moving traffic, is covered by a semi-transparent red wall. In addition, a hands-on symbol is displayed above the road, along with the text "please take over" (in German). Starting from the ego-vehicle, a possible safe trajectory to the right lane is suggested by a green turn arrow. The clear right lane is also indicated by a green check mark at the bottom right of the windshield. In both designs with MRM (design 2 and design 3), the emergency braking can be overridden and it does not start until it is detected that the driver is not responding to the TOR. Figure 9 shows a tree or state-transition diagram of the three designs.

$N = 20$ subjects participated in the study (45% female). The age of the participants ranged from 18 to 54 years ($M = 28.90$ years, $SD = 12.57$ years). The results of the Karolinska Sleepiness Scale (KSS) as well as the Sofi scale, which measure the fatigue of test subjects, did not differ significantly between the takeover design groups. Subjects were randomly assigned to the use cases intersection and highway and to the designs, resulting in each subject experiencing one design and both use

Fig. 8 Simulator study of confidence horizon designs in two use cases ($N = 20$, Snapshots from the gaze scene video, blue dot represents drivers gaze)

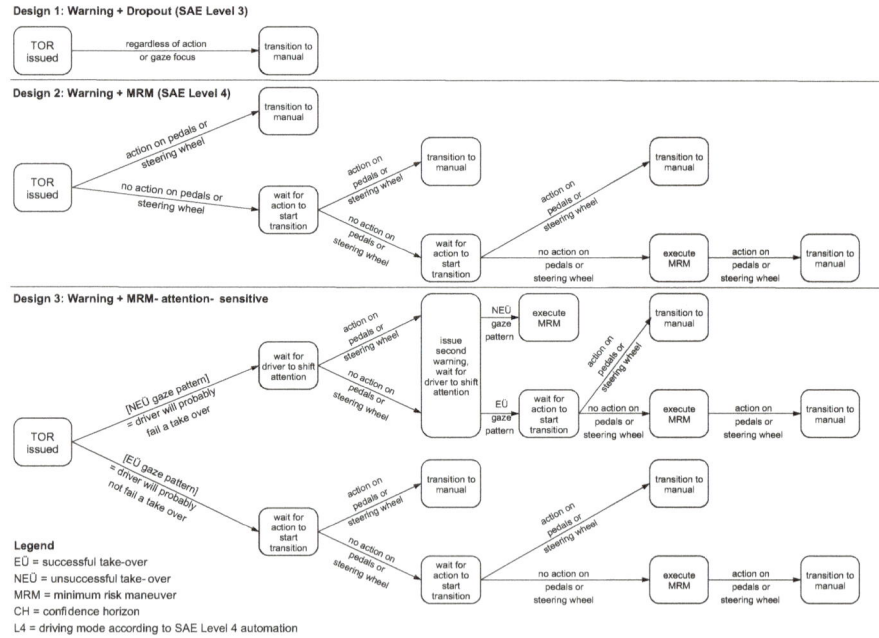

Fig. 9 Tree representation of the three designs of cooperation patterns

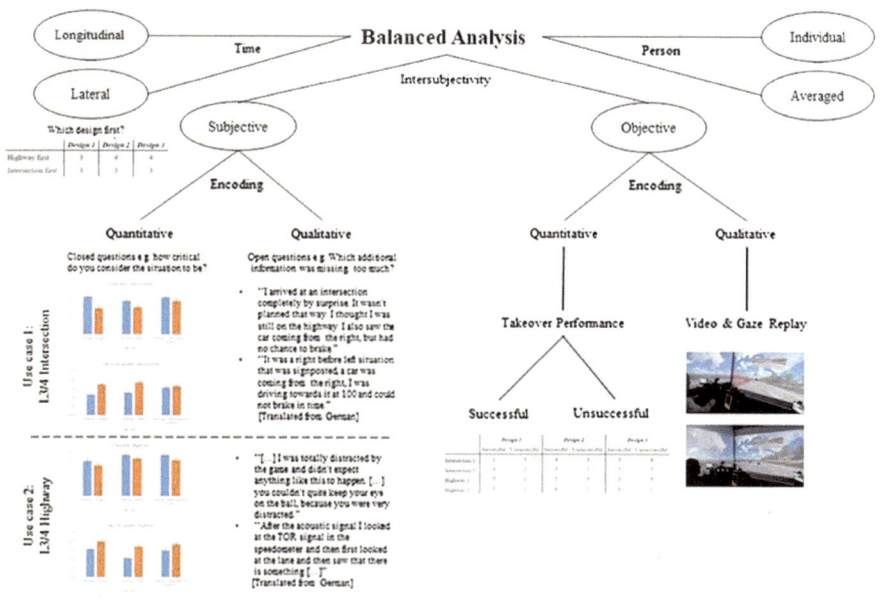

Fig. 10 Principle of balanced analysis in the example of the driving simulator study

cases. The distribution of subjects was carefully balanced so that, as far as possible, there were an equal number of subjects in each design and in each possible use case sequence combination. $n = 6$ were assigned to design 1, $n = 7$ to design 2 and $n = 7$ to design 3. All subjects experienced each use case twice. The first use case trial is referred to as t_1 and the second trail as t_2.

6 Results and Discussion

The evaluation was carried out in accordance with the principle of balanced analysis, which combines and balances subjective with objective, quantitative with qualitative, individual with averaged, and time-longitudinal with time lateral perspectives (see Fig. 10, e.g. Flemisch et al. [12]).

The subjective data are further subdivided into results from the closed and open questions (quantitative vs. qualitative). An extraction of the objective results is shown in Table 1. Here, the takeover success by design and use case is presented.

Not surprisingly, the results reveal that across all designs and situations, subjects took over more successfully at t_2 than at t_1. Contrary to the hypothesis that subjects in design 3 were fundamentally more successful in taking over the driving task than in designs 1 and 2, it appeared that design 3 performed better than design 1 only

Table 1 Takeover success by design and use case

	Design 1		Design 2		Design 3	
	Successful	Unsuccessful	Successful	Unsuccessful	Successful	Unsuccessful
Intersection t_1	1	5	4	3	3	4
Intersection t_2	4	2	5	2	5	2
Highway t_1	5	1	2	5	2	5
Highway t_2	5	1	6	1	6	1

in the intersection use case. In the highway use case, however, the results were inverse, indicating an effect of the cooperation design, or of the experimental design. However, these influencing effects need to be investigated in more detail to avoid potential side effects of the more complex attention-sensitive design, and realize the true potential of the concepts, already seen in the results in one of the two use cases, in the future for all use cases.

Analysis of data related to driver ability in both use cases and all designs was conducted based on aggregated data sets, as shown exemplary in Fig. 11. The data

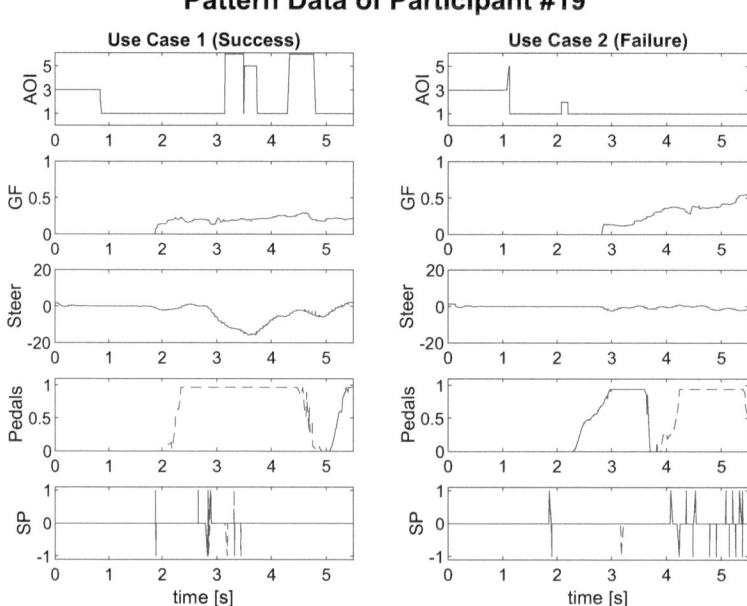

Fig. 11 Example of a data set for a single participant. (AOI = gaze area of interest, 1 = front view, 2 = instrument cluster, 3 = center stack, 4 = mirror left, 5 = mirror right, 6 = rear mirror; GF = sum of normalized grip force activation; Steer = steering angle [deg]; Pedals = normalized pedal activation (straight line: accelerator, dashed line: brake); SP = change of seat pressure focus point (straight line: longitudinal coordinate, dashed line: lateral coordinate))

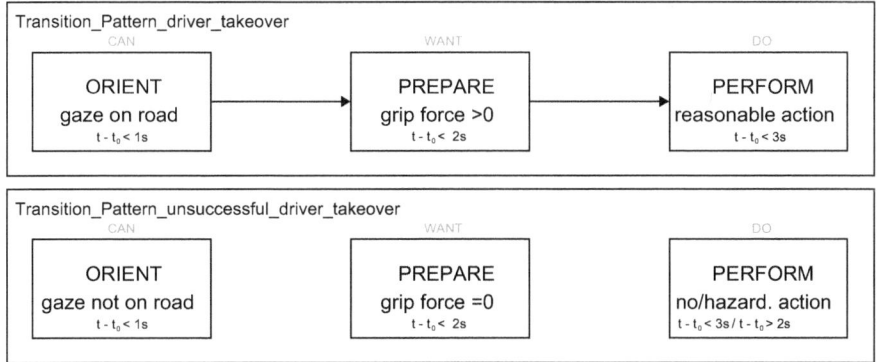

Fig. 12 Top: Structure of the successful takeover pattern. Drivers follow the pattern of orient, prepare, then perform. Time annotations of each interaction block shows the time window in which the event has to occur for the takeover to be still successful in both use cases, with t = observation time and t_0 = time point that TOR is issued by the automation. Bottom: Structure of the unsuccessful pattern. The driver fails to fulfill the subpatterns in the given time frame. Activation of subpatterns does not necessarily follow an order in this case

set consists of gaze AOI (area of interest) data, grip force on the steering wheel, steering angle, pedal activation and seat and seat back pressure. Data sets were evaluated to find a most universal pattern, which describes the ability or inability of the human driver to takeover control after the TOR was issued.

Regarding the ability of the driver, results indicate a possible detection of the inability to take over. Gaze behavior shows, that only 11.7% of successful drivers did take a look at any mirror more than once and tend to have a stable gaze on the road, which tends to lead to a successful takeover, however, it does not guarantee it.

While the initial driver gaze gives a hint on the early orientation behavior of drivers, its analysis also leads to the conclusion, that a successful takeover is not describable by driver gaze alone, hence more data points (c.f. Fig. 11) were added to the analysis.

The combination of gaze, grip force and driver input (pedals and/or steering wheel) leads to a first model of a pattern for the successful[2] control transition to the driver after the TOR was issued by the automation. Figure 12 displays the successful (Fig. 12 top) and unsuccessful (Fig. 12 bottom) pattern found. 87% of all successful drivers followed the successful transition pattern, while 95% of all unsuccessful drivers followed the unsuccessful pattern, which hints towards a better performance of the unsuccessful pattern. Focusing on the orientation and preparation stages of the pattern alone, still 82% of both successful and unsuccessful transitions are being detected.

---

[2] Successful means in this context that the driver took over and resolved the situation without causing a crash of the ego vehicle or other vehicles.

This analysis and first pattern model give an orientation on how to implement the human part of the confidence horizon, however, the transfer from post-processing to an online detection of the confidence horizon still has to be made. A more detailed report on the analysis and found pattern will be published in the near future [26].

The subjective, qualitative results from the balanced analysis provided a variety of indications for possible causes as well as further adaptation options for the HMI. For example, several subjects from all designs ($n = 6$) stated that they would like to see a TOR notice on the tablet. Furthermore, a clearer description of the hazard situation via a voice output instead of just a sound was desired ($n = 4$). The participants' statements on perceived criticality, subjectively perceived takeover quality, and stress did not differ significantly between the designs, which is probably due to a small sample size. A detailed evaluation of the results and recommendations for the further development of the confidence horizon concept will be published in the near future.

7 Conclusion and Outlook

The initial concept of confidence horizon, in conjunctions with new ideas of diagnostic take over requests (described in more detail in the chapter by Herzberger et al.), helped us to open up a new direction of attention and ability sensitive design of automated and cooperative systems. The concept can support design and development teams in cooperative vehicle automation, but also in other domains where machines and humans cooperate, to dynamically balance abilities of agents, and to design and engineer the transitions of control in a more transparent way compared to the traditional "on/off"-thinking. With design explorations and experiments, some of which were described here, we were able to cut through a vast design and use space at least in the driving simulator, and to identify the most prominent dimensions of the vast space of possibilities. Even if we are far from really mastering this new space of attention- and ability-based transitions, the chances are good that in close cooperation with other research projects e.g., from the DFG priority program CoInCar, the first design patterns can already be transferred to real vehicles and products. Equally important, we have paved the ground for more research which will be necessary to fully master this design and use space of transitions, as an important aspect of cooperatively interacting vehicles and human machine cooperation.

Acknowledgements This publication was funded within the Priority Programme 1835 "Cooperative Interacting Automobiles (CoInCar)" of the German Science Foundation (DFG).

References

1. Alexander, C., Ishikawa, S., Silverstein, M.: A Pattern Language: Towns, Buildings, Construction, Center for Environmental Structure Series, vol. 2. Oxford University Press, New York, NY (1977)
2. Baltzer, M.C.A.: Interaktionsmuster der kooperativen Bewegungsführung von Fahrzeugen. Dissertation, Shaker Verlag and Dissertation, RWTH Aachen University, 2020, Aachen, DOI 40345 (2021). https://publications.rwth-aachen.de/record/818952
3. Baltzer, M.C.A.: Interaktionsmuster der kooperativen Bewegungsführung von Fahrzeugen: Lehr- und Forschungsgebiet Systemergonomie/Lehrstuhl und Institut für Arbeitswissenschaft. Dissertation, Shaker Verlag and Dissertation, RWTH Aachen University, 2020, Aachen, DOI 40345 (2021). https://publications.rwth-aachen.de/record/818952
4. Baltzer, M.C.A., Altendorf, E., Meier, S., Flemisch, F.: Mediating Interaction between Human and automation during the arbitration processes in cooperative guidance and control of highly automated vehicles: base concept and first study. In: Ahram, T., Karwowski, W., Marek, T. (eds.) Proceedings of the 5th International Conference on Applied Human Factors and Ergonomics AHFE 2014, AHFE International, Kraków, Poland, pp. 2107–2118 (2014)
5. Borchers, J.O.: A pattern approach to interaction design. In: Boyarski, D., Kellogg, W.A. (eds.) Proceedings of the Conference on Designing Interactive Systems Processes, Practices, Methods, and Techniques—DIS '00, pp. 369–378. ACM Press, New York, New York, USA (2000). https://doi.org/10.1145/347642.347795
6. Flemisch, F.: Pointillistische Analyse der visuellen und nicht-visuellen Interaktionsressourcen am Beispiel Pilot-Assistentensystem. Ph.D. thesis, Universität der Bundeswehr München (2001)
7. Flemisch, F., Herzberger, N., Usai, M., Baltzer, M., Schwalm, M., Voß, G., Krems, J., Quante, L., Trommler, D., Strelau, N., Burger, C., Stiller, C.: Cooperative Hub for Cooperative Research on Cooperatively Interacting Vehicles: Use Cases, Design and Interaction Patterns. Springer (in Press)
8. Flemisch, F., Adams, C.A., Conway, S.R., Goodrich, K.H., Palmer, M.T., Schutte, P.C.: The H-Metaphor as a guideline for vehicle automation and interaction (2003). https://ntrs.nasa.gov/citations/20040031835
9. Flemisch, F., Schieben, A., Kelsch, J., Löper, C.: Automation spectrum, inner/outer compatibility and other potentially useful human factors concepts for assistance and automation. In: de Waard, D., Flemischm, F., Lorenz, B., Oberheid, H., Brookhuis, K.A. (eds.) Human Factors for assistance and automation. Shaker Publishing (2008). https://elib.dlr.de/57625
10. Flemisch, F., Heesen, M., Hesse, T., Kelsch, J., Schieben, A., Beller, J.: Towards a dynamic balance between humans and automation: authority, ability, responsibility and control in shared and cooperative control situations. Cogn., Technol. Work. 14(1), 3–18 (2012). https://doi.org/10.1007/s10111-011-0191-6
11. Flemisch, F., Altendorf, E., Canpolat, Y., Weßel, G., Baltzer, M.C.A., López Hernández, D., Herzberger, N.D., Voß, G., Schwalm, M., Schutte, P.: Uncanny and unsafe valley of assistance and automation: first sketch and application to vehicle automation. In: Advances in Ergonomic Design of Systems, Products and Processes, pp. 319–334. Springer, Berlin, Heidelberg (2017). https://doi.org/10.1007/978-3-662-53305-5_23
12. Flemisch, F., Preutenborbeck, M., Baltzer, M., Wasser, J., Meyer, R., Herzberger, N., Bloch, M., Usai, M., Lopez, D.: Towards a balanced analysis for a more intelligent human systems integration. In: Advances in Intelligent Systems and Computing, pp. 31–37. Springer, Cham (2021). https://doi.org/10.1007/978-3-030-68017-6_5
13. Flemisch, F., Preutenborbeck, M., Baltzer, M.C.A., Wasser, J., Kehl, C., Grünwald, R., Pastuszka, H.M., Dahlmann, A.: Human systems exploration for ideation and innovation in potentially disruptive defense and security systems. In: Advanced Sciences and Technologies for Security Applications, pp. 79–117. Springer International Publishing, Cham (2022). https://doi.org/10.1007/978-3-031-06636-8_5

14. Flemisch, F., Usai, M., Herzberger, N.D., Baltzer, M.C.A., Hernandez, D.L., Pacaux-Lemoine, M.P.: Human-machine patterns for system design, cooperation and interaction in socio-cyber-physical systems: introduction and general overview. In: 2022 IEEE International Conference on Systems, Man, and Cybernetics (SMC), pp. 1278–1283. IEEE (2022). https://doi.org/10.1109/SMC53654.2022.9945181

15. Flemisch, F., Usai, M., Wessel, G., Herzberger, N.: Human system patterns for interaction and cooperation of automated vehicles and humans. at - Automatisierungstechnik **71**(4), 278–287 (2023). https://doi.org/10.1515/auto-2022-0160

16. Gamma, E., Helm, R., Johnson, R., Vlissides, J.: Design patterns: abstraction and reuse of object-oriented design. In: European Conference on Object-Oriented Programming, pp. 406–431. Springer, Berlin, Heidelberg (1993). https://doi.org/10.1007/3-540-47910-4_21

17. Guo, H., Zhang, Y., Cai, S., Chen, X.: Effects of level 3 automated vehicle drivers' fatigue on their take-over behaviour: a literature review. J. Adv. Transp. **2021**, 1–12 (2021). https://doi.org/10.1155/2021/8632685

18. Herzberger, N., Usai, M., Schwalm, M., Flemisch, F.: Cooperation Between Vehicle and Driver: Predicting the Driver's Takeover Capability in Cooperative Automated Driving Based on Orientation Patterns (in Press)

19. Herzberger, N.D., Eckstein, L., Schwalm, M.: Detection of missing takeover capability by the orientation reaction to a takeover request. In: 27th Aachen Colloquium Automobile and Engine Technology 2018, pp. 1231–1240 (2018)

20. Herzberger, N.D., Usai, M., Flemisch, F.: Confidence horizon for a dynamic balance between drivers and vehicle automation: first sketch and application. Hum. Factors Transp. (2022). AHFE International. https://doi.org/10.54941/ahfe1002431

21. Löper, C., Kelsch, J., Flemisch, F.: Kooperative, manöverbasierte Automation und Arbitrierung als Bausteine für hochautomatisiertes Fahren. In: AAET—Automatisierungs-, Assistenzsysteme und eingebettete Systeme für Transportmittel, Net_372work, GZVB, Braunschweig (2008)

22. López Hernández, D., Vorst, D., Baltzer, M.C.A., Bielecki, K., Flemisch, F.: Parts of a whole: first sketch of a block approach for interaction pattern elements in cooperative systems. In: Mařík, V. (ed.) International Conference on Systems, Man, and Cybernetics. IEEE (2022)

23. Rhede, J., Wäller, C., Oel, P.: Der FAS Warnbaukasten. Strategie fuer die systematische Entwicklung und Ausgabe von HMI-Warnungen. In: 6. VDI-Tagung Der Fahrer im 21. Jahrhundert, VDI Verlag, Düsseldorf, VDI-Berichte (2011). https://trid.trb.org/view/1217567

24. SAE: SAE International Standard J3016: Taxonomy and Definitions for Terms Related to Driving Automation Systems for On-Road Motor Vehicles (2021)

25. Sebanz, N., Bekkering, H., Knoblich, G.: Joint action: bodies and minds moving together. Trends Cogn. Sci. **10**(2), 70–76 (2006). https://doi.org/10.1016/j.tics.2005.12.009

26. Usai, M., Herzberger, N., Flemisch, F.: Understanding human ability and intention to improve cooperative automated driving takeovers following a pattern approach, submitted to IEEE SMC 2023 (2023)

27. Winkler, S., Werneke, J., Vollrath, M.: Timing of early warning stages in a multi stage collision warning system: drivers' evaluation depending on situational influences. Transport. Res. F: Traffic Psychol. Behav. **36**, 57–68 (2016). https://doi.org/10.1016/j.trf.2015.11.001

28. de Winter, J.C.F., Happee, R., Martens, M.H., Stanton, N.A.: Effects of adaptive cruise control and highly automated driving on workload and situation awareness: a review of the empirical evidence. Transport. Res. F: Traffic Psychol. Behav. **27**, 196–217 (2014). https://doi.org/10.1016/j.trf.2014.06.016

Cooperation Behavior of Drivers at Inner City Deadlock-Situations

Nadine-Rebecca Strelau, Jonas Imbsweiler, Gloria Pöhler,
Hannes Weinreuter, Michael Heizmann, and Barbara Deml

Abstract In urban traffic, there are several situations in which the right of way is not regulated. For automated vehicles in mixed traffic to show behavior that is considered acceptable by all parties, the cooperation behavior of drivers in these situations must be understood. An observational study identified several behaviors in these situations at equal narrow passages and T-intersections that can be classified as offensive and defensive. These behaviors were tested in an experiment whether they can communicate the intention to drive or to stop. Drivers respond to defensive behaviors of the cooperation partner by continuing to drive, and stopping when the behavior is offensive. In the equal narrow passage, drivers felt safest when they drove first, whereas at the T-intersection, drivers felt safest when the cooperation partner drove first. In further experiments, it was shown that at T-intersections the entry position has an influence on whether drivers drive first or stop. Pedestrians or other traffic do not have an influence on the behavior. However, if drivers follow a vehicle that is driving ahead of them, they drive first through the deadlock situation. Recommendations for the behavior of automated vehicles in these situations are derived from the findings of the studies.

1 Introduction

The introduction of highly automated vehicles in the coming years holds great potential for road safety [44]. Nevertheless, potential problems can also arise, especially in mixed traffic of manual and automated vehicles. This is particularly critical in inner-city traffic, for example at intersections, where there is a higher risk of accidents [38]. These critical situations include situations that are not clearly regulated by road traffic regulations. Here, the behavior of other road users must be predicted in each case in order to then cooperate adequately. These deadlock situations occur, for example, at equal narrow passages or T-intersections with a certain constellation of road users (Fig. 1). Here, none of the drivers has the right of way and the situation

N.-R. Strelau (✉) · J. Imbsweiler · G. Pöhler · H. Weinreuter · M. Heizmann · B. Deml
Karlsruhe Institute of Technology, Karlsruhe, Germany
e-mail: nadine-rebecca.strelau@kit.edu

© The Author(s) 2024
C. Stiller et al. (eds.), *Cooperatively Interacting Vehicles*,
https://doi.org/10.1007/978-3-031-60494-2_19

545

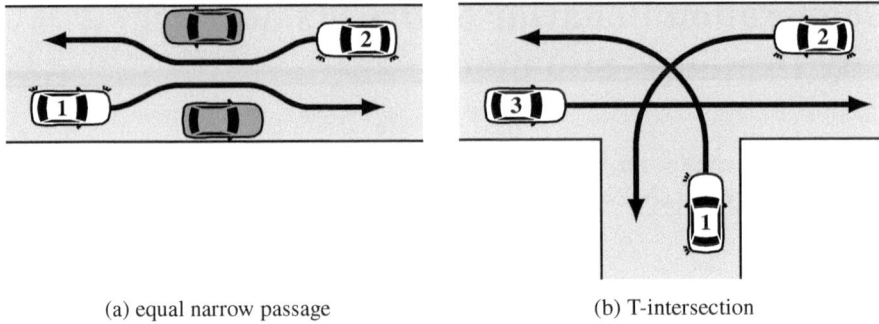

(a) equal narrow passage   (b) T-intersection

Fig. 1 Examples of deadlock situations

must be resolved by cooperative behavior. In this case, an automatic vehicle guidance must also be able to solve the situation cooperatively and recognize the intention of the manual drivers. At first glance, the safest solution would be for the automated car to stop and allow the other road users to drive first [9]. However, this behavior is not necessarily the most comfortable and accepted behavior. For example, if another road user is driving defensively and wants to yield the right of way, it would cause a complete stillstand. This undesirable and possibly unexpected behavior could degrade the acceptance of automated vehicles by both the passenger and interaction partner. In order to cooperate successfully, it is necessary to correctly classify the behavioral decisions of other road users and to be able to predict the intention of the other driver. If the behavior of the other person is not correctly anticipated, this can lead to unsuccessful cooperation, which in turn can lead to conflict [42]. In deadlock situations, it is therefore necessary for both manual drivers and automatic vehicle guidance to recognize the intention of the other in order to resolve this situation successfully. The intention of the other can be recognized by communication signals. Therefore, the first goal of our project was to understand the communication of drivers in the two presented deadlock situations and thus to be able to predict the intention of the drivers. Since one of our findings was that the complexity of the situation could affect cooperation behavior, as a second step, the influence of complexity in the sense of the presence of other traffic participants on cooperation behavior was investigated. Based on the findings, recommendations for the behavior of automated vehicles can be derived.

The chapter is structured as follows: First, an overview of the theoretical background of communication in road traffic is given. Then, two studies are described that examine communication in deadlock situations. Furthermore, the theoretical background on the influence of other road users on driving behavior in the context of complexity and the conducted studies are described. Finally, recommendations are given for the behavior of automated vehicles in deadlock situations.

2 Communication in Road Traffic

Communication is a necessary requirement for successful cooperation [29]. According to Hoc [15], cooperation occurs when two agents interfere on goals or resources and try to manage these interferences to facilitate each individual goal. Because communication is also necessary for cooperation, no cooperation takes place when two road users meet briefly on a road from the oncoming direction, for example. The same resource is shared, but communication does not take place [23]. In contrast, the two deadlock situations to be investigated fulfill the requirements of a cooperative situation. All road users need to use the same part of the road and need to communicate in some way to resolve this situation without conflict by agreeing who will drive first. Drivers must effectively communicate with each other so that they can understand and predict each other's intentions and actions. If this is not successful, the situation cannot be understood correctly and thus the drivers cannot react appropriately in the situation [23]. Furthermore, the way in which drivers communicate with each other can have an influence on their decisions to act [49]. Since the communication of the intentions is essential in these cooperative situations, it is important to understand how drivers communicate.

In general, communication is understood as the exchange of information. A sender transmits a signal or message to a receiver who is intended to get this information [36]. The challenge in transmitting the information is to ensure that the signal sent arrives correctly at the receiver [35]. A characteristic of human communication is that both verbal and nonverbal channels can be used. Nonverbal communication through gestures and body language can be used for very fast communication [29] and can also initiate, coordinate, or be used to avoid cooperation [5]. However, these general findings of communication cannot be fully applied to communication in road traffic. Road traffic is a volatile system in which situations can be dynamic and very complex. In order to cope with the complexity of situations in road traffic, road users apply schemata in which behavior is concluded from road user characteristics or they signal the expected action to other road users. A different strategy is to wait in a situation and first gather as much information as possible about it or to follow the actions of other road users [30]. Another limitation in road traffic is the limited options for communication, which can lead to misunderstandings [41], as communication is limited to the nonverbal level and is also anonymous [30]. Furthermore, drivers cannot escape the situation. This means that any actions at any time can be interpreted by others as a communication signal. It can thus lead to both intentional and unintentional communication in road traffic [3]. In the context of everyday communication, the axiom of Watzlawick, Bavelas, and Jackson [46] is applicable here, which states that one cannot not communicate. Since there is always communication, it can further be divided into explicit and implicit communication. This division can also be assigned to the different driving tasks according to Geiser [13]. Here, the driving task is divided into three subtasks. The primary driving task includes navigation as well as maneuvering and stabilizing the vehicle. The secondary driving task supports the primary driving task and mainly serves to inform other road users and to react to environmental conditions. This includes, for example, using the indicator or the headlight flasher. The tertiary driving task is independent of the actual driving task and serves primarily to increase comfort by, for example, operating the air conditioning or radio. It can be concluded that the implicit communication through the driven trajectories falls under the primary driving task, while the explicit communication belongs to the secondary driving task. According to this categorization, implicit communication always takes place, while explicit communication is always just an additional form of communication that can support implicit communication.

Communication in road traffic can only take place non-verbally and this can be a challenge. Different areas of non-verbal communication can be distinguished [1, 9, 33]: Facial expression and eye contact, gestures and body movements, voice and manner of expression, spatial behavior, and technical signals. Merten [30] proposes the thesis that without eye contact no communication can take place in road traffic. However, Witzlack, Beggiato, and Krems [48] showed that eye contact is often overestimated in driver-pedestrian interactions. Eye contact only served as confirmation and is thus not a necessary requirement. Moreover, in mixed traffic, eye contact would not be helpful under certain conditions, for example, when the driver of the automated vehicle is looking at the traffic but is not involved in the driving task [9]. Among the most commonly used gestures are gestures indicating to other road users that they should slow down, that they can or should drive, and that the driver's own right of way is being yielded [9].

Kitazaki and Myhre [25] showed that at intersections, using a hand gesture combined with the vehicle behavior showed larger effects on the drivers' anticipation of intention and therefore decision compared with the vehicle behavior alone. Possible technical signals that drivers can use include the turn signal, horn, headlight flashers, and hazard warning lights [33]. Ba, Zhang, Reimer, Yang, and Salvendy [2] investigated these explicit signals (with the exception of the headlight flasher) for different traffic situations. They found that drivers prefer when the other driver uses an explicit signal. However, even without an explicit signal, subjects can recognize the intention of drivers. Lee and Sheppard [26] showed subjects both pictures and videos of a vehicle approaching an intersection that would either continue straight or turn. The vehicles used a valid or invalid turn signal. Even though the subjects were better to judge the behavior of the vehicle when it gave a valid signal, in most cases they were also able to correctly judge the behavior despite the invalid signal. Thus, the explicit signal is helpful to estimate the intention of a driver, but not necessary. This is also supported by the finding that the intention could be better estimated in the videos than in the pictures. Drivers therefore also use the dynamic behavior of the vehicle, such as the braking behavior, to estimate the intention.

When looking at all these explicit signals, it becomes clear that they cannot be used alone, but only in combination with other signals. This is especially true for spatial behavior, since the driver is moving on the road at all times and thus the driven trajectory can be interpreted as a communication signal at any time. The driven trajectory from lateral and longitudinal driving behavior is considered to be implicit communication. In some situations, this is even more meaningful than the explicit signals [48]. Especially longitudinal behavior, i.e. approaching, is used for intention detection. For example, when changing lanes, deceleration behavior, speed reduction, and reaction speed in particular are used as indicators of cooperative behavior [24]. At intersections, when the other driver maintains speed or accelerates, drivers expect the other to proceed through the intersection. In contrast, when slowing down, the vehicle is expected to yield and stop [4].

Implicit behavior is especially interesting for automatic vehicle guidance to predict the intention of other road users. One reason for this is that it is technically easier to interpret implicit behavior rather than explicit signals, which may not be used consistently, especially in situations that are not clearly defined, such as deadlock situations. In mixed traffic, it is crucial that both automated vehicle guidance and human drivers are able to recognize each other's intentions. This is especially true in deadlock situations, where the intention must be anticipated in order for the situation to be resolved. Since intentions can be communicated via both explicit and implicit signals, an automatic vehicle guidance system must be able to interpret both in order to react appropriately. At the same time, it should also be able to use the signals itself to display behavior that the human cooperation partner expects. There has not yet been sufficient research on how intentions are communicated and which combination of possible signals is used in deadlock situations. In a first step, the use of communication signals at intersections and equal narrow passages was investigated.

2.1 Observational Study

The aim of the observational study [17] was to identify cooperative behavior, to classify it into offensive and defensive behavior, and to derive behavioral sequences from the behavioral patterns. For this purpose, a T-intersection and an equal narrow passage in Karlsruhe (Germany) were each observed for five hours by two and three trained observers, respectively. The behavior of the cooperating drivers were recorded: the order of arrival and departure, the direction of driving (right, left, straight ahead), driving behavior (acceleration, deceleration, stopping, maintaining speed), and explicit signals (turn signal, horn, gesture). Analysis of individual gestures as well as the recording of drivers' gaze direction was omitted, as this is difficult to observe and technical aids could not be used for data protection reasons. The observations of the individual observers were combined afterwards in order to extract behavior sequences for the individual situations.

A total of 33 events with 12 different traffic situations could be observed at the T-intersection. The results of the observation showed that explicit communication plays a minor role. In fact, 71 implicit signals were observed in contrast to only 32 explicit signals. Of these explicit signals, the indicator was mostly used to indicate turning. Of the behaviors, defensive behaviors such as stopping and braking were more frequently exhibited than offensive behaviors such as accelerating or maintaining speed. In particular, left-turning was associated with a defensive behavior pattern, while right-turning showed more offensive behavior. For the deadlock situation at the T-intersection, six different situations could be observed. These could be classified into defensive and offensive behavior patterns and were used in the following experiment (see Sect. 2.2).

At the equal narrow passage, 40 events could be registered. As at the T-intersection, the observation showed that explicit communication takes a minor role compared to implicit communication. The most frequently used explicit signal was the headlight flasher. When drivers arrived first at the narrow passage but drove second, defensive behaviors were mainly exhibited and they stopped. Conversely, drivers who arrived second at the equal narrow passage but drove first could be observed to exhibit mainly offensive behaviors such as maintaining speed and accelerating. Our findings from the T-junction and equal narrow passage suggest that implicit communication plays an important role in deadlock situations in order to be able to recognize the intentions of the other drivers. Furthermore, behavioral sequences can be classified well into offensive and defensive behaviors.

2.2 Experiments

The behaviors identified in the observational study were further examined in two experiments to test whether they are suitable for conveying intention and whether it is possible to determine from the behaviors whether drivers want to drive or stop in deadlock situations. For this, subjects drove a test vehicle on a traffic training area through a deadlock situation at an equal narrow passage [19] and a T-intersection [20] (Fig. 2). The cooperation vehicles in these situations were driven by one respectively two instructed examiners, who followed predefined behavior scripts. These behaviors were intended to convey the intentions to drive or not drive and to represent offensive or defensive behaviors. For the defensive behavior, the examiner was to stop and let the subjects drive first. For the offensive behavior, on the other hand, the examiners were asked to drive through the equal narrow passage or T-intersection first, if the subject's behavior allowed. For both the equal narrow passage and the T-intersection, six different situations were presented, each with three offensive and three defensive

Fig. 2 Deadlock-Situation at the T-intersection during the experiment at the traffic training area

Table 1 Approaching behaviors of the examiners at the T-intersection [20]. The number of the examiner describes the position in the T-Intersection (see Fig. 1)

Situation	Behavior classification	Behavior
1	Defensive	Examiner 3 decelerates, stops, uses flasher Examiner 1 or 2 indicates and stops
2	Defensive	Examiner 3 decelerates and stops Examiner 1 or 2 indicates and stops
3	Defensive	Examiner 1 or 2 decelerates and indicates Examiner 1 or 2 decelerates, indicates, uses flasher
4	Offensive	Examiner 3 maintains speed Examiner 1 or 2 indicates and decelerates
5	Offensive	Examiner 3 decelerates Examiner 1 or 2 indicates and decelerates
6	Offensive	Examiner 1 or 2 decelerates and indicates Examiner 1 or 2 indicates, decelerates, uses gesture

behaviors. Each situation was driven through twice, resulting in a total of 12 runs through the intersection or equal narrow passage for each subject. For the equal narrow passage, the defensive behaviors of the examiners were: 1. stopping distinctly, 2. braking to a speed of 15km/h and using the flasher, 3. stopping distinctly and using the flasher. The offensive behaviors were: 1. maintaining speed, 2. accelerating, 3. braking to 15km/h and continuing to drive toward the equal narrow passage. The behaviors of the examiners for the T-intersection are shown in Table 1.

After driving through all situations, subjects were shown video clips of their driving and asked to rate how confident they were to drive first or second, how high they perceived the risk of an accident and the willingness of the involved drivers to cooperate. During the drive, the CAN bus data of the test vehicle were also recorded, as well as the eye movements of the subjects using an eye tracker. The results of these data can be found in [18, 21]. In total, the experiments lasted approximately 90 min. For the equal narrow passage, 22 subjects (21 males, average age $M = 23.91$, $SD = 2.10$) were surveyed, for the T-intersection 20 subjects (18 males, average age $M = 23.35$, $SD = 3.51$).

2.2.1 Results and Implications for the Situation "Equal Narrow Passage"

For the equal narrow passages, defensive behaviors by the examiners resulted in a significantly higher probability of subjects driving first rather than stopping. For defensive behaviors 1 and 2, the subjects had a probability of 83% to drive first through the equal narrow passage, for behavior 3 the value was 75%. For the offensive behaviors, the probabilities of driving first were significantly lower at 31% (behavior pattern 4), 35% (behavior pattern 5), and 9% (behavior pattern 6). Furthermore, for the

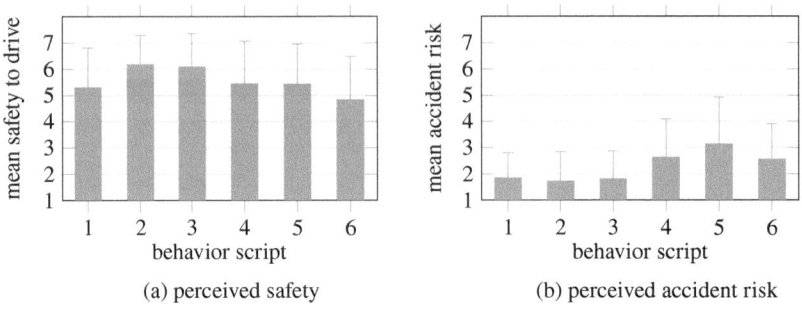

Fig. 3 Perceived accident risk and safety to drive for the different behavior scripts at the equal narrow passage

different behaviors, there were significant differences in how confident the subjects were in driving first or second ($F(1, 3.426) = 4.42$, $p < .05$). They felt the safest when the cooperative vehicle braked and flashed its lights (defensive behavior). In contrast, they felt least safe when the oncoming vehicle slowed down from 30 to 15 km/h (offensive behavior) (see also Fig. 3a). The perceived accident risk also differed significantly between the different behaviors ($F(1, 3.221) = 6.942$, $p < .001$) (Fig. 3b). In particular, when the cooperation vehicle accelerated before the equal narrow passage, the accident risk was perceived to be significantly higher compared to the defensive behaviors. For the perceived willingness to cooperate, there were significant differences between the behaviors, $F(1, 5) = 14.096$, $p < .001$. Defensive behaviors of the cooperation partner were perceived to be more cooperative than offensive behaviors. Perceived willingness to cooperate further influenced whether subjects would drive first or yield the right-of-way to their counterparts. When perceived willingness to cooperate is considered very low, the probability of driving first decreases significantly even for the defensive behaviors. For the offensive behaviors, the probability of driving first increases accordingly if the behavior is perceived as cooperative.

Overall, the results of the study show that all six behavior patterns produce the expected behavior in the cooperation partner and are therefore also suitable for producing a desired behavior in a certain situation. An important requirement is that the behavior is perceived as cooperative. This works very well with behavior pattern 2, for example, braking with the headlight flasher. Here, the behavior is perceived as cooperative, the drivers drive first with a high probability, and are very confident in their decision to do so. Thus, this behavior seems to be suitable for automatic vehicle guidance at an equal narrow passage. Behavior 5, acceleration, on the other hand, also shows the expected behavior: the drivers stop and let the other person drive first. With this decision, they also feel safe. However, this behavior is not perceived as cooperative. Since automatic vehicle guidance should probably be perceived as cooperative in order to be accepted, such behavior would likely be inappropriate for automatic vehicle guidance.

2.2.2   Results and Implications for the Situation "T-Intersection"

As in the case of the equal narrow passage, the different behavior patterns of the
cooperation partners showed a significant influence on whether the subjects drove
through the T-intersection first or gave way to one of the other two vehicles. For
defensive behaviors 1 and 3, the probabilities of driving first were relatively high
at 58% and 74%, respectively. Defensive behavior 2, on the other hand, failed to
produce the expected behavior. Here, the probability of driving first was only 5%. The
offensive behaviors showed lower probabilities to drive first with 56% (behavior 5),
8% (behavior 4) and 30% (behavior 6). The increased probability of behavior 6 may
have been due to the fact that some subjects thought they were driving on a main road.
Thus, with the exception of one behavior, a desired behavior can be achieved among
drivers at a T-intersection, similar to the equal narrow passage. Subjective evaluations
of the situation also showed differences between the different behavior patterns. For
the offensive behaviors, subjects were significantly more confident with their decision
to drive than for the defensive behaviors ($\chi^2(5) = 621.776$, $p < .001$). Additionally,
the order in which subjects drove through the intersection had an influence. When
subjects drove second or third, they were significantly more confident in this decision
than when they drove first. There was also a significant difference between behaviors
for perceived cooperativeness ($\chi^2(5) = 5.190$, $p < .001$). The offensive behaviors
were rated less cooperative than the defensive behaviors. The accident risk, on the
other hand, is estimated to be the same for all behaviors. A possible reason for this
could be that in many cases the vehicles were in a standstill, thus minimizing the
objective risk of accidents. In addition, this experiment aimed to investigate whether
the subjects themselves gave explicit signals. Explicit signals were used in only 42
of the 240 cases. It is noticeable that half of them were given to the right cooperating
partner, i.e. the one who has the right of way over the subject anyway according to
the road traffic regulations. This could be an indication that the deadlock situation
is sometimes not correctly understood and that accordingly no adequate strategy is
used to resolve the situation.

2.2.3   Conclusion from the Experiments

In summary, both at the equal narrow passage and at the T-intersection, certain offensive as well as defensive behaviors influence the behavior of drivers, i.e. whether they
stop in a deadlock situation or drive first. There is an interesting difference in terms
of perceived safety for driving: subjects rate their confidence of driving at the equal
narrow passage higher when the cooperation partner shows defensive behavior. In
contrast, at the T-intersection, confidence of driving is rated higher when the cooperation partners show offensive behavior. In other words, it can be concluded that
drivers prefer to drive first themselves at the equal narrow passage, while they prefer
to give way to other vehicles at the T-intersection. One difference between the two
situations is their complexity. At the equal narrow passage, only one lane needs to be
considered and there is only one cooperation partner, making it a relatively simple sit-

uation. In contrast, at the T-intersection, two lanes and two cooperation partners must be considered. Therefore, it is apparent that in more complex deadlock situations, drivers prefer not to drive first. Following this logic, drivers should show different cooperation behavior within the same situation, which differs only in its complexity. This was further investigated using the deadlock situation at the T-intersection.

3 Complexity and Driving Behavior

In order to examine the effects of different aspects of complexity on cooperation behavior at deadlock situations, it is useful to look at the definition of complexity in the context of road traffic as well as its effects on drivers' behavior. The influence of complexity on driver behavior and workload has been studied for a variety of situations, but not for cooperation behavior at deadlock situations. Additionally, there is no precise consistent definition or operationalization of complexity among these studies. A basic classification of traffic situations in terms of their complexity was established by Fastenmeier [10]. According to him, the task complexity of traffic situations results from the demands on information processing and vehicle handling. Faure, Lobjois, and Benguigui [11] used this classification to measure the subjective and objective mental workload of drivers. They classified driving on the highway with low demands on both information processing and vehicle handling, driving in rural environment with high demands on vehicle handling and low demands on information processing, and an urban environment that is visually rich with buildings, street furniture, traffic lights, intersections and roundabouts with high demands on both information processing and vehicle handling. Driving on the highway showed the lowest mental workload according to both the subjective ratings of the participants as well as eyetracking and steering wheel parameters. The results for rural and urban environment were not quite as expected. Although the subjective workload was higher for the urban environment, the eyetracking and steering wheel parameters indicated a higher workload for the rural environment. A possible explanation according to the authors is that there were only few intersections or roundabouts on the urban roads and thus the demands on vehicle handling were low. In contrast, on the rural roads there were many sharp curves, which resulted in very high demands on vehicle handling. Similarly, Oviedo-Trespalacios, Haque, King, and Washington [31] found that sharp bends on roads increase task demands and drivers therefore adapt their speed more compared to straight roads. Another reason given by Faure et al. [11] is that there were no other road users in their urban environment and therefore little information processing was required. This is in contrast to the results of Oviedo-Trespalacios et al. [31], who showed that in urban areas a greater speed adaptation takes place due to higher demands compared to suburban areas, even if in both no other traffic is present.

Jahn, Oehme, Krems, and Gelau [22] also used Fastenmeier's [10] classification, but interpreted it differently from Faure et al. [11]. Like Faure et al. [11], they classified highway as a situation with low demands on information processing and vehicle handling. In contrast to Faure et al. [11], however, they classified a rural environment and urban environment without interactions with other traffic with low requirements on both dimensions. According to them, only city centers and complex intersections are situations with high requirements for information processing and vehicle handling. They also found the expected differences in mental workload between these two differently defined complexity groups. However, since Faure et al. [11] also found differences between highway and rural environment, this division is apparently not sufficient. Patten, Kircher, Östlund, Nilsson, and Svenson [32] used another group in addition to the two groups of low and high demands: they defined situations with high demands on information processing and low demands on vehicle handling and vice versa, such as intersections regulated by traffic lights or by road signs where the driver has the right of way. Compared to the situations with low demands in both categories, drivers who do not drive much showed significantly longer reaction times for a peripheral detection task and thus higher workload in these medium situations, but not drivers with high mileage. The latter, on the other hand, showed worse performance in the high demand situations compared to the medium demand situations. Driver experience thus also appears to play a role in how different situations affect drivers. Overall, these studies with the different classifications of complexity and results show that the classification according to Fastenmeier [10] into different traffic situations is not sufficient in that way and can only give a first indication of the complexity. Instead, the exact specific conditions within these situations must also be defined, as these can have a direct influence on driver behavior.

Törnros and Bolling [45] showed this for the urban environment. They found that reaction time of drivers in a complex urban environment is higher than in a medium or low complex urban environment. The high complex urban environment was thereby described with buildings on both sides, pedestrian tracks, car and pedestrian crossings, parked cars and busses. The medium and low complex urban environments, on the other hand, featured only some traffic, parked busses and were residential areas. Drivers also show lower speed and higher subjective ratings of mental workload when driving on streets where buildings and shops are located directly to the sidewalk compared to streets where the buildings are set far back from the road [34]. The same can be shown for areas where cars are parked at the roadside compared to streets with no parking spaces or empty parking spaces [8]. The amount of visual information a driver pays attention to seems to have an influence on behavior. At intersections with more visual information (vehicles, pedestrians, stores, construction site) drivers reduce their speed. On the highway with a lot of visual information (advertisements, billboards, buildings, highway furniture), drivers do not estimate their subjective mental workload higher than for stretches of road with little visual information, but do decrease their speed here nonetheless [16], indicating that there is an effect of those visual information on drivers.

Other road users themselves have an effect on drivers' behavior and workload. For example, traffic congestion causes drivers to behave more aggressively in the section after the congestion than if they had not driven through any congestion [27]. Individual road users also influence drivers. When a vehicle is in front of their own vehicle, drivers adjust their speed more than when they are free to drive on the road [31]. In turn, overtaking a vehicle in front also leads to an increased workload compared to driving freely [6]. Traffic density also has a negative impact on the workload of driving. This is true for both driving on the highway [43] and at intersections [28].

To the authors knowledge, up to now, the influence of complexity on cooperation behavior and especially on deadlock situations has not been studied. Since the studies presented so far only give an indication of the influence of complexity on general driving behavior and, as described, do not provide a comprehensive description or definition of complexity, the task-capability model of Fuller [12] was considered. According to this model, driving behavior can be explained in terms of task difficulty. This is composed of the relative proportion of task demands on the driver's capability. The task demands can have both information input and response output character. The incoming information of the task demands includes a variety of factors such as operational characteristics of the vehicle, route choice, environmental characteristics and other road users. Task demands in the sense of output factors are the driver's own behavior, i.e. speed and trajectory. The capability includes knowledge and skills arisen from training and experience, the mental representation of the situation as well as physiological characteristics like information processing capacity or reaction time. The task difficulty then results from the task demands and the capability, for which in turn each driver has an individual range that they accept. If this threshold is exceeded, compensatory actions are taken to reduce the task difficulty. This is usually done by reducing the speed, i.e. an output function of the task demands. Applied to cooperation behavior at deadlock situations, this would imply that compensatory actions should be taken in more difficult deadlock situations and that drivers should therefore stop rather than proceed through the intersection or equal narrow passage first.

The environmental factors according to Fuller [12] are not broken down in detail. Since this can be essential for describing a situation as described above, the classification of visual clutter according to Edquist [7] was further considered. This concept is closely related to that of task difficulty, but offers a further breakdown of the relevant factors. The visual clutter is divided into objects that must be attended for safe driving and objects that distract from safe driving. The latter, together with the background complexity, are called built clutter and refer, for example, to stores, advertising or infrastructure such as light poles. The objects that must be observed for safe driving can be further subdivided according to this taxonomy. Road markings, traffic signs or signals are referred to as designed clutter. The situational clutter consists of vehicles, cyclists and pedestrians. For the deadlock situation in the present study, the influence of other vehicles and pedestrians on the perceived visual clutter as well as the perceived difficulty was investigated. In addition, the position from which the intersection is entered was considered. At intersections, it has an impact on the workload whether drivers drive straight ahead or turn. This difference was implemented in the methodology of the experiment described above for communication at deadlock situations [20]. Yet, it was not distinguished from where the drivers approached the intersections. However, this position also has an influence on the driving behavior

Fig. 4 The screenshot of a video used in study 1 shows the approach from entry position left with a vehicle in front of the ego vehicle

[14], which is why the entry position to the intersection was additionally taken into account in the following experiments.

3.1 Experiments

To investigate the influence of entry position of the T-intersection and complexity in a controlled setting, two online studies [39, 40] were conducted with 30 and 34 subjects, respectively. The subjects were shown videos of the approach to T-intersections from the driver's perspective, which were created using the driving simulation software SILAB 6.5 (see Figs. 4 and 5). All three possible entry positions to a T-intersection were shown. The two cooperation vehicles as well as the own vehicle decelerated before the intersection and came to a stop at the intersection. One second before this, the videos were cut off so that the situation had not yet been resolved. The subjects were then asked to state how likely they themselves would then be to drive through the intersection first in this situation. In addition, they were asked to rate the perceived difficulty and visual clutter. In study 1 [39], it was varied whether a vehicle passed through the intersection in front of the ego vehicle, whether other vehicles were traveling behind the cooperation vehicles at the intersection, and whether other traffic was seen during the approach to the intersection. Study 2 [40] examined the influence of pedestrians. For this, the number of pedestrians walking on the sidewalk was varied (none, 20, 80). In addition, it was varied whether the pedestrians were walking close to the front of the house or street as well as a barrier separating the sidewalk from the street.

3.1.1 Results

Across all situations, the probability of driving first was relatively low at 24% in both study 1 and study 2 (Fig. 6). The entry position had a significant effect on

(a) many pedestrians with a barrier

(b) few pedestrians without a barrier

Fig. 5 The screenshots of videos used in study 2 show the approach from the left for two different scenarios

whether subjects would drive first through the T-intersection or not (study 1: $\chi^2(2) = 88.14$, $p < .001$; study 2: $F(2, 776) = 64.35$, $p < .001$). When drivers approach from below, the intention to drive first is lowest. Study 2 additionally showed a significant difference between entry positions left and right. The lower probability when approaching from below could be an indication that the deadlock situation was not recognized correctly and that the straight-through road was possibly interpreted as a priority road. This was also the case for some subjects in the previously described study at the traffic training site [20]. In addition, Björklund and Åberg [4] were able to show that this main road effect exists at intersections (however, they did not consider deadlock situations). If one were to assume such a main road effect exists also in deadlock situations, one would expect that drivers from entry position left would drive first, since this would be in accordance with the right-of-way rule of a main road. However, this is not evident in the data from either study. In study 1, there are no differences between the entry positions left and right. In study 2, there is even a higher probability of driving first from entry position right compared to left, i.e. an opposite behavior to the right-of-way rule on a main road. Overall, therefore, there seems to be no accurate understanding of the deadlock situation. An automated vehicle must therefore be aware that manual drivers may think they have the right of way at a deadlock situation. The prediction of the intention through the displayed behavior then plays a special role.

The perceived difficulty of the situation does not differ between the three positions. For the assessment of the visual clutter, however, there are significant differences for the different entry positions (study 1: $\chi^2(2) = 13.461$, $p = .001$; study 2: $\chi^2(2) = 13.58$, $p = .001$). When approaching from below, visual clutter is rated

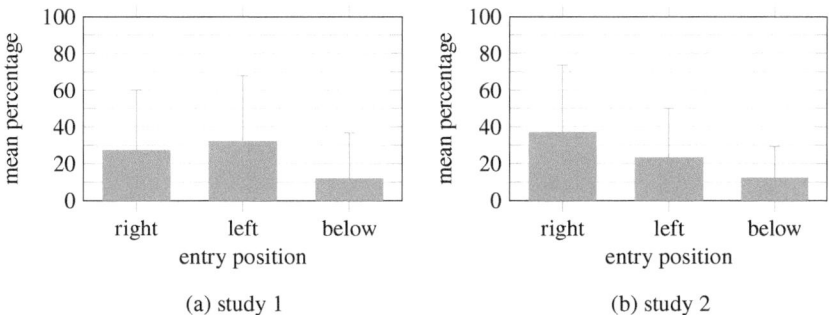

Fig. 6 Probability to drive first through the T-intersection for study 1 and 2

higher than from the other positions; in study 2, visual clutter is also rated higher when approaching from the right compared to the left. This is against the expectation according to Edquist (2008) that drivers should show a lower probability to drive first in situations with high visual clutter. The entry positions below and right were both rated with higher visual clutter than the position on the left. Nevertheless, for these two positions different tendencies for the probability to drive first can be seen: For the position below, a lower probability of driving first is shown, whereas for the position on the right, a higher probability of driving first is shown than for the position on the left. Again, a lack of or incorrect understanding of the situation could be an explanation. Overall, it can be concluded from these results that both perceived difficulty and visual clutter do not seem to be adequate to explain or predict cooperative behavior in deadlock situations.

Situations with no pedestrians (27%) showed on a descriptive level a slightly higher probability to drive first than those with few (24%) or many pedestrians (23%), but this difference was not significant. The zone in which pedestrians walked as well as the barrier did not affect the subjects' indicated behavior. As with the different entry positions, there were no differences in perceived difficulty. For the visual clutter, however, differences could be observed. The presence of many pedestrians increased the perceived visual clutter of the situation in contrast to situations with no pedestrians. However, since the pedestrians had no influence on the cooperation behavior, it can also be assumed that there is no correlation between behavior and visual clutter in deadlock situations.

A vehicle passing through the intersection before one's own vehicle increased the probability that drivers themselves would be the first to pass through the intersection. Further traffic, on the other hand, had no influence. When considering additionally the perceived difficulty and visual clutter, an interesting pattern emerges. A vehicle ahead significantly increases both the difficulty and visual clutter of the situation. In this more difficult and more cluttered situation, the probability of driving through the intersection first increases. The effect of position, on the other hand, showed the opposite pattern. Here, situations that were rated higher in visual clutter showed a lower probability of being the first to drive through the intersection. These results support two findings: First, the concepts of visual clutter and difficulty are not effective in explaining or predicting drivers' cooperation behavior in deadlock situations at T-intersections. Since the complexity of intersections without deadlock can contribute to predicting certain behaviors [47], this seems to be exclusive to deadlock situations. This in turn supports the second finding: deadlock situations do not seem to be comprehended sufficiently or there seems to be an uncertainty about how to proceed in such situations. In an uncertain situation, where one is not clear how to behave, the behavior of others is imitated [37]. This is precisely what can be observed in the deadlock situation: In a situation that is perceived as more difficult, subjects follow the vehicle in front, thus imitating the behavior of another road user.

4 Conclusion

The aim of the work was to better understand the cooperation behavior of drivers in deadlock situations in inner city traffic, so that automated vehicles can show cooperation behavior in mixed traffic of automated and manual vehicles, which is similar to that of humans and thus is accepted by all involved parties. To identify the communication behavior that drivers exhibit in these situations, an observational study was conducted, and these observations were then further investigated in an experiment. In addition, building on the findings from the experiment, video studies were conducted in which the influence of complexity, in this case other traffic and pedestrians, as well as the entry position of the T-intersection was investigated.

The observational study showed that drivers communicate at intersections and equal narrow passages primarily through implicit signals, that is, driving behavior. Explicit communication, on the other hand, plays only a minor role. Of the explicit signals, the headlight flasher was most often used as a sign of defensive behavior, especially at the equal narrow passage. For deadlock situations at both T-intersections and equal narrow passages, several behaviors could be observed that can be classified into offensive and defensive behaviors. These were tested in an experiment to find out whether these behaviors can communicate the intention to drive or stop, and if drivers adjust their behavior accordingly. In both situations, participants showed a higher probability of stopping when the cooperating partner showed offensive behavior. In contrast, when the cooperating partner showed defensive behavior, they were more likely to drive first. Thus, in deadlock situations, drivers are able to recognize the intention of the other person based on his or her behavior and behave accordingly.

A difference between the T-intersection and the equal narrow passage can be seen in the subjective evaluation. At the equal narrow passage, drivers feel safest when the cooperating vehicle shows defensive behavior, and they themselves can drive first and least safe when the cooperating vehicle shows offensive behavior. At the T-intersection, on the other hand, the opposite picture emerges. For the offensive behaviors, subjects were significantly more confident with their decision to drive when the cooperation partner showed offensive behaviors and participants therefore drove as second or third. Thus, for different deadlock situations, there also seem to be different expectations and behaviors. Since the two situations differ in the number of vehicles and lanes involved, it seems that the complexity of the situation has an influence on the cooperation behavior. For the simpler of the two situations, the equal narrow passage, it is therefore easier to make recommendations for the behavior of an automated vehicle.

Drivers prefer to drive first in deadlock situations at equal narrow passages. Accordingly, an automated vehicle should rather show defensive behavior in order to give the manual driver the opportunity to drive first. To show defensive behavior and the intention to stop, a use of the headlight flasher or a clear stop seems to be most suitable for the narrow passage. For the T-intersection, the recommendations for the behavior of automated vehicles cannot be derived quite as clearly. Since drivers tend to prefer to drive second or third here, automated vehicle guidance should tend to show offensive behavior and drive through the intersection first. Since there is no clear explicit signal for offensive behavior, driving behavior must be used here to indicate to the manual driver that the automated vehicle will proceed. For this purpose, it is most suitable to maintain the speed.

At the T-intersection, however, other aspects must be considered as well, as the findings from the video studies indicate. The entry position to the intersection can influence whether drivers stop or not. When approaching from below, most drivers would stop. Automated vehicle guidance encountering a manual driver from the entry position below in a deadlock situation should proceed through the intersection first. One possible reason why drivers from the position below do not drive first is that the deadlock situation is not recognized as such. This is especially important when a manual driver approaches from the positions on the left or right. Here, drivers show a higher probability of driving first themselves. Therefore, the automated vehicle guidance system must expect that manual drivers may want to drive first, because they may assume that they have the right of way in this situation. In its intention detection, the automated vehicle guidance system must therefore recognize whether drivers are approaching the intersection as if they have the right of way or whether there is uncertainty in their behavior that suggests they have recognized the deadlock situation. In the latter case, the automated vehicle should then drive first. Other traffic with which the vehicles involved in the deadlock situation do not interact, as well as pedestrians, have no influence on the cooperation behavior and can be ignored for the behavior decision. However, if a vehicle passes through the intersection before the manual vehicle, it can be expected that this manual vehicle will also pass through the intersection first in the deadlock situation. In this case, an automated vehicle should then exhibit defensive behavior, preferably by coming to a distinctive stop, and yield the right of way to the manual driver.

Overall, the studies provided initial insights into the cooperation behavior of drivers in two different deadlock situations. These provide indications of how an automated vehicle should behave in these situations in order to resolve the situation to everyone's satisfaction.

Acknowledgements This project has been funded within the priority program 1835 "Cooperatively Interacting Automobiles" by the *German Research Foundation (DFG)*.

References

1. Argyle, M., Schmidt, C.: Körpersprache & Kommunikation: Das Handbuch zur nonverbalen Kommunikation. 9. Junfernmann, Paderborn (2005)
2. Ba, Y., Zhang, W., Reimer, B., Yang, Y., Salvendy, G.: The effect of communicational signals on drivers' subjective appraisal and visual attention during interactive driving scenarios. Behav. Inf. Technol. **34**(11), 1107–1118 (2015). https://doi.org/10.1080/0144929X.2015.1056547
3. Bauer, T., Risser, R., Soche, P., Teske, W., Vaughan, C.: Kommunikation im Straßenverkehr: Literaturstudie aus juridischem, kommunikationstheoretischem und psychologischen Blickwinkel. Kuratorium für Verkehrssicherheit, Wien, Zwischenbericht (1980)
4. Björklund, G.M., Åberg, L.: Driver behaviour in intersections: formal and informal traffic rules. Transport. Res. F: Traffic Psychol. Behav. **8**(3), 239–253 (2005). https://doi.org/10.1016/j.trf.2005.04.006
5. Burgoon, J., Floyd, K., Guerrero, L.: Nonverbal communication theories of interaction adaption. In: Berger, C.R. (ed.) The Handbook of Communication Science, pp. 93–108. SAGE, Thousand Oaks, Californiad (2010)
6. Cantin, V., Lavallière, M., Simoneau, M., Teasdale, N.: Mental workload when driving in a simulator: effects of age and driving complexity. Accid. Anal. Prev. **41**(4), 763–771 (2009). https://doi.org/10.1016/j.aap.2009.03.019
7. Edquist, J.: The effects of visual clutter on driving performance. Dissertation (2008). https://scholar.archive.org/work/hv3zorovnjhlzlrscarwu653u4/access/wayback/, https://au-east.erc.monash.edu.au/fpfiles/7363204/monash7883.pdf
8. Edquist, J., Rudin-Brown, C.M., Lenné, M.G.: The effects of on-street parking and road environment visual complexity on travel speed and reaction time. Accid. Anal. Prev. **45**, 759–765 (2012). https://doi.org/10.1016/j.aap.2011.10.001
9. Färber, B.: Kommunikationsprobleme zwischen autonomen fahrzeugen und menschlichen fahrern. In: M. Maurer (ed.) Autonomes Fahren, pp. 127–146. Springer Vieweg, Berlin and Heidelberg (2015). https://doi.org/10.1007/978-3-662-45854-9_7
10. Fastenmeier, W. (ed.): Autofahrer und Verkehrssituation: Neue Wege zur Bewertung von Sicherheit und Zuverlässigkeit moderner Straßenverkehrssysteme, Mensch, Fahrzeug, Umwelt, vol. 33. Dt. Psychologen-Verl. and Verl. TÜV Rheinland, Bonn and Köln (1995)
11. Faure, V., Lobjois, R., Benguigui, N.: The effects of driving environment complexity and dual tasking on drivers' mental workload and eye blink behavior. Transport. Res. F: Traffic Psychol. Behav. **40**, 78–90 (2016). https://doi.org/10.1016/j.trf.2016.04.007
12. Fuller, R.: Driver control theory: from task difficulty homeostasis to risk allostasis. In: B.E. Porter (ed.) Handbook of Traffic Psychology. Academic Press (2011). https://www.sciencedirect.com/science/article/pii/b9780123819840100025
13. Geiser, G.: Mensch-maschine-kommunikation im kraftfahrzeug. Automobiltechnische Zeitschrift (2), 77–84 (1985). https://trid.trb.org/view/1033305
14. Hancock, P.A., Wulf, G., Thom, D., Fassnacht, P.: Driver workload during differing driving maneuvers. Accid. Anal. Prev. **22**(3), 281–290 (1990). https://doi.org/10.1016/0001-4575(90)90019-h
15. Hoc, J.M.: Towards a cognitive approach to human-machine cooperation in dynamic situations. Int. J. Hum Comput Stud. **54**(4), 509–540 (2001). https://doi.org/10.1006/ijhc.2000.0454
16. Horberry, T., Anderson, J., Regan, M.A., Triggs, T.J., Brown, J.: Driver distraction: the effects of concurrent in-vehicle tasks, road environment complexity and age on driving performance. Accid. Anal. Prev. **38**(1), 185–191 (2006). https://doi.org/10.1016/j.aap.2005.09.007
17. Imbsweiler, J.: Kooperation im Straßenverkehr in innerstädtischen Pattsituationen. Dissertation (2019). https://core.ac.uk/download/pdf/210569723.pdf
18. Imbsweiler, J., Linstedt, K., Palyafári, R., Weinreuter, H., Puente León, F., Deml, B.: Quasi-experimentelle untersuchung des blickverhaltens und der fahrparameter von autofahrern in engstellen. Zeitschrift für Arbeitswissenschaft **71**(4), 242–251 (2017). https://doi.org/10.1007/s41449-017-0083-6. https://link.springer.com/article/10.1007/s41449-017-0083-6

19. Imbsweiler, J., Palyafári, R., Puente León, F., Deml, B.: Untersuchung des entscheidungsverhaltens in kooperativenverkehrssituationen am beispiel einer engstelle. at - Automatisierungstechnik (7), 477–488 (2017). https://www.degruyter.com/document/doi/10.1515/auto-2016-0127/html

20. Imbsweiler, J., Ruesch, M., Weinreuter, H., Puente León, F., Deml, B.: Cooperation behaviour of road users in T-intersections during deadlock situations. Transport. Res. F: Traffic Psychol. Behav. **58**, 665–677 (2018). https://doi.org/10.1016/j.trf.2018.07.006

21. Imbsweiler, J., Wolf, E., Linstedt, K., Hess, J., Deml, B.: Relevant eye-tracking parameters within short cooperative traffic scenarios. In: D. de Waard, F. Di Nocera, D. Coelho, J. Edworthy, K. Brookhuis, F. Ferlazzo, T. Franke, and A. (ed.) Proceedings of the Human Factors and Ergonomics Society Europe Chapter 2017 Annual Conference, pp. 2333–4959 (2017)

22. Jahn, G., Oehme, A., Krems, J.F., Gelau, C.: Peripheral detection as a workload measure in driving: effects of traffic complexity and route guidance system use in a driving study. Transport. Res. F: Traffic Psychol. Behav. **8**(3), 255–275 (2005). https://doi.org/10.1016/j.trf.2005.04.009

23. Juhlin, O.: Traffic behaviour as social interaction: implications for design of artificial drivers. Sosiologisk Arbok 1–20 (2001)

24. Kauffmann, N., Winkler, F., Vollrath, M.: What makes an automated vehicle a good driver? In: Proceedings of the 2018 CHI Conference on Human Factors in Computing Systems. ACM, New York, NY, USA (2018). https://doi.org/10.1145/3173574.3173742

25. Kitazaki, S., Myhre, N.J.: Effects of non-verbal communication cues on decisions and confidence of drivers at an uncontrolled intersection. In: Driving Assessment Conference, vol. 8, issue 2015, pp. 113–119 (2015). https://doi.org/10.17077/drivingassessment.1559. https://pubs.lib.uiowa.edu/driving/article/id/28583/print/

26. Lee, Y.M., Sheppard, E.: The effect of motion and signalling on drivers' ability to predict intentions of other road users. Accid. Anal. Prev. **95**, 202–208 (2016). https://doi.org/10.1016/j.aap.2016.07.011

27. Li, G., Lai, W., Sui, X., Li, X., Qu, X., Zhang, T., Li, Y.: Influence of traffic congestion on driver behavior in post-congestion driving. Accid. Anal. Prev. **141**, 105508 (2020). https://doi.org/10.1016/j.aap.2020.105508

28. Manawadu, U.E., Kawano, T., Murata, S., Kamezaki, M., Muramatsu, J., Sugano, S.: Multiclass classification of driver perceived workload using long short-term memory based recurrent neural network. In: 2018 IEEE Intelligent Vehicles Symposium (IV). IEEE (2018). https://doi.org/10.1109/ivs.2018.8500410

29. Merten, K.: Kommunikation: Eine Begriffs-und Prozessanalyse. Westdeutscher Verlag, Opladen (1977)

30. Merten, K.: Kommunikationsprozes im straßenverkehr, klassifikation und beurteilung von verkehrssituationen. Unfall- und Sicherheitsforschung, pp. 115–126 (1977)

31. Oviedo-Trespalacios, O., Haque, M.M., King, M., Washington, S.: Effects of road infrastructure and traffic complexity in speed adaptation behaviour of distracted drivers. Accid. Anal. Prev. **101**, 67–77 (2017). https://doi.org/10.1016/j.aap.2017.01.018

32. Patten, C.J., Kircher, A., Östlund, J., Nilsson, L., Svenson, O.: Driver experience and cognitive workload in different traffic environments. Accid. Anal. Prev. **38**(5), 887–894 (2006). https://doi.org/10.1016/j.aap.2006.02.014

33. Risser, R.: Behavior in traffic conflict situations. Accid. Anal. Prev. **17**(2), 179–197 (1985). https://doi.org/10.1016/0001-4575(85)90020-x

34. Rudin-Brown, C.M., Edquist, J., Lenné, M.G.: Effects of driving experience and sensation-seeking on drivers' adaptation to road environment complexity. Saf. Sci. **62**, 121–129 (2014). https://doi.org/10.1016/j.ssci.2013.08.012

35. Schulz, W.: Politische Kommunikation: theoretische Ansätze und Ergebnisse empirischer Forschung, 2., vollst. überarb. und erw. aufl. edn. VS Verl. für Sozialwiss, Wiesbaden (2008). https://doi.org/10.1007/978-3-531-91830-3

36. Shannon, C., Weaver, W.: The Mathematical Theory of Communication. University of Illinois Press, Urbana, IL, USA (1949)

37. Song, G., Ma, Q., Wu, F., Li, L.: The psychological explanation of conformity. Soc. Behav. Pers. **8**(40), 1365–1372 (2012). https://doi.org/10.2224/sbp.2012.40.8.1365. https://www.ingentaconnect.com/content/sbp/sbp/2012/00000040/00000008/art00015

38. Statistisches Bundesamt: Straßenverkehrsunfälle nach unfallkategorie, ortslage (07.07.2022). https://www.destatis.de/DE/Themen/Gesellschaft-Umwelt/Verkehrsunfaelle/Tabellen/polizeilich-erfasste-unfaelle.html

39. Strelau, N.R., Weinreuter, H., Heizmann, M., Deml, B.: Verklemmungs-situationen im stadtverkehr—der einfluss von verkehrsaufkommen auf die wahrgenommene schwierigkeit und das kooperationsverhalten an t-kreuzungen. In: GfA (ed.) Tagungsband 67. GfA-Frühjahrskongress Arbeit HUMAINE gestalten (2021)

40. Strelau, N.R., Weinreuter, H., Heizmann, M., Deml, B.: Der einfluss von fußgängern auf die wahrgenommene übersichtlichkeit und das kooperationsverhalten von autofahrern an t-kreuzungen. In: GfA (ed.) Tagungsband 68. GfA-Frühjahrskongress Technologie und Bildung in hybriden Arbeitswelten (2022)

41. Sucha, M.: Road users' strategies and communication: driver-pedestrian interaction. Transport Research Arena (TRA) 2014 Proceedings (2014)

42. Svensson, A., Hydén, C.: Estimating the severity of safety related behaviour. Accid. Anal. Prev. **38**(2), 379–385 (2006). https://doi.org/10.1016/j.aap.2005.10.009

43. Teh, E., Jamson, S., Carsten, O., Jamson, H.: Temporal fluctuations in driving demand: the effect of traffic complexity on subjective measures of workload and driving performance. Transport. Res. F: Traffic Psychol. Behav. **22**, 207–217 (2014). https://doi.org/10.1016/j.trf.2013.12.005

44. The Royal Society for the Prevention of Accidents: Road Safety Factsheet: Autonomous Vehicles. Birmingham (2021)

45. Törnros, J., Bolling, A.: Mobile phone use—effects of conversation on mental workload and driving speed in rural and urban environments. Transport. Res. F: Traffic Psychol. Behav. **9**(4), 298–306 (2006). https://doi.org/10.1016/j.trf.2006.01.008

46. Watzlawick, P., Bavelas, J.B., Jackson, D.D.: Menschliche Kommunikation: Formen, Störungen, Paradoxien, 13., unveränderte auflage edn. Klassiker der Psychologie. Hogrefe, Bern (2017). https://doi.org/10.1024/85745-000

47. Weinreuter, H., Strelau, N.R., Qiu, K., Jiang, Y., Deml, B., Heizmann, M.: Intersection complexity and its influence on human drivers. IEEE Access **10**, 74059–74070 (2022). https://doi.org/10.1109/ACCESS.2022.3189017

48. Witzlack, C., Beggiato, M., Krems, J.: Interaktionssequenzen zwischen fahrzeugen und fußgängern im parkplatzszenario als grundlage für kooperativ interagierende automatisierung. In: VDI (ed.) Fahrerassistenz und Integrierte Sicherheit, VDI Berichte, pp. 323–336. VDI Verlag (2016). https://doi.org/10.51202/9783181022887-323

49. Zaidel, D.M.: A modeling perspective on the culture of driving. Accid. Anal. Prev. **24**(6), 585–597 (1992). https://doi.org/10.1016/0001-4575(92)90011-7

Measuring and Describing Cooperation Between Road Users—Results from CoMove

Laura Quante, Tanja Stoll, Martin Baumann, Andor Diera, Noèmi Földes-Cappellotto, Meike Jipp, and Caroline Schießl

Abstract Safe and efficient traffic requires that road users interact and cooperate with each other. Especially in situations which are not explicitly regulated, and the right of way is not clearly defined, it is of great importance that road users are able to communicate their own intentions and understand the communication and cooperation behaviour of the other involved road users. When automated vehicles enter the current traffic system, their ability to fit into the system, that is their ability to communicate and cooperate, will determine their success. Therefore, the development of cooperatively interacting, automated vehicles requires detailed knowledge about human cooperation behaviour in traffic, which can only be obtained using appropriate methods and measures. By focusing on road narrowings and lane changing, this chapter gives an overview on how to measure cooperation between road users, considering methods for data collection, subjective and objective measures of cooperation as well as behaviour modeling, to support the systematic research on

L. Quante (✉) · C. Schießl
Institute of Transportation Systems, German Aerospace Center (DLR E.V.), Lilienthalplatz 7, 38108 Braunschweig, Germany
e-mail: laura.quante@dlr.de

C. Schießl
e-mail: caroline.schiessl@dlr.de

T. Stoll · M. Baumann · A. Diera · N. Földes-Cappellotto
Department of Human Factors, Institute of Psychology and Education, Ulm University, Albert-Einstein-Allee 43, Ulm, Germany
e-mail: tanja.stoll@uni-ulm.de

M. Baumann
e-mail: martin.baumann@uni-ulm.de

A. Diera
e-mail: andor.diera@uni-ulm.de

N. Földes-Cappellotto
e-mail: noemi.foeldes@uni-ulm.de

M. Jipp
Institute of Transport Research, German Aerospace Center (DLR E.V.), Rudower Chaussee 7, 12489 Berlin, Germany
e-mail: meike.jipp@dlr.de

565
C. Stiller et al. (eds.), *Cooperatively Interacting Vehicles*,
https://doi.org/10.1007/978-3-031-60494-2_20

cooperation in road traffic. This overview is extended by findings from studies conducted within CoInCar, including results on factors influencing human behaviour in cooperative situations, either in a manual or an automated setting, and initial findings from modeling the cognitive processes underlying cooperative driving behaviour.

1 Introduction

Given the technological development in recent years and the expected further development, the introduction of automated agents, such as highly or fully automated vehicles, into the traffic system seems to be within reach in the near future. According to previously published roadmaps of many car manufacturers, suppliers and other stakeholders, driving highly automated in certain driving contexts was expected to occur in the next decade [19]. With reference to the defined automation levels by the Society of Automotive Engineers (SAE) [35], this means that the vehicle automation is able to perform the longitudinal and lateral control, to monitor the environment and to bring the vehicle into a safe state if necessary in any given situation, not requiring the human driver as a back-up for the automation if the vehicle is in the high automation level. The efforts behind these developments are motivated by the promise that the technology will bring benefits regarding traffic safety and efficiency, convenience for drivers, mobility for different types of road users etc.

However, these promised benefits can only be achieved if these automated vehicles are not only able to drive safely but also possess the required communicative and cognitive capabilities to cooperate with the human road users effectively and efficiently. The reason for this is that the traffic system can be seen as a social system in which different types of road users, such as drivers, pedestrians, or bicyclists, try to safely, efficiently and comfortably achieve their goals. This in turn requires that either possible goal conflicts in the future are detected and prevented by an appropriate adaptation of behaviour or, if road users are already in a conflict situation, that negotiation strategies are available to solve the conflict. In many situations, this requires that different individuals cooperate with each other to achieve their own goals. These interactions between them have essential influences on traffic safety, such as the interaction or communication between driver and pedestrian [63].

Accordingly, in Germany, for example, the road traffic regulations explicitly call on road users to show mutual consideration (§1, StVO) and to communicate (§ 11, StVO). Both aspects are especially important in situations which are not explicitly regulated, and the right of way is not clearly defined, e.g., two drivers simultaneously arriving at a two-sided road narrowing from opposite directions. To resolve this situation safely and efficiently, the two drivers have to be considerate of each other and cooperate by communicating who will pass the bottleneck first. Additionally, road users can cooperate with other road users by facilitating their intended maneuvers, e.g., drivers on highways adapting their speed to facilitate a lane change for other vehicles. Therefore, it is of great importance that road users are able to understand the behaviour of the other involved road users and infer their intentions, i.e., that they can correctly interpret the signals of other road users, and send clear signals themselves.

As a new participant in the traffic system, automated vehicles need to be able to participate in these interaction and cooperation processes with human road users, in order to integrate themselves smoothly into the traffic system, negotiate conflicting action plans and adapt efficiently to the traffic situation. For this it is necessary that automated vehicles are able to understand and predict others' states and behaviour, that they are able to establish and maintain a shared situation representation with their interacting human partners [6, 23], and that these vehicles are able to change their own action plans and/or trigger their cooperating human partners to adapt their action plans in the light of changing situational characteristics and demands [14, 90].

Consequently, introduction of cooperatively interacting vehicles demonstrates the relevance of understanding cooperation in traffic. When automated vehicles enter the current traffic system, their ability to fit into the traffic system will determine their success [78], that is their ability to communicate and cooperate. Current automated vehicles lack an understanding of human behaviour in traffic, making conservative and defensive behaviour necessary for safe operation, which is associated with reduced traffic flow and higher accident involvement [77, 80]. Therefore, the development of cooperatively interacting, automated vehicles requires detailed knowledge about human cooperation behaviour in traffic, which can only be obtained using appropriate methods and measures. For example, we must understand how human drivers communicate their intentions, e.g., via movement patterns or explicit signals, how human drivers understand these signals, in which contexts which signals are used, what kind of situational characteristics trigger cooperative behaviour, especially when cooperation is optional, and what cooperative behaviour looks like.

The aim of our work within the project CoMove as part of the priority program "Cooperatively Interacting Vehicles" (CoInCar) of the German Research Foundation was, on the one hand, to find measures for the systematic description and evaluation of cooperation and, on the other hand, to identify situational factors which influence human behaviour in cooperative situations in order to develop a comprehension- and decision-based model of driver-vehicle cooperation. The present chapter gives an overview on how to measure cooperation in traffic, considering potential methods for data collection (Sect. 2), subjective and objective measures of cooperation (Sect. 3) as well as behaviour modeling (Sect. 4). This overview is complemented by selected findings and results of studies conducted as part of CoMove. In this chapter, we focus on two concrete scenarios that are prototypical for two classes of cooperation situations: lane changing situations as an example of cooperation situations where not cooperating is an option, and two-sided road narrowings where cooperation is necessary to solve the deadlock in the situation.

The term *cooperation* requires a domain-specific definition since interaction and cooperation in traffic differ from other forms of social interaction. For example, drivers will most likely never meet again [45]. In this chapter, cooperation is understood as a specific form of interaction between two or more (human) road users whose actions interfere with each other (space-sharing conflict; [57]), and who adjust their behaviour to support each other and to solve the potential conflict [30, 54]. This cooperation is achieved by means of communication [30], which includes explicit (e.g., horn, indicator, hand gesture) and implicit signals (e.g., deceleration, acceleration) [17, 59]. Cooperative situations are characterized as being often non-symmetrical, meaning one of the involved road users has to make a compromise [58].

2   Methods for Data Collection

Appropriate data collection methods are a prerequisite for studying cooperation. This section describes exemplary studies that have investigated cooperation in road traffic in order to give a brief overview of potential data collection methods, mentioning some important advantages and disadvantages of the different methods. Traffic observations, video-based online and laboratory experiments as well as test track and simulator studies are considered. Special attention is given to coupled simulator studies, as they represent a promising but under-researched method.

2.1   *Traffic Observations*

On-site observations are an inexpensive and simple tool to study cooperative behaviour, but are limited in that not every detail can be recorded, such as the duration and exact sequence of signals. Imbsweiler et al. and Rettenmaier et al. [31, 71], for example, conducted on-site observations of drivers at narrow passages. Observation protocols included lateral and longitudinal behaviour (e.g., swerving to the side of the road, accelerating, stopping), use of horn, indicators, headlights, and hand gestures, and order of arrival and departure. These protocols were used to examine implicit and explicit communication signals and to derive prototypical offensive and defensive approaching behaviours, which were evaluated in further studies (e.g., [34, 70]).

Traffic observations using video cameras that are either permanently installed or set up for a specific period of time (see Fig. 1 for an example) allow a much more detailed analysis of traffic behaviour. Schuler et al. [76], in comparison to [31, 71], based their analysis on video data, and were able to determine not only the frequency and sequence of communication signals, but also the spatial occurrence of the signals with respect to the distance to the narrow passage. Quante et al. and Zhang et al. [67, 94] demonstrate that additional trajectory data is useful in pre-selecting relevant interactions (e.g., based on surrogate safety measures such as time to collision) and allows the inclusion of precise measures of, for example, drivers' velocity and relative position, in the analysis.

Traffic observations have in common that the observers take an external perspective on the situations and thus the subjective experience of the observed road users is usually missing. In addition, the observed behaviour usually shows a large variance, so that only limited conclusions can be drawn about factors influencing behaviour.

Fig. 1 DLR's application platform for intelligent mobility (AIM) research junction (right) and mobile installations (left)

2.2 Online Experiments

One possible tool to efficiently study the reactions of a large number of human drivers to highly controlled traffic scenarios in a standardized manner are online experiments (e.g., [49, 59]). Online experiments are generally easy to implement and time- and cost-efficient and allow both the collection of specific reaction data, such as decision data, and even reaction times, as well as subjective data, such as ratings. However, they partly lack realism since participants only passively experience a situation sitting in front of a computer screen. One such example is [59]: In two video-based online experiments, Miller and colleagues presented videos of vehicles approaching a narrow passage in which participants took the perspective of the driver approaching from the opposite direction. The approaching behaviour was systematically varied with respect to the longitudinal and lateral vehicle movements as well as the timing of a given movement. Participants were asked to rate the other driver's intention, the intention's explicitness and the cooperativeness of the observed behaviour.

2.3 Laboratory Studies

Similar to online studies, laboratory-based studies using video material allow a high level of standardization of the investigated situations, since the exact same situation can be presented to the participants. It allows the measurement of reaction time and eye-tracking parameters. Compared to online studies, the number of participants is smaller. However, it is easier to control that participants fully focus on the study. In CoMove, these kinds of studies were carried out to identify situational factors that influence the driver's behaviour in lane changing situations both when changing lanes on the highway and when a car merges onto the highway from an on-ramp [85–87], which were then further investigated in a driving simulator setting [83, 84] (see Sect. 3.3).

2.4 Test Track and Simulator Studies

Test track and simulator studies, in comparison to online and laboratory video-based studies, allow both subjective experience and objective behaviour to be studied in a more realistic, yet controlled and standardized (but costlier) setting. Imbsweiler et al. [34], for example, conducted a study on a test track in which participants encountered another driver at a narrow passage. The other driver was trained to approach the narrow passage in six predefined ways. After every encounter, participants rated the other driver's cooperativeness, the degree of cooperation in the given situation and the participants' confidence to pass first or second. Rettenmaier et al. [70] performed a similar study in a driving simulator: Participants repeatedly encountered a vehicle at a narrow passage, which showed nine different approaching behaviours. With the goal of designing movements for autonomous vehicles, participants' driving behaviour and subjective ratings were used to evaluate the implemented approaching behaviours. In [83], the selected behaviour in lane changing situations when another vehicle wants to overtake a slower truck were investigated. Participants were told to support their automated vehicle to find the best decision. In [84], the automated vehicle was carrying out the decision on its own and participants were asked if they agree with their vehicle's decision.

When comparing test track and simulator studies, test track studies offer the advantage of interaction with real road users. This, however, may bear a certain risk for the participants and is therefore not always ethical. However, simulated road users are usually only human-like to a limited extent, which can be problematic when studying cooperation. A promising alternative are coupled simulators, in which two or more participants move and interact in the same simulated environment [24, 60, 61, 64, 75].

2.5 Coupled Simulator Studies

Coupled simulators have been used to study, for example, traffic safety (e.g., [26]), driver-assistance systems (e.g., [66]), and automated vehicles (e.g., [24]), not only considering drivers but also pedestrians (e.g., [52]), cyclists (e.g., [53]), and motorcyclists (e.g., [92]). With respect to cooperation, only a limited number of published studies exists. For example, [25, 28] investigated cooperative behaviour in a lane-change scenario in a multi-driver simulator. In both studies, the driver on the right lane had to perform a lane change to the left lane because of a braking lead vehicle, interfering with the driver in the left lane. In this scenario, [28] studied the effects of the availability of a left lane, indicator use and the brake strength of the lead vehicle on the occurrence of cooperative behaviour. Friedrich et al. [25], in addition, manipulated participants' belief about whether the other driver was a human driver or simulated.

2.5.1 Exemplary Coupled Simulator Study

Within CoMove, a coupled driving simulator study was conducted to describe and evaluate cooperative behaviour of two drivers encountering narrow passages (S6 in Table 1). Two scenarios were implemented: In one scenario, drivers came from opposite directions and had to pass a narrow passage caused by traffic beacons on both sides of the road (two-sided bottleneck; Fig. 2A). In the other scenario, drivers came from the same direction but on separate lanes. The right lane was blocked by an excavator (one-sided bottleneck; Fig. 2B), so a lane change was necessary. The two drivers were driving through a city on a fixed route, which took them past the two narrow passages eight times each. They were synchronized by traffic lights to ensure they would encounter each other in the two scenarios.

The coupled driving simulator MoSAIC (Modular and Scalable Application-Platform for ITS Components) at the Institute of Transportation Systems at the German Aerospace Center (Braunschweig, Germany) was used. For the study, two fixed-based driving simulators were coupled. Each driving simulator was equipped with a steering wheel, accelerator and brake pedal. Three monitors created a 180° view (see Fig. 2C). The driving simulators were placed on the right and left side of the room, separated by a third simulator which was not used during the study. Participants could use the headlights, indicators, and horn. The sound of the engine and horn was transmitted via headphones. The environment was designed using Unreal Engine (Version 4.24) and Trian3DBuilder. In-house software was used to connect and control the driving simulators.

Twenty-two participants, i.e. eleven pairs, took part in the study (16 male, 6 female). The mean age was 27.3 years ($SD = 6.6$ years). After receiving general information and driving a ten-minute training session, participants encountered each other 16 times in the two scenarios. After every encounter they stopped and answered several questions about the interaction. At the end of the study, they filled out different questionnaires regarding demographic information, driving behaviour, personality, and their experience with the simulator. In addition to subjective data, driving data (e.g., velocity, acceleration, pedal positions) and videos of the experimental drive from the drivers' perspective were recorded. Participants knew that they were interacting with each other in the simulation.

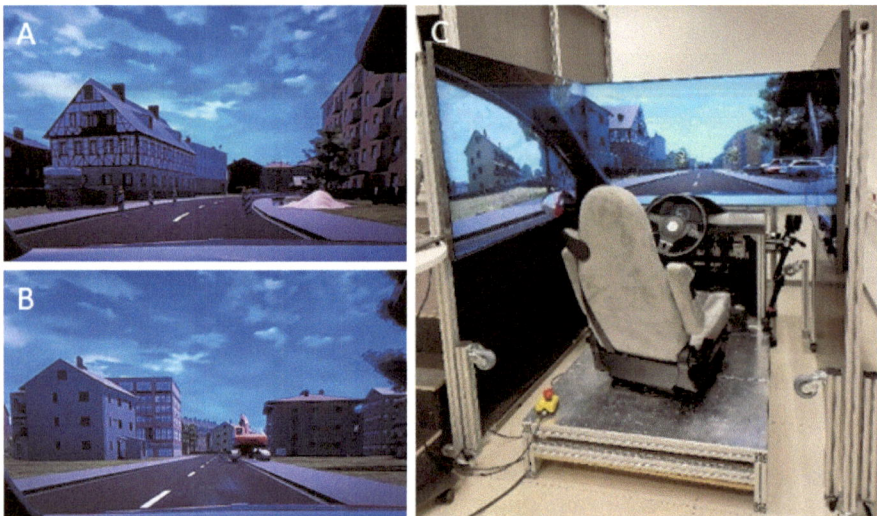

Fig. 2 The cooperative scenarios of the simulator study, two-sided bottleneck (**A**) and one-sided bottleneck (**B**), which were implemented in DLR's coupled simulator AIM MoSAIC (**C**)

Besides the aim to describe and evaluate participants' cooperation behaviour (not published yet), part of the analysis was also to answer whether cooperative behaviour could be provoked within the study and whether it felt realistic. Focusing on the two-sided bottleneck, in more than 80% of the encounters, drivers stated that they cooperated with the other driver. On average, the driving and communication behaviour of the other driver as well as the interaction with the other driver were rated as realistic. When asked what aspects of the interaction behaviour were not realistic, 14 participants mentioned limited communication, noting the absence of hand signals and missing eye contact.

2.5.2 Implications for Future Studies

Based on these findings, future studies on cooperation in a coupled driving simulator should investigate whether performing and perceiving hand gestures and seeing the other driver's head orientation increases the feeling of realistic interaction and cooperation. It should also be considered that drivers might behave more defensively and cooperatively when knowingly interacting with another human driver (see also [25]). At this point, it is important to note that neither the influence of limited communication possibilities, the awareness of interacting with real humans, nor other factors such as the repeated encounter of the same participants, the spatial proximity in the laboratory, the degree of familiarity of the participants, (lack of) sympathy, and participants' characteristics (e.g., age, gender) have been systematically investigated so far. Accordingly, there is a great need for research to ensure that cooperative behaviour shown in a coupled simulator is equivalent to cooperative behaviour shown on the road.

3 Measures of Cooperation

In order to study cooperation systematically and gain a detailed understanding of how and when road users cooperate, the construct of cooperation has to be operationalized, i.e., cooperation must be made measurable. This requires, on the one hand, a precise definition of cooperation and its aspects and, on the other hand, measures that can reliably capture these different aspects of cooperation. Within CoMove, we have therefore conducted several empirical studies to methodically advance the systematic and scientific assessment of cooperation (see Table 1 for an overview).

This section first presents one of these studies which was conducted with the goal of better understanding the construct of cooperation and identifying aspects of cooperation in road narrowings and lane changing situations, followed by an overview of existing objective and subjective measures of cooperation. Finally, it is complemented by a detailed outline of experiments investigating factors influencing cooperation in lane changing scenarios in order to give a practical example of how to measure cooperation.

3.1 *Identifying Aspects of Cooperation*

Focused interviews were conducted and qualitatively analysed to identify potential criteria and metrics for the description and evaluation of cooperation (see also [69]). It was focused on two cooperative scenarios: drivers encountering each other from opposite directions at a narrow passage and a lane change with surrounding traffic. Twelve traffic researchers (5 male, 7 male) were interviewed. They were between 26 to 37 years old ($M=30.08$ years, $SD=3.87$ years) and owned a driver's licence for at least eight years ($M=11.67$ years; $SD=3.37$ years). Interviewees were presented with short videos of traffic encounters. For the lane change, the video material was recorded from within a vehicle driving either on a highway or in an urban environment. One camera was directed to the front, the other camera was directed to the back (see Fig. 3B). For the narrow passage, videos were recorded from two perspectives at a road narrowing in Braunschweig, Germany (see Fig. 3A) via two portable sensor poles which are part of DLR's Application Platform for Intelligent Mobility Mobile Traffic Acquisition [46] (see Fig. 1).

Encounters between two or more drivers were extracted and rated with respect to their degree of interaction by two raters. Videos with identical ratings and different degrees of interaction were chosen for the interview study, resulting in 29 videos of encounters at the narrow passage and 51 videos of lane changes (of which 12 were recorded in an urban environment). After answering demographic questions, every

Table 1 Empirical studies conducted within CoMove to improve the systematic assessment of cooperation

No.	Method	Scenario	Goal	Refs
S1	Focussed interviews	NP, LC	Identification of potential indicators/metrics of cooperation	[69]
S2	Video-based online experiment	NP	Influence of arrival order and time delay on evaluation of cooperation	
S3	Video-based online experiment	NP	Influence of arrival order, time delay, perspective and stopping distance on evaluation of cooperation	
S4	Traffic observation	NP	Identification of metrics to quantify arrival order	[67]
S5	Online survey	NP	Development of questionnaire for the subjective evaluation of cooperation (CoopQ)	[68]
S6	Coupled simulator	NP, LC	Further development of CoopQ; exploratory analysis of cooperation in coupled simulator setting	
S7	Coupled simulator	NP	Influence of arrival order and approaching behaviour of oncoming driver on evaluation of cooperation and cooperation behaviour	
S8	Online survey		German validation of the Prosocial and Aggressive Driving Inventory (PADI)	[82]

NP = narrow passage; LC = lane change

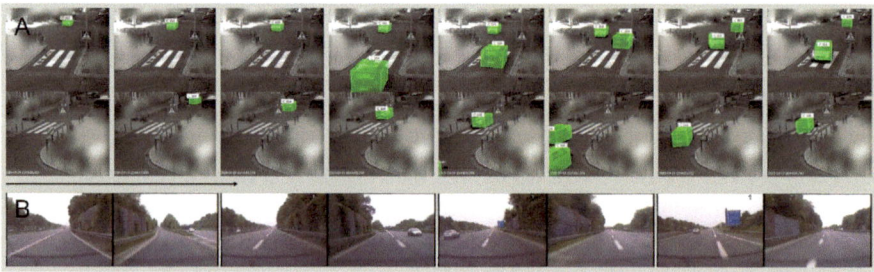

Fig. 3 Examples of video material used in the interview study. **A** shows a narrow passage scenario, **B** shows a lane change

participant sequentially watched 37 videos while commenting aloud on the drivers' behaviour. The instruction was given as follows (translated from German):

"Please comment aloud on the videos by describing and evaluating the behaviour of the drivers. The following questions serve as a guide: How do you evaluate the behaviour of the drivers? What do you base your evaluation on? Did the drivers communicate with each other? If so, who communicated what? And how did the other react? On what do you base this? In what order did they communicate?"

The order of blocks (lane change vs. narrow passage) and trials were randomized between participants. Four interviews were conducted in person, eight were conducted via Skype for Business calls due to Covid-19 restrictions. Participants' answers were recorded, transcribed and analysed via MAXQDA Analytics Pro 2020. As an example, a participant's comment on a narrow passage video is presented below (translated from German):

"Here we see that two vehicles are approaching the bottleneck at relatively the same time and therefore also meet in the bottleneck or shortly before it. And you can already see that the vehicles have to brake heavily in any case, or at least one of them, namely the 734, is really hitting the brakes. The, what is it, the T5 or whatever, it's speeding through it quite recklessly, I would say. So, he says I'm the stronger one and, yes, okay, admittedly, because he's faster, he's also the first to get into the bottleneck. Well, how do I evaluate the behaviour of the drivers? Well, one of them [...] drives defensively. That's the one who brakes, of course. The T5 is driving [...] offensively, [...], also drives much faster".

Codes were developed in an iterative process and organized into four categories: description of behaviour, interpretation of behaviour, factors influencing behaviour, and evaluation of behaviour. The results for the categories description and evaluation of behaviour are summarized in Tables 2 and 3. Aspects mentioned by at least half the interviewees are listed.

Particularly relevant for the description and evaluation of the narrow passage scenario seem to be (1) the time delay with which drivers arrive at the narrow passage, and (2) the arrival and departure order (who arrives first and who passes the narrow passage first). For the lane change, (1) the distance between vehicles, (2) the speed difference of vehicles, and (3) the necessity of a lane change seem to be most relevant.

3.2   Operationalizing Cooperation

Once it has been worked out which aspects of cooperation are to be investigated, these aspects need to be operationalized to be either experimentally manipulated

Table 2 Aspects used to *describe* drivers' behaviour (number of interviewees who mentioned a given aspect)

Individual behaviour	Narrow passage		Lane change	
Longitudinal behaviour	Wait Stop (standstill) Decelerate Constant Accelerate	(12)	Decelerate Constant Accelerate	(12)
Lateral behaviour	Movement to the right side	(12)	To/from left/middle/right lane Fast/slow lane change	(12)
Velocity	Fast Slow	(12)	Fast Slow	(12)
Start of behaviour	Early Late	(10)	Suddenly Early Late Simultaneously	(12)
Explicit communication	Headlight flash Indicator	(4)	Headlight flash Indicator	(12)
Relative behaviour	Narrow passage		Lane change	
Time/space distance	Successive Simultaneous	(12)	–	
Arriving/leaving vehicle	Same Different	(10)	–	
Distance between vehicles	–		Large Small	(12)
Speed difference	–		Large Small	(11)

as independent variables or analyzed as dependent measures. Based on the definition of cooperation given in Sect. 1 and the findings from the focussed interviews (Sect. 3.1), at least three aspects of cooperation could be derived: the temporal and spatial proximity of road users, costs and benefits of a cooperative situation, and the dynamics between interacting road users. Section 3.2.1 provides an overview of potential objective measures to assess these three aspects. Since cooperation in road traffic is not a clear-cut phenomenon but depends on the subjective evaluation of the involved road users, subjective measures of cooperation are presented in Sect. 3.2.2.

Table 3 Aspects used to *evaluate* drivers' behaviour (number of interviewees who mentioned a given aspect)

Aspect	Narrow passage		Lane change	
Defensiveness	Defensive, passive, calm, considerate, careful, cautious, cooperative	(12)	Polite, considerate, defensive, anticipatory, cooperative	(9)
Clarity	Predictable, explicit, unambiguous, certain, clear, anticipated, expected	(11)	Clear, ambiguous, unexpected	(8)
Offensiveness	Offensive, aggressive, inconsiderate, careless, dynamic, brash, brazen, impatient, uncooperative	(10)	Ruthless, aggressive, egoistic, impudent, brash, reckless, uncooperative	(7)
Criticality	Critical, safe, risky, dangerous, unproblematic	(10)	Critical, uncritical, dangerous, safe, unsafe, unproblematic	(12)
Efficiency	Efficient, well timed, fast, flowing, time saving, obstructive	(8)	Flow, time saving, obstructive	(10)
Necessity	–		Necessary, unnecessary, causeless, needless	(7)

3.2.1 Objective Description of Cooperative Behaviour in Traffic

A space-sharing conflict [57], i.e., a certain temporal and spatial proximity of road users, is a precondition for cooperation. The measurement of temporal and spatial proximity is particularly relevant with respect to traffic safety. Well-known surrogate measures of safety have been described [36], for example, the time to collision (TTC) or the post encroachment time (PET), which can provide information on the criticality of a situation and thus, on the presence of a space-sharing conflict. In addition to temporal and spatial proximity, cooperation is associated with facilitated goal achievement and, in the case of non-symmetrical cooperation, with the postponement of one's own goals. Therefore, it might be of interest to measure the costs and benefits of cooperating road users, for example, in terms of safety, efficiency or comfort. Safety, as described above, can be assessed by surrogate measures of safety, for example TTC , PET, deceleration to safety time or conflict severity [36]. Efficiency, in turn, can be measured by, for example, passing time (e.g., [70]), journey time and standard deviation of speed (e.g., [94]). Driving style, which can be described in terms of acceleration, jerk, quickness, and lane deviation, among others, has a major influence on experienced comfort [7]. Düring and Pascheka [20], for example, took the idea of costs and benefits and determined the type of cooperation, namely altruistic, rational, and egoistic, by comparing the utility of a maneuver (calculated by cost functions for both agents involved) with a reference behaviour.

A major challenge is to describe the dynamics between interacting road users, since there is usually not only one stimulus (e.g., a headlight flash) and one reaction (e.g., acceleration), but an interplay of multiple stimuli and reactions evolving over time. So far, mainly scenario-specific approaches exist. Hidas [29], for example, used the gaps between following and leading vehicles before and after a lane change to infer whether behavioural interference occurred between road users and to classify a lane change as free, forced or cooperative. For a convoy of vehicles, the length of the convoy and the standard deviation of the lateral position of the vehicles have been used to describe the interactions between the convoy's vehicles [60, 61, 64]. Oeltze and Schießl [64], in addition, used the distances between vehicles to describe the adaptive behaviour of vehicles within a convoy in response to a driver assistance system. We adapted this approach in order to estimate the arrival order of drivers at a narrow passage [67]. In this study, trajectory data was used to calculate both the distance and time to arrival (TTA) to the narrow passage for both drivers. Next, the difference in distances/TTAs was calculated such that this difference was positive (negative) for one driver if he/she was closer (more distant) to the narrow passage than the other driver. The arrival order was then defined based on the minimum of this difference over a given space. The relationship of arrival order and cooperation for the narrow passage scenario has been investigated in further studies but results have not yet been published (S2, S3 and S7 in Table 1).

A more general approach is to correlate time series data, e.g., drivers' velocity, to capture the dependency of two road users and thus their interaction behaviour [51, 52, 61]. Furthermore, different visual descriptions have been used to study the interplay of road users, such as interaction plots (e.g., [61, 92]), time-space-diagrams (e.g., [61]), and sequence diagrams (e.g., [91]). Figure 4 depicts two encounters from the coupled simulator study (S6 in Table 1), illustrating interaction plots, time-space diagrams and cross correlations. The upper interaction plot in Fig. 4 illustrates an encounter, in which the two drivers arrive simultaneously at the narrow passage (i.e., they are the same distance away from the bottleneck for a longer period of time; see also the upper time-space diagram) until the driver passing second (red) stops while the other driver (green) passes the bottleneck. In contrast, the lower interaction plot visualizes an encounter, in which the two drivers arrive after each other, i.e., the driver passing first (green) is always closer to the narrow passage than the driver passing second (red). The cross-correlation plots show that for the upper encounter

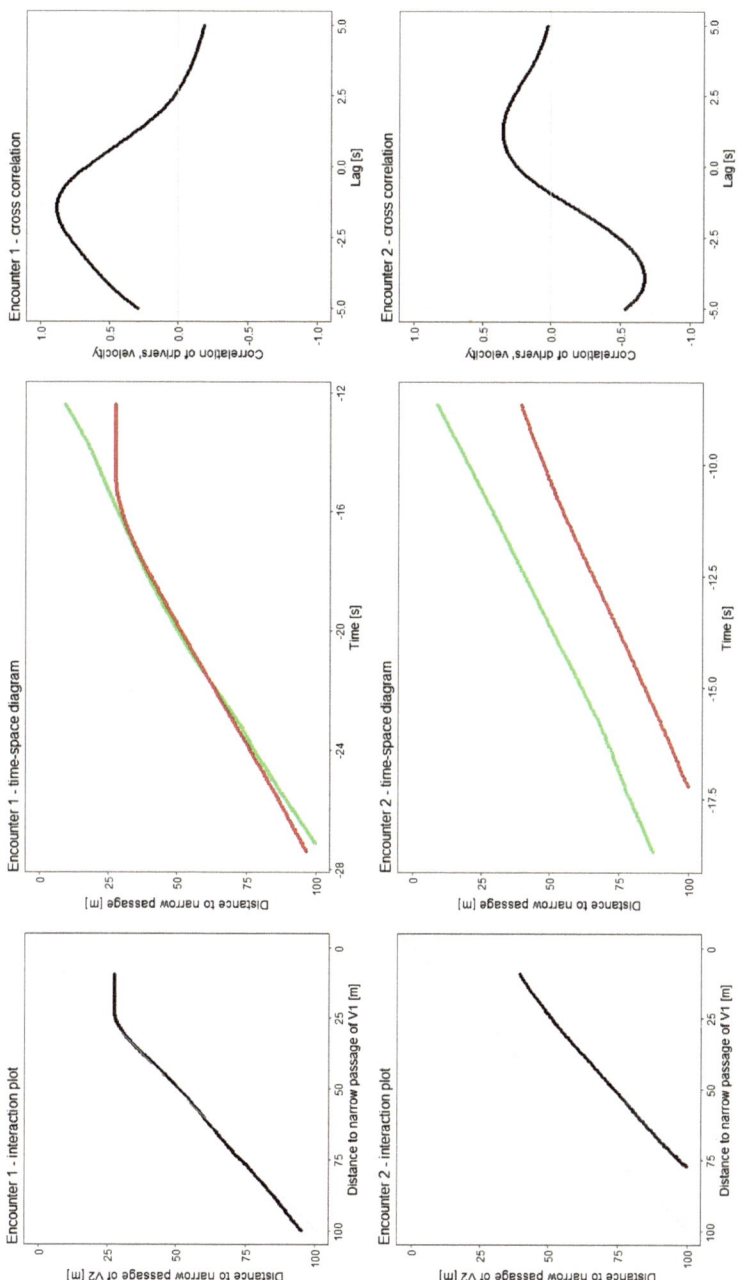

Fig. 4 Interaction plots, time-space diagrams and cross correlations of two exemplary encounters, in which two drivers encounter a two-sided narrow passage from opposite directions. The data comes from the coupled simulator study described in Sect. 2.5.1

drivers' velocities show a high positive correlation (i.e., velocities are similar) when the time series are shifted by -1.25 s, whereas for the lower encounter, the highest correlation is negative (i.e., velocities are in reverse) for a lag of -3.75 s. The lag for which the correlation maximizes might allow to identify the leading and following driver [51].

It must be emphasized that so far there are only isolated cooperation-specific objective measures. Most of the measures originate from other research areas, so that the relationship with cooperation has yet to be investigated. Thus, on the one hand, measures to objectively capture cooperation are still missing, and on the other hand, existing measures' reliability and validity for the construct cooperation still have to be proven.

3.2.2 Subjective Evaluation of Cooperation: PADI and CoopQ

To complement objective measures, cooperative behaviour in traffic also needs to be captured from a subjective perspective, for example by self-reporting questionnaires. The Prosocial and Aggressive Driving Inventory (PADI; [27]) is based on two scales, addressing both prosocial and aggressive driving behaviour, which distinguishes this questionnaire from other self-reporting driving questionnaire that only focus on one of the two (e.g., The Positive Driver Behaviour Scale by [65]). It is based on the assumption that driving behaviour is a stable and enduring characteristic of drivers [27]. Since prior studies linked unsafe driving behaviour to dimensions of the Five-Factor Model of personality, [15, 27, 89] correlated the PADI scales with the Big Five traits. Since no comparable questionnaire existed in German before, the questionnaire was translated into German and successfully validated during the course of this project [82] and used in different studies [81]. Based on [11], PADI was translated into German by two independent translators. These two versions were compared and combined into a third version which was translated back into English by a third translator. Finally, the third version was compared with the initial English version regarding the meaning of each single item. After some minor wording adaptation, a final version was used for the questionnaire's validation. In an online study, $N = 291$ filled in the PADI, NEO-Five-Factor Inventory (NEO-FFI) [10] and driving-related questions. A confirmatory factor analysis, a principal component analysis with varimax rotation and a logistic regression supported the structure of the original questionnaire. Only one item had to be excluded in the German version. The German version of PADI consists of 28 items: 16 measuring prosocial driving (e.g., "drive more cautiously to accommodate people or vehicles on the side of the road (e.g., slow down, move over)") and twelve measuring aggressive driving (e.g., "speed up when another vehicle tries to overtake me") [82].

The subjective evaluation of cooperation in a specific encounter, on the other hand, has mainly been measured uni-dimensional using multilevel response scales. Kauffmann et al. [37], for example, asked their participants to rate how cooperative another driver was on a scale from $0 = $ not at all to $15 = $ very cooperative. Similarly, participants in [59] rated another driver's cooperativeness on a 7-point rating scale

(1 = not cooperative at all, 7 = completely cooperative). In [32], the willingness to cooperate (of themselves and another driver) and the intensity of cooperation in a given situation were each judged on 7-point rating scales. In contrast, [96] assessed the subjective perception of cooperation in more detail by addressing the dimensions satisfaction, relaxation, accordance, and trust. Their participants rated twelve adjective pairs (e.g., frustrating/satisfying, delaying/time-saving) on 6-point forced choice semantic differential scales, with four items related to the experienced situation, three to the participant him-/herself, three to the other driver, and two to the situation impact.

Building on [96], a questionnaire has been designed within CoMove to assess the subjective evaluation of cooperation in a traffic encounter in even more detail (see also [68]). The questionnaire was developed to answer the following questions: (A) Could a given encounter between road users be considered cooperation? (B) Did road users cooperate successfully? For question A, 39 statements reflecting different aspects of cooperation were formulated, for example "The drivers competed with each other". Based on different definitions of cooperation, aspects like altruism, coordination, communication, competition, goal orientation, reciprocity, dependence, interference, mutual agreement, negotiation, costs and benefits were considered [8, 18, 20, 22, 30, 39]. To answer question B, 40 adjective pairs were identified, which reflect common motives in road traffic, for example safety and efficiency [8, 79, 88]. Based on an online survey with 123 participants, the number of items was reduced. By means of descriptive statistics, item analysis and factor analysis, ten items and 22 pairs of adjectives were selected for a first version of the cooperation questionnaire (CoopQ; Table 4). The CoopQ questionnaire was used in studies S2, S3, S6, and S7 (Table 1). An evaluation of the questionnaire's reliability and validity has still to be published.

3.3 Situational Characteristics Influencing Cooperative Lane Changing

3.3.1 A Model of Recognition-Primed Decision Making in Cooperative Situations

Within CoMove, we adapted the recognition-primed decision (RPD) model by [41, 42] to investigate how situational factors influence the behaviour in the cooperative situation of a lane change [81]. The RPD model is a naturalistic decision-making model. According to this model, rather than trying to find the best possible option in a decision situation, the first workable option is selected by the decision maker. The RPD model was developed to describe decision making for experienced agents under complex and uncertain conditions facing personal consequences of their actions and having to react fast. As these assumptions fit to the dynamics of traffic situations and the context of drivers' decision making in traffic the model has been selected to describe decisions made by drivers.

Table 4 Selected items and adjective pairs for the first version of CoopQ

Part A—Could a given encounter between road users be considered cooperation?

The drivers wanted to occupy the same space at the same time

The drivers have adapted to each other

The drivers cooperated

The drivers competed with each other

The drivers acted amicably

(At least one driver/Driver X) showed the other driver consideration

(At least one driver/Driver X) acted selfishly

(At least one driver/Driver X) were at an advantage because of the situation

(At least one driver/Driver X) were at a disadvantage because of the situation

(At least one driver/Driver X) acted with foresight

7-point rating scale ("does not apply at all" to "applies perfectly")

Part B—Did road users cooperate successfully?

Beneficial/obstructive	Supportive/hindering
Relieving/burdening	Enjoyable/unpleasant
Satisfying/frustrating	Pleasant/unpleasant
Relaxed/stressful	Calm/aggressive
Effective/ineffective	Efficient/inefficient
Goal-oriented/unplanned	Coordinated/uncoordinated
Harmonized/not harmonized	Consensual/non-consensual
Fair/unfair	Controlled/uncontrolled
Safe/unsafe	Harmless/dangerous
Risk-free/risky	Understandable/misleading
Unambiguous/ambiguous	Necessary/unnecessary

5-point forced choice semantic differential scale, e.g. efficient ☐ ☐ ☐ ☐ ☐ inefficient

Following the model, well experienced drivers do not have to weigh several alternatives but match the current situation with patterns they already know [43]. The most relevant clues are highlighted by the patterns, which also provide expectations, identify plausible goals and suggest typical forms of reactions in the specific situation [42]. If the situation contains similarities to an already known situation, the most typical alternative is mentally simulated regarding its outcomes in the context of the current situation. If the mental simulation is successful, the intended action will be carried out. Otherwise, this action will be modified mentally until the expected outcome can be reached. Our adapted version of the model is presented in Fig. 5.

In [33], we combined the RPD model with a questionnaire survey inspired by [9], which investigated yielding behaviour in crossings. They surveyed the usage of formal and informal traffic rules and showed that traffic decisions can be portrayed using questionnaires. We used their approach in a questionnaire study on cooperative

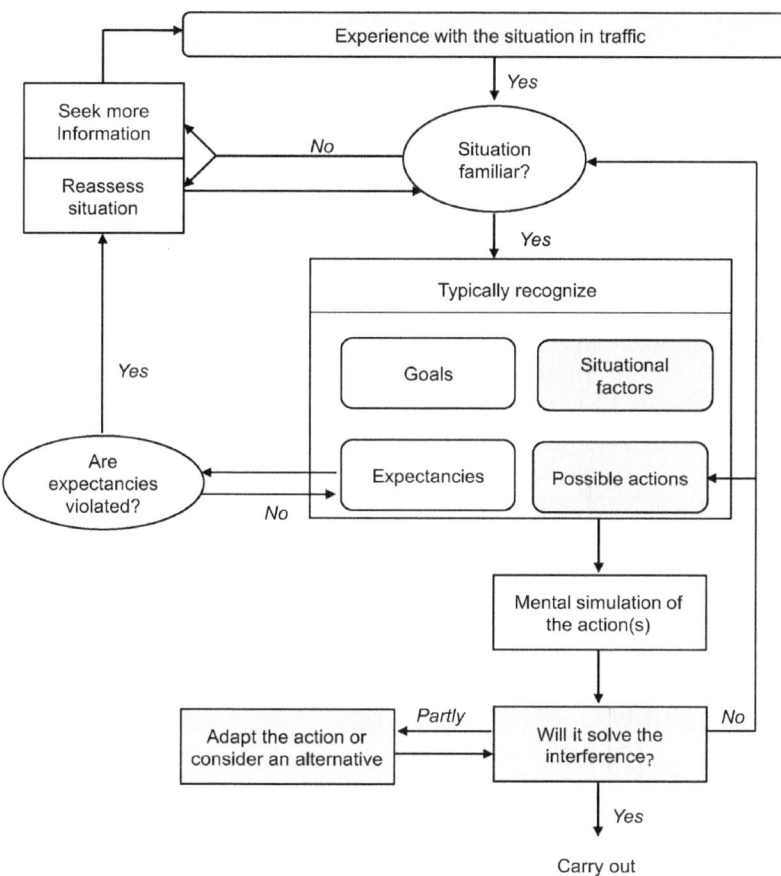

Fig. 5 Adapted decision making model from [81] after the original model from [41]; grey shading represents changes from the original. If experienced drivers encounter a familiar situation, they typically recognize goals, possible actions, situational factors and expectancies. If there are no violations of expectations this knowledge structure allows to simulate actions and their effects to decide whether these actions will solve the current interference. If such an action is found, it is carried out

traffic situations with $N = 281$ participants, which was analysed with the Natural Decision Making approach. This approach made it possible to categorize individual communication signals into offensive or defensive signals and thus make predictions about the intention of the driver.

When a cooperative situation occurs, both partners have to detect the interference and figure out if the situation is familiar. Therefore, the process shown in Fig. 5 has to be carried out by both partners and will be influenced by the actions of the cooperating partner. Moreover, since the cooperative situations are dynamic, it might have to be assessed multiple times depending on the changing conditions due to the behaviour of the interfering partners or other involved agents (e.g., drivers in additional lanes or behind the agents). In a lane-change scenario, the driver who is asked for cooperation has to assess the situation as familiar or not. Since lane changing is a common driving situation, drivers are expected to have experience with this situation and to be able to recognize the situation.

The mental simulation is influenced by expectancies, different situational factors, the goals as well as the possible actions that could be carried out [41]. If a mental simulation indicates that a certain action would solve the interference this action is chosen and carried out. During the simulation process, it might be detected that a certain action only partly solves the interference, which implies that the action has to be adapted or reconsidered. It might be necessary to decelerate even stronger than originally simulated or to adapt the behaviour in another way.

Meanwhile, the driver that wants to change lanes has to indicate the intention and wait until the behaviour of the driver being asked for cooperation indicates that the cooperation being asked for is accepted and the corresponding actions are or will be carried out. This action has to be interpreted correctly by the partner asking for cooperation which has to react to that. Nevertheless, this is the optimal process. At every step of the process, errors might occur. If an interference is not detected or one of the involved drivers finds him-/herself in an unfamiliar situation, the process might be very different and the cooperative situation might not be solved successfully. It could be that the driver that wants to change lanes does not wait, which forces the other driver to step back (e.g., open a gap) even though this driver decided not to accept cooperation.

To better understand which situational characteristics facilitate drivers' recognition of an interference situation and under which circumstances which actions are preferred to solve the situation, we carried out several studies [81, 83–87] investigating how situational factors influence the processes in cooperative situations on the example of lane changing on highways. An overview over these studies – which will be discussed in the following – is given in Table 5.

In these studies, participants were taking the perspective of a vehicle on the faster, left lane on the highway (egocar). On the slower right lane, a vehicle (lane changer) was driving behind a slower commercial vehicle (see Fig. 6a). This scenario describes a discretionary lane change meaning a vehicle merges into another lane to maintain the desired speed level [2] but being not obliged to change the lane. In contrast to that, in [87] we investigated a mandatory lane change [2] in the form of an on-ramp scenario: The egocar drives on the highway when approaching an on-ramp from which another vehicle (lane changer) wants to merge onto the highway (see Fig. 6b).

3.3.2 Investigating Influencing Factors on Preferred Actions in Cooperative Situations

Scope of Action and Situation Criticality

Table 5 Empirical studies conducted within CoMove to investigate the effect of situational characteristics on drivers' recognition of and behavioural preferences in cooperative situations

No.	Method	Scenario	Goal	Refs
S9	Video-based laboratory experiment	Highway LC (2 lanes)	Investigating influence of scope of action and criticality for the lane-changing vehicle on the preferred action.	[85]
S10	Video-based laboratory experiment	Highway LC (2 vs. 3 lanes)	Investigating influence of scope of action, criticality for the lane-changing vehicle, indicator usage and additional lane on the preferred action	[86]
S11	Video-based laboratory experiment	On-ramp (2 vs. 3 lanes)	Investigating the influence of criticality for lane changer vehicle, indicator usage, additional lane and traffic signs on the preferred action	[87]
S12	Driving simulator	LC (2 lanes)	Investigating influence of scope of action, criticality for the lane-changing vehicle and indicator usage on the selected action.	[83]
S13	Driving simulator	LC (2 lanes)	Investigating the acceptance of automated behaviour depending on scope of action, criticality for the lane changing vehicle, indicator usage and the maneuver which was carried out by the automation	[84]
S14	Driving simulator	LC (2 lanes)	Investigating the influence of highlighting relevant situational factors on the preferred action	[81]

LC = lane change

In S9 [85], a video-based study, the preferred behaviour of the driver in the faster lane was investigated based on manipulating the scope of action (here called distance in time and space) and the situation's criticality for the other driver (lane changer) in the slower vehicle. The participants took the perspective of a driver in the faster left lane (egocar) while the simulation-controlled lane changer driving in the right lane approached a slower truck. An experimental setting using videos allowed strict control over these influential factors at a fixed decision point, on which certain lev-

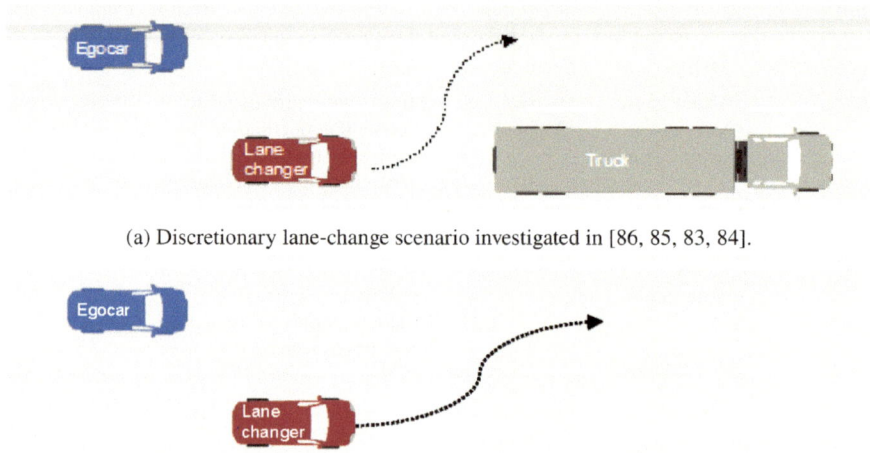

(a) Discretionary lane-change scenario investigated in [86, 85, 83, 84].

(b) Mandatory lane-change scenario investigated in [87].

Fig. 6 These two sketches show the principal scenarios investigated in the different experiments. The participants were always situated in the car on the left lane (blue car called "Egocar") from where they observed the development of the situation. The red car (called "Lane changer") was always controlled by a simulation and represented a vehicle that might intend to change the lane. In Fig. 6a the situation of a discretionary lane change of the "Lane changer" is depicted, in Fig. 6b a mandatory lane change situation of the "Lane changer" is shown. The captions of the figures list the experiments where each of the situations was applied

els of situation's criticality for the lane changer and scope of action were reached. Each of the 43 participants assessed 81 lane-change scenarios regarding the preferred behaviour (accelerate, decelerate or maintain speed) and the situation's criticality (on a five-point rating scale) when the video stopped. The scope of action was operationalized in two ways: first by TTC and second by headway distance (HW) between the egocar and the lane changer. The situation's criticality for the lane changer was operationalized by TTC between the lane changer and the truck and the HW between the lane changer and truck. TTCs were manipulated on three levels, ranging from a critical TTC (2 s) to ambivalent (4 s) and an uncritical TTC (6 s). The HWs between these vehicles were set at 6 m, 13 m, and 20 m. One aim of the exploratory study was to identify the best operationalizations for the scope of action and situation's criticality for the lane changer. The study design was a 3x3x3x3 within-subjects design with a randomized presentation of each scenario. For the egocar, the speed was between 87 km/h and 152 km/h. The truck's speed was kept at 80 km/h which is the maximum allowed speed level for trucks on German highways. The speed of the lane changer varied between 84 km/h and 116 km/h. These ranges result from the combination of TTC and HW manipulation. All vehicles kept their speed constant during the video.

Results showed that the preferred behaviour was almost equally distributed between the options accelerate, decelerate, and maintain speed. A generalized linear mixed model was used to analyze the effects of TTCs and HWs. It showed that the TTC between egocar and lane changer was a significant predictor for the preferred behaviour: By increasing this TTC (higher scope of action), the frequency of decelerating increased, and the frequency of maintaining speed decreased. In contrast, if the TTC between changer and truck increased (a less critical situation for the changer), the frequency for decelerating decreased, and the frequency for accelerating and maintaining speed increased. HW between egocar and changer showed to be a significant factor for accelerating but not for decelerating, while the HW between changer and truck was a significant factor for decelerating but not for accelerating. When the participants decided to decelerate the mean criticality rating was highest and lowest when they decided to maintain speed. This study [85] indicates that the situational factors criticality for the lane changer and the scope of action are both relevant for the participants to adapt their decision in accordance with the interference in this lane change situation. Furthermore, the results showed that the manipulation of TTC was the more useful manipulation for the scope of action and criticality for the lane changer compared to HW, even though it might be more difficult to estimate [40].

Following [85], a second video-based study was carried out [86] (S10). In [85], the options regarding the preferred behaviour for the participants were limited to a longitudinal adaptation of vehicle speed due to the two-lane paradigm. In [86], the possibility for lateral adaptation by changing to a third lane as an additional option was investigated. Changing lanes is considered less costly compared to decelerating [28]. We expected participants to prefer the less costly alternative and, therefore, to change lanes if possible. In [85], the lane changer used the indicator in all presented situations. Since we expected that the communication of intention has an effect on the preferred behaviour, we also manipulated the communication of intention in the form of indicator usage in [86].

The scenario was similar to the one of [85]. 51 participants had to assess different lane-change scenarios. Four factors were manipulated: The scope of action by the distance between egocar and lane changer (TTC; 2s vs. 6s), the availability of a lane change to the left (2 vs. 3 lanes; lane change possible yes vs. no), the criticality for the lane changer (TTC; 2s vs. 6s), and the way the intention to change lanes was indicated. The latter was manipulated on four levels: (1) no indicator usage, (2) brake lights, (3) indicator usage, and (4) indicator and an additional arrow over the vehicle as an augmented-reality (AR) display. With this fourth way of indication, we investigated if the indicator is salient enough or if a more salient stimulus is needed. Additionally, by flashing brake lights, we investigated how another form of communication influences the participants' behaviour because braking lights might be interpreted such that the vehicle is slowing down in order to stay behind the truck. Every combination of the four factors was shown twice as a repeated measurement. To hide this repetition the surrounding landscape was changed between the repetitions. The presented highway was a three-lane highway, but in half of the scenarios, the left lane was closed due to a construction site, creating a two-lane scenario. If the left lane was free, participants had the additional option to change lanes to solve the interference in the situation.

The results of S10 [86], displayed in Fig. 8 (left), showed a clear behavioural preference for a lane change to the left to solve the cooperative situation when an additional lane was available which was in line with our expectation. In the two-lane condition, without the possibility to change lanes for the participants, participants preferred maintaining speed over decelerating and accelerating. Moreover, the factors situation's criticality, scope of action, and the indicator usage affected the preferred behaviour. If a third lane was available, a lower scope of action increased the preference to change lanes compared to maintaining speed but not for accelerating or decelerating. In the two-lane condition, indicating the intention to change lanes increased the preference for lane changing to the left, decelerating, and even accelerating compared to maintaining speed.

An additional arrow had only a small additional effect compared to a regular indicator due to a ceiling effect. When the braking lights were turned on or when no signal was presented, the preference was to maintain the current speed. In these cases, no need for cooperation was detected by the participants.

The main difference to S9 [85], the introduction of an additional option on how to solve the cooperative situation and the indication of the intention to change lanes were both influential on the preferred behaviour. The results indicate that indicator lights seem to be salient enough. Therefore, a missing adaptation in behaviour can not be explained by the low salience of an intention signal.

Discretionary Versus Mandatory Lane Changes

The aim of S11 [87] was to investigate the behaviour in an on-ramp situation (see Fig. 6b), which is categorized as a mandatory lane change [2]. On-ramps are considered as challenging situations that are a cause of traffic jams [3, 55]. We aimed to investigate if the factors indication of the intention (blinking) to change lanes and the situation's criticality for the lane changer are influencing the preferred behaviour as they did in a discretionary lane change situation on highways [85, 86]. Based on earlier studies, we expected more adaptation of the behaviour (any behaviour which is not to maintain the current speed) (1) if the situation's criticality for the lane changer is high, (2) if the intention to change lanes is communicated explicitly, and (3) if the third lane was available. Additionally, a traffic sign explaining merging traffic was shown. We expected more adaptation in behaviour if this sign was presented at the beginning of the scenario since it should help drivers identify the interference. Moreover, we expected a two-lane scenario to be perceived as more critical than a three-lane scenario.

The videos were presented to the same 51 participants as in [86]. Again, the perspective was the one of a vehicle driving on the highway while, this time in front of them, another vehicle wanted to merge onto the highway coming from an on-ramp. As in S10 [86], we manipulated the way the merging vehicle indicated its intention to merge onto the highway on the same four levels ((1) no indicator usage, (2) braking lights turn on, (3) indicators usage, and (4) indicator and an additional arrow over the vehicle as an AR display). The situation's criticality for the merging vehicle was operationalized by manipulating the TTC towards the end of the on-ramp (2 s vs. 6 s). Additionally, we manipulated the availability of unoccupied lanes. In half the scenarios, the left lane was free, while in the other half of the scenarios, this lane was blocked by a construction site. As in S10 [86], this leads to the additional option "change lanes" in these scenarios. Every factor combination was presented twice. Therefore, each participant experienced 64 lane-change scenarios.

Results show that when having the possibility to change lanes to the left, this was the preferred behaviour (see Fig. 8). When this option was not given, the preference was to decelerate (see Fig. 7). The situation's criticality for the lane changer was considered in the two-lane condition both for maintaining speed versus decelerating as well as accelerating vs. decelerating, increasing the preference for decelerating if the criticality for the lane changer was raised.

In the three-lane condition only the odds of accelerating compared to changing lanes increased if the criticality for the lane changer decreased. The indicator usage or the indicator usage with an additional AR arrow decreased the odds of maintaining speed or accelerating compared to decelerating in the two-lane condition, which was not found in the three-lane condition. When braking lights of the other vehicle were turned on, the odds of accelerating compared to decelerating increased significantly in the two-lane condition and also in the three-lane condition in comparison to lane change. The traffic sign explaining merging traffic only had an effect in the two-lane condition for maintaining speed compared to decelerating, in the form that the odds for maintaining speed were lower when the sign was presented. In all other cases, it was not significant. The participants assessed the two-lane scenario as more critical than the three-lane condition.

Comparable to [86], participants preferred the less costly alternative of changing lanes if possible [28]. This human behaviour corresponds with the maneuver automated vehicles most likely will carry out in these situations [12].

The effect of a traffic sign explaining merging traffic was small. However, a carry-over effect might have influenced the results since participants might have remembered the traffic sign even in the scenarios in which the sign was not presented at the beginning of the video. Therefore, further studies should investigate whether this sign or other signs facilitate the recognition of an oncoming interference using a between-subjects design.

The results indicate that participants considered the mandatory character of the lane change due to the end of the on-ramp in their preferred behaviour.

Since [86, 87] were carried out together, it was possible to investigate whether more behavioural adaptation is shown when the lane changer intends to carry out a mandatory lane change compared to a situation in which the lane changer intends to carry out a discretionary lane change [81]. As Fig. 7 shows, the preferred action in the discretionary lane change condition with two lanes was to maintain the speed, while in the mandatory lane change condition, the preferred option was to decelerate. The proportion of accelerating was marginally lower in the mandatory condition compared to the discretionary lane change condition. In the three-lane condition, the preferred choice was to change lanes both in the discretionary lane change and in the mandatory scenario (Fig. 8). However, in the mandatory scenario, the other behaviour options were rarely selected, while in the discretionary lane-change scenario, maintaining speed was selected in 35 % of the cases.

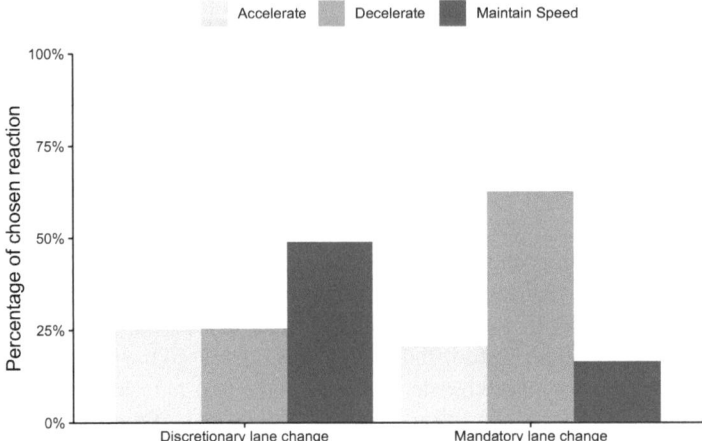

Fig. 7 Percentage of chosen actions in discretionary (left) from [86] and mandatory (right) from [87] lane change situations

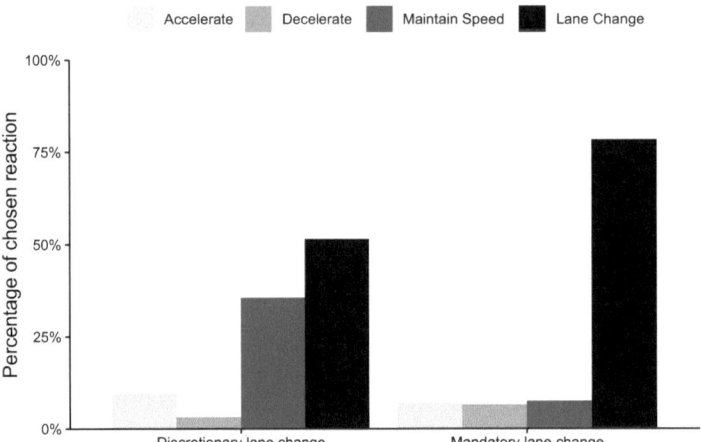

Fig. 8 Percentage of chosen actions in situations with three lanes either for discretionary lane chances (left) or mandatory lane changes (right)

The results support the hypothesis that more behavioural adaptation is shown when the lane changer intends to carry out a mandatory lane change compared to a situation in which the lane changer intends to carry out a discretionary lane change for the two-lane condition. The difference between discretionary lane change and mandatory lane change was significant both for accelerate vs. maintain speed, and decelerate vs. maintain speed. The differences between discretionary and mandatory lane change were significant for accelerate versus maintain speed, decelerate vs. maintain speed and change lanes versus maintain speed.

If a mandatory lane change was intended, the willingness to adapt the behaviour increased in this study. This is in line with the research by [2]. However, the behaviour carried out in actual traffic might differ from the first intention participants show in this study.

Besides the intention to facilitate a necessary lane change, the results might be influenced by the experience participants made in this situation. Zheng [95] assumes that the decision making for the driver changing lanes is different in a mandatory lane change compared to a discretionary change. Their research indicates that drivers carrying out a mandatory lane change are willing to take higher risks than drivers in a discretionary lane change situation. Further studies investigating these lane changes will be needed, also focusing on the agent intending to change lanes regarding the decision making when and how to change lanes.

In three studies [85–87] the focus was on the preferred behaviour at a particular moment rather than the behaviour the participants would show when experiencing this situation as a driver supported by an automation. However, the behaviour they would carry out might differ from the preferred behaviour of an automated vehicle they are sitting in. Therefore, the next step was to investigate the effect of the influencing factors when the situation was experienced in the driving simulator [83].

3.3.3 Preferred Actions of Automated Vehicles in Cooperative Situations

To achieve comparability to the previous studies [85–87], the same levels of TTCs were set to investigate the scope of action and the criticality for the lane changer. To do so, automated driving was introduced, since individual differences in speed, acceleration, and deceleration under manual driving would have affected comparability between participants as well as to the earlier results.

Based on the results of the video studies, we expected an adaptation in behaviour (deceleration or acceleration) if (1) the scope of action was high compared to when it was low, (2) if the criticality for the lane changer was high compared to when it was low, and (3) if the lane changer used the indicator to communicate the intention to change lanes compared to no communication of intentions.

In a driving simulator study, 32 participants experienced 36 lane-change scenarios on a two-lane highway. Three factors were manipulated: The scope of action (TTC 2 vs. 6 s), the criticality for the lane changer (TTC 2, 4 and 6 s), and indicators usage (yes vs. no). Every scenario was presented twice. The landscape around the highway was altered in the scenario so that the participants did not notice the repetition.

Participants were instructed that their automated vehicle would ask for their preference in specific situations and would carry out the selected maneuver automatically. This preference had to be provided at a specific point in time, indicated by an acoustic signal. The specific point varied based on the manipulated factors. Participants had to answer within 2 s, and if they did not decide, the car would continue with the current speed, and the lane changer would stay behind the truck. No change in behaviour was shown by the lane changer if participants selected maintaining speed or accelerating. When decelerating was selected, the lane changer would change the lane to the left and overtake the truck and return back to the right lane after the maneuver.

Results showed that maintaining speed was the maneuver that was selected the most, followed by decelerating and accelerating. As expected, the indicator usage had a significant effect on the selected behaviour: If the lane changer used the indicator to indicate the intention to change lanes, the preferred behaviour was to decelerate. In line with the hypothesis, the adaptation in selected behaviour was significantly influenced by the criticality for the lane changer. When the TTC was low, decelerating and accelerating were selected more often compared to when TTC was high.

Additionally, an interaction between criticality for the lane changer and indicator usage occurred: More adaptation of the behaviour was shown both for decelerating and accelerating when the scope of action increased. The additional effect that the indicator was stronger in situations with a low criticality for the lane changer might indicate that participants need this additional information that the lane changer has the intention to merge into the faster lane when the situation is uncritical, while in situations with a high criticality for the changer, the short distance to the truck in front might need less additional explanation through the indicator. Therefore, additional possibilities to communicate one's intentions were focused on in the final study [81].

Moreover, it is relevant to know how the actively chosen behaviour differs from the acceptance of a behaviour carried out by an automation since future technology might work like that, which was addressed in study [84].

3.3.4 Investigating Acceptance of Automated Maneuvers in Cooperative Situations

In [84], we investigated the acceptance of the performed maneuver by an automated vehicle in the same lane-change scenario as used in [83]. Different to the study described above (see Sect. 3.3.3), here the automation suggested a maneuver and participants were asked whether they would accept the suggestion.

Research related to driving style and comfort indicates that participants would accept behaviour carried out by an automation that differs from their manually performed driving behaviour [4, 93]. Therefore, it is necessary to know if a comparable pattern can be found in cooperative situations and how situational factors that influence the preference of certain behaviour influence the acceptance when an automated vehicle is in control.

It was investigated if drivers prefer the automation to adapt its behaviour over not adapting the behaviour when another vehicle wants to change lanes. Additionally to the studies before, the influence of being cognitively distracted by a secondary task on this preference was investigated. Higher acceptance for the automated behaviour was expected (1) when the behaviour was an adaptation (decelerating or accelerating) and the lane changer used its indicator, (2) when the scope of action was small and (3) when the participants were cognitively distracted. That third hypothesis was based on the assumption that distracted participants would have less situation awareness and agree with any decision made by the automation. Additionally, it was expected that higher criticality for the lane changer increases the acceptance of a behavioural adaptation.

In a driving simulator study, 20 participants experienced 48 lane changes in an automated vehicle. The vehicle was either accelerating, maintaining speed, or decelerating when it approached the lane changer driving behind the truck. The scenarios were the same as in the study before [83], with the same manipulations of indicator usage, the scope of action, and criticality for the lane changer. After each scenario, participants had to decide if they accepted (yes vs. no) the automated vehicle's behaviour. In half the scenarios, participants were distracted by an auditory one-back task.

Overall, the acceptance rate was generally high (74%). The highest acceptance was shown for decelerating, resulting to be a significant factor. Acceptance for maintaining speed or accelerating decreased when the lane changer was using the indicator compared to decelerating. We expected a higher acceptance both for decelerating and accelerating compared to maintaining speed, when the indicator was turned on. However, the effect was only significant for decelerating compared to maintaining speed. Therefore, this hypothesis had to be rejected. In line with the hypothesis, a small scope of action resulted in a significantly higher acceptance compared to a large scope of action. Also, the indicator usage significantly affected the acceptance. Being engaged in a secondary task had no significant effect on the acceptance, which contradicts our hypothesis.

Participants, sitting in a simulated automated vehicle, showed a clear preference that the automation adapts the behaviour when another vehicle plans to change lanes with the highest acceptance rate for decelerating. They preferred a more defensive driving style than the participants in the studies before when being the driver: In comparison to [83], they accepted maneuvers performed by the automation that differed from what participants would have preferred when asked before the maneuver. Indicator usage influenced the acceptance rate. The general high acceptance of the automated behaviour dropped, if the automation accelerated or maintained the speed.

The results show that with a higher scope of action, the preference or selection of deceleration increased while accelerating or maintaining speed decreased. Decreasing the speed allows the lane changer to merge in front of the egocar while accelerating would close the gap. Additionally, the less critical the situation was for the lane changer, the more maintaining speed was selected. Therefore, it can be assumed that participants considered the criticality of the situation. Moreover, also the necessity of the lane change is taken into account as the results showed: Participants' selection of decelerating was higher in the on-ramp scenario when the lane change was necessary compared to the lane change on the highway when the lane change was discretionary. When given the additional option to change lanes to a third lane on the left, participants preferred this option, which is generally seen as a less costly alternative [28].

Regarding communication, participants expected the lane changer to communicate the intention to change lanes. If no intention was communicated, they kept their speed constant [83, 86]. An additional more salient indicator was not necessary [86]. In line with that, participants expected their automated vehicle to behave accordingly: If the lane changer used the indicators and the automated vehicles reacted with maintaining speed or acceleration, the acceptance was lower compared to the automated reaction of deceleration [84].

Further research will be needed but the results imply that participants prefer cooperative automated vehicles.

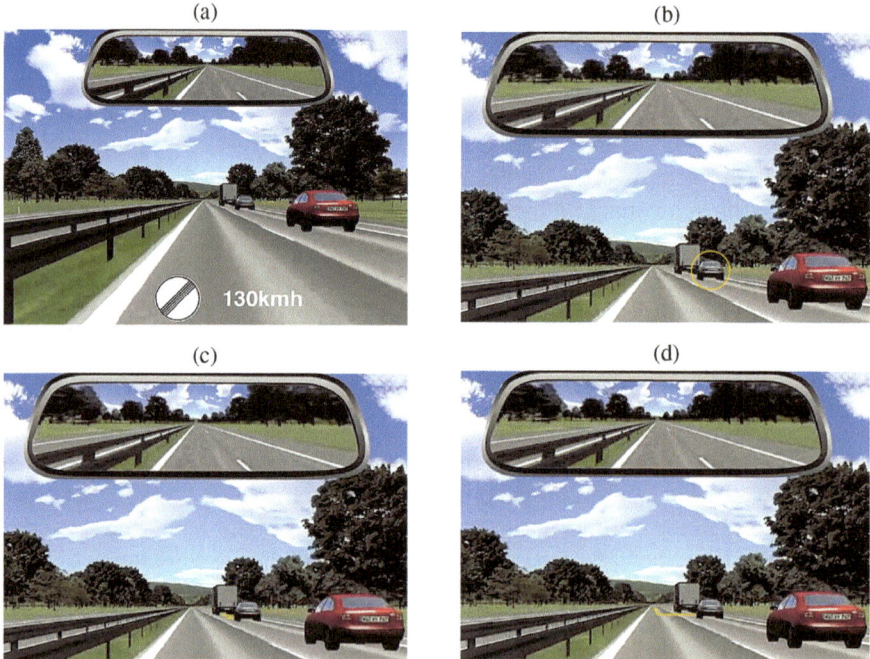

Fig. 9 Different presented information: **a** no additional information, **b** identifying the lane changer, **c** distance to the slower vehicle, **d** the planned trajectory. Note: The current speed was presented in all four conditions but was not presented here for better visualization. Original from [81]

3.3.5 Supporting Situational Awareness by Highlighting Situational Factors in Cooperative Situations

In the final study [81], situational factors were highlighted using an AR display to investigate how highlighting aspects of the situation influences the preferred action (see Fig. 9). Results of prior studies [38, 48] show that AR displays could be a suitable way of supporting human drivers in cooperative situations. Corresponding with the Situation Awareness model by Endsley [23], we highlighted either: (1) The vehicle that wants to change lanes, which should help the driver identifying the vehicle that wants to cooperate. This corresponds to the perception (Level 1) in Endsley's model. (2) The reason for the lane changer to change lanes, which in this case was the decreasing distance between the vehicle and the vehicle ahead. This should support the comprehension level (Level 2) in Endsley's situation awareness model. (3) The planned trajectory was shown, which should support the projection level of situation awareness (Level 3).

In the experiment, 29 participants took part. All participants held a valid driving license for at least one year. In contrast to studies so far, there were two vehicles following the truck. Between the scenarios, it varied which vehicle (Car A or Car B) wants to change lanes. The intention was to better test the HMI since one of the elements intends to improve the detection of a potential lane-changing vehicle.

The results from [81] indicate that highlighting relevant situational factors to support the driver in the understanding of being in a cooperative situation is a promising approach. Compared to not showing any additional information, highlighted relevant factors increased deceleration, which is seen as an increase in the willingness to support solving the cooperative situation. When highlighting the planned trajectory for the vehicle asking to change lanes, the preference to decelerate and open a gap was the preferred choice (Fig. 10).

3.3.6 Summary of the Empirical Studies on Factors Influencing Action Preferences in Cooperative Situations

The aim of these six studies was to deepen the understanding of the influence of situational factors on the action selected in cooperative situations under the theoretical framework of decision-making. For that reason different influencing factors were manipulated: The criticality for the changing vehicle, the scope of action, and indicating the intention to change lanes. These were investigated regarding their influence on the preferred action via video studies (S9, S10 and S11 in Table 5) and on the selected action (S12 in Table 5) as well as accepted action (S13 in Table 5). In the final study (S14 in Table 5), these influential factors were highlighted in an AR display to investigate how drawing attention towards these factors influences the preferred action.

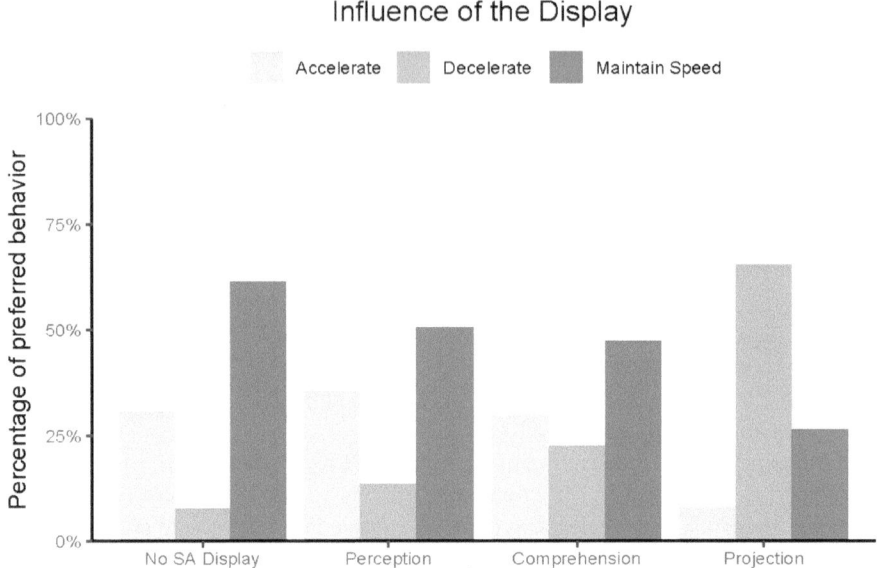

Fig. 10 Percentage of preferred automation behaviour with HMI supporting different levels of situation comprehension. Results from [81]

4   Modeling Cooperation Behaviour

In addition to the need for valid forms of measuring cooperative behaviour and for a firm empirical base on the factors that influence it, there is a need for theories of cooperative behaviour that are able to integrate the various empirical findings and that allow for generalization across various concrete scenarios. As driving is a task that involves many psychological processes, from perception to decision making and action execution, aspects of driving can only be described and explained by considering the interplay of these processes in a given situation and how it produces a given behaviour.

As described above, one of the theoretical concepts that is highly relevant to the field of cooperation in traffic is the understanding of the current situation and its future development [6, 21, 23]. Successful cooperation requires a shared understanding of the current situation and its requirements. This includes also a mutual understanding of the interaction partner's goals, state and action plans [14, 44] to enable appropriate communication and plan negotiation between the partners [44] as seen in our experimental studies described above [83, 84, 86] and referred to in the cooperation model based on recognition primed decision making [33] (see Fig. 5).

Comprehension based models of situation understanding [6, 21] describe in detail the psychological processes and structures underlying the construction of a mental representation of a dynamic situation and thus provide a suitable basis for modeling

the cognitive processes involved in the construction of a shared situation representation in cooperation in traffic. In general, this construction includes the perception
of the relevant situation elements, the understanding of these elements in relation to
the overall situation and the prediction of the future development of the situation on
the basis of learned expectations or active projections of the driver [23]. That means,
the construction of this mental representation involves many cognitive processes and
structures of the human information processing system. It is the interplay between
perception, attention, long-term memory, working memory, evaluation processes and
decision making that leads to a given representation in a given situation which is then
the basis of, as in our experimental situation, the decision whether to accelerate, to
decelerate or to maintain speed in the lane change situation depicted in Fig. 6a.

Consequently, to model the process of comprehending a dynamic situation and
of maintaining and updating it requires to model the interplay of all these processes
in a dynamic scenario. One possible framework that can be used for modeling this
interplay are cognitive architectures based on Alan Newell's concept of unified theories of cognition [62]. Cognitive architectures represent a theory of how the brain
achieves cognition. They are implemented as computational modeling platforms for
cognitive tasks that enable the creation of models that can be used to explain and predict task performance. They have been used by researchers of artificial intelligence
and cognitive psychology for many decades [47]. Many of these architectures are
extensively used in modeling complex cognitive tasks involved in driving situations
[73], piloting air-crafts [13], and air-traffic control [50].

4.1 ACT-R: A Unified Theory of Cognition

One of the most prominent candidates of unified theories of cognition is ACT-R [1].
ACT-R (which stands for Adaptive Control of Thought—Rational) underwent multiple revisions since its original publication, and has spawned over 1.100 publications
in the field of cognitive science [72]. It is grounded on a firm empirical basis of basic
psychological research on cognitive processes that play a major role in theories of
how humans comprehend dynamic situations and was already successfully applied
to model some aspects of the driving task [73, 74]. Therefore, we decided to use
this architecture to develop a computational cognitive model of decision making in
cooperative lane change situations such as those depicted in Fig. 6a.

The computational implementation of the ACT-R theory is a production-system
architecture, that models knowledge either as declarative or procedural knowledge.
Declarative knowledge is knowledge we are aware of and we are generally able
to verbalize. An example would be "Berlin is the capital of Germany". In ACT-
R declarative knowledge is represented as a set of chunks of factual information.
Each chunk consists of a collection of pairs of attributes and values. This declarative
knowledge represents the knowledge that a person is assumed to have when she/he
performs a task or solves a problem. An example of such a possible chunk of ACT-R's
declarative knowledge database is shown in Fig. 11.

Procedural knowledge represents knowledge that controls behaviour but of which we are often not conscious. Examples are how we produce language or ride a bike. Procedural knowledge is represented as condition-action production rules in ACT-R as shown in Fig. 12. The condition side of a production rule specifies a set of features that has to be true for the rule to be selected and executed. The action side of the production rule consists of a set of actions that the model performs if the rule is selected, i.e. "fired". It has to be noted that Figs. 11 and 12 represent an informal natural language description of chunks and production rules to provide an overview of what the concepts mean. "Real" ACT-R chunks and production rules have to be more precise and specific to be actually implemented in ACT-R.

Chunks and production rules are the basic building blocks of any ACT-R model that is executed by the ACT-R cognitive architecture. This architecture consists of a set of modules and each module implements a specific cognitive function. Information is exchanged between the modules via buffers. Each module includes any number of buffers and each buffer can hold one chunk at a time. The information held in a buffer is accessible for other modules to be read or modified. The chunks in the buffers of the modules consequently represent that set of information that is immediately accessible to the ACT-R model. The declarative module holds all chunks of the ACT-R model, that is it represents the model's declarative memory. It has one buffer, the so called retrieval buffer. The declarative module reacts to requests to the module by searching through its set of chunks to find a chunk that matches the request. This chunk is then placed into the retrieval buffer.

The production rules of an ACT-R model are held in the procedural module. The procedural module does not have a buffer and it does not react to requests. The procedural module continuously checks the buffers of the other modules for patterns that match the conditions of some of the productions it holds. If it finds such a pattern the matching production rule is fired and the actions defined on the action side of the production rule are executed. Such actions usually modify the contents of buffers of

```
Car-023
    lane right
    lead-car yes
```

Fig. 11 An example of an ACT-R chunk that represents that there is a car on the right lane that has a lead car in front

```
IF the goal is to drive safe
    and you are driving on the left lane
    and there is a car in front of you on the right lane
    and this car is fast approaching the lead car on the right lane
THEN create the goal to provide space for this car on your lane
```

Fig. 12 An example of an ACT-R production rule that might represent a piece of procedural knowledge as part of a driving strategy that prioritizes safe driving

one or more modules. The procedural and declarative module are accompanied by a number of other modules, such as a visual module to process information from the visual field, a manual module to control motor skills, or a goal module to oversee and control objectives and intentions.

The internal structure and rules of the whole architecture are inspired by cognitive neuroscience and make ACT-R capable of validating experiments in cognitive science by matching human reaction times and predicting error rates and strategy choices [16, 56].

4.2 Modeling Decision Making in Cooperative Traffic Situations in ACT-R

As the starting point for an ACT-R model of cooperation in traffic we used the adapted decision making model of [81] depicted in Fig. 5 in combination with the basic ideas of theories of situation comprehension [6, 21]. In this chapter we can only sketch the main components of the assumed information processing steps of the first version of the ACT-R model of cooperation in traffic. These steps are depicted in Fig. 13.

As an empirical basis for the model development data and the experimental paradigm from previous studies with human participants [81, 83–87] were used. We designed a traffic simulation scenario in ACT-R identical to the one shown in Fig. 14. The ACT-R model is given the task to decide whether to accelerate, decelerate or maintain speed in dynamic scenarios with differing TTC values between the vehicles. In the visual field of the model, three visual objects are present: (a) the blue box representing the lane changer car, (b) the red block representing the slow truck, and (c) a virtual near point on the road in front of the egocar based on [73]'s model of gaze behaviour while driving. The model continuously moves its visual attention between the visual objects to gather relevant information about the situation. Visual information such as the indicator of the lane changer and the visual cues used for estimating the TTC values between the different vehicles are crucial in selecting the appropriate action for the given traffic situation. As can be seen in Fig. 13, perceiving these situational characteristics triggers a memory retrieval process by which a chunk representing a previously experienced comparable situation is retrieved. This corresponds to the recognition of familiar situations as described in the decision making model in Fig. 5. The retrieved chunk contains information about the possible development of the situation in the near future, and about the criticality that has been experienced in similar situations in the past. Consequently, based on this memory retrieval the model assesses the situation in terms of its criticality and whether an action has to be carried out to avoid a safety critical situation or whether no action is necessary. If the model comes to the conclusion that an action is required it selects the action that has led to a successful solution in the past and that is therefore connected to the retrieved situation representation. This process corresponds to the recognition based decision-making process as described in Fig. 5.

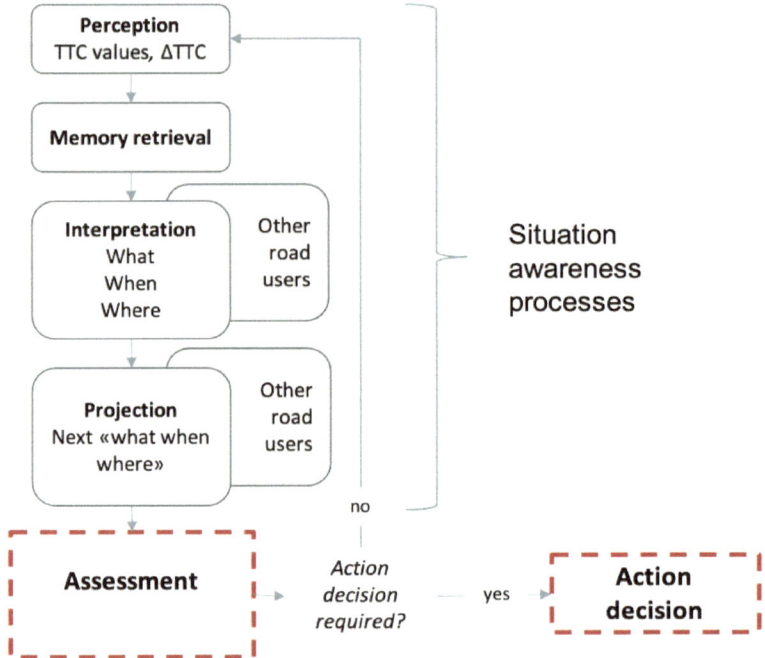

Fig. 13 Flow chart of the information processing steps of the first version of the ACT-R model for cooperation in traffic as described in [5]

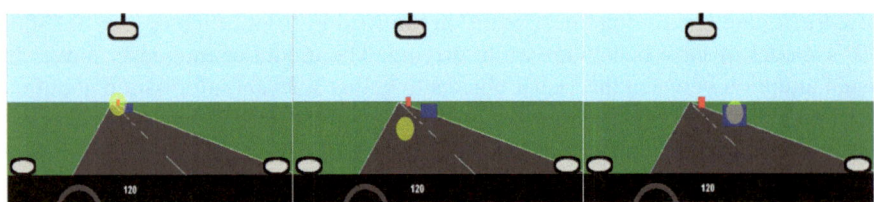

Fig. 14 The traffic scenario in the ACT-R simulation at different time stamps. The red rectangle represents the truck, the blue rectangle depicts the lane changer, and the yellow circle illustrates the location of the visual attention

To date, however, the necessary understanding of the role and underlying processes of criticality estimation and its integration into an understanding of the situation represented in a situational model of the driving situation has yet to be clarified to be able to model it precisely in ACT-R. The preliminary experiments conducted to validate the first model versions clearly show that the perception of the indicator status of the lane changer and the TTC values are crucial for the model's decision making. This is, of course, in line with the experimental data collected in the series of experiments by Stoll and colleagues [81–87]. But the simulation data based on these first model versions also show that these situational characteristics are not sufficient to explain the observed behaviour. It became clear that, for example, the integration of assumptions about the other driver's perception and interpretation of the current situation and its projection into the near future into the situation representation of the driver on the left lane is necessary to be able to describe the observed behaviour of human drivers in such situations. In Fig. 13, this is represented by the grey boxes "Other road users" next to the "Interpretation" and "Projection" box. Filling these gaps and identifying general causal mechanisms underlying cooperation behaviour in a variety of driving situations are the goals of future studies on cooperation in traffic. Although the current studies in ACT-R are inconclusive to date, modeling the underlying psychological processes in ACT-R is a promising method to advance research on cooperation in traffic and to validate experimental results because it provides quantitative predictions of driver behaviour that enable robust hypothesis testing.

5 Conclusion

By focusing on road narrowings and lane changing, this chapter gives an overview on methods for data collection, subjective and objective measures of cooperation, factors influencing cooperation behaviour and behavioural and computational cognitive modeling to support the systematic research on cooperation in traffic. This overview is based on findings collected in studies conducted within CoInCar in the project CoMove. In this respect, the results on factors influencing human behaviour in cooperative situations, either in a manual or an automated setting, and initial findings from modeling the cognitive processes underlying cooperative driving behaviour are worth highlighting. This chapter illustrates that there is still a great need for research on cooperation in road traffic, which includes not only the manifold thematic research questions but also the methodological approach.

So far, there are only a small number of (scenario-specific) measures of cooperation and no standardized and established procedures to assess and measure cooperative behaviour. Coupled simulators are a promising method to investigate cooperative situations, but need further research on their behavioural validity. With respect to subjective measures of cooperation, the German version of the Prosocial and Aggressive Driving Inventory [27, 82] was successfully validated and an additional multidimensional questionnaire for the subjective evaluation of cooperation [68] was developed within the project. Both questionnaires now allow to assess cooperative traffic behaviour either in general or within a specific traffic situation. In addition, objective measures allow to quantify different aspects of cooperation, for example the temporal and spatial proximity, costs and benefits, and the dynamic interplay of road users, but their reliability and validity have yet to be confirmed. In particular the last aspect, the dynamic interplay, requires further measures in order to systematically describe cooperation. In this regard, within CoMove we have been able to develop measures, at least for road narrowings, that allow us to quantify the arrival order of drivers [67].

In a series of highly controlled experiments various situational characteristics have been identified that influence drivers' decision making in dynamic cooperative situations. The results indicate that, besides others, the perceived criticality of the situation, the available scope of action, and the assumed planned actions of the inter-action partner asking for cooperation clearly influence the decision-making process of the driver being asked for cooperation. The empirical results also show that there is a clear overlap between actions chosen by drivers when driving manually and which actions of an automation are preferred in cooperative situations. But this overlap is not complete and the origin of the differences has still to be explained in order to create automation behaviour that is accepted by human drivers.

The results of the first versions of an ACT-R model of cooperative behaviour in traffic show that considering the situational characteristics of a cooperative situation is important but definitely not sufficient to explain human cooperation behaviour in traffic. Clearly, the driver's assumptions about the situation understanding of the cooperating partners and their likely goals and action plans are strongly influencing the driver's own decision making in these situations. This becomes very clear when one considers the effects of situational elements that provide information about the future development of the situation or the cooperating partners' intentions. The first one was investigated via a HMI that highlights situational elements that should support different levels of situation understanding. The results clearly showed that it is the support of the projection that has the greatest effect on triggering cooperative behaviour. In the same sense, basically all empirical studies, in which indicator usage was manipulated, showed a significant effect on action selection of the indicator. This supports the importance of understanding the cooperating partner's intention for the selection of one's own behaviour in cooperative situations. Building a computational cognitive model of the underlying psychological processes is associated with many both theoretical and technical challenges, but offers a great potential both for the theoretical progress and for integrating behaviour into technical systems, such as the vehicle automation, that is accepted and trusted by humans as it considers human-like behaviour preferences.

In much more cases than we as psychologists and human factors specialists would wish, we have to admit that there is not enough empirical knowledge available, the available knowledge is too vague or our theories are not precise enough to provide the basis for the precise statements that are required when building such computational models. But exactly this requirement for precise theories make the existing gaps visible and help to fill this gaps by dedicated and theory-driven experimental studies. And the prize to win with such modeling is that the huge amount of empirical results that was collected in the past and will be collected in the future becomes available for the integration into technical systems as this knowledge is condensed and transformed into computational models that can be used by engineers and computer scientists to inform their automation algorithms. The vision is that this leads to technical systems, such as cooperatively interacting vehicles, that then behave in way that is accepted and trusted by human road users as they possess an executable knowledge about human preferences, goals and needs.

Acknowledgements This project has been funded within the priority program 1835 "Cooperatively Interacting Automobiles" by the German Research Foundation (DFG).

References

1. Anderson, J.R., Matessa, M., Lebiere, C.: ACT-R: a theory of higher level cognition and its relation to visual attention. In: Human–Computer Interaction, vol. 12.4, pp. 439–462 (1997). https://doi.org/10.1207/s15327051hci1204_5
2. Balal, E., et al.: Analysis of discretionary lane changing parameters on freeways. Int. J. Trans. Sci. Technol. **3**.3, 277–296 (2014). https://doi.org/10.1260/2046-0430.3.3.277
3. Baselt, D., et al.: Merging lanes—fairness through communication. Veh. commun. **1**.2, 97–104 (2014). ISSN: 2214-2096. https://doi.org/10.1016/j.vehcom.2014.05.005
4. Basu, C., et al.: Do you want your autonomous car to drive like you? In: Proceedings of the 2017 ACM/IEEE International Conference on Human-Robot Interaction, pp. 417–425. IEEE (2017). https://doi.org/10.1145/2909824.3020250
5. Baumann, M., Földes-Cappellotto, N., Stoll, T.: Challenges in modelling situation awareness in cooperative traffic scenarios. Paper presented at the International Conference of Traffic and Transport Psychology 2022. Gothenburg, Aug. 2022
6. Baumann, M., Krems, J.F.: A comprehension based cognitive model of situation awareness. In: Lecture Notes in Computer Science (including subseries Lecture Notes in Artificial Intelligence and Lecture Notes in Bioinformatics), vol. 5620, pp. 192–201. LNCS (2009). ISSN: 03029743. ISBN: 978-3-642-02808-3. https://doi.org/10.1007/978-3-642-02809-0_21. http://link.springer.com/10.1007/978-3-642-02809-0_21
7. Bellem, H., et al.: Objective metrics of comfort: developing a driving style for highly automated vehicles. Transp. Res. F: Traffic Psychol. Behav. **41**, 45–54 (2016). https://doi.org/10.1016/j.trf.2016.05.005
8. Benmimoun, A., Neunzig, D., Maag, C.: Effizienzsteigerung durch professionelles/partnerschaftliches Verhalten im Straßenverkehr. FAT-Schriftreihe Nr. 181. Frankfurt/Main: Forschungsvereinigung Automobiltechnik e.V. (2004)
9. Björklund, G.M., Åberg, L.: Driver behaviour in intersections: formal and informal traffic rules. Transp. Res. Part F: Traffic Psychol. Behav. **8**.3, 239–253 (2005). ISSN: 13698478. https://doi.org/10.1016/j.trf.2005.04.006
10. Borkenau, P., Ostendorf, F.: NEO-Fünf-Faktoren Inventar: Nach Costa u. McCrae; NEO-FFI. Hogrefe, Verlag f. Psychologie (2008)
11. Bullinger, M., et al.: Translating health status questionnaires and evaluating their quality: the IQOLA project approach. J. Clin. Epidemiol. **51**.11, 913–923 (1998). https://doi.org/10.1016/S0895-4356(98)00082-1
12. Burger, C., et al.: Rating cooperative driving: a scheme for behavior assessment. In: *2017 IEEE 20th International Conference on Intelligent Transportation Systems (ITSC)*, pp. 1–6 (2017). ISBN: 2153-0017. https://doi.org/10.1109/ITSC.2017.8317794
13. Chen, H., et al.: Developing an improved ACT-R model for pilot situation awareness measurement. IEEE Access **9**, 122113–122124 (2021). https://doi.org/10.1109/ACCESS.2021.3108438
14. Christoffersen, K., Woods, D.D.: How to make automated systems team players. In: Advances in Human Performance and Cognitive Engineering Research, vol. 2, pp. 1–12. Elsevier (2002). ISBN: 0762308648. ISSN: 14793601. https://doi.org/10.1016/S1479-3601(02)02003-9. http://www.emeraldinsight.com/journals.htm?articleid=1781588&show=abstract

15. Dahlen, E.R., White, R.P.: The Big Five factors, sensation seeking, and driving anger in the prediction of unsafe driving. In: Personality and Individual Differences, vol. 41.5, pp. 903–915 (2006). https://doi.org/10.1016/j.paid.2006.03.016

16. Daily, L.Z., Lovett, M.C., Reder, L.M.: Modeling individual differences in working memory performance: a source activation account. Cogn. Sci. **25.3**, 315–353 (2001). https://doi.org/10.1207/s15516709cog2503_1

17. De Ceunynck, T., et al.: Road safety differences between priority-controlled intersections and right-hand priority intersections: behavioral analysis of vehicle–vehicle interactions. Transp. Res. Rec. **2365.1**, 39–48 (2013). https://doi.org/10.3141/2365-06

18. Deutsch, M.: A theory of co-operation and competition. In: Hum. Relat. **2.2**, 129–152 (1949). https://doi.org/10.1177/001872674900200204

19. Dokic, J., Müller, B., Meyer, G.: European Roadmap Smart Systems for Automated Driving, pp. 1–39 (2015). https://doi.org/10.1017/CBO9781107415324.004. pmid: 25246403

20. Düring, M., Pascheka, P.: Cooperative decentralized decision making for conflict resolution among autonomous agents. In: IEEE International Symposium on Innovations in Intelligent Systems and Applications (INISTA) Proceedings, vol. 2014, pp. 154–161. IEEE (2014). https://doi.org/10.1109/INISTA.2014.6873612

21. Durso, F.T., Rawson, K.A., Girotto, S.: Comprehension and situation awareness. In: Durso, F.T. et al. (eds.), Handbook of Applied Cognition, 2nd ed., pp. 163–193. Wiley, Chicester (2007). https://doi.org/10.1002/9780470713181.ch7

22. Ellinghaus, D: Rücksichtslosigkeit und Partnerschaft. Eine sozialpsychologische Untersuchung über den Umgang unter Kraftfahrern im Straßenverkehr. In: UNIROYAL VERKEHRSUNTERSUCHUNGEN, vol. 12 (1986)

23. Endsley, M.R.: Toward a theory of situation awareness in dynamic systems. Hum. Factors **37.1**, 32–64 (1995). https://doi.org/10.1518/001872095779049543

24. Feierle, A. et al.: Multi-vehicle simulation in urban automated driving: technical implementation and added benefit. In: Information, vol. 11.5, p. 272 (2020). https://doi.org/10.3390/info11050272

25. Friedrich, M., et al.: Validation of the MoSAIC-Driving Simulator–Investigating the impact of a human driver on cooperative driving behavior in an experimental simulation setup. In: Proceedings of the Human Factors and Ergonomics Society Annual Meeting, vol. 57. 1, pp. 2052–2056. SAGE Publications Sage CA, Los Angeles, CA (2013). https://doi.org/10.1177/1541931213571458

26. Hancock, P.A., De Ridder, S.N.: Behavioural accident avoidance science: understanding response in collision incipient conditions. Ergonomics **46.12**, 1111–1135 (2003). https://doi.org/10.1080/0014013031000136386

27. Harris, P.B., et al.: The prosocial and aggressive driving inventory (PADI): a self-report measure of safe and unsafe driving behaviors. Accid. Anal. Prev. **72**, 1–8 (2014). https://doi.org/10.1016/j.aap.2014.05.023

28. Heesen, M. et al.: Investigation of cooperative driving behaviour during lane change in a multi-driver simulation environment. In: Human Factors and Ergonomics Society (HFES) Europe Chapter Conference Touluse, pp. 305–318 (2012)

29. Hidas, P.: Modelling vehicle interactions in microscopic simulation of merging and weaving. Transp. Res. Part C: Emerg. Technol. **13.1**, 37–62 (2005). https://doi.org/10.1016/j.trc.2004.12.003

30. Hoc, J.-M.: Towards a cognitive approach to human–machine cooperation in dynamic situations. Int. J. Hum.-Comput. Stud. **54.4**, 509–540 (2001). https://doi.org/10.1006/ijhc.2000.0454

31. Imbsweiler, J., et al.: Entwicklung einer Beobachtungsmethode von Verhaltensströmen in kooperativen Situationen im innerstädtischen Verkehr. In: Proceedings of the 32. VDI/VW-Gemeinschaftstagung, Fahrerassistenz und automatisiertes Fahren, Wolfsburg, Germany, pp. 8–9 (2016). https://doi.org/10.51202/9783181022887-439

32. Imbsweiler, J., et al.: Quasi-experimentelle Untersuchung des Blickverhaltens und der Fahrparameter von Autofahrern in Engstellen. In: Zeitschrift für Arbeitswissenschaft, vol. 71.4, pp. 242–251 (2017). https://doi.org/10.1007/s41449-017-0083-6

33. Imbsweiler, J., et al.: Insight into cooperation processes for traffic scenarios: modelling with naturalistic decision making. In: Cognition, Technology and Work, vol. 20.4, pp. 621–635. Publisher: Springer London (2018). ISBN: 0123456789. ISSN: 14355566. https://doi.org/10.1007/s10111-018-0518-7. http://link.springer.com/10.1007/s10111-018-0518-7

34. Imbsweiler, J., et al.: Relevant eye-tracking parameters within short cooperative traffic scenarios. In: Proceedings of the Human Factors and Ergonomics Society Europe Chapter 2017 Annual Conference (2018)

35. SAE International.: Taxonomy and Definitions for Terms Related to Driving Automation Systems for On-Road Motor Vehicles. SAE, June 2018

36. Johnsson, C., Laureshyn, A., De Ceunynck, T.: In search of surrogate safety indicators for vulnerable road users: a review of surrogate safety indicators. Transp. Rev. **38.6**, 765–785 (2018). https://doi.org/10.1080/01441647.2018.1442888

37. Kauffmann, N., et al.: Learning the "Language" of road users-how shall a self-driving car convey its intention to cooperate to other human drivers? In: International Conference on Applied Human Factors and Ergonomics. Springer, pp. 53–63. (2017). https://doi.org/10.1007/978-3-319-60366-7_6

38. Kelsch, J., Dziennus, M., Köster, F.: Cooperative lane change assistant: background, implementation and evaluation. In: AAET 2015. ITS Niedersachsen, pp. 65–85 (2015). https://elib.dlr.de/95232/

39. Khamis, A.M., Kamel, M.S., Salichs, M.A.: Cooperation: concepts and general typology. In: 2006 IEEE International Conference on Systems, Man and Cybernetics, vol. 2, pp. 1499–1505. IEEE (2006). https://doi.org/10.1109/ICSMC.2006.384929

40. Kiefer, R., Flannagan, C.A., Jerome, C.: Time-to-collision judgments under realistic driving conditions. Hum. Factors: J. Hum. Factors Ergon. Soc. **48.2**, 334–345 (2006). ISSN: 00187208. https://doi.org/10.1518/001872006777724499

41. Klein, G.A.: Recognition-primed decisions. In: Rouse, W.B. (ed.) Advances in man-machine systems research, pp. 47–92. JAI, Greenwich, CT (1989)

42. Klein, G.A.: "Naturalistic decision making". In: *Human Factors* 50.3 (2008), pp. 456–460. https://doi.org/10.1518/001872008X288385

43. Klein, G.A.: "The recognition-primed decision (RPD) model: Looking back, looking forward". In: *Naturalistic decision making*. Ed. by Caroline E. Zsambok and Gary A. Klein. Expertise Research and applications. New York and London: Routledge, 2009. ISBN: 080581874X

44. Klein, G.A. et al.: "Ten challenges for making automation a "team player" in joint human-agent activity". In: *IEEE Intelligent Systems* 19.6 (2004). ISBN: 1541-1672, pp. 91–95. ISSN: 15411672. https://doi.org/10.1109/MIS.2004.74

45. Klein, G.A. et al.: "Common Ground and Coordination in Joint Activity". In: *Organizational simulation*. Ed. by William B. Rouse. Wiley series in systems engineering and management. Hoboken, NJ: Wiley, 2005, pp. 139–184. ISBN: 9780471739449. https://doi.org/10.1002/0471739448.ch6

46. Knake-Langhorst, S., Gimm, K.: AIM Mobile Traffic Acquisition: instrument toolbox for detection and assessment of traffic behavior. J. Large-Scale Res. Facilities JLSRF **2**, A74–A74 (2016). https://doi.org/10.17815/jlsrf-2-123

47. Kotseruba, I., Tsotsos, J.K.: 40 years of cognitive architectures: core cognitive abilities and practical applications'. In: *Artificial Intelligence Review* 53.1 (2020), pp. 17–94. https://doi.org/10.1007/s10462-018-9646-y

48. Kraft, A.-K.: Kooperation zwischen Verkehrsteilnehmern. Entwicklung und Evaluation von HMI-Konzepten zur Unterstützung kooperativen Fahrens. Doctoral Dissertation, Ulm University (2021)

49. Lau, M., Jipp, M., Oehl, M.: One solution fits all? Evaluating different communication strategies of a light-based external human-machine interface for differently sized automated vehicles from a pedestrian's perspective. Accident Anal. Prevention **171**, 106641 (2022). https://doi.org/10.1016/j.aap.2022.106641

50. Lebiere, C., Anderson, J.R., Bothell, D.: Multi-tasking and cognitive workload in an ACT-R model of a simplified air traffic control task (2001)

33. Imbsweiler, J., et al.: Insight into cooperation processes for traffic scenarios: modelling with naturalistic decision making. In: Cognition, Technology and Work, vol. 20.4, pp. 621–635. Publisher: Springer London (2018). ISBN: 0123456789. ISSN: 14355566. https://doi.org/10. 1007/s10111-018-0518-7. http://link.springer.com/10.1007/s10111-018-0518-7

34. Imbsweiler, J., et al.: Relevant eye-tracking parameters within short cooperative traffic scenarios. In: Proceedings of the Human Factors and Ergonomics Society Europe Chapter 2017 Annual Conference (2018)

35. SAE International.: Taxonomy and Definitions for Terms Related to Driving Automation Systems for On-Road Motor Vehicles. SAE, June 2018

36. Johnsson, C., Laureshyn, A., De Ceunynck, T.: In search of surrogate safety indicators for vulnerable road users: a review of surrogate safety indicators. Transp. Rev. **38**.6, 765–785 (2018). https://doi.org/10.1080/01441647.2018.1442888

37. Kauffmann, N., et al.: Learning the "Language" of road users-how shall a self-driving car convey its intention to cooperate to other human drivers? In: International Conference on Applied Human Factors and Ergonomics. Springer, pp. 53–63. (2017). https://doi.org/10.1007/ 978-3-319-60366-7_6

38. Kelsch, J., Dziennus, M., Köster, F.: Cooperative lane change assistant: background, implementation and evaluation. In: AAET 2015. ITS Niedersachsen, pp. 65–85 (2015). https://elib. dlr.de/95232/

39. Khamis, A.M., Kamel, M.S., Salichs, M.A.: Cooperation: concepts and general typology. In: 2006 IEEE International Conference on Systems, Man and Cybernetics, vol. 2, pp. 1499–1505. IEEE (2006). https://doi.org/10.1109/ICSMC.2006.384929

40. Kiefer, R., Flannagan, C.A., Jerome, C.: Time-to-collision judgments under realistic driving conditions. Hum. Factors: J. Hum. Factors Ergon. Soc. **48**.2, 334–345 (2006). ISSN: 00187208. https://doi.org/10.1518/001872006777724499

41. Klein, G.A.: Recognition-primed decisions. In: Rouse, W.B. (ed.) Advances in man-machine systems research, pp. 47–92. JAI, Greenwich, CT (1989)

42. Klein, G.A.: "Naturalistic decision making". In: *Human Factors* 50.3 (2008), pp. 456–460. https://doi.org/10.1518/001872008X288385

43. Klein, G.A.: "The recognition-primed decision (RPD) model: Looking back, looking forward". In: *Naturalistic decision making*. Ed. by Caroline E. Zsambok and Gary A. Klein. Expertise Research and applications. New York and London: Routledge, 2009. ISBN: 080581874X

44. Klein, G.A. et al.: "Ten challenges for making automation a "team player" in joint human-agent activity". In: *IEEE Intelligent Systems* 19.6 (2004). ISBN: 1541-1672, pp. 91–95. ISSN: 15411672. https://doi.org/10.1109/MIS.2004.74

45. Klein, G.A. et al.: "Common Ground and Coordination in Joint Activity". In: *Organizational simulation*. Ed. by William B. Rouse. Wiley series in systems engineering and management. Hoboken, NJ: Wiley, 2005, pp. 139–184. ISBN: 9780471739449. https://doi.org/10.1002/ 0471739448.ch6

46. Knake-Langhorst, S., Gimm, K.: AIM Mobile Traffic Acquisition: instrument toolbox for detection and assessment of traffic behavior. J. Large-Scale Res. Facilities JLSRF **2**, A74–A74 (2016). https://doi.org/10.17815/jlsrf-2-123

47. Kotseruba, I., Tsotsos, J.K.: 40 years of cognitive architectures: core cognitive abilities and practical applications'. In: *Artificial Intelligence Review* 53.1 (2020), pp. 17–94. https://doi. org/10.1007/s10462-018-9646-y

48. Kraft, A.-K.: Kooperation zwischen Verkehrsteilnehmern. Entwicklung und Evaluation von HMI-Konzepten zur Unterstützung kooperativen Fahrens. Doctoral Dissertation, Ulm University (2021)

49. Lau, M., Jipp, M., Oehl, M.: One solution fits all? Evaluating different communication strategies of a light-based external human-machine interface for differently sized automated vehicles from a pedestrian's perspective. Accident Anal. Prevention **171**, 106641 (2022). https://doi.org/10. 1016/j.aap.2022.106641

50. Lebiere, C., Anderson, J.R., Bothell, D.: Multi-tasking and cognitive workload in an ACT-R model of a simplified air traffic control task (2001)

51. Lehsing, C., Kracke, A., Bengler, K.: Urban perception-a cross-correlation approach to quantify the social interaction in a multiple simulator setting. In: IEEE 18th International Conference on Intelligent Transportation Systems, vol. 2015, pp. 1014–1021. IEEE (2015). https://doi.org/10.1109/ITSC.2015.169

52. Lehsing, C., et al.: Effects of simulated mild vision loss on gaze, driving and interaction behaviors in pedestrian crossing situations. Accident Anal. Prevention **125**, 138–151 (2019). https://doi.org/10.1016/j.aap.2019.01.026

53. Lindner, J., et al.: A coupled driving simulator to investigate the interaction between bicycles and automated vehicles. In: 2022 IEEE 25th International Conference on Intelligent Transportation Systems (ITSC), vol. 2022, pp. 1335–1341. IEEE. https://doi.org/10.1109/ITSC55140.2022.9922400

54. Löper, C., Kelsch, J., Flemisch, F.O.: Kooperative, manöverbasierte Automation und Arbitrierung als Bausteine für hochautomatisiertes Fahren (2008)

55. Marczak, F., Daamen, W., Buisson, C.: Merging behaviour: empirical comparison between two sites and new theory development. In: Transportation Research Part C: Emerging Technologies, vol. 36, pp. 530–546 (2013). ISSN: 0968-090X. https://doi.org/10.1016/j.trc.2013.07.007

56. Marewski, J.N., Mehlhorn, K.: Using the ACT-R architecture to specify 39 quantitative process models of decision making. In: Judgment and Decision Making (2011). https://doi.org/10.1017/S1930297500002473

57. Markkula, G., et al.: Defining interactions: a conceptual framework for understanding interactive behaviour in human and automated road traffic. In: Theoretical Issues in Ergonomics Science, vol. 21.6, pp. 728–752 (2020). https://doi.org/10.1080/1463922X.2020.1736686

58. Martin, J.: Organizational culture: mapping the terrain. Sage Publ. (2001). https://doi.org/10.4135/9781483328478

59. Miller, L., et al.: Implicit intention communication as a design opportunity for automated vehicles: understanding drivers' interpretation of vehicle trajectory at narrow passages. In: Accident Analysis and Prevention, vol. 173, p. 106691 (2022). https://doi.org/10.1016/j.aap.2022.106691

60. Mühlbacher, D., et al.: The multi-driver simulator–a new concept of driving simulation for the analysis of interactions between several drivers. In: Human Centred Automation, pp. 147–158 (2011)

61. Mühlbacher, D., et al.: Multi-road user simulation: methodological considerations from study planning to data analysis. In: UR:BAN Human Factors in Traffic, pp. 403–418. Springer (2018). https://doi.org/10.1007/978-3-658-15418-9_23

62. Newell, A.: Unified Theories of Cognition. Harvard University Press, Cambridge, MA (1990)

63. Obeid, H., et al.: Analyzing driver-pedestrian interaction in a mixed-street environment using a driving simulator. In: Accident Analysis and Prevention, vol. 108, pp. 56–65, July 2017. ISSN: 00014575. https://doi.org/10.1016/j.aap.2017.08.005. pmid: 1623688. http://dx.doi.org/10.1016/j.aap.2017.08.005

64. Oeltze, K., Schießl, C.: Benefits and challenges of multi-driver simulator studies. In: IET Intelligent Transport Systems, vol. 9.6, pp. 618–625 (2015). https://doi.org/10.1049/iet-its.2014.0210

65. Özkan, T., Lajunen, T.: A new addition to DBQ: positive driver behaviours scale. In: Transportation Research Part F: Traffic Psychology and Behaviour, vol. 8.4, pp. 355–368 (2005). https://doi.org/10.1016/j.trf.2005.04.018

66. Preuk, K., et al.: Does assisted driving behavior lead to safety-critical encounters with unequipped vehicles' drivers? In: Accident Analysis and Prevention, vol. 95, pp. 149–156 (2016). https://doi.org/10.1016/j.aap.2016.07.003

67. Quante, L., Gimm, K., Schießl, C.: Trajectory-based traffic observation of cooperation at a road narrowing: Implications for autonomous driving. In: at-Automatisierungstechnik, vol. 71.4, pp. 249–258 (2023). https://doi.org/10.1515/auto-2023-0003

68. Quante, L., Schießl, C.: CoopQ: Questionnaire for measuring the subjective evaluation of cooperation in road traffic encounters. Poster presented at Tagung experimentell arbeitender Psychologen (TeaP) (2022). https://elib.dlr.de/185877/. Accessed: 17 May 2023

69. Quante, L., Schießl, C.: Criteria for the evaluation of interaction behaviour of drivers in a bottleneck scenario. Poster presented at HFES Europe Chapter Annual Meeting. https://elib.dlr.de/186210/. Accessed 17 May 2023

70. Rettenmaier, M., Dinkel, S., Bengler, K.: Communication via motion–Suitability of automated vehicle movements to negotiate the right of way in road bottleneck scenarios. Appl. Ergon. **95**, 103438 (2021). https://doi.org/10.1016/j.apergo.2021.103438

71. Rettenmaier, M., Witzig, C.R., Bengler, K.: Interaction at the bottleneck–a traffic observation. In: International Conference on Human Systems Engineering and Design: Future Trends and Applications, pp. 243–249. Springer (2019). https://doi.org/10.1007/978-3-030-27928-8_37

72. Ritter, F.E., Tehranchi, F., Oury, J.D.: ACT-R: A cognitive architecture for modeling cognition. In: Wiley Interdisciplinary Reviews: Cognitive Science, vol. 10.3, p. e1488 (2019). https://doi.org/10.1002/wcs.1488

73. Salvucci, D.D.: Modeling driver behavior in a cognitive architecture. In: Human Factors, vol. 48.2, pp. 362–380 (2006). https://doi.org/10.1518/001872006777724417

74. Scharfe-Scherf, M.S.L., Wiese, S., Russwinkel, N.: A cognitive model to anticipate variations of situation awareness and attention for the takeover in highly automated driving. en. In: Information, vol. 13.9, p. 418, Sept. 2022. ISSN: 2078-2489. https://doi.org/10.3390/info13090418. https://www.mdpi.com/2078-2489/13/9/ 418 (visited on 04/19/2023)

75. Schindler, J., Koster, F.: A model-based approach for performing successful multi-driver scenarios. In: Proceedings of the DSC 2016 Europe, pp. 93–97 (2016)

76. Schuler, K., et al.: Communication between drivers in a road bottleneck scenario. In: Plant, K., Praetorius, G. (eds.), Human Factors in Transportation. AHFE (2022) International Conference. AHFE Open Access, vol. 60 (2022). https://doi.org/10.54941/ahfe1002461

77. Schwarting, W., et al.: Social behavior for autonomous vehicles. In: Proceedings of the National Academy of Sciences, vol. 116.50, p. 24972 (2019). https://doi.org/10.1073/pnas.1820676116

78. Sharma, A., et al.: Assessing traffic disturbance, efficiency, and safety of the mixed traffic flow of connected vehicles and traditional vehicles by considering human factors. In: Transportation Research Part C: Emerging Technologies, vol. 124, p. 102934 (2021). https://doi.org/10.1016/j.trc.2020.102934

79. Steg, L.: Car use: lust and must. Instrumental, symbolic and affective motives for car use. In: Transportation Research Part A: Policy and Practice, vol. 39.2–3, pp. 147–162 (2005). https://doi.org/10.1016/j.tra.2004.07.001

80. Steward, J.: Why People Keep Rear-Ending Self-Driving Cars Human drivers (and one cyclist) have rear-ended self-driving cars 28 times this year in California—accounting for nearly two-thirds of robocar crashes. In: Wired, vol. 1 (2018). https://www.wired.com/story/self-driving-carcrashes-rear-endings-why-charts-statistics/

81. Stoll, T.: Cooperation in Traffic: Influence of Situational Factors in Cooperative Situations. Ph.D. thesis. Ulm University, Ulm, Germany (2022)

82. Stoll, T., Lanzer, M., Baumann, M.: German validation of the prosocial and aggressive driving inventory (PADI). In: Driving Assesment Conference, vol. 10. University of Iowa (2019). https://doi.org/10.17077/drivingassessment.1697

83. Stoll, T., Lanzer, M., Baumann, M.: Situational influencing factors on understanding cooperative actions in automated driving. Transp. Res. F: Traffic Psychol. Behav. **70**, 223–234 (2020). https://doi.org/10.1016/j.trf.2020.03.006

84. Stoll, T., Mühl, K., Baumann, M.: Do drivers accept cooperative behavior of their automated vehicle on highways? Transp. Res. F: Traffic Psychol. Behav. **77**, 236–245 (2021). https://doi.org/10.1016/j.trf.2020.12.002

85. Stoll, T., Müller, F., Baumann, M.: When cooperation is needed: the effect of spatial and time distance and criticality on willingness to cooperate. In: Cognition, Technology & Work, vol. 21.1, pp. 21–31 (2019). https://doi.org/10.1007/s10111-018-0523-x

86. Stoll, T., Strelau, N.-R., Baumann, M.: Social interactions in traffic: the effect of external factors. In: Proceedings of the Human Factors and Ergonomics Society Annual Meeting, vol. 62, pp. 97–101. 1. Sage Publications Sage CA, Los Angeles, CA (2018). https://doi.org/10.1177/1541931218621022

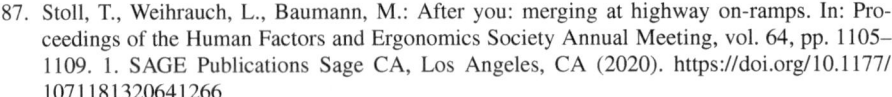

87. Stoll, T., Weihrauch, L., Baumann, M.: After you: merging at highway on-ramps. In: Proceedings of the Human Factors and Ergonomics Society Annual Meeting, vol. 64, pp. 1105–1109. 1. SAGE Publications Sage CA, Los Angeles, CA (2020). https://doi.org/10.1177/1071181320641266

88. Summala, H.: Towards understanding motivational and emotional factors in driver behaviour: comfort through satisficing. In: Modelling Driver Behaviour in Automotive Environments, pp. 189–207. Springer (2007). https://doi.org/10.1007/978-1-84628-618-6_11

89. Taubman-Ben-Ari, O., Mikulincer, M., Gillath, O.: The multidimensional driving style inventory—scale construct and validation. In: Accident Analysis and Prevention, vol. 36.3, pp. 323–332 (2004). https://doi.org/10.1016/S0001-4575(03)00010-1

90. Walch, M., et al.: Cooperative overtaking: overcoming automated vehicles' obstructed sensor range via driver help. In: 11th International Conference on Automotive User Interfaces and Interactive Vehicular Applications (AutomotiveUI 2019) (Level 5 2019), pp. 144–155 (2019). https://doi.org/10.1145/3342197.3344531. http://dl.acm.org/citation.cfm?doid=3342197.3344531

91. Wilbrink, M., et al.: Preliminary interaction strategies for the interACT Automated Vehicles. Technical report interACT D.4.1. interACT project (2018)

92. Will, S.: A new approach to investigate powered two wheelers' interactions with passenger car drivers: the motorcycle–car multi-driver simulation. In: UR:BAN Human Factors in Traffic. Springer, pp. 393–402 (2018). https://doi.org/10.1007/978-3-658-15418-9_22

93. Yusof, N.M., et al.: The exploration of autonomous vehicle driving styles: preferred longitudinal, lateral, and vertical accelerations. In: AutomotiveUI '16: Proceedings of the 8th International Conference on Automotive User Interfaces and Interactive Vehicular Applications, pp. 245–252 (2016). https://doi.org/10.1145/3003715.3005455

94. Zhang, M., Dotzauer, M., Schießl, C.: Analysis of implicit communication of motorists and cyclists in intersection using video and trajectory data. In: Frontiers in Psychology, vol. 13 (2022). https://doi.org/10.3389/fpsyg.2022.864488

95. Zheng, Z.: Recent developments and research needs in modeling lane changing. In: Transportation Research Part B: Methodological, vol. 60, pp. 16–32 (2014). ISSN: 0191-2615. https://doi.org/10.1016/j.trb.2Q13.11.009. https://www.sciencedirect.com/science/article/pii/S019126151300218X

96. Zimmermann, M., Fahrmeier, L., Klaus, J., Bengler: A roland for an oliver? Subjective perception of cooperation during conditionally automated driving. In: International Conference on Collaboration Technologies and Systems (CTS), vol. 2015, pp. 57–63. IEEE (2015). https://doi.org/10.1109/CTS.2015.7210400